Mineral Resources, Economics and the Environment

Written for students and professionals, this revised textbook surveys the mineral industry from a geological, environmental, and economic perspective. Thoroughly updated, the text equips readers with the skills they need to contribute to the energy and mineral questions currently facing society, including issues regarding oil pipelines, nuclear power plants, water availability, resource tax policy, and new mining locations.

Key features

- A new chapter on metals used in the technology industry is included as well as separate chapters on mineral economics and environmental geochemistry.

- Topics of special interest are highlighted in boxes, technical terms are highlighted when first used, and references are included to allow students to delve more deeply into areas of interest.

- Carefully designed figures simplify difficult concepts and show the location of important deposits and trade patterns, emphasizing the true global nature of mineral resources.

Stephen E. Kesler is an Emeritus Professor in the Earth and Environmental Sciences department at the University of Michigan, and a leading expert in the field of mineral resources. He has taught economic geology for almost 50 years, and worked with numerous exploration, mining, and energy companies worldwide. His research interests include geology and geochemistry of ore deposits, and mineral exploration and economics.

Adam C. Simon is an Associate Professor in the Earth and Environmental Sciences department at the University of Michigan, specializing in economic geology, igneous petrology, and geochemistry. He combines field, laboratory, and experimental work to investigate the physical and chemical evolution of magmatic systems and the formation of ore deposits.

Mineral Resources, Economics and the Environment

Stephen E. Kesler
University of Michigan, Ann Arbor

Adam C. Simon
University of Michigan, Ann Arbor

CAMBRIDGE
UNIVERSITY PRESS

CAMBRIDGE
UNIVERSITY PRESS

University Printing House, Cambridge CB2 8BS, United Kingdom

Cambridge University Press is part of the University of Cambridge.

It furthers the University's mission by disseminating knowledge in the pursuit of education, learning and research at the highest international levels of excellence.

www.cambridge.org
Information on this title: www.cambridge.org/9781107074910

© Stephen E. Kesler and Adam C. Simon 2015

First published 2015
3rd printing 2019

Printed in Singapore by Markono Print Media Pte Ltd

A catalogue record for this publication is available from the British Library

Library of Congress Cataloguing in Publication data
Kesler, Stephen E.
Mineral resources, economics and the environment / Steve Kesler, University of Michigan, Ann Arbor, Adam Simon, University of Michigan, Ann Arbor.
   pages   cm
ISBN 978-1-107-07491-0 (hardback)
1. Mines and mineral resources.   2. Mineral industries.   3. Mines and mineral resources – Environmental aspects.   I. Simon, Adam F., 1965–   II. Title.
TN145.K44   2015
333.8′5–dc23   2015014692

ISBN 978-1-107-07491-0 Hardback

Additional resources for this publication at www.cambridge.org/kesler

To Elias, Kai, and Torsten – the next generation of mineral consumers

Steve Kesler

To Alicia, Abigail, James, Laura, and Ethan – for everything

Adam Simon

CONTENTS

The color plates appear between pages 212 and 213.

PREFACE

As we move into the twenty-first century, mineral supplies have become a truly global concern. For most of human history, developed countries have consumed far more than their per capita share of world mineral production. Everyone talked about the day when the rest of the world might want its share of the pie, but it was largely an abstract notion. Now, they are at the door. In fact, China has become the world's largest consumer of mineral resources, and India is not far behind. This momentous change poses two threats. First, there is the possibility that we will run out of the minerals even sooner than we thought. Second, there is the increased pollution caused by their extraction and consumption, which has already destroyed the environment in some areas.

These threats have generated a wide range of opinions about mineral resources and the environment. At one end of the spectrum are those who advocate a dramatic reduction in new mineral production with recycling and conservation providing for the future. At the other end are those who feel that vigorous exploration will always find new minerals and that they can be produced safely with minimal attention to the environment. Both camps are on perilous ground. Many mineral commodities, such as oil and fertilizers, cannot be recycled and the growing demand from developing countries will consume any minerals that are conserved by developed countries. To make matters worse, numerous studies have shown that Earth's storehouse of mineral deposits is indeed finite and that substitutes for important mineral commodities are scarce. Finally, we cannot ignore the environmental catastrophes that have been caused by past mineral production or the impending problems likely to be caused by increasing global mineral consumption.

Unless we are willing to make a dramatic reduction in our standard of living, however, we must find a way to produce and consume the enormous volumes of minerals that we need without significant degradation of the environment. In other words, we must find a middle ground in these arguments, and this means compromise. Unfortunately, compromise is impossible if the parties involved do not understand the problem. That is where this book comes in. It provides an introduction to the geologic, engineering, economic, and environmental factors that govern the production and consumption of mineral resources. This sort of comprehensive information is required if we are to understand all sides of an argument and, hopefully, find a solution.

The book is intended largely for use as a college text, although it can also be used as a primer for anyone with an interest in mineral resources. Mineral professionals who seek a broader view of their field will also find it useful. Because this audience has such a wide range of backgrounds, an effort has been made to make the book a self-contained document, in which all terms and concepts are explained. A basic high school education is all that is needed to read this book. Introductory material on geology, chemistry, engineering, economics, and accounting have been included, along with a glossary of terms, which appear in bold in the text on first mention. Appendices with information on elements, minerals, rocks, mineral commodities, units of weight and measure (including useful conversion factors), and mineral reserves and resources, have also been included, as have references to recent literature. In keeping with their wide use throughout the world, metric (SI) units are used as much as possible in this book, including the term tonne for metric ton, although the (US) short ton, or more simply ton, and (British) long ton are used in some cases where data were reported in these units. Other units, such as flasks and troy ounces, are also employed where dictated by convention.

This book deals with controversial subjects and we have expressed opinions about some of them. We have tried to do this on a case-by-case basis, without following any specific agenda or point of view. It is encouraging in that respect that the book has been cited as too "industry oriented" by some and too "environmentally oriented" by others. Hopefully, each camp will find much that is familiar and friendly, but also much that challenges assumptions and encourages factual debate intended to solve problems and produce a consensus. We will all find many areas in which more data are needed before final decisions can be reached.

Although this book has two authors, it is the product of many minds. We are very grateful to the numerous geologists, mining and petroleum engineers, metallurgists, mineral economists, and other professionals who have allowed access to their projects or operations over the years, and to the many environmentalists who have discussed their research and concerns with us. We are equally grateful to the many students, particularly those in GS/ES380 at the University of Michigan, who have been a constant source of new information and challenging questions. We are grateful to Dale Austin and Marc Gellote for invaluable assistance with the figures, to Hannah Sherman for help with the references, and to Zoë Lewin for especially careful review of the entire manuscript.

CHAPTER

1 Introduction

1.1 Our mineral resource crisis

We are facing a global mineral resource crisis. In fact, we have two of them. First, Earth has a finite supply of minerals for a population that is growing faster than at any time in history (Figure 1.1). Second, mineral consumption is growing even faster than the population. Until recently, we were deeply concerned that most minerals were used in more developed countries (MDCs) with smaller consumption in less developed countries (LDCs) (Table 1.1). Although MDCs account for only 13% of world population, they consume 40% of world oil, 34% of world copper, 28% of world aluminum, 23% of world coal, and 21% of world steel, far more than their share. Now, the MDCs have been joined by China, which alone consumes 49% of world coal, 46% of world steel, 43% of world aluminum, 34% of world copper, and 11% of world oil, also far above its 20% share of world population. Demand is also increasing from India and other large LDCs as global affluence grows.

This creates a dilemma. Although we need more minerals to supply civilization, we are becoming increasingly aware that their production and use are polluting the planet. Effects that were once local in scale have become truly global, with mineral consumption implicated strongly in problems ranging from global warming and acid rain to destruction of the **ozone** layer and pollution of groundwater. Just when we need to expand mineral production, there is concern that Earth is reaching its limit of mineral-related pollution.

We cannot ignore this crisis. Our civilization is based on mineral resources. Most of the equipment that supports a modern life style is made of metals and powered by energy from fossil fuels. The machines that we have developed to

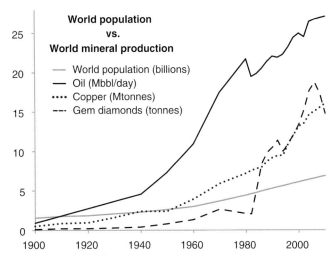

Figure 1.1 Change in world population since 1960 compared to the increased production of oil, copper, and gem diamonds (based on data of the US Geological Survey)

transition us into a renewable energy future are also made entirely of mined materials. Our dependence on minerals pervades society and managing their flow is a major challenge to society (Figure 1.2). Large-scale production of food for growing populations depends on mineral fertilizers, the buildings in which we live and work are made almost entirely of mineral material, and even the gems and gold that we use for adornment and to support global trade come from minerals. Although some might seek a return to Walden Pond to free them from mineral dependency, most of Earth's 7 billion inhabitants are actively seeking the comforts that mineral consumption can provide. If global population and affluence continue to grow as rapidly as many estimates suggest, the pressure to find and produce minerals will be enormous.

Table 1.1 High-income countries, termed more developed countries (MDCs) in this book, listed in order of decreasing per capita gross national income (GNI) in US dollars. This list is based on data for 2012 from the World Bank and does not include data for the following countries that have been listed as high-income in previous years: Andorra, Bahrain, Bermuda, Israel, Kuwait, Liechtenstein, Libya, Macao, Monaco, New Zealand, Oman, Qatar, Saudi Arabia, and San Marino. All other countries are referred to in this book as less developed countries (LDCs).

Norway	98,860	Germany	44,010	Slovak Republic	17,180
Switzerland	82,730	France	41,750	Estonia	15,830
Luxembourg	76,960	Ireland	38,970	Barbados	15,080
Denmark	59,770	Iceland	38,710	Trinidad-Tobago	14,400
Australia	59,570	United Kingdom	38,250	Chile	14,280
Sweden	56,210	Italy	33840	Latvia	14,200
Canada	50,970	Spain	30,110	Lithuania	13,920
United States	50,120	Cyprus	26,000	Equatorial Guinea	13,560
Netherlands	48,250	Greece	23260	Uruguay	13,510
Austria	48,160	Slovenia	22,810	St. Kitts and Nevis	13,330
Japan	47,870	Korea, Rep.	22,670	Croatia	13,290
Singapore	47,210	Portugal	20,580	Russian Federation	12,700
Finland	46,940	Malta	19,760	Poland	12,660
Belgium	44,990	Czech Republic	18,130	Antigua-Barbuda	12,640

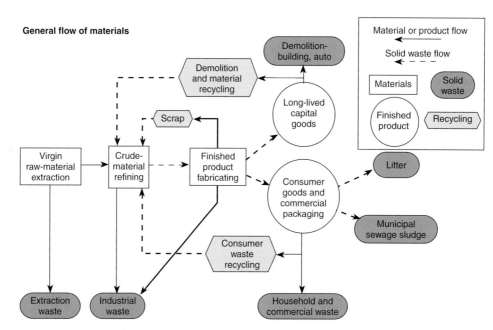

Figure 1.2 Flow of mineral materials through the US economy showing the role of both waste and recycling

Although the magnitude of our growing demand is easy to see, we have become dangerously complacent about it. This would have been unimaginable to the authors of *Limits to Growth* (Meadows *et al.*, 1972), who alerted the world in 1972 to its finite mineral supplies and soaring consumption. The collision between these forces had been developing for almost a century as world living standards improved. Between 1900 and 1973, world oil consumption grew by more than 7%

annually, with each succeeding decade using about as much oil as had been consumed throughout all previous history. World oil supplies were said to be on their way to exhaustion by the turn of the century (Bartlett, 1980a). With steel, aluminum, coal, and other commodities following similar trends, it appeared that we were about to witness the end of a brief mineral-using era in the history of civilization (Petersen and Maxwell, 1979).

However, this did not happen. In the mid 1970s, world mineral consumption slowed just as *Limits to Growth* was published. At the same time, exploration, stimulated by predicted mineral shortages, fanned out across the globe, dramatically increasing reserves for most mineral commodities. In fact, production increased so much that it created a glut of minerals on world markets. Thus, just when we were supposed to feel the cold breath of shortages and rising prices, the world saw an excess of mineral supplies and plummeting prices.

Unfortunately, the respite was brief. As can be seen in Figure 1.1, production curves resumed their climb by the early 1980s and since then production has continued to rise with short interruptions for economic downturns. Interestingly, the urgency expressed by *Limits to Growth* did not resurface as production began to rise again. Instead, it was replaced by a new concern about the environment.

Only a short time ago, our mineral supplies were determined largely by geologic, engineering, and economic factors. Their relation to Earth's mineral endowment was usually depicted as shown in Figure 1.3. Here it can be seen that the most important part of the mineral endowment consists of **reserves**, material that has been identified

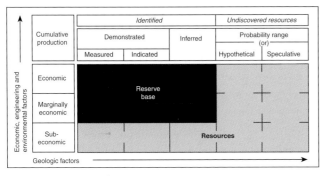

Figure 1.3 Mineral resource classification of the US Geological Survey. The horizontal axis of the diagram represents the level of geological knowledge about deposits, possible deposits, and even undiscovered deposits. The horizontal axis conflates all other information, which affects economic, engineering, and environmental factors that determine whether a deposit might be extracted economically.

geologically and that can be extracted at a profit at the present time. **Resources** include reserves plus any undiscovered deposits, regardless of economic or engineering factors. But, addition of environmental factors to the vertical axis of this diagram has made the situation much more complex. Now, we must ask, not only whether the deposit can be extracted at a profit, but can we also do it in a way that does not compromise the quality of our planet. Environmental costs impact the economic axis of Figure 1.3, thereby controlling the overall profitability of extraction. Just as importantly, however, and more difficult to show in the diagram, are government regulations and public opinion. Today, extraction of mineral deposits in most MDCs and many LDCs must be approved by environmental regulators and accepted by the public, regardless of their economic and engineering merits. The social license to find and operate mineral deposits has become a major constraint on our ability to supply society with minerals (Thompson and Boutilier, 2011).

Thus, the nature and extent of our global mineral endowment is no longer controlled strictly by market forces and administered by mineral professionals who make decisions on the basis of geologic, engineering, and economic factors. Instead, it is in the hands of a broader constituency with a more complex agenda focused largely on the environment, but with additional concerns about distribution of wealth. Addition of this new constituency threatens to push the challenge of supplying society with minerals into the realm of wicked problems, those in which there is a lack of certainty about how actions are related to outcomes and where there is much debate about the relative values of constraints (Metlay and Sarewitz, 2012; Freeman and Highsmith, 2014).

As more and more of us express opinions about mineral deposits, we incur an obligation to understand the factors that control their distribution, extraction, and use. That is what this book is about. We will start with a brief review of the four major factors that control mineral availability.

1.2 Factors controlling mineral availability

1.2.1 Geologic factors

Our mineral supplies come from **mineral deposits**, which are concentrations of elements or minerals that formed by geologic processes. Where something can be recovered at a profit from these concentrations, they are referred to as **ore deposits**. Mineral deposits can be divided into four main

> ## BOX 1.1 | NIMBY – THE "NOT-IN-MY-BACKYARD" SYNDROME
>
> Many mineral deposits are in "inconvenient" places, including heavily settled regions, and production from them is often resisted by local residents. Other activities, such as half-way houses for persons released from prison to garbage dumps, are also resisted, and the practice has become known as the "not-in-my-backyard" (NIMBY) syndrome. However, if we need the minerals, they must be produced somewhere. This brings up the question of whether the NIMBY approach, whether by individuals and governments, is fair to others. Hydraulic fracturing (fracking) provides a good example of the problem. In 2014, the state of New York banned fracking, spurred in part by environmental problems at early gas production wells. Similar anti-fracking moves have been made by some towns in the United States and even by the French parliament. We will learn about fracking later in the book, but for the moment consider the ramifications of this decision. New York is a major consumer of natural gas, and a large proportion of its supply comes from adjacent states where fracking is applied. If fracking is too risky for residents of New York, why would they want to subject Pennsylvanians to that risk? A similar question might be asked of people who expect to use copper mined in other countries with lower levels of environmental regulation. Unless we find a way to get our minerals from an uninhabited asteroid or planet, we will ultimately have to face the moral dilemma posed by the NIMBY syndrome.

groups. The most basic group comprises soil and water, which lack the excitement of gold and oil, but have been essential to civilization from its beginning. **Energy resources** can be divided into the **fossil fuels**, including **crude oil**, **natural gas**, **coal**, **oil shale**, and **tar sand**; the nuclear fuels, including uranium and thorium; and geothermal power. As interesting as they are for the future, wind, tidal, and solar power are not derived from minerals and, along with hydro-electric power, have been omitted from this discussion in order that we can concentrate on minerals, as the title suggests. **Metal resources** range from structural metals such as iron, aluminum, and copper, to ornamental and economic metals such as gold and platinum, and the technological metals such as lithium and rare earths. **Industrial mineral resources**, the least widely known of the four groups, include more than 30 commodities such as salt, potash, and sand, which are critical to our modern agricultural, chemical, and construction industries.

The essential resources, soil and water, require special consideration in our discussion of mineral resources. Our interest in most of the other mineral resources discussed here deals with the balance between the benefits that we derive from them and the environmental damage that they cause. In contrast, soil and water have become the main dumping grounds for most of the wastes that are produced by modern society, including those related to mineral resources. Thus, the essential resources become the context in which we assess the environmental cost–benefit ratios of other mineral resources. Rather than being the focus of a single chapter, then, their role in world mineral extraction and use must be discussed throughout the text.

As we will see throughout this book, there is a close relation between the type of mineral resource found in an area and its geologic setting. Just as common sense tells us not to look for oil in the crater of a volcano, study of Earth has taught us to look for minerals in favorable geologic environments. As population pressures place more demand on land, geologic controls on the distribution of mineral deposits will become increasingly important in land-use decisions.

1.2.2 Engineering factors

Engineering factors affect mineral availability in two ways, technical and economic. Technical constraints are imposed when we simply cannot do something regardless of desire or funding. An example is extraction of iron from Earth's **core**, which is too deep and hot to be reached by any mining method. Economic factors constrain mineral availability only when we judge the cost of a project to be too great. We could build the necessary equipment to mine the Moon, for instance, but the cost of the equipment and the mining expedition would far exceed any benefit that the minerals might afford us.

Engineering considerations place important limits on our ability to extract minerals from Earth. Mining does not extend below about 2.3 km in most areas and the gold mines of South

BOX 1.2 | ARE MINERAL RESOURCES SUSTAINABLE?

Mineral deposits have two geologic characteristics that make them a real challenge to modern civilization. First, almost all of them are **non-renewable resources**; they form by geologic processes that are much slower than the rate at which we exploit them. Whereas balanced harvesting of fishery and forest resources might allow them to last essentially forever, there is little likelihood that we will be able to grow mineral deposits at a rate equal to our consumption of them. Recent estimates suggest that we are consuming gold about 17,000 times faster than it is being concentrated in deposits (Kesler and Wilkinson, 2009). This means that the term sustainability cannot be applied in its strictest sense to mineral resources. Second, mineral deposits have a place value. We cannot decide where to extract them; Nature made that decision for us when the deposits were formed. The only decision that we can make is whether to extract the resource or leave it in the ground.

Figure 1.4 (a) The German ultra-deep borehole project was undertaken to provide information on geologic conditions at depth in Earth's crust. Two separate holes, which were drilled from the station shown here, reached a total depth of 4 km. Temperatures at the bottom of the holes were 120 °C and pressures were 40 megapascals, conditions that are extremely challenging for drilling equipment (photograph courtesy of KTB-Archive, GFZ Potsdam). (b) The Finiston Pit (also known as the Super Pit) at Kalgoorlie, Western Australia is one of the largest open pit mines in the world, measuring 3.5 km long, 1.5 km wide, and 0.6 km deep. The pit moves about 15 million tonnes of rock annually containing about 20,000 kg of gold. Waste rock removed to reach the gold ore is placed on the gray waste-rock dumps to the right of the mine and pulverized ore after processing to remove gold is placed in the white tailings ponds in the upper right. See color plate section.

Africa, the deepest in the world, reach depths only to about 3.7 km. Wells extend to deeper levels; some oil and gas production comes from depths of about 8 km and experimental wells extend to 12 km (Figure 1.4a). However, there is little likelihood that significant production will come from these depths in the near future simply because few rocks at these levels have holes from which fluids can be pumped. Additional engineering constraints are imposed by the need to process most raw minerals to produce forms that can be used in industry and by the need to handle wastes efficiently and effectively.

1.2.3 Environmental factors

Environmental concerns about mineral resources focus on two main problems. The first to be recognized was pollution associated with mineral production (Figure 1.4b). Mining and mineral processing wastes are ten times greater by volume than municipal waste, and by far the largest amount of waste generated in the economic cycle (Hudson-Edwards *et al.*, 2011). The study of older mineral extraction sites has shown that elements and compounds were dispersed into the environment around them for distances of many kilometers. In an effort to prevent future calamities of this type, laws and regulations have been developed to control the generation and disposal of waste products from mineral exploration and production. The cost of compliance with these regulations has increased enormously and has become a growing factor in determining whether a mineral deposit can be extracted profitably. Only recently, have we begun to explore ways to reuse these wastes (Bian *et al.*, 2014).

We have been slower to recognize the importance of wastes associated with mineral consumption, but are making up for lost time. These wastes are more widely dispersed and it has required longer periods of observation and better analytical techniques to demonstrate that the soil, water, and air around us are changing in response to our activities (Figure 1.5). This recognition has produced legislation to remove lead from gasoline, to decrease the amount of SO_2 emitted from smelters, and to limit the release of salt and fertilizers from storage areas, important changes that improve environmental quality but add to the cost of using minerals.

1.2.4 Economic factors

Economic factors that control mineral production include those on the supply side, which are largely engineering and environmental costs related to extraction and processing, and those on the demand side, which include commodity prices, taxation, land tenure, and other legal policies of the host government. Although the balance among these forces can be considered from many political and economic perspectives, it is impossible to avoid the fact that the cost of producing a mineral must be borne by the deposit from which it comes or, in some special cases, by some other segment of the host economy.

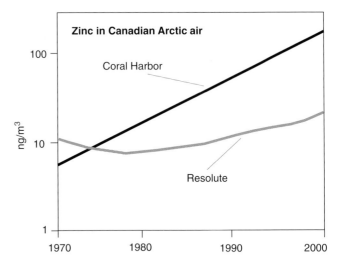

Figure 1.5 Although most airborne pollutants have decreased in concentration over the last few decades in response to environmental clean-up, some continue to increase. Shown here is the change in zinc content in air at Resolute and Coral Harbor in Arctic Canada, which has increased. Note that the scale for this diagram is logarithmic, indicating an enormous, and as yet poorly understood, increase in airborne zinc at these remote locations (based on data in Li and Cornett, 2011).

BOX 1.3 | **THE RIGHT TO MINERAL RESOURCES**

The globalization of environmental concerns presents complex ethical problems that we have just begun to face. Just what right does any country have to pollute the atmosphere and ocean, when that pollution affects other countries? MDCs are at least trying to limit damaging emissions, but many LDCs continue to be major polluters. A related problem is the tendency of MDCs to "export" pollution by importing raw and sometimes even processed minerals from LDCs with fewer environmental regulations. In a world with finite resources and growing demand, the decision not to exploit one deposit requires that another be exploited to supply world demand. What might have happened, for instance, if Kuwait had responded to the environmental damage of Iraqi sabotage during the 1991 war by limiting oil production to just enough for its own energy needs? Would the MDCs have accepted that, and increased domestic exploration and production, or do they expect environmental sacrifices from supplier countries, which they are not willing to make themselves? Finally, what about states and nations whose increased environmental awareness leads them to forbid specific mineral production activities, such as has happened with the ban on fracking for oil and gas production in New York? Do these entities have a right to expect others to supply their mineral needs, or should they be excluded from commerce in that commodity? As demand increases these questions might well become more than tantalizing thought experiments.

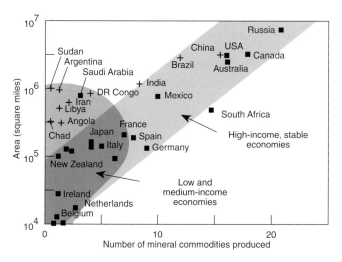

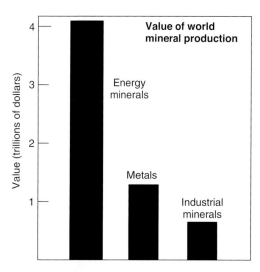

Figure 1.6 Relation between land area and number of minerals produced by various countries showing a good relation for high-income, stable economies and a poor relation for low and medium-income countries with less stable economies. This distinction between countries is similar to the LDC–MDC distinction used in this book and shows that countries with large land areas (and consequent variable geology) and stable fiscal and operational regulations are more likely to host operating mineral deposits.

Figure 1.7 Value of world production for the three main classes of mineral products. Recycled material is not included. Steel and cement are the only processed mineral products included here and the exclusion of these would cause the metals and industrial minerals totals to drop to $0.8 billion and $0.4 billion, respectively (compiled from data of the US Geological Survey and International Energy Agency).

In a free market, costs and prices are usually part of a global system, which places similar constraints on all countries. However, legal, tax, and environmental regulations differ from country to country. The overall importance of these factors to mineral availability is shown by the positive correlation between the number of minerals produced and land area for countries with high-income, stable economies (Figure 1.6). This correlation supports the notion that large areas of Earth are more likely to have lots of important mineral resources than small areas. A similar correlation is not seen for low- and middle-income countries. In view of the relatively weak environmental regulations in most LDCs, the lack of a correlation in these countries probably reflects a more uncertain legal and tax framework, which discourages investment (Govett and Govett, 1977). For this reason, we have included chapters on land tenure and on mineral economics and taxation in the book.

1.3 Minerals and global economic patterns

The impact of minerals on the global economy is enormous. World fuel and metal production are worth about $4.2 and $1.3 trillion, respectively, and industrial mineral production is worth about $550 billion (Figure 1.7). A good indication of the role of mineral production in economic activity in any country can be obtained by comparing the value of mineral production and gross domestic product (GDP). As can be seen in Table 1.2, raw mineral production makes up only a few percent of GDP in MDCs such as the United States, the Netherlands, and Sweden, but reaches 7 to 12% in others, including Australia and Canada. Norway holds the crown among MDCs with mineral production making up more than 35% of the GDP. Such unusually high percentages are more common in some LDCs including Papua New Guinea and Zambia, which are major copper producers, and the Persian Gulf countries that supply most of the world's oil. It is a mistake to conclude that countries are unimportant mineral producers just because raw mineral production makes up a small percentage of the GDP, however. The United States, for instance, is the leading world producer of many mineral commodities with a total value of more than $520 billion and, according to the US Geological Survey, the value added to the US economy by major industries that consume these minerals is about $2.44 trillion.

Classical theory holds that economic activity depends on domestic mineral resource availability (Hewett, 1929). According to this scheme raw-mineral exports occur early in a nation's development, as mineral deposits are discovered (Figure 1.8a). Profits from these exports are used to build an industrial infrastructure, which supports growing exports of

Table 1.2 Approximate value of energy and mining production in major producing countries (in billions of $US). Compiled from data of the World Bank, International Energy Agency, and International Council on Mining and Metals.

Energy production			Mined mineral production		
Country	*Value*	*% of GDP*	*Country*	*Value*	*% of GDP*
Russia	$534	20.85%	Australia	$72	7.80%
United States	$499	3.11%	China	$69	1.20%
China	$486	3.87%	Brazil	$47	2.30%
Saudi Arabia	$401	46.80%	Chile	$31	14.70%
Canada	$167	11.15%	Russia	$29	1.90%
Iran	$164	15.41%	South Africa	$27	7.50%
United Arab Emirates,	$125	46.04%	India	$26	1.50%
Venezuela	$114	28.93%	United States	$23	0.20%
Kuwait	$109	99.67%	Peru	$19	12.00%
Qatar	$107	50.37%	Canada	$14	0.90%
Iraq	$104	45.17%	Indonesia	$12	1.70%
Norway	$94	35.53%	Ukraine	$9.3	6.70%
Indonesia	$93	7.24%	Mexico	$8.4	0.80%
Australia	$68	7.07%	Kazakhstan	$7.3	4.90%
India	$54	1.12%	Iran	$4.4	1.30%
South Africa	$23	4.13%	Philippines	$4.2	2.10%
Netherlands	$19	2.57%	Sweden	$4.0	0.90%
Germany	$18	0.01%	Ghana	$3.9	12.70%
Poland	$13	0.01%	Zambia	$3.8	23.80%
Kazakhstan	$11	4.53%	Papua New Guinea	$3.2	33.40%

goods manufactured from domestic raw materials. As mineral reserves dwindle, imports rise to support continued manufacturing. Many LDCs, such as Zambia and the Democratic Republic of Congo, have been bogged down at the start of this evolution and their national budgets and overall welfare are highly dependent on raw-mineral prices. Because these prices vary unpredictably, these countries cannot control their revenues, a factor that limits stable development. This situation is a universal sore spot, with almost all countries wishing to sell more finished goods and less raw minerals. Even Canada, which occupies an enviable position in a global context, agonizes about its role as "hewer of wood and drawer of water" for the world.

It can therefore be seen that classical mineral economic theory predicts disaster for countries that lack raw minerals

to support manufacturing and exports. But things have changed. Japan has a strong positive balance of trade in spite of an enormous annual deficit in mineral imports (Figure 1.8b). Lower wages and higher domestic productivity are commonly cited reasons for Japan's success. Just as important, and less widely recognized, have been the Japanese raw-material trade policies. During the last two decades Japan has invested in mineral extraction projects throughout the world. Most of these investments have involved agreements to buy some or all of the production, thus assuring an orderly supply of minerals.

A more modern view of global mineral trade is shown in Figure 1.9 using iron and steel as an example. Note that Japan and Korea, both major exporters of manufactured goods, are heavily dependent on imported raw material. The European

BOX 1.4 | **THE GLOBAL FOOTPRINT OF A SMARTPHONE**

In 2015, 70% of the world's population, almost 5 billion people, owned a mobile phone, with nearly 2 billion of these being smartphones that function as handheld computers. This is a dramatic increase from none in 1990. The technology embedded in a smartphone exceeds that in the Apollo Guidance Computer used in 1969 to send humans to the Moon. That computer weighed 70 pounds, cost $150,000, and had a total storage capacity of 4 thousand bytes of information. Compare this to an Apple iPhone that weighs less than 4 ounces, costs only a few hundred dollars and comes standard with a storage capacity of 64 billion bytes of data. This remarkable technology comes with a huge natural-resource footprint. Among the more than 40 elements used are aluminum, potassium, and silicon for the ion-strengthened glass screen; carbon, cobalt, and lithium for the batteries; indium and tin to conduct electricity in the transparent touch screen; nickel for the microphone; lead and tin used as solder; antimony, arsenic, boron, phosphorus, and silicon in various semiconductors and chips; oil for the plastic housing; bromine in the plastic for fire retardation; copper, gold, and silver in the wiring; tantalum for the capacitors; the rare-earth elements gadolinium, neodymium, and praseodymium for the magnet, neodymium, dysprosium, and terbium to reduce vibration, and dysprosium, gadolinium, europium, lanthanum, terbium, praseodymium, and yttrium to produce colors. That is roughly one-half of all naturally occurring elements. Mining all of these resources consumes vast quantities of energy, as does shipping them and the finished products around the world. Almost 90% of the rare earths are mined in China, lithium is mined in Chile, cobalt in the Democratic Republic of Congo, aluminum in Australia, phosphorus in Morocco, nickel in Canada, and oil is extracted by using hydraulic fracturing to stimulate permeability in unconventional shale reservoirs. Smartphones truly have a global environmental footprint. And in the United States the average user buys a new phone every 2 years.

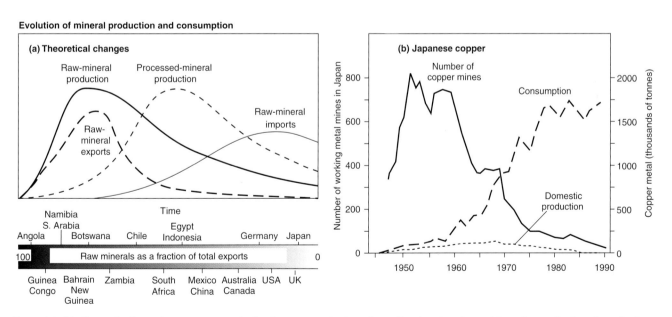

Figure 1.8 (a) Classical relation between economic development and mineral supplies showing the position of several mineral-producing countries as indicated by the proportion of minerals in their total exports. (b) Change in copper mining and production in Japan from 1940 to 1990 showing increased consumption despite decreased domestic production (based on Ishihara, 1992).

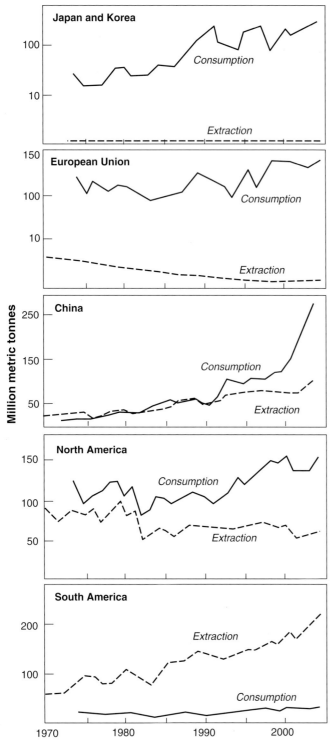

Figure 1.9 Relation between consumption and extraction for iron and steel in various parts of the world, showing the high dependency of the European Union and Japan–Korea on imports, with lesser dependence in China and North America and a large export market for South America (compiled from Rogich and Matos, 2008)

Union is in only slightly better shape. China, despite its major role as a raw-mineral importer, is able to supply a much larger proportion of its needs domestically, in part because of its larger size and greater geologic diversity. North America comes even closer to self-sufficiency, and South America is a major source of raw materials. Thus, the pattern of mineral use is global, with LDCs supplying mineral raw materials to MDCs that manufacture goods and export them (Graedel and Cao, 2010).

Some feel that the great increase in the global trade of minerals has weakened the concept of **strategic minerals**, which holds that the security of a country depends on its mineral supplies, particularly those that are necessary for defense needs. However, the 1991 Iraq war and its successors in the Middle East have shown just how hard MDCs will fight for access to mineral supplies, suggesting that the strategic minerals concept has not died away. Global mineral trade has, however, eroded the power of mineral cartels by promoting market transparency, in which production and consumption data are shared by producers throughout the world.

1.4 The new era of world minerals

Mineral resource availability is entering a new era, one in which traditional geologic, engineering, and economic constraints are joined and often trumped by environmental considerations. Dealing with these many factors and the uncertainties that they involve, while moving ahead to supply the next generation with minerals, will require compromises based on a full understanding of the issues. As a first step in this direction, this book explores the ramifications and interrelations of geologic, engineering, economic, and environmental constraints on global mineral resources. We hope it makes you a better decision maker as we approach these major problems.

CHAPTER

2 Origin of mineral deposits

A cubic kilometer of average rock contains millions of dollars worth of mineral commodities (Table 2.1). Although this might make you think that we could supply our mineral needs from average rock, this notion is dispelled by a look at the costs involved. A tonne of average rock contains only a few cents worth of gold versus the several dollars that you would have to spend to extract the gold. Unfortunately, you cannot get around this constraint by extracting other metals from the same tonne of average rock because most mineral commodities require different processes, each with its own high cost. To help keep things in perspective, remember that it takes several dollars just to buy a tonne of gravel, which takes essentially no processing. That is why we have to depend on **mineral deposits** for our needs. Mineral deposits are simply places where Earth has concentrated one or more mineral commodities. Where the degree of concentration is high enough for profitable extraction, we have an **ore deposit**.

With all of these factors to consider, you might lose sight of the fundamental fact that we can only produce minerals from mineral deposits. So, we begin with a review of the geologic setting and formation of mineral deposits.

2.1 Geologic framework of Earth and mineral deposits

Earth consists of four global-scale divisions: the **atmosphere**, **hydrosphere**, **biosphere**, and lithosphere (Figure 2.1). Mineral deposits are part of the **lithosphere**, which is made up largely of rocks and minerals. **Minerals**, which control the distribution of elements in Earth, are naturally occurring solids with a characteristic crystal structure and definite chemical composition. They are divided into groups on the basis of their chemical compositions (Appendix 2, Table A2.1). Although about 3,800 minerals are known, only about 30 minerals make up almost all common rocks, another 50 or so account for most metal ores, and about 100 comprise the industrial minerals (Appendix 2).

Rocks consist of grains of minerals that hold together well enough to be thrown across a room. They are divided into groups on the basis of their origin and composition

Table 2.1 Value of selected mineral commodities in a cubic kilometer of average upper crustal rock and in 1 tonne of rock based on 2014 prices. The total value of just these commodities in a cubic kilometer of average rock is more than $200 billion!

Element	Value in 1 km^3 of crust ($million)	Value in 1 tonne of crust ($)	Average crustal abundance (ppm)
Aluminum	172,000	140.00	81,500
Iron	37,000	30.00	39,200
Manganese	2,100	1.70	774
Nickel	1,100	0.91	47
Copper	233	0.19	28
Zinc	170	0.14	67
Gold	83	0.068	0.0015
Tin	59	0.05	2.1
Lead	43	0.035	17
Molybdenum	42	0.034	1.1
Platinum	3.1	0.003	0.0005

BOX 2.1 | MINERAL DEPOSITS VS. ORE DEPOSITS

The distinction between ore deposits and mineral deposits is a dynamic function of economic, engineering, political, and environmental factors. For example, the need to use catalytic converters to clean automobile exhaust changed platinum-bearing mineral deposits around Rustenburg, South Africa, into profitable ore deposits. Similarly, increased oil prices in the 1970s provided the key to large-scale mining of the extensive tar sands of Alberta, and subsequent price increases in the 2000s transformed Canada into a major oil producer. On the other side of the coin, mining of lithium from the large **pegmatite** deposits at Kings Mountain, North Carolina, came to a halt when production began from the large brine deposits in the Atacama Desert of Chile, simply because it is much easier to get lithium from a liquid than from a mineral.

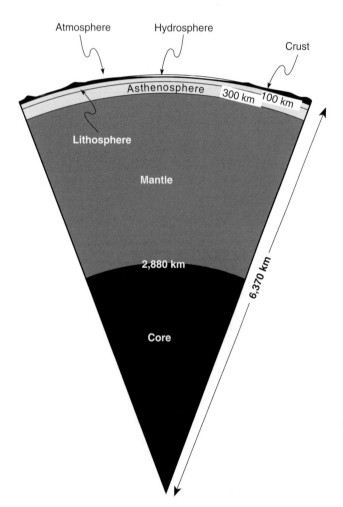

Figure 2.1 Distribution of the lithosphere, hydrosphere, and atmosphere (the biosphere is too small to be shown at this scale) and major divisions of the solid Earth (core, mantle, asthenosphere, and lithosphere). Note that the term lithosphere as used by environmental geochemists refers to the entire solid part of the planet rather than the outer 100-km-thick plate that geologists call the lithosphere.

(Appendix 1; Table A2.2). **Igneous rocks** form by crystallization or solidification of molten rock called **magma**, which is less dense than rock and will rise through the lithosphere. Where magma reaches the Earth's surface it is called **lava** and forms **extrusive rocks**, including volcanoes. Where magma solidifies below the surface, it forms **intrusive rocks**, including large bodies known as **stocks** and **batholiths** and tabular bodies known as **sills** and **dikes**. Most **igneous rocks** consist largely of silicon and aluminum, but they are divided into smaller groups on the basis of their mineral and chemical compositions, as discussed later. **Sedimentary rocks** consist of material that was deposited by water, wind, or ice. They can consist of clasts or fragments of pre-existing rocks, of minerals precipitated from water, or organic matter. **Metamorphic rocks** form when other rocks are buried in the crust where high heat and pressure recrystallize original minerals, forming new minerals and textures.

The solid Earth, or lithosphere, is divided into the core, mantle, and crust (Figure 2.1). The **core**, which consists largely of iron, is one big mineral deposit. Unfortunately, engineering constraints prevent us from extracting it, although cores of failed planets come to us in meteorites. The **mantle**, which forms a thick shell around the core, consists of **ultramafic** rocks that are strongly enriched in magnesium; they contain locally important concentrations of chromium, cobalt, and nickel. Even the mantle is too deep to be within reach of mining, although we do extract ore from it where mantle rocks have been moved to the surface by plate tectonics, as discussed later.

Most of our mineral deposits are in the **crust**, which covers the mantle and contains the greatest variety of Earth's rocks. It is divided into two parts: **ocean crust**, which is 5 to 10 km thick and consists of **mafic** igneous rocks, largely **basalt**, that is enriched in iron, calcium, and magnesium; and **continental**

crust, which is 20 to 70 km thick and consists of **felsic** igneous rocks that are enriched in sodium and potassium, metamorphic rocks, and a covering layer of sedimentary rocks. Even the ocean crust is difficult to reach from an engineering standpoint, and we extract mineral resources from it mostly where it has been moved onto the continents by faulting. It is the continental crust that hosts the vast majority of our mineral resources.

BOX 2.2 | LOOK IT UP

Throughout the book, we have tried to explain most terms and concepts that might be new to you. If we have failed, have a look in the glossary. And, if you want to get a little more information on the composition and classification of common rocks and minerals, the main ore minerals for each element, or the most important weights and measures that we deal with in the book, you can look at the appendices. We make specific reference to the appendices in this chapter, but will not do so in the remaining chapters and will trust you to *look it up*.

BOX 2.3 | PLATE TECTONICS

Earth's upper part is divided into large lithospheric plates (Figure Box 2.3.1) that move about by the process known as **plate tectonics**. The plates are about 100 km thick and consist of ocean and/or continental crust and some of the underlying mantle (Figure Box 2.3.2). They are underlain by plastic mantle known as **asthenosphere**. The plates meet in three types of **margins**. **Divergent margins** form where plates move apart, such as at **mid-ocean ridges** where rising mantle melts to produce mafic basalt magma that flows onto the ocean floor creating new ocean crust. Divergent margins beneath continental crust create **rifts** such as the Dead Sea, which enlarge to form new oceans such as the Red Sea. **Convergent margins** form where plates move together. Most convergent margins include **subduction zones**, where the ocean crust sinks back into the mantle causing the production of **intermediate** to felsic magmas that form stocks and batholiths in the crust and volcanoes at the surface. At some convergent margins, neither plate sinks and one rides onto the other along **obduction zones**. **Transform margins** form where two plates move horizontally past each other along features like the San Andreas fault in California.

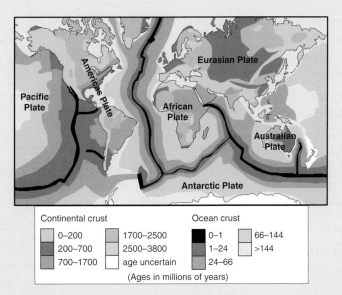

Figure Box 2.3.1 Distribution of lithospheric plates. Note that continental crust is much older than ocean crust and the ages of ocean crust increase away from spreading ridges.

BOX 2.3 | (CONT.)

Plate tectonics and mineral deposit environments

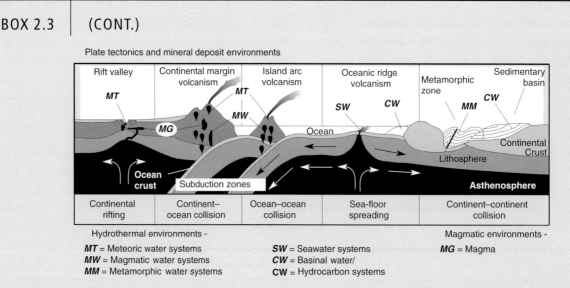

Figure Box 2.3.2 Plate-tectonic environments, showing their relation to ore-forming magmatic and hydrothermal processes discussed in this chapter and shown in Figure 2.12

The continents are buoyant features that float on the mantle, whereas denser ocean crust sinks into the mantle at subduction zones. In fact, the oldest known ocean crust is only about 200 million years (Ma) old, whereas the oldest rocks on the continents are about 4 billion years old (Figure 2.2). Even these old continental rocks are significantly younger than the 4.65 billion year (Ga) age of Earth because the oldest rocks have been destroyed by erosion or covered by younger ones. Thus, even on the continents, rocks of the four major divisions of Earth history, the **Hadean**, **Archean**, **Proterozoic**, and **Phanerozoic** eons (Appendix 1, Table A1.3), become progressively less abundant with increasing age. In most cases, rocks of Hadean and Archean age form stable cores for the continents, which are known as **cratons** or **shields**, and Proterozoic and Phanerozoic rocks are found largely in deformed belts surrounding them.

Plate tectonics moves the lithospheric plates at the agonizingly slow rate of a few centimeters per year. This rate, in turn, controls magma formation, erosion, and sedimentation, all of which form mineral deposits (Sawkins, 1990). The resulting rate at which individual mineral deposits form is much slower than the rate at which we extract them, which is why we refer to mineral deposits as **non-renewable resources**. This distinguishes them from **renewable resources** such as forests and fisheries. It also highlights the fact that, in the long run, mineral resources are not really sustainable, at least not in the sense that we can use them at a rate equal to the rate at which Earth forms them.

2.2 Geologic characteristics of mineral deposits

Ore deposits represent work that nature does for us. For instance, Earth's crust contains an average of about 55 ppm (parts per million) of copper, whereas copper ore deposits must contain about 5,000 ppm (0.5%) copper before we can mine them. Thus, geologic processes need to concentrate the average copper content of the crust by about 100 times to make a copper ore deposit that we can use (Table 2.1). We then use industrial processes to convert copper ore into pure copper metal, an increase of about 200 times. By a fortunate natural coincidence, elements and materials that we use in large amounts need less natural concentration than those that we use in small amounts. Thus, we are likely to have larger deposits of mineral commodities that we use in large amounts. As long as energy costs remain high, the relation between work that we can afford to do and work that we expect nature to do will control the lower limit of natural concentrations that we can exploit, and this puts very real limits on our global mineral resources.

Most ore deposits contain an **ore mineral** in which the element or substance of interest has been concentrated. For instance, the common zinc ore mineral, sphalerite (ZnS), contains about 64% (640,000 ppm) zinc. In rocks that contain sphalerite, the production of zinc is greatly simplified because grains of sphalerite (Figure 2.2a) can be separated from the ore to produce a smaller volume of material for further treatment to obtain zinc metal. Zinc does not form sphalerite everywhere, however. Most of the Earth's zinc **substitutes** at very low (ppm) concentrations for iron in silicate minerals that make up most of the crust. Ore deposits form only where geologic processes liberate this zinc and combine it with sulfur to form sphalerite. Formation of most mineral deposits requires this sort of concentration, first into an ore mineral and then into a large enough concentration of the ore mineral to be of economic interest. In some cases, the concentrated material is not actually a mineral. For instance, crude oil and natural gas deposits form when **organic material** dispersed through sedimentary rock accumulates in concentrations that we can exploit (Figure 2.2b) and groundwater accumulates in zones call aquifers.

Figure 2.2 (a) The dark crystals of galena grew on top of smaller crystals of white dolomite. Whereas the dolomite contains almost no zinc, sphalerite contains as much as 64% by weight and is our principal ore mineral for zinc. (b) The dark material is solidified crude oil known as **bitumen**, which is lining a cavity in the white limestone. See color plate section.

The importance of ore minerals or compounds is well illustrated by molybdenum and germanium, which have approximately the same average concentrations in the crust, about 1.5 ppm. In spite of this, the world molybdenum production of 270,000 tonnes is over 2,000 times greater than that for germanium. This difference reflects the fact that molybdenum forms the common ore mineral, molybdenite (MoS_2), whereas germanium forms no common minerals.

Ore minerals such as sphalerite and molybdenite are rarely found in massive clumps that can be extracted directly. Instead, they are found in mixtures with minerals that have no commercial value in that setting, known as **gangue minerals**. For instance, quartz (SiO_2) is a common gangue mineral in many metal deposits and is discarded as waste, although pure deposits of quartz are valuable for glass sand and other uses. The concentration of the element or compound of interest in an ore is referred to as the **grade** of a deposit, and the minimum concentration needed to extract the ore at a profit is known as the **cut-off grade**. Many zinc deposits have an average grade of about 5% zinc, almost all of which is in sphalerite. Grades for natural gas, crude oil, and other fluids are not usually quoted as percentages. Instead, they are given as the amount of fluid that can be recovered from a given volume of rock.

In addition to grade, a mineral deposit must attain a minimum size. No matter how high its grade, a deposit must contain enough ore to pay for the equipment, labor, and costs of extraction. The size of solid mineral deposits is almost always given as the mass of ore, which is specified in metric tons (tonnes) and the size of fluid deposits is given as the volume of fluid, with the understanding that it can be recovered from a given volume of rock.

Although we refer to processes that form mineral deposits as ore-forming processes, many deposits are not actually ores. In the following discussion, we divide ore-forming processes into those that take place at or near Earth's surface and those that take place at depth (Table 2.2).

2.3 Ore-forming processes

2.3.1 Surface and near-surface ore-forming processes

Weathering and soils
Weathering is the geologic term for the wide range of processes that take place in the **critical zone**, where rocks and minerals that formed at depth equilibrate with water, air, and plants near the Earth's surface (Lin, 2010). Weathering

Table 2.2 Geologic processes that form mineral deposits, with examples of deposits formed by each process and elements concentrated in them.

Type of process	Types of deposits formed and minerals concentrated surface processes
Surface processes	
Weathering	Laterite deposits – nickel, bauxite, gold, clay soil
Physical sedimentation	
Flowing water	Placer deposits – gold, platinum, diamond, ilmenite, rutile, zircon, sand, gravel
Wind	Dune deposits – sand
Chemical sedimentation	
Precipitation from or in water	Evaporite deposits – halite, sylvite, borax, trona
	Chemical deposits – iron, manganese
Organic sedimentation	
Organic activity or accumulation	Hydrocarbon deposits – oil, natural gas, coal
	Other deposits – sulfur, phosphate
Subsurface processes	
Involving water	Groundwater and related deposits – uranium, sulfur
	Basinal brines – Mississippi Valley-type (MVT), sedimentary exhalative (SEDEX)
	Seawater – volcanogenic massive sulfide, SEDEX
	Magmatic water – porphyry copper–molybdenum, skarn
	Metamorphic water – gold, copper
Involving magmas	Crystal segregation – chromium, vanadium
	Immiscible magma separation – nickel, copper, cobalt, platinum-group elements

produces the **regolith**, a layer of partly decomposed rock and soil that covers most of the land surface from depths of a few centimeters to several hundred meters. Some civil engineers refer to essentially all of the regolith as **soil**. Farmers refer to soil as the upper part of the regolith that will support marketable plant life. Most geologists and environmentalists extend this definition to take in the upper part of the regolith throughout the world, including materials somewhat deeper than the soil of agriculture. This layer contains most of the minerals that are produced by weathering and it bears the brunt of environmental pollution (Foth, 1984; Sparks, 2002; Hillel, 2007).

In warm, humid climates, soil forms largely by chemical and biological weathering involving the dissolution of common rock-forming minerals. Dissolved elements such as sodium, potassium, calcium, and magnesium are carried by streams to the ocean. Because most common rock-forming minerals are not very soluble, weathering depends heavily on acid water and on biological activity (Chorover *et al.*, 2007). Acid rain is generated by dissolution of atmospheric gases such as CO_2 and SO_2, as discussed in the next chapter, and acidity is increased in the soil by bacterial activity that boosts

CO_2 concentrations to levels of several percent versus only 0.03% in the atmosphere; by decomposition of pyrite and other sulfide minerals to form sulfuric acid; and by acids formed during the decay of organic matter. Some rock-forming elements, such as iron and carbon, are oxidized during weathering, further contributing to the decomposition of minerals containing them, and plant roots create special chemical environments that enhance mineral dissolution.

Physical weathering involves processes that actually break the rock into pieces. Root growth has this effect, as does ice, which has a volume 9% greater than an equivalent amount of water. Growth of salt and calcite crystals when water evaporates from cracks can also fracture a rock. Diurnal (daily) and seasonal temperature changes cause expansion and contraction that gradually promote fracturing, as does the removal of overlying sediment by erosion. These effects dominate only in arid and Arctic environments, where chemical processes are limited by the absence of moisture or by low temperatures.

Weathering and soil formation actually form ore deposits of two main types. First, these processes can remove soluble constituents leaving behind relatively insoluble elements and minerals of value in residual deposits. The most obvious

BOX 2.4 | CLASSIFICATION OF SOILS

Soils can be classified on the basis of their vertical zonation. In humid areas where population concentrations are greatest and environmental effects of greatest interest, the generalized soil profile shown in Figure Box 2.4.1 is a useful start. It consists of an upper O horizon rich in organic material from vegetation, which grades downward into an A horizon dominated by silicate and oxide minerals formed by weathering of the original rock. Acid water leaches material from the lower part of the A horizon, which is sometimes given the separate name E horizon. These dissolved constituents move downward and are deposited in the B horizon, which becomes enriched in soluble material. Underlying all of these is the C horizon, which consists of variably weathered bedrock. In many areas of deep weathering, the C horizon consists of intensely weathered material known as saprolite, which retains the original textures of the parent rock. In more arid regions and in areas of restricted drainage such as swamps this zonation is not as well developed.

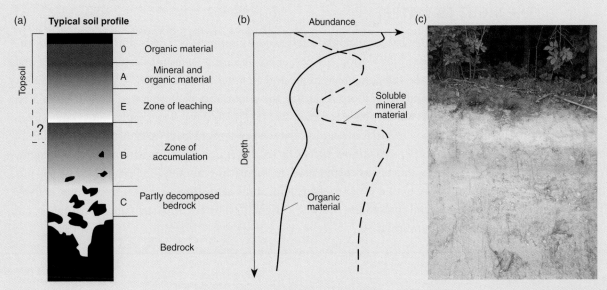

Figure Box 2.4.1 Typical soil profile (a) with the related abundances of soluble mineral and organic material (b) of a well-drained soil in a humid temperate region showing zones in which soluble mineral and organic matter are leached and accumulated. (c) Photo of soil profile showing organic matter at the top (gray), leached A zone (white) and zone of accumulation (reddish-brown). See color plate section.

example is grains of gold in rocks; the rocks decompose but the gold remains in the residuum. Similar processes form concentrations of barite, ilmenite, and other less-soluble minerals when their enclosing rocks dissolve and disintegrate. Redistribution of soluble constituents in the soil also creates mineral deposits such as **laterites** in which aluminum, iron, nickel, or other elements accumulate in the lower part of the B or the C horizons. Laterites often make up the upper part of saprolite zones, and represent the parts of the weathering zone where original rock textures have been totally destroyed.

Soil is actually our most important mineral resource. It is the basis for agriculture as well as the principal host of many microorganisms that carry out chemical reactions essential to life. It is also subject to many assaults, of which erosion is the most dangerous. By its very nature, soil is not coherent or lithified like rock, and this makes it easy to erode. Soil is held in place largely by plant roots and where they are removed, soil that has taken thousands of years to form can be removed in a few hours (Dotterweich, 2013). Humans are more important to global erosion and destruction of soils than natural processes. Human activity has lowered continental land surfaces by about 6 cm, an enormous amount in view of the short time that we have been on Earth (Wilkinson and McElroy, 2007). Soil is also subject to **desertification**, which occurs naturally on the margins of deserts, but is also caused by

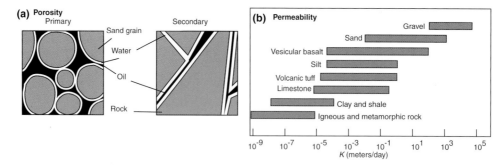

Figure 2.3 (a) Typical primary and secondary porosity, showing the relation between water and oil in pores. (b) Hydraulic conductivity *K* (permeability to water) of various common rock types under near-surface conditions (modified from data of Bear 1972).

BOX 2.5 | THE RIGHT TO WATER

Water is the only mineral commodity that is currently referred to by some as a "right." On July 28, 2010, the United Nations General Assembly passed Resolution A/RES/64/292 declaring "safe and clean drinking water and sanitation a human right essential to the full enjoyment of life and all other human rights." Although this is only the latest in a long line of documents and declarations regarding the importance of clean water, almost 900 million people remain without safe drinking water today (Saltman, 2014). Possible ways to supply water range from collection, purification, and distribution of surface water to expanded systems of local wells using groundwater. In many cases, groundwater is the less expensive option, especially in arid regions or areas with limited funding for water distribution systems. Groundwater is also being used in increasing amounts for irrigation, which pits humans against themselves as they try to find water to drink and, at the same time, water to grow crops. Is there really enough water; can a finite mineral commodity really be a "right"? These are important questions and part of the answer lies in the geology of groundwater.

---

overgrazing, improper cultivation, and mining without proper reclamation. Salinization of soils is a problem in areas of excessive irrigation, where soluble salts concentrate near the surface as the water moves upward by capillary action and is evaporated. Major changes in soil chemistry can also be caused by the removal of vegetation, which produces lateralization, particularly the formation of iron-rich laterite soils where vegetation is removed from tropical areas. Finally, soil is the locus of most of the pollution that affects the lithosphere, as discussed in the next chapter.

Groundwater

Water is the mineral resource that we must have on a daily basis, and therefore the most essential. It is so readily available in many parts of the world that it is easy to forget its critical importance to life. Much of our water demand is hidden. The United Nations estimates that about 70% of water is used for agriculture, another 20% for industry and only 10% for domestic use. Water use has tripled since the mid 1900s and is currently increasing at the rate of 64 km^3 annually.

Most of the upper part of the continental crust is filled with water that is fresh or nearly so (Freeze and Cherry, 1979; Driscoll, 1986). The storage and movement of this water is controlled by porosity and permeability (Figure 2.3a). **Porosity** includes all types of holes in a rock, whether they are coeval with the rock or formed later. Primary porosity, which is coeval with the rock, is most common in sedimentary rocks such as **sandstone** and **conglomerate** that consist of rounded grains with as much as 30% effective porosity. Shales are usually **impermeable** because their small grains are too closely packed. Secondary porosity, which was imposed on the rock after it formed, is created largely by faulting, fracturing (which is discussed later), and chemical reactions. In many igneous and metamorphic rocks, fault zones are the only areas with high porosity and permeability.

Chemically corrosive water creates pores in otherwise impermeable rocks, such as in limestones where descending rainwaters form caves. Hydraulic fracturing or fracking, which will be discussed in later chapters, creates secondary porosity that allows production of oil and gas. The other factor that controls movement of fluids is **permeability**,

Table 2.3 Brief summary of Darcy's law, the relation that explains the movement of fluids through rocks. It says, in general, that the amount of fluid that will move through a rock is directly proportional to permeability and the pressure forcing the fluid to flow.

Darcy's law

$q = -K^* dh/dl$

Definition of terms used in Darcy's law

q = volume of fluid per unit of area per unit time

K = fluid conductivity

dh/dl = head gradient

$K = C \times F \times R$

C = constant

F = fluid term = density or viscosity of fluid

R = rock term = permeability
(proportional to square of average diameter of grains)

which describes the degree to which pores connect together to form pathways for migrating fluids (Figure 2.3b).

Rocks with high porosity and permeability that contain water are **aquifers** (Figure 2.4). The flow of underground fluids is described by Darcy's law (Table 2.3), which says that the amount of fluid that will move through a porous rock per unit of time is directly proportional to head gradient and fluid conductivity. The head gradient reflects the elevation of fluid along the flow path and any additional pressure forcing it along. Fluid conductivity describes both the rock and the fluid, and is known as **hydraulic conductivity** if the fluid is water. The rock is described by its permeability and the fluid by its density and viscosity.

Groundwater comes from **recharge**, the fraction of precipitation that flows into the ground. This water flows downward through an unsaturated zone, known as the vadose zone, to the **water table**, beneath which rock is saturated and can constitute an aquifer. Groundwater aquifers range from regolith and sediments just below the surface to sedimentary layers or fracture zones that extend to depths of many kilometers. Recharge can also come from seawater or waters made saline by evaporation in desert areas. Saline and freshwaters do not mix easily and less-dense freshwater usually floats on more dense saline water (Figure 2.4).

Where groundwater aquifers are undisturbed, they reach an equilibrium between the amount of recharge and the amount of water exiting to the surface at springs, lakes, and streams. Our use of groundwater has disrupted this

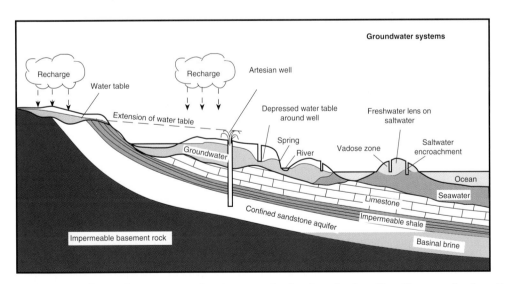

Figure 2.4 Schematic illustration of groundwater systems showing unconfined and confined aquifers. The unconfined aquifer consists of a layer of glacial deposits that covers the entire surface. Note that the water table mimics surface topography and forms lakes, springs, and rivers where it meets the surface. Freshwater floats on saltwater where the aquifer extends beneath sea level. A confined aquifer, which is recharged in the hills to the left, supports an artesian (flowing) well. As long as the water table in the confined aquifer is above the top of the well, it will continue to flow.

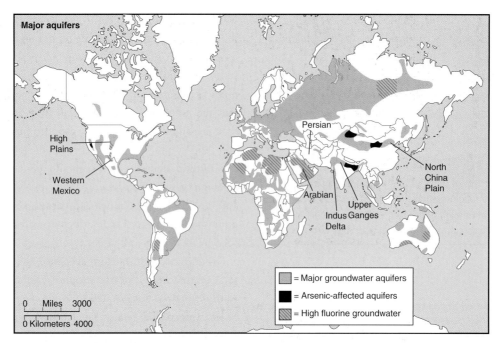

Figure 2.5 Global distribution of major groundwater aquifers showing named aquifers with the highest stress related to overuse, and those with arsenic and fluorine contamination (based on data of Kemper *et al.*, 2005; Amini *et al.*, 2008a, b; Gleeson *et al.*, 2012)

equilibrium, leading to problems in some areas. Most commonly, excessive withdrawal lowers the local water table, sometimes drying up shallow wells in the vicinity. If it occurs over a large area, it can cause regional water-table declines and groundwater depletion (Figure 2.5). In a recent study, the US Geological Survey estimated that groundwater depletion in the United States totaled about 1,000 km³ between 1900 and 2008 (Konikow, 2010). Global depletion is even worse, having increased by more than 100% between 1960 and 2000 to about 283 km³/year, an amount estimated to be about 2% of global recharge (Wada *et al.*, 2010).

In coastal regions and on ocean islands, excessive withdrawal causes **saltwater encroachment**, a process by which seawater invades and contaminates the freshwater aquifer. This can even occur relatively far inland, as in the eastern part of San Francisco Bay, where excessive withdrawals have been made for agriculture and pipeline export to the southern parts of the state. In areas like southern Florida, saltwater encroachment is aggravated by ocean-access canals for pleasure boats, which permit influx of seawater from above. Irrigation also exacerbates saltwater encroachment and salinization when evaporation increases the salt content of irrigation waters that recharge local aquifers. Excessive withdrawal also causes subsidence of the land surface (Figure 2.6). **Subsidence** of only a few inches can cause big problems by greatly enlarging areas subject to flooding and by reversing the slope of sewer and water pipes. In areas underlain by

shallow cave systems, such as central Florida, excessive withdrawal of groundwater removes support for rock, causing sinkholes.

Water shortages are widespread around the world but are most acute in northern Saudi Arabia, northwestern Mexico, Iran, and the Upper Ganges River area (Standish-Lee *et al.*, 2005; Gleeson *et al.*, 2012). Where surface water is not available at all, especially in Kuwait and Saudi Arabia, water supplies are obtained by desalinization of seawater and saline groundwater (Murtaugh, 2006; Green, 2013). Desalinization, although very effective, costs up to ten times more than current water rates in the United States and can only be used in countries with excess energy (Younos, 2005).

An obvious solution to water shortages is importation. Long-distance water transfers, whether surface or groundwater, have become more and more common. The pioneer in all of this is southern California, which imports water from the northern part of the state, as well as the Colorado River (Jacobs, 2011). Most residents of wetter parts of the world are not enthusiastic about such long-distance water transfers. The logic of this stance is puzzling in view of our willingness to transfer oil, gas, and other fluids around the world. No one says that all of us have to live near the oil wells or the milk production. The same logic probably applies to water, as long as we can afford to move it to market and it can be done without disrupting natural flow patterns and recharge.

Figure 2.6 Removal of groundwater resulted in 5.5 m of subsidence north of Luke Air Force Base near Glendale, AZ, between 1957 and 1991. H. H. Schumann, who provides scale for the photo, is 1.88 m tall (photo courtesy of H. H. Schumann and Associates).

Sedimentary ore-forming processes

Sedimentary processes, all of which happen at and near Earth's surface, can be divided into three types. **Clastic sedimentation** deposits grains of rocks and minerals. The most obvious mineral deposits that form this way are stream gravels and beach or dune sands. The best known mineral deposits resulting from clastic sedimentation are *placer deposits*, which form where heavy minerals are dropped by water or wind that passes an obstruction or diversion (Figure 2.7). The most familiar placer deposits contain gold, which has a specific gravity about six times greater than average rock. Other placer deposits contain diamond, magnetite, and zircon. Placer deposits that have been buried and preserved in older sedimentary rocks are known as paleoplacers.

Chemical sedimentation is an important ore-forming process only where the influx of clastic sediment is very limited, such as in isolated arms of the sea, inland lakes, or the open ocean far from land. The most widespread ores that form by chemical sedimentation are **evaporites**, which consist of minerals deposited by evaporation of water. **Marine** evaporites, which contain elements that are abundant in seawater, are our main source of salt, gypsum, potassium, and bromine. Non-marine evaporites form in arid regions of the continents where water flows into enclosed valleys with intermittent saline lakes and evaporates to form salt pans known as **salars** or **playas**. The composition of lake-hosted, or **lacustrine**, evaporites usually reflects the composition of rock in the local drainage basin, and includes a complex array of sodium, magnesium, boron, and lithium minerals. Pore spaces in buried evaporite deposits and their host clastic sediments contain aqueous solutions more saline than seawater, which are known as **brines**, and which contain high concentrations of similar elements.

Organic sedimentation also forms mineral deposits. Coral reefs, which consist largely of the calcium carbonate minerals calcite and aragonite, are mined for limestone. Skeletal material in other organisms consists of the calcium phosphate mineral, apatite, which is also mined. Some mineral deposits form by modification of organic material in sediments. For instance, chemical changes associated with burial beneath younger sediment transform organic matter rich in carbon and hydrogen into coal, oil, and natural gas. Bacteria facilitate these changes, usually because they derive energy from them. For instance, some bacteria convert dissolved sulfate ions (SO_4^{-2}) into H_2S gas, which is later reduced to the elemental form, known as native sulfur. Many such changes take place below Earth's surface and merge with the subsurface processes discussed below.

2.3.2 Subsurface ore-forming processes

Subsurface processes that form mineral deposits involve aqueous, hydrocarbon, and magmatic fluids. Aqueous fluids consist of water and dissolved solids and gases, including freshwater, steam, and brine. Hydrocarbon fluids consist of crude oil and natural gas, and magmatic fluids include magmas of all compositions. Aqueous and magmatic fluids form mineral deposits in two ways; either they precipitate ore minerals from aqueous solution or crystallize them from magma, or they form fluid concentrations that constitute ore in their own right, such as groundwater, oil, and gas accumulations, and some special metal-rich magmas.

BOX 2.6 | **GROUNDWATER MINING**

Excessive withdrawal constitutes groundwater mining. The Ogallala aquifer in the High Plains area of the United States (Figure 2.5) contained at least 4,000 trillion liters of water before large-scale farming began in the late 1800s (Gutentag *et al.*, 1984). Since that time, about 330 km^3 of fossil groundwater, which accumulated over the last 13,000 years, has been withdrawn. At present withdrawal rates, groundwater resources in the southern third of the area will be exhausted by 2040 (Scanlon *et al.*, 2012). The continued fall of the water table has put some farmers out of business because of the increased cost of pumping water from these great depths and has caused the US Internal Revenue Service to allow a "cost-in-water" depletion allowance, an approach similar to the depletion allowance for conventional minerals that is discussed in Chapter 6.

A similar situation is underway at the foot of the Himalaya Mountains in the Indian states of Rajasthan, Punjab, and Haryana, which are home to over 100 million people and large-scale agricultural production. Recent studies show that water withdrawals greatly exceed recharge causing groundwater levels to drop at a rate of about 4 cm per year (Rodell *et al.*, 2009). Continued production at this rate will deplete groundwater sources within a few decades. Largely in response to groundwater mining of this type, many states and smaller jurisdictions have systems to control withdrawals from large aquifers.

BOX 2.7 | **THE LOS ANGELES WATER SYSTEM**

The Los Angeles Department of Water and Power (LADWP) provides 180 billion gallons (680 million liters) of water to 3.8 million people and is the largest municipal water system in the United States. When the city was first established, its water was supplied by a private company, which was taken over by the city in 1898. Since then, the system has grown to draw water from most of central and southern California as well as the Colorado River. The first major water transfer came via the Los Angeles Aqueduct, built in 1913, which collects surface and groundwater in the Owens Valley on the east side of the Sierra Nevada. This was followed by the 390-km Colorado River Aqueduct and the 1,130-km California Aqueduct, which draws water from the western Sierra Nevada and the Sacramento River. The LADWP system includes 114 storage tanks and reservoirs, 78 pumping stations and 11,600 km of pipe. In 2013, 43% of its water came from the California Aqueduct, 37% from the Los Angeles Aqueduct, 11% from groundwater in the greater Los Angeles area, 8% from the Colorado River, and 1% from recycled water.

Discussion of these subsurface fluids and their ore-forming processes requires a brief review of terminology. **Groundwater**, for instance, really means all water in the ground, but it is widely applied to near-surface water that is consumed for municipal, agricultural, and other uses, as is done here. **Hydrothermal solutions**, which are hot groundwaters, are usually distinguished as a separate type of fluid. The distinction between groundwater and hydrothermal solutions is reflected in the terms **supergene**, which refers to water that moves downward through the crust, and **hypogene**, which refers to water that moves upward. In general, hypogene water is of hydrothermal origin, whereas supergene water is rain and snow on its way down into the crust. Finally, magmas and their role in formation of mineral deposits will be discussed separately. Although this division is imperfect and imposes categories where nature has not, it allows us to discuss crustal fluids in a way that reflects their relation to mineral resources.

Hydrothermal ore-forming systems

Hydrothermal systems form ore deposits by precipitating minerals from **hydrothermal solutions**, which consist largely of water with variable amounts of CO_2, H_2S, NaCl, silica, and carbonate (Kesler, 2005). Much of what we know about hydrothermal solutions comes from the study of **fluid inclusions**, which are small imperfections in crystals that trap the fluid from which they were growing (Figure 2.8a). These tell

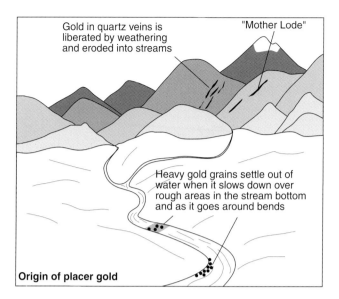

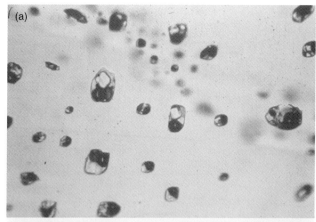

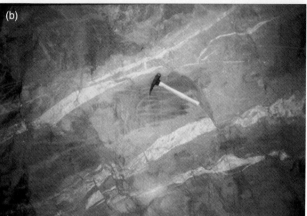

Figure 2.7 Placer deposits are concentrations of heavy clastic grains in stream, lake, or ocean sediment. Gold placer deposits form when grains of gold are liberated from quartz veins by weathering and erosion carries them downstream to be deposited where the water slows.

us that hydrothermal solutions range in temperature from less than 100 °C to at least 600 °C and that many of them contain small but important amounts of dissolved metals such as copper or gold. Hydrothermal solutions react with surrounding rock as they flow, causing changes in the mineral composition known as **hydrothermal alteration** (Figure 2.8b). The widespread distribution of hydrothermal alteration and mineral deposits proves that hydrothermal systems were abundant in the crust.

Hydrothermal flow systems form ore deposits because they pass large volumes of fluid through relatively small zones where the dissolved metals or other minerals are concentrated relative to their background concentration in the crust. Deposition of ore minerals from hydrothermal solutions can take place by open-space filling or replacement. Open-space filling refers to the precipitation of minerals in existing porosity. If it is flowing through a fracture, the new minerals can form a **vein**. Where rock is highly fractured and consists of many intersecting veins and veinlets, it is known as a **stockwork**. Replacement refers to the process by which new minerals actually take the place of preexisting minerals almost atom for atom. This process involves many materials in addition to ores. In petrified wood, carbon and hydrogen have been replaced in this way by opal, chalcedony, or some other form of silica.

Hydrothermal fluid systems form in very distinct geologic and plate-tectonic environments (Figure Box 2.3.1) and can be

Figure 2.8 (a) Fluid inclusions inside a crystal of transparent quartz from the Granisle porphyry copper deposit, British Columbia, as seen through a microscope. Each inclusion is about 30 micrometers in diameter and they contain transparent saline water, a vapor bubble (dark) and several daughter minerals including hematite (red) and halite (clear cubes). The inclusion was originally trapped as a liquid, but during cooling the liquid shrank to form the vapor bubble and deposit the daughter minerals. If the quartz crystal is reheated under the microscope the vapor bubble will shrink and the daughter mineral will dissolve, providing information on the temperature at which the liquid was trapped in the inclusion (photograph by the authors). (b) Hydrothermal alteration (light zones) in wallrock around quartz veins (white) at the Mt. Charlotte gold deposit, Kalgoorlie, Australia (photograph by the authors). See color plate section.

divided into meteoric, seawater, basinal, magmatic, and metamorphic types (White, 1974). Meteoric hydrothermal systems are made up of water from rain or snow that recharged into the crust and went deep enough to become heated. Meteoric flow systems usually involve water that percolates downward over a wide zone, is heated, and becomes buoyant enough to rise along more focused zones (Figure 2.9). These systems are best developed in volcanic areas, where shallow intrusions underlying the volcanoes provide heat to drive convection of

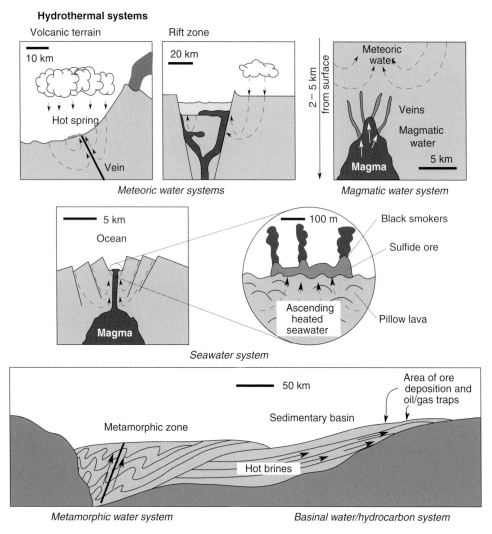

Figure 2.9 Schematic illustration of geologic environments for major types of hydrothermal systems. The relation between these systems and larger plate-tectonic environments is shown in Figure Box 2.3.2.

the water, and they are frequently associated with geothermal systems (Norton, 1978; Heasler *et al.*, 2009). Where the intrusion is near the surface, meteoric systems often form hot springs (Figure 2.10a) or episodic fountains of hot water known as **geysers**. Temperatures in such systems reach a maximum of about 350 °C at depths of a kilometer or so, but many are only 100–200 °C. In areas with no volcanic activity, even cooler meteoric systems form along large faults where descending water becomes buoyant by the natural increase in heat downward in the crust, known as the **geothermal gradient** (see Figure 4.16). The most common type of mineral deposits formed by meteoric hydrothermal systems are epithermal Au–Ag deposits and roll-front U deposits.

Seawater hydrothermal systems are the submarine equivalent of meteoric hydrothermal systems in the subaerial environment (Figure 2.9). They usually form along mid-ocean

ridges at divergent plate margins where seawater flows downward into hot rocks surrounding basalt magma chambers (Cathles, 1981; Hannington *et al.*, 2005; Tivey, 2007). Heated seawater returns to the surface to form spectacular hot-spring vents that have been observed by research submarines (Figure 2.10b). Water that leaves the vents at temperatures as high as 350 °C is prevented from boiling by high pressures at the 3-km water depths. The hot-spring water contains dissolved metals and H_2S, which react to precipitate minute particles of metal sulfide as the vent water mixes with cold seawater, a process that has earned them the name **black smokers**. Although seawater hydrothermal systems can form veins, they usually form sediments called exhalative deposits. The most common are volcanogenic massive sulfide (VMS) deposits that consist of layers of massive Fe, Cu, Zn, and other metal sulfides on the seafloor. By reacting with ocean crust,

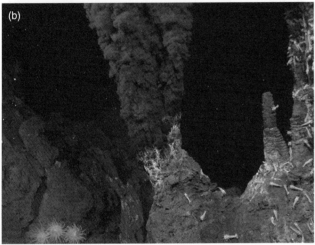

Figure 2.10 Hydrothermal systems. (a) The famous Pamukkale travertine (calcite) deposits in Turkey were produced by warm, meteoric (~35 °C) water flowing from springs that have been active for at least the last 55,400 years. The Greco-Roman city of Hierapolis, which was built at the springs, has been partly enveloped by younger deposits. Epithermal precious-metal deposits in the same area of Turkey are probably deeper extensions of systems such as this one (photograph courtesy of David Stenger). (b) Submarine hydrothermal "black smoker" vent surrounded by shrimp (white) at the Mid-Cayman rise in the Caribbean Sea. Water expelled from this vent at a temperature of 400 °C mixes with cold seawater to precipitate fine-grained black particles of metal sulfide that form the "smoke" (photo by Chris German, WHOI/NASA/ROV Jason © 2012 Woods Hole Oceanographic Institution). See color plate section.

seawater hydrothermal systems have also controlled the composition of seawater through time (Kump, 2008).

Basinal fluid systems (Figure 2.9) include all fluids in or derived from sedimentary basins such as the thick prism of sediment accumulating off the Gulf coast of Louisiana and Texas (Cathles and Adams, 2005). Seawater that was trapped when the sediments were deposited is known as connate

water. Basins also contain **meteoric water** that flows in from above and some water is released when clays and other hydrous minerals undergo low-temperature metamorphism caused by burial of the sediment. Changes in organic material during burial also release oil and gas. During burial, water moves through sediments, becoming formation water, a non-specific term for water in sedimentary rocks. Formation water that is very saline is known as basinal brine (Hanor, 1987). Basinal waters frequently circulate to deep enough levels to reach temperatures of 200 °C and the brines, in particular, are potent solvents because they contain abundant chloride that combines with metals to form soluble complex ions. The flow of basinal fluids takes place largely in response to gravity and the geological evolution of the basin. Compaction of sediment during burial expels basinal fluids, and tilting and deformation of the basin causes them to flow out the low side, often pushed by recharge from the high side. During this outward fluid migration, some oil and gas accumulates in traps and some flow to the surface to form the tar pits discussed in Chapter 7. Outward migrating basinal brines also form Mississippi Valley-type (MVT) Pb–Zn deposits where they intersect suitable chemical traps and SEDEX Pb–Zn deposits where they are expelled onto the seafloor.

Magmatic hydrothermal systems consist of fluids that separate from a magma as it cools and crystallizes at depths of several kilometers (Figure 2.9). All magmas contain dissolved water, as well as gases such as CO_2, SO_2, H_2S, and HCl. When the magma crystallizes, most of these gases remain in the magma, gradually exerting greater pressure as their concentration increases. When this pressure exceeds that of the surrounding rocks, the water and gases exsolve from the magma to become magmatic hydrothermal solutions with temperatures of 600 °C or more. These solutions are enriched in chloride, sulfur, and metals, which also concentrate in residual magmas (Holland, 1972; Candela and Piccoli, 2005; Simon and Ripley, 2011). Magmatic hydrothermal systems are best developed in and around felsic intrusions (which dissolve more water than mafic intrusions) and their most common product is porphyry Cu deposits. Where felsic magmas intrude limestone, reactions at and near the contact form skarn, a rock consisting of calcium silicates with metal oxide and sulfide minerals.

Metamorphic hydrothermal systems consist of fluids that were expelled from rocks during prograde metamorphism (Figure 2.9). As temperature and pressure increase during metamorphism, minerals such as clay, mica, and calcite, react to form minerals like feldspar, releasing H_2O and CO_2

BOX 2.8 | GEOCHEMISTRY OF HYDROTHERMAL SOLUTIONS

It might have occurred to you in reading about the types of hydrothermal systems that water is water. How can one water molecule be distinguished from another and so how can we say where the water came from? The best way to do this is by analysis of the isotopic composition of the water, which involves measurement of the relative abundances of isotopes of hydrogen and oxygen that make up water. Water in different geologic environments has characteristic hydrogen and oxygen isotopic compositions (Figure Box 2.8.1). In fossil hydrothermal systems where the water is long gone, information can be obtained from fluid inclusions or minerals that contain hydrogen and oxygen from the parent hydrothermal solution. These measurements have provided important insights into the nature of hydrothermal systems, but they are not foolproof. Complications result because isotopic compositions of waters of different types overlap and, worse still, because the isotopic composition of many minerals is changed by reaction with waters that pass by after the deposits form. Thus, efforts are still underway to sort out the distribution and history of fluid movement in Earth's crust.

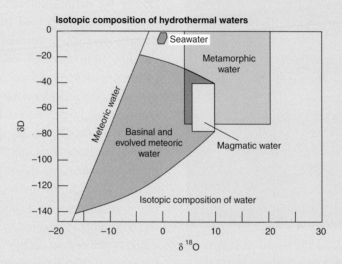

Figure Box 2.8.1 Oxygen and hydrogen isotopic compositions of some important types of hydrothermal systems. $\delta^{18}O$ indicates the ratio of ^{18}O to ^{16}O in the sample versus that in a standard and δD indicates the same ratio of ^{1}H to ^{2}H (deuterium). These isotopes are stable (that is, they are not produced by radioactive decay), but their relative abundances in water and other phases are affected by physical and chemical reactions involving oxygen and hydrogen.

(Fyfe, 1978; Elmer et al., 2006). Sulfide minerals break down to release H_2S and metals, and organic matter releases methane and other organic molecules (Pitcairn et al., 2006). The resulting fluid formed by these reactions is stored in the rock under pressure, and is released during episodic faulting events (Sibson et al., 1988). The abundant CO_2 in upward migrating metamorphic fluids commonly causes formation of carbonate minerals in overlying rocks. Fluids of this type are thought to form orogenic gold deposits, which are quartz veins that occupy continent-scale faults along convergent margins.

You might also have wondered why a hydrothermal solution that had dissolved metals or other elements and transported them for a great distance would deposit them. That is, what causes hydrothermal fluids to actually form

hydrothermal mineral deposits? Most minerals become less soluble with decreasing temperature; thus, they should precipitate as the solutions rise from deep, hot levels of the crust toward the cooler surface environment. However, cooling occurs gradually over a long flow path and usually produces widespread, dilute mineral concentrations rather than rich ore deposits. More localized chemical changes that produce concentrated ore deposition occur when the hydrothermal solution boils, mixes with another type of water, or reacts with rocks such as limestone (Figure 2.11).

Magmatic ore-forming systems

Although some igneous rocks such as crushed stone and dimension stone are mineral deposits in their own right,

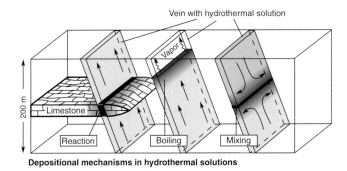

Depositional mechanisms in hydrothermal solutions

Figure 2.11 Processes that deposit ore minerals from hydrothermal solutions include reaction between solution and chemically reactive wallrock such as limestone, boiling that releases gases from the hydrothermal solution, and mixing with descending, cool meteoric waters.

most magmas must do more than simply crystallize to form mineral concentrations that are of economic interest. The processes that form these concentrations usually involve either **crystal fractionation** or magmatic **immiscibility** (Figure 2.12).

Crystal fractionation involves the selective concentration of a particular mineral during magma crystallization. Magmas crystallize over a range of temperatures, with different minerals crystallizing in different parts of this range. If early crystallized minerals sink to the bottom of the magma chamber before others crystallize, they might form a type of "magmatic sediment" known as a **cumulate** layer. Processes of this type are particularly important in large mafic intrusions known as **layered igneous complexes**. The largest of these is the Bushveld Complex in South Africa, a major source of chromium, platinum, and vanadium.

Another way to concentrate elements from a magma is to have it split into two separate, immiscible magmas of different composition at some point in the crystallization history. In all cases, the main magma is a relatively typical silicate magma such as forms most igneous rocks. The other magma, however, can be rich in metal sulfides or metal oxides. This process is known to happen in mafic and ultramafic magmas, where it produces immiscible magmas consisting largely of iron–nickel–copper sulfides with platinum-group elements or iron–titanium oxides (Arndt *et al.*, 2005). Experiments have shown that it can also happen in some felsic magmas, which generate an immiscible iron oxide-rich magma (Naslund, 1983), although field evidence for this process is debated. Immiscible magmas are heavier than the silicate magma and would usually sink to the bottom of their magma chamber.

A few magmas contain enough valuable minerals to be extracted in bulk. **Kimberlites** and **lamproites**, some of which are the main source of the world's diamonds, are potassium-

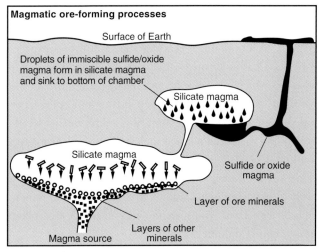

Figure 2.12 Schematic diagram showing magmatic processes that form mineral deposits by crystal fractionation and immiscibility. (All processes probably do not take place in the same magma.)

rich, ultramafic fragmental rocks that originated in the mantle. **Carbonatites**, which are sources of vanadium, columbium, and tantalum, are igneous rocks that consist almost entirely of calcite and other carbonate and phosphate minerals. Both of these unusual igneous rocks are usually found in old cratons.

2.3.3 Distribution of mineral deposits in time and space

Mineral deposits are sensitive indicators of changes in Earth history. For instance, during Archean time the upper mantle was hotter and partial melting took place at higher temperatures, extracting unusual ultramafic magmas that formed rocks known as **komatiite**. Komatiite magmas were rich in sulfur and nickel and formed immiscible magmatic sulfide magmas more readily than the basalt magmas that come from the mantle today. By late-Archean and early-Proterozoic time, sizeable continental masses had developed. Erosion of them produced large sedimentary basins with shallow shelf areas. Where these basins are preserved, they contain large chemical and clastic sedimentary deposits, including much of Earth's iron, gold, and uranium ores. These deposits also reflect the changing composition of the atmosphere because they apparently did not form after the Great Oxidation Event when the oxygen content of the atmosphere became high enough to change the geochemical behavior of iron and uranium at the surface (Holland, 2005).

By Phanerozoic time, the continents had grown large enough to host even bigger sedimentary basins, which contained limestone and evaporite minerals. The appearance of marine organic life enhanced the probability of forming crude oil and natural gas and the appearance of vascular land plants enabled the formation of coal deposits. Hot brines expelled from these basins formed Mississippi Valley-type lead–zinc deposits, which are rare in pre-Phanerozoic rocks. The increased concentration of oxygen in the atmosphere permitted more effective dissolution of uranium during weathering of the rocks and allowed the creation of roll-front-type uranium deposits related to groundwater movement. Finally, numerous collisional margin volcanic arcs developed on the newly formed continents, leading to the formation of mineral deposits associated with meteoric and magmatic hydrothermal systems (Goldfarb *et al.*, 2001).

The resulting pattern of ore deposits through time includes concentrations in space that are known as metallogenic provinces and concentrations in time (Figure 2.13) that are known as metallogenic epochs (Wilkinson and Kesler, 2009). Both concentrations are closely linked to the supercontinent cycle in which continents merged to form large masses and then broke apart again (Barley and Groves, 1992; Kerrich *et al.*, 2005).

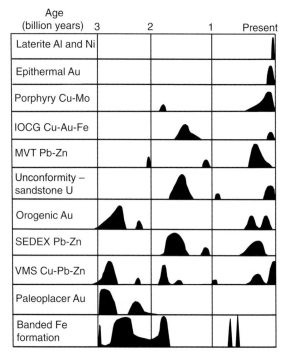

Figure 2.13 Formation and preservation of important types of mineral deposits through geologic time. Deposits at the top of this figure formed at progressively deeper levels in the crust and therefore older deposits have been preserved from erosion. Deposits at the bottom of the figure formed in the ocean and were preserved in part by younger sediments. IOCG = iron oxide–copper–gold (Modified from Kesler and Wilkinson *et al.*, 2009; Bekker *et al.*, 2010; Huston *et al.*, 2010; Goldfarb *et al.*, 2010).

Conclusions

It should be apparent from this discussion that our mineral endowment is fixed by geologic factors. In other words, the mineral endowment of any country is a function of its geologic setting. The good news from this is that we can use geologic relations to predict the type of undiscovered mineral deposits that might be present in an area. The bad news is that no amount of government action or popular effort will change these geologic constraints. The ray of hope in all of this is that we do not yet know everything about the geology of the world or about every type of mineral deposit that it contains. Thus, if we continue to study Earth's geology and explore for mineral concentrations, we can hope that there will be some pleasant surprises that will help us with the growing problem of dwindling resources.

CHAPTER

3 Environmental geochemistry and mineral resources

Most environmental debates require an understanding of the fundamental principles of geochemistry that govern the movement of pollutants. These principles control environmental decisions in the same way that geological principles control the distribution of mineral deposits. Failure to understand them creates the risk of advocating impossible solutions, such as disposing of radioactive waste by burning it. Thus, before beginning our discussion of mineral resources, we must review the chemical controls that determine their environmental effects.

3.1 Principles of environmental geochemistry

Just in case you have forgotten some of your high school science, here is a quick review. The fundamental building blocks of Earth are **atoms**, which consist of a positively charged nucleus surrounded by a cloud of negatively charged electrons. The nucleus is made up of **neutrons** with no charge and **protons** with a positive charge equal to that of an electron. Different atoms, each with a specific number of protons, make up the **elements**, which are arranged in the periodic table (Figure 3.1). Elements lose or gain electrons to become ions. Those with fewer electrons than protons have a net positive charge and are known as **cations**; those with more electrons than protons have a net negative charge and are known as **anions**. Ions that lose electrons are said to be oxidized, whereas those that gain electrons are reduced. The valence or **oxidation** state of the ion refers to the number of electrons that it has lost or gained. Elements with different numbers of neutrons but the same number of protons are known as **isotopes**. They are commonly referred to by their

number of protons plus neutrons, or mass number, and are written in the form ^{235}U.

Atoms form **chemical bonds** with other atoms to produce **compounds**, substances with a definite chemical composition. A group of atoms is known as a **molecule** if it has no net charge, such as CO_2, or a **complex ion** if it has a net charge, such as NH_4^+ or CO_3^{2-}. Compounds are classified as organic if they contain carbon in association with hydrogen, oxygen, nitrogen, sulfur, phosphorous, chlorine, or fluorine, or inorganic if they do not contain carbon. Compounds that form during processes occurring in Earth are considered natural and those that are made in the laboratory are synthetic.

Matter undergoes physical, chemical, and nuclear changes, all of which can be described by an equation with reactants on one side and products on the other. When undergoing physical changes, a compound retains the same chemical composition and simply changes from one state to another, as from liquid to vapor when water is boiled. When undergoing chemical changes, reactants change composition to form products with new chemical compositions, as when methane oxidizes (burns) in oxygen to form carbon dioxide and water, or when pyrite reacts with air-saturated water to form acid, sulfate, and ferric iron ions in solution. Physical and chemical changes are subject to the law of conservation of matter, which states that these reactions do not create or destroy matter.

Some substances are **stable** in their physical and chemical environment and others are **unstable**. In natural systems, it is not possible to determine whether a substance is stable without considering all aspects of its physical and chemical environment. For instance, liquid water is stable at atmospheric pressure and surface temperature, but converts to vapor above

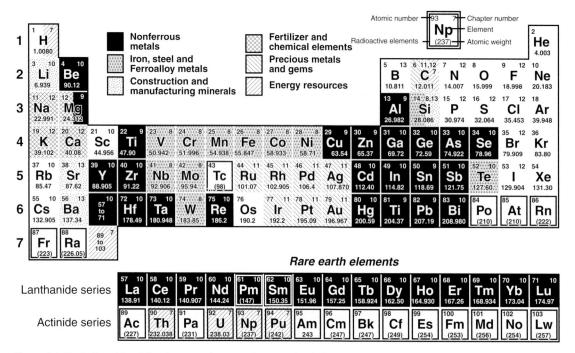

Figure 3.1 Periodic table of the elements showing elements divided into groups discussed in the text

---

BOX 3.1 | **THE STATE OF MATTER**

The degree of combination of atoms, molecules, and ions determines the state of matter. **Gases** contain atoms or molecules with little mutual attraction. **Liquids** consist of loosely bonded clusters of atoms, molecules, or compounds with no long-range frameworks. **Solids** consist of the same basic building blocks that are linked by stronger bonds that form ordered structures, and crystalline solids have a long-range atomic structure (see Figure 3.2). Solidified liquids such as glass lack a well-defined atomic structure and are referred to as **amorphous**. The margins of all solids consist of broken chemical bonds that can combine with ions or molecules in surrounding liquids and gases, a process known as **adsorption**, and that is where a lot of environmental problems and solutions take place.

---

a temperature of 100 °C. The iron sulfide mineral, pyrite, is stable in an argon atmosphere because it will not react with argon. But, in contact with water and oxygen, pyrite reacts to form acid water containing dissolved iron. Definitions of stability are complicated by the rate or **kinetics** of these changes. Although some reactions proceed at a very rapid rate, many do not, and some never produce stable compounds. Compounds that persist even though they are not stable are described as **metastable**. Diamond and most organic compounds, including those in your body, are metastable at Earth's surface.

Nuclear changes, which involve the nucleus of atoms, convert one isotope to another isotope. Reactions of this type include natural radioactivity, nuclear fission, and nuclear fusion. All nuclear changes obey the law of conservation of matter and energy, which states that the sum of matter and energy in the reaction is constant. Of the 2,000 or so known isotopes, only 266 are stable. The rest are naturally **radioactive** and have a finite, measurable probability of forming a different isotope by radioactive decay. The fixed period of time required for half of the amount of the radioactive isotope to decay is known as the **half-life**. Half-lives

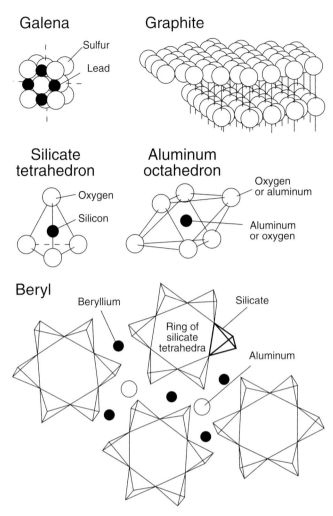

Galena

Sulfur

Lead

Graphite

Silicate tetrahedron

Oxygen

Silicon

Aluminum octahedron

Oxygen or aluminum

Aluminum or oxygen

Beryl

Beryllium

Ring of silicate tetrahedra

Silicate

Aluminum

Figure 3.2 Crystal structure of galena (PbS) and graphite (C). Also shown are the silicate tetrahedron and aluminum octahedron, which are the basic crystal units of many silicate minerals. Beryl ($Be_3Al_2Si_6O_{18}$), a silicate mineral, consists of rings of six silicate tetrahedra that are linked by beryllium and aluminum.

vary from milliseconds to billions of years for the different isotopes.

Products of radioactive decay include **alpha particles**, which are the nuclei of helium atoms, **beta particles**, which are electrons or positrons, and gamma rays, which have very high-energy radiation, with an even shorter wavelength than X-rays. These decay products are called **ionizing radiation** because they dislodge electrons from atoms in surrounding compounds, damaging their structures and producing reactive ions. Nuclear **fission**, the basis for presently used nuclear power, takes place when isotopes of elements with large mass numbers, such as ^{235}U, are struck by neutrons and split into isotopes of lower mass number, more neutrons, and energy. Most of the isotopes created by nuclear fission are radioactive and undergo further decay, creating our problems with

nuclear waste disposal. In nuclear **fusion** light isotopes combine to form a heavier isotope plus energy, as when hydrogen and tritium combine to form helium. This reaction takes place in stars, but controlled fusion remains an elusive goal for civilization (Rafelski and Jones, 1987; Waldrop, 2014).

3.2 Geochemical reservoirs – their nature and composition

Earth can be divided into regions of relatively similar average composition, which are known as **reservoirs** (Horne, 1978). The largest of these are the atmosphere, hydrosphere, biosphere, and lithosphere, which have very different compositions (Table 3.1), and which can be further divided into smaller reservoirs such as mineral deposits and human bodies. Environmental geochemistry deals with the chemical interaction between the biosphere and any of the other reservoirs, and the environmental importance of a substance is determined by its effect on life. Access of a substance to the biosphere is controlled largely by its ability to be concentrated in the lithosphere, particularly in the soil, to dissolve in the hydrosphere, or to evaporate into the atmosphere. The capacity of substances to move from one reservoir to another is termed mobility, and the degree to which substances concentrate in living organisms is known as **bioaccumulation**.

3.2.1 Lithosphere

The lithosphere is where it all started. The original Earth probably consisted only of the lithosphere. The atmosphere and hydrosphere formed later from water and other gases released by magmas and impacted on the surface by comets, and the biosphere probably began with the formation of organic compounds in the early hydrosphere. The composition of all of these reservoirs has continued to evolve with time. The lithosphere, for instance, separated into the core, mantle, and crust, all of which continue to undergo changes induced by plate tectonics.

The most important part of the lithosphere from the standpoint of environmental geochemistry is **soil**, which is in intimate contact with the other reservoirs. Soil is a **heterogeneous mixture** of water, minerals, and living and dead organic matter. Soil minerals include quartz, calcite, iron–manganese oxides, micas, and clay minerals. Of these, clay minerals are most important in terms of both abundance and reactivity. Clay-mineral structures are based on the **silicate tetrahedron**, a pyramid-shaped unit consisting of a silicon ion bonded to four oxygen ions, and a similar octahedral unit with a central

Table 3.1 Ten most abundant elements in Earth's major reservoirs. The composition given for the biosphere includes water in living organisms; the composition for the lithosphere is that of the crust. In addition to the single-element gases tabulated here, the atmosphere also contains 0.25% H_2O, 1.79 ppmv CH_4, 0.325 ppmv NO, 400 ppmv CO_2, 0.002% SO_2, 0.02 ppmv NO_2, 0.00006% H_2S and 0.04 ppmv O_3.

Atmosphere		Hydrosphere		Biosphere		Lithosphere	
(All abundances in parts per million (ppm))							
Nitrogen	780,840	Oxygen	857,000	Oxygen	523,400	Oxygen	464,000
Oxygen	209,500	Hydrogen	108,000	Carbon	302,800	Silicon	282,000
Argon	9,300	Chlorine	19,000	Hydrogen	67,500	Aluminum	82,300
Carbon	90.05	Sodium	10,500	Nitrogen	5,000	Iron	56,000
Neon	18.18	Magnesium	1,350	Calcium	3,700	Calcium	41,000
Krypton	1.14	Sulfur	885	Potassium	2,300	Sodium	24,000
Helium	5.24	Calcium	400	Silicon	1,200	Magnesium	23,000
Hydrogen	0.55	Potassium	380	Magnesium	980	Potassium	21,000
Xenon	0.09	Bromine	65	Sulfur	710	Titanium	5,700
Sulfur	0.5	Carbon	28	Aluminum	555	Hydrogen	1,400

aluminum or magnesium ion and six oxygen or hydroxyl (OH^-) ions. These units combine to form a wide variety of crystal structures on which the many silicate minerals are based, including the sheets that make up the clay minerals (Figure 3.3).

Clay minerals are divided into three major groups, the kaolinite group, which contains alternating sheets of tetrahedra and octahedra, the smectite group (including the common clay mineral montmorillonite), which contains an octahedral sheet sandwiched between two tetrahedral sheets, and the illite group, which also contains an octahedral sheet sandwiched between two tetrahedral sheets (Figure 3.3). The composition of kaolinite is essentially constant $[Si_4]Al_4O_{10}(OH)_8 \cdot nH_2O$ (where n = 0 or 4). The composition of smectite can vary tremendously, with different ions substituting for silicon and aluminum, and a general formula $[(Ca,Na,H)(Al,Mg,Fe,Zn)_2 (Si,Al)_4O_210(OH)_2 \cdot xH_2O]$, where x represents variable amounts of water. Illite, which contains more potassium, has the general formula $[(K,H)Al_2(Si,Al)_4O_{10}(OH)_2 \cdot xH_2O]$.

Exchange and adsorption involve competition for available sites on a mineral. The winners are determined by size and bonding characteristics, and by their abundance. In soil water, for instance, an ion with a large dissolved concentration has a better chance of finding a place on a soil mineral than one that is dissolved in small amounts. The hydrogen ion (H^+) competes very effectively because it is small, and it takes up most of the cation exchangeable sites on minerals in acid solutions. In

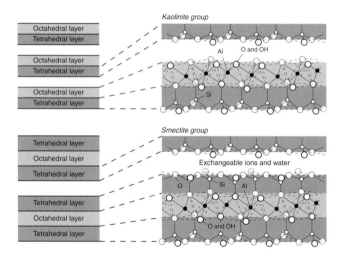

Figure 3.3 Crystal structure of the kaolinite and smectite groups (the smectite group and the illite group have almost identical structures). Exchangeable cations and water are found largely between the sheets, with montmorillonite containing much larger amounts.

neutral and alkaline soils, the competition is open to more ions and, in an interesting twist, small, highly charged metal ions have a better chance than large ions such as potassium that have a low charge. Thus, metal ions concentrate on soils even when they are present in relatively low concentrations in natural water, a factor that can create pollution problems or be used to solve them.

BOX 3.2	THE ENVIRONMENTAL IMPORTANCE OF ADSORPTION

Many clay minerals have poorly developed crystal structures and some iron oxides and manganese oxides are almost amorphous. Thus, they have a large surface area consisting of many broken bonds, which give them an enormous capacity to adsorb elements and compounds from their surroundings, a process that is the key to pollution of the lithosphere. In smectite clays, for instance, substitution of Al^{+3} for Si^{+4} in a silicate tetrahedron creates an additional negative charge that can be satisfied by substitution of another positively charged ion such as sodium (Na^+) in the structure. The new ion might be adsorbed on the surface of the mineral or it might substitute in its crystal structure. Substitution or adsorption sites can exchange one ion for another as concentrations of the ions change in the surrounding water, a process known as **ion exchange**. Clays of the smectite group are known as **swelling clays** because they adsorb water between their layers, with a resultant increase in volume (Buol *et al.*, 1989). This is why the clay vermiculite is used as a packing material inside boxes when liquids are shipped.

3.2.2 Hydrosphere

The hydrosphere consists of water that is not chemically bound in minerals, along with everything that the water can dissolve (Westall and Stumm, 1980; Broecker, 1983). Water is one of the few compounds that forms a stable solid, liquid, and vapor under surface conditions. Changes between these states form the basis for the water cycle, the pattern of **evaporation** and **condensation** that allows us to pour our filth into water and have Nature clean it up for us (Schnoor, 2014). About 96.5% of the hydrosphere consists of seawater, which contains about 3.5% (by weight) dissolved material (Table 3.1). The ocean is a major reservoir of elements, and the total ocean abundance of metals such as Mg, K, Li, Ba, Mo, Ni, Zn, U, V, Ti, Cu, Au, and Th exceeds land-based resources (Bardi, 2010). Recovery of most of these scarce elements directly from seawater is difficult because of the large amount of energy required, although selective **adsorption** methods offer promise for the future. Most of the rest of the hydrosphere is in glaciers, ice caps, and permanent snow (1.74%), fresh and saline groundwater (0.76% and 0.93%, respectively), rivers (0.0001%), fresh and saline lakes (0.007% and 0.006%, respectively), ground ice and permafrost (0.022%), soil moisture (0.001%), swamp water (0.0008%), and the atmosphere (0.001%) (USGS Water Science School, 2014).

Water is an extremely effective solvent because it consists of two hydrogen atoms that extend outward from one side of an oxygen atom. This configuration creates a molecule with a positive charge on one side and a negative charge on the other, which allows it to attract charged compounds into solution. Water also dissociates to produce ionic H^+ and OH^-, which determine its pH (the negative logarithm of the H^+ concentration). At normal surface temperatures, water has equally effective concentrations of H^+ and OH^- at a pH of 7. Natural rainwater is more acid than this because it dissolves carbon, sulfur, and nitrogen gases from the atmosphere, as discussed below. In general, acid solutions are more effective solvents than neutral solutions and most natural materials become more soluble as temperature increases.

The mobility of a substance in the hydrosphere is controlled by its **solubility**, which is the amount that can be dissolved. Water that cannot dissolve any more of a substance is said to be **saturated** in that substance. In neutral water at room temperature, saturated water dissolves up to 264,000 ppm halite (NaCl) compared to only about 20 ppm quartz (SiO_2). Many ionic substances dissociate into more than one ion when they dissolve, making their solubility dependent on the concentrations of these ions in solution. For instance, halite dissolves largely by dissociating into Na^+ and Cl^- ions, and its solubility will therefore be greater in pure water than in water with dissolved sylvite (KCl), which already contains Cl^-. Water also contains a wide range of organic and inorganic **suspended material**, largely in the form of **colloids**, which are particles with dimensions of about 1 to 1,000 nanometers. Colloids remain suspended because they have highly charged surfaces and electrostatic repulsion prevents them from aggregating into larger grains that would sink. Colloids are more abundant in freshwater; in seawater, dissolved ions such as sodium and chloride ions nullify their surface charges, causing them to aggregate into larger grains, a process known as **flocculation**.

In general, the hydrosphere is oxidizing except where special conditions have consumed all available oxygen. This takes place most commonly where water bodies become stratified. In temperate areas, lakes more than about 10 m deep are

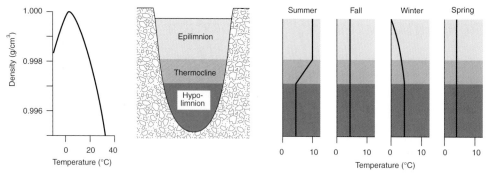

Figure 3.4 Thermal stratification in dimictic lakes, showing the change in the density of water with temperature and its relation to seasonal variation of thermal stratification. Note that stratification occurs because water is most dense at 4 °C.

stratified throughout most of the year, but undergo two periods of mixing or overturn in spring and fall (Figure 3.4). While they are stratified, the upper part of the lake, known as the **epilimnion**, mixes with the overlying atmosphere to maintain a high concentration of oxygen. The underlying **hypolimnion** is isolated from the atmosphere, however, and becomes more reducing with time. During overturn, reduced material from the hypolimnion is oxidized by contact with the atmosphere (Horne, 1978). The overall evolution of the hydrosphere also reflects a gradual change from reducing to oxidizing conditions. Elements with more than one natural oxidation state, such as sulfur and iron, were reduced in the early ocean, whereas they are oxidized in the modern ocean (Holland, 2006).

3.2.3 Atmosphere

The atmosphere consists of gases (Table 3.1) and particulate matter known as the **atmospheric aerosol** (Eiden, 1990; Andrae and Crutzen, 1997). The concentration of gases in the atmosphere is determined by their vapor pressures and abundances, and the main atmospheric gases have high vapor pressures (Figure 3.5a). Water has a low vapor pressure and is an important part of the atmosphere only because it is an abundant liquid at Earth's surface. Organic compounds with high vapor pressures, such as ethane (C_2H_6), propane (C_3H_8), acetylene (C_2H_2) and ethylene (C_2H_2), are not naturally common at Earth's surface and are correspondingly rare in the atmosphere, although we have begun to change that by extraction and use of fossil fuels, as discussed below (Hewitt, 1998; Boynard et al., 2014). Since the Industrial Revolution, fugitive emissions during production, processing, and combustion of coal, oil, and natural gas have increased atmospheric concentrations of methane and ethane, the two most abundant hydrocarbons in the atmosphere (Dewulf and Van Langenhove, 1995; Simpson et al., 2012).

The atmospheric aerosol includes liquid particles called mist or fog, solid particles from combustion known as **smoke**, and wind-blown rocks and minerals known as dust (Figure 3.5b). **Smog** consists of atmospheric gases that were condensed into particles by photochemical processes based on light energy. SO_2, for instance, reacts with calcium to form calcium sulfate either as **dry deposition** (aerosol particles) or **wet deposition** (acid from dissolved gas). Pollen, spores, bacteria, and other organic material and condensates can be important locally, and volcanic eruptions are very important periodically (Rampino et al., 1988; Li et al., 2012). The aerosol is dominated by large dust particles in unpopulated continental areas, and by smaller combustion and gas condensate particles in populated areas (Figure 3.5b; Boynard et al., 2014). Over the ocean, the aerosol is largely ocean spray, which diminishes landward from coasts. The aerosol is removed by precipitation and sedimentation, with particles larger than about 2 micrometers in diameter forming most nuclei for raindrops, and smaller particles having a longer residence time in the atmosphere.

The size of aerosol particles plays an important role in their health effects. Particulate matter < 10 micrometers (PM_{10}) can pass into the bronchi, and particulate matter < 2.5 micrometers ($PM_{2.5}$) can pass through the lungs and into the circulatory system (Perrino, 2010). Increased concentrations of PM_{10} and $PM_{2.5}$ are correlated with an increased incidence of respiratory and cardiovascular disease (Dockery and Stone, 2007; Peréz et al., 2009). Globally, anthropogenic sources including vehicle and power-plant emissions, industrial processes, and domestic heating with biomass represent about 10% of atmospheric particulate matter. Sulfate aerosols from combustion of fossil fuels are of particular concern because they increase the number of cloud droplets that are smaller than natural condensation nuclei. Sulfate particles also reflect radiation, making clouds

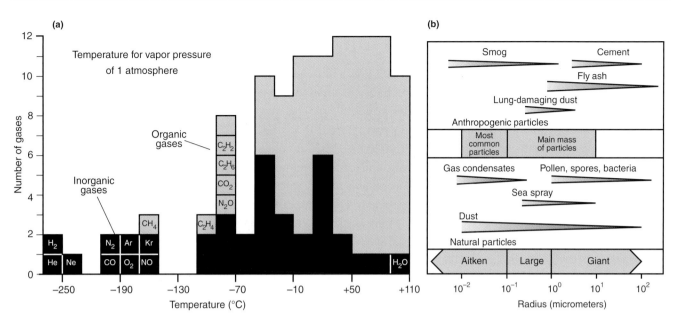

Figure 3.5 (a) Histogram of temperatures at which common gases have vapor pressure of 1 atmosphere. Inorganic gases, which make up most of the atmosphere, reach this vapor pressure at low temperatures. (b) Size classifications and sources of some atmospheric aerosol particles (the width of the bar indicates the relative number of particles). The minimum size of the atmospheric aerosol is limited by coagulation of particles with radii less than 0.01 micrometers. The maximum particle size of about 200 micrometers is limited by gravitational settling. The main mass of the atmospheric aerosol is contained in the larger particles. Modified from Horne (1978), Winchester (1980), and Warneck (1988).

BOX 3.3 | METHANE IN THE ATMOSPHERE

The concentration of methane increased from approximately 700 parts per billion (ppb) in the 1800s to 1,800 ppb at the beginning of the twenty-first century. Methane is a significant concern because it has a stronger effect, relative to carbon dioxide, on retention of longwave radiation in the atmosphere. Since the 1980s, the growth rate of fugitive methane and ethane in the atmosphere declined from about 1–1.2% per year in 1984 to 0.11% per year by 2010 (Aydin *et al.*, 2011; Simpson *et al.*, 2012). This decline in growth rate of emissions does not mean that hydrocarbon concentrations of the atmosphere are decreasing. They are not. Rather, the growth rate appears to be reaching a plateau because of improved engineering of air/fuel mixtures and higher combustion temperatures in engines, as well as the capture of co-produced natural gas during crude-oil extraction (Aydin *et al.*, 2011). The largest global anthropogenic point source of methane remains agricultural livestock, which release about 10% more methane to the atmosphere than fuel combustion and fugitive hydrocarbons combined (DeWulf and Van Langenhove, 2003; EPA, 2014).

---

more reflective, which has implications for climate forcing (Boucher, 1995).

3.2.4 Biosphere

The biosphere consists of both living and dead organic matter that is rich in carbon, hydrogen, and other light elements (Table 3.1). Living organic material can be divided into skeletal material, body fluids, and soft tissue or biomaterial.

Skeletal material in animals consists of calcite or aragonite, which are different crystal forms of calcium carbonate, various forms of silica, and the calcium phosphate mineral, apatite. Most apatite found in skeletal material, including that which makes up human teeth and bones, contains fluorine and is known as fluorapatite. Although many plants do not have true skeletal material, vascular land plants are supported by **lignin**, a complex hydrocarbon consisting of ring structures with attached OH$^-$ ions. The **body fluid** in plants and animals

---

BOX 3.4 | THE GREAT OXIDATION EVENT

The early atmosphere was more reducing than it is now and lacked abundant oxygen (Holland, 2006). For almost the first two billion years (Ga) of Earth history the atmosphere contained as little as 0.02% free oxygen, compared with 21% in the modern atmosphere (Lyons *et al.*, 2014). The concentration of oxygen in the atmosphere increased rather rapidly at about 2.4 Ga, an event referred to as the Great Oxidation Event (Holland, 2005). Oxygen levels in Earth's early atmosphere probably increased by the loss of hydrogen from water vapor in the upper atmosphere. Then, in the late Archean and early Proterozoic time, oxygen levels in the atmosphere fluctuated for several hundred million years as photosynthetic microbes, which produce free oxygen, struggled to overtake oxygen-consuming reduced gases such as hydrogen, carbon, and sulfur in the atmosphere. By about 2.1 Ga, the concentration of oxygen in the atmosphere rose to at least several percent. The proliferation of respiring land plants in early Paleozoic time is thought to mark the point at which atmospheric oxygen neared its present concentration (Holland, 2006; Planavsky *et al.*, 2014). This increase in atmospheric oxygen was critical for the evolution of life on Earth, and also impacted weathering in the oceans and on land. Sulfide minerals that had been stable on Earth's surface for almost two billion years were attacked by free oxygen, and the sulfide was oxidized to sulfate that ultimately made its way to the oceans.

---

is largely water with dissolved inorganic and organic substances, including hemoglobin and chlorophyll as discussed below, and the biomaterial includes a wide range of organic molecules, such as cellulose, as well as organs (Buol *et al.*, 1989).

Living plants and animals have a tremendous effect on the chemical and physical nature of soil. As much as a full body-weight of soil passes through earthworms each day, enhancing soil porosity and permeability. The cylinder of soil surrounding plant roots, which is known as the **rhizosphere**, is much more acidic than adjacent soil and has a higher concentration of microorganisms, allowing the plant to take up elements that would otherwise remain undissolved in the lithosphere (Cloud, 1983; Foth, 1984). The ability of plants to concentrate metals, a process known as phytoextraction or phytoremediation, has even been used to remove pollutants from contaminated soils (Garbisu and Alkorta, 2001; Sessitsch *et al.*, 2013). For example, sunflower plants helped extract radioactive cesium and strontium from soil after the Chernobyl nuclear accident (Adler, 1996), and willow can extract cadmium and zinc from contaminated soil (Greger and Landberg, 1999).

Dead organic material in soils comes largely from plants and consists of **sugars**, **proteins**, and **cellulose**, which are relatively easily degraded by bacteria; and fats, waxes, and lignin, materials that do not degrade easily. Degradation of this material yields humus, which is divided into two groups. The humic group, which dominates, is an amorphous, black to brown substance, part of which can be dissolved in acidic or basic solutions and is known as **fulvic acid**. Part is soluble only in basic solution and is known as **humic acid**. The remaining material that will not dissolve is called humin. The less abundant non-humic group consists of smaller, sugar-like molecules, including polysaccharides $[C_n(H_2O)_m]$ (Brady, 1990). Partly decomposed organic material known as **kerogen** is the raw material from which coal, crude oil, and natural gas form and is also an active ion-exchange medium similar to clay minerals and iron and manganese oxides (Dorfner, 1990).

Biomaterial and body fluids incorporate phosphorus, sulfur, and nitrogen, as well as many **trace metals**, such as fluorine, strontium, and lead. Some of these elements are essential for life, whereas others can be damaging (Table 3.2). Some are essential in small amounts, but toxic in large amounts (Goldhaber, 2003). Metals combine with organic matter largely in compounds called **chelates**, which are complex ions with one pair of atoms that can bond with a metal ion. The blood chelate molecule, **hemoglobin**, in which each of four Fe^{+2} ions is bonded to four nitrogen atoms, a protein, and an oxygen atom, carries oxygen during respiration in animals. Chlorophyll, which is essential for photosynthesis in plants, has a similar structure involving Mg^{+2}. Synthetic chelate compounds such as ethylenediaminetetraacetic acid (EDTA) are used as scavengers to remove heavy metals from the human body because they form unusually stable complex ions involving six separate bonds with a single metal ion. Metals can be incorporated into living organic material by methylation, in which they combine with methyl ions (CH_3^+), a reaction that is facilitated by microorganisms (Ehrlich, 1997; Mason, 2013). Microbial methylation of

Table 3.2 Biological importance of selected elements (after Horne, 1978; Zumdahl, 1986). As indicated in the second part of this table, some elements that are necessary to life are toxic in higher concentrations.

Element	Biological importance
Elements necessary for life	
Cr	Aids insulin, which controls blood sugar
Mn	Necessary for some enzymatic reactions
Fe	Metal in blood molecule hemoglobin
Co	Component in vitamin B_{12}
Ni	Component of some enzymes
Cu	Component of some enzymes, produces pigment
Zn	Component of insulin and enzymes

Elements inimical to life (oral toxicity to small animals)

Very toxic	Moderately toxic	Slightly toxic	Harmless
As	Cd	Al	Cs
Pu	Cu	Mo	Na
Se	Hg	Ta	I
Te	Pb	W	Rb
Tl	Sb	Zn	Ca
	V	R	K

mercury is of particular concern because monomethylmercury is a neurotoxin that bioaccumulates in the food chain (Blum *et al.*, 2013); it is commonly concentrated in reduced, methanogenic environments such as rice paddies (Gilmour, 2013). This is of special concern because rice supplies about 20% of the world's per capita human energy, and about 30% each of dietary energy and protein in less developed countries (LDCs) (IRRI, 2001; Sautter *et al.*, 2006).

Elements that exist in more than one oxidation state under surface conditions are important energy sources for life. Bacteria use many of these elements and can greatly enhance the rate and completion of **oxidation** and **reduction** reactions (Quantin *et al.*, 2001). Sulfur, which has several stable and metastable oxidation states from +6 to −2, is widely used by bacteria. *Desulfovibrio, Desulfotomaculum,* and *Desulfuromonas*

convert dissolved sulfate SO_4^{-2} to H_2S. *Thiobacillus thiooxidans* and related strains convert dissolved H_2S to SO_4^{-2}. *Ferrobacillus* converts Fe^{+2} to Fe^{+3}. These reactions facilitate rock weathering, and have the potential to recover metals from ores or waste streams of chemical and pharmaceutical industries (Krebs *et al.*, 1997; Frankham, 2010).

3.3 Geochemical cycles and the dynamic nature of global reservoirs

3.3.1 Geochemical cycles

Substances are continually moving from one reservoir on Earth to another, and among smaller reservoirs within the four large, global reservoirs. The amount of a substance that moves in a given period of time is known as the **flux**, its average stay in each reservoir is known as **residence time**, and its overall pattern of movement from reservoir to reservoir is known as its **geochemical cycle**. Reservoirs that release a substance are known as **sources**; those that capture or concentrate it are **sinks**. One task of environmental geochemistry is to distinguish natural sources and sinks from those created by humans, which are known as **anthropogenic** sources and sinks, and to determine whether anthropogenic contributions disturb the cycle (Broecker, 1985; Holton, 1990; Blum *et al.*, 2013).

The global sulfur cycle (Figure 3.6) illustrates these principles as well as the challenges facing environmental chemistry and regulation (Zenhder and Zinder, 1980; Chin *et al.*, 2000; Sievert *et al.*, 2007; Johnston, 2011). The most important reservoirs of sulfur are the lithosphere and hydrosphere, both of which contain large amounts of sulfur with long residence times. Sulfur in the lithosphere is largely in the form of the sulfide mineral pyrite and the sulfate mineral gypsum, some of which form mineral deposits. Sulfur in the hydrosphere is present largely as dissolved sulfate in the oceans, which comes from weathered sulfide and sulfate minerals. Sulfur returns to the lithosphere by evaporation of seawater to form gypsum and other sulfate minerals or by precipitation of metal sulfides. Sulfur enters the atmosphere as SO_2, H_2S, and other gases through volcanic eruptions, weathering of sulfide minerals, and dispersal of sea spray (Table 3.3). It also enters the atmosphere from anthropogenic sources including burning of fossil fuels containing sulfur-bearing minerals and organic compounds, where it is a potential cause of acid rain, as discussed below.

In order to determine the role of anthropogenic sources in the sulfur cycle, we must first recognize and measure the natural sources. This is a major challenge, whether it is done on a global or local scale. Many natural sources (and sinks) are not well understood and emission rates from others are difficult to measure (Malinconico, 1987; Stoiber *et al.*, 1987). For volcanoes on land, such as Mount Pinatubo in the Philippines,

Table 3.3 Reactions that decompose sulfide minerals to produce acid mine drainage and natural geochemical dispersion halos. Me = metal.

A. Oxidation of pyrite in contact with air-saturated water to form acid, sulfate, and ferric iron ions in solution:
$4FeS_2 + 2H_2O + 15O_2 \rightarrow 4H^+ + 8(SO_4)^{-2} + 4Fe^{+2}$
(Fe^{+2} is then oxidized to Fe^{+3} by bacteria)

B. Deposition of hydrated iron oxide to produce more acid:
$Fe^{+3} + 3H_2O \rightarrow Fe(OH)_3 + 3H^+$

C. Dissolution and oxidation of sphalerite:
$ZnS + 2O_2 \rightarrow Zn^{+2} + (SO_4)^{-2}$

D. Leaching of metals from clay minerals by acid water:
$Clay\text{-}Me^{+2} + H^+ \rightarrow Clay\text{-}H^+ + Me^{+2}$

E. Neutralization of acid water:
$CaCO_3 + 2H^+ \rightarrow Ca^{+2} + H_2O + CO_2$

activity is intermittent and unpredictable, although we can at least observe and attempt to measure the eruptions, but submarine volcanoes along the mid-ocean ridge are much more difficult to observe and measure (Figure 3.7). The entire volume of water in the oceans circulates through the hot, fractured rocks of the mid-ocean ridges within only a few million years and has a profound effect on sulfur emission rates. Similar problems beset all environmental studies, making it very important to obtain reliable measurements before constructing a geochemical cycle.

The construction of a geochemical cycle allows us to evaluate the relative importance of natural and anthropogenic contributions, and the degree to which human activity has perturbed the natural system. In developing such models, it is important to recognize that pollution can be both natural and anthropogenic.

3.3.2 Natural contamination

The chemical relation between rocks, soils, water, plants, and even gases is the province of landscape geochemistry

Global sulfur cycle

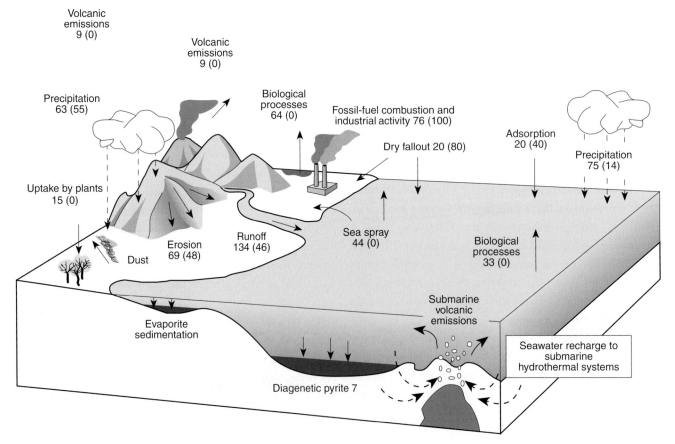

Figure 3.6 Generalized global sulfur cycle, showing major sources, sinks, and fluxes (amounts in millions of tonnes annually). Numbers in parentheses are percentage of emissions contributed by anthropogenic activity. Modified from Zenhder and Zinder, 1980; Adams *et al.*, 1981, Stoiber *et al.*, 1987).

Figure 3.7 Natural and anthropogenic sulfur sources. (a) Mutnovsky volcano (seen from above) consists of a large crater and lake with fumaroles releasing 620 °C gases consisting of H_2O, CO_2, SO_2, H_2S, HCl, HF, and H_2. The fumaroles release about 4×10^5 tonnes of SO_2

BOX 3.5 | MINERAL DEPOSITS AND NATURAL POLLUTION

Mineral deposits are important local sources of natural contamination, which commonly forms patterns that are used in geochemical exploration, as discussed in the next chapter. In the vicinity of metal deposits, it is common to observe natural metal concentrations at the surface, such as the copper-rich bogs of Coed-y-Brenin, Wales (see Figure 3.8a). The rhizosphere surrounding plant roots facilitates the uptake of metals from mineral deposits into trees, which are very enriched in metals around some deposits. This effect can produce natural stress, which limits tree growth and is sometimes manifested by early loss of leaves in the fall (Canney *et al.*, 1979). Animals, also, can show variations in composition related to mineral deposits. Sea urchins living near an unmined lead–zinc deposit contain higher concentrations of iron (from pyrite) and zinc than urchins at a greater distance, and fish livers in lakes in British Columbia have been shown to be enriched in metals near mineral deposits (Bohn, 1979).

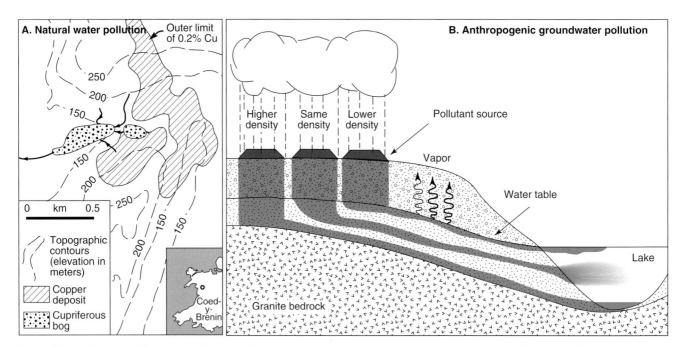

Figure 3.8 (a) Copper-rich bogs formed by natural weathering of the Coed-y-Brenin copper deposit, Wales, showing outline of the deposit as indicated by the area containing soil with more than 0.2% copper (after Andrews and Fuge, 1986). (b) Schematic illustration of contaminant plumes in groundwater formed by liquids with different densities. Typical low-density liquids include gasoline and oil; high-density liquids are those with high dissolved solid contents.

Caption for Figure 3.7 (cont.)

annually. Some of the sulfur is deposited in vents and fumeroles on the volcano (such as the 2-m-high chimney in (b)) but most of it goes into the atmosphere. The eruption of Mount Pinatubo, Philippines, in 1991, released about 2,000 times more SO_2 in only a few hours. (c) Treatment of nickel–copper ores at Sudbury, Ontario, have been significant sources of sulfur to the atmosphere. Since mining began, roasting and smelting have released about 10^8 tonnes of sulfur (up to 2×10^6 tonnes annually), contributing to the lack of vegetation as seen in the foreground of the photo. Improvements began in the early 1970s with construction of the 380-m superstack, which disperses emissions over a much broader area. The later addition of sulfur capture systems has cut annual emissions by at least 90% (photos by the authors). See color plate section.

(Fortescue, 1992; Smith, 2015). Where these relations affect public health, they become the concern of **geomedicine** (Warren, 1989; Davenhall, 2012). Health problems can be caused by deficiencies or enrichments of a natural substance and they can be related to regional-scale rock units or smaller features such as mineral deposits. It is widely known that different types of rocks have different average minor- and trace-element contents. Ultramafic rocks, for instance, lack potassium and phosphorus that encourage plant growth, and rarely support good crops. Rocks that lack adequate trace metals such as copper, cobalt, molybdenum, and zinc can also cause problems (Allcroft, 1956; Thornton, 1983). Sedimentary rocks that are enriched in selenium cause blind staggers in cattle that graze on the land and deformities in birds that live in surface water fed by drainage from the rocks (Anderson *et al.*, 1961; Presser and Barnes, 1985).

Rocks have their greatest effect on life by influencing the composition of water. In early times, **goiter**, a thyroid disease caused by iodine deficiency, was a common problem in populations far from the sea where salt spray did not contaminate streams and lakes, and rocks lacked iodine. Some granitic igneous and metamorphic rocks, as well as phosphate sediment, are enriched in uranium, thorium, and their radioactive decay product radon, which concentrate in groundwater supplies and present health concerns (Aieta *et al.*, 1987; Wanty *et al.*, 1991; Murphy, 2011). Al-Jarallah *et al.* (2004) reported a 90% correlation between radon emissions and radium concentration in granite building stone in Saudia Arabia. The absence of selenium in soil and water has been implicated in the high incidence of stroke in the southern United States and China, and high concentrations of calcium and magnesium in groundwater have been related to cardiovascular disease (Barringer, 1992; Pocock *et al.*, 1985; Wei *et al.*, 2004; Sengupta, 2013).

The relation between health and regional geochemical variations can be determined only through geochemical surveys in which the average composition of the soil is determined at closely spaced sites throughout a large area. Studies of this type show surprising patterns, such as the curious barium-enriched zones that cross Finland (Koljonen *et al.*, 1989). In the Guangdon province of China, elevated concentrations of arsenic in soil and groundwater are related directly to weathering of arsenic-bearing limestone and sandstone, and not to anthropogenic inputs (Zhang *et al.*, 2006). The trace-element content of natural soils is not always a function of the underlying rocks. In areas where soil development is slow, a large part of the soil might be from wind-borne dust, often from thousands of kilometers away, bringing with it trace elements that are not of local origin. This is the case for soil in parts of Florida, where dust particles can be traced to the Sahara Desert in northern Africa (Prospero, 1999). Wind-delivery of Saharan dust increases during the summer months, and it must be monitored to meet air quality standards established by the Clean Air Act (Prospero, 1999).

3.3.3 Anthropogenic contamination

Anthropogenic contamination has overprinted natural contamination creating the complex geochemical pattern that affects us today (Nriagu, 1989). We are currently removing metals and other mineral compounds from Earth at a rate that is approximately equal to the rate at which they are liberated by weathering of rocks and other natural processes (Pacyna and Pacyna, 2001). Comparison of the global cycles of aluminum, chromium, copper, iron, lead, nickel, silver, and zinc indicates that anthropogenic processes mobilize these metals at an annual rate approximately equal to their mobility among natural reservoirs (Rauch and Pacyna, 2009). Natural mobility of metals is dominated by water transport, biomass growth and death, volcanoes, and wind, and includes production and subduction of crust, sediment erosion, and ocean deposition. Anthropogenic mobility is caused largely by fossil-fuel combustion, agricultural biomass burning, metal production and fabrication, and metal disposal, and anthropogenic contributions are estimated to have doubled the natural flux of these metals (Milliman and Syvitski, 1992). Anthropogenic processes that release metals from rocks and soil include mining as well as construction of infrastructure such as roads and buildings (Syvitski *et al.*, 2005; Wilkinson and McElroy, 2007). Although old mines, oil fields, smelters, and refineries are important sources of anthropogenic emissions, newer facilities built in response to environmental regulations, combined with economic and sociopolitical factors that incentivize companies to reduce waste, have resulted in statistically significant reductions in anthropogenic emissions of metals (Chander, 1992). Lead is a prime example of this. It was added to gasoline starting in the 1930s to improve the octane rating of fuel. By the early 1970s, anthropogenic emissions of lead to the atmosphere were 345 times higher than natural emissions (Lantzy and MacKenzie, 1979; Nriagu, 1989). The observation that the inhalation of lead causes cognitive delay in children stimulated the US Environmental Protection Agency to phase out lead as a fuel additive starting in the 1970s, and a complete ban was required by 1996 (Newell and Rogers, 2003). Elimination of lead additives resulted in a 95% decrease in anthropogenic emissions (Rauch and Pacyna, 2009).

Currently, it is estimated that natural mobilization of aluminum and iron exceed anthropogenic mobility. However, anthropogenic mobilization of copper is about ten times more than natural mobilization, slightly less than that for carbon, nitrogen, and sulfur, and about five times for phosphorus (Falkowski *et al.*, 2000).

Although atmospheric pollution is most obvious, water pollution is a growing threat. Worldwide anthropogenic metal emissions that reach the oceans are largest for arsenic, cadmium, cobalt, chromium, copper, mercury, manganese, nickel, lead, and zinc, with most contributions coming from manufacturing and domestic wastewaters followed by atmospheric fallout (Nriagu, 1990b; Mason, 2013). Contamination of surface water by fertilizers and road salt is an inevitable result of the requirement that these materials be used in highly soluble forms. Groundwater aquifers have been damaged by drainage from these applications, as well as from point sources such as salt storage areas, sewage-treatment plants, and storage facilities for crude oil and refined liquids. Contaminated groundwater commonly forms a plume that originates at the point of input and gradually enlarges down the flow path, with its ultimate form determined by the relative densities of contaminated water and groundwater (Figure 3.8b). Because hydrocarbons are lighter than water, they remain close to the surface invading sewers and basements where they cause explosions. Most modern industrial facilities use impermeable underlayers in order to prevent groundwater contamination, but individuals continue to dump wastes at the surface, creating a layer of contamination at the top of the water table.

Recent high-sensitivity geochemical studies have shown that anthropogenic contamination patterns are truly global in scale and that they have affected the composition of the biosphere. Global effects are most obvious in the atmosphere, where mobility is highest, and they include NO_x, CO_2, and SO_2, largely from combustion of fossil fuels, and volatile organic compounds (VOCs), such as methane (CH_4), from oil and gas production and agricultural livestock. Global **dispersion** of metals is also a concern; the mercury isotope composition of fish caught near Hawaii is from coal combusted in China and India (Blum *et al.*, 2013). Sources for particulate emissions include soot, smoke, road dust, and chemical conversion of anthropogenic gases. The effects of these contributions are seen in the concentration of contaminants where population is greatest and in the close correspondence between visibility and SO_2 emissions (Figure 3.9). In the 1990s, atmospheric sulfur particulate matter of anthropogenic origin was responsible for 60–90% of visibility reduction

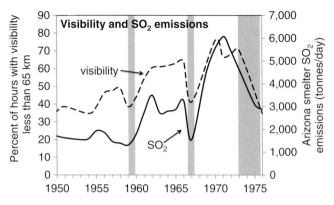

Figure 3.9 Relation between atmospheric visibility in Arizona and SO_2 emissions from copper smelters between 1950 and 1975. Vertical gray bars represent global recessions, which resulted in decreases in copper production. Emission controls and smelter closings have greatly diminished this effect.

in the eastern United States and 30% in the western United States (Malm, 1999). It is also possible to fingerprint the source of atmospheric sulfur particles. For example, sulfate collected at the south rim of the Grand Canyon can be traced by analyzing the ratio of sulfate to spherical aluminosilicate particles, and the concentrations of selenium, arsenic, lead, and bromine in air samples (Eatough *et al.*, 2000, Watson *et al.*, 2002). In the summer of 1992, 33% of the sulfate was derived from southern California, 14% was traced to copper smelting in southern Arizona, and 23% was traced to three coal-fired power plants in Arizona and neighboring Nevada (Malm, 1999). Two aspects of anthropogenic contamination, acid pollution and global change, have become household terms and require special attention.

3.3.4 Acid pollution

The most important pollutant of the hydrosphere is acid (H^+) from rain and mine drainage. **Acid mine drainage** results from the decomposition of pyrite, usually catalyzed by bacteria, to produce iron hydroxide and dissolved H^+ and SO_4^{-2} (Taylor *et al.*, 1984; Evangelou and Zhang, 1995; Johnson, 2003) (Table 3.3). About 95% of the acid mine drainage in the United States comes from coal and metal mine waste, much of which contains pyrite (Baker, 1975; Powell, 1988). Acid water formed by dissolution of pyrite dissolves more pyrite, thus accentuating the effect. As acid water moves downstream, it mixes with less acid water causing dissolved iron to precipitate as oxides and hydroxides, and this produces more acid. The multiplicative effects of these reactions are so great that oxidation of a single molecule of pyrite can yield

Table 3.4 Generalized reactions that produce (A, B, C) and consume (D, E, F) acid rain. Reactions shown here are "net reactions" that show reactants and final products, but omit intermediate products. Because limestone is more soluble, this reaction is much more important than the silicate reaction. Both reactions are essentially the same as reactions that take place during normal rock weathering.

A. Production of carbonic acid:
$$CO_2 + H_2O \rightarrow H_2CO_3$$

B. Production of sulfuric acid:
$$SO_2 + H_2O + \tfrac{1}{2}O_2 \rightarrow H_2SO_4$$

C. Production of nitric acid:
$$NO + \tfrac{1}{2}H_2O + \tfrac{3}{4}O_2 \rightarrow HNO_3$$

D. Limestone dissolution
$$CaCO_3 + H^+ \rightarrow Ca^{+2} + (HCO_3)^-$$

E. Silicate dissolution (feldspar)
$$2KAlSi_3O_8 + 2H^+ + H_2O \rightarrow 2K^+ + Al_2Si_2O_5(OH)_4 + 4SiO_2$$

F. Lime dissolution
$$CaO + H^+ \rightarrow Ca^{+2} + OH^-$$

four H^+ ions. Acid water produced by these reactions dissolves other sulfide minerals as well as rock-forming minerals, and leaches elements adsorbed on the surfaces of clays and other poorly crystallized minerals, thus increasing the trace-element content of streams (Johnson and Thornton, 1987; Likens *et al.*, 1996). Water from waste piles recharges groundwater systems, where it can form plumes of acid water (Zaihua and Daoxian, 1991; Naicker *et al.*, 2003; Figure 3.8b).

Acid rain is the product of reactions between atmospheric water and CO_2, SO_2, and NO_x (Table 3.4). Although natural levels of CO_2 in the atmosphere would give precipitation a pH of about 5.7, it is even more acid than this in eastern North America, western Europe, and eastern China, reaching levels as low as 2.9 (Figure 3.10). The close correlation of this acidity with areas of sulfur and nitrate deposition (Figure 3.11) is interpreted to indicate that anthropogenic SO_2 and NO_x are the main sources of acid. The effects of acid rain on surface water and soils is governed in part by the presence of underlying limestone, which consumes acid to produce dissolved

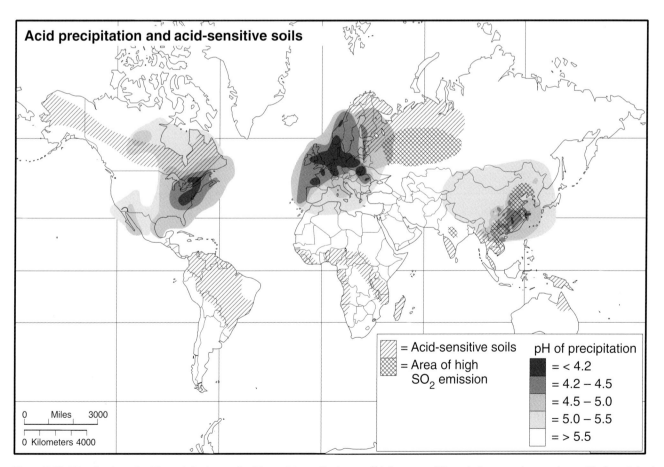

Figure 3.10 Distribution of acid precipitation and acid-sensitive soils. Areas of high current SO_2 emissions are shown where pH of precipitation is not available. Contours in North America are for pH of precipitation in 2013 (see Figure 3.11).

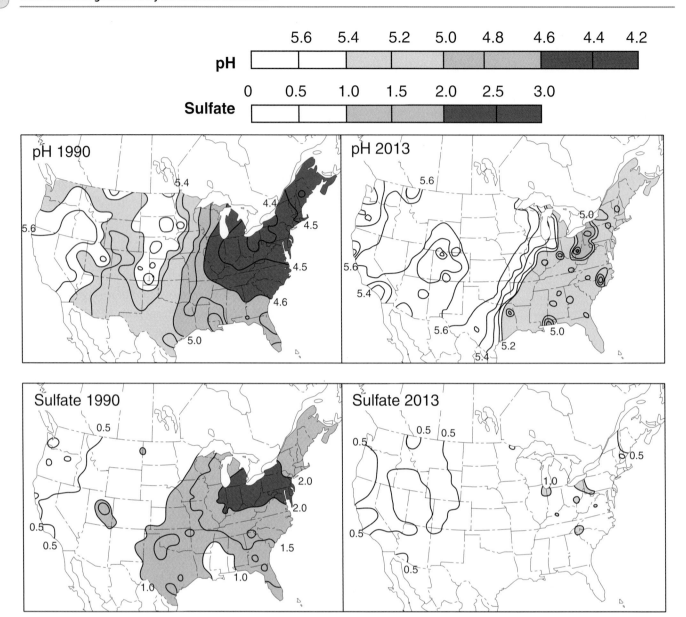

Figure 3.11 Change in pH and wet sulfate deposition (in kilograms/hectare) of precipitation in North America between 1990 and 2013 (from the National Atmospheric Deposition Program)

calcium and bicarbonate. Intrusive and metamorphic rocks, which consist largely of silicate minerals, also react with acid, but at a much slower rate. Thus, lakes in areas underlain by limestone are usually less acidic than those underlain by silicate rocks, even when they receive the same acid precipitation.

The distribution and history of acid precipitation has long been a focus of environmental concern. Lakes underlain by silicate rocks in the Adirondacks and southern parts of the Canadian and Baltic Shields are surprisingly acid (Figure 3.12). In the Adirondacks and the Upper Peninsula of Michigan, about 10% of the lakes larger than 4 hectares have a pH of less than 5 (Cook *et al.*, 1990; Irving, 1991).

Diatoms discussed in Chapter 13 show that this acidity has developed within the last hundred years or so, and it appears to have caused large declines in fish populations, probably through increases in dissolved aluminum and other metals that are more soluble at low pH (Longhurst, 1989; NAPAP, 1991). Parallel declines have been observed in the health of forest trees, particularly red spruce.

3.3.5 Global change

Global change is our acknowledgement that Earth did not stop changing just because humans showed up. We have always recognized that short-term changes take place, including

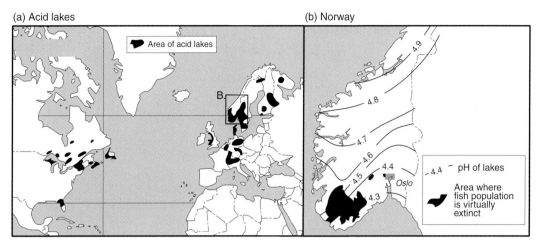

Figure 3.12 (a) Areas of acidic lakes in North America and northern Europe. (b) Relation between lake acidity (indicated by pH contours) and fish mortality in an area of Norway that is underlain by silicate rocks that react slowly with acid water. Compiled from GEMS (1989), Taugbol (1986), and State of Norway (2014).

BOX 3.6 | THE HUBBARD BROOK EXPERIMENT

One of the challenges in recognizing a cause–effect relationship for anthropogenic emissions and acid rain is the timeframe over which data are collected. The long-term study at the Hubbard Brook Experimental Forest in New Hampshire provided important insight. It measured the acidity of precipitation continuously from 1964 to 2000, along with the concentrations of atmospheric sulfate and nitrate (Likens *et al.*, 2005). Acidity and sulfate concentrations were strongly correlated (0.80) over this time period but acidity and nitrate were poorly correlated (0.22). At first glance this suggests that sulfate is the main cause of acidity. However, the nitrate concentrations were relatively constant from 1964 to 1990, but over the period from 1991 to 2000 they were more variable and showed a strong correlation with acidity (0.69). Thus, both agents appear to have contributed to acid.

---

volcanic eruptions, earthquakes, and floods, all of which regularly disrupt life. We have also known for many years that Earth underwent longer-term changes, such as the glacial advances that have swept over North America and Eurasia intermittently for the last million years, and the unusually hot climate that prevailed about 100 million years ago in the Cretaceous time. It now appears that anthropogenic emissions can affect these longer-term changes, and the greatest present concern centers on the possibility that they are causing surface temperatures to increase dramatically (IPCC, 2013). Recent analyses suggest that the average global temperature integrated over all land and ocean surfaces increased by 0.85 °C (1.58 °F) between 1880 and 2012 (IPCC, 2013).

The increased average global temperature is related to carbon dioxide, methane, water vapor, nitrous oxide, chlorofluorocarbons, and other gases, which are known as **greenhouse gases** because they absorb infrared radiation that would otherwise radiate from Earth into space, thus heating the surface of the planet. Historical records and analyses of gases trapped in ice caps show that CO_2 and CH_4 concentrations in the atmosphere have increased significantly over the last few centuries, and have fluctuated significantly over the past 800,000 years (Figure 3.13). The possibility that gas variations control temperature is supported by the close correlation through time between the composition of bubbles of air trapped in the polar ice caps and world ocean temperatures indicated by isotopic measurements (Barnola, 1987).

Although the composition of the atmosphere has clearly evolved through geologic time, the recent increase is thought to be more abrupt than normal, further supporting its suggested relation to anthropogenic emissions. There is no question that we are emitting more of these gases (Figure 3.14). Compilations of global CO_2 emissions suggest that they tripled between 1950 and 2013 to almost a billion tonnes annually (IPCC, 2013). In the United States, 38% of CO_2 emissions are generated by electricity production, 32% by

(a)

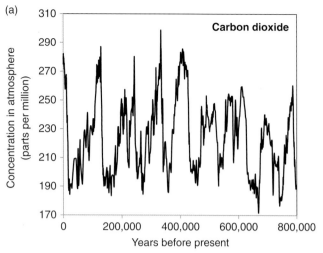

(b)

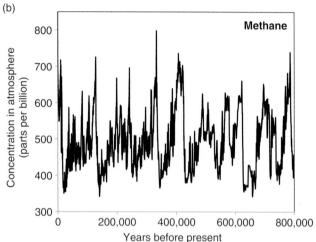

(c)

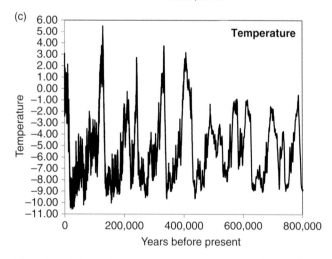

Figure 3.13 Relation between concentrations of CO_2 (a) and CH_4 (b) in ice core and (c) temperature (°C) of the atmosphere at the time of ice formation for the past 800,000 years in Antarctica as determined from stable isotope compositions of hydrogen and oxygen. Temperature data are corrected for seawater isotopic composition (after Bintanja *et al.*, 2005), and for ice-sheet elevation (after Parrenin *et al.*, 2007 on the EDC3 age scale). (Gas concentrations from Lüthi *et al.*, 2008; temperature data from Jouzel *et al.*, 2007).

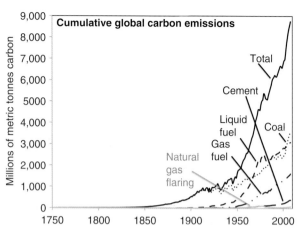

Figure 3.14 History of CO_2 emissions to the atmosphere from 1750 to 2005 from specific sources. (From Boden *et al.*, 2013; original data sources include: 1959–2010 estimates for fossil-fuel combustion are from the Carbon Dioxide Information Analysis Center (CDIAC) at Oak Ridge National Laboratory; 2011, 2012, and 2013 estimates are based on energy statistics published by BP Statistical Review of World Energy reports; emissions from cement production were estimated by CDIAC based on cement production data from the United States Geological Survey. Reported uncertainty is about ±5% for a ±1 sigma confidence level.)

transportation, 14% by industrial processes, 9% by residential and commercial building, and 6% by processes that do not involve fossil fuels (EPA, 2014). Globally, about 87% of CO_2 emissions come from combustion of fossil fuels, with 43% of these from coal, 36% from oil, 20% from natural gas, and 1% from gas flaring (Le Quéré, 2013). Land-use changes are responsible for 9% of global CO_2 emissions, and cement production emits 4%. The United States was the largest source of CO_2 emissions throughout the nineteenth century, as it was for SO_2. However, China eclipsed the United States in emissions of both of these gases by the early twenty-first century (Figure 3.14, 3.15a).

Another way to assess emissions is to compare those related to domestic consumption to those that are effectively transferred via international trade. This has been done for CO_2, and the results indicate that the United States and the European Union are the largest emitters of CO_2 when both domestic emissions and life-cycle emissions attributed to product consumption are included (Figure 3.15b). Emissions are beginning to decline, however. After tripling between 1900 and 1975, SO_2 emissions in the United States began to decline in the late 1970s because of legislative action and pollution controls on power plants and industrial facilities (Davies and Mazurek, 1998). As China and other LDCs transition to consumer-based

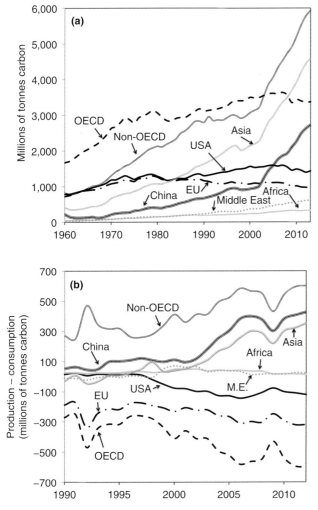

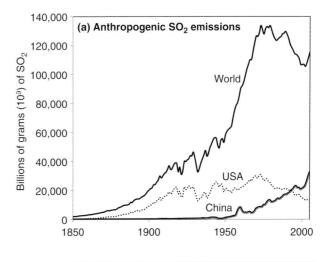

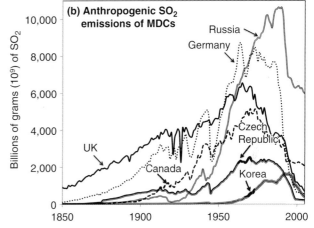

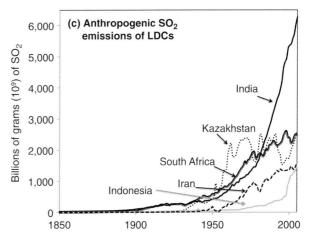

Figure 3.15 (a) Increase in atmospheric emissions of CO_2 from 1960 to 2012 from fossil fuels and cement production for specific regions (OECD = Organisation for Economic Co-operation and Development). (b) Difference between total CO_2 emitted from all sources in a region and CO_2 emissions that are embedded in products traded to that region for the period 1990 to 2012 (data updated from Peters *et al.*, 2011).

Figure 3.16 Anthropogenic atmospheric emissions of SO_2 for: (a) the entire world, the United States and China; (b) more developed countries (MDCs); and (c) LDCs. Note that the ordinate scale changes in each panel. Data from Smith *et al.* (2011).

economies, similar pollution control strategies will need to be implemented (Figure 3.16).

Preliminary calculations using projected anthropogenic greenhouse-gas emission rates suggest that they could cause global land-surface and sea-surface temperatures to increase by up to 4 °C by 2100 (IPCC, 2013). A temperature increase of this magnitude would equal the entire temperature change of the last major glacial advance and could melt enough of the polar ice caps to increase global sea level by 28 to 98 cm, flooding large coastal cities (IPCC, 2013). The ocean is a sink for both heat and atmospheric CO_2. Since 1971, the upper 75 m of the oceans have warmed by an average of 0.11 °C per

decade, and the increased concentration of dissolved CO_2 has made the oceans more acidic (IPCC, 2013).

If global change is placed on the doorstep of modern civilization, it will require that the use of fossil fuels be curtailed and substitutes found, including significant increases in energy from renewable sources and possibly nuclear power (Hoffert *et al.*, 2002; Ngo and Natowitz, 2009; Towler, 2014). Most of our mineral extraction and processing methods will have to change, as will our current mineral-use patterns, which are based on extensive global trade. In the meantime, we must curtail offending emissions, which will require assessment and remediation, as discussed next.

3.4 Assessing and remedying the environmental impact of pollution

Most modern environmental studies begin with surveys of the abundance of the contaminant, followed by an assessment of its environmental effects. If contaminant levels need to be reduced, remediation is undertaken. Any survey of contaminant abundances actually involves two separate steps, sample selection and sample analysis. This sequence of efforts provides the only hard facts on which to base environmental assessments, predictions of future conditions, and regulations. Although these studies can be tedious and time-consuming, they are absolutely essential to responsible management of our environment.

3.4.1 Sample selection and sample statistics

Regardless of the imaginative nature of any hypothesis or the quality of the analyses carried out to test it, a study is only as good as the samples on which it is based. Samples must faithfully represent the material that is being investigated, and there must be enough of them to illustrate any compositional variations in the study group. The selection of proper samples depends on a full knowledge of the science of the material under investigation. To obtain representative samples of lake sediment, it is necessary to understand limnology and geology; to obtain samples from humans, a knowledge of biology and anatomy is needed. There are many examples of confusion caused by improper sample material or sampling methods. Early ice cores used for the study of global lead dispersion were collected with equipment that contained traces of lead, producing spurious high values (Ng and Patterson, 1981; Boutron and Patterson, 1983). Similarly, early analyses for mercury were contaminated by air and dust, giving results that were too high (Porcella, 1990). Asbestos samples used in some early studies contained several types of fibrous minerals and yielded a confused relation between exposure and symptom. It can even be important to analyze material from the right part of a sample. As can be seen in Figure 3.17, lead contributed by atmospheric fallout from leaded gasoline and coal-fired power plants is concentrated in the upper part of the soil horizon, whereas lead from natural soil formation is concentrated in the B horizon of the soil (Graney *et al.*, 1995; Chiaradia *et al.*, 1997). Sampling the wrong zone can confuse efforts to distinguish between anthropogenic and natural pollution.

Attention must also be given to sample statistics, which deal with the characteristics of groups of samples called **populations**. The first question posed by statisticians is whether a population of samples truly represents the group from which it came. In the interests of cost and convenience, we need to know the minimum number of samples that must be collected to assemble a representative population. In order to answer this, statisticians need to know the type of distribution that the sample population shows. Populations can be described by

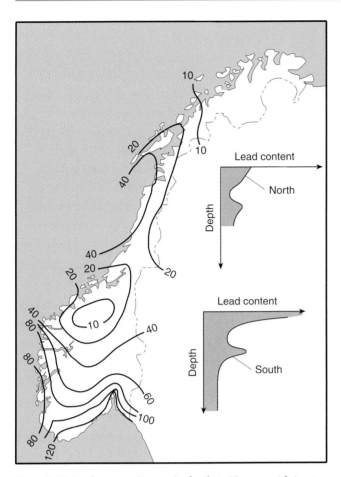

Figure 3.17 Lead content (in ppm) of soils in Norway with insets showing the position of lead enrichment in soil profiles. Soil in northern Norway has normal E/A (leached) and B (enriched) zones, whereas the profile from southern Norway has enrichment of lead in the O zone reflecting atmospheric fallout largely from coal-fired electric power generation in Europe to the south (modified from Steinnes, 1987).

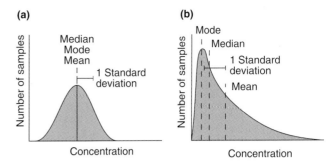

Figure 3.18 Frequency diagrams showing the relative values of mean, mode, and median for (a) normal (Gaussian) and (b) skewed populations

their **mean** or average value, **mode** or most common value, **median** or middle value, and **standard deviation**, the square root of the sum of the squares of the difference between all individual values and the population mean. These descriptors can be used to distinguish between two distributions that are important in mineral-resource and environmental studies. The population labeled **normal** in Figure 3.18 has a mode, median, and mean that are identical, and the values form a bell-shaped distribution. In the other type of population, here called **skewed**, the mode and median values are significantly lower than the mean. Many environmental and geological populations, such as the concentrations of copper in soil, the grades of gold deposits, or the sizes of oil fields, are skewed, with a few large values and many small ones.

The mean is much more representative of a **normal population** than it is of a skewed population. It follows that means

are useful only if the type of population they represent is known and if the population truly represents the material or geochemical reservoir being investigated. Although the statistical basis for most environmental and geological studies is adequate, the conclusions from these studies are sometimes confused in media reports by careless use of these concepts.

3.4.2 Analytical preparation and methods

Once a group of samples of rocks, minerals, water, plants, or air has been collected, the next step is the measurement of the parameter of interest, usually by chemical analysis (Marr and Cresser, 1983). These measurements are used by the exploration geologist to search for ore, by the miner to determine that the ore is being recovered effectively, by the environmentalist to determine the dispersal of polluting substances, and by the public health professional to determine the toxicity of a substance. Often, they are compared to allowable amounts that have been specified by government agencies (Table 3.5).

Analytical measurements involve two important steps: preparation of the sample, and the actual analysis. For most atmosphere and water samples, the only sample preparation needed is filtration, which separates fluid and particle fractions so that they can be analyzed separately. For solid samples such as rocks, soils, or trees, matters are more difficult because elements and compounds can be present in different minerals and other forms. For instance, lead in stream sediments might be found as adsorbed ions on the surface of fine-grained clay minerals, iron oxides, or kerogen, in trace amounts substituting for potassium in silicate minerals, or as clastic grains of the lead mineral galena. To determine the total lead content of the sample, a total analysis must be performed, such as by completely dissolving the sample and analyzing the resulting solution. If, on the other hand, we want to estimate the amount of lead that was introduced by waters flowing through a sample,

> ## BOX 3.8 | ANOMALOUS SAMPLES
>
> Many of the analyses carried out in mineral exploration and environmental studies are designed to determine whether any samples are **anomalous**. Anomalous samples are those that differ from most samples in the population under consideration. Samples usually differ in having a higher concentration of some element or substance, although negative anomalies are possible. Samples that represent the mean composition of the population are often referred to as **background**. There is no single, widely accepted way to designate a sample as anomalous and great care must be taken in using this term. At a very simplistic level, anomalies from a single survey can be defined as those values that differ from the survey mean by 1, 2, or 3 standard deviations. More sophisticated techniques include the use of multivariate analysis and geostatistics (discussed in Chapter 4). Comparison to surveys in other areas is also necessary to determine whether all of the samples in a survey are anomalous with respect to global averages. Finally, it is important to look for patterns of variation among samples.

some means must be found for removing lead from the appropriate part of the sample, a technique known as a partial analysis. One way to do this is to leach the sample with a weak acid that removes only adsorbed lead without decomposing lead-bearing minerals such as galena or feldspar that would have been in the sample before water flowed through it.

Modern analyses of properly prepared samples are made by an amazing array of methods, and anyone carrying out a geochemical study is immediately faced with the need to select the proper analytical method for the study at hand. Although it is best to compare analytical methods on the basis of the chemical principles they employ, a quick comparison can be based on three measures of their capability. The **detection limit** refers to the minimum concentration that can be measured. Detection limits for different elements or compounds vary tremendously, even for a single analytical method. Complicating matters is the fact that different elements and compounds are usually present in a single sample in concentrations that differ by many orders of magnitude. Even when you are only analyzing one element with one method, things can be complicated because samples in a large group can have a wider range of concentrations than can be measured by that method. In some cases, this might require that the same element be analyzed by two different methods, each with their own limitations or complications. It is also common for the same samples to be analyzed by more than one independent laboratory, and sometimes the laboratory is not told the history of a sample; this is referred to as quality-control blind analysis (Glenn and Hathaway, 1979).

Analytical methods also differ in their **precision**, which is the ability to obtain the same result if the same sample is analyzed again. Precision is commonly expressed as the deviation from an average value for a sample that has been analyzed repeatedly throughout the duration of the study. **Accuracy**, which is the ability to obtain the correct answer for a sample of known composition, is based on analysis of a **standard sample**, in which the concentration of the component of interest has been agreed upon by a large number of analytical laboratories. It is possible for analyses to be precise without being accurate and vice versa. Because mineral-resource-related samples have so much economic and environmental importance, it is critically important that they be reported along with information on the analytical method used and its precision and accuracy. Ideally, analyses of standard samples should be provided as well.

As the detection limit, precision, and accuracy of analytical methods improve, we are able to see variations that were not detectable previously. One famous instance of this was the development of the electron-capture detector for gas chromatographs, which allowed measurement of low concentrations of chlorofluorocarbons (CFCs) in air and led to recognition that they were accumulating in the atmosphere (Lovelock, 1971).

3.4.3 Risk and dosage in environmental chemistry

Many environmental studies related to mineral resources seek to determine whether specific elements, compounds, or materials are damaging to life. The term **toxic**, which is frequently used in this context, refers to any substance that causes undesired effects such as alterations in DNA, birth defects, illness, cancer, or death (Doull *et al.*, 1980). In most environmental studies, toxicity is measured by exposing a number of individuals of a specific organism to a known concentration of the compound of interest, and then observing how many of the

Table 3.5 Maximum allowable concentrations of pollutants of interest to mineral resources (from Environmental Protection Agency and US Public Health Service Drinking Water Standards). Primary drinking-water regulations are legally enforceable standards for which all public water systems must comply; secondary drinking-water regulations are non-enforceable guidelines. Data from the US Environmental Protection Agency.

Primary drinking-water regulations				Gases in atmosphere
Substance dissolved	Maximum contaminant level (mg/liter)	Potential health effects of long-term exposure above the maximum contaminant level		Average amount with time period for averaging
Arsenic	0.010	Skin damage; circulatory problems; cancers	CO_2	Primary standard: 35 ppm – 1 hour (not to be exceeded more than once per year) Primary standard: 9 ppm – 8 hours (not to be exceeded more than once per year)
Barium	2.0	Increased blood pressure	SO_2	Primary standard: 75 ppb – 1 hour (99th percentile of 1-hour concentration averaged over 3 years) Secondary standard: 0.5 ppm – 3-hour mean (not to be exceeded more than once per year)
Cadmium	0.005	Kidney damage	NO_2	Primary standard: 100 ppb – 1 hour, 98th percentile, averaged over 3 years; Primary and secondary standard, 53 ppb, annual mean
Chromium	0.1	Allergic dermatitis	TSP*	$PM_{2.5}$ – primary standard: 12 micrograms/m^3, annual mean averaged over 3 years Secondary standard: 15 micrograms/m^3 annual mean, averaged over 3 years Primary and secondary standard: 35 micrograms/m^3 – 24 hours, 98th percentile averaged over 3 years PM_{10} – primary and secondary standard: 150 micrograms/m^3, 24 hours, not to be exceeded more than once per year on average over 3 years
Copper	1.3	Gastrointestinal problems; liver or kidney damage;	Ozone	Primary and secondary standard: 0.075 ppm, 8 hours, annual fourth-highest daily maximum 8-hour concentration, averaged over 3 years
Fluorine	4.0	Bone disease; mottle teeth	Pb	Primary and secondary standard: 0.15 micrograms/m^3, rolling 3-month average
Lead	0.015	Physical and cognitive delay in children; kidney problems; high blood pressure		
Mercury	0.002	Kidney damage		
Selenium	0.05	Fingernail or hair loss; circulatory problems; peripheral neuropathy		
Thallium	0.002	Hair loss; kidney, intestine, liver problems		

Secondary drinking-water regulations

Substance dissolved elements	Secondary standard (mg/liter)	Potential health effects of long-term exposure
Chloride	250	Not observed
Iron	0.3	Cirrhosis; liver cancer; cardiac arrhythmias
Manganese	0.05	Cognitive delays in children
Silver	0.01	Skin and eye discoloration
Sulfate	250	Dehydration; diarrhea
Zinc	5.0	Stomach cramps, nausea
pH	6.5–8.5	
Dissolved solids	500	

* TSP = total suspended particles. Some states recommend lower levels for some substances. For example, the Minnesota Department of Health issued a guidance that infants who consume tap water (mixed with baby formula) consume water with no more 100 ppb manganese.

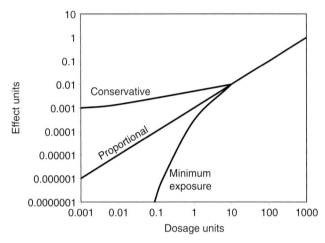

Figure 3.19 Three possible relations between dosage and effect. The conservative curve is based on the assumption that a constant amount of damage is caused below some lower limit. The proportional curve assumes that there is a constant relation between dosage and effect over the entire range of dosages. The minimum exposure curve assumes that exposure below a specified minimum has no detectable effect.

organisms die in a specific time period. By convention, the results are reported as the concentration necessary to produce 50% mortality, a factor known as LD_{50}. Toxicity measurements of this type can be made on any living organism, although aquatic organisms are used most commonly. There is no universal scale of toxicity because different organisms respond differently to the same compound. Furthermore, these measurements assess only the mortality that occurs in a short time span, usually 96 hours, and do not take into account long-term effects such as increased risk of cancer.

Longer-term studies of human populations are often based on comparisons to standard mortality ratios, which show the proportion of a population expected to die of a specific exposure. The effects of materials such as chrysotile asbestos can be investigated by determining whether groups that have been exposed to it exhibit a higher mortality ratio than the general population. In the absence of total control of all aspects of the

life of all subjects for years or decades, it is difficult to isolate specific causal factors. The fact that lower dosages require longer periods of time to manifest symptoms aggravates this problem because it is necessary to continue studies over decades, thus increasing the variation in exposure to other factors.

One method designed to cope with these complications is the extrapolation technique, which involves determination of the effect per unit of high dosage and extrapolation of this relation to lower dosages to give the effect typical of natural settings. The simplest assumption, represented by the proportional curve in Figure 3.19, accepts a linear relation between dose and effect. As one decreases, the other does so also, at a constant rate. A frequently quoted pitfall in this method is illustrated by its application to estimate fatalities associated with crossing a river. If 100 out of 1000 of the people attempting to cross a 10-meter deep river die by drowning, then 10 out of 1000 people should die trying to cross a river 1 meter deep, and 1 out of 1000 people should die attempting to cross a river only 0.1 meter deep. This analogy is obviously not correct and it suggests that there is a minimum threshold value below which no damage is experienced (minimum exposure curve, Figure 3.19).

Dose–effect relations are even more complex for some compounds and elements. For instance, underexposure to elements that are essential to life, such as zinc and copper, can be just as debilitating as overexposure. In contrast, it seems clear that even small levels of exposure to metals such as mercury and lead have negative health effects for children and pregnant women (Needleman, 2004; Trasande *et al.*, 2005; Patrick, 2006).

3.4.4 Remediation

Once an anthropogenic emission has been identified as damaging, means must be found to stop it or clean it up, efforts known generally as remediation. Most regulatory agencies require that pollutants in water and air leaving mineral and

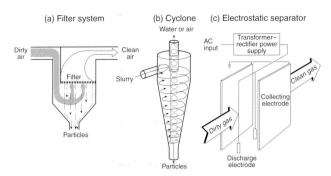

Figure 3.20 Methods of separating particles from liquids or air. In filter systems (a), particles are removed when they fail to pass through small holes. In the cyclone (b), the fluid–particle mixture moves down along the walls of the cone where particles fall out and the fluid rises to exit through the top. In the electrostatic separator (c), particles are charged by electrons in a high-voltage field and attracted to one of the poles where the charge is neutralized, causing them to fall to the bottom of the chamber.

industrial facilities be at or below recommended standards (Table 3.5), even if they did not meet those standards entering the property. (This requirement has had unlikely consequences, such as the construction of a shelter to protect mine waste in northern Michigan from contamination by fallout of airborne mercury from outside the property.) Treatment of some type is commonly necessary to meet these standards. The simplest methods involve physical separation of particulate material. The least expensive and most common method for air and water samples uses a **cyclone** (Figure 3.20b). Greater efficiency can be obtained at additional cost with a filtration system that removes smaller particles, and very small particles can be separated from air by more expensive electrostatic methods that ionize particles causing them to be attracted to an electrode (Figure 3.20c). Even after they have been recovered, the disposal of particles is a major problem. Although biological separation methods have not been widely used yet, genetic engineering promises to open some very exciting possibilities, including plants that take up several percent of their weight in metals, and bacteria that consume crude oil.

Chemical separation methods are much more varied and complex. In liquids, they can involve exchange of some less harmful substance for the pollutant, precipitation of the pollutant in solid form, or adsorption of the pollutant on an active surface. **Ion exchange** is the most commonly used process, with typical water softeners being the prime example. These units operate by exchanging the offending calcium in **hard water** for sodium in zeolite minerals to produce sodium in solution and calcium in the zeolite. Natural and synthetic

zeolites that operate in the same way are being developed to remove metals and other ions from solution. Precipitation and exchange are actually closely related. Acid mine drainage, for instance, can be neutralized by the addition of calcite, limestone, or lime to the solution. If the concentrations of dissolved calcium and sulfate in the resulting solution are low enough, this reaction is essentially an exchange of calcium in the rock for hydrogen ions in the water. On the other hand, if the concentrations of calcium and sulfate in the solution become high enough, gypsum or anhydrite precipitate, forming a sludge that must be removed from solution. Substances can also be removed from solution by adsorption on **activated charcoal**, a form of carbon that has been treated to increase its ion-exchange capacity. For dissolved ions, this is actually the same as exchange, because it is necessary to maintain the electrical balance of the solution (that is, equal total charges of cations and anions). Thus, most of the commonly used water-cleaning methods simply replace undesirable dissolved constituents with those that are less undesirable; if you want to remove all dissolved material, you must distill the sample and dispose of the precipitate, a very costly process.

Chemical separation processes for gases are different. Gases are not electrically charged, but can be adsorbed onto active surfaces such as charcoal and molecular sieves. They can also be removed by reaction with other chemicals, either in the gaseous state or after being dissolved in water. At present, air cleaning systems are used largely to remove SO_2 from exhaust gases, a process known as **flue-gas desulfurization (FGD)**. When the concentration of SO_2 in the exhaust gas is high, as in many modern metal sulfide smelters, it can be reacted with water and oxygen in an acid plant to make sulfuric acid (H_2SO_4) (Table 3.6). If the SO_2 concentration is low, it is usually treated in a **scrubber**, where it reacts with a finely dispersed compound such as calcite or lime to form calcium or sodium sulfite. In most cases, this is done by spraying a water slurry of the compound through the flue gas, although "dry" scrubbing can be carried out by spraying a finely divided powder of water and a reactive mineral. Concentrations of SO_2 are so low in many coal-burning electric utilities that neither process works very efficiently, and large amounts of SO_2 still escape from these sources. Scrubbers can be regenerative or non-regenerative, depending on whether the chemical process used to trap SO_2 can be reversed. Most non-regenerative scrubbers produce large volumes of waste that must be disposed of and are environmental contaminants in their own right. Other gases can be removed by similar systems.

Table 3.6 Common reactions used to remove SO_2 from flue gases, including those that produce sulfuric acid (acid plant) and those that precipitate calcium or sodium sulfite (scrubber). Both systems are most efficient on gas with a high concentration of SO_2, making them more useful in cleaning gases from smelter rather those from electric utilities, which have lower SO_2 concentrations (modified from Elliot and Schwieger, 1985 and Manahan, 1990). Efficiencies of SO_2 removal range from 50 to 98% (US Environmental Protection Agency). Eighty-five percent of FGD systems operating in the United States are wet systems, which remove 98% of SO_2.

Process	Chemical reaction	Consequences
Sulfuric acid reaction	$SO_2 + H_2O + \frac{1}{2}O_2 \rightarrow H_2SO_4$	
Sulfate and sulfite precipitating reactions		
Lime slurry	$Ca(OH)_2 + SO_2 \rightarrow CaSO_3 + H_2O$	Large volume of waste $CaSO_3$
Limestone slurry	$CaCO_3 + SO_2 \rightarrow CaSO_3 + CO_2$	Less efficient than $Ca(OH)_2$
Magnesium hydroxide slurry	$Mg(OH)_2 + SO_2 \rightarrow MgSO_3 + 2H_2O$	Can regenerate $Mg(OH)_2$

Nitrogen from air that is converted to nitrogen oxides (NO_x) during combustion is regulated as part of the Clean Air Act Amendments of 1990. As of 2012, NO_x emissions from coal-fired power plants had decreased by about 65% from 1995 levels (EPA, 2014).

New regulations for CO_2 emissions, and the possibility of a carbon tax, are the latest efforts to decrease anthropogenic contributions to the atmosphere. These proposed standards seek a 30% reduction in carbon emissions by the year 2030 from 2005 levels. The US Environmental Protection Agency proposal allows states to establish their own mechanisms for complying with the regulation. If the successful history of compliance with SO_2 and NO_x emissions standards, both of which decreased significantly following regulatory action, is a guide then it is likely that strategies for carbon capture and storage (CCS) will be successfully implemented. As of 2014, CCS implemented at a coal-fired power plant in Saskatchewan, Canada, resulted in a 90% reduction of CO_2 emissions (Danko, 2014). The implementation of CCS raises new challenges though because the captured carbon must be stored somewhere, a process known as geologic **carbon sequestration** that is an active area of research (Benson and Cole, 2008).

Conclusions

Environmental geochemistry is essentially the study of global and local geochemical cycles for substances that move through the biosphere. Although the biosphere is not the major reservoir for any important elements, many substances affect life through it. Hence, environmental geochemistry takes a somewhat distorted view of Earth processes, but one that is critical to our continued survival on the planet. Many of the cycles studied for environmental reasons, such as that of salt, involve natural, geologic materials. Others, such as that for chlorofluorocarbons and hydrofluorocarbons, involve synthetic materials. Still others, ranging from the global cycle of zirconium to the cosmic cycle of hydrogen, are studied by geochemists and cosmochemists, but are not of major interest to environmental geochemists because they do not impinge significantly on the biosphere.

Studies involving environmental geochemistry have several goals. First, the reservoir of interest must be recognized and its average composition for the element or compound of interest must be determined. Second, the form of the element or compound in related reservoirs must be specified. Third, reactions that transfer the element or compound from reservoir to reservoir must be determined. Finally, the impact of all parts of this cycle on the biosphere must be understood. When this is done completely, it is possible to distinguish between natural and anthropogenic inputs to environmental systems.

The results of these studies are commonly used to formulate legislation and regulations. Environmental legislation has occupied a large part of legislative time in many parts of the world. It is up to us to be certain that the studies on which this legislation is based are carried out and interpreted in ways that are consistent with geologic and chemical principles.

CHAPTER

4 Mineral exploration and production

Our global appetite for mineral resources is so great that even large mineral deposits are exhausted rapidly, making it necessary to search constantly for new ones. No example of this is more dramatic than the Prudhoe Bay oil field in Alaska, which was discovered in 1967. With an original recoverable oil reserve of 9.45 billion barrels (Bbbl), Prudhoe Bay was the largest oil field in North America and the 18th largest in the world (Carmalt and St. John, 1986). Before Prudhoe Bay was discovered, the largest field in the United States was East Texas, which ranked 56th in the world with "only" 5.6 Bbbl. The discovery of Prudhoe Bay increased US oil reserves by about 30%. Your first reaction in 1967 might have been that we could relax; this huge field would take care of our needs forever. Not so. At the 1967 US consumption rate of 6 Bbbl annually, we could have exhausted Prudhoe Bay in less than 2 years. It lasted longer only because it joined many other already-producing fields and was produced at a much slower rate. Even so, US oil production peaked in 1989 and declined steadily until shale-oil production started in 2007.

This is a distressingly short period of time to exhaust something that nature took thousands or millions of years to form. That is why we call mineral deposits **non-renewable resources**. They differ greatly from **renewable resources** such as trees and fish that can be replenished naturally in periods similar to our lifetimes. It follows that we must think of Earth as having a fixed inventory of minerals to supply our needs.

You might say that there is no point in looking for deposits if even the large ones will be exhausted so rapidly. This argument is no more logical than it would be to stop fixing meals because each one is eaten as soon as it is prepared. So, let's look into what mineral exploration entails.

Mineral exploration is a business in which failure is the norm and success is the exception (Slichter, 1960; Woodall, 1984, 1988, 1992). Success requires enough capital to withstand a long series of failures. Most major oil companies spend billions of dollars each year in exploration, and most mining companies spend hundreds of millions. In 2014, even after a major decline in commodity prices, about $11 billion

---

BOX 4.1 | SUSTAINABILITY – GEOLOGISTS VERSUS ECONOMISTS

Does Earth really have a finite supply of mineral deposits? Many economists do not think so. They maintain that increased oil prices and new technology will make it possible to exploit deposits that are too low grade for today's market. In fact, this is exactly what has happened in the last few years with the development of shale-oil fields in which fracking of source rocks yields oil. Geologists agree that changes of this type are important, but they note that the volume of the crust is limited and that this places an ultimate limit on the conventional and even the new deposits that we can find. So, although economists might win in the short run, geologists are right in the long run. In the meantime, we have to find the deposits that remain.

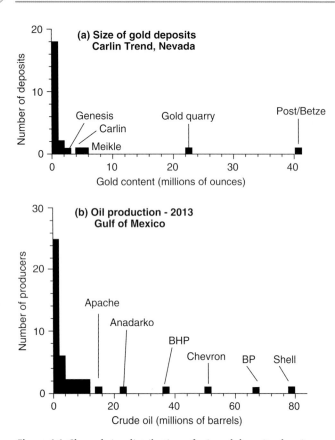

Figure 4.1 Skewed size distribution of mineral deposits showing how a few deposits can contain a large fraction of total mineral reserves. (a) The Carlin Trend in Nevada is the largest gold-mining region in the United States. (b) The Gulf of Mexico is the largest oil-producing province in the United States.

was spent in exploration for non-ferrous metal deposits around the world and the major oil companies spent about $660 billion on exploration (including production costs) (Steel, 2014; OGJ, 2015). The combined effects of inflation and the increasing difficulty of discovery will require that these expenditures increase significantly if we are to maintain our mineral reserves at present levels.

Things are made worse by the fact that mineral deposits come in a wide range of sizes (Singer, 2013), forming a skewed distribution with many small ones and only a few large ones (Figure 4.1). Size distributions of this type tell us that, even if we discover a deposit, it is much more likely to be small. Unfortunately, many small deposits cannot be extracted at a profit. Thus, most discoveries are "geologic successes" but economic failures.

With all of these complications, success in the mineral business is understandably elusive and anything that can be done to enhance it is most welcome. This is where prospectors, geologists, geochemists, and geophysicists come in. Some

of the concepts and methods used in their search are discussed later in this chapter.

4.1 Mineral exploration methods

Although the basic function of mineral exploration is to find new mineral deposits to replace those that become exhausted, it is not directed evenly at all mineral commodities. This is particularly true in market economies, where exploration is driven by the desire to make a profit and therefore focuses on commodities with increasing demand or price. Figure 4.2, which shows the relation between mineral reserves and price increases, provides a good indication of global exploration targets. As can be seen here, reserves of oil, gold, and silver are low and prices have increased since 1960. In contrast, large reserves of commodities such as magnesium, sodium carbonate, phosphate, and potash discourage price increases and exploration. Base and structural metals, which are shown by black symbols, are exploration targets because of their low reserve/production ratios. Whatever the target, mineral exploration can be undertaken in several different ways.

4.1.1 Prospecting and random exploration

Mineral exploration has not always been a science. Many good deposits, such as the Spindeltop oil field, were found by **prospectors** with nothing more than hope and determination. A modern manifestation of this approach is the proposal that exploration be carried out by **random drilling** (Drew, 1967; 1990). Geologic clues to the location of deposits will not be ignored, but they will not be the primary guides in the search. The number of exploration drill holes required by this approach and their distribution will be determined by the nature and size of the deposit being sought and the degree of statistical certainty desired for the search.

Although no truly random method has been used, a modified version led to the discovery of zinc in central Tennessee (Callahan, 1977). Inspiration for the search came from the eastern part of Tennessee near Knoxville, where large zinc deposits fill parts of a buried cave system in Ordovician-age limestone and dolomite known as the Knox Group (Figure 4.3). To the west around Nashville, Knox Group rocks were buried by hundreds of meters of younger sediments and drilling that concealed possible deposits. The favorable area was huge, over 18,000 km^2, and the geologic clues at the surface were simply not good enough to select the best place to drill. So, the area was tested by drilling holes randomly, based largely on where land owners agreed to

---

BOX 4.2 | MINERAL EXPLORATION, PERSONAL INVESTMENTS, AND YOU

This is a good place to insert a word of caution. Mineral exploration has led to some very large fortunes, including those of the Rockefeller, Carnegie, Mellon, and Getty families. Maybe that could happen to you someday.

Before investing in a mineral opportunity, however, evaluate the quality of available geologic information and the environmental, legal and tax framework in which the project would operate. Pay attention to the technical experience of those involved in the project, and be wary of volunteers. Some people are simply not very good at what they are doing and some are downright dishonest. Things were so bad during the early days of the American West that Mark Twain is reported to have said that a mine was a hole in the ground with a liar on top.

Salting, the practice of scattering ore minerals around a worthless property, still happens. You can do it yourself by rubbing your gold ring onto a rock, leaving a trace of gold that would make the rock seem attractive to an uninformed investor. By far the most spectacular salting effort in recent history was the Bre-X scandal in which small amounts of gold were added to exploration drill samples from the Busang deposit in Indonesia. These produced high gold values when the samples were assayed and led to estimates of a very large gold resource in the deposit.

Before the fraud was discovered, Bre-X stock had reached a market value of $6 billion all of which vanished. Litigation related to the case continued for almost a decade but did not result in a conviction (Goold and Willis, 1997). However, it did result in formulation of the Canadian National Instrument 43–101, which provides strict guidelines for handling of exploration samples and reporting of data. Similar guidelines, including the Joint Ore Reserves Committee Code in Australia and Reporting of Mineral Resources and Mineral Reserves in South Africa went a long way toward removing the loopholes through which frauds of this type could wriggle.

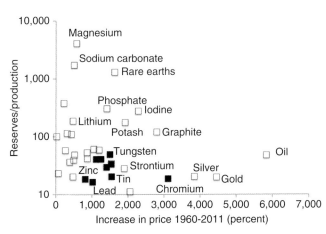

Figure 4.2 Mineral resources classified according to their attractiveness as exploration targets. Individual boxes represent specific commodities, only a few of which are labeled because of space limitations. The vertical axis shows the ratio of reserves to production – large numbers mean large reserves are already available. The horizontal axis shows the increase in price from 1960 to 2011. Note that oil, gold, and silver, which are the most sought after commodities today, have had the greatest price increase and have a low ratio of reserves to production. Dark symbols show that most base metals, which are also exploration targets, have low reserve-to-production ratios.

exploration. Luckily, the 79th hole intersected ore and led to the development of a major zinc-producing district.

The Knox Group extends northward beneath much of Kentucky (Figure 4.3). Would you explore for another big deposit in this area? It would not be an easy job. The favorable region measures over 5,000 km^2 and a large deposit might have an area of only about 1 km^2, only about 0.2% of the favorable area. Obviously, it would help to have additional information to guide exploration. That is where geology, geochemistry, and geophysics come in.

4.1.2 Geological exploration

Geological exploration works because most mineral deposits form in specific geological environments by processes that leave evidence in the rock (Ohle and Bates, 1981). By studying deposits that have already been discovered, geologists learn about these environments and processes and then look for them elsewhere.

Discovery of the Las Torres silver deposits in Guanajuato, Mexico, resulted from this sort of study (Gross, 1975). Mining of vein deposits at Guanajuato started in 1548 and had produced over 31,000 tonnes of silver and 125 tonnes of gold by the late 1960s, when most veins seemed to be exhausted.

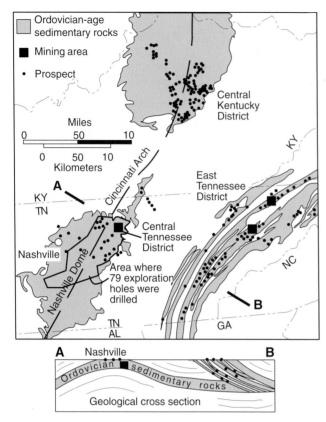

Figure 4.3 Geologic map and cross section showing the distribution of Mississippi Valley-type (MVT) zinc deposits and prospects in Knox Group sedimentary rocks in the southeastern United States (compiled from Callahan, 1977 and Clark, 1989)

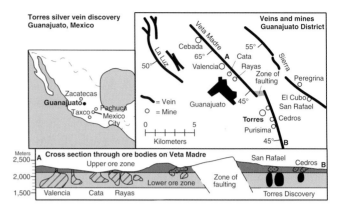

Figure 4.4 Geologic map of the Guanajuato district, Mexico and the Veta Madre and the area between A and B shown in the cross section at the bottom. In this section, ore zones fall into a lower zone (light shading) containing the Valencia, Cata, and Rayas deposits and an upper zone (dark shading) containing the San Rafael and Cedros deposits, which overlie the Torres discovery in the lower zone (modified from Gross, 1975).

However, reexamination of the main vein, the Veta Madre, showed that it contained both upper and lower ore zones and that only the upper zone had been mined in the southern part of the vein (Figure 4.4). Deep drilling beneath the upper zone in this area discovered the Las Torres deposit and revived mining at Guanajuato. Between 2009 and 2013, the Veta Madre produced almost 6 million ounces of silver and about 45,000 ounces of gold, and had been explored to depths of almost 700 m (Great Panther Silver Ltd.).

Basic geologic information is a critical part of mineral exploration, and much of it comes from maps and reports prepared by government agencies such as the US Geological Survey, the Geological Survey of Canada, the Sernageomin in Chile, and the Australian Geological Survey Organization. Hamilton (1990) estimated that government surveys were important aids in the discovery of more than half of the mines found in the United States, Canada, and Australia up to that time.

4.1.3 Geochemical exploration

Geochemical exploration depends on the fact that nature works just as hard to destroy mineral deposits as it does to

create them. At Earth's surface, weathering disperses the components of mineral deposits into the surrounding water, soil, vegetation, and air to create chemically enriched zones known as geochemical anomalies (Rose *et al.*, 1979; Levinson, 1983). The Montico copper–zinc deposit in the Dominican Republic provides a good example of geochemical dispersion (Figure 4.5). Montico is at the top of a hill in humid, tropical forest and grassland. It consists of copper, zinc, and iron sulfides in a matrix of quartz, that have been weathered to depths of at least 30 m by reactions similar to those that generate acid mine drainage (Table 3.4). As a result, the weathered top of Montico consists of quartz grains, rusty colored limonite, and acidic water containing dissolved copper and zinc like those shown in Figure 4.6.

Almost as quickly as they formed, these weathering products were dispersed into the surrounding surface environment. Quartz and limonite grains became part of the regolith, which moved downslope by slump, landslide, and sheetwash into arroyos (stream valleys), where it was carried downstream as sediment. Limonite in the stream, which gave the Arroyo Colorado its name, became a helpful exploration guide. (A more famous example is the Rio Tinto, which drains the large copper deposits near Huelva in Spain, and for which the Rio Tinto Corporation is named.) Soluble elements such as zinc were carried away in groundwater and partially adsorbed on clay and iron oxide minerals, the water and adsorbed elements entered the arroyo yielding high copper and zinc values in sediment far downstream (Figure 4.5).

BOX 4.3 | HYDROTHERMAL ALTERATION

Geological exploration is often aided by **hydrothermal alteration**, which forms when hot water reacts with adjacent rocks, changing their mineral and chemical composition. Altered rocks usually contain minerals in a zonal arrangement that can be used to vector toward the center of the mineral system. A classic example of this vectoring happened at the San Manuel deposit in Arizona (Lowell and Guilbert, 1970). As the San Manuel deposit was mined, it became apparent that it was actually only half of a deposit; the other half seemed to have been cut off by a fault (Figure Box 4.3.1). Early exploration holes drilled to look for this lost half of San Manuel went to depths of 630 meters, but stopped when copper was not found. Review of these holes with hydrothermal alteration in mind, however, suggested that they had entered the outer alteration zones around a porphyry copper deposit, and subsequent deepening of the holes resulted in the discovery of the Kalamazoo ore body, San Manuel's other half (Lowell, 1968).

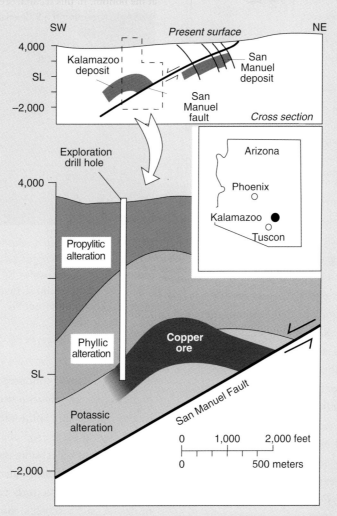

Figure Box 4.3.1 Kalamazoo–San Manuel area, Arizona, showing how zones of hydrothermal alteration around the Kalamazoo porphyry copper deposit were used to guide deep exploration. Potassic, phyllic, and propyllitic alteration refer to types of alteration containing potassium feldspar, muscovite, and chlorite, respectively. SL = sea level (compiled from Lowell, 1968 and Chaffee, 1982).

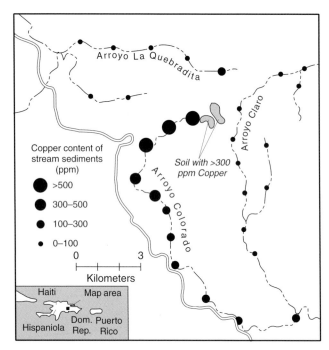

Figure 4.5 Copper (Cu) content of stream sediments and soils around the Montico deposit, Dominican Republic. Soil containing more than 300 ppm Cu covers a large area over the deposit and stream sediments draining this area contain over 500 ppm Cu near the deposit and as much as 250 ppm many kilometers downstream.

Geochemical exploration surveys also use other natural features, such as plants, whose roots reach deep into the underlying regolith, as well as lakes and swamps (Figure 4.7a). In areas where glaciers have covered the bedrock with sand and gravel, these can be sampled for traces of ore scraped from the underlying rock. Surveys are also based on gases such as methane and ethane that leak upward from oil and gas fields (Figure 4.7b). Radioactive decay of uranium produces helium and radon that can percolate upward through pores in the soil, and the weathering of metal sulfides forms mercury and sulfur gases that can be used in surveys.

Regional geochemical surveys help locate these areas, as well as stimulating exploration and aiding strategic planning. They have been used by the United Nations to encourage mineral exploration in less-developed countries (LDCs) (Guy-Bray, 1989). More-developed countries (MDCs) have used geochemical surveys, such as the US National Uranium Resource Evaluation (NURE) program, to locate areas favorable for mineral deposits, and similar studies have long been used to assess areas proposed for wilderness status (Corn, 2008).

4.1.4 Geophysical exploration

Geophysical exploration involves the search for deposits by measuring physical properties of rocks, such as magnetic

Figure 4.6 (a) Weathered veins in the Pueblo Viejo gold–copper deposit have released copper, some of which has precipitated as blue-green carbonates and oxides. (b) This pool of water near the weathered veins contains so much dissolved copper that the knife blade was coated by copper that was deposited from the water. See color plate section.

intensity, electrical conductivity, radioactivity, and the speed of shock (seismic) waves passing through them (Dobrin, 1976; Kearey *et al.*, 2002). Some of these measurements can actually detect the presence of the desired element or mineral, although most reflect only the general nature of buried rock.

Geophysical exploration for metal deposits measures magnetic, electrical, and radioactivity properties. Under favorable circumstances, measurements made at the surface can detect metal ore bodies at depths of a hundred meters. Many of these geophysical measurements can be made from the air, thus increasing the speed of surveys and minimizing their environmental impact (Figure 4.8). Some minerals, such as magnetite and pyrrhotite, are *magnetic* and can be detected by measuring the direction and intensity of Earth's magnetic field (Figure 4.9a). Many metallic minerals

Geochemical exploration

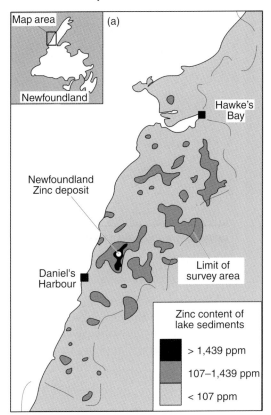

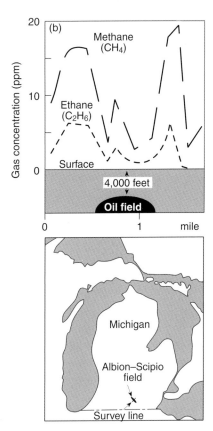

Figure 4.7 Natural geochemical pollution around mineral deposits. (a) Zinc content of lake sediments around Daniel's Harbour zinc deposits in northwestern Newfoundland (modified from Davenport *et al.*, 1975). (b) Methane and ethane contents of soil gas over the Albion–Scipio oil field, Michigan. Soil gas contents are thought to be highest on the edges of the field because rocks immediately above the field are less permeable (modified from Burns, 1986).

BOX 4.4 | GEOCHEMISTRY AND THE PRISTINE EARTH

Geochemical exploration is living proof that Earth is not "pristine." The natural landscape contains many natural geochemical dispersion patterns that would be classified as highly polluted if they were of anthropogenic origin. In 1793, George Vancouver, the navigator for Captain Cook, reported floating balls of tar and oil seeps in the Santa Barbara Channel offshore from southern California, an area that later became a major oil producer. Arsenic is widespread naturally in many sedimentary rocks and has been a major contaminant of water and food in areas of India, Bangladesh, and China (Garelick *et al.*, 2008).

---

conduct electricity better than the surrounding rocks (Hohmann and Ward, 1981), a phenomenon that can be used to carry out a wide array of *electrical* and *electromagnetic* measurements (Figure 4.9b). Surveys of this type have been a critical element in the exploration of large areas in Canada, Scandinavia, and Russia, where glacial deposits obscure most underlying rock (Figure 4.10a). Radioactivity can be used to detect uranium, thorium, and potassium, all of which have naturally radioactive isotopes. Original **Geiger counters**, which detected alpha radiation, have been superseded by **scintillometers** and gamma-ray spectrometers, which measure gamma radiation. Gravity methods, which measure the strength of Earth's gravitational field, can reflect differences in the type of rocks present in large regions, but are not commonly sensitive enough to be used directly to locate mineral deposits.

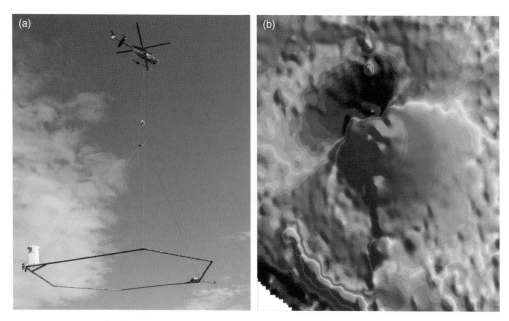

Figure 4.8 (a) Helicopter in Alaska towing a large wire loop that measures electromagnetic fields in rock below (photograph courtesy of Kirsten Ulsig, SkyTEM Surveys APS). (b) Magnetic anomaly over a possible diamond-hosting kimberlite pipe. The anomaly consists of a zone of high magnetic intensity (red-purple) and a zone of low magnetic intensity (blue). A line of smaller anomalies that trends directly north and south from the kimberlite is caused by a related kimberlite dike (image courtesy of Brooke Clements and Jennifer Pell, Peregrine Diamonds Ltd.). See color plate section.

Geophysical exploration

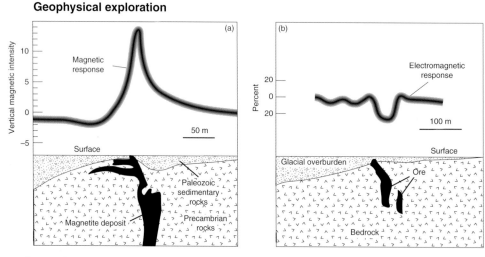

Figure 4.9 Geophysical anomalies associated with mineral deposits. (a) Magnetic anomaly over iron ore body at Iron Mountain, Missouri. The clarity of magnetic anomalies depends on the shape of the ore body and its orientation relative to Earth's magnetic field (from Leney, 1964). (b) Electromagnetic anomaly (in phase MaxMin-II horizontal loop at 888 Hz) over Montcalm township nickel discovery in Ontario (adaptation of Figures 6 & 7 of *Geophysics of the Montcalm Township Copper–Nickel Discovery* by Douglas C. Fraser, published in the CIM Bulletin, January 1978. Printed with permission of the Canadian Institute of Mining, Metallurgy and Petroleum).

Exploration for oil and gas is done largely by **seismic** methods, in which shock waves from an explosion or a vibrating source are sent into the ground from a land-based unit or a ship (Figure 4.11). By determining how long it takes these shock waves to reach detectors around the area, it is possible to construct an image of the sedimentary layers in the subsurface and to delineate features that might contain oil or gas.

Older seismic surveys produced cross sections showing subsurface geologic relations in two dimensions (Figure 4.11b), but modern methods yield a three-dimensional image of the underlying rock that has significantly improved exploration success. Geophysical measurements can also be made down the length of a drill hole. This method, which is known as **well-logging**, includes measurements of radioactivity, electrical

Figure 4.10 (a) Gravels carried into the area by glaciers cover many parts of North America making mineral exploration difficult. In this photo, from the Soudan mine in Minnesota, a layer of gravels about 10 m thick covers iron ore. Geophysical methods are useful in exploring country of this type because they can "see through" the gravels to determine the character of underlying rocks. (b) The final stage of mineral exploration involves drilling to obtain samples of buried rock, either chips or cylinders known as core. The core samples shown here are cut by dark veins of sphalerite. See color plate section.

resistivity, and other properties that provide information on porosity and lithology of the rock (Darling, 2006).

4.1.5 Measuring the size and quality of deposits

Most deposits are found by drilling, which is carried out with a drill rig that consists of a superstructure with pumps and a motor that rotates a "string" of steel drill pipe into the ground. Drill rigs come in a wide range of sizes from portable to ship-mounted (Figure 4.12). In all of them, the drill pipe comes in segments that are screwed together. At the bottom of the pipe is a drill bit that is impregnated with diamond chips that are hard enough to cut through any type of rock. Rock fragments cut by the drill bit are washed to the surface by air, water, or a suspension of minerals and chemicals known as **drilling mud**, which is pumped down through the hollow center of the pipe. Drill bits can be designed to cut the rock into tiny pieces, called **cuttings**, or a cylindrical core of rock (Figure 4.10b), which provides more detailed information on the nature of the buried rocks. The most famous type of drill bit, consisting of three cone-shaped cutting heads and known as the Tricone, was first marketed by the Hughes Tool Company and formed the basis for the fortune accumulated by Howard Hughes.

Over the last decade there have been major improvements in directed and horizontal drilling (Samuel and Liu, 2009). Directed drilling uses special wedges and motors to make the drill hole go in a desired direction. Contrary to popular notion, drill holes do not go straight down, even if they start as vertical holes at the surface. Instead, their path is influenced by heterogeneities in the rock and by spinning of the drill rods, causing some holes to wander long distances from their intended targets. Directed drilling was developed to prevent this, and it is now used to direct holes toward locations that are hard to reach from above. This method can be used to drill a large number of holes from a single platform, thereby decreasing disturbance of the surface. Horizontal drilling is a variation on this procedure in which a vertical hole is turned so that it proceeds horizontally. This method has been widely applied in oil and gas exploration, especially where sedimentary layers are horizontal, because it permits the hole to intersect a much greater length of the target rock. **Hydraulic fracturing** or **fracking** is widely used in horizontal oil and gas wells to enhance porosity, as discussed in Chapter 7.

Offshore drilling has been one of the most active areas of innovation (Leffler *et al.*, 2003). It started in 1896, when rigs were placed on piers to drill the Summerland field in southern California. In 1938, true offshore drilling began with the discovery of the Creole field in 4.5 m of water about 2.4 km off the Louisiana coast. Since then, drilling has reached distances of several hundred kilometers from shore and water depths of almost 3,000 m. Drill rigs with jack-up legs can work in water as deep as 130 m, but deeper water requires special drilling ships (Figure 4.12b). During deep-water drilling, the ship must be kept on location above the hole by constant positioning adjustments. These techniques were perfected during the Deep Sea Drilling Project (DSDP), which explored the deep ocean floor to clarify the history of plate tectonics from 1968 to 1983. This program drilled over 325 km of holes into the ocean crust in water up to 7,000 m deep. Production of oil or gas from deep water requires stationary equipment, usually a fixed-leg or floating platform that is tethered to the

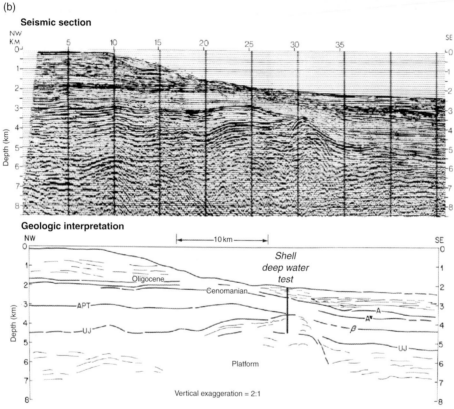

Figure 4.11 (a) Marine seismic survey ship, which uses collapsing air bubbles to send shock waves (that are harmless to marine life) through the water and into the underlying rocks. Returning waves are detected by measuring devices known as hydrophones, which are towed behind the ship (courtesy of Society of Exploration Geophysics). (b) Seismic survey across the Baltimore Canyon offshore from the eastern United States showing the edge of the carbonate platform that has been the target of exploration drilling for oil (from Klitgord and Watkins, 1984). See color plate section.

seafloor. One of the deepest producing platforms in the world is the Shell Oil Perdido, a $3 billion floating platform extracting oil from water depths of up to 2,925 m in the Gulf of Mexico (Mouawad and Meier, 2010; Shell Oil Company).

Drilling is very expensive. Drill holes used to explore metal deposits cost up to $200 per meter of hole, and a single deposit can require thousands of meters of drilling. Resulting costs, exclusive of administration and other support, range up to

Figure 4.12 Drill rigs. (a) Small "back-pack" drill rig in use on a Mexican hillside. Drill rods are rotated down the hole cutting a core of rock, while drilling mud is pumped in from the top to wash away rock cuttings (photograph by the authors). (b) The *D/V Chikyu* holds the world marine record of 7,740 m for depth of hole drilled below sea level. The ship, which is operated by the Japan Agency for Marine–Earth Science and Technology (JAMSTEC), is designed to drill through the ocean crust into the underlying mantle (photograph courtesy of JAMSTEC). See color plate section.

BOX 4.5 | OFFSHORE EXPLORATION IN ICEBERG ALLEY

As exploration and production have moved into Arctic areas, it has become necessary to deal with floating ice. In water up to 15 m deep along the northern Alaska coast, production has come from sand–gravel-constructed islands that can resist ice. In deeper water, production must be done from large, stable platforms that can resist ice. In one such area, "Iceberg Alley" along the coast of Newfoundland, Labrador, and Greenland, oil from the 2-Bbbl Hibernia field is produced from a 600,000-tonne structure that rests on the ocean floor and is designed to withstand impacts from icebergs. Farther north, drill ships operating offshore from Greenland are protected by vessels that monitor and tow icebergs when necessary. As climates in the area warm, the number and size of icebergs is expected to increase.

$20 million per deposit. Costs for individual holes are much higher in the oil and gas business. The average cost of a well drilled in the United States in 2007 was about $4 million, up from less than $1 million as recently at 2003. Exploration wells in offshore and frontier areas cost much more, averaging about four times more expensive than onshore wells. Keep in mind that these costs apply to both successful and unsuccessful exploration drilling programs. Over 70% of all oil and gas holes in the United States are dry and must be paid for by revenue from the other holes.

4.1.6 Reserve estimation and feasibility analyses

If drilling indicates that a deposit is present, it is necessary to undertake a more formal reserve estimate using the drill-hole data (Cronquist, 2001; Sinclair and Blackwell, 2006). In the simplest approach, holes are drilled on a regular grid or randomly. The depth and grade of ore or reservoir quality are determined in each hole and a judgment is made about continuity from hole to hole. The challenge in reserve estimation is to obtain an estimate that is highly accurate and precise at a reasonable cost. The uncertainty of these measurements can be minimized by drilling holes closer together, but this is expensive and the most successful drilling programs delineate the most ore or oil for the smallest amount of drilling.

Geostatistics, which started in the gold mines of South Africa, has been most successful in achieving this goal (David, 1977; Hohn, 1998; Chilés and Delfiner, 2012). The approach works because measurements of ore grade or reservoir quality in any part of a deposit are related to measurements of the same type in other parts of the same deposit, with the relation decreasing in strength as the distance separating the two points increases. Thus, information from one hole can predict ore or reservoir quality in adjacent holes. By determining the distance at which this relation breaks down, it is possible to specify the minimum spacing of drill holes needed to get a reliable estimate of the reserve in a deposit. This relation also allows bankers and environmental regulators to judge whether a deposit or area have been adequately evaluated.

When the reserve estimation is complete, a deposit can be divided into proven, probable, and possible reserves. These terms are commonly used for ore in a single deposit, whereas the generally similar terms, measured, indicated, and inferred reserves, are used to discuss estimates for a larger region containing many deposits. Reserve estimates form the basis for a **feasibility analysis** to determine whether the deposit can be exploited economically (Wellmer, 1989; Harris, 1990; Kasriel and Wood, 2013). First, engineers determine the rate and cost of extraction, as well as costs for processing, transportation, and administration. Estimates must also be made of costs related to environmental monitoring and reclamation, including the amount and nature of any bonds that are required, and of taxes and royalties and other applicable charges. Finally, it is necessary to estimate future prices for the commodity of interest.

Once these estimates are on hand, the cost of extracting the resource and selling it can be compared to estimates of the future prices of the commodity to determine the potential profitability of the operation. If estimated costs are significantly less than the estimated value of future production, the deposit will probably be put into production, a process known as **development**. This involves the construction of drilling and production platforms and pipelines for oil and gas or a beneficiation plant and tailings disposal areas for hard minerals, as discussed later. If the deposit does not look economically attractive at this stage, it will be abandoned or held for possible later reconsideration. Many projects are stalled at this stage because they do not meet requirements for economic production. If the project will be developed it might require outside financing to help with costs. Although funds are sometimes raised by public stock offerings, most financing is accomplished by bank loans. Some banks and investment houses employ or retain mineral advisors to review the technical quality of data supporting loan requests.

4.1.7 Environmental effects of mineral exploration

Geological, geochemical, and geophysical exploration have a limited impact on the environment, usually confined to access roads, survey lines, and sample trenches, all of which require government permits. More damage is caused in wetlands if canals or raised roads are used for access, especially in delta areas of the Mississippi, Niger, and other major rivers (Getschow and Petzinger, 1984; Kadafa, 2012). In the United States, dewatering of abandoned, flooded mines for further exploration requires a permit from the National Pollutant Discharge Elimination System, as well as relevant state agencies.

Drilling is of more concern environmentally because it requires larger-scale access. It also has the potential to introduce drilling fluids into the ground and release natural fluids, such as brines, oil, natural gas, and H_2S, to the surface. These risks are greatest in oil and gas exploration, where reservoir characteristics are unknown and pressurized fluids might be encountered unexpectedly. Fluid escape can be minimized by the use of **drilling mud**, which consists of a suspension of fine-grained barite and bentonite clay in water, with or without organic fluids. The bulk density of this slurry is high enough to wash rock cuttings made by the drill bit to the surface and to help prevent the escape of pressurized fluids. Thousands of gallons of drilling mud can be used at a single well; it is recycled unless it is lost when the hole enters a cave system or other large void. Barite is stable in oxidizing water and does not cause environmental problems, but organic fluids in drill muds are of concern (Miller *et al.*, 1980a; Miller and Pesaran, 1980b; Neff *et al.*, 2000; Contreras-León, 2013).

The main safety from blowouts due to unexpected pressures during exploration drilling is provided by a **blow-out preventer**. In the event of a sudden increase in pressure, this valve shuts the hole automatically. If it fails, oil or brine can be caught in ponds enclosed by dikes that surround wells on land. Where pressurized oil or gas reaches the daylight surface, it forms a **blowout** or **gusher**. In the worst case, the oil and gas are ignited by heat or sparks from drilling equipment to become a tower of fire. Once the fire is extinguished, some blowouts can be stopped by capping the well whereas others require the injection of mud and cement from a new hole drilled to intersect the blowout well at depth. Blowouts can cause much more damage in offshore settings because oil spills into the ocean.

BOX 4.6 | **DEEPWATER HORIZON OIL SPILL**

The Deepwater Horizon (Macondo) oil spill in the Gulf of Mexico provided a catastrophic example of what can happen when a blowout preventer fails (Wassel, 2012). The spill coated about 675 km of Louisiana coast, damaging vegetation that prevented shoreline erosion, weakening populations of organisms from snails to dolphins, and requiring over $4.4 billion of fines and clean-up costs (Cornwall, 2015). The total amount of oil spilled, most recently estimated to be 4.9 Mbbl, is only about 0.015% of global annual consumption, underscoring the rarity of oil spills, but their enormous potential for damage. Oil spills are discussed further in Section 7.1.2.

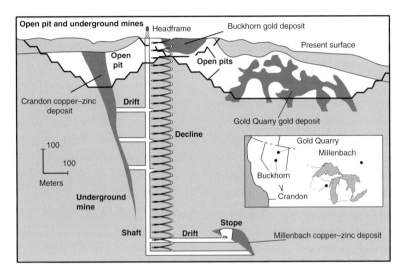

Figure 4.13 Cross section showing depth of four actual metal deposits and the way in which they can be mined by open pit or underground methods.

4.2 Mineral extraction methods

Once a mineral deposit has been found and judged to be economically attractive, it must be extracted if we are to benefit from it. This can be done by mining, in which rock is removed from the ground or by use of well systems that remove only a fluid, leaving the rock behind.

4.2.1 Mining

There are two main types of mines, and the one that is used depends on the depth to be mined (Hartman, 1987). In general, deposits within a hundred meters of the surface are extracted from **open pit mines** (also known as **open cast mines**) and those at greater depth come from **underground mines** (Figure 4.13). In general, the cost and difficulty of mining increase with depth, and extremely deep ore deposits are not mineable at a profit by any method. Curiously, open pits account for over 90% of the ore mined in the United States but only about 50% of the ore mined in the rest of the

world. There is no geological reason for this, nor are there tougher environmental regulations in LDCs, which might encourage underground mining. Instead, the controlling factor is economic and relates to average wage scales in LDCs. Because underground mining requires more worker-hours per tonne of ore produced, wages are a larger proportion of mining costs. Thus, in areas with lower wages, underground mining and other labor-intensive ore concentrating processes can be used on deposits that could not be mined in the United States and other MDCs (Figure 4.14).

A typical open pit mine removes ore and **overburden**, the worthless rock that overlies ore. The ratio between the volumes of overburden and ore, known as the **stripping ratio**, rarely exceeds ten and commonly is less than five. Volumes of rock that are moved by these operations are truly huge. The Grasberg mine in Indonesia, one of the largest open pit mines in the world and one with a relatively low stripping ratio, removes over 250,000 tonnes of ore and 150,000 tonnes of waste each day. Walls of open pit mines are sloped to prevent rock from falling to the bottom where people are working, which makes the pit much wider at the top than at the bottom (Figure 4.13). When commodity prices are low, some mines will economize by mining less barren rock and making pit walls steeper, a practice that can lead to failure of the pit wall (Figure 4.15). Other factors, such as faults that produce landslides, can make even the best pits susceptible to failures, as happened in 2012 at the Bingham mine just outside Salt Lake City in Utah.

Dredging, a special type of open pit mining, is used where the inflow of water is too high to be pumped out economically and the ore is sufficiently unconsolidated to be dug without blasting. It is almost always used on placer deposits, as shown in the titanium-deposit section in Chapter 9. The dredge digs its way forward, bringing along its own lake (Figure 9.9). Mined material is processed on board and gangue minerals are disposed of behind the advancing dredge. **Hydraulic mining** is a cross between dredging and conventional mining in which a jet of water is used to disaggregate the rock and wash it into a processing facility. **Strip mining** is also a special open pit method that is used on shallow, flat-lying ore bodies such as coal (Atwood, 1975). In spite of the unsavory connotation of its name, strip mining is actually one of the most environmentally acceptable forms of mining because it fills the pit and restores the land surface to its original form, as discussed further in the section on coal.

A typical underground mine is entered through a vertical **shaft** or a horizontal **adit** (Hartman, 1987). Tunnels or drifts enter the deposit at different levels, and ore is extracted from holes called **stopes**. It is taken to the surface by trucks along a spiral ramp or in large buckets up a shaft (Figure 4.13). Mines in flat-lying, relatively soft ores such as coal and potash use continuous mining, in which a machine advances steadily, breaking up the ore and sending it by conveyor to a collection point. Tabular and flat-lying ores are mined by **room-and-pillar** and **longwall** methods such as those discussed for coal mining. Steeply dipping ore bodies are mined by more selective methods such as **cut-and-fill mining** or by bulk mining methods such as **block caving**. Geologic factors complicating underground mining include weak rock that caves into stopes and dilutes the ore, and groundwater that flows in too rapidly to be pumped away economically (Cook, 1982).

In practice, not many open pit mines reach depths of more than 200 m, although giant pits are becoming more common. Examples include the Escondida and Chuquicamata pits in Chile, with depths of about 700 m and 800 m, respectively, and the Superpit in Kalgoorlie, Western Australia, with a depth of almost 600 m. The depth of underground mines is limited by the geothermal gradient, Earth's downward increase in temperature (Figure 4.16). Geothermal gradients are high in areas of active volcanism and relatively low in areas of old, **Precambrian** crust. For instance, the Limon gold mine along the Central American volcanic arc in western Nicaragua is only a few hundred meters deep but contains 75 °C water. At depths of more than 1,000 m, some mines experience **rock bursts**,

Figure 4.14 Mineral production in LDCs. (a) Overview of the KingKing porphyry copper deposit on Mindanao Island, Philippines. The upper part of this deposit has been excavated by hand by thousands of unemployed people to recover gold from the weathered ore. Small houses scattered around the deposit provide scale. (b) Young miner panning gold from weathered KingKing ore (photographs by the authors). See color plate section.

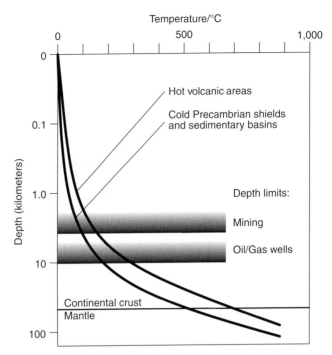

Figure 4.16 Geothermal gradients showing how Earth's temperature increases downward in hot volcanic areas and in colder rocks in Precambrian shield areas, and comparing them to the lower limit of mines and oil or gas wells. Note that the depth scale is logarithmic in order to show all depths.

Figure 4.15 (a) The Escondida porphyry copper mine in Chile, with annual production of more than 1 million tonnes of metal, is the world's largest copper producer. The mine reaches a depth of almost 700 m and measures 4 km by 3 km at the surface. About 1.4 million tonnes of ore and waste are removed from the mine each day using more than 150 large trucks that move up the inclined ramp at the back of the pit (photograph courtesy of BHP Billiton Ltd). (b) Side of the main pit at Grasberg, Irian Jaya in Indonesia, showing a landslide that formed where the pit wall collapsed. Pit-wall failures of this type are not uncommon and are caused by faults, water-saturated ground, and oversteep pit walls. The truck at the bottom center of the pit provides scale (photograph courtesy of Karr McCurdy). See color plate section.

which are small explosions that take place as rocks in newly opened, underground workings decompress due to the release of confining pressure (Durrheim, 2010). Shaft depths are also limited by the maximum length of cable that will not break from its own weight, usually about 2,600 m. Levels deeper than that must be accessed through internal shafts, which greatly increases the time required for workers to reach their stations. These factors combine to limit underground mining to about 2 km in most areas, although the deepest operating mine in the world, the TauTona gold mine near Carletonville, South Africa, had reached a depth of 3.9 km in 2012.

4.2.2 Well systems and pumping

Pumping and well systems are used to recover groundwater, crude oil, natural gas, and some solid minerals that can be dissolved or melted. Most well systems are devoted to oil and gas extraction and an entire discipline known as reservoir engineering involves the determination of the best distribution and type of wells for most efficient extraction (Lyons, 2009). Although reservoir rocks for oil and gas are porous and permeable, it is common to enhance their permeability in the area of producing wells by **fracking**, the process of injecting fluid and sand or other granular material at high pressure to fracture the rock. **Drill stem tests** are used to determine how much oil and gas can be extracted from a specific zone of rock in the well by isolating it with pads and measuring the flow rate that it produces. If these tests are successful, pumping facilities are installed, a process known as **well completion**, and the exploration well becomes a producer.

Production from many onshore fields is achieved by connecting to a pipeline, but wells in frontier areas must sometimes wait for years for pipelines to reach them. In the

Figure 4.17 Offshore oil production in increasingly deep water. (a) The artificial ice-proof island that produces from the Endicott oil field near Prudhoe Bay on the north coast of Alaska was constructed in the mid 1980s for over $1 billion. The 445-acre (18-hectare) island has close spacing of wells designed to minimize its area and environmental impact. Initial oil production at a rate of about 100,000 bbl/day has declined to about 10,000 bbl/day today, and total production so far amounts to about 460 Mbbl of oil (photo courtesy of BP America). (b) The Monopod platform, in 20 m of water in Cook Inlet, Alaska, began operation in 1967. Its single leg, which is less vulnerable to floating ice, has 32 wells that reach the entire oil field. With enhanced recovery methods that were applied in 2014, it produces about 2,900 bbl/day (photo courtesy of Chevron Corporation). (c) The Jasmine platform, at a depth of 81 m in the central North Sea, consists of a production unit with 24 slots for holes into the field and a living-quarters unit. The platform, with gross capacity of 140,000 bbl/day of oil equivalent (oil and gas), began production in 2013 (photo courtesy of ConocoPhillips). (d) In deep water where platforms do not work well, floating platform storage and offloading (FPSO) systems are used. The FPSO Aseng, shown here, began production from the Aseng field offshore Equatorial Guinea in November, 2011, at a water depth of about 1,000 m. The unit produces 80,000 bbl/day using injection of 150,000 bbl of water and has a storage capacity of 1.695 Mbbl (photo courtesy of Glencore Ltd.). See color plate section.

meantime, oil can be pumped into storage tanks and taken by truck or train to a pipeline connection or refinery as is the case for much of new shale-oil production from North Dakota. Offshore production takes place from platforms and islands (Figure 4.17) that are among the engineering wonders of the world (El-Reedy, 2012). Water depths in which these platforms operate have increased steadily over the last decade and are approaching 3,000 m. Production from deeper water comes from subsea wellhead systems.

Solution mining, the process by which solid minerals are dissolved or melted and then pumped from a well system, is the most environmentally acceptable form of mining because no overlying rock has to be removed. Evaporite minerals, such as halite and sylvite (potash), are uniquely suited to this process because they are easily dissolved in water. A variant of solution mining, known as the **Frasch process**, is used to mine native sulfur, which melts at 115 °C and can be flushed to the surface by superheated steam. Some metal oxide and carbonate minerals can be dissolved by sulfuric acid, uranium oxide can be dissolved by acidic or alkaline solutions, and gold can be dissolved by cyanide solutions. For these solutions to be effective mining agents, they must react only with the

BOX 4.7 | **FRONTIER EXPLORATION AND PRODUCTION**

As nearby deposits are exhausted, production must move into frontier areas. In fact, resource-based communities have been the traditional way that many frontier areas were settled. A few decades ago, some Canadians thought that a complex of mining and oil production communities would stretch across the Arctic. A few were established, notably Nanisivik and Pine Point in the Northwest Territories (Figure 5.7), but they were largely abandoned after the mines were exhausted. The more modern approach involves commuters. For instance, workers at the Lupin gold mine in the Northwest Territories lived in Edmonton and other cities to the south, or in established northern communities, and worked "fly-in" shifts several weeks long followed by extended vacation time. Similar fly-in work schedules are used at some of the large iron mines in Western Australia. This arrangement produces a happier, more stable work force, as well as eliminating the need for a large infrastructure at the mine, with attendant environmental problems. It also does away with **ghost towns**, the legacy of exhausted mineral deposits.

desired mineral rather than enclosing rock. An additional challenge confronted by solution mining is channeling, in which leach solutions find a few easy paths through the ore, never contacting the rest of it.

Mining will probably change greatly in the future, spurred by the need to reduce labor costs and the desire to produce materials with less environmental damage. Solution mining will almost certainly expand as better solvents are developed and ways are found to fracture rock more evenly. Emphasis will be on the conversion of raw material to a marketable form as close to the deposit as possible. An early start on this will probably be made with *in situ* combustion, in which combustible ores are burned underground to produce a gas or liquid that can be pumped to the surface (Hammond and Baron, 1976). Where physical removal of ore must still be done, it will undoubtedly involve larger-scale, continuous mining equipment that will increase processing volumes but decrease unit costs.

4.2.3 Environmental effects of mineral extraction

Mines and quarries of all types occupy about 0.3% of Earth's surface, roughly similar to the area occupied by railroads and reservoirs and much smaller than the areas occupied by logging operations (1.8%), cropland (12.8%), and permanent meadows and pastures (23.4%) (Hooke and Martin-Duque, 2012). Because the important construction materials stone, clay, and sand and gravel make up such a large proportion of mined material, MDCs have more than their share of land disturbed by mining. About 60% of disturbed areas is used for excavation and the remaining 40% is used for the disposal of overburden and similar wastes, which account for about 40% of the solid wastes generated annually in the United States.

The most important statute regulating surface mining in the United States is the Surface Mining Control and Reclamation Act of 1977 (SMCRA), which pertains largely to coal mines and is administered by the Office of Surface Mining and related state agencies. Mining wastes are regulated in the United States by the Environmental Protection Agency (EPA) and related state agencies under provisions of the Clean Water Act, Clean Air Act, Comprehensive Environmental Response, Compensation and Liability Act, and Toxic Substance Control Act, among others. A small amount of mining wastes is regulated under Subtitle C of the Resource Conservation and Recovery Act of 1986 (RCRA), which addresses hazardous waste disposal, but most are deemed non-hazardous. Finally, the National Environmental Policy Act requires that an environmental impact statement be prepared before any large mineral operation is started, and federal regulations require that any operation on land that will disturb more than 5 acres (12.35 hectares) must present and have approved plans for operation and reclamation, and deposit a bond to assure that reclamation will take place.

In the United Kingdom, environmental aspects of mineral production are regulated under the Environment Protection Act of 1990. In Canada, environmental regulation is carried out largely by the provincial governments. The earliest and most comprehensive of these laws were Bill 56 in British Columbia and Bill 71 in Ontario (Champigny and Abbott, 1992). In Australia, mining is regulated by individual states (Fitzgerald, 2002). Reclamation in most jurisdictions includes removal of buildings, restoration of the land surface to an acceptable contour, and alleviation of acid mine drainage caused by weathering of rock and unmined ore (Johnson and Paone, 1982; Carlson and Swisher, 1987). In the United States, the SMCRA requires that strip mines be restored to their original contours, with no slopes greater than 20°. It is not usually possible to fill other types of mines because they proceed deeper and deeper in the same spot, although some operations fill earlier pits with waste as they move to new pits. Aggregate quarries near towns have been used widely for disposal of construction wastes and a few metal mines have been used for municipal waste, as was done at Rustenburg and Pretoria, South Africa. Even where the pits are not filled, regulations require that pit walls and overburden dumps be reshaped and revegetated in a manner consistent with local topography. Mine reclamation can be quite successful; Butchart Gardens, one of the main tourist attractions in Victoria, British Columbia, was once a rock quarry, as was Queen Elizabeth Park in Vancouver. Many homes and recreation areas in MDCs surround partly flooded gravel pits that have become lakes. Most governments require that a reclamation plan be approved and a bond-securing completion of the plan be paid before a mine can begin operation. Where an operator plans to abandon a mine, a bond is usually required to cover future environmental problems such as the failure of the waste confinement system, seepage of mine water into nearby streams, or collapse of mine walls.

Although reclamation requirements are part of the law for active mines, many mines were abandoned before these laws were in place. In the United States, most abandoned mines that pose a threat to the environment have been identified and are listed in the Superfund National Priorities List administered by the EPA. As of 2014, this list contained 130 mines and processing facilities, most of which had been reclaimed. The EPA has estimated that the cost of reclaiming abandoned mine lands will be at least $35 billion, which has spurred a search for responsible parties that might help foot the bill. The net for potentially responsible parties is very wide and includes: (1) retroactive responsibility, in which parties can be held liable for actions taken before enactment of the Superfund in 1980, (2) joint and several responsibility, which holds that anyone associated with the property can be responsible for the entire clean-up regardless of their actual contribution to the problem, and (3) the fact that a potentially responsible party can be held negligent even if it was operating by industry standards. Although this approach to finding funds for reclamation is logical from a legal point of view, it discourages ownership of brownfields lands where mining has taken place. This is bad because many new mines are found close to old ones and because it ignores society's collective responsibility for the original mess. A more equitable solution would be to reclaim abandoned properties with funds from severance taxes, as is done with coal. A final consideration is the need to dispose of mine wastes in a way that allows the next generation to reach them. After all, today's waste is likely to be tomorrow's ore.

Table 4.1 Lung diseases (pneumoconiosis) associated with mineral dusts and fumes (compiled from Logue, 1991; Cleveland Clinic, 2015).

Disease	Related mineral material
Anthracosis	Carbon (soot)
Asbestosis	Asbestos
Bauxite lung disease	Bauxite fumes
Berylliosis	Beryllium minerals and fumes
Cadmium pneumonitis	Cadmium fumes
Coal workers' pneumoconiosis	Coal
Lung cancer	Asbestos
Mesothelioma	Amphibole asbestos
Silicosis	Crystalline silica
Silicosiderosis	Hematite

Health and safety considerations are also a major aspect of current mining regulations. Although fatalities do occur because of rock collapse and equipment failure, underground mine fires are the most serious problem. Most fires begin in wood supports, but they often spread to flammable rock such as coal. Even metal sulfides will burn if the fire is hot enough. Fatalities from fires are usually caused by carbon monoxide (CO), a highly toxic gas. Since the Sunshine silver mine fire in Idaho, US mines have been required to provide all personnel with a portable unit to covert CO to CO_2. Dust is actually of wider concern because it is the source of so many lung and system ailments (Table 4.1). The biggest problem is crystalline silica, including quartz, which causes the lung disease **silicosis**. Silicosis was not understood until well into the twentieth century, and earlier outbreaks were caused by poor dust control in quartz-rich mines. Dust and other particles are also of concern in uranium mines because they contain adsorbed radioactive isotopes that become lodged in the lung, where radiation damage can cause lung cancer. **Black lung disease** (also known as **coal workers' pneumoconiosis (CWP)**) is a related disease observed in coal miners.

The main environmental problems related to oil and gas extraction are the escape of underground fluids and land subsidence. Large-scale fluid escape from wells and local distribution systems is amazingly rare. About 15 Bbbl of oil have been produced from offshore wells in the United States since 1960 and, excluding the Deepwater Horizon (Macondo) accident, only 567,863 bbl have been spilled. Even if we include the 4.9 Mbbl estimated by the US Geological Survey to have been lost from Macondo, the spillage rate rises to only 0.034%. Only two wells of the thousands that have been drilled during that time have blown out, and one of these was Macondo.

Subsidence of the land surface, which is a problem over some shallow oil and gas fields and groundwater aquifers, can be alleviated by injection of replacement fluids. The most troublesome fluid that comes from oil and gas wells is brine. Although some brines contain high enough concentrations of sodium, potassium, bromine, or related elements to be of commercial interest, most are worthless and must be disposed of, usually by injection back into the sedimentary **strata** from which they came. The early practice of collecting the brine in unlined ponds called oil pits near producing wells is no longer permitted in MDCs because it contaminates local surface and groundwater. Instead, brines must be held in lined ponds or in tanks, and then injected into the ground. Brines have elicited additional concern because of their dissolved metals, organic compounds, and anomalous amounts of radium, a radioactive decay product of uranium and thorium. Accumulation of radium in oil pits, pipes, and other facilities has created local radioactive hazards. Reinjection of brines has greatly minimized this problem, although old brine pits remain a major legacy environmental problem in many areas (Schneider, 1990; Rowan *et al.*, 2011; Silverstein, 2013).

4.3 Processing of mineral resources

4.3.1 Processing of metal ores

Production of metals from ore is usually a two-step process. The first step involves **beneficiation**, which physically separates the ore in a **concentrate** containing the ore mineral and **tailings** consisting largely of gangue (waste) minerals. The concentrate is processed chemically to separate the desired metal from the ore mineral, a process known as **smelting**. In most aluminum ores, as well as some uranium, vanadium, gold, and other ores, concentrates are not produced and the metal is recovered directly from the ore (Cummings and Given, 1973).

The first step in beneficiation is usually crushing and grinding the ore to separate grains of ore mineral from grains of gangue minerals. This is carried out at the mine site in order to avoid the cost of transporting worthless rock to a distant location. Coarse-grained ore minerals can sometimes be separated by handpicking, but most ores are so fine-grained that they must be pulverized into a powder and separated by more complex methods based on the different physical properties of the ore and gangue minerals. Grains of magnetic minerals such as magnetite and pyrrhotite can be separated with an

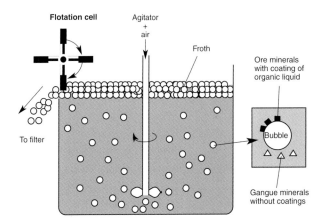

Figure 4.18 Cross section of a froth flotation cell showing how ore minerals adhere to bubbles and float to the surface of the cell. Waste mineral grains do not adhere to the bubbles because of their different surface properties and consequently remain at the bottom of the cell.

electromagnet. Density differences can also be used because most ore minerals are heavier than gangue minerals. In **froth flotation**, the most widely applied beneficiation method, a hydrophobic organic liquid coats desired minerals, which then attach to bubbles passing through the suspension, causing the grains to float. Unwanted minerals, which are not coated by the organic liquid, do not attach to bubbles and sink (Figure 4.18). Another method uses a sensor to test grains falling off the end of a conveyor and a jet of air to blow away grains that do not meet predetermined criteria.

Separation of metal from the concentrate is usually done in smelters (Figure 4.19) using **pyrometallurgy**, in which the melted concentrate separates into two immiscible liquids (Gilchrest, 1980). One liquid contains the metal and the other, which is called **slag**, contains the waste elements. The metal-bearing liquid is heavier than slag and sinks to the bottom of the mixture where it can be removed (Table 4.2). Smelting of iron ores in a **blast furnace** yields a metal liquid (**pig iron**), which is transferred to a second furnace for further processing to make **steel**. Conventional smelting of metal sulfide concentrates involved a two-step process in which waste minerals were separated into slag and sulfur was removed as SO_2, but newer smelting methods use a single, continuous process to limit energy consumption and the escape of gases. These include the use of **direct reduction** for producing iron and **continuous smelting** for producing copper and other base metals. Metal from a smelter must be purified by **refining**, to recover traces of gold, silver, and minor metals that followed the principal metal through the smelting process.

Metals can also be liberated from concentrates by **hydrometallurgy**, which relies on a caustic solution or solvent to

Metal reduction facilities

(a) Blast furnace   (b) Smelter – Noranda continuous system

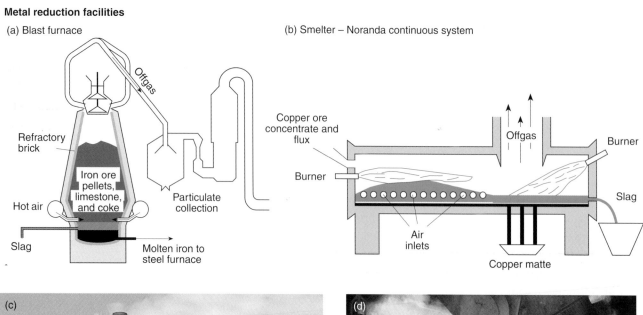

Figure 4.19 Metal ore reduction facilities. (a) Blast furnace used to convert iron ore concentrates (usually pellets) to pig iron, which is then transferred to steel furnaces. (b) Smelter used to convert metal sulfide concentrates to molten sulfide (matte), which is then transferred to **converters** to remove remaining sulfur as SO_2. (c) Blast furnace at Bremen, Germany (courtesy of ArcelorMittal). (d) Molten matte at the Alto Norte smelter, Chile (courtesy of Glencore plc.). See color plate section.

leach metal from the ore mineral. The metal is recovered from solution by precipitation (usually as a compound containing the metal) or by an electrolytic process known as **electrowinning** in which pure metal is precipitated onto the cathode of an electrochemical cell. This process is applied widely in many types of metal recovery systems, where it is often referred to as solvent extraction-electrowinning (SX-EW). Some hydrometallurgical processes can be carried out on piles or heaps of ore under the open sky. This process, which is known as **heap leaching**, requires less investment and is effective only on some gold, silver, copper, and

uranium–vanadium ores, as discussed in Chapter 11. Although present hydrometallurgical processes are based on inorganic chemistry, future methods will probably utilize microbes (Brierley and Brierley, 2001; Gahan *et al.*, 2012).

4.3.2 Environmental aspects of metal ore processing

Almost all beneficiation plants at mines in MDCs must recycle their water, even in areas as remote as the Red Dog mine in northwestern Alaska. Surface water flowing through the

Table 4.2 Chemical reactions that take place during pyrometallurgical smelting of iron and base metal ores.

Iron smelting reactions

$$Fe_2O_3 \;+\; 3CO \;\longrightarrow\; 2Fe \;+\; 3CO_2$$
hematite   pig iron

$$Al_2O_3 \;+\; SiO_2 \;+\; CaCO_3 \;\longrightarrow\; Ca\text{-}Al\text{-}Si\text{-}silicate \;+\; CO_2$$
gangue minerals limestone   slag

Copper smelting reactions

$$CuFeS_2 \;+\; O_2 \;\longrightarrow\; Cu\text{-}Fe\text{-}S \;+\; SO_2$$
chalcopyrite air   matte

$$Al_2O_3 \;+\; SiO_2 \;+\; CaCO_3 \;\longrightarrow\; Ca\text{-}Al\text{-}Si\text{-}silicate$$
gangue minerals limestone   slag

$$Cu\text{-}Fe\text{-}S \;+\; O_2 \;+\; SiO_2 \;\longrightarrow\; Cu \;+\; Fe\text{-}silicate \;+\; SO_2$$
matte air silica copper slag

BOX 4.8 | ENERGY AND METAL PRODUCTION

Energy is a big part of material production (Gutkowski *et al.*, 2013). Life-cycle assessment (LCA) methods are very useful in comparing energy requirements for the production of various metals and their impact on the environment. They can be viewed in two ways. Figure Box 4.8.1a shows that, in general, production of light metals such as aluminum and titanium require much more energy per kilogram. For the base metals, the environmental impact of producing a kilogram of metal by pyrometallurgical methods is considerably lower than for hydrometallurgical methods. Perhaps surprisingly, steel and lead show up with the lowest environmental impact per kilogram of metal (because their ore minerals are easiest to melt). However, when we take the total amount of each metal that is produced into consideration, things change a lot. Figure Box 4.8.1b shows that global production of steel has a much larger total environmental impact than the other metals, followed by aluminum, zinc, and pyrometallurgical copper.

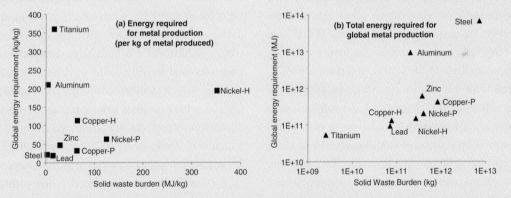

Figure Box 4.8.1 Environmental impact of metal production based on the LCA of total energy required for production (global energy requirement) and solid waste produced (solid waste burden) (a) per kilogram of production and (b) for total global production. P and H for copper and nickel refer to pyrometallurgical and hydrometallurgical methods, respectively (compiled from data of Norgate *et al.*, 2007; Rankin, 2012).

property must meet local water quality standards when it leaves the property, even if it did not meet them when it came onto the property and was not used in the process. Escape of process water into groundwater aquifers must also be limited, and leach pads, overburden dumps, and tailings piles containing cyanide and other caustic or toxic solutions must be shielded by impermeable barriers to prevent the downward movement of these waters. As discussed in the section on gold, cyanide solutions require special handling.

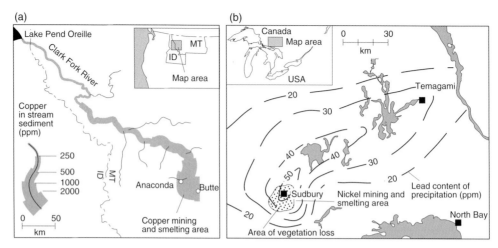

Figure 4.20 Legacy pollution around old mineral operations. (a) Copper content of stream sediment in Clark Fork River draining Butte, Montana, USA (after Axtmann and Luoma, 1991). (b) Asymmetric smelter plume at Sudbury, Ontario, reflecting prevailing winds from the west, showing lead content of precipitation prior to installation of pollution control equipment. From Semkin and Kramer (1976).

Although mining and beneficiation operations are clean, older plants and their wastes are major sources of **legacy pollution** (Castilla, 1983). Butte, Montana, for example, was the largest copper mine in the United States from the 1880s until the 1940s, producing about 450 million tonnes of copper ore. Before installation of tailings ponds in the 1950s, about 100 million tonnes of tailings and smelter wastes were dumped into tributaries of the Clark Fork River, which drains into Lake Pend Oreille, 550 km downstream (Figure 4.20a). Stream sediment near Butte contains over 100 times more copper than average background values and anomalous copper values extend all the way to Lake Pend Oreille (Axtmann and Luoma, 1991). Each rain or increase in the flow of the streams stirs up the sediment, making new metal available to the system, and there is enough metal in the streams for this process to continue for hundreds of years. The only real solution is the removal and stabilization of the contaminated sediment, which is underway in parts of the drainage (see http://www.cfwep.org). Additional problems come from the large open pit at Butte that is gradually filling with very acid water that will overflow in the late 2010s (Gammons and Duaime, 2006). In Peru, a similar problem developed when water from the Rio San Juan, which drains the large Cerro de Pasco mining area, was diverted into Lake Junín to supply seasonal flow for hydroelectric power (Rodbell et al., 2014).

Smelting is of great environmental concern because it produces gases and dust. Early concern in Pittsburgh and other steel-making cities focused on dust emissions, a problem that was largely eliminated by compacting the ore mineral concentrate into pellets as discussed in the Section 8.1.2 on iron deposits. Sulfur dioxide from metal sulfide smelters has been of concern because of its contribution to acid rain, as discussed in the previous chapter. The effects of smelter-induced acid precipitation are most obvious in humid areas because there is more vegetation to kill and because the soil is naturally more acid (in contrast to the alkaline soils of many arid areas). The Sudbury area of Ontario, the world's largest source of nickel, is a case in point. A century of smelting left Sudbury surrounded by 100 km^2 of barren land and another 360 km^2 of stunted birch-maple woodland (Winterhalder, 1988). Soil near the smelters had a pH of almost 3 and that of nearby lakes was as low as 4. Metals such as lead, mercury, arsenic, and cadmium, which vaporize at the high temperatures used to smelt ore, were dispersed over an enormous area encircling the smelters (Figure 4.20b). All this has changed, however. In response to environmental legislation, US smelters now recover almost all of their SO_2 and other emissions. At Sudbury, annual SO_2 emissions have declined from 2 million tonnes in 1977 to only 30,000 tonnes projected for 2019; vegetation and lake recovery is widespread (Keller et al., 2004; Tollinsky, 2015). The decline in US smelter emissions is particularly impressive when compared to the steady trend of coal-fired power-plant emissions, the major anthropogenic source of SO_2.

The effort to clean smelter emissions has centered on the SO_2 content of flue gas. Gases from reverberatory furnaces, the most common first step in most old copper smelters, have low SO_2 concentrations that were difficult to clean. Modern smelters, which have been designed to emit gases with higher SO_2 concentrations, have recovery rates exceeding 90%, and 1.4 million tonnes of SO_2 that used to escape into the environment is now recovered each year from smelters. Smelters under construction now will recover more than 99% of their sulfur emissions, effectively removing smelters from the list of anthropogenic sources of SO_2. Facilities that melt scrap metal for recycling, which are known as *secondary smelters* to distinguish them from *primary smelters* that process ore concentrates, are not important sources of sulfur emissions. However, without proper exhaust-gas cleaning, they can emit metals that vaporize at low temperatures, such as lead.

4.3.3 Processing of fossil fuels

Fossil fuels consist of carbon and hydrogen with small amounts of sulfur, oxygen, and nitrogen. They form molecules that range from CH_4 with only five atoms, to others with thousands of atoms. Under near-surface conditions, these molecules form vapors (natural gas), liquids (crude oil), and solids (coal, asphalt, bitumen, oil shale, tar sands, gilsonite). Natural gas is made up of methane (CH_4), by far the most abundant component, along with small and variable amounts of other natural gases including ethane (C_2H_6), propane (C_3H_8), butane (C_4H_{10}), hydrogen sulfide (H_2S), helium (He), carbon dioxide (CO_2), and nitrogen (N_2). Pentane (C_5H_{12}) and heavier molecules, which can be present as vapors in natural gas at depth in the crust, will condense at the surface and are commonly removed at the wellhead to form **natural gas liquids**.

Natural gas is processed to remove ethane, propane, butane, and related gases, which constitute **liquefied petroleum gases**, as well as the non-hydrocarbon gases, which can be sufficiently abundant to constitute resources in their own right. Removal of essentially all H_2S is of particular concern because it is toxic and reacts with moisture in pipelines to create highly corrosive sulfuric acid.

Processing of crude oil is carried out in refineries, which separate the hydrocarbon molecules by molecular weight and modify them further to produce hundreds of products, including asphalt, fuel oil, gasoline, jet fuel, lubricating oil, naphtha, paraffin, petroleum coke, petroleum jelly, wax, and white spirit, as well as feedstock for petrochemical manufacture. Although individual refineries differ according to the type of oil they process and the types of product that they make, they usually require three steps (Figure 4.21). The first step is **distillation** in which much of the crude oil is vaporized by heating. The vapor is passed to an atmospheric pressure distillation tower where it cools, with lighter molecules, such as gasoline, condensing high in the tower and heavier molecules, such as fuel oil, condensing at lower levels. Unvaporized heavy oil is fed to a vacuum distillation tower, where it too is vaporized and condensed into fractions. In the second step, **cracking**, heavy molecules from distillation are heated under pressure and broken down into

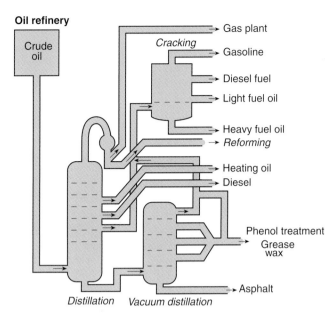

Figure 4.21 Schematic diagram of an oil refinery showing distillation/vacuum distillation, cracking, and reforming stages

smaller molecules that can be used in gasoline and other light hydrocarbon products. The third step, **reforming**, is a generally similar process in which the actual molecular structure of each product is changed to make it more acceptable to today's markets (Leffler, 2012). The balance of products from refineries depends both on the type of crude oil feed and the type of equipment, a factor that creates some problems when demand changes rapidly. In 2014, there were 143 refineries operating in the United States, and no new ones had been built since 1978, although this might change if shale-oil production continues to increase (Lefebvre, 2014).

Processing of the solid fossil fuels varies widely depending on the desired product. The simplest form of processing for coal involves cleaning to remove non-combustible ash and sources of sulfur such as pyrite. Coal can also be heated and reacted with steam and other gases in a process known as **coal gasification**, which produces an impure natural gas that can be reformed into a wide range of products including gasoline. Other solid and semisolid hydrocarbons can also be treated to yield oil and gas, largely by heating to drive off the lower-molecular-weight fraction, which can undergo further conventional refining (Hammond and Baron, 1976).

4.3.4 Environmental aspects of fossil-fuel processing

The most important environmental effects of fossil-fuel processing are caused by the escape of hydrocarbons from refineries. Unlike smelters, which can be located near mines, oil

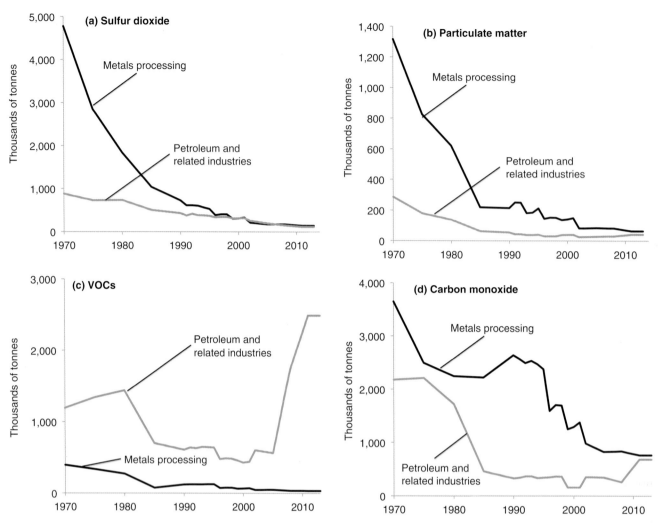

Figure 4.22 Change through time of emissions of (a) sulfur dioxide, (b) particles, (c) VOCs, and (d) carbon monoxide by the petroleum and mining industries. The large change in VOC emissions is related to expanded production of shale gas (from EPA, 2015).

refineries are commonly located where markets are strongest or near ports where transport is easiest, which places them in and near cities. At refineries, hydrocarbons can be lost as **volatile organic compounds** (VOCs), including important greenhouse gases such as methane, or in wastewater, crude oil, and synthetic organic liquids. Most environmental attention has been given to wastewaters and refinery sludge, which contain organic compounds and heavy metals that can be toxic (Kent, 1994; Hua *et al.*, 2012). Escape of these liquids from early refineries has created contaminated groundwater plumes that vary in form depending on the degree to which the organic pollutant dissolves in water. Modern refineries now recycle wastewater and are isolated from groundwater aquifers by impermeable underlayers, but many older installations are still being cleaned (Kapoor *et al.*, 1992; Chestnut, 1990).

Conclusions

Estimates of environmental impact are made by many regulatory agencies and they provide useful mileposts to assess the performance of the mineral industry. Data for atmospheric pollution in the United States show that metal and SO_2 emissions from metal production have decreased from 4.775 million tonnes in 1970 to 0.145 million tonnes in 2013, and that particulate emissions have decreased from 1.3 million tonnes in 1970 to 0.063 million tonnes in 2013. For petroleum refining, VOC emissions have decreased from 266,000 to

41,000 tonnes (Figure 4.22). It is particularly important for the average citizen to hear this news. In the same way that small investors learn about a rising stock market just in time to buy at the top, average citizens often get the news about pollution just as the problem is being solved. We are only now becoming fully aware of the incredible legacy of the pollution from early mining and oil and gas production, and the tremendous cost that we must bear to clean it up. As enormously important as these problems are, we must not let them cause us to lose sight of the fact that modern mineral extraction is much cleaner.

Things are not perfect; accidents will continue to happen, and emissions will exceed acceptable levels locally. But, with improved methods, and vigilance on the part of operating personnel, regulatory agencies, and the general public, we can hope to extract Earth's remaining resources in a manner more consistent with our obligation to respect the planet.

CHAPTER

5 Mineral law and land access

Land is the currency of mineral exploration and production. Without land there is nowhere to look and nothing to produce. This simple constraint is frequently forgotten in our growing concern over future mineral supplies. The fact that mineral deposits must be extracted where they are found (Figure 5.1) is an extreme example of location theory, the branch of spatial economics that focuses on factors controlling the location of economic activity. Whereas most other economic activity has some flexibility with respect to location, mineral extraction has none. This requires a rational approach to land ownership and use at all levels from local to continent-wide. In this chapter, we review forms of land tenure and mineral law, as well as the policies that control implementation of these laws.

5.1 Types of land and mineral ownership

Laws governing mineral ownership take two approaches (Ely, 1964). **Regalian** legal systems hold that minerals are owned by the state, regardless of surface land ownership. Because Spain operated under this system as the New World was explored, most of the western hemisphere outside Canada and the United States adopted variations on this system (Prieto, 1973). English mining law, which was influenced by early Roman practice, holds that minerals are owned by the owner of the surface, and it became the basis for mineral law in the United States, Canada, and other parts of the original Commonwealth.

In the English system, the simplest form of land ownership, fee simple, involves control of the surface and everything on or below it. It is possible, however, to divide land ownership into separate parts, which are referred to as rights. Common

Figure 5.1 Living near mineral production. (a) The Terry Peak ski area in the Black Hills of South Dakota looked down on three gold mines that produced more than 60 million ounces of gold and were a major factor in the local economy. At the base of the hill is the Golden Reward open pit mine, in the far right distance is the Gilt Edge open pit mine, and barely visible in the far left distance is the headframe for the giant Homestake underground mine (photograph by Paul A. Bailly). (b) The city of Butte, Montana, is directly adjacent to the Berkeley open pit and overlies the extensive underground operations that mined veins (photograph by the authors). See color plate section.

> ## BOX 5.1 | LAND LAW, FORTUNES, AND PHILANTHROPY
>
> The difference between the English and regalian legal systems has had a big impact on economic development in the New World. Most of the early personal fortunes in the United States, for instance, were based on the natural resources, especially minerals. Included in these were the Rockefellers, Mellons, Carnegies, and Guggenheims, all of whom created important corporations that continue in one form or another today. These families made major philanthropic donations to cultural and historic institutions that have enriched American life. In countries with government ownership of minerals, it was more difficult to develop such fortunes and related philanthropic activity has been more limited.

divisions include **surface rights**, **water rights**, **timber rights**, and **mineral rights**, which constitute ownership of these parts of the land. These rights reflect the variety of interests that people and businesses take in land, as well as their willingness to permit multiple use of the same parcel of land, an aspect of land tenure that will undoubtedly become more common as population pressures increase.

The possibility of divided land ownership can come as a surprise to land owners if rights were severed in the past. In older mineral-producing regions such as the northern Appalachian coal fields, for instance, far-sighted companies bought mineral rights early in the 1900s, but did not mine the ground until much later. The reverse situation, in which several owners claim a single mineral, can also occur. For instance, who owns methane that contaminates some coal mines? This methane was historically considered a mining hazard, but is now responsible for about 10% of US natural gas production. Is this coal-bed methane owned by the owner of the coal, by the property owner, or by a third party? State Supreme Courts in Virginia and Kansas in 2004 and 2006, respectively, decided that land owners who previously sold the mineral rights to coal beneath their property retained legal ownership of methane within the coal seam (Owens, 2012). Similarly, the US Supreme Court decided that coal-bed methane beneath land owned by the Southern Ute Tribe in Colorado is to be treated as a split estate, with ownership separate from solid coal (Kennedy, 1999). This is not the rule everywhere, however; the Supreme Court of Pennsylvania has determined that the owner of the coal owns the coal-bed methane.

There is even the question of how to divide surface from subsurface. Most jurisdictions consider surface rights to stop below the immediate land surface. However, Texas courts have allowed the owner of surface rights to claim minerals down to a depth of 61 m if their removal would result in destruction of the present surface (Aston, 1993). This curious twist of logic reflects the long history of oil production in Texas, which does not usually result in significant disturbance of the land.

The rights of overlapping governments, such as states and federal governments, to ownership of public land has been handled in different ways. In Canada, public lands and mineral rights were given to the provinces at the time of confederation. In the United States, in contrast, public lands and their mineral rights reverted to federal control as states joined the union. The only important exception made by Congress was Texas, which was permitted to retain title to its public lands when it was annexed in 1845. This moment of weakness cost the Federal Treasury billions of dollars in later oil and gas royalty payments that went to Texas.

Ownership of water and water rights is a separate issue in many countries, particularly arid lands (Walston, 1986; Williams *et al.*, 1990; Lee *et al.*, 2005). In the eastern United States and other countries with abundant surface water, laws are based on the doctrine of **riparian rights**, which holds that the right to use water on the land comes with ownership of the land and is maintained even if the water is not used. In the western United States and other arid regions, water law applies the concept of **appropriation rights**, which grants access to water on the basis of the time it was first used and the purpose for which it is used. For example, the Colorado River Compact signed in 1922 by Arizona, California, Colorado, Nevada, New Mexico, and Utah established maximum annual allotments of water from the Colorado River that could be used by the signatories according their population at the time (Boime, 2002). In 1922, the maximum legally allowed water use among all seven states was established assuming an annual flow totaling 16.4 million acre feet, which exceeds the average water volume of 14.5 million acre feet that has prevailed during the late twentieth and early twenty-first centuries, and this has led to legal challenges in western states with fast-growing populations (National

Research Council, 2007; Jacobs, 2011). Similar compacts allocate water rights on rivers that cross geopolitical boundaries, such as the Connecticut, Delaware, Red, and Potomac rivers in the United States, the Columbia River in Canada and the United States, and the Colorado River in Mexico and the United States.

Groundwater law developed separately from surface water law because of the mistaken impression that the two water reservoirs were not related. The first attempt to regulate groundwater use originated in England in the early nineteenth century, as described in the English law case of *Acton versus Blundell* (1843), which gave a land owner the right to withdraw as much groundwater as desired without concern for potential adverse consequences for neighboring land owners. This policy, known as the Rule of Capture or Absolute Ownership, remains in practice in Connecticut, Georgia, Indiana, Louisiana, Maine, Massachusetts, Rhode Island, and Texas (Great Lakes Protection Fund, 2007). Most other US states adhere to a Reasonable Use Doctrine, which states that land owners can pump groundwater from beneath their property as long as it does not adversely effect groundwater availability on contiguous properties (Joshi, 2005). In Michigan, the state legislature passed a bill in 2006 that regulates new groundwater use exceeding 100,000 gallons per day, which will likely lead to legal challenges owing to the interpretation of the word *new*. Although most states accept that groundwater ownership accompanies land ownership, Alaska, Colorado, Idaho, Kansas, Montana, Nevada, New Mexico, North Dakota, Oregon, South Dakota, Utah, Washington, and Wyoming have declared that they own groundwater. This ownership claim is usually manifested in agencies that determine how and by whom water can be pumped from the ground; it does not mean that groundwater is reserved for state use.

5.2 Land ownership and law in the United States

In most countries, the government is responsible for surveying the land and providing a regional location system on which land ownership is based. In the United States, locations are based on three survey systems (Figure 5.2). Land in the original US colonies and some of the older states was divided into irregular tracts based on the old English system of metes and bounds (Muhn and Stuart, 1988; PLS, 1988). Divisions between parcels were indicated by landmarks such as trees and fences, many of which have disappeared. Land in the southern and southwestern states, which was originally under Spanish influence, was divided even less systematically. Present land boundaries in many of these states have been relocated by modern surveys, but persist by tradition and agreement in a few areas. Land west of Pennsylvania was surveyed into 2.59 km^2 sections under the provisions of the Land Ordinance Act of 1875. These sections were part of a north–south, east–west grid of townships and ranges into which almost all of the central and western United States was divided. The resulting system includes townships that measure 6 miles (9.7 km) on a side and contain 36 sections.

Control over the 2.3 billion acres (9.3 million km^2) of land that makes up the United States is about equally divided among the federal government, state, and local governments, and private groups and individuals. Early in US history, the federal government owned another 1.287 billion acres (5.2 million km^2) that have since been transferred to state and

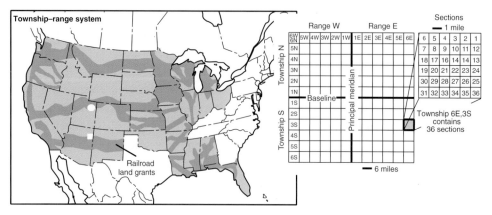

Figure 5.2 Areas of the United States surveyed by the township–range system; inset shows the relation between townships, ranges, and sections. Note that the term township refers to both the east–west tiers of divisions and the divisions themselves. Also shown are areas covered by railroad land grants, which consisted of alternative sections that created a "checkerboard" land-ownership pattern that persists today.

Table 5.1 Disposition of public land in the United States, 1781–2012. ANSCA: Alaska Native Claims Settlement Act (Data from BLM, 2013a).

Type	Acres
Disposition by methods not classified elsewhere	303,500,000
Granted or sold to homesteaders	287,500,000
Granted to States for:	
Schools	77,630,000
Reclamation of swampland	64,920,000
Railroads	37,130,000
Miscellaneous institutions	21,700,000
Unclassified	117,600,000
Canals and rivers	6,100,000
Construction of wagon roads	3,400,000
Granted to railroad corporations	94,400,000
Granted to veterans as military bounties	61,000,000
Confirmed as private land claims	34,000,000
Sold under timber and stone law	13,900,000
Granted or sold under timber culture law	10,900,000
Sold under desert land law	10,700,000
Granted to State of Alaska and ANSCA:	
State conveyances	99,100,000
Native conveyances	43,700,000
Total	1,287,180,000

local governments, private groups, Native American tribes, and individuals (BLM, 2013a; Gorte *et al.*, 2012). The Homestead Act of 1862, the Desert Land Act of 1877, and related laws, which encouraged settling of the West, accounted for 287 million acres (1.16 million km^2) of these transfers. Larger blocks of land, totaling 328 million acres (1.33 million km^2), were granted to the states for construction of schools, railroads, canals, and other improvements (Table 5.1). Between 1850 and 1870, another 94.4 million acres (380,000 km^2) were transferred to companies as incentives to build railroads into the West. These grants commonly took the form of odd-numbered sections of land in 20-mile (32-km)-wide strips centered on the rail line. Known mineral lands were specifically excluded from these grants, but many deposits have since been discovered on these lands, making western railroads into important mineral producers (Gorte *et al.*, 2012).

Even after all of these transfers, the federal government controls and manages most of the land in the western states, ranging from a high of 85% in Nevada to a low of 29% in Washington (Table 5.2). This land includes national parks, monuments, and other scenic and recreational areas, as well as wilderness areas, where mineral exploration is largely prohibited. It also includes vast areas of national forest, grasslands, and other land, where mineral entry is permitted and which supply the bulk of domestic mineral production.

5.2.1 US federal land laws

Access to federal land for mineral exploration and production is governed by a host of laws and regulations. Of most importance to exploration are the laws that deal with the discovery–claim and leasing systems, as discussed next.

Discovery–claim system

United States federal mineral law is based on the experience of prospectors, such as those in the California gold rush of 1849. In the system that they developed, the discoverer of a deposit had the right to claim ownership in it. This concept became federal law as the General Mining Law of 1872, which is a combination and amplification of the earlier Lode Mining Law of 1866 and Placer Mining Law of 1870 (Ely, 1964). These laws divided mineral deposits into two geological types; those in actual bedrock were called **lodes** and those in river and stream gravels were called **placers**. The 1872 law allows any citizen or domestic corporation to stake an unlimited number of claims. Claims must measure 600 by 1500 ft (183 by 457 m) for lode deposits and 20 acres (0.08 km^2) for placer deposits. Their location must be marked on the ground, and they must be registered with the US Bureau of Land Management (BLM). Mining claims can be located under the 1872 law in Alaska, Arkansas, Arizona, California, Colorado, Florida, Idaho, Louisiana, Mississippi, Montana, Nebraska, Nevada, New Mexico, North Dakota, Oregon, South Dakota, Utah, Washington, and Wyoming. Mineral access on federal land in the other states is administered under the Mineral Leasing Act of 1920, which is discussed later.

According to the original law, claims could be held in perpetuity by the holder doing annual assessment work worth $100 on each claim. This was changed in 2015 to an annual rental of $155 for **lode claims**, and $155 for the first 20 acres up to a maximum of $1,240 for 160 acres of a placer claim. The annual rent eliminates the requirement of maintenance work, although claim owners can apply for a waiver if

Table 5.2 States with important federal land ownership, showing value of non-fuel mineral production. The top 20 states listed here account for 90% of federal land holdings and 69% of US non-fuel mineral production. (Land area data from Gorte et al., 2012. Non-fuel mineral production data from National Mining Association, 2015.)

State	Federally owned land total area (km^2)	Percent of area	Non-fuel mineral production (2013, $ billion)	Percent of total US non-fuel mineral production (2013)
Nevada	242,657	85.1	9.66	7.5
Utah	141,776	66.5	4.18	5.4
Alaska	913,976	61.8	3.51	4.5
Idaho	132,073	61.7	1.20	1.5
Oregon	132,192	53.0	0.35	0.5
Wyoming	121,582	48.2	2.35	3.0
California	193,430	47.7	3.51	4.5
Arizona	124,406	42.3	8.06	10.4
Colorado	97,473	36.2	2.32	3.0
New Mexico	109,272	34.7	1.87	2.4
Washington	49,266	29.2	0.89	1.1
Florida	18,360	13.1	2.99	3.9
Michigan	14,722	10.0	2.41	3.1
Arkansas	12,796	9.4	1.03	1.3
Virginia	9,543	9.2	1.11	1.4
North Carolina	9,821	7.7	1.30	1.7
Minnesota	14,039	6.8	4.71	6.1
South Dakota	10,709	5.4	0.31	0.4
Missouri	6780	3.8	2.48	3.2

they own fewer than ten claims and demonstrate that at least $100 of work is performed in good faith to develop the claim. The original law also permitted the holder of a claim to purchase the title to its surface and mineral rights from the federal government, a process known as **patenting**. Original patenting requirements included completion of $500 worth of improvements on the claim, public notification of intent, and payment of $2.50/acre for a placer claim and $5/acre for a lode claim. Early abuses of the patent option to obtain land for other purposes, including vacation homes, were prevented by stipulating that claims could be patented only if they contain mineral deposits that are good enough to merit further work by a "prudent" person and that the minerals must be "marketable." The US Supreme Court has issued decisions in several cases (e.g. *Castle* versus *Womble*, 1894; *Chrisman* versus *Miller*, 1905; *Layman* versus *Li*, 1929; *US* versus *Coleman*,

1968) that followed an economic approach to these concepts, requiring that the claimant demonstrate reasonable intent to sell the minerals and that the sale of the minerals be profitable (Sokoloski, 1996). Recently, environmental groups in Colorado filed mineral claims on federal land that they have no intent to mine (Whitney, 2005). Such actions seem to violate the prudent man and marketability concepts embedded in the 1872 law, and have re-focused debate on the claim system.

Periodic efforts have been made to clear the books of old claims. The most recent of these was part of the Federal Land Policy and Management Act of 1976 (FLPMA), which required that all active mining claims be registered with the BLM by October, 1979. This requirement is estimated to have removed from the books inactive claims covering at least 1.2 million acres (486 km^2).

BOX 5.2 | IMPROVING THE MINING LAW

The General Mining Law of 1872 is not without faults. Because it was designed to encourage settlement of the land and development of its mineral resources, the law made access to the public land for minerals relatively easy, gave miners priority over other land uses, and required only exploration or mining work with no payment of fees to hold the land. It also made no provision for environmental safeguards. It was intended to protect small deposits that could be covered by a few claims, and even included the concept of **extralateral rights** for lode claims (Figure Box 5.2.1) permitting the miner to follow the vein if it extended downward off the claim. Modern exploration for larger, deeply buried deposits requires many claims covering ground that has been identified as having high potential for a deposit by methods discussed in Chapter 3. The ground must be held while detailed exploration is carried out to evaluate the economic potential of the discovery.

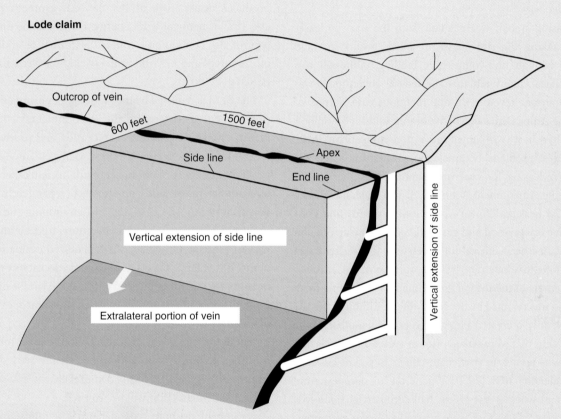

Figure Box 5.2.1 Lode mining claim. Note that the miner has the right to pursue veins off the claim (extralateral rights) if their uppermost part or vertex is on the original claim. This practice is not consistent with the irregular nature of many veins and led to major legal battles in Butte, Montana, and other western mining districts (modified from US Forest Service, 1995).

Recent attempts to revise the law have focused on the need to charge an annual rent on claims, to impose a royalty on production, and to have more stringent environmental regulations, although none have succeeded. In 2007, for instance, the US House of Representatives passed the Hardrock Mining and Reclamation Act that would have permanently banned patents for mineral claims, required royalties of 4% of gross revenue on existing hardrock mining operations, and 8% for future hardrock mining (Anonymous, 2007). The US Senate, facing strong opposition from mineral-producing states such as Nevada, refused to discuss the bill and it died legislatively. It is important that changes be made with careful comparison of their benefits, including increased tax revenues and less environmental disruption, and their costs, including decreased employment, increased import reliance, and declining balance of trade.

Mineral leasing system

The 1872 law was not well suited to exploration for oil, coal, gravel, and other minerals that form large or concealed deposits. This led to laws that established leasing as a second system of mineral entry on federal land. The first of these, the Mineral Leasing Act of 1920, originally applied to coal, oil, oil shale, natural gas, sodium minerals (evaporites), and phosphate rock. It was later amended to include sulfur in New Mexico and Louisiana, as well as potassium evaporites and geothermal resources. Under the 1920 law, leases can be acquired by either competitive or non-competitive bidding, with competitive bidding required on land with known mineral deposits.

The leasing system was plagued from the start by leaseholders making big discoveries or quick profits on land obtained through non-competitive bidding. Although this was not illegal, many felt that the federal government was losing important revenue. In one infamous case, a tract of land in the Amos Draw area of Wyoming sold for $1 million shortly after it was obtained in a non-competitive lease. Because these problems became apparent first for coal, non-competitive leasing for coal was eliminated by the Federal Coal Leasing Amendment Act of 1976. The Federal Onshore Oil and Gas Leasing Reform Act of 1987 provided that non-competitive leases of oil and gas land could be made only on tracts that had been offered unsuccessfully for leasing in an earlier competitive sale. The leases require payment of royalties and rentals to the federal government, and the title to all leased land was retained by the government. Although the law works well now, it created a long-lived political football in the form of claims that were staked prior to 1920 for minerals that must now be leased. In particular, efforts to patent oil-shale claims under the 1872 law have gone on for decades, amid debate about whether the claims had commercial value to a prudent person when they were originally filed or even now.

The leasing system was modified by the Materials Act of 1947, which pertains to most construction and industrial minerals. This law permits operators to sell materials from federal land under permits obtained by non-competitive or competitive bids, depending on the volume of planned production. The Acquired Lands Act of 1947 authorizes leasing of acquired lands, which are those obtained by purchase, condemnation, donation, or exchange.

Administration of US mineral laws

Management of mineral resources on US federal land is the responsibility of the BLM (OTA, 1978; Muhn and Stuart, 1984). The BLM monitors the location and status of all claims on federal land and is responsible for all subsurface mineral activities. As of 2014, the BLM has under its supervision 5.3 billion barrels (Bbbl) of oil reserves, 1.95 trillion m^3 of natural gas reserves, 1.23 trillion barrels of oil in oil shale, all of the nation's major tar sand, ~300 million tonnes of coal, 35% of the nation's uranium, as well as world-class deposits of phosphate, sodium, potash, lead, and zinc, and most of the nation's undiscovered but geologically predictable domestic supplies of aluminum, antimony, beryllium, bismuth, cadmium, chromium, cobalt, copper, fluorspar, lead, manganese, mercury, molybdenum, nickel, platinum-group metals, silver, tungsten, and vanadium. The BLM administers land that produces nearly 50% of the nation's geothermal energy, and 15% of national wind energy. Federal lands under BLM control also contain an estimated 30.5 Bbbl of undiscovered oil resources and 6.5 trillion m^3 of undiscovered natural gas (USDOI, 2008).

The BLM is required by the National Environmental Policy Act of 1970 to determine the environmental effect of all activities, including mineral exploration and production, on federal land. (The National Forest Management Act of 1976 provides that the surface impact of mining and other activities on national forest lands be managed separately by the US Forest Service.) The first step in determining the potential impact of a proposed mineral operation impact is an environmental assessment, which is a generalized evaluation of the nature of the land. This is used to guide decisions on all types of land use, including mineral entry. If the land is opened for mineral entry, and part of it is leased or claimed, exploration plans must be reviewed and approved. If the activities have potential for significant disturbance to the land, further permitting is required. If the proposed activities are judged to be sufficiently serious, a more comprehensive environmental impact statement (EIS) will be required.

Leasing of onshore federal land is administered by the BLM, which oversees the mineral estate on 258 million surface acres of public land and 700 million acres of subsurface public and private land (BLM, 2013a). The BLM also provides technical support for the management of 57 million acres of Native American trust lands (BLM, 2013a). Leasing of approximately 1.8 billion acres of offshore federal land is administered by the Bureau of Ocean Energy Management (BOEM) (Figure 5.3). Offshore lands are divided into tracts that can be as large as 5,000 acres (approximately 23 km^2) (BOEM, 2011a). These agencies identify tracts of land and offer them for competitive bids, which must contain a work plan, including an environmental assessment that has to be approved by the leasing agency. Annual rental for onshore oil

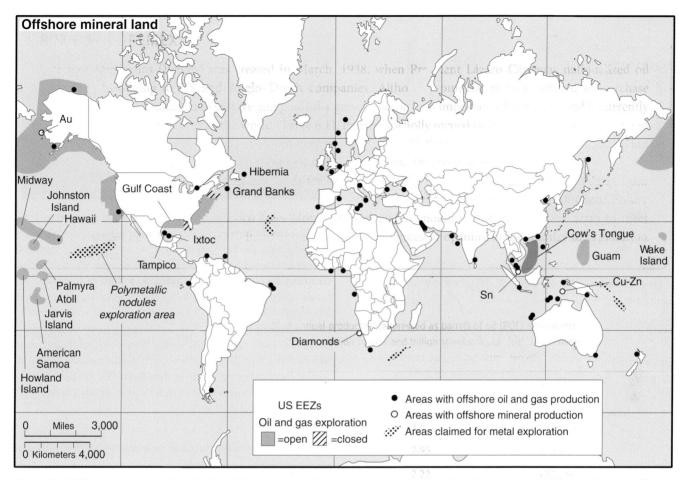

Figure 5.3 Offshore areas claimed by the United States as exclusive economic zones (EEZs). Shaded areas in the EEZs were closed to offshore leasing for oil and gas exploration in 2012. Regions of the world with offshore oil, gas, or mineral production are also shown.

and gas leases is $1.50/acre for the first 5 years and $2/acre thereafter, with a royalty of 12½% of the value of any future production (Vann, 2012). Offshore oil and gas leases historically required royalties of 16⅔% in shallow water and 12½% in deep water, defined as depths greater than 200 m. In 2008, the royalty payment requirement was increased to 18¾ percent for all leases at all water depths. Rental fees are assessed based on water depth, and range from $7/acre for water less than 200 m deep, to $11/acre for water greater than 2,000 m deep (BLM, 2012; BOEM, 2014). The initial duration of offshore leases is also tied to water depths, with 5-year leases for water depths less than 800 m, 7-years leases for depths of 800 to less than 1,600 m, and 10 years for depths greater than 1,600 m (BOEM, 2011a; BLM, 2012). If wells are started in a given lease block, rental rates increase in years 6, 7, and 8, and remain fixed thereafter for the life of a producing well (BLM, 2012). Terms for other minerals differ, but have approximately the same economic effect. Bids are compared on the basis of a work plan and the amount of **bonus bid** that is offered. This payment as well as the first year's rent are due on signing the lease, and the bonus bid is not returnable if the lease turns out to contain no oil or gas.

By far the largest number of federal leases are for oil and gas exploration and production (Figure 5.4). As of 2013, there were almost 36 million acres of offshore land leased for oil and gas exploration, with activity on about 7 million acres (BOEM, 2014). Onshore competitive federal land leases considered in effect for oil and gas exploration and production totaled about 22 million acres as of 2013, with an additional 15 million acres having been leased non-competitively prior to the Federal Onshore Oil and Gas Reform Act of 1987 (BLM, 2013a). This is a decrease from nearly 70 million onshore leased acres in 1988 (BLM, 2013b). The number of producing onshore acres over the same time period decreased only slightly, from 12.9 to 12.6 million acres (BLM, 2013a). The story is the same for federal land leases for coal, which declined from 730,247 acres in 1990 to 484,017 in 2012. Active leased federal lands for all other solid materials, including resources such as phosphate, sodium, metals, sand, and gravel totaled about 200,000 acres (BLM, 2013a).

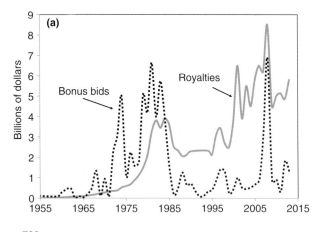

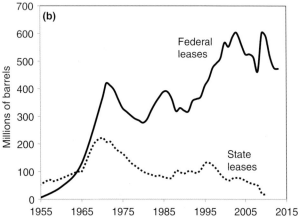

Figure 5.4 Oil and gas revenues from production from federal land in the United States. (a) Relation between bonus bid and royalty revenue from offshore US federal land oil and gas leases. (b) Change in oil production from state and federal offshore leases. Decline in production from state leases reflects the exhaustion of fields in shallow waters (2014 data from the Office of Natural Resources Revenue, ONRR, see http://statistics.onrr.gov/ReportTool.aspx).

Figure 5.5 Oil fields in the Santa Barbara Channel, just offshore from the city of Santa Barbara, California, are an important source of US oil production. The Dos Cuadras field, which is served by Platforms A, B, and C shown here, is in federal offshore land, just beyond the state limit. The field is at a depth of 2,000 to 3,500 feet and has produced more than 260 Mbbl of oil since it was discovered in 1968. The field is outside the 3-mile limit and therefore is administered by the US federal government. A spill of almost 100,000 barrels from this field in 1969 was a catalyst in the US environmental movement. Although this was a large spill, it is much smaller than the natural oil seepage that has been going on in the Channel for the last few hundred thousand years (photograph courtesy of Doc Searls).

Federal lands yield a large proportion of US national production (Table 5.2), and are an important source of non-tax revenue for the US government. Bonus bids were historically a very important part of these revenues, both because of their large size and because they are paid regardless of whether the tract produced minerals. Originally, the nature of the bidding process led to large overbids and the associated winner's curse, with winning bids on average 45% higher than the next-highest competing bid (Megill and Wightman, 1984; Cramton, 2006). The value of winning bids has decreased, however, from an average winning bid, in constant 1982 dollars, of $15.07 million ($2,224/acre) for the period 1954 to 1982, to $0.62 million for the period 1998 to 2006 ($263/acre) (Haile *et al.*, 2010). In 2012, bonus bids totaled $2.77 billion for competitive offshore oil and gas leases, $226 million for onshore oil and gas leases, and $1.9 billion for coal (ONRR, 2013).

If oil and gas are produced from leased acreage, royalty and rent payments are made to the federal government (Figure 5.4a). Between 1954 and 2013, $81 billion in bonus bids, $2.5 billion in lease rents, and $75 billion in royalties were paid for US federal leases, almost all of that from offshore oil and gas (Figure 5.4b; BOEM, 2014). The relative importance of royalty income and bonus bids has changed with time (Figure 5.4a).

Companies now perform high-resolution seismic surveys before submission of bids, which greatly increases their ability to select smaller areas with higher potential for discovery. For example, almost all offshore bids today are made after seismic surveys are completed, compared with fewer than five percent of bids in 1989 (Haile *et al.*, 2010). Bonus bids declined and remained relatively steady through the beginning of the twenty-first century, largely because of a moratorium on drilling in new offshore areas. This moratorium was lifted by executive order in 2008, which opened almost 500 million acres of new offshore land to drilling, including offshore Virginia, and resulted in a dramatic increase in bonus-bid revenue. In 2012, the moratorium was re-imposed eliminating new drilling along the entire Atlantic and Pacific coasts (Figure 5.5), as well as large parts of the Gulf of Mexico. This caused a 40% decrease in bonus-bid payments during the period 2011–2013 compared with 2008–2010. In 2011, nearly

BOX 5.3 | **WHERE DO LEASE REVENUES GO?**

Revenues from the leasing system in the United States are distributed to the states, the federal Treasury, and several specific funds (Table Box 5.3.1; GAO, 2012). The Land and Water Conservation Fund Act of 1965 commits a portion of offshore lease revenue for purchase of park and recreation lands. Additional offshore lease revenues support state purchase of land for historic preservation under the National Historic Preservation Act of 1986. Forty percent of onshore revenue goes to a fund to support reclamation of mined land by state governments in 17 western states, excluding Alaska (GAO, 2012). In addition, states receive a proportion of revenues from onshore mineral production on federal land within their boundaries, and revenues from Native American lands are distributed directly to the tribes and organizations with leased land. Fifty percent of onshore lease revenue is paid to the state in which the lease is located, except for Alaska, which receives 90% of onshore revenue (Malm, 2013).

Table Box 5.3.1 Disbursement of federal mineral royalty revenue for fiscal year 2013. Data from the US Office of Natural Resources Revenue.

Fund	Amount
American Indian tribes	$932,956,397.19
Historic Preservation Fund	$150,000,000.00
Land and Water Conservation Fund	$895,580,482.27
Reclamation Fund	$1,592,181,337.45
State share: offshore	$40,939,426.68
State share: onshore	$1,964,029,388.52
US Treasury	$8,648,688,422.06
Total	**$14,224,375,454.17**

For offshore leases within 3 miles (4.8 km) of the coastline, 27% of all royalty revenue is paid to the state that hosts production. Leases more than 3 miles offshore of Texas, Mississippi, and Louisiana pay 37½% royalties to the states. In 2013, about $930 million was paid to Native American tribes, $895 million went to the Land and Water Conservation Fund, $8 billion to the US Treasury, and more than $2 billion in revenue was disbursed to 35 states. The largest payments went to Wyoming ($933 million), New Mexico ($479 million), Utah ($138 million), Colorado ($130 million), California ($102 million), North Dakota ($90 million), Montana ($36 million), Louisiana ($27 million), Alaska ($19 million), and Texas ($17 million). Among states east of the Mississippi, Michigan got $330,884, Virginia got $45,215, Ohio got $284,145, and North Carolina got $158 (Malm, 2013; BLM, 2013c; WyoFile, 2013).

93% of payments to the federal government were in the form of royalties from oil and gas production on federal onshore and offshore land (GAO, 2012).

It is not clear how much longer we will be able to enjoy the largesse of the federal leasing system. Its main source of funds, bonus bids from offshore leasing, has decreased significantly. The Gulf of Mexico is essentially the only area of active, large-scale leasing, and most shallow areas of the Gulf have been leased. Production in these shallow fields is declining rapidly (Figure 5.4). In 1995, the US Congress passed the Outer Continental Shelf Deepwater Royalty Relief Act to stimulate deep-water drilling, with the hope that this would yield greater bonus-bid and royalty payments. Over the 20-year period ending in 2012, oil production from deep- (1,000 to 5,000 feet) and ultra-deep (>5,000 feet)-water depths rapidly outpaced shallow-water (<1,000 feet) production in spite of the fact that the development of the production infrastructure in deep-water tracts is, on average, twice as expensive as in shallow-water tracts. Over the same period, revenues from deep-water production were almost four times higher than in shallow water, and profits per tract averaged $3.76 million in deep water compared with average losses of $1.64 million in shallow water (Haile *et al.*, 2010). However, more than 90% of federal offshore land is off limits to oil and gas exploration activity.

5.2.2 State and local mineral laws in the United States

Discovery–location systems are not applied on land held by US states, and access to this land (if it is open to mineral entry) is by some form of leasing. Private land of all types is also leased, with most attention focused on large landholders such as ranchers and railroad, timber, or paper companies. Lease agreements are simplest for oil and gas exploration, where a more or less standard lease form is widely accepted. Leases for other types of minerals are not standard and must be negotiated separately for almost all cases. The main sticking point in these negotiations is the level of the royalty percentage, caused by land owners with unrealistic expectations about prevailing practice. Several US states have "mineral lapse" laws that terminate mineral interests that have been undeveloped for 20 years.

States have made important contributions to mineral law in two areas related to oil and gas (Ely, 1964). Interest in **tidewater land** was generated by the need to apportion royalty and tax revenue from offshore production. The Submerged Lands Act of 1953 divided land between the states and federal government. States bordering the Atlantic and Pacific oceans, as well as Louisiana, Mississippi, and Alabama obtained land within 3 miles (4.8 km) of shore. For Texas and Florida, the boundary was set at 3 marine leagues (10.35 miles; 16.65 km) from shore to reflect original Spanish conventions. Leasing regulations for these federal offshore lands were spelled out in the accompanying Outer Continental Shelf Lands Act of 1953. The key word in these statues was *coast*, a term that turned out to require clarification. In some areas, such as southern Louisiana, the ocean shallowed and freshened gradually into swamp and in many other areas the shoreline was eroding landward or growing seaward. Offshore islands further confused the situation. Attempts to make these dynamic geologic processes conform to acceptable legal terminology have occupied years of geologic mapping and court time.

Unitization of oil and gas fields was also largely a state legal issue, mostly because much early oil production was on private and state land. Unitization refers to the process by which production from an oil or gas field is administered as a single unit, regardless of the number of land parcels and owners involved in the field. Before unitization, land owners drilled as many oil wells as possible on their land in order to prevent oil under their land from being pumped from wells on adjacent land. This practice, which placed a premium on closely spaced wells and rapid pumping, greatly decreased the amount of recoverable oil in a field and was soon recognized as a wasteful practice. In its place, each owner was assigned a proportion of the production that was anticipated from the entire field. The field was then produced with the well spacing and pumping rate that was judged most favorable to maximize recovery. States took on the job of enforcing unitization, beginning with the misnamed Texas Railroad Commission (Prindle, 1981). The interest that these groups had in production rates gradually grew to take in the relation between production rates and prices, especially since oil prices had a strong effect on state royalty revenues. For many years, oil production rates determined by the Texas Railroad Commission had a strong impact on world oil prices.

Most states and many local jurisdictions also have environmental agencies, resulting in overlapping jurisdictions that can cause confusion in determining which agency has the appropriate authority. Most counties or municipalities zone land, and some of these zoning regulations deal with exploration activities such as drilling. Almost all projects that can be carried out under the zoning laws must be assessed for environmental impact. These agencies require reclamation, although many restrict such requirements to operations that move more than 1,000 yd^3 (765 m^3) of material per acre. An area of growing legal interest is the degree of control that state and local agencies can exert on the use of federal land in their jurisdictions. Recent Supreme Court decisions indicate that states may control federally approved mineral activities on federal land if operators do not comply with reasonable state environmental regulations.

5.3 Mineral ownership and land laws in other countries

5.3.1 North America

In Canada, mineral rights accompanied surface rights until the early 1900s when they were declared government property. About 90% of Canadian mineral rights are currently government owned, divided about 40% and 50% between federal and provincial governments, respectively (Peeling *et al.*, 1992). The Canadian federal government controls public lands and minerals in the Northwest and Nunavut Territories, although some control has been ceded to aboriginal groups, as discussed later. Individual provinces retain title to their public lands and set their own mineral laws. In 2009, the province of Ontario withdrew Crown rights to mineral resources under privately held surface land where no claims or leases had previously been recorded.

Prospecting licenses are required for exploration in the Northwest Territories, British Columbia, Manitoba, Ontario, Quebec, New Brunswick, and Nova Scotia, and all provinces require that the prospector has a license in order to stake a claim. In some cases, staking can be carried out online rather than on the ground. Individual mining claims usually measure 16 to 25 hectares in area, with shape and size restrictions. Additional regulations allow the area of the claim or claims to be much larger, including the practice of having individuals stake the claims and then transfer them to a company. There is no competitive bidding, and claims remain active as long as there is work (and expire after 2 years if no work is done). A mining lease, usually for 20 to 21 years, is required before mining can begin. The traditional "free-entry system" for mineral exploration has been modified in Ontario to require consultation with indigenous groups and to limit exploration in populated areas. Mining royalties range from a low of 5% of net for small mines in Ontario to 17% of net in Manitoba, and these recover an average of about 4% of the gross value of the mineral produced (Mining Watch Canada, 2012).

Oil and gas exploration on land is controlled by the individual provinces and offshore by joint federal–provincial agreements such as the Canada–Newfoundland Atlantic Accord Implementation Act, which resulted from the discovery of the Hibernia field offshore from Newfoundland. Lease applications in the Arctic are given in three stages, starting with competitive bidding for an exploration license for 9 years, followed by a significant discovery license for more detailed evaluation, and finally a production license for 25 years, automatically renewable if production is achieved. Royalty payments on production start at 1% but rise to the greater of 30% of net or 5% of gross after project payout. Annual royalty payments in Alberta, where most production is located, average about $1 billion, with less than 10% of this coming from oil sands.

In Mexico, all minerals are owned by the federal government, regardless of surface ownership. Oil and natural-gas exploration and production were operated for much of the twentieth century as a monopoly by the Mexican state corporation, Petroleos Mexicanos (Pemex). To attract much-needed international capital to stimulate new onshore and offshore oil and gas exploration and production, Mexico enacted a constitutional amendment in 2013 that ended government control of oil and gas. Salt as well as uranium and other nuclear fuel minerals are reserved for the government. Most other minerals can be produced by individuals and Mexican corporations. The stipulation that mineral deposits must be controlled by Mexican citizens or corporations was a major impediment to mineral exploration in Mexico for many years. Mexican joint-venture partners lacked adequate capitalization and foreign investors were unwilling to forfeit control over projects. In 1996, legislative amendments to the Constitution's Foreign Investment Law allowed foreign entities to control mining activities. The changes to mineral, oil, and gas regulations led to a resurgence in international investment, and as of 2010 more than 700 foreign entities spent $7 billion on mineral exploration in Mexico (Oancea, 2011).

Ground control for exploration and mining is accomplished by staking claims, which are called **concessions**. Exploration concessions are valid for 6 years and are not renewable, whereas extraction concessions are valid for 50 years and renewable once (Sanchez-Mejorada, 2000). Concession sizes are measured in hectares (100 by 100 m), must be astronomically oriented east to west and north to south, and are not limited in number. Annual rent is charged on concessions and assessment work is required.

5.3.2 South America

Mineral ownership in Brazil rests with the state, although mineral law and administration are in a state of flux. Anticipated changes will bring mining concessions with a duration of 40 years, possible extension for another 20 years and the creation of a National Council for Mining Policy to advise the president on mineral affairs. Current low royalties are anticipated to rise to as much as 4% of gross (Visconti and Fernandes, 2015).

Mining in Chile is controlled by the Chilean Mining Code of 1983 and regulated by the Ministry of Mines. Exploration is carried out under renewable 2-year concessions, and are granted without competition. Concessions for lithium and oil and gas are treated separately but follow generally similar rules. In addition to income taxes, large producers of copper (including Codelco, the national mining company) pay taxes at a rate of 5 to 14% of taxable income.

5.3.3 Europe

In the United Kingdom, the crown holds title to gold, silver, coal, oil, and gas, and individuals and corporations can own other minerals. Exploration and production leases can be obtained for gold and silver from the Crown Estate Mineral Agent and for coal, oil, and gas from the Department of Energy and Climate Change (DECC). Although the owner of land is entitled to all minerals beneath the land, this title is secure only if registered before October 13, 2013. In the Duchies of Lancaster and Cornwall, which occupy an area of about 730 km^2, mineral rights are held by the Sovereign and eldest son, respectively. Recent interest in the old mines of Cornwall, which have a wide suite of elements including some technology metals, have led to reassertion of these rights (Carter, 2013). Northern Ireland does not follow this format, however. In 1969, mineral titles were vested in the Department of Enterprise, Trade and Investment, which is in charge of granting exploration and mining licenses (Eldridge and Boileau, 2014). Oil and gas exploration and production are controlled by the Petroleum Act of 1998, which permits the Secretary of State to grant exploration and production rights through competitive bidding. Most oil and gas activities were located offshore until the development of fracking to produce gas from shales. The widespread presence of favorable subsurface shale on land has led to a strong government initiative to smooth the way for mineral entry in more populated areas (Carrington, 2014). This effort has been complicated by efforts of the European Union to draft strong environmental protection laws (Ross, 2013).

The Republic of Ireland was one of the first countries to benefit from significant changes in mineral law, including the Minerals Development Act of 1940, which allowed free land entry, and the Finance Act of 1956, which provided tax advantages for new mining companies (as discussed further in Chapter 6). This resulted in the discovery of at least five major zinc deposits including Tara at the old city of Navan, the largest zinc mine in Europe. Currently, prospecting licenses are granted through competitive bidding that takes place four times a year.

In an interesting twist to the land access problem, the Brown Coal Act of 1950 in Germany allows companies to use the right of **eminent domain** for access to coal on private land. In the United States and Canada, this right, which allows purchase of the land at a fair price for a specific, non-governmental use, is commonly available only to railroad, pipeline, power-line, and hydroelectric companies. The recent increase in coal mining in Germany, following the decision to phase out nuclear power, has increased focus on the legal ability for coal producers to relocate surface occupants for the purpose of mining coal (Friederici, 2013). This law, which reflects a strong motivation to minimize dependence on imported resources, could become a model for legislation in other countries as domestic resources dwindle.

In Spain, all minerals belong to the Crown according to the Mining Law of July 21, 1973 and the Hydrocarbon Law of October 7, 1998. Exploration and production activity is administered at the regional rather than federal level under concessions with durations of a few years for exploration and up to 30 years for mining. The tax and royalty system provides incentives to mineral activity (Nicoletopolous, 2013). In direct contrast, although relatively new mining laws have been promulgated in France, fracking has been banned at the national level and the only recent metal-exploration effort was stopped by widespread protests.

5.3.4 Australia and New Zealand

In Australia, mineral entry is regulated by individual states and territories under laws ranging from the 1971 Mining Act of South Australia to the 2012 Mineral Titles Act of the Northern Territory. A license, which is required to conduct exploration, is granted based on the applicant's work plan and financing, and is for a period of 5 years. All minerals are treated similarly with the exception of uranium in Western Australia, which requires a radiation management plan. Rights of indigenous Australians are protected by the Native Title Act of 1993, which requires that they be included in negotiations for use of land. Royalties apply to most mineral production, with a special rate of 30% for iron ore and coal operations with resource profits of over $50 million (Bogdanich and Purtill, 2015).

New Zealand has seen special controversy over ownership of their extensive geothermal resources. In 1840, when the Maori people signed the Treaty of Waitangi with the British, they retained ownership of fishing and sacred places. As it turns out, many of these sacred places were geothermal areas. The significance of this situation to ownership of New Zealand's geothermal power plants and deposits must be clarified by litigation. It could become even more important if the Maori claim ownership of extinct geothermal systems, many of which contain gold deposits.

5.3.5 Other countries

In Japan, where as much as 70% of the land has been used only for timber and is thus potential exploration country, surface rights are privately owned but all mineral rights are owned by the federal government. The federal government grants claims or concessions, although exploration on them can be greatly complicated by the need to negotiate land access with surface owners.

In South Africa, most land is privately owned, and historically the surface owner also controlled the rights to subsurface mineral deposits. However, in 2004, the Mineral and Petroleum Resources Development Act abolished prior laws related to mineral rights and transferred the rights to undiscovered mineral deposits to the state (van der Vyver, 2002). Ongoing fears of nationalization of the mining industry in South Africa have impeded international investment (SACSIS, 2014).

Mineral land access in less-developed countries (LDCs) is usually by concessions that must be negotiated with the government, although the process is not always straightforward. Some countries have mineral laws on the books, but incumbent governments are often not in agreement with the laws and do not facilitate operations under them. Reasons for these disagreements include valid environmental and economic concerns and the lack of technical and clerical staff to process applications, but they can also reflect the desire for extra payments for approving applications. In many countries that lack modern banking infrastructure for electronic transactions, it is common for business transactions to be done by using paper money. This has led to criticism that some cash payments do not find their way into the treasury of the country (Ensign and Matthews, 2014). In an effort to avoid this problem some concession applicants offer funds for public improvement projects that are carried out or monitored by independent groups, in lieu of bonus bids (Rodriguez et al., 2013).

The World Bank has spent more than $1.5 billion to implement mining-sector reform in about two dozen mineral-rich LDCs, including the Democratic Republic of Congo, Mongolia, and Uganda (World Bank, 2013). These projects aim to build partnerships among international mining companies that have the capital to conduct exploration and extraction operations, local and national government agencies that are eager for tax revenues, and non-government organizations (NGOs) that want revenues to be used for education and healthcare. Where concessions are obtained, they are often set up with low rentals in the first few years and increased payments after that. The increased payments are often linked to "drop-out" clauses requiring that a proportion of the land be released each year. This approach stimulates concession holders to search diligently and prevents hoarding of mineral concessions.

5.4 Exploration versus exploitation concessions

A final element of concern to explorationists in many countries is the link between exploration and extraction. Although this link is a fundamental part of the law in many countries, giving discoverers the right to extract mineral deposits that they find, some countries require completely new negotiations after a discovery. In many LDCs, these negotiations focus on economic factors, especially the division of profits. This problem was most severe during the 1970s and early 1980s when mineral exploration was widespread, and metal prices and economic nationalism were both on the rise.

The huge Cerro Colorado porphyry copper deposit in Panama is a good example of the casualties of these times. Cerro Colorado was discovered in the 1960s and exploration of the deposit continued through the early 1970s with expenditures totaling about $25 million. Negotiations over the economic framework for production of the deposit broke down during the later stages of exploration and the property was confiscated with partial repayment of exploration expenses to the discovering group. Cerro Colorado was then leased to a third group, which carried out further exploration but decided not to put the property into production because falling copper prices limited its economic appeal under the ownership and tax structure advocated by the government. In 2011, Cerro Colorado was still not in production even though copper prices had increased dramatically, and it remains a testament to the debate that pits unemployment and industrial development against sovereign control and profit sharing (Simms and Moolji, 2011). Perhaps the worst example of these problems is Oyu Tolgoi, the giant copper mine in Mongolia, which has been subject to repeated negotiations about tax and ownership structure long after it was put into production (Flynn, 2014).

In some more-developed countries (MDCs), notably the United States, this link is most likely to be disturbed by environmental issues. Because of high levels of public participation and conflicting interests, it has become difficult to agree on all environmental aspects of proposed mineral projects and some have failed on the basis of these complications. For instance, the US Environmental Protection Agency recently denied the application for the Pebble deposit in Alaska, one of the largest copper deposits in the world. Although the high level of concern about environmental issues is necessary and desirable, the failure to reach a compromise permitting production is not encouraging for domestic mineral supplies in MDCs. The recently permitted Eagle

---

BOX 5.4 | **OIL IN THE SOUTH CHINA SEA**

One of the most aggressive challenges to the Law of the Sea has come from China, which has claimed a large area in the South China Sea, which has the potential to host large oil and gas deposits. The area, which is known as the Cow's Tongue (Figure 5.3), lies between Vietnam on the west and the Philippines on the east, and it extends as far south as Malaysia and Kalimantan. This large sea claim by China is based on Chinese claims to the Spratly Islands and several other islands in the area, which are also claimed by the other countries. The claim is motivated by China's need to assure its supplies of oil and have better navigational access to the Pacific Ocean.

---

copper–nickel mine in Michigan's Upper Peninsula is a positive step toward balancing environmental concerns with societal demand for mineral resources.

5.5 Law of the sea

A world mineral law would be an ideal solution to our vexing land access problems for mineral exploration. Although such a law for land-based minerals is unlikely, it was drafted for marine mineral resources as part of the 1982 United Nations Law of the Sea (Morell, 1992; Schmidt, 1989; United Nations, 1982). The Law of the Sea holds that ocean mineral resources are the "common heritage of mankind." It puts them under an International Seabed Authority (ISA), which would license mining projects, receive royalties from production, and disburse receipts to all members of the United Nations. Companies are exempt from royalties during the first 5 years of production, which allows exploration and infrastructure development costs to be recovered. A royalty of 1% of the value of production is paid in year 6, and an additional 1% annually until the 12th year when royalty payments are capped at 7% annually (Groves, 2011). The law also stipulates that the ISA must receive all technical data and that it can choose to develop areas being explored by private groups.

Concerns over these provisions prevented the United States, Canada, Belgium, Germany (then West Germany), Italy, the United Kingdom, and Japan from signing the Law of the Sea when it was drafted. Between 1982 and 1993, the LOS was ratified by only 45 countries, all of which were LDCs except Iceland. At least 60 countries had to ratify the treaty for it to become international law, and MDCs initially refused. In response to the failure of the LOS to address the interests of MDCs, the United States proclaimed in March, 1983, that the ocean area between 3 and 200 miles (4.8 and 320 km) offshore from its territory was part of an **exclusive economic zone** (EEZ) under US control (Figure 5.3) and many other coastal nations followed suit. The US EEZ has an area of 3.9 billion acres

(15.8 million km^2), much larger than its land surface, and contains a wide variety of mineral deposits (Cronan, 1992; Ferrero *et al.*, 2013).

In 1994, the ISA passed an amendment that recognizes that all coastal states have sovereign rights to their own EEZs. As a result, 106 countries, including Canada, Japan, and the European Union, signed the treaty, although the United States held back over concerns of loss of sovereign control and royalty revenue from oil and gas production on the continental shelf beyond 200 nautical miles (Groves, 2011; Cover, 2012). The absence of the United States as a member to the treaty does not seem to be hindering ocean-floor exploration. The ISA has contracts with nearly two dozen companies that are actively exploring copper–gold deposits along the Clarion–Clipperton fracture zone of the Mid-Atlantic Ridge and the Southwest and Central Indian ridges, as well as cobalt-rich ferromanganese crusts in the western Pacific. As land-based reserves dwindle, it seems certain that marine minerals will assume center stage in our thinking (Earney, 1990; OTA, 1987; Borgese, 1985; Ferrero *et al.*, 2013).

5.6 Land access and land policy

Many jurisdictions have mineral laws on their books but control access to land and production by their land policies (Culhane, 1981; Foss, 1987; Hargrove, 1989). These policies have come to rival geology as controls on world mineral reserves. The most important policy matters of current interest are land withdrawals and land classification.

5.6.1 Land withdrawals and wilderness areas

Land withdrawal, the prohibition of mineral, grazing, agricultural, or related activities on public land, has become an important part of land policy (Vann, 2012). In the United States, it began with the creation of Yellowstone National Park in 1872, an event that marked the end of a century in which

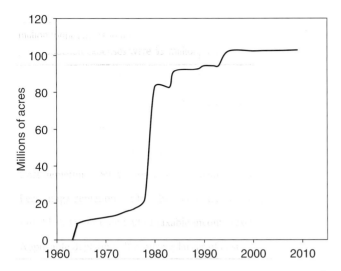

Figure 5.6 Historical growth of acreage placed in the US National Wilderness Preservation System. The large increase from 1978 to 1980 consists largely of Alaskan land (data for 2014 from the Bureau of Land Management).

federal land policy consisted largely of giving away land to settle the country. In 1906, Theodore Roosevelt withdrew more than 66 million acres (267,000 km^2) from homesteading and leasing under existing agricultural laws. Passage of the Pickett Act, or General Withdrawal Act, in 1910 empowered the president to withdraw federal lands for classification and to stop mineral entry to these lands if it was in the public interest. These laws were not motivated by environmental concerns. Rather, they reflected a desire to set aside land containing oil, coal, and fertilizer minerals thought to be vital to national security. (The major source of US fertilizer minerals at that time was Germany.) The Naval Petroleum Reserves, areas set aside to preserve oil for the US fleet, were started at this time, as was federal control over oil-shale lands.

Federal land withdrawals reached a new level with passage of the Wilderness Act of 1964, which set aside 9.1 million acres (36,800 km^2) of federal land to be preserved as wilderness. Through this act and the FLPMA, the Department of Interior was charged with evaluating all federally owned roadless areas larger than 5,000 acres (20.2 km^2) for possible inclusion in a National Wilderness Preservation System. The wilderness system grew rapidly and by 2014 it contained about 110 million acres (444,400 km^2), more than 12 times larger than the original 9.1 million acres (Figure 5.6). Wilderness lands come from areas controlled by the Bureau of Land Management (BLM), Forest Service, Fish and Wildlife Service, and Park Service, which have been responsible for making recommendations to Congress on which lands to include in the wilderness system. Over the life of the system,

Congress has actually approved about 25% more wilderness land than the agencies have recommended. Including federal land withdrawn for other purposes, more than 90% of offshore federal land, and 50–60% of onshore federal land is closed to mineral, oil, and gas exploration and production. In 2012 alone, almost 1.4 million acres of federal land were withdrawn by the BLM and US Forest Service (BLM, 2013a).

Unfortunately, most of the land that has been withdrawn from exploration has not been adequately explored. The Wilderness Act recognized the possibility that unknown minerals might be present on this land, and the FLPMA required preparation of management framework plans that include geological resource inventories for the land for which BLM has oversight of the mineral estate (Kornze, 2013). These studies have been largely the responsibility of the US Geological Survey and the BLM, whose work has been of a reconnaissance nature, largely involving geological mapping, and geochemical and geophysical surveys. The average cost of these surveys has been about $4/acre, considerably less than it would cost to mow a lawn of that size. Full evaluation requires drilling, which was not done for cost and policy reasons. Thus, there is little doubt that potentially economic deposits have been included in withdrawn areas, including the wilderness system. Present administration of the wilderness system gives essentially no attention to the mineral potential of land after it is included in the system. In 2014, of the 400 presentations at the National Wilderness Conference held to commemorate the 50th anniversary of the wilderness system, only one paper mentioned minerals, and it was only in the context of historical mineral prospecting in the Boundary Water Canoe Area that is now designated wilderness land (Duffy, 2014).

Even offshore areas are subject to withdrawals. In 1970, a Presidential decision withdrew most of the Pacific offshore, as well as parts of Florida and the Grand Banks (Shabecoff, 1990). The 1970 withdrawal contained provisions for buying back leases that had already been made in parts of the areas, something that had never been done before. As noted earlier, approximately 90% of offshore United States areas have been withdrawn from oil and gas exploration, including Atlantic and Pacific offshore until at least 2017 and the Gulf of Mexico off the coast of Florida until at least 2022. These offshore lands are estimated to contain undiscovered technically recoverable resources of 88.6 Bbbl oil and 11.3 m^3 of natural gas (BOEM, 2011b).

Pressure also continues for a permanent withdrawal of the Arctic National Wildlife Refuge (ANWR) from oil exploration (Figure 5.7). The ANWR has an area of 19 million acres (76,900 km^2) and its geology is very similar to that of the

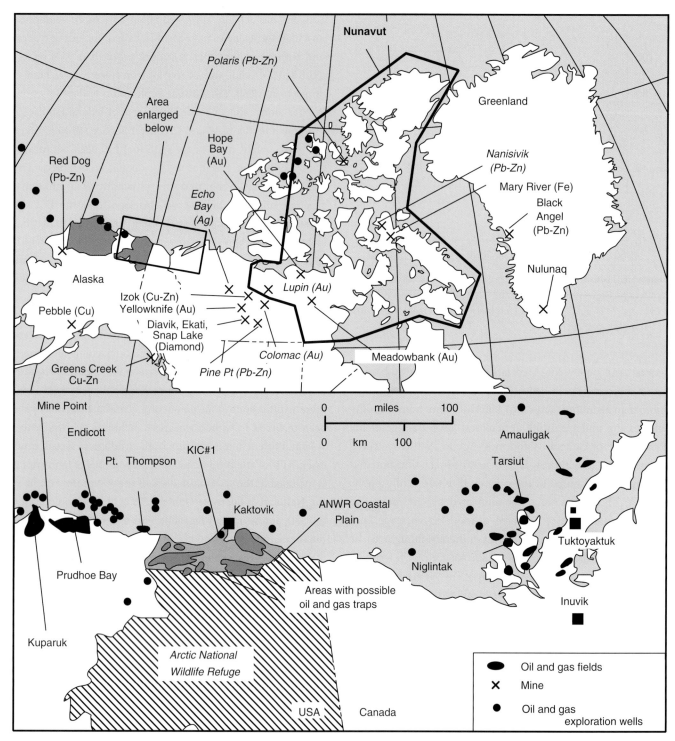

Figure 5.7 Northern North America showing Canadian aboriginal territory of Nunavut and areas of major mineral and oil and gas activity in the high Arctic. Inactive and exhausted mines and currently inactive oil fields are shown in italics; important Pb–Zn and W mines in the Yukon Territory are not shown because of lack of space. The detailed map shows the relation between the Arctic National Wildlife Refuge (ANWR) and oil and gas fields in the Prudhoe Bay–MacKenzie Delta area, along with possible oil and gas traps indicated by seismic surveys in ANWR. Exploration well KIC#1 was drilled on Native American land (compiled from US Geological Survey Professional Paper 1850; Hilyard, 2010; Bishop *et al.*, 2011).

BOX 5.5 | **LAND WITHDRAWAL VERSUS PROPERTY RIGHTS**

Public concern about mineral activities in scenic or populated areas has led to various initiatives to withdraw privately owned land from mineral entry. At the local level, this includes efforts by some municipalities to ban oil drilling. These efforts run afoul of property rights, which are firmly embedded in law in most areas. By banning mineral entry, the municipality prevents owners of land in the area from producing their minerals and therefore from the attendant economic benefit. Courts have consistently ruled against these bans and, in some cases, have levied heavy monetary judgments against the government responsible for the ban.

Prudhoe Bay area to the west, suggesting that it could contain similar oil resources (USGS, 1989). Based on 1,400 miles of seismic surveys carried out in part of the ANWR, the US Geological Survey estimated that there is a 95% chance that this area could contain 5.7 Bbbl of economically recoverable oil (USGS, 1999; Bird and Houseknecht, 2008), which would make ANWR one of the largest producing areas in the United States. In 1987, after completing a comprehensive assessment of potential impact on wildlife, the BLM recommended that approximately 1.5 million acres of the ANWR coastal plain be opened to oil and gas exploration. Forty-two percent of the ANWR is already designated federal wilderness, and slightly more than 90% is completely off limits to development. In spite of these compelling observations, there is strong sentiment that exploration in the ANWR should not be allowed (NRDC, 2011). Much of the concern relates to the welfare of the large Porcupine caribou herd that uses the ANWR as its summer feeding ground. In 2005, the US Senate, by a 51 to 49 vote, attempted to include opening the ANWR coastal plain to exploration as part of the federal budget, but the House of Representatives removed the provision from the budget reconciliation bill (Blum, 2005; Weisman, 2005). In 2010, the US Fish and Wildlife Service, an agency within the Department of Interior, began assessing the ANWR coastal plain for possible inclusion in the wilderness system, and in 2014 released a draft Comprehensive Conservation Plan for the ANWR. The draft contained six different plans for long-term management of the ANWR and received 94,061 public comments.

Most other countries have not been as aggressive in withdrawing wilderness. Canada has no national wilderness law, although it does have an extensive system of national and provincial parks, ecological reserves, conservation areas, and wildlife and management areas (McNamee, 1990; Fluker, 2009). Although wilderness laws have been enacted by some provinces, only British Columbia and Newfoundland have proposed significant withdrawals. Nevertheless, almost 1 million km^2 of Canada's 10 million km^2 land surface is designated as provincial or national protected areas not open to mining activities, and grassroots campaigns are underway to enlarge this (Boyd, 2001; Parks Canada, 2009; Environment Canada, 2011). In an action directly linked to minerals, the British Columbia government turned the area containing the very large Windy Craggy copper deposit into a park (Thompson and Boutilier, 2011). The impact of such movements on Canadian economic activity would be considerably greater than that in the United States, in view of the greater proportion of its gross national product that is contributed by mineral production.

Antarctica is the ultimate land withdrawal (Zumberge, 1979; Riding, 1991). This continent is jointly administered by members of the Antarctic Treaty, which when originally signed in 1959 included the United States, the Soviet Union, France, Britain, Italy, Belgium, Germany, Poland, Norway, Japan, China, India, South Africa, Australia, New Zealand, Chile, Argentina, Brazil, and Uruguay. Antarctica contains many potentially important mineral deposits (de Wit, 1985; de Wit and Kruger, 1990), including the Dufek layered igneous complex, the closest possible analog to the huge Bushveld mafic igneous complex that hosts the world's largest reserves of platinum, chromium, and vanadium (Figure 5.8). Pressure to explore these deposits has been dampened by distance and harsh operating conditions, although it may increase as more readily accessible deposits are exhausted. To forestall this, the Antarctic Treaty group agreed in 1991 to ban mineral and oil exploration for at least 50 years. By 2010, the treaty had been signed by an additional 17 nations.

5.6.2 Land classification

Land withdrawals are a first big step toward national land classification. Whether we like it or not, all land is likely to be classified according to preferred use in the near future. To do this intelligently, maps must be prepared for all areas showing

Figure 5.8 The Dufek layered igneous complex in Antarctica, as seen from a passing aircraft, consists of layered mafic rocks similar to those in the Bushveld mafic igneous complex and could host important mineral deposits (photo by S. B. Mukasa, University of New Hampshire).

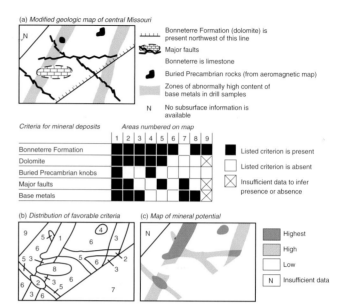

Figure 5.9 The system of geologic land classification advocated by the US Geological Survey begins with the preparation of a map (a) showing geologic features related to potential mineral deposits, in this case Mississippi Valley-type (MVT) lead–zinc deposits in the Viburnum Trend of southern Missouri. As tabulated below the map, these deposits are found in the Bonneterre Formation where it has been altered from limestone to dolomite, usually around older, Precambrian rocks or along faults. Deposits are also surrounded by traces of base metals in the rock. Second map (b) shows the distribution of rocks with these favorable characteristics, and third map (c) uses these zones to classify the area in terms of the probability of containing MVT deposits (from Shawe, 1981).

how the land is used presently and special factors or attributes that might affect classification decisions. This is not a new idea. Most municipalities consider minerals in their zoning regulations, usually focusing on the preservation of conveniently located deposits of sand, gravel, and stone for local construction. Land classification in larger regions requires attention to all types of potential mineral resources. Studies done by the US Geological Survey through their Mineral and Energy Resources Programs set the standard for such efforts (Shawe, 1981; Ferrero *et al.*, 2013). Their work provides information on the type of mineral concentrations that are present in an area, the geologic and geochemical environments in which these deposits are found, and the potential that undiscovered deposits might be present. The information is provided in maps (Figure 5.9) that can be compared by non-technical people to maps for other potential land uses, thus facilitating informed land classification (Nielsen, 2011).

The most important land-classification effort in history started in Alaska in the late 1960s (Singer and Ovenshine, 1979). When Alaska joined the Union in 1959, it was given the right to select about 103 million acres (413,000 km²) of federal land by 1984. Interestingly for the state, it went about this selection process rather slowly, focusing primarily on agricultural land. Then came the 1967 discovery of the Prudhoe Bay oil field on land that the state had selected. A lease sale in 1969 for oil exploration around Prudhoe Bay area yielded over $900 million dollars in bonus bids to the state. Suddenly, mineral potential became the focus of state land-classification efforts. In addition to claims by the state, the federal government was faced with long-standing claims by Alaskan native

people. These claims were recognized by the Alaska Native Claims Settlements Act of 1971, which provided 44 million acres (178,000 km²) to native Alaskans along with $962.5 million, of which $462.5 million was paid immediately. The act also provided for the formation of regional corporations controlled by native Alaskans, with shares in the corporation awarded to all native residents in each region. In the face of competing claims from these groups, land distribution in Alaska was stopped by the federal government in 1978 so that an enormous program of land classification was undertaken to guide further decisions. As of 2014, about 43.7 million acres (0.18 million km²) had been conveyed to native groups and another 99.1 million acres (0.40 km²) had gone to the state (BLM, 2013a).

The Alaskan land-distribution effort shows the growing importance of aboriginal land claims and their emergence as a major factor in land classification and land access. After a century of quiet, many aboriginal groups are asking the courts to support their claims to land. Some claims are based simply on prior occupancy, others on religious significance, and still

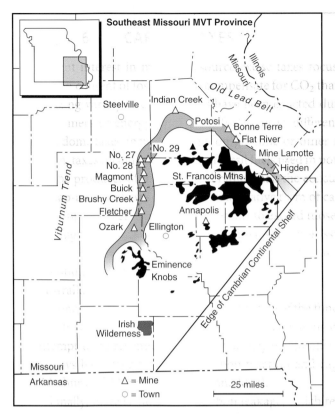

Figure 5.10 Schematic geologic map of southeastern Missouri showing the location of important MVT lead–zinc mines of the Old Lead Belt and Viburnum Trend. Old Lead Belt mines are exhausted and flooded, and explored only by local scuba divers. Lead–zinc reserves remain only in the Viburnum Trend and possibly in its southern extension (light gray shaded zone), which continues into the Irish Wilderness (compiled in part from Larsen, 1977).

others on treaties that were signed and forgotten or ignored by settlers (Albers, 2015; Aboriginal Affairs and Northern Development Canada, 2015). In a settlement of this type, the Canadian federal government granted the Gwich'in tribes 6,162 km² of land and mineral rights in the Mackenzie River delta, an area of oil exploration interest. In 1993, the Nunavut Land Claims Agreement established a 2 million km² area of the eastern Northwest Territories as a new Inuit-administered territory known as Nunavut (Figure 5.7). This agreement also awarded Nunavut's 17,500 Inuit residents $1.173 billion over a 14-year-period, and ownership of about 18% of the territory, including mineral rights to 2% of the land, with the rest remaining in federal hands (Farnsworth, 1992).

Not everyone approaches land classification in the same way. A few simply ignore geologic or other information. Some conservation groups actually target areas with mineral potential for withdrawal. The idea here is that withdrawal of ground with high mineral potential will cut down future environmental

Figure 5.11 (a) White arrows show the location of four man-made islands immediately offshore from Long Beach, California, that were constructed with gravel and are used for production of oil from the Wilmington field. The islands, which are named after US astronauts, Grissom, Freeman, White, and Chaffee, have produced 930 Mbbl of oil through 2014, with a revenue sharing agreement signed in 2012 that gives 49% of revenue to the oil producers, 49% to the state of California, and 2% to the city of Long Beach (Belk, 2012). (b) Since the upper photo was taken, development of Long Beach harbor has put pleasure-boat docks very close to the islands, and the islands themselves have been modified with drill rigs enclosed in structures that look like high-rise residential complexes as seen in this photo of Island Grissom (photos from US Department of Energy and "Donielle from Los Angeles, USA" (oil island uploaded by PDTillman) [CC-BY-SA-2.0 (http://creativecommons.org/licenses/by-sa/2.0)], via Wikimedia Commons). See color plate section.

problems by locking up land in wilderness areas or wildlife refuges. A case in point is the Irish Wilderness area in Missouri. This 16,500-acre (67 km²) area is near the southern end of the Viburnum Trend (Figure 5.10), which accounts for about 90%

of lead and 20% of zinc production in the United States. Mining in the Viburnum Trend, which has been a major employer in the area since 1970, takes place along an almost continuous line of MVT lead–zinc deposits (discussed in Chapter 9), at least 70 km long and 0.6 km wide, buried at a depth of about 300 m. Almost no evidence of these deposits can be detected at the surface, and they were discovered by deep drilling based on geologic concepts.

Limited drilling around the Irish Wilderness showed that it appeared to be the southern continuation of the Viburnum Trend and had high potential for mineable ore. Although this information was provided to all interested parties during classification decisions, the area was included in the wilderness system. Ironically, the Irish Wilderness is not wilderness or even unique countryside. It was logged extensively early in the 1900s, and was described then as: "The pines were gone. Wildlife, deprived of food and cover, also was gone. The soil, stripped of its protective trees and even of much of its grasses, eroded and washed into the streams and choked them with gravel" (Barnes, 1983, p. D-1). By the time it entered the wilderness system, the land was covered by scrub forest similar to that found throughout the area. In this case and in many others, wilderness classification was assigned with little consideration of local employment or long-term mineral requirements.

Conclusions

It is interesting to speculate on how a visitor from Mars might view our present land laws and policies. Our Martian might ask where it is that we expect to get minerals for the next generation if we withdraw all distant and wilderness land from exploration. The only land left will be in more settled areas of the world, where the **NIMBY** ("not in my back yard") syndrome is most strongly entrenched. The visitor might note that most of us lack information on the distribution of mineral operations in our own backyard. Oil exploration and production has taken place in the Los Angeles area for almost 100 years with little attention from the general public, often from wells housed in large structures that look like office buildings (Figure 5.11). Sand and gravel, crushed stone, and other construction materials are mined near most towns from operations that are similar in size to large metal mines. Landfills for municipal waste are even more common and make up the largest "mining" activity in most MDCs.

Hopefully, our Martian visitor would conclude from this that we have coexisted with mineral extraction for centuries and that we will find a way to continue. Taking the longer-term view, our Martian might realize that land can be used for different purposes at different times, and that mineral extraction is particularly well suited for this sort of sequential land use if the land is reclaimed properly. Having seen our strong interest in environmental issues, the visitor could return to Mars confident that we would meet the challenge of producing minerals for our next generations.

CHAPTER

6 Mineral economics

6.1 History and structure of the mineral industry

A mineral deposit must be an ore deposit before society can use it. In other words, it must have the potential to yield a profit. But, what is a profit? It could be defined narrowly as income remaining after deduction of expenses and taxes for a single mineral deposit. But a broader view that includes other deposits or activities is also possible. The definition that prevails depends largely on the organization and the economic framework in which it operates. Many different types of organizations have been involved with mineral deposits and their definition of profit has varied considerably.

6.1.1 Historical perspective

During the Industrial Revolution, most minerals were produced by individuals or small groups funded by wealthy backers. As the scale of operations grew, small producers became large corporations funded by loans from banks, governments or other institutions, or sale of **shares** in the corporation (also known as **stock**). Where shares were sold to a few people and not widely traded, the operation was *privately owned*. Where shares were sold to the general public, a *publicly owned corporation* was created and its shares were traded on a **stock exchange**. *Government-owned corporations* were those in which the government held all the shares. These were not publicly owned because people could not buy a share in the company, although citizenship allowed them to benefit indirectly from the corporation's activities.

Consolidation became necessary when mining of isolated deposits showed that they were, in fact, parts of a single large deposit. The classic example of this happened in the diamond deposit at Kimberley, South Africa, which was discovered in 1871 (Wheatcroft, 1985). Early arrivals to Kimberley simply staked a claim on the deposit and began digging. At the height of activity, 30,000 people were digging on adjacent small parcels of land covering the kimberlite pipe that contained the diamonds. Gradually, adjacent claims were joined to form groups, and groups combined with neighboring groups to form small syndicates. By the mid 1870s, the original 3,600 mines had condensed into 98 groups, and by 1889, two groups, De Beers Mines headed by Cecil Rhodes and Kimberley Mines headed by Barney Barnato, vied for total control. With payment of over 5 million pounds, control was gained by De Beers, which went on to become the world's leading diamond producer (van Zyl, 1988). Meanwhile, mining grew into a deep open pit and finally an underground mine (Figure 6.1). The fortune that Rhodes earned from this venture and its sequels funded many philanthropies, including Rhodes Scholarships to Oxford University, an institution that had the foresight to let Cecil Rhodes attend university intermittently for several years as he commuted to and from South Africa.

As the scale of world mineral production grew, corporations became vertically integrated. Producers of crude oil moved into refining, and then into marketing of gasoline and other products. Metal ore miners began to smelt ore and fabricate bars, plates, wires, and other metal products. As raw-material needs grew and deposits were exhausted more quickly, it became necessary to explore for new deposits. This led corporations into countries where geologic conditions were favorable for mineral discoveries even though there was no significant domestic demand for the minerals, and it

Figure 6.1 (a) The Kimberley mine in South Africa, began as small diggings that gradually grew to form the deep pit shown here, which continues to depth as the underground mine shown in Figure 11.18. The dark, high-rise building in the background is Harry Oppenheimer House, where diamonds from South Africa and Botswana are sorted and graded (photograph by the authors). (b) The Chuquicamata mine in northern Chile is one of several mines owned by US companies that were nationalized by the Chilean government in 1970 to form the state-owned corporation, Codelco. This mine, one of the largest in the world, measures 4.3 km long, 3 km wide and is almost 1 km deep. (c) This shovel is one of the tiny dots at the bottom of the Chuquicamata pit in the middle picture. The person standing below the bucket of the shovel, and another just beside it, show just how large this machine is (photographs by Antonio Arribas). See color plate section.

produced a new type of corporation, the multinational, which extracted minerals in one country, moved them to other countries for processing, and sold them wherever the demand was greatest (Jacoby, 1974; Sampson, 1975; Thoburn, 1981; Mikdashi, 1986; Mikesell and Whitney, 1987).

With the growth in mineral production came an increased awareness of its importance to national welfare and the advent of government-owned corporations (Grayson, 1981; Radetsky, 1985; Mikesell, 1987; Victor *et al.*, 2012). Driving forces for government involvement included the security of mineral supplies and a desire for more of the profits from domestic economic activity. The first such company was created in 1908 when discovery of oil in the Persian Gulf resulted in the establishment of Anglo-Persian Oil, the predecessor of British Petroleum (BP). It was designed to prevent the oil from falling into the hands of corporations from other countries, a situation that would have made the British Royal Navy dependent on foreign firms for its oil supplies. Some oil companies, such as Compagnie Francaise des Petroles (CFP) in France and Azienda Generale Italiana Petroli (Agip) in Italy, grew from small government investments in oil-production organizations. Others, such as Statoil, the national oil company of Norway, were created by government decree and assigned an interest in all domestic concessions that were granted to other companies. Still others, such as Petro-Canada, were started by the transfer of government-owned

stock in existing companies (Syncrude and Panarctic Oils) to a new company.

6.1.2 Nationalization and expropriation

The trend toward government participation was strongest in less-developed countries (LDCs), where the definition of profit and its distribution between operating company and host government became a major sticking point (Dasgupta and Heal, 1980; Nwoke, 1987). Lack of funds prevented most LDCs from purchasing an equity interest in mineral operations by conventional means, and this led to the use of **nationalization** or **expropriation**, the forced transfer of all or part of the operation to the government, usually without adequate compensation. A pioneer in such efforts was Mexico, which nationalized its oil industry in 1938, to form the giant government-owned Petróleos Mexicanos Corporation (Grayson, 1981). This was followed in 1951 by nationalization of British oil interests in Iran, and in 1973 Saudi Arabia began its takeover of Aramco. The transition to government ownership of non-fuel minerals reached an apex in the late 1970s and early 1980s, with large fractions of world iron ore, bauxite, copper, tin, and phosphate production being put into government corporations. Codelco, the Chilean national copper corporation, was formed during this period by nationalization of copper mines in the country (Figure 6.1 b and c).

BOX 6.1 | PEMEX

Petróleos Mexicanos (Pemex) was created in March, 1938, when President Lázaro Cárdenas nationalized oil operations belonging to US and Anglo–Dutch companies. Although countries initially refused to purchase Mexican oil, Pemex managed to survive and gradually grew to become an important oil exporter, and is currently the eighth largest oil producer in the world (Table 6.1). Pemex is wholly owned by the Mexican government and pays taxes that supply a large fraction of government revenues. In 2009, Pemex paid taxes and duties of $12 billion on sales of $23.7 billion, far higher than most private oil companies. This high level of taxation is typical of many government-owned corporations that use funds for social needs. The lack of funds has prevented Pemex from investing in new technology and accounts in large part for the gradual decline in its production. Future exploration and production in Mexico faces large technical and economic challenges. Various ways to meet these challenges have been suggested, including privatization of Pemex or opening of the country to exploration by foreign firms.

Table 6.1 Largest oil companies in the world, based on 2013 annual production (expressed as barrels of oil (BOE) equivalent). Compiled from Helman (2013). Reserves for oil and gas are shown in billion barrels and trillion standard cubic feet, respectively. Note that most of the large companies are government-owned, a fact that sheds a different light on the term "big oil."

Company	Government ownership (%)	Production (BOE)	Reserves
Saudi Aramco	100	4.6	260, 284
Gazprom	50.1	2.95	4500, 19
NIOC (Iran)	100	2.22	137, 29
ExxonMobil		1.93	12.8, 74.1
Rosneft	69.5	1.67	16.7, –
Royal Dutch/Shell		1.46	6.2, 42.8
Petrochina	86.3	1.43	11, 67.6
Pemex	100	1.31	11.4, 12.7
Chevron		1.28	6.5, 29.2
Kuwait Petroleum Corporation	100	1.24	111, 1.8
BP		1.13	10, 40.3
Total (France)		0.94	5.7, 30.9
Petrobras (Brazil)	64	0.91	10.9, 12.4
ADNOC (United Arab Emirates)	100	0.87	137, –
Lukoil		0.84	13.4, 23.5
Ministry of Oil (Iraq)	100	0.98	115, –
Sonatrach (Algeria)	100	1.30	10, 140
PDVSA (Venezuela)	100	1.40	130, 195
Statoil (Norway)	67	0.70	2.1, 17.7

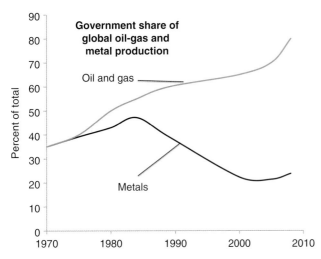

Figure 6.2 Change in government control of global metals vs. oil and gas companies (compiled from reports of the World Bank).

Although some early takeovers were carried out with compensation, many others were not, leading to long legal and political battles (KCC, 1971; Girvan, 1972).

Increased government control of mineral production changed the face of the world mineral industry. In 1970, multinationals controlled about 70% of world oil and gas production. By 1980, their share had declined to about 50%. As of 2013, 15 of the top 20 oil producers in the world are either completely or partially government-owned and they control about 80% of world oil reserves (Table 6.1). The potential longevity of the national oil companies is even more impressive. On the basis of 2013 reserves, the seven largest multinationals can continue to produce at current rates for 11 years, whereas the seven largest national oil companies have a time frame that is more than ten times greater (Kretzchmar *et al.*, 2010). In the hard-mineral sector, government control increased into the 1980s but then began to decrease (Figure 6.2). The main reasons for this reversal in trend were the highly cyclic nature of mineral prices, which made profitable operation difficult during times of low prices, as well as poor government management and re-investment. This led to privatization, which dominated the 1990s and 2000s.

6.1.3 Privatization and resource nationalism

Large-scale privatization began in the 1990s when the British Coal Corporation announced that it would privatize after 45 years of declining results under government control. In Africa, production by the large government copper company Zambian Consolidated Copper Mines (ZCCM) declined

steadily until it was privatized in 2000. The largest wave of privatization followed the breakup of the Soviet Union, when numerous state-owned mineral operations were privatized entirely or, more commonly, with the government retaining a controlling interest. Among the new companies to emerge from this transition are Rusal, currently the world's largest aluminum producer, and the giant oil and gas companies Gazprom and Rosneft (Table 6.1).

Most privatization efforts were driven by the fact that government-run organizations were generally not as efficient as those in the private sector and were more susceptible to internal corruption. An example of the problem is Petrobras, the giant Brazilian oil company that is 64% owned by the government. The company has teamed with outside expertise to make major discoveries of deep oil under salt layers along its Atlantic margin. However, the pace of production has not matched expenditures, leaving the company mired in debt and plagued by scandal about internal corruption (Romero and Thomas, 2014). These pressures remain in many government-owned organizations and continue to drive privatization.

As privatization matured in the late 2000s, resource nationalism appeared as a backlash (Johnson, 2007; Mares, 2010; Gardner, 2013). Resource nationalism includes a wide range of strategies that allow governments to derive greater benefits from their natural resources. Expropriation and nationalization remain part of the tool kit. For example, in 2007, Venezuela directed four companies with operations in the oil-rich Orinoco River basin to relinquish majority control or be nationalized. Exxon and ConocoPhillips left the country while Statoil and BP remained. In one of the strangest deals, in 2012 Argentina forced Repsol, the Spanish multinational oil company, to sell 51% of the shares of its YPF subsidiary for approximately $5 million. YPF, which was originally an Argentine government-owned oil company, was sold to Repsol during privatization efforts in 1999. Other approaches that increase government control include changes in the tax regime, development of new exploration and production laws, joint government–corporate ownership and/or control, and requirements for greater degrees of processing in the host country. In the mining sector, increased government participation and mining taxes have been instituted in Bolivia, Argentina, Democratic Republic of Congo, Ghana, and Mozambique, and restrictions on exports of unprocessed ores were put in place in Indonesia, Gabon, and Tanzania (EY, 2013).

Ironically, some of the national oil and mineral companies that were set up to assure control of domestic resources have now become multinationals themselves with exploration and production interests outside their own boundaries. Both Statoil of Norway and Codelco of Chile are involved in numerous projects around the world, some of which have become subject to the new wave of resource nationalism. Even China, which retains close control of mineral operations in its own country, has felt the pressure as it invests in other countries. Chinese government-owned companies purchased partial interests in large multinational mining companies including the British–Australian Rio Tinto and Teck of Canada, and complete ownership of iron, copper, and zinc deposits in Peru, Korea, and Canada, respectively (Bank, 2011), which generated discussion about the appropriate level of foreign investment in the United States, Canada, and Australia (Wilson, 2011). In Africa, Chinese investments in oil and mineral projects have by-passed international restrictions on foreign investment (Burgess and Beilstein, 2013).

Resource nationalism has given rise to two challenges. Most obviously, resource-poor countries are offering tax relief or lower overall operating costs in an effort to lure mineral processing and other down-stream activities. Second, corporations are attempting to move profits to more tax-friendly venues through transfer pricing, the process by which materials move from one corporate division to another in the corporation (Ashley, 2014).

6.1.4 Tomorrow's business structure

The cycle of expropriation, privatization, resource nationalism, and growth of sovereign wealth funds has produced governments, corporations, and funds with a more mature and focused understanding of their mutual goals. All are trying to maximize the gain from their mineral assets without killing the golden goose or growing too fat from its production. Although simple transfer of ownership is still happening, the trend is toward cooperative efforts and overlapping ownership interests for more complex mineral deposits where exploration and production expertise is essential. BP, for instance, sold its interest in a Russian joint venture to Rosneft in return for a 19.75% interest in the company, two seats on the board of directors, and probably some help with the project. Similarly, Exxon, swapped a partial interest in many of its North American oil and gas projects for a partial interest in Russian Arctic exploration and production by Rosneft (Horn, 2014).

Most projects are also guided by the **Equator Principles**, which provide a framework for due diligence to support responsible decisions regarding risky investments. Unlike

earlier guidelines, Equator Principles give attention to social and community issues as well as financial aspects of proposed investments.

Greater cooperation between government and private corporations has led to complications in foreign policy. For instance, proposed sanctions against Russian oil and gas production in response to Ukrainian issues will likely hurt Exxon, Shell, and BP, which have interests in Russian energy companies (Krauss, 2014). The blending of interests, however, has led to a more unified accounting system for recognition of profits in the industry, the subject that we take up next.

6.2 Profits in the mineral industry

6.2.1 Differing views of mineral profits

With its complex history, it is no surprise that the mineral industry has wrestled with the definition of profit. Consider the Real del Monte silver–gold mines near Pachuca, Mexico. These veins have been mined since the 1500s (Randall, 1972) and by 1973 had produced more than 20 billion ounces of silver, over 6% of the total world supplies to that time. By 1974, however, the mines were not profitable and they were sold to Fomento Minero, a branch of the Mexican government, for $3.5 million. A major reason for Fomento's purchase was the Mexican government's need to provide continued employment to thousands of miners, rather than to have them drift around the country looking for work, and perhaps contributing to social unrest. The decision was probably a good one; as the price of precious metals rose in the late 1970s and 1980s, new veins were found, good profits were made, and many jobs were saved. Mines in the district continued to operate until 2005 and, as of 2014, some veins were still being explored.

Even where the government does not buy the mineral property, it can affect profits with subsidies. During the 1960s, gold producers were especially needful of help because government action had held the world price at $35 per ounce since 1934. As costs rose, many mines were forced to close, a problem that was most acute in the big gold-mining areas of Canada and South Africa. In Canada, the Emergency Gold Mining Assistance Act subsidized operations between 1948 and 1973. Over 300 million Canadian dollars was paid on production of 61.8 million ounces of gold, a subsidy of almost 5 Canadian dollars per ounce at a time that gold sold for only $35 per ounce (EMR, 1973). A similar subsidy offered by the South African government was particularly necessary because many of the mines were joined underground. Thus,

BOX 6.2 | SOVEREIGN WEALTH FUNDS

As governments have gained a greater share of the profits from mineral production, there has been a move to make these gains sustainable. This actually began in Texas with the formation of the Permanent School Fund (current value $30 billion) in 1854, followed by the Permanent University Fund ($15 billion) in 1876. A second wave of funds based on mineral income began in 1958 with the New Mexico State Investment Council ($17 billion), and continued in the 1970s with the Wyoming Mineral Trust (1974, $5.6 billion) and the Alaska Permanent Fund (1976, $49 billion), in 1985 with the Alabama Trust Fund ($2.5 billion), and finally in 2011 with the North Dakota Legacy Fund (2011, $1.7 billion). In Canada, the Alberta Heritage Fund ($16 billion) was started in 1976.

Similar funds, which began in the 1970s around the world, have come to be known as **sovereign wealth funds**. Of the 74 funds active at present, 44 derive all or most of their revenues from oil and gas production, including the largest of these, the Government Pension Fund of Norway, worth $838 billion in 2014 (Table 6.2). Another seven funds, including two in Chile, and one each in Western Australia, Botswana, Kiribati, Mongolia, and Wyoming derive their funds from production of other minerals including coal, copper, diamonds, and phosphate. Total holdings in 2014 of the 74 sovereign wealth funds amounted to about $6.4 trillion of which $3.8 billion came from oil and gas production.

Table 6.2 The 15 largest sovereign wealth funds (out of a total of 74 funds with $6.4 trillion in assets) showing assets as of March, 2014 (from information supplied by Sovereign Wealth Fund Profiles – http://www.swfinstitute.org/fund-rankings/).

Country	Sovereign wealth fund name	Assets $billion	Start	Source of funds
Norway	Government Pension Fund	838	1990	Oil
UAE – Abu Dhabi	Abu Dhabi Investment Authority	773	1976	Oil
Saudi Arabia	SAMA Foreign Holdings	676	n/a	Oil
China	China Investment Corporation	575	2007	Variable
Kuwait	Kuwait Investment Authority	410	1953	Oil
China–Hong Kong	Hong Kong Monetary Authority	327	1993	Variable
Singapore	Government of Singapore Investment Corporation	320	1981	Variable
Singapore	Temasek Holdings	173	1974	Variable
Qatar	Qatar Investment Authority	170	2005	Oil and gas
China	National Social Security Fund	161	2000	Variable
Australia	Australia Future Fund	89	2006	Variable
Russia	National Welfare Fund	88	2008	Oil
Russia	Reserve Fund	87	2008	Oil
Kazakhstan	Samruk-Kazyna JSC	77	2008	Variable
Algeria	Revenue Regulation Fund	77	2000	Oil and gas

Many sovereign wealth funds have grown so large that they are the ultimate source of funds during difficult times, such as the financial downturn of 2008 (Beck and Fidora, 2008; Selfin *et al.*, 2011). Their very presence generates additional concerns about the use of new-found mineral and other wealth, known as **Dutch disease** or the **resource curse**. This affliction occurs when an event causes a large flow of foreign currency into a country,

BOX 6.2 | (CONT.)

generating inflation, a decline in manufacturing and other domestic economic activity, and a decrease in exports (Corden and Neary, 1982; Buiter and Purvis, 1983; van Wijnbergen, 1984). This is usually caused by the discovery and development of a large mineral resource, such as happened with the discovery of the Groningen gas field in the Netherlands. One of the goals of the sovereign wealth funds, then, is to derive the benefit of the mineral wealth while avoiding the negative effects.

groundwater filling of abandoned gold mines would have increased pumping costs at adjacent operating mines, dragging them into bankruptcy.

Several other considerations might allow continued operation of mineral deposits under conditions that are not profitable in the strict sense of the word. **Strategic minerals**, which are usually defined as minerals that are essential for national defense, often receive special treatment from governments (van Rensburg, 1986). Some governments maintain **stockpiles** of strategic minerals and oil. Although the original intent of most such stockpiles was defense, sales and purchases can be used to influence commodity prices and sales can also generate revenue to the government in periods of budget deficit. Sales of gold from government stockpiles during the high-price period of the 2000s served this purpose in many countries.

The need for foreign exchange is another consideration affecting government attitudes toward mineral production. The sale of minerals on international markets provides funds for other foreign purchases. During its tenure as owner and operator of the copper mines, the government-controlled Zambian Consolidated Copper Mines provided 85% of the foreign-exchange earnings of the entire country. Currently, oil and related products account for about two-thirds of Russian export earnings and finance over half of the federal budget (Krauss, 2014). Some countries barter minerals for other commodities; for many decades, Cuba traded nickel for Soviet oil. An extreme example of the "minerals for foreign exchange" concept is the use of "conflict minerals" such as diamonds and **coltan** (columbium–tantalum, as discussed in later chapters), which provide income to insurgent groups.

Even in publicly or privately owned corporations, the extraction of a mineral deposit might not be controlled exclusively by its immediate profitability. The most obvious case occurs when the price of a commodity decreases to a level at which production is unprofitable, but management is convinced that the price will rise in the future. Under those conditions, it can be less expensive to continue operations at a loss until prices improve rather than shut down temporarily.

6.2.2 Accounting methods and mineral profits

In free-market economies, the profit of a mineral operation is calculated by subtracting costs from revenue. The exact steps in the calculation and the specific costs that are allowed depend on prevailing accounting principles, but are generally similar worldwide (Gentry and O'Neil, 1984; Vogely, 1985; Wellmer, 1989). The results of this calculation provide a measure of the benefits derived from the operation, and are also the basis for determining any taxes or royalties that it might owe, as discussed later in this section.

Publicly owned companies publish an annual report outlining their activities and financial results, including a balance sheet that compares assets and liabilities, as well as a **statement of income** and expenses. In the United States, more detailed information on corporations, often with additional geological information, is available from 10-K reports, which are submitted to the US Securities and Exchange Commission, and similar information is available in most other countries with active share markets.

Table 6.3 shows condensed financial information for four large mineral companies, ExxonMobil, BP, BHP Billiton (which began as the Australian company called Broken Hill Proprietary), and Rio Tinto. ExxonMobil and BP are almost exclusively in the energy business, whereas BHP and Rio Tinto are largely mining companies. Total **revenues** for 2013 for these companies range from a high of almost $440 billion for ExxonMobil to a low of about $51 billion for Rio Tinto. These are huge numbers and would place ExxonMobil at about number 25 and Rio Tinto at about number 75 in a list of national gross domestic profits. You will sometimes hear these numbers used to describe the companies – as in, "ExxonMobil, the $440 billion company." This is misleading because it gives no idea of how profitable the company is. The

Table 6.3 Condensed statement of earnings for 2013 for major mineral companies. All amounts in millions of US dollars.

	ExxonMobil	BP	BHP	Rio Tinto
Total revenues	438,255	396,212	70,098	51,171
Deductions	380,544	365,991	52,226	43,741
Exploration cost	1,976	3,441	1,124	948
Operating income	57,711	30,221	17,872	7,430
Tax	24,263	6,463	6,797	2,426
Profit after tax	33,448	23,758	11,075	5,004
Earnings per share	7.37	1.24	2.04	1.94
Dividends per share	2.46	2.19	1.16	1.78
Percentage returned to shareholders	33.4%	176.6%	56.9%	91.8%
Cash flow from operations	10.16	6.60	6.84	17.57
Free cash flow	2.54	–1.08	–1.51	1.12
Total equity	180,495	130,407	38,997	23,080
Equity per share	40.85	41.09	14.61	12.42
Operating profit/ shareholders equity	32.0%	23.2%	45.8%	32.2%

basic question is, "did they make a profit on all of this money?" To answer that question, we have to go determine their taxable income and tax.

As you can see in Table 6.3, each company had large *deductions* from revenues, ranging from about $381 billion for ExxonMobil to almost $44 billion for Rio Tinto. These deductions included a wide variety of costs such as: the purchase of crude oil or mineral concentrates from other mines or wells; wages and salaries for employees; fuel and energy for operations; all sorts of supplies, maintenance, and repair; depreciation of tangible assets such as trucks; and **amortization** of intangible assets such as loans or goodwill. They also include the cost of exploration for new mineral deposits, which we show separately in the table. In some countries, the loss in value of the mineral deposit itself, known as depletion, was also deducted, as discussed later in this chapter.

Deduction of internal costs from revenue leaves the **operating profit** of the corporation, which is subject to tax, and the amount that remains after this is the after-tax profit. It probably occurs to you at this point that claiming extra deductions

is a good way to decrease taxes. Although tax codes try to be specific, there is always room for discussion and the fine points of what might be deductible continues to be worked out in courts all over the world. There is no room for frivolous accounting, however.

After-tax profit can be expressed in terms of the total number of shares that are outstanding for each company. For our four corporations, the per-share profit ranged from a high of $7.37 for ExxonMobil to a low of $1.24 for BP.

Once after-tax profit has been determined, it is time to decide what to do with this profit. Strictly speaking, it belongs to the owners of the corporation (the shareholders), so they could demand that it be paid to them in the form of **dividends**. This matters to all of us because we all derive benefits from corporate dividends. Even if you do not own shares in a corporation personally, other groups that you depend on, such as your bank, credit union, or pension fund do own shares, and dividends are a major source of their income. Note in Table 6.3 that ExxonMobil and BHP paid dividends that were only 33% and 57% of after-tax profit, respectively. The remaining profit was retained for investment in new equipment or new projects, which is the only way that the business can grow. In contrast, Rio Tinto paid out almost all (92%) of its after-tax profit in dividends and BP paid more than (177%) its after-tax profit. Such high levels of dividend payment are unusual and they reflect special situations. BP was still paying for the 2010 Deepwater Horizon oil spill in the Gulf of Mexico and Rio Tinto was paying down a debt incurred for the purchase of Alcan Aluminum in 2007.

So, where does the money come from to pay dividends that exceed after-tax profit? One possibility is to borrow money and pay it out as dividends, especially when interest rates are very low. The more reasonable way it to take advantage of **cash flow**, which is the actual amount of cash that flows through the corporation. Cash flow and operating profit differ because some of the items that are deducted from income to determine operating profit are not actual expenditures. Depreciation is an example; making a depreciation deduction of 10% of the value of a truck each year leaves that amount of money in the corporation as cash flow. Cash flow can be separated into cash flow from operations and cash flow from financial transactions, with the sum of these being free cash flow. As you can see in Table 6.3, Rio Tinto and BP had large cash flows from operations that supported their dividends, although BP had a negative free cash flow.

The amount of money left to pay dividends depends, of course, on how much was taken in taxes. Taxes paid by our four corporations ranged from about 21% to 42% of operating

BOX 6.3 | INTERNAL AND EXTERNAL COSTS

We should pause a moment here and distinguish between internal and external costs. Internal costs are those that are deducted to determine taxable income. As such, they are part of the cost of the mineral that is produced. External costs are those that are associated with production (or consumption) of the product but that are paid by society. Black lung disease contracted by coal miners was an external cost of coal until a fund was set up by the US government and a levy was collected on each short ton of coal mined. Today, CO_2 is an external cost of fossil fuels and society is wrestling with ways, such as cap-and-trade or a carbon tax, to internalize this cost. **Life-cycle assessment**, which quantifies the energy and materials required and the emissions released, provides a way to recognize these costs, even if they are not internalized (Hesselbach and Herrmann, 2011).

BOX 6.4 | THE LORNEX STORY – A TALE OF TAXES AND PROFITS

Lornex Mining Corporation provides an example of the interplay between profits and taxes in the mineral industry. In the early 1970s, when the Lornex story begins, it had just started a porphyry copper mine in British Columbia, and was a subsidiary of Rio Tinto. Now, it is part of the Highland Valley Complex, a group of closely adjacent operations that make up the largest copper mine in Canada (Figure Box 6.4.1), and is a division of Teck Corporation, Canada's largest mining company.

Figure Box 6.4.1 Highland Valley Copper Operation in British Columbia, showing the Lornex pit, beneficiation plant, and waste rock piles. Note the revegetated waste rock pile on the left behind the red domes. (Photograph courtesy of Teck Resources Ltd.) See color plate section.

In 1975, Lornex reported revenues of $C51 million and operating expenses of $C30.8 million. Additional deductions from revenues included a depreciation deduction of $C6 million for tangible property such as trucks, ore beneficiation equipment and computers, and an amortization deduction of $C6 million for interest on money borrowed to construct the mine and beneficiation plant. This left an operating profit of $C8 million. Taxes on this income amounted to $C7.5 million, leaving after-tax profit of $C626,000. So, the Lornex mine paid 87% of its operating profit in tax, leaving only 13% for the owners.

This unusual level of tax was caused by a peculiar overlap of new taxes and commodity prices. First, the price of copper jumped from an average of $0.46/lb in 1973 to $0.90/lb in 1974 and then plummeted back to $0.54 in 1975, the year of the unusual taxes. The huge price rise in 1974 affected most metals and resulted in large profits for most producers, including Lornex. Thus, governments throughout the world were faced with high commodity prices,

BOX 6.4 | (CONT.)

large corporate profits, and a generally disgruntled citizenry that did not like what had happened to the prices they were paying for mineral products and resented the profits that were being reported to them by newscasters.

The British Columbia government responded to these high prices and profits by levying an extra royalty that was retroactive to the start of 1974. Unfortunately, the royalty hit after prices had declined, and it was so large that it caused a decline in Canadian federal tax revenues. In the ensuing debate about which government was entitled to tax revenues from the company, Revenue Canada disallowed the British Columbian royalty as a deduction in calculation of federal income tax. The squabble was eventually resolved, but changes in tax rates and tax policy continue to be a major factor in determining whether mineral deposits are ore deposits all over the world.

income and averaged 33% (Table 6.3). The exact percentage depends, of course, on how many deductions each company had, but the average of 33% is just about the same as the personal tax rate in many more-developed countries (MDCs). Those who argue for increased tax rates on corporations should keep in mind that dividends from corporations are taxable income to individuals in most MDCs. So, corporate income is actually doubly taxed already. Now, take a look at Box 6.4, and note that in 1975, the Lornex mine paid 87% of its operating profit in tax. This shows that taxes are critically important to the profitability of a mineral operation. So, our next step in understanding profits is to understand the various taxes that can be applied to mineral operations.

6.3 Mineral taxation and mineral profits

The original purpose of taxation was to generate income to support government activities, but lawmakers have not been able to resist using taxes as instruments of policy. They justify this because taxing less desirable activities might promote desirable ones. Mineral supplies are subject to many conflicting interests, including conservation, exploration, and inexpensive products, making them obvious targets of policy oriented taxes and a consequent wide variety of taxes.

6.3.1 Income taxes, depletion allowances, and windfall profit taxes

The most common tax on mineral operations is income tax. It is usually expressed as a rate or percentage of operating profit. The amount of tax paid is controlled by the deductions that are taken in determining taxable income. Because it involves the extraction of non-renewable resources, the mineral industry is sometimes afforded a special deduction known as the **depletion allowance** (O'Neil, 1974; Gentry and O'Neil, 1984;

Shulman, 2011). Unlike a clothing manufacturer, for instance, which could operate indefinitely by buying cloth and replacing sewing equipment with funds set aside from the depreciation deduction, a mine or oil well must stop production when the reserve of ore or oil is exhausted. The depletion allowance is a special form of depreciation, which permits mineral producers to deduct a portion of income from tax each year to compensate for the loss of this asset.

The problem comes, of course, in determining the magnitude of the depletion allowance deduction. Present practice in the United States uses two methods, **cost depletion** and **percentage depletion** (Table 6.4). Percentage depletion was introduced in response to excessive litigation resulting from disagreement between producers and the Internal Revenue Service over the calculation of cost depletion. Producers are required to calculate both types of depletion and to use the larger of them in calculating taxable income.

Even at the lowest rate, the depletion allowance can shield a significant amount of revenue from taxation (Salzarulo, 1997), and this has made it a contentious aspect of the tax code. Supporters argue that it is a logical extension of the depreciation concept and that it is required to make mineral production competitive with other investment opportunities. Critics say that it is a loophole through which revenue escapes taxation. Opponents are particularly disturbed when these deductions are used for purposes other than obtaining a new deposit to replace the exhausted one, although that practice is legal, just as depreciation deductions taken by a business need not be used to buy new equipment.

Adverse opinion about the depletion allowance reached a crescendo in the United States during the OPEC oil embargo of 1973 and the Iranian revolution of 1979, when the rapid rise in oil prices increased oil prices and oil company profits. In response to public concern about these profits, the depletion allowance for most oil production was disallowed and the

Table 6.4 Comparison of cost and percentage depletion deductions in the US Federal tax code for a typical small copper deposit that produced 11 million tonnes of ore in the tax year, with revenues of $52 million and taxable income of $4.5 million. The purchase cost of the deposit was $1.5 million, pre-production expenses were $3 million, annual exploration costs were $1.5 million, and remaining reserve is 100 million tonnes.

Cost depletion = prorated value of each tonne of ore times number of tonnes produced that year.

Deposit value = cost of deposit + pre-production

Exploration expenditures = $1,500,000 + $3,000,000 = $4,500,000

Value of tonne = deposit value/number of tonnes = $4,500,000/100,000,000 tonnes = $0.045/tonne

Cost depletion = $0.045/tonne × 11,000,000 tonnes = $495,000

Percentage depletion = allowable percentage multiplied by revenue, not to exceed 50% of taxable income = $52,000,000 × 15% = $7,800,000

But this is larger than half of taxable income (excluding depletion allowance) = $4,500,000 × 50% = $2,250,000

Applicable depletion allowance = larger of cost and percentage depletion = $2,250,000

Table 6.5 Percentage depletion rate for various minerals as used for US federal tax calculation (from Internal Revenue Service).

Type of mineral deposit	Rate
Sulfur, uranium, and, if from deposits in the United States, asbestos, lead ore, zinc ore, nickel ore, and mica	22%
Gold, silver, copper, iron ore, and certain oil shale, if from deposits in the United States	15%
Borax, granite, limestone, marble, mollusk shells, potash, slate, soapstone, and carbon dioxide produced from a well	14%
Coal, lignite, and sodium chloride	10%
Clay and shale used or sold for use in making sewer pipes or bricks or used or sold for use as sintered or burned lightweight aggregates	7½%
Clay used or sold for use in making drainage and roofing tiles, flower pots, and kindred products, and gravel, sand, and stone (other than stone used or sold for use by a mine owner or operator as dimension or ornamental stone)	5%

Windfall Profits Tax was established (Goodman, 1992). In spite of its misleading name, the Windfall Profits Tax is an excise tax, as discussed in a later part of this chapter. Depletion remains for small oil producers and for most other mineral commodities (Table 6.5).

Other governments approach the depletion concept differently. Until 1989, the Canadian federal tax code included the concept of **earned depletion** (Parsons, 1981), in which corporations were able to deduct four-thirds of exploration and other qualifying expenditures. Although this deduction was an improvement over a pure depletion allowance because it focused the deduction on mineral-related expenditures, it was discontinued. Because of the importance of mineral production to the Canadian economy, other deductions have been allowed from time to time, including flow-through financing, in which individuals can deduct the cost of corporate exploration expenditures directly from current income (Hasselback, 2013).

Mineral prices are highly cyclical and this makes profits cyclical (Figure 6.3). Operations that start at a low point in the cycle can be very profitable, but those that start at high points in the cycle when investment capital is easiest to obtain often lose money (Harris and Kesler, 1996) During periods of high profits, it is common to hear calls for taxes on "windfall profits," which are supposed to be profits gained solely from the increase in price of the commodity with no effort from the corporation. A true windfall profits tax would tax positive operating income, not revenue. Many so-called excess profits taxes did not do this; instead they charged a flat tax on revenue, and strictly speaking were excise taxes.

The most famous excise tax in recent US history is the inaptly named Windfall Profits Tax. This tax, which was signed into law on April 2, 1980, was an outgrowth of the national preoccupation over large profits made by oil companies in the late 1970s when oil prices rose to over $30/bbl. The main thrust of the tax was to place a 70% tax on the sales price of oil above a base price of $12.81/bbl (Table 6.6). Formulation of this tax occupied Congress for several months, and negotiations over provisions of the tax were

BOX 6.5 | **TAX HOLIDAYS AND OTHER TAX RELIEF**

Tax holidays have been used as a means to encourage exploration and development in areas with high unemployment. When Ireland passed legislation in 1956, exempting mineral production from income and corporate taxes, an exploration boom found several important mines, including Tara near the town of Navan, the largest zinc producer in Europe. By 1973, mining was well established in Ireland and taxes had risen to world norms. The same approach was taken in Australia where gold mining, the most important productive activity in the arid, western part of the country, was granted freedom from taxation. Gold mining increased tremendously in the area, due in part to this incentive and in part to the increased world price of gold, and the law was later repealed, although immediate deduction of exploration expenses has been part of the tax code for many years. Both of these successful tax holidays were originally passed without a termination date. In contrast, tax holidays with specific terms have been notably unsuccessful. Prior to 1973, the Canadian federal tax code permitted a 3-year, tax-free production period for new mines. Unfortunately, this led to wasteful extraction practices such as removal of only the highest grade ore during the tax holiday in order to maximize profit. This made it harder to mine remaining low-grade ore and caused mines to close prematurely.

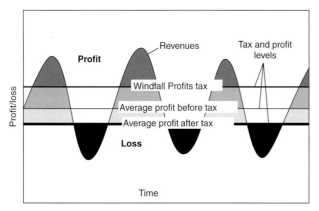

Figure 6.3 Cyclical nature of mineral profits showing how operations cycle from profit to loss through time. If we assume that the only tax paid is a windfall profits tax, then we can model the impact of the tax with the three lines. The line labeled "Average profit before tax" shows what the profit of the operation would be if it simply averaged its years of profit and loss. It would make a lot of money in the good years and lose money in the bad years. The line labeled "Windfall Profits Tax" shows how this tax would take profits above the line. The effect of taking the high profits is to lower the average profitability of the operation to the new line labeled "Average profit after tax." Note that the only way to return the operation to its previous profit level is to add a "windfall loss support" program. In fact, variants of these were used to support auto and bank businesses during the 2008 economic downturn.

based largely on economic projections that it would bring in $227 billion by 1990. Not surprisingly, the economic projection was wildly wrong; by the late 1980s, the price of oil had fallen to $12/bbl and the tax was generating essentially no income. In August, 1988, it was repealed, after having generated $77 billion in tax revenue (Crow, 1991).

In retrospect, most people agree that the Windfall Profits Tax was a failure. It was not very successful as a revenue-generating vehicle. Much worse, it consumed important national energies at a time when they were needed to address the real problem of how to lessen dependence on foreign oil. The obvious solution was to find oil closer to home, a challenge that had to be met by increased exploration, which required money. It was ironic that the main move taken by Congress in response to this national emergency was to decrease oil-company profits, thereby cutting funds available for exploration (Figure 6.4). In the end, this move simply weakened the multinational oil companies and strengthened the national oil companies, which control oil reserves and oil prices today.

Australia is heavily dependent on mineral production and has also experimented with windfall profits taxes. As mineral prices rose during the 2000s, many royalties and other taxes did not keep pace with profits to the mineral companies. This led to efforts to develop a tax system that would provide the government with a greater share of profits during high-price periods. The first proposal was the Resource Super Profits Tax, which charged 40% of taxable income and did not really focus on excess income. It was replaced by the Minerals Resource Rent Tax, which charged 30% of profits above $75 million. This tax did focus on income, but used a fixed ceiling rather than a percentage. It was repeated in late 2014.

6.3.2 Ad valorem taxes – severance taxes, royalties, and carbon taxes

In addition to taxing income, it is possible to tax the value of assets. For individuals, this might be a tax on a home or other

Table 6.6 Provisions of the Windfall Profits Tax. Oil was divided into three tiers, each of which had a base price above which all income was taxed at the rate specified (Gelb, 1983).

Base price ($/bbl)	Tax rate	Comments
Tier 1 Oil – Oil in production before 1979		
12.81	70%	50% tax on first 1,000 bbl/day of production by crude producers with no refining capacity
Tier 2 Oil – Stripper oil (oil produced from old wells in amounts of less that 10 bbl/day)		
15.20	60%	30% tax on first 1,000 bbl/day of production by crude producers with no refining capacity
Tier 3 Oil – Oil discovered after 1978, heavy oil, and oil recovered by tertiary extraction methods		
16.55	30%	Base price adjusted 2% annually in addition to inflation adjustment

Additional specifications:
(1) Total tax not to exceed 90% of new income from property
(2) Tax deductible as business expense for income-tax purposes
(3) Amount subject to tax can be reduced by most state severance tax payments
(4) Tax to phase out over 33-month period beginning in January, 1988, if it raised $227.3 billion, or by January, 1991 otherwise

Exemptions:
(1) Alaskan oil except that from Sadlerochit field, which was taxed as pre-1979 oil
(2) Oil from properties owned by Native Americans and state and local governments
(3) Oil from properties owned by non-profit medical or educational institutions or by churches that dedicated proceeds to these purposes

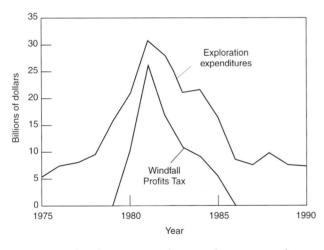

Figure 6.4 Oil exploration expenditures and payments to the Windfall Profits Tax in the United States compiled from data of the American Petroleum Institute. The rise in tax revenues in the early 1980s reflected higher oil prices and the parallel rise in exploration expenditures reflected the industry's anticipation of higher prices in the late 1980s and 1990. Because the high prices did not materialize, both tax revenues and exploration expenditures fell.

personal property. For corporations, it might be a tax on the physical plant where material is processed. In the case of mineral companies, the mineral reserve can be taxed. **Reserve taxes** are usually expressed as a percentage of the value of the mineral reserve. Proponents of the reserve tax view it as a logical extension of the **property-tax** concept.

Opponents counter that value is realized only when minerals are produced and sold, and that changing economic conditions might convert ore to waste before it is mined. They also argue that reserve taxes limit recoverable reserves because they apply equally to both high-grade and low-grade ores, making it proportionally less economic to produce low-grade ores. Sticklers even point out that it is not logical to disallow the depletion allowance on the one hand, as is done for most oil, thereby indicating that the asset has no value, and impose a reserve tax on the other hand, which accepts that a deposit has value. This logical conundrum has not daunted legislators in the United States, largely because property and reserve taxes are levied by state and municipal governments, whereas the depletion allowance is part of the federal tax code.

Severance taxes are also specific to the resource industries and are applied to each unit of production, as are sales taxes. Severance taxes are similar to royalties (Bourne, 1989; Olson and Klecky, 1989), which are commonly granted to the owner of land from which minerals are produced. Severance taxes, however, are levied by the host government. So, in the case of production from private land in a state or province with a severance tax, the producer would pay a severance tax to the government as well as a royalty to the land owner. In theory, these taxes are supposed to compensate the owner of the land (and the host government) for any loss in value caused by removal of the minerals. Severance taxes yielded almost

Table 6.7 (A) Severance tax revenues for 2012 for the ten US states with largest revenue, showing the percent that these revenues constitute of total state revenue and the main source of revenue (from the National Council of State Legislatures).

State	Severence tax revenue (millions of US dollars)	Approximate percent of total revenue	Source of revenue
Alaska	5,787	90	Oil and gas
Texas	3,656	15	Oil and gas
North Dakota	3,187	75	Oil and gas
Wyoming	968	50	Oil, gas, and coal
Louisiana	886	14	Oil and gas
Oklahoma	849	17	Oil and gas
New Mexico	768	25	Oil and gas
West Virginia	626	20	Oil, gas, and coal
Kentucky	346	10	Oil, gas, and coal
Montana	305	25	Oil, gas, coal, and metals

(B) Royalty, mining taxes, and related payment to Canadian provincial governments for the fiscal year 2012 (from Toms and McIlveen, 2011)

Province	Amount (millions of Canadian dollars)
Alberta	1,215
Saskatchewan	860
Quebec	207
Newfoundland and Labrador	178
Ontario	110
Northwest Territories (Nunavut)	56
Manitoba	21
New Brunswick	20
Nova Scotia	1
Yukon	0.2

$19 billion to the 35 states that levied this tax in 2012. Most revenues come from taxes on oil and gas production (Table 6.7), and most of these go to Alaska and Texas. Canadian provincial royalties generally range from 1 to 2% of the gross value of mineral produced, although several provinces offer tax holidays for the first years of production as well as other incentives for mineral development (Natural Resources Canada, 2014). For the fiscal year 2012 these taxes yielded about $C3.3 billion (Table 6.7B).

Most federal and provincial or state mining laws include royalties to the host government. One of the few that does not is the US Mining Law of 1872, a feature that will likely change if the law is ever revised. Royalties are also important sources of income to private individuals and corporations. Royalties are almost always the same for oil and gas, one-eighth of the value of production, but differ greatly for most hard minerals.

Two aspects of severance taxes merit special attention. First, there is the issue of whether severance taxes restrain

---

BOX 6.6 | CARBON TAXES

Present interest in mineral-resource excise taxes focuses on **carbon taxes**, which would impose a levy on the carbon content of fossil fuels to compensate for CO_2 that they generate and the global warming that they cause. An amazing variety of carbon taxes have been enacted during the last 20 years. Many of them are directed toward consumers of energy from fossil fuels, but in different ways. In Ireland, Norway, Costa Rica, and the United Kingdom, taxes focus primarily on gasoline or other motor fuels, whereas in Switzerland, France, Korea, and Japan, taxes include other types of fossil fuels used in power generation and industrial activity. In India, taxes focus on the producer of the fossil fuel rather than the consumer.

Cap-and-trade systems offer an alterative form of carbon taxation (Cleetus, 2010). In these systems, emissions are capped at a certain level and permits to exceed those limits are auctioned. This provides flexibility to polluters, although the allocation of initial permits is a major issue (Bushnell and Yihsu, 2012). Also, the system does not provide a continuous revenue stream to governments; their only revenue comes from the initial sale of the pollution permits. Cap-and-trade systems are in operation in New Zealand, the European Union, and parts of Australia.

Concern has been expressed about the fate of the funds generated by these systems. In some cases, the funds are put directly to efforts to minimize carbon emissions but in others they simply go into the general fund. Some attempt to be revenue neutral in that they would offset expenditures in other sectors of the economy, although surveys have found that both carbon taxation and cap-and-trade are generally regressive and poorly designed (Mathur and Morris, 2014; Robson, 2014).

Finally, there is the issue of **carbon leakage**, which refers to the degree to which carbon emissions simply move to jurisdictions without cap-and-trade or other limiting systems. As we noted earlier in this section, taxes can be used either for revenue generation or implementation of policy. In efforts to limit carbon emissions, we have an extremely complex policy initiative and one that is likely to take some time to perfect.

---

interstate trade (Stinson, 1982). The issue was brought to a head by the very high severance taxes imposed on coal by Wyoming and Montana. Low-sulfur coal in these states is in high demand to cut SO_2 emissions from power plants in the eastern United States. Because there are no other domestic sources of low-sulfur coal, these states can impose large severance taxes without fear of losing sales. The issue of just how large these taxes can be without imposing an unfair burden on consumers in other states has been brought to the courts, which have ruled that it should be decided by legislative action.

Second, there is the degree to which severance taxes actually reimburse citizens for a loss of what is sometimes referred to as their natural (or national) heritage. The idea here is that removal of the minerals diminishes future employment and economic opportunities in the area. Some governments put part of their severance tax revenues into funds that will be used to relieve these problems in the future. Minnesota allots 50% of its severance tax revenue to a fund for use in the northern part of the state where iron ore is produced, and other states use some of the revenues for environmental clean-up. The sovereign wealth funds discussed earlier are a response to this issue.

6.3.3 Tariffs and export–import quotas

Governments can also affect mineral profits through tariffs or **import quotas**, which are often used to protect a weak or inefficient domestic industry from foreign competition. The motivation for such protection has commonly been the desire to maintain employment levels in a large industry such as coal or steel, or the need to preserve domestic productive capacity for a strategic mineral commodity such as oil. Tariffs imposed by the US government are divided into countries with the designation of **permanent normal trade relations status**, which replaced the most-favored-nation status in 1998. The only countries that are permanently without this designation are Cuba and North Korea, although other countries move on and off the list according to prevailing issues and foreign policy.

Import quotas have been employed to protect the US steel and oil industries. During the 1990s, steel imports were controlled by voluntary restraint agreements (VRAs) that specified the fraction of the US market that could be supplied by each country or economic region. For instance, in 1990 these agreements allowed imported steel to supply 20% of the US market, with 7% of the market coming from the European Community, 5.3% from Japan, and the remaining 7.7% from South Korea, Brazil, Mexico, Australia, and other countries. They were replaced in 1993 by proposals for steep tariffs on steel imports from 19 countries, most of which were alleged to have provided unfair support to their steel industries. Since that time, tariff disputes have plagued the US steel industry, the latest concerning the import of pipe from Korea (Levinsohn, 2014).

The longest-running import quota effort is the attempt by the US government to control oil imports. In contrast to efforts on behalf of steel, which involved cooperation among nations, the oil import effort was largely a unilateral move. It began in 1955 when oil companies were asked to limit their growing imports of cheap foreign oil to 1954 levels, in order to protect the domestic industry. This failed, and in 1957 the Voluntary Oil Import Program was proclaimed because oil importation was deemed to be a threat to national security. The Secretary of the Interior was authorized to establish permissible import levels and allocate quotas. Lack of cooperation led to the Mandatory Oil Import Program in 1959. To make a long story short, this also failed to control imports in a way that furthered the interests of the country, the consumer, or the industry. Refineries in the northeastern United States, where cheap oil was usually imported, were not receiving sufficient crude oil. Foreign producers attempted to sneak oil into the United States via Canada as "strategically secure" Canadian oil. Consumers were denied the cheaper gasoline that could be refined from imported oil. Just as the system was about to collapse, the world oil situation changed with the emergence of the Organization of the Petroleum Exporting Countries (OPEC) and the start of oil embargoes.

The US export–import saga for oil reached a head in 2014 with the appearance of two new issues. First among them was the controversy over whether to import oil from the tar-sand operations in Alberta via the proposed Keystone XL pipeline. Despite approval by appropriate government agencies, the pipeline is opposed by environmental groups that wish to limit the use of oil. The second issue focuses on the possibility of exporting oil and gas. This is astounding for a country that was dreadfully short of crude oil only a decade ago, but it has gained traction because the production of shale oil has greatly increased domestic supplies and because delivery of oil and,

especially, gas to Europe would loosen their dependence on Russian gas. Resolution of these two issues will determine the degree to which North America attains energy independence, at least for oil and gas, as well as the economic integration of Russia into Europe.

It comes as no surprise that most companies will opt to operate in a country with low taxes. In a recent survey, effective tax rates for a model copper mine were shown to vary from a low of about 28% in Sweden to a high of about 63% in Uzbekistan (Mitchell, 2009). Just as important as the level of taxation is its predictability: whether or not the country can be relied upon to levy fair taxes with few abrupt changes. As we have seen, even the best countries can have their bad days, and many of these are caused by abrupt changes in the prices of mineral commodities.

6.4 Mineral commodity prices

Mineral commodity prices change because of changes in supply and demand of the commodity. Recall from your basic economics course that the law of demand states that higher prices lead to lower demand, whereas the law of supply states that higher prices lead to higher supply. When supply and demand are equal, they are at **equilibrium** and all users receive adequate supplies at a price they are willing to pay. Unfortunately, equilibrium is easily disturbed, particularly for mineral commodities.

An imbalance of supply and demand, or disequilibrium, can have a wide variety of causes (MacAvoy, 1988). Many commodities come from only a few locations and closing of one of these because of an accident, war, or exhaustion of the deposit can have a large impact on supply. Other sites are usually producing at maximum rates and cannot increase to make up for the disruption. Conversely, if a new supplier does come on line, it might cause disequilibrium by producing more than is needed. Other events that can cause disequilibrium include technological changes that permit substitution of one commodity for another, environmental legislation that requires or forbids use of a commodity, or governmental actions including price supports, tariffs, or import restrictions.

If supply and demand are delicately balanced, only small perturbations in either supply or demand can cause large changes in price. You can imagine the importance of **marginal demand** of this type by estimating the price of air in an air-tight room holding 100 people for an hour but containing only enough air for 99 of them. This is particularly true for mineral commodities, where only a few percentage points

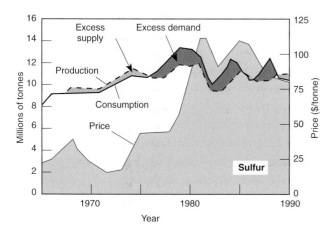

Figure 6.5 Changes in price and supply of sulfur in the United States from 1962 to 1990 showing how small changes in production and consumption caused large changes in price (compiled from US Geological Survey Minerals Yearbooks)

difference in supply–demand relations can cause large changes in price (Figure 6.5).

6.4.1 Supply–demand disequilibrium and coping mechanisms

Supply–demand disequilibrium and the resulting changes in commodity prices make life difficult for producers and consumers alike. Both would like to know the price of the commodity and plan accordingly. To minimize the impact of price fluctuations, **commodity futures** markets provide a forum in which buyers and sellers establish future prices. These markets trade contracts for the delivery of specified quantities of commodities at a certain date and price. Copper contracts on the New York Mercantile Exchange are for 25,000 lb of the metal for delivery on any of the next 23 months and on any March, May, July, September, and December for the next 60 months. Contracts can be closed before they expire by selling those that were originally bought or vice versa. Thus, no one has to deliver or take delivery of the commodity. This permits speculators with no use for the commodity to participate in the market, thereby increasing the number of potential buyers and sellers and improving the liquidity of the market. To further increase liquidity, most exchanges permit contracts to be bought and sold with a small payment, a practice referred to as trading on margin.

By examining the futures markets, producers and consumers can determine the consensus regarding future prices for mineral commodities. Commodities exhibiting **contango** have future contracts that are more expensive than the cash, or present, price. Those with cheaper future contracts are said to exhibit **backwardation**.

6.4.2 Price controls and cartels

The important control that mineral prices exert on supply and demand, and on profits, has led to numerous efforts at price control by individuals, corporations, and governments. These efforts have been motivated by a desire for increased profits, protection from high prices, and prevention of losses in a declining market (Prain, 1975). When control involves a single individual or organization, it is known as a monopoly; when a group is involved, it is known as a **cartel**.

Modern mineral monopolies are rare simply because of the large scale of mineral world markets. One such effort involved the Hunt brothers, who tried to corner the world silver market in 1979 (Hurt, 1981). They purchased more than $1 billion in silver in the form of futures and other contracts, helping the price rise from $9/ounce to almost $50/ounce. When they could not put up additional funds as collateral for the funds borrowed to buy the silver, they had to sell some of their contracts, causing prices to collapse and driving them into bankruptcy. In a similarly ill-fated venture, the government of Malaysia lost hundreds of millions of dollars in 1981 and 1982 during a clandestine attempt to control the price of tin on world markets (Pura, 1986).

Cartels, which are more common, include both corporations and governments, with government cartels usually being more powerful. Cartel activity in the United States was limited by anti-trust legislation of the Sherman and Clayton Acts of 1890, which had a big effect on the structure of the domestic mineral industry. The first major action against a mineral corporation under the law came in 1911, when the Standard Oil Trust was broken up to form Amoco, Chevron, Exxon, Mobil, and Standard of Indiana. Legal action was also brought by the US federal government against US Steel and Alcoa, the dominant producers in the steel and aluminum industries, although with less effect.

Government efforts at price control by edict have usually been given palliative names such as mineral price stabilization. The US National Bituminous Coal Commission, which operated between 1937 and 1943, had the power to set coal prices and was intended to stabilize the weakened coal industry. The Texas Railroad Commission, which has controlled the production of oil in Texas since the mid 1930s, was set up ostensibly to prevent damage to the natural drive of oil fields by overproduction. In fact, it was able to influence world oil supplies because of the large percentage of production coming from Texas during this time. In some cases, government price-stabilization groups in the same country have come into conflict with one another. During the 1970s in the

BOX 6.7 | THE PRODUCER'S CLUB

One of the strangest market-control efforts in recent times concerns the Producer's Club, a group of uranium-producing companies that was formed in the early 1970s. This group, which was encouraged by their host governments, attempted to place a floor on uranium prices as government purchases for nuclear weapons declined. Uranium markets were made even smaller for non-US producers when imports to the US market were curtailed to support domestic producers. At the same time, nuclear-reactor manufacturers began to compete for the electric-power-generating market. In order to induce utilities to use their reactors, the Westinghouse Corporation offered to supply reactor fuel at low prices. By late 1974, Westinghouse had contracted to deliver about 80 million lb of low-priced uranium to its reactor customers, but held contracts to buy only about 15 million lb. This short position, which amounted to about 2 full years of US uranium production, caused upward pressure on uranium prices at the same time that the overall uranium market picked up because of forecast demand for reactor fuel. The resulting price rise quadrupled the price of U_3O_8 to about \$40/lb, leaving Westinghouse with a potential loss of \$1 to \$2 billion if it lived up to its contracts. Instead, it defaulted and began a protracted legal war against the Producer's Club, charging that they fixed the price of uranium and the utilities that bought the reactors countersued to get their contracted cheap uranium (Taylor and Yokell, 1979; Gray, 1982).

United Kingdom, the National Coal Board and British Gas Council, which were in charge of production and prices in their respective commodities, engaged in direct competition for energy expenditures by the public.

One of the largest price-stabilization efforts ever undertaken was the quixotic attempt by the US government to control domestic oil prices. This saga began in 1971 when President Nixon froze all prices and wages in the economy under the auspices of the Economic Stabilization Act of 1970. Controls on oil were replaced briefly by voluntary guidelines in 1973, but these failed almost immediately in the face of the OPEC oil embargo and resulting price rises that began in late 1973. To combat this, buyer–seller agreements were frozen in December 1, 1973, and the Federal Energy Office was given the responsibility of controlling oil prices. In the ensuing confusion, prices were determined by endless bickering over byzantine formulae that required producers of "old," cheap oil to pay entitlements to rival companies with higher priced crude, simply for the right to refine their own crude oil. By late 1975, payments totaled over \$1 billion and the system collapsed. The Energy Policy and Conservation Act of 1975 then divided oil into three price control tiers, but failed to cope with the strain of price changes associated with the Iranian revolution of 1979. Going from bad to worse, Congress replaced price controls with the Windfall Profits Tax in 1980 (Ghosh, 1983; Chester, 1983).

Although the US government has discouraged cartels (other than its own efforts at price control), governments in other countries have been more accommodating. During the last century, cartels or monopolies have been organized with varying success in aluminum, bauxite, coal, copper, diamonds, lead, mercury, nickel, nitrate, crude oil, potash, silver, steel, sulfur, tin, tungsten, uranium, and zinc. Of these, only the Central Selling Organization (CSO) of De Beers, which deals in diamonds, maintained long-term control over its market. Even OPEC, which was the most powerful economic force in the world during the late 1970s and early 1980s, failed to control oil prices through the crude oil glut of the late 1980s. As new oil supplies came on stream from non-OPEC nations including the United Kingdom, Canada, Norway, and Mexico, and as internal disagreements limited its action, OPEC has devolved into an important player on the world oil scene (Johany, 1980; Al-Chalabi, 1986; Kohl, 1991; Parra, 2010; Danielson, 2013).

To succeed, a cartel must control the supply of the commodity and the commodity itself must be essential and non-substitutable. In addition, members of the cartel must be well-financed and willing to withstand periods of low demand, and they must have similar internal cost structures so that low commodity prices do not cause competition among members that are still making a profit and others that are already losing money. A mutual belief or value system is also helpful to hold groups together during trying times.

Most recent cartels have lacked at least one of the key elements, and the future of classical cartels appears dim indeed.

Present efforts to stabilize prices focus more on market transparency, in which producing and consuming countries share information on production, consumption, and other aspects of a commodity, making it easier for producers and consumers to make plans (Cammarota, 1992). For most commodities, this takes place largely through organizations known variously as "study groups" or "councils", such as the International Lead and Zinc Study Group and the World Gold Council.

It should be obvious from this short review that prices have a life of their own. Efforts to control prices for any real length of time have rarely been successful. For this reason, prices will probably continue to change dramatically and along with them profits will change. This leads us back to the question of how to divide profits from mineral production.

6.5 Distribution of profits

The distribution of profits is the central question in mineral production and taxation. How do we divide the pie in a way that attracts producers into the business, gives owners and host governments satisfactory revenue, and provides a product that is priced acceptably for the consumer? The key player in this list is the producer; if they are not attracted into the business, there is nothing to tax and nothing to consume. In terms of global mineral supplies, that means that the mineral exploration and production business has to be attractive enough for shareholders in a corporation to invest in it, rather than putting their money and time into the computer or drug business, for instance.

6.5.1 Return on investment

Shareholders who invest in any corporation do so with the hope of making their money grow. This growth, which is referred to as the **return on investment**, is the most important parameter controlling whether an investment will be made. In simplest terms, it can be expressed as the percentage of the investment that is returned as interest or dividends to the investor each year. For bank deposits and government bonds, the rate of return is determined by the prevailing interest rate. Although interest rates vary considerably depending on the state of the host economy and the threat of inflation or deflation, annual rates on long-term (30-year) government bonds in the United States and Canada have averaged about 5% since 1960. That is the rate of return against which all businesses compete.

For most corporations, one common measure of return on investment is obtained by dividing operating profit by the **shareholders' equity**, which is the amount of money that the shareholders have invested in the corporation. The shareholders' equity is essentially the difference between the assets (cash, funds owed to the corporation, goods that have not yet been sold (inventories), and other supplies) and liabilities (debt or taxes that have not yet been paid). In the case of Lornex, the shareholders' equity at the end of 1974 was $64 million, but the profit after tax and royalties (and before income tax on dividends received by shareholders) was only about 1% ($600 thousand/$64 million), far less than the rate of return on a bank deposit or government bond.

Something was obviously wrong. It is common knowledge that people put money into banks or bonds partly for safety and ease of access, and that they accept a lower return on their investment for the peace of mind. We also know that investments with a higher risk of failure must pay a higher rate of return in order to attract investment. The mineral business is certainly risky, so it should pay a higher rate of return than banks and bonds. Proof of this is seen in the rates of 23–46% for the four corporations that we have been following in Table 6.3.

The general objective that most governments have in taxing is to capture as much of the profit as possible, leaving only enough to attract new investment. The amount or fraction of profit that meets these criteria is known as **economic rent**, and it is indeed an elusive quantity. Any combination of income tax, severance tax, royalty, or even government equity can be used to approximate economic rent. In general, corporations are most interested in repaying their investment and achieving an acceptable return on their investment in as short a period as possible. Some governments allow this by way of depreciation and depletion deductions and attempt to maximize their return over a longer period. This is sometimes done with a resource rent tax that is imposed after corporate goals have been attained. Australia imposed a tax of this type on oil production in 1990 and has since applied it intermittently to mining companies.

6.5.2 Valuation of deposits

One of the most important numbers used in decisions about the economic future of a mineral deposit is its **discounted present value**. This value was likely the basis for negotiations in 2014 that led Glencore Xstrata to sell its interest in the large Las Bambas copper deposit in Peru to a consortium led by China Minmetals Corporation for $7 billion. The discounted present value of a deposit is obtained by summing the present values of after-tax cash flow for the deposit for each future year of operation using the relation:

$$P = S/(1 + i)^n$$

where P is the present value of the deposit, S is the after-tax cash flow from operation of the deposit in the nth year, and i is the **discount rate**, which is the expected rate of return on the investment.

This calculation reflects the fact that a dollar received today is worth more than one received tomorrow, and it requires that certain estimates be made about the future of the deposit. The value of S for any year can be determined from information on the cost of production and the price of the commodity, which are gathered in a feasibility study. This also involves an estimate of the period over which the deposit will be operated, which depends largely on the scale of extraction, the size of the deposit, and the amount of money to be invested in the first place. The discount rate to be used in the calculation depends on the risk involved in the business. For money to be invested in the bank or a bond, prevailing and projected future interest rates would be used. For a business project, it is necessary to arrive at a new discount rate that reflects the risks of the project. The risks might be technical (maybe oil will not flow at the predicted rate), economic (maybe the price of oil will change), or governmental (maybe the tax rate will go up, or the operation will be nationalized). Arriving at a discount rate to represent these factors is obviously difficult, but it must be done.

As its name implies, the present-value calculation provides an estimate of the present cash value of anticipated future profits from a mineral deposit. Figure 6.6 shows how important the choice of a discount rate can be to calculation of the present value. The higher the discount rate, the smaller the present value of the deposit. In other words, knowledgeable buyers will pay more for a project with a low anticipated risk than they will for one with a high anticipated risk, and this sort of project might be taxed at a higher rate. It is also apparent from the calculation that the earliest years in the operation of a deposit have the greatest effect on present value. Thus, the present value of a large deposit that will last for 100 years might be only twice as great as one that will last for 10 years, although governments might place a higher value on this "far-out" cash flow than would corporations.

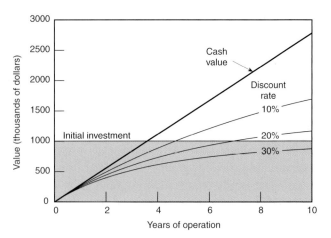

Figure 6.6 Present value of a hypothetical mineral deposit with an initial investment of $1 million (spent in exploration and development of the deposit) and annual after-tax profit of $275,000, showing the time needed to pay back the initial investment for different discount rates. If no discount rate is used (that is, the prevailing interest rate is assumed to be 0 percent), the cash value reaches the initial investment in slightly less than 4 years.

Present-value calculations are most useful in comparisons among different deposits, whether it is a corporation attempting to decide which deposit to extract, or a government trying to decide how large a tax or environmental cost to impose on a group of deposits. As shown in Figure 6.6, the time necessary to repay an initial investment of $1 million required to put a hypothetical deposit into operation, known as the **payback period**, would be about 5 years at a discount rate of 10% and 7 years at a discount rate of 20%. At a 30% discount rate, however, the deposit would not recoup its initial investment even within 10 years. Any similar deposit with lower payback periods would obviously be more attractive unless other factors intervened.

The reverse of this relation concerns **lead time**, the period between discovery and production (Wellmer, 1992). Obviously, a deposit must be put into production as soon as possible in order to minimize the payback period. An exploration investment of $1 million can be paid back the next year for only $1.1 million at a discount rate of 10%. If development of the deposit is delayed for 10 years after the exploration expenditure, the amount required to repay the original investment would be $2.7 million. This is a very real amount for the investor whose money is tied up in the deposit, because this money could otherwise be earning interest in a bank. Although geological and engineering complications often delay the development of mineral properties, the most common culprits are disagreements about economic rent or environmental control.

Conclusions

Mineral economics may be the great, final battleground between modern civilization and the natural order. We have not been following Mother Nature's system and it is unclear just how much longer we will able to flout her authority. As originally planned, Earth concentrated minerals into deposits, we explored for them and then extracted the best ones, perhaps moving on to poorer ones as time passed. By erecting an economic structure that renders some rich deposits worthless through high taxes and encourages extraction of poor ones through subsidies and tariffs, we have confused the system.

As minerals become more difficult to find and produce, we need to establish national mineral policies that recognize the strong linkage between mineral supplies and economic environment. These policies must provide mineral producers with a secure and reasonable return on investment, governments with a fair share of profits derived from their soil, and citizens with a clean environment. While we wait for these miracles to occur, we should work to encourage more accurate use of economic terms in discussions of mineral resources. We have shown here that there is a huge difference between the revenues of a mineral operation and any profits that might be obtained from it, and both are very different from the gross value of minerals in the deposit. In the end, it is only the profit that counts. In spite of this, mineral companies persist in describing their discoveries in terms of gross value, and newscasters and politicians perpetuate the error. This exaggerates the perception of profits that are available from deposits, which encourages debate over division of fictional spoils between government and owners, thus delaying decisions to develop mineral deposits. It would be naive to think that such debates penalize only the corporation; higher costs are passed along to the consumer. Thus, we must strive to make the period of time between discovery and development of mineral deposits as short as possible without jeopardizing our important economic and environmental goals. This requires an informed public; one that does not have to spend years catching up on basic geologic, economic, and environmental concepts in order to express an informed opinion.

CHAPTER

7 Energy mineral resources

The value of annual world energy mineral production in 2014 was almost $4 trillion, an amount that exceeded the gross domestic product (GDP) of every country in the world except China, Japan, and the United States. Energy production makes up almost 4.5% of world GDP and as much as 40 to 90% in major oil-producing counties including Russia, Iran, Nigeria, and Saudi Arabia (Table 7.1). In countries with larger and more diversified economies such as the United States and United Kingdom, the value of energy production ranges from 2 to 5% of GDP.

The importance of energy extends far beyond these numbers, however, because most other activities depend on energy, as shown by the strong correlation between energy production and per capita gross national product (GNP) for most countries (Figure 7.1). Energy production, particularly oil, has also formed the basis for many of the world's great fortunes, including the Sultan of Brunei, the Rockefeller, Mellon, and Getty families and the royal families of Saudi Arabia, United Arab Emirates (UEA), and Qatar, which have supported a wide range of philanthropic, political, and religious efforts (Yergin, 1990, 2011). The emergence of **sovereign wealth funds**, most of which are based on revenues from oil production, are a further indication of the degree to which energy production impacts global events.

World energy has been derived from minerals since the late 1800s, when coal began to surpass wood as the main source (Figure 7.1e). Coal gave way to oil in the mid 1900s and natural gas and nuclear energy began to grow in importance in the latter half of the twentieth century. Since 1960, global natural-gas use has grown at a relatively high rate, whereas uranium has stalled after its rapid start.

7.1 Fossil fuels

Coal, oil, natural gas, oil shale, and tar sand are called **fossil fuels** because they are derived from fossil plant and animal remains that have been preserved in rocks (Fulkerson *et al.*, 1990). These remains, commonly referred to as organic matter, are changed by geologic processes into various forms that have many names (Figure 7.2). Ultimately, most organic matter decomposes by oxidation to form CO_2 and H_2O, which are recycled into the hydrosphere and atmosphere. A small amount of organic matter is shielded from the atmosphere by burial beneath other sediments where it is preserved to become fossil fuels, an important endowment of solar energy that Earth has set aside for later use. This accident of geologic history was probably the most important single factor in the development of our present industrial civilization. Had it not occurred, we would have had no widely available fuel to bridge the gap between wood and more advanced, non-fossil fuels, and the Industrial Revolution might have been delayed indefinitely.

7.1.1 Coal

Coal powered the Industrial Revolution and dominated global energy output through the middle of the nineteenth century (Figure 7.1). Consumption of coal for electricity production increased by 300% between 1970 and 2010, and it still supplies 30% of the global primary energy demand (slightly behind oil at 33%) and 41% of global electricity (World Coal Association, 2014; IEA, 2012). Annual world coal production is worth almost $800 billion and many studies suggest that it will continue to be a primary fuel for electric power generation (Humphreys and Sherlock, 2013).

Table 7.1 Major producers and reserves for oil, natural gas, coal, and uranium (from BP Statistical Review of World Energy, June 2014 for oil and gas; the World Coal Association for coal; and the World Nuclear Association for uranium).

Commodity	Country	Production	Reserves
Oil		(Billions of barrels per year)	(Billions of barrels)
	Saudi Arabia	4.21	265.9
	Russia	3.94	93.0
	United States	3.65	44.2
Natural gas		(Billion cubic meters)	(Billions cubic meters)
	United States	687.6	930
	Russia	604.8	31,300
	Iran	166.6	33,800
Coal		(Millions of short tons)	(Millions of short tons)
	China	3,561	114,500
	United States	904	237,925
	India	613	60,600
Uranium		(Tonnes Uranium)	(Tonnes Uranium)
	Kazakhstan	22,500	629,100
	Canada	8,999	467,700
	Australia	6,991	1,660,000

Geology and global distribution of coal

Coal is plentiful because it forms by simple geologic processes. It is a sedimentary rock made up largely of altered plant remains such as **cellulose** and **lignin**. Although some coals contain fossil bark, leaves, and wood, the fossils in most coal have been obliterated, leaving a hard, black rock that will burn. Under the microscope, coals can be seen to consist of grains called **macerals**, which include: **vitrinite**, the remains of wood or bark; **liptinite** or **exinite**, the remains of spores, algae, resins, and needles and leaf cuticles; and **inertinite**, a complex mixture of fungal remains, oxidized wood or bark, and other altered plant materials (Hessley *et al.*, 1986; Schobert, 1987; Kendall *et al.*, 2010).

Coal begins as **peat**, an accumulation of partly decomposed, brownish plant remains. As peat is buried beneath new sediment, the increasing temperature and pressure cause it to release water and other gases, gradually increasing the proportion of carbon, a process known as **coalification** (Table 7.2). As coalification proceeds coal increases in **rank** from **lignite**, through **subbituminous** and **bituminous coal**, to **anthracite**. The thermal stability of lignite and high-rank bituminous coal indicates that the maximum temperatures that they could have experienced were about 200 °C and 300 °C, respectively. Anthracite is found in rocks that have been folded and subjected to higher temperatures, such as those typical of low-grade metamorphism (Ward, 1984).

Peat accumulates in swamps, which form in flood plains and deltas of rivers, coastal barrier island systems, and poorly drained glacial regions. Most of the world's large coal deposits formed in temperate swamps that contained fresh or, at most, brackish water. The swamps were largely along flat coastal plains and continental interiors, where surrounding land had been leveled by erosion and could not contribute much clastic sediment. Modern equivalents of this environment might include the Dismal Swamp of Virginia and North Carolina, and the coastal swamps of Sumatra (Figure 7.3). Large coal-forming swamps required stable sedimentary conditions over large areas and for long periods of time in order to accumulate

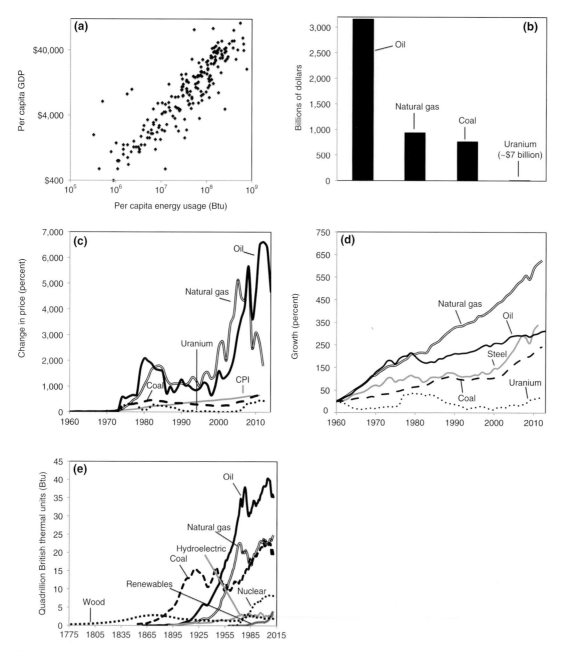

Figure 7.1 (a) Positive correlation between per capita energy usage and GDP. (b) Value of world energy mineral production in 2013. (c) Change in prices of energy minerals since 1960, also showing the change in the consumer price index (CPI). (d) Growth in production of energy minerals since 1960, compared to growth in steel production. (e) Growth in consumption of energy by source since 1775 (compiled using data from the US Energy Information Agency; BP World Energy Outlook; International Energy Agency OECD iLibrary, World Nuclear Association; Organization of the Petroleum Exporting Countries Annual Statistics Bulletin).

thick layers of peat. Peat layers decrease in thickness by almost ten times as they become coal. So, to make a 10-m-thick coal layer, or **seam** as the miners call them, a peat layer about 100 m thick had to accumulate. These peat layers were also very extensive; the famous Pittsburgh seam covers about 52,000 km² (Figure 7.3), and it is only one of about 50 coal seams that have been mined in the Appalachian basin.

The distribution of coal-forming swamps through geologic time has been a function of Earth's plate-tectonic, climatic, and life history. During plate-tectonic movement, mountains formed with large rivers draining into swampy deltas, and during warm periods these deltas had lots of vegetation. None of this mattered, of course, until vascular land plants developed in Late Silurian time. Thus, pre-Silurian sedimentary basins, notably in South America and Africa, lack coal (Figure 7.4). After Silurian time, coal formation took place largely during Late Carboniferous (Pennsylvanian) and Permian time and intermittently during Jurassic through Neogene time (Fettweiss, 1979). During the latter half of Carboniferous time, Earth's polar areas were covered beneath ice caps that advanced and retreated, affecting global sea level and forming some of the largest coal deposits on the planet.

Coal production and related environmental concerns

Coal can be mined from both underground and open pit mines (Kendall *et al.*, 2010). Most open pit coal mines are **strip mines**, which are actually the best form of mining because it is possible to fill the pit and reclaim the land surface. Coal is amenable to strip mining in many areas where it forms flat layers that are covered by only a few tens of meters of overburden. In the first step of strip mining, soil is removed and stockpiled for later replacement, then overburden is removed separately. In the second step, the underlying coal layer is removed, and in the final step the overburden is replaced, the soil is placed on top of it, and the surface is reclaimed (Figures 7.5, 7.6). Overburden removal, coal mining, and overburden replacement are repeated back and forth across strips of land, giving the method its name. Most strip mines are large-scale, expensive operations; a single dragline, the large shovel commonly used to remove overburden, can cost tens of millions of dollars (Figure 7.6). **Contour mining** is a special form of strip mining that is used around the sides of hills where flat-lying coal layers are too thin to merit removal of overburden from the entire hill (Figure 7.5b). Contour mining requires smaller investments in equipment and permits recovery of coal seams that cannot be extracted

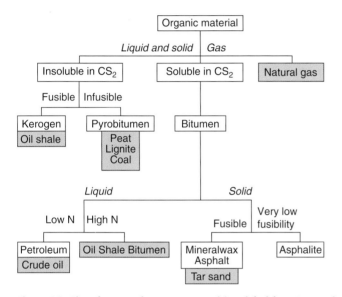

Figure 7.2 Classification of organic material (modified from Yen and Chilingarian, 1979)

BOX 7.1 | HOW RAPID WAS THE BOOM IN FOSSIL-FUEL USE?

Wood was the major fuel during the United States Civil War, which lasted from 1861 to 1865, and whale oil was the primary means to illuminate homes and offices at night. During the Civil War, the Union and Confederate armies together felled 400,000 acres of trees each year and used the wood for heat, as well as to improve visibility of the landscape. At about the same time, Abraham Gesner, a Canadian geologist, figured out how to produce liquid kerosene from coal, and Benjamin Silliman, Jr., a chemistry professor at Yale University, figured out how to distill crude oil from Edwin Drake's well in Pennsylvania to make products such as kerosene, paint, lubricants, and paraffin for candles. Almost overnight kerosene replaced whale oil, and because it was so much less expensive it was possible to lengthen the day by using artificial light. A few decades later, in 1893, US President Grover Cleveland amazed attendees at the Chicago World's Fair when he turned a switch that caused electricity to illuminate 100,000 light bulbs that had been installed by George Westinghouse using Nikola Tesla's new invention of alternating current. Only a few years later, Henry Ford began mass production of the automobile. In the 1850s the US and UK navies transitioned from wind-powered sails to coal-fired boilers, and then during World War I to oil-fired boilers. These events, coupled with population increase and an abundance of coal, oil, and natural gas in the United States, set society on a its path of dependence on fossil fuels. Today, residents of the United States consume almost 100 quadrillion British thermal units (Btu) of energy every year, almost one-fifth of the 550 quadrillion Btu consumed worldwide and a dramatic 3,000% increase from our days of burning wood and whale oil.

Table 7.2 Composition and energy content of coals of different rank (compiled after Schobert, 1987 and Meyers, 1981).

Rank of coal	C	H	Volatile	Fixed carbon	Calorific value	
	(Weight percent)				(Btu/lb)	(MJ/kg)
Lignite	73.0–78.0	5.2–5.6	45–50	50–55	<8,300	<19.31
Subbituminous	78.0–82.5	5.2–5.6	40–45	55–60	8,300–11,500	19.31–26.75
Bituminous						
High-volatile	82.5–87.0	5.0–5.6	31–40	60–70	11,500–14,000	26.75–32.56
Medium-volatile	87.0–92.0	4.6–5.2	22–31	70–80	>14,000	>32.56
Low-volatile	91.0–92.0	4.2–4.6	14–22	80–85	>14,000	>32.56
Anthracite	92.0–98.0	2.9–3.8	2–14	85–98	>14,000	>32.56

Coal basins

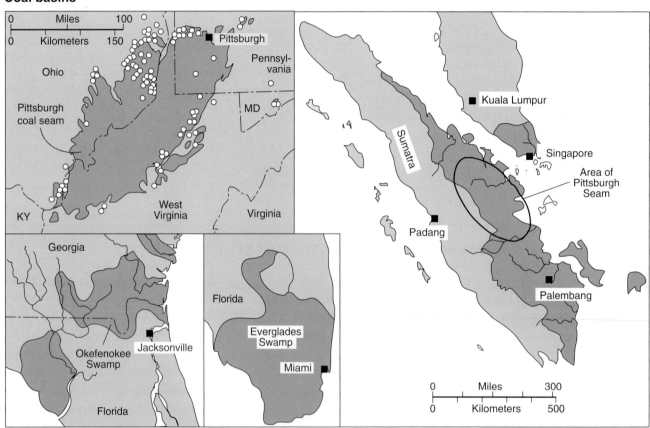

Figure 7.3 Comparison of the Pittsburgh coal seam with modern swamps containing peat. Distribution of test holes in the Pittsburgh seam shows the level of information about the quality of this seam (modified from Garbini and Schweinfurth, 1986).

economically by strip mining. In some areas, the scale of contour mining has increased to the point that entire mountain tops are removed and put into adjacent valleys. This "mountaintop mining" has allowed much higher recoveries of coal, but at the disturbance of even greater areas of land (Copeland, 2013; Lutz *et al.*, 2013).

Early strip and contour mines were major environmental offenders. In old strip mines, overburden was not graded flat after it was replaced, leaving ugly, hummocky topography and no vegetation. Old contour mines were simply left as open cuts along hillsides that continued to erode long after they were abandoned. Pyrite, a common contaminant in coal, was

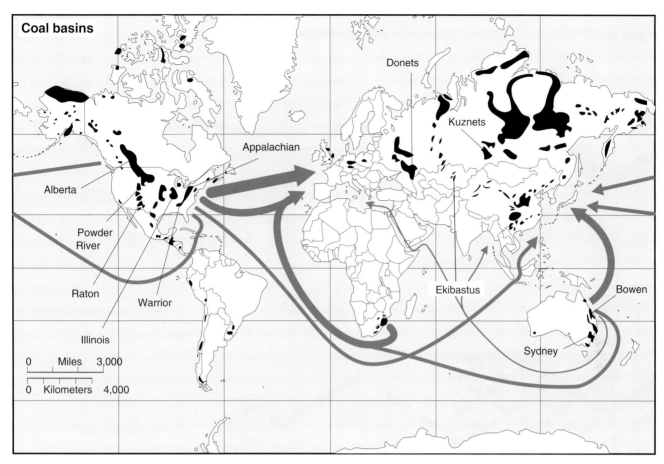

Figure 7.4 Location of coal basins of the world (modified from Fettweis, 1979), including names for important producing basins. Major lines of marine coal trade are also shown (from International Energy Annual, US Department of Energy).

exposed in these old mines and it oxidized (Table 3.3) to create sulfuric acid that entered surrounding drainages (Figure 7.7). The Surface and Mining Control Reclamation Act of 1977 (SMCRA) was a response to these abuses in the United States. This regulation is enforced vigorously; 1,194 km^2 of high-priority, hazardous land was reclaimed between 1978 and 2012, and safety and environmental hazards were remediated on 1,275 km^2 of abandoned mine land (Table 7.3). Congress amended SMCRA in 2006 to mandate that 83% of coal reclamation fees be made available annually to states for high-priority reclamation sites, and reclamation at the remaining 5,200 coal-related, high-priority abandoned mines is anticipated to be complete by 2021. Similar statutes have required remediation in most more-developed countries (MDCs).

Groundwater quality is an important concern in reclaimed mines. Although original aquifers are destroyed by mining, broken rock that is used to fill the mine is commonly more porous and permeable, and actually increases water storage capacity. The increased surface area of this broken rock and its

exposure to weathering during mining allows groundwater to dissolve elements such as calcium, sulfur (as sulfate), and some metals, which are enriched in coals, although this can be minimized by addition of lime to make the groundwater less acid. Uranium is also concentrated in some coals and disturbances related to mining can release radon (Johnson and Paone, 1982; Hossner, 1988; NRC, 1990). Improved efforts at strip mining, however, have minimized most problems, and reclaimed land is usually suitable for other uses (Figure 7.6).

Underground coal mines are used to mine coal that is too deep for surface mining. Older **room-and-pillar mines** remove rooms of ore leaving walls or pillars of ore between them to support the overlying rock (Figure 7.5c). Although some pillars are eventually removed, most are left to prevent collapse that would damage the mine. More modern underground mines use continuous or **longwall mining** equipment consisting of large rotating cutters that break coal away from the layer and drop it onto a conveyer belt that carries it out of the mine (Figure 7.5d). The entire coal layer is removed and

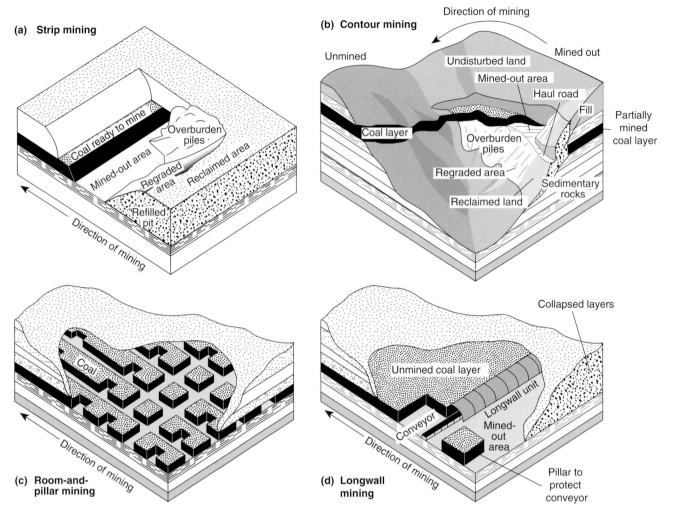

Figure 7.5 Coal mining methods, with unmined coal layer on left and mined area on right in all diagrams. (a) Strip mining. Unmined coal seam has been stripped of overburden, which has been put in spoil piles that are regraded and revegetated to produce reclaimed land after coal is removed. (b) Contour mining. Unmined coal seam on hillside is mined from **bench** (center), which is then reclaimed. (c) Room-and-pillar mining. Unmined coal seam has been cut into a series of rooms and surrounding pillars that are left to hold up the roof (back) of the mine. (d) Longwall mining. Unmined coal seam is mined by the longwall method, which removes the entire layer, allowing complete collapse of overlying rock.

the mine opening collapses behind the mining equipment as it moves forward (Figure 7.8). Longwall mining has generated concern about methane emissions, groundwater degradation, and even earthquakes caused by abrupt subsidence (Booth, 2003; Karacan, 2008; Gutierrez et al., 2010; Marcak and Mutke, 2013). It increases mine productivity and coal recovery, however, and is clearly the way of the future for underground mining (Mark, 2007).

Coal mining is one of the most rigorous lines of work in the world, especially underground mines. **Black lung disease** or **coal worker pneumoconiosis** (CWP), which results from inhalation of coal dust, is a major health problem in coal-producing countries (Hilts, 1990; UMW, 2013). Present US federal regulations require that dust levels in mines be kept

below 2 mg/m^3 of air, and dust levels are monitored in more than 2,000 mines across the country. Nevertheless, about three-quarters of the mine workers who died in the Upper Big Branch mine accident in West Virginia in 2010 were found to have signs of CWP (James, 2011). Congress enacted the Black Lung Disability Trust Fund (BLDTF) in 1977 to provide benefits to eligible miners. The Trust requires US coal producers to pay an excise tax of $1.10 per ton of underground mined coal and 55 cents per ton of surface mined coal. The Trust was restructured in the US Emergency Economic Stabilization Act of 2008, which extended the BLDTF through 2018.

Fires and explosions caused by methane and coal dust are also of concern. Methane-rich gas intersected during mining

Figure 7.6 Surface coal mining methods. The Freedom lignite mine near Beulah, North Dakota, supplies about 16 million tons of coal each year to the basin Electric Power Cooperative. (a) Mining is carried out with a dragline (background) that removes the light-colored sediments that cover the horizontal coal layer piling it on the left side of the photo. The black coal layer in the bottom of the pit is removed by shovels and loaders (small equipment in center of photo). (b) After mining, the dragline replaces the waste rock and the area is graded to near-original contours (including wetlands), covered with topsoil and revegetated, as shown here. Additional information on the operation can be seen at http://www.basinelec tric.com/About-Us/Organization/Subsidiaries/Dakota-Coal-Company/ (photos courtesy of Bill Suter, Coteau Properties). See color plate section.

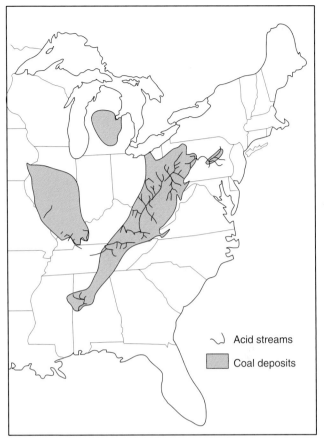

Figure 7.7 Distribution of streams in the USA made acid by coal mining prior to present mining regulations (modified from CEQ, 1989)

can flow into mine ventilation systems, where it builds up rapidly and can cause explosions before the mine can be evacuated. In addition to the cost in lives, explosions start fires that can destroy mine supports and equipment and even the coal, itself. The famous Centralia mine fire in Pennsylvania has burned for decades and may not go out until the supply of air or coal is exhausted. In the western United States, outcropping coal seams have been struck by lightning and burned to great depths destroying significant coal reserves (Heffern and Coats, 2004). Underground mines also have **subsidence** problems that are of greatest concern where surface and mineral rights have been divided (Bisc, 1981). In general, mining that causes subsidence will be restricted to areas overlain by agricultural land, where simple grading will correct the problem, although some mining proceeds under farm dwellings and some subdivisions have been built over abandoned mines. Most longwall mines collapse within a few months, however, permitting companies and land owners to agree on compensation relatively rapidly.

The cost of reclamation in US coal mines that were started after 1977 is secured by a bond put up by the mine operator. This bond is not released until reclamation has been approved, commonly in 5 to 10 years. Mines abandoned before 1977 are reclaimed by the Abandoned Mine Land Reclamation Fund (AMLRF), which was set up by SMCRA. The AMLRF legislation was amended in 1990 to provide funding for mines abandoned after 1977. Funding for AMLRF comes from production fees of 35 cents/ton of

Table 7.3 Disposition of the AMLRF in important coal-producing states and Native American lands, through the life of the program (Office of Surface Mining Reclamation and Enforcement, US Department of Interior, 2014). The amount collected represents all collections by both states and federal governments.The amount awarded represents the funds returned to states for environmental reclamation. Most reclamation funds are spent on mines abandoned before 1977. Lower amounts spent in some western states reflect extensive reclamation undertaken during present coal mining.

State	Amount collected ($)	Amount awarded to states and tribes ($)
Alabama	180,232,294	68,201,538
Alaska	16,135,747	5,894,025
Arkansas	998,231	429,379
Colorado	187,723,191	60,688,993
Illinois	333,225,238	131,197,800
Indiana	350,438,481	122,948,586
Iowa	2,416,182	1,179,798
Kansas	7,068,352	3,082,416
Kentucky	1,091,195,033	399,311,876
Louisiana	9,038,206	1,819,847
Maryland	29,943,731	10,259,603
Mississippi	27,271,757	787,691
Missouri	27,434,351	12,506,367
Montana	389,310,508	108,293,881
New Mexico	137,101,477	46,093,945
North Dakota	93,115,736	31,655,868
Ohio	262,732,019	102,868,947
Oklahoma	27,567,208	11,278,770
Pennsylvania	536,498,280	200,428,499
Tennessee	2,878,896	1,158,763
Texas	156,708,167	44,385,257
Utah	103,272,175	33,955,118
Virginia	230,862,251	83,979,776
West Virginia	1,001,357,026	340,523,652
Wyoming	3,057,793,498	582,148,121
Crow Tribe	50,869,221	12,155,093
Hopi Tribe	47,459,341	9,242,732
Navajo Nation	228,784,525	66,106,635

Figure 7.8 Underground longwall mine using a shearer loader. Note the cutting head at left and the hydraulic roof support canopy above the operator at right (courtesy of Bob Quinn, Eickhoff Corp.).

surface coal, 15 cents/ton for underground coal, and 10 cents/ton for lignite, which are paid by all active mines. As of 2012, $10.1 billion had been paid into this fund (Office of Surface Mining, 2015). Half of the funds are returned to their state of origin, where they must be applied to reclamation, and the other half goes to high-priority reclamation, regardless of location (Table 7.3). Since 1977, the AMLRF has distributed $7.6 billion in grants to states and tribes (Office of Surface Mining, 2015).

Coal mining has been a leader in mechanization, a trend that correlates with an increase in the importance of surface mining, a decrease in coal-related employment and an increase in productivity (Figure 7.9). The slow increase in productivity that characterized early mechanization in most parts of the world was increased by efforts to comply with mine safety regulations passed in the late 1960s. For instance, productivity increased from 1.93 to 6.99 tons/miner/hour from 1980 to 2000. Improvements were greatest in open pit mines; in 2011, Wyoming, which has almost entirely open pits, was producing at a rate of 30 tons/miner/hour, compared with West Virginia with largely underground mines, which produced at a rate of 2 tons/miner/hour. Over the same period mechanization in Germany increased production from 7.5 tonnes/miner/hour to 14 tonnes/miner/hour (according to the Research Fund for Coal and Steel). In spite of decreased total output, average hourly productivity in the United Kingdom improved from 0.08 tonnes/miner/hour in 1913 to 1 tonne/miner/hour in 2012 (Mitchell, 1988; Greasley, 1990; UK Department of Energy and Climate (see https://www.gov.uk/government/collections/coal-statistics).

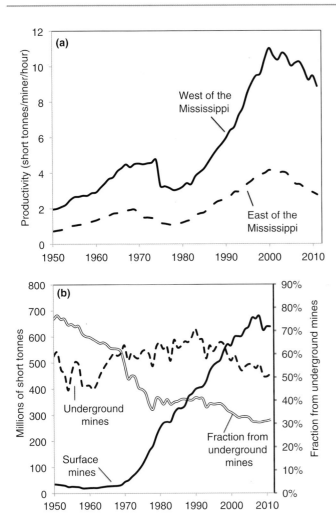

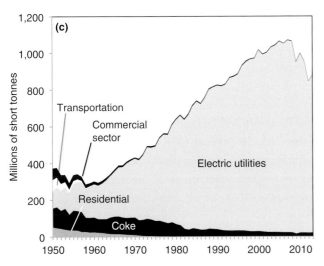

Figure 7.9 Patterns of coal production and use. (a) Change in productivity of coal mining in the eastern United States compared with the western United States. (b) Increase in production of coal mining from surface mines compared with underground mines, and the decrease of underground mines. (c) Change in US coal markets since 1950 showing the growth in the consumption of thermal coal for electric power generation (data from US Energy Information Administration).

The situation is worse in less-developed countries (LDCs) that lack mechanized coal production. India, which generates 60% of its electricity from coal produced in open pit mines, averaged only 0.4 tonnes/miner/hour in 2011. China is the world's largest producer of coal, producing 3,576 million tonnes in 2011 (IEA, 2012), largely from underground mines. Approximately 6 million miners work in Chinese underground mines and more than 1,000 of them are killed annually in mine accidents (Homer, 2009). Coal productivity was 0.15 tonnes/miner/hour at federally owned mines where mechanization is common, and 0.04 to 0.02 tonnes/miner/hour at state- and locally-owned mines, where artisanal mining is practiced (Liu *et al.*, 2007).

Coal use and quality

Coal started the 1900s as a fuel for home heating, manufacturing, and rail transport, and the main feedstock for the dye and chemicals industry. Now, its markets in the United States and other MDCs are **thermal coal** for electric power generation and **coking coal** to make coke for steel production (Figure 7.9c). In the United States, about 93% of coal goes for electric power, with the rest going for steel production. Even in China, which has enormous steel production, as discussed in the next chapter, the figure is 85%. In Europe, coal use for electric power generation ranges from about 5% in France where nuclear power is dominant to over 80% in Poland (World Bank, 2015).

Concerns over coal quality are growing in response to recognition that it is the dirtiest of the fossil fuels (Table 7.4). The most important factors affecting coal quality are ash, sulphur, and trace elements (Krishnan and Hellwig, 1982; Shephard *et al.*, 1985; Cecil and Dulong, 1986). Ash, the residue that remains after burning, can make up as much as 30% of coal. It consists of clastic sediment such as clay minerals and quartz that were part of the original sediment and were accidentally included with the coal during mining. During combustion, heavy **bottom ash** settles out and lighter **fly ash** goes out with flue gases, where about 90% is recovered by particulate recovery systems. Some coal ash is used in construction materials, although most is disposed of as a slurry in tailings ponds. One such pond at the Kingston coal-fired power plant in Tennessee spilled about 1 million gallons of slurry into the Clinch River, calling attention to the need for safe disposal and effective oversight of these operations (Seramur *et al.*, 2012).

Coal-fired power plants are annually responsible for 67% of SO_2 emissions in the United States and similar amounts in most MDCs. Sulfur occurs in coal in three main forms (Arora *et al.*, 1980). Most obvious are pyrite and other sulfide minerals, which are often visible in coal as small, shiny grains. Another form, organically bound sulfur, is invisible because it is part of the complex organic molecules that make up coal. Sulfate minerals such as gypsum are a third, considerably less important, type of sulfur. The distinction between high-sulfur and low-sulfur coals is made at sulfur contents of 1.5% (by weight), although coal can contain as much as 12% sulfur. Sulfur enters coal as sulfate dissolved in seawater that floods swamps. Coals in eastern and central North America, which are high in sulfur (Figure 7.10), formed in swamps that suffered periodic incursions of seawater, whereas those in western North America, which are low in sulfur, formed largely in freshwater. During combustion, sulfur forms SO_2, which reacts with atmospheric moisture to create acid rain (Table 3.4). Nitrogen from coal combustion also forms gases (NO_x) that contribute to acid rain. Some of this nitrogen is in the coal molecules, although most of it comes from air in which the coal burns.

Coals are also enriched in trace elements such as mercury, nickel, chromium, and arsenic, most of which are combined at the atomic scale as chelated organic molecules (Smith *et al.*, 1980; Vourvopoulos, 1987). When coal is combusted, at temperatures of about 1,500 °C, many of its trace metals vaporize. Some condense on fly ash or form aerosol particles that are caught on their way out of the exhaust systems, but others escape to form fallout around power plants. This metal-bearing fallout is easily leached by acid rain. So, a thin film of ash and aerosol particles around coal-burning plants could release trace elements to water, plants, and animals (Grisafe *et al.*, 1988; Blum *et al.*, 2013). According to the US EPA about two-thirds of global mercury emissions are caused by coal combustion, and coal-fired power plants account for 50% of mercury, 22% of chromium, 28% of nickel, and 62% of arsenic emissions in the United States. In the case of mercury, implementation of the Clean Air Act Amendments in 1990 resulted in reduction of total coal-combustion mercury emissions

Table 7.4 Relative contributions to atmospheric pollution in 2014 by coal, oil, and natural-gas electric power generation in the United States (data from the US Energy Information Administration and the US Environmental Protection Agency).

Fuel type	Carbon dioxide	Sulfur dioxide	Nitrogen oxides
Coal	972.73	5.91	2.72
Oil	793.18	5.45	1.82
Natural gas	554.55	0.045	0.77

All amounts are in tons of gas per million kW-hr of electric power. Data for oil are the average for distillate oil (No. 2) and residual oil (No. 6). Data for coal are the average for bituminous, subbituminous, and lignite.

BOX 7.2   WHAT HAPPENS TO COAL ASH?

The solid material that remains after combustion of coal is referred to as **coal combustion residuals** (CCRs), or coal ash. About 100 million tons of CCRs are generated annually from US coal-fired power plants. Because CCRs contain toxic metals such as arsenic, cadmium, and mercury, the US Environmental Protection Agency (EPA) regulates their disposal as part of the Resource Conservation and Recovery Act. In late 2014, the US EPA proposed new rules to strengthen existing regulations for CCRs. The rules aim to prevent CCR dust from entering the atmosphere and leaking into groundwater aquifers, and they require coal-fired power-plant operators to keep, and make public, detailed records of the fate of CCRs. The rules require periodic assessments of the structural stability of surface impoundments, and preparation of emergency hazard plans that can be implemented in the case of leakage or impoundment failure.

The proposed rules apply only to disposal of CCRs, and do not address beneficial use of CCRs encapsulated in building materials (individual states are largely responsible for regulating beneficial uses). Fly ash is used in a variety of products such as cement, bricks, and grout, and flue-gas-desulfurization (FGD) material is used as synthetic gypsum. These uses prevent about 50% of CCRs from entering landfills, and reduce the need to mine raw materials, which reduces CO_2 emissions. The US Green Building Council's Leadership in Energy & Environmental Design (LEED) program classifies the use of fly-ash cement and FGD-gypsum as post-industrial recycled material. In fact, the drywall used to construct US EPA headquarters in Washington, DC contains FGD-gypsum.

from 59 to 53 tons per year from 1990 to 2005; a decrease of 10%. In 2011, the US EPA proposed the Mercury and Air Toxics Standards that require a 90% reduction in mercury emissions. Emission of mercury and other heavy metals are more severe in LDCs where many coal-fired plants lack even particulate collection systems. As noted later, sizeable numbers of fatalities in China have been attributed to arsenic poisoning of soils and foods caused by emissions from plants of this type.

Clean coal technology, the response of industry to these challenges, focuses largely on minimizing sulfur emissions, although efforts are now expanding to other gases and metals. Sulfur emissions can be minimized by removing sulfur from

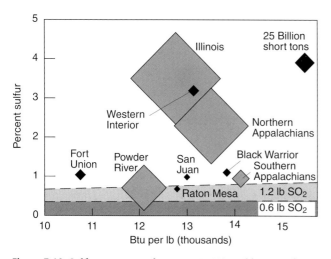

Figure 7.10 Sulfur content and reserves in US coal basins. The reserve of each basin is shown by the size of the symbol. Coal falling below the line labeled 1.2 lb meets emission standards for SO$_2$ (expressed in lb of SO$_2$ per million Btu) for the year 2000. Note that the Powder River Basin is the only large, low-sulfur coal reserve (from Garbini and Schweinfurth, 1986; EIA, 1993).

the coal before combustion or by recovering sulfur gas from the exhaust after combustion (Eliot, 1978; Osborne, 1988). Coal can be cleaned before burning by standard beneficiation methods, involving grinding, washing, and flotation, although these processes do not remove organically-bound sulfur, nitrogen, or other trace elements. Cleaning after burning centers on particulate recovery systems and scrubbers. Most scrubbers in use on power plants today are non-regenerative and spray a wet slurry of powdered lime or limestone (containing calcium, or magnesium oxide) to react with SO$_2$ gas to form calcium sulfite, a process called **flue-gas desulfurization** (FGD). Implementation of pollution controls at modern coal-fired power plants accounts for roughly half the cost of plant construction, occupies about half of the space on the power-plant grounds, consumes between 2 to 4% of the power from the plant, and removes between 80 and 90% of SO$_2$ (Victor *et al.*, 2009). They also produce about three tonnes of sulfate sludge for every tonne of SO$_2$ that they remove from flue gas, producing a huge disposal problem similar to that of coal-mine sludge (Luxbacher *et al.*, 1992). It is possible to recover the calcium sulfite and use it to manufacture synthetic gypsum.

Compliance with the US Clean Air Act and subsequent regulations such as the Cross State Air Pollution Rule (CSAPR) and Mercury and Air Toxics Standards (MATS) adds between 20% and 80% to the cost of energy generation (Interagency Force on Carbon Capture and Storage, 2010). In 2010, US coal plants with operating FGD scrubbers generated 58% of the total coal-sourced electricity, while emitting 27% of coal-sourced SO$_2$, indicating that that scrubbers do significantly reduce SO$_2$ emissions. Sulfur emissions from US power plants decreased from 14.5 to 3.6 million tonnes from 1990 and 2012, even though coal use increased from

---

BOX 7.3 | GLOBAL IMPACT OF MERCURY FROM COAL COMBUSTION

Combustion of coal releases mercury that is present in coal. Some of this mercury reaches the atmosphere where it returns to Earth via precipitation. In oxygen-poor environments anaerobic organisms use this mercury to produce the **methylmercury cation**, CH$_3$Hg$^+$, which is ingested by aquatic species. It then **bioaccumulates** up the food chain where it can be consumed by humans. Elevated levels of methylmercury impair neurological development, and various health agencies recommend against consumption of certain fish and shellfish, especially for pregnant women and children. We used to think that mercury emissions posed problems only for the local area around coal-fired power plants; however, we now know that mercury emissions are global in scale. Fish caught in the Pacific Ocean near Hawaii contain methylmercury produced by microbes in the ocean. Atmospheric circulation patterns suggest that the source of the mercury is coal-fired power plants located in Asia several thousand miles away.

820 to 900 million tonnes (EIA, 2015a). Even so, coal-fired power plants remain the major source of atmospheric sulfur pollution, a fact that is driving the search for improved scrubber and carbon-capture-and-storage (CCS) technologies.

Monitoring of CO_2 emissions from US coal plants was required starting in 1995, and the total annual emissions of CO_2 increased from 2.1 to 2.2 billion tonnes from 1995 to 2009 (EIA, 2015b). In 2007, the US Supreme Court ruled in Massachusetts versus Environmental Protection Agency that the EPA was required to regulate emissions of CO_2 if it is a threat to human health. Following this ruling, the EPA proposed emission standards for CO_2 that limit coal plants to 1,100 lb/MW-hr. Compliance with the standard for CO_2 requires coal plants to capture and store 20–40% of produced CO_2 (Plummer, 2013). A coal plant in Kemper County, Mississippi, constructed to meet the new EPA requirements takes advantage of Integrated Gasification Combined Cycle (IGCC) technology, which should capture 65% of produced CO_2 and 90–95% of the mercury. IGCC technology converts coal to synthetic gas, and removes CO_2 and other impurities before combusting the gas to produce electricity. Capital costs for IGCC plants are estimated to be 20–25% higher than traditional coal plants (Hutchison, 2009). The captured CO_2 must still be disposed of in some way.

A final way that coal can be "cleaned" is by *in situ* underground coal gasification (UGC), in which coal is burned, while still underground, to produce combustible gases (Berkowitz and Brown, 1977). This is similar to IGCC, but does not require that the coal be mined and transported to a power plant. Rather, this process involves strategic positioning of injection and production wells into a coal seam, where gasifying agents such as air, oxygen, steam/air, and steam/oxygen are used to ignite coal and produce synthetic gas (Perkins and Sahajwallaa, 2007). Underground coal gasification allows combustible gas to be recovered from conventionally unrecoverable, hence uneconomic, coal seams; thus extending the lifetime of coal reserves. It also offers the environmental benefit of leaving by-product ash underground, while taking advantage of scrubber and CCS technologies similar to those used in IGCC (Self *et al.*, 2012). Coupling UGC with on-site, geological CCS, where CO_2 is stored in porosity generated during the UGC process, is viewed as a cost effective means for compliance with CO_2 emission regulations (Roddy and Gonzalez, 2010). *In situ* extraction and energy production processes are being field tested in the United States, Australia, Canada, China, South Africa, New Zealand, Pakistan, Europe, and Uzbekistan, and are currently producing gas in several of these locations (Couch, 2009).

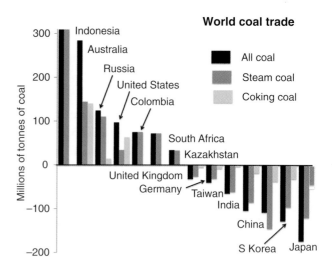

Figure 7.11 World coal trade for 2013. The amount shown for each country is the difference between coal consumed and coal produced for three categories: all coal, steam coal (for electric power generation), and coking coal (for steel making). Exporters plot above the line and importers plot below the line. Data from the US Energy Information Agency.

Coal reserves, resources, trade, and transport

World coal recoverable reserves at the end of 2014 were 861 billion tonnes, and the global reserve to production ratio was 110 years. Global conventionally unrecoverable coal, some of which might be used for UGC, are about 16.3 trillion tonnes (Couch, 2009). Coal reserves exist in about 70 countries and are huge in relation to other fossil-fuel reserves and global energy needs. The growing demand for coal has pushed exploration into many areas of the world that have little record of coal production. For example, in Mongolia, 80 coal deposits are being developed to supply markets in China, and similar activity is taking place in Mozambique, Kazakhstan, and Brazil. Coal reserves might increase even more if coal layers formerly deemed uneconomic because of depth considerations can be accessed by UCG.

Exports and imports account for only about 15% of total world coal production, which was 7.8 billion tonnes in 2013; this is quite an increase from 4.7 billion tonnes in 1990. International trade involves flow from Indonesia, Australia, Russia, the United States, Colombia, and South Africa to China, Japan, South Korea, India, Taiwan, Germany, and the United Kingdom (Figure 7.11). The growth in coal production is dominated by the Asia–Pacific region, where increases in 2012 of 3.5% and 9% in China and Indonesia, respectively, helped offset an overall decline of 7.5% in US production. China and India have emerged as major coal producers, although both countries consume all their domestic coal and rank globally among the top coal importers. China increased its share of global coal consumption from 10% in 1965 to 50% in 2011. Production from western Europe has declined steadily in the face of declining reserves and increasing environmental regulations, although Germany has seen an increase due to its decision to eliminate nuclear power. Russia, which had stagnated in coal production during the collapse of the Soviet Union, is now among the top ten global coal producers, increasing coal production from 220 to 315 million tonnes from 1990 to 2013.

Coal makes an important contribution to the trade balance in Australia, where exports increased by 50% over the decade ending in 2010. Coal production employs 200,000 people directly and indirectly, and accounts for 3.5% of Australia's GDP (Australia Coal Association, 2015). The health of the Australian coal industry is largely contingent on coal consumption in China, Japan, Korea, India, and Taiwan, which collectively receive 88% of Australia's coal exports. Indonesia exports up to 80% of its produced coal, and demand from countries such as China and India helped Indonesia become the world's largest exporter of coal in 2013. Coal exports account for 4.5% of Indonesia's GDP and 11% of export trade. The future of coal trade in India is likely to change because 300 million residents, one-fourth of the country's population, currently lack electricity; domestic coal is abundant and easily accessible, and is likely to supply much of this energy in the future. Significant coal trade is also driven by coal quality. Historically, the most important quality-based trade has been inside North America, where large amounts of low-sulfur coal from the Powder River Basin in Wyoming are used in eastern power plants. The low cost of coal production in the western United States, coupled with better coal quality, was viewed as a way to meet Clean Air Act requirements. This resulted in increasing unemployment in the high-sulfur, coal-mining areas of the eastern United States. Increasing unemployment is particularly challenging in communities that exist almost solely because of coal mining. For example, in Harlan, Kentucky, a town nicknamed the "cul-de-sac of coal," the number of coal-mining jobs decreased by 48% as coal-fired power plants either converted to natural gas, as discussed below, or increased the amount of low-sulfur coal from western US states (Maher and McGinty, 2013).

Rail transport dominates coal trading in the United States and China. Coal transport in the United States is accomplished with large **unit trains** with three to five locomotives, and 100 cars carrying nearly 100 tonnes each, which move directly from western mines to eastern power plants. In China, in 2011, trains transported twice the mass of coal than was consumed in the United States. The international coal market is dominated by seaborne transport, which accounts for 94% of global coal transport. Ocean vessels have capacities of 40 to 80 thousand tonnes of coal, and transportation costs can account for 80–90% of the final price of coal (Schernikau, 2010). Coal can also be transported by **slurry pipelines**, in which a mixture of equal-parts pulverized coal and water is pumped through a pipeline. This method is particularly suited to coal, which is less dense than most minerals, and coal pipelines were used in England over 100 years ago. Such transport in the United States is unlikely, however, because it would require movement of water from already water-poor areas such as the upper Colorado River valley (Karr, 1983). In fact, the longest coal-slurry pipeline in the world, which carried 5 million tonnes of coal annually for 439 km from Arizona to Nevada, was shut down in 2006 because it used about 4.5 billion liters of water per year.

A few decades ago, the outlook for coal consumption was very strong. Now, it is not as clear because of a combination of environmental concerns and alterative fuels. In the United States, regulations allow coal-fired power plants to phase in emission controls over several years; however, the cost associated with retrofitting existing power plants with pollution controls is expected to result in many plants being decommissioned. The number of coal plants in the United States decreased from 633 to 589 from 2002 to 2011. In the United Kingdom, increased European regulations are expected to cause an almost complete end to coal-fired electric power generation by about 2030. Denmark has pledged to eliminate all coal-fired plants by 2025. Although the impetus for these changes was originally largely environmental, it has now become economic. The availability of relatively cheap natural gas makes it even less likely that power plants can be retrofitted with suitable scrubbers economically (Tierney, 2012). There is also increasingly effective social pressure to transition to wind and solar energy. The real question from a global coal perspective, though, is whether LDCs will institute similar environmental regulations, thus pushing them toward other fuels for power generation. If so, coal's day in the sun might end long before its reserves are exhausted.

7.1.2 Crude oil and natural gas

Crude oil dominates global economics. With annual world production worth about $3.1 trillion at 2012 prices, oil is the most valuable basic commodity in world history. Natural gas, with current world production valued at about $1 trillion, is in second place as a result of new discoveries and new ways to

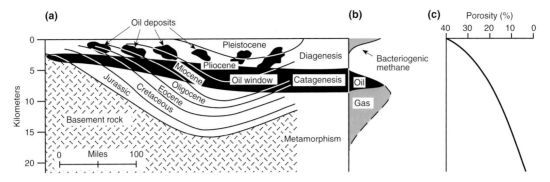

Figure 7.12 Formation of oil in a typical sedimentary basin. (a) Distribution of sedimentary formations, showing the oil window in which temperatures are high enough to form oil and gas from kerogen in source rocks. (b) Diagram showing the relation between natural gas and oil formation during burial. (c) Generalized decrease in porosity of sediments due to compaction and mineralogical changes caused by burial.

produce it. These new developments have expanded into new areas, putting more of our population in contact with oil and gas production, and raising environmental and societal concerns (Healy, 2013). At the same time, they have made oil and natural gas surprisingly abundant and cheap. So, just at the time when we expected them to become scarce and expensive, oil and gas are dominating the world energy picture and putting economic pressure on our efforts to transition to an energy economy based on renewables.

Geology and origin of oil and gas

Crude oil and natural gas consist largely of chains and other forms of carbon and hydrogen atoms, known as hydrocarbons. In natural gas, which consists largely of methane (CH_4) with smaller amounts of ethane (C_2H_6) and propane (C_3H_8), the chains are very short. In crude oil, the chains contain 4 to 30 carbon atoms and are more complex. Crude oil and natural gas are derived from fats and other lipids in marine algae and other aquatic plants that were buried with sediment (Hunt, 1979; Link, 1982; Buryakovsky et al., 2005). Conversion of lipid-rich organic matter to hydrocarbon fluids begins immediately after sedimentation, even in landfills, when microbes begin to alter organic matter and produce methane (Figure 7.12). As organic matter is leached and altered by groundwater, it becomes **kerogen**, a relatively insoluble form of organic matter consisting of molecules much larger than those in oil or gas. With deeper burial, temperature increases and kerogen decomposes to form crude oil and natural gas, leaving a complex, solid residue. Oil generation usually begins when buried organic matter reaches temperatures of 50–60 °C, the lower limit of the **oil window**. By temperatures of about 100 °C, kerogen has released most of its oil, and further reactions produce largely methane gas (Figure 7.12). This process that forms oil and gas is known as **catagenesis**.

Sediments with abundant organic matter that might form oil and gas are called **source rocks**. Although the best source rocks are marine shales and other fine-grained clastic sediments, oil and gas were not historically recovered from these rocks. Instead, oil and gas were extracted from **reservoir rocks**, which have adequate porosity and permeability (Figure 2.3) for production. This is because source rocks such as shale contain oil and gas, but their permeability is so low that oil and gas cannot escape rapidly enough to be produced economically. This situation changed during the last decade, however, when the combination of hydraulic fracturing and horizontal drilling made it possible to extract oil and gas from shale source rocks, as discussed below. Today, oil and gas are produced from conventional reservoirs where, over geological time, hydrocarbon fluids seep out of shale and migrate into reservoir rocks, and from unconventional sources, where fluids can be removed after hydraulic fracturing increases rock permeability.

Oil and gas that escape from source rock are less dense than surrounding pore water and flow toward the surface. Some reaches the surface where natural leakage of oil and gas is widespread, including the famous La Brea tar pits of Los Angeles. Oil and gas deposits exist because some hydrocarbons are trapped before reaching the surface. **Hydrocarbon traps**, which are reservoir rocks with an **impermeable** cap that impedes upward migration, can be divided into structural and stratigraphic types (Figure 7.13). Structural traps form where the shape of the trap or its ability to hold fluid are due to folding or faulting of the host rocks. Typical structural traps consist of anticlines, with alternating layers of reservoir rock and impermeable sediment, or more complex features related to faults or the margins of impermeable salt domes (Figure 7.14). Stratigraphic traps form where variations in porosity or permeability of the sedimentary sequence create

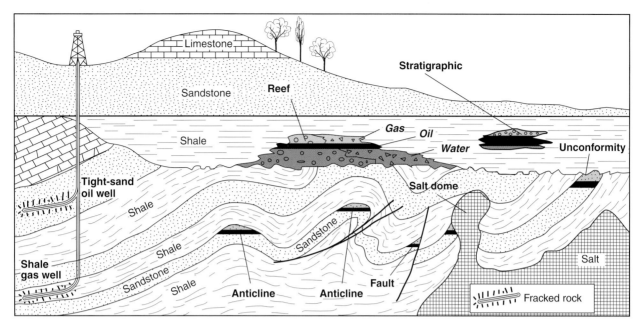

Figure 7.13 Conventional oil and gas traps and unconventional production. The two sedimentary sequences shown here are separated by an unconformity that formed when the earlier sequence was folded and eroded. The lower, folded sequence contains conventional anticline and fault traps, as well as traps beside salt domes and beneath unconformities. The upper, flat-lying sequence contains conventional stratigraphic traps, including a carbonate reef with reef **talus**, and a sand lens. All of the traps are surrounded or overlain by impermeable shale. The well on the far left produces unconventional oil and gas from tight sand and shale that required hydraulic fracturing. Strictly speaking, these are not reservoirs in the sense that oil or gas have migrated to them and been trapped; they are essentially source rock that is being tapped to yield oil and gas.

isolated reservoirs. If oil and gas are present in the same trap, they tend to separate, with gas floating on oil, which floats on water.

There are almost 800 sedimentary basins on Earth but less than half of them produce oil (Figure 7.15). The age of sediments is not as great a limiting factor as it was for coal. Marine aquatic plants from which oil and gas could be generated formed long before Silurian time, and Precambrian oil is known, though scarce. The composition of sediments is more important and requires a combination of source rock, maturation, migration, and traps. In highly productive basins such as the North Sea and Middle East, all factors converged to maximize hydrocarbon generation and preservation. In the North Sea, important factors were the presence of a rich Jurassic-age source rock and a late stage of deep burial related to rifting, which increased heat flow into the basin facilitating **maturation** of the organic material (Prather, 1991). In the Middle East, source rocks were unusually rich, releasing large amounts of oil close to large reservoirs consisting of fractured limestones that had been folded into long, continuous anticlinal traps covered by impermeable evaporites and shales. Seepage, along with uplift and erosion of oil fields, probably limits the hydrocarbon endowment of most basins, as shown by the fact that younger basins contain more oil (Tissot and

Welte, 1978; Hunt, 1979). Finally, if meteoric water invades a basin during uplift, it can flow through the reservoirs removing much of the oil (Szatmari, 1992).

Oil and gas production and related environmental concerns

Hydrocarbons are commonly produced from drill holes that penetrate traps (Koederitz et al., 1989). In a simple reservoir with layers of water, oil, and gas (Figure 7.13), pressure from expansion of gas and buoyancy provide a natural drive forcing fluids up the well. Where fluids do not reach the surface naturally, they must be pumped. If oil is pumped too rapidly, water and gas, which flow more easily through porous rock, will stream into the reservoir, isolating zones of oil. Reservoir engineering involves the determination of the optimum type and distribution of wells needed to maximize production without causing this type of damage. Producing wells were originally vertical, but the use of directional and horizontal drilling allows wells to intersect vertical fractures and permeability that might otherwise be missed. Directional wells require about twice the initial capital expenditure, but the production rate is about 3.2 times higher than for vertical wells (Halker Consulting, 2013). Because oil and gas extraction rates govern royalty returns, governments sometimes

Figure 7.14 (a) Production on land. The anticlinal structure at Elk Basin, Wyoming, is a typical simple oil trap. Sedimentary layers dipping away in both directions from the center of the photo show the crest of the anticline. Oil is produced from the buried part of the crust by pumping units. (b) Offshore production. The Hibernia field, 315 km east–southeast of St. John's. Newfoundland, is the largest oil deposit along the east coast of North America. The field contains an estimated 2.1 billion bbl (Bbbl) of oil in place, of which about 700 million are recoverable. It was discovered in 1979 and put into production in 1997 with the platform shown here, which sits on the seafloor (80 m deep). A fleet of tugs protects the platform from passing icebergs calving off Greenland (photo courtesy of Suncor Energy). See color plate section.

regulate pumping rates, as was done by the curiously named Texas Railroad Commission, which is discussed in Chapter 6. Before production begins, the reservoir rock is commonly fractured to enhance the rate at which fluids can flow to the drill hole. This process, which is known as hydraulic fracturing (fracking), is discussed further in Section 7.1.4.

Secondary production methods, which are used to increase oil recovery, were originally applied late in the life of a field but they are now used from the start of production. The most common methods, **gas injection** and **water flooding** (Figure 7.16), use gas and water from earlier production in the same field to maintain reservoir pressure. In the early days of oil production, accompanying natural gas was burned (flared) rather than being injected, although this has decreased, in part due to public awareness caused by improved satellite imagery (Lavelle, 2013; Ebrahim and Friedrichs, 2013). Still, oil production in 2011 resulted in flaring 138 billion cubic meters of natural gas, equal to about 4% of global natural gas production (EIA, 2012). Even after the best conventional primary and secondary production methods, as much as 50% of the oil remains in the ground. Some of this oil can be recovered by tertiary or enhanced methods (Department of Energy, 2015). **Enhanced oil recovery** (EOR) uses methods such as combustion of oil and gas on the margin of a field or injection of CO_2 or alkaline solutions to decrease the viscosity of oil and increase the pressure forcing it to flow from the reservoir.

Some tax codes encourage EOR to maximize recovery and increase employment. The only way to improve on EOR is oil mining, in which reservoir rock is removed, disaggregated, and washed clean of oil (Herkenhoff, 1972). As discussed below, mining is already done for some heavy oil and tar sands. It has been carried out on some shallow oil fields, but widespread oil mining will be difficult because so many fields amenable to the method are in populated areas of Pennsylvania, Ohio, and California.

Production in most onshore fields in populated areas requires simple pumps and other facilities, and hydrocarbons reach markets through pipelines (Figure 7.17). Frontier wells that lack pipeline connections send oil by truck, rail, or tanker to a pipeline connection or a refinery. Offshore production takes place from very expensive platforms that are being improved constantly (Figure 7.18a). In 1982, the deepest water platform in the United States, Shell Cognac in the Gulf of Mexico, sat on rigid legs in 311 m of water. By 1990, Conoco's Joliet, also in the Gulf, was held in place in 533 m of water by metal legs that were under tension rather than supporting the platform. In 2012, Shell's Perdido platform was held in place, floating in water that was 2,438 m deep, by nine 3.2-km-long mooring lines. Perdido set a world record for the deepest well drilled below the seafloor, reaching a depth of 2,925 m and produces 100,000 bbl/day from three separate oil fields. In the mature Gulf of Mexico basin, deepwater drilling increased from 6% to 79% of total effort between 1985 and 2012 and now accounts for almost half of total oil production (Figure 7.18) (BSEE). Globally, deepwater production is dominated by the United States, Brazil, Angola, and Nigeria, although it is increasing in Norway, India, Australia, and Egypt (Sandrea and Sandrea, 2010; Trefis Team, 2012).

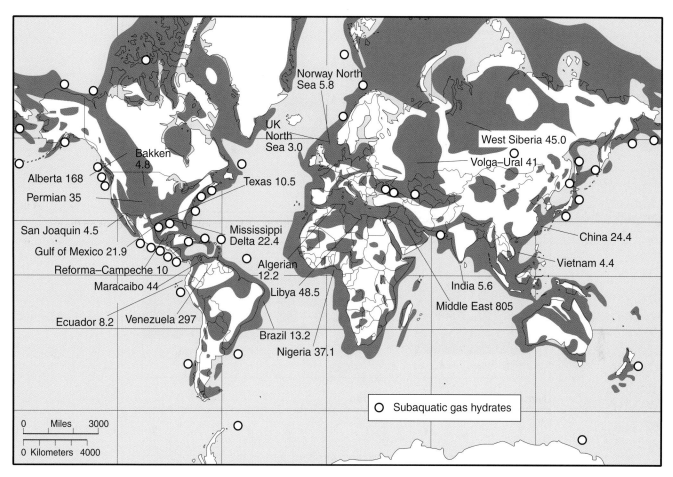

Figure 7.15 Distribution of sedimentary basins showing the location of major oil and gas fields (reserves shown for oil in Bbbl). Note that many basins extend offshore into the continental shelf and slope. Known areas of gas hydrate are also shown (after Klemme, 1980; Nehring, 1982; Kvenvolden *et al.*, 1993; additional data from the US Energy Information Administration).

BOX 7.4 | ENHANCED OIL RECOVERY

The Denver Unit of the Wasson oil field in West Texas is the largest CO_2 enhanced oil recovery project in the world (DOE, 2012). This field, which contained more than 2 Bbbl of oil, underwent primary production from 1938 to 1965, with a peak in 1945 and a gradual decline after that. Initiation of water flooding (secondary production) in 1965 caused production rates to rise again. However, the rate of water injection soon matched the rate of water produced, and oil production slowed by the late 1970s. EOR using CO_2 injection was implemented in 1983, and production increased again. The rate of oil production in 2008 was 31,500 bbl/day, of which 26,850 bbl/day is attributed to the use of CO_2 flooding.

Environmental problems related to oil exploration and production include subsidence, wastewater discharge, and oil spills (Williams, 1991; Gossen and Velichkina, 2006). Subsidence, which is caused by collapse of the reservoir rock as fluids are withdrawn from the pore space, was more significant in early oil fields. Between 1926 and 1967, for instance, production of 1.2 Bbbl of oil from the Wilmington field in Long Beach, California, formed a conical depression at the surface that had an area of 52 km^2 and a maximum depth of 8.8 m (Figure 7.19). Production-related subsidence in deltas, such as the Louisiana gulf coast, is aggravated by levies constructed to limit flooding, which prevent river sediment from replenishing the land (Getschow and Petzinger, 1984). Water spills can release brines that contaminate local surface and groundwater,

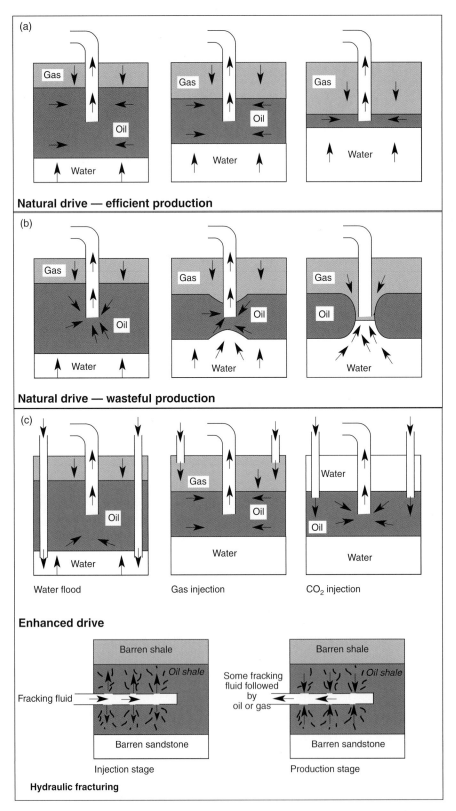

Figure 7.16 Efficient (a), wasteful (b), and enhanced (c) conventional oil production methods. Pumping rate for efficient production must not exceed the rate at which oil can flow into well. Enhanced methods include water flooding, gas injection, and CO_2 injection. Panel (c) shows the two stages of hydraulic fracturing.

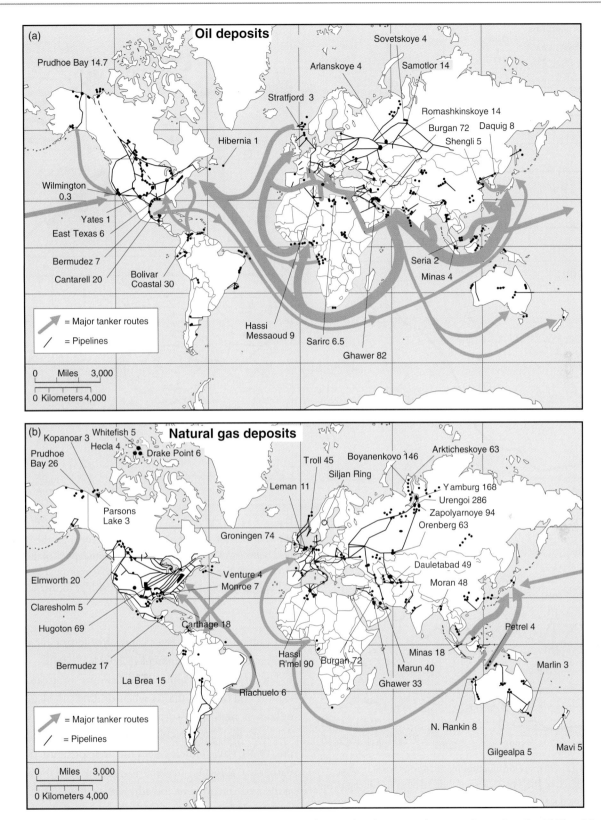

Figure 7.17 Distribution of major oil (a) and gas (b) fields showing pipelines and tanker routes (reserves shown for oil in Bbbl and for gas in trillion cubic feet (Tcf); after Oil and Gas Journal; Carmalt and St. John, 1986). Detail on parts of California and Middle East in Figures 7.19 and 7.21. Dashed pipeline in the top map shows the location of proposed MacKenzie Valley and Keystone pipelines.

(a) Offshore platforms

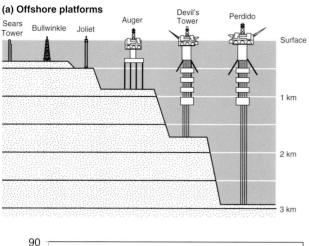

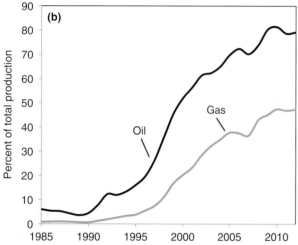

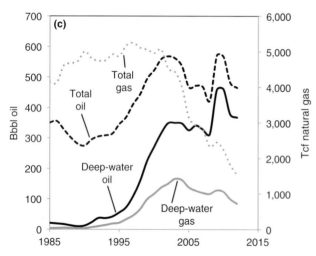

Figure 7.18 (a) Offshore production platforms for work in deep water compared to the 442-m Sears (Willis) Tower in Chicago. The Auger platform stands on rigid legs, whereas the Perdido platform is attached to the floating platform and to a base that is fixed to the sea floor. (b) Change in offshore deep-water production relative to total production in the Gulf of Mexico. (c) Growth in deep-water oil and gas production, enabled by advances in geophysical exploration and engineering of drilling platforms (data from the Bureau of Ocean Energy Management, US Department of Interior).

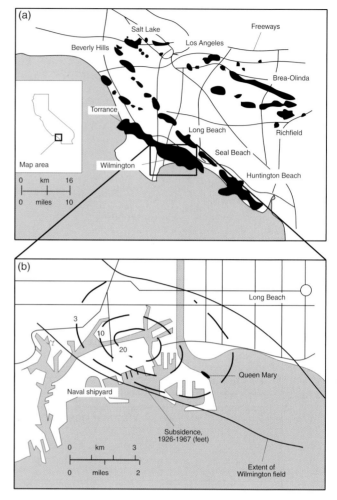

Figure 7.19 (a) Oil fields in the Los Angeles basin. (b) Contours showing location of maximum land subsidence over the Wilmington field between 1926 and 1967 (modified from reports of the California Division of Oil and Gas and Mayuga, 1970, reprinted by permission of the AAPG whose permission is required for further use).

and radium in some brines causes low-level radioactive contamination of well equipment (Schneider, 1990).

The risk of oil spills and blowouts is greatest during exploration because that is when drilling might intersect an unknown, highly pressurized reservoir. However, only two of the more than 50,000 wells drilled in the United States since the 1960s have blown out. One well offshore from Santa Barbara, California, in 1969 lost about 77,000 bbl through faults and fractures in the adjacent seafloor rather than the actual well. It is no coincidence that this well was in a zone of about 2,000 natural oil seeps, which extends along the Santa Barbara Channel into the La Brea tar pits of Los Angeles (Link, 1952) (Figure 7.19). The other was the Deepwater Horizon well in the Gulf of Mexico, which released 4.9 Mbbl (National Response Team Report, 2011). This blowout took

Table 7.5 Oil spills from tankers (global data) and production/exploration facilities (US data) since the mid 1960s (data from the Bureau of Ocean Energy Management, US Department of Interior; and the International Tanker Owners Pollution Federation). Spills listed in order of declining size (in bbl of oil). For perspective, daily crude oil consumption in the United States is about 19 Mbbl.

Production/exploration spills			
Spill/cause	Date	Location	Amount of oil
Well blowout	1910	Lakeview Gusher #1, CA (onshore)	9,000,000
Well blowout	2010	Deepwater Horizon, Gulf of Mexico	4,900,000
Well blowout	1979	Ixtoc #1, Mexico	3,500,000
Anchor damage to offshore pipeline	1967	West Delta, LA	160,638
Well blowout	1969	Santa Barbara Channel, CA	80,000
Well blowout	1970	South Timbalier, LA	53,000
Well blowout	1970	Main Pass, LA	30,000
Anchor damage to offshore pipeline	1974	Eugene Island, LA	19,833
Tanker spill	Date	Location	Amount of oil
Atlantic Empress/ Aegean Captain	1979	Off Trinidad	2,110,000
ABT Summer	1991	700 nautical miles off Angola	1,912,000
Castillor de Dellver	1983	Off Saldanha Bay, South Africa	1,850,000
Amoco Cadiz	1976	Off Brittany, France	1,640,000
Haven	1991	Off Genoa, Italy	1,059,000
Odyssey	1988	700 nautical miles off Nova Scotia, Canada	971,000
Torrey Canyon	1967	Off Scilly Isles, UK	875,000
Sea Star	1972	Gulf of Oman	846,000
Irenes Serenade	1980	Navarino Bay, Greece	735,000
Urquiola	1976	La Coruna, Spain	735,000
Exxon Valdez	1989	Prince William Sound, Alaska	240,000

87 days to contain, and released about five times more oil than the worst tanker-related oil spill in history (Table 7.5). All spills of more than 1 bbl of oil must be reported to the National Response Center. From 1964 through 2009, for the approximately 27.6 Bbbl of oil produced from US offshore leases, only about 0.002% (570,000 bbl) was lost by spills (Anderson *et al.*, 2012). Adding the 2010 Deepwater Horizon spill increases this to about 0.02% of total offshore oil production.

Other sources of oil pollution include tank cleaning operations and oil-refinery waste discharges, which release about the same amount of oil to the marine environment as does offshore oil production (Gossen and Velichkina, 2006). Natural oil seeps in the Santa Barbara Channel release an estimated 14,600 to 244,500 bbl/yr, and have been a source of beach pollution for centuries (Estes *et al.*, 1985). Natural hydrocarbon leaks of this type are widespread and are an important part of global hydrocarbon pollution (Prior *et al.*, 1989). These are similar in magnitude to losses from production, and they are minimized if production in a field decreases overall reservoir fluid pressures, thus limiting natural leakage.

Reclamation of abandoned oil fields is a matter of ongoing concern. Surface installations must be removed and wells need to be plugged so that they do not leak to the surface or to other fluid-bearing zones. A typical oil well penetrates numerous layers of sediment, including some with salty brines or drinkable water. Mixing of these fluids is prevented during operation of the well by lining the hole with a **casing** of steel pipe that is perforated only at the level of the oil- and gas-bearing layers. After the well is abandoned, it must be filled with cement. Some older wells remain open, however, and programs to shut them down properly are in place in most states. In Texas, an estimated 116,500 abandoned wells require plugging and the Oil Field Cleanup Fund (OFCF) requires oil and gas producers to pay a permit fee of $100 to $200 per new well to fund the plugging of abandoned wells.

Oil and gas markets and quality

Oil is used largely to make gasoline and other fuels, whereas natural gas is used largely for home and industrial heating and electric power generation. As with coal, the key quality factor for both crude oil and natural gas is sulfur, which occurs as atoms in crude-oil molecules and as hydrogen sulfide (H_2S) in natural gas. Sulfur contents can range up to several percent, and are most serious in natural gas. Natural gas with more than 1 ppm (0.0001%) H_2S is called **sour gas**; that with less than 4 ppm at standard temperature and pressure is called **sweet gas**. The permissible amount is so low because H_2S

reacts with air and moisture to form sulfuric acid, which corrodes metal pipes, causing leaks. Thus, long before acid rain became a concern, sulfur was removed from natural gas, and major sour-gas producers such as those in Alberta were important sources of by-product sulfur. The main source for sulfur in oil and gas is seawater sulfate, which is reduced to H_2S during formation of the oil and gas. If metals are present in surrounding waters, H_2S reacts with them to deposit metal sulfides such as pyrite, thereby removing sulfur from the natural gas. Where metals are absent, as in many limestone reservoirs, H_2S contents of natural gas can reach high levels. In addition to H_2S, natural gas can contain important amounts of carbon dioxide, nitrogen, argon, and helium. Helium found in some gas fields in the midcontinent and Rocky Mountain regions of the United States is derived from radioactive decay of uranium in underlying Precambrian granitic rocks (Leachman, 1988).

The dominant consideration in crude-oil quality is its suitability for making gasoline, the most important oil product. **Light oil**, which yields more gasoline, commands higher prices than does **heavy oil**. Refining costs for gasoline have long been a driving force in the economics of oil, with a constant battle between cost and environmental considerations (Marshall, 1989). Early concern centered on tetraethyl lead [$Pb(C_2H_5)_4$], which was introduced in the 1920s as the inexpensive way to minimize premature combustion, which created a knock in engines. Tetraethyl lead use has declined since the mid 1970s because it is a major contributor to global lead pollution, as discussed in Section 9.2.2. It also coats the surface of catalytic converters used to clean automobile exhaust from **unleaded gasoline**. It has been replaced by gasoline formulations that do the same thing, but cost more to produce. The main problem with gasoline is incomplete combustion, which produces CO

and NO_x that contribute to smog. Incomplete combustion occurs because of scarcity of oxygen in the combustion chamber, and it can be alleviated by increasing the oxygen content of the gasoline. In the United States, the Clean Air Act Amendments of 1990 required that all gasoline contain at least 2% oxygen additive. This requirement was changed by the Renewable Fuel Standards (RFS) provision of the Energy Policy Act of 2005, which requires that refiners use ethanol (C_2H_5OH) synthesized from sugar cane, corn, and other plants as the oxygenating agent in gasoline. In 2007, RFS provisions were modified to require that 9 billion gallons of ethanol be added to the US gas supply in 2008, increasing to 36 billion gallons in 2022 (EPA-RFS, 2015). As of 2013, the amount of ethanol added to US gasoline was about 10% by volume. Brazil, which has been even more aggressive, currently requires 25% ethanol in gasoline. Benefits of combusting ethanol include significantly reduced emissions of CO, NO_x, sulfate, and volatile organic compounds (VOCs) compared to gasoline, as well as reduced crude-oil imports (Hower, 2013; Fridell *et al.*, 2014). On the other side of the coin, ethanol contains 33% less energy than gasoline (DOE, 2011), and its use in gasoline removes corn and other plants from the food supply, and increases corn acreage and related nitrate and phosphorus fertilizer pollution (Carter *et al.*, 2013; Thompson, 2013).

Oil and gas reserves

Global reserves for oil in conventional reservoirs are estimated to be almost 1.7 trillion barrels (Tbbl) and world natural gas reserves are about 6,800 trillion cubic feet (Tcf). As of 2015, there were about 4,000 actively producing oil fields globally, each producing at least 20,000 bbl/day. There are about 500 giant oil and gas fields, defined as containing greater than 500 Mbbl of recoverable oil or 3 Tcf gas and

BOX 7.6   HOW MUCH ENERGY DOES IT TAKE TO PRODUCE ENERGY?

All of the energy resources used by society require energy to produce, and this is energy that is not available to the consumer. This includes not only energy used to produce oil, natural gas, corn-ethanol, but also energy to mine copper and rare-earth metals for turbines to produce electricity. The concept of "energy return on energy invested" (EROEI) compares the amount of energy necessary to extract a particular energy resource to the amount of energy actually available to society (Inman, 2013). For liquid fuels, gasoline refined from crude oil pumped from conventional reservoirs such as those in the Middle East returns 16 units of energy for every 1 unit required to produce the gasoline. Ethanol from sugarcane returns 9 units, soy biodiesel returns 5.5, oil from the Alberta tar sands returns 5 units, heavy oil from California returns 4, and ethanol from corn returns only 1.4. To understand the "at-the-pump" energy available to consumers from each of these resources, the distance driven by a car can be calculated for each fuel by multiplying the fuel economy of a car (miles per gallon) and the EROEI for each fuel, and then dividing this by the energy density of each fuel, expressed in gigajoules per gallon (Inman, 2013). Assuming 1 gigajoule of energy invested in the production of each fuel, a car can drive 3,600 miles using gasoline from conventional oil; 2,000 miles using ethanol from sugar cane; 1,400 miles using biodiesel from soy; 1,100 miles using gasoline from tar-sands oil; 900 miles using gasoline from heavy oil; and 300 miles using ethanol from corn. The same EROEI concept applied to electricity indicates that hydroelectric returns 40 units of energy for each unit invested, wind returns 20 units, coal 18 units, natural gas 7 units, photovoltaic solar 6 units, and nuclear 5 units.

producing 100,000 bbl/day for more than one year (Figure 7.20)(Simmons, 2002). The giant oil fields account for about 60% of global production and contain about half of global reserves, and the 20 largest fields account for about one-quarter of global production (Halbouty, 2001; Mann *et al.*, 2007; Höök *et al.*, 2009). The rate of discovery of giant oil fields continues to decline (Figure 7.20). The 120 giant fields discovered between 1960 and 1969 are about equal to the total number of new fields discovered between 1970 and 2010. These giant oil and gas fields cluster in the Middle East, the Gulf of Mexico, Alaska, western Siberia, offshore Brazil, Venezuela, Nigeria, and the mid and western United States (Mann *et al.*, 2007).

The two largest oil fields, Ghawer in Saudi Arabia and Burgan in Kuwait (Figure 7.21; Table 7.6), account for about 10% of world reserves. According to the Brisith Petroleum Statistical Review of World Energy, Venezuela has the largest proven reserves by country, with 297 Bbbl, followed by Saudi Arabia with 266 Bbbl, Canada with 173 Bbbl (mostly in tar sands as discussed later), and Iran with 154 Mbbl. The Middle East as a region contains an estimated 807 Bbbl, or 48% of global reserves. Oil reserves in the United States were 44 Bbbl at the end of 2014, an increase of about 5 Bbbl from 1992 estimates, largely in the Bakken shale-oil field in North Dakota and deep-water fields in the Gulf of Mexico (Figure 7.15).

World gas reserves are more evenly distributed globally (Figure 7.17). Russia and Iran each have about 18% of global reserves, Qatar has 13%, and no other country has more than 5%. US gas reserves increased from 165 to 300 Tcf from 1991 to 2012, driven by horizontal drilling and hydraulic fracturing of tight-gas reservoirs. Gas reserves in the United States amount to only 13 times annual production, slightly more than for oil. While the United States capitalized early on the shale-gas boom, with shale-gas reserves of about 650 Tcf, shale-gas reserves in Algeria (700 Tcf), Argentina (800 Tcf) and China (1100 Tcf) are larger and are certain to play an important role in future global gas production.

Reserves depend strongly on the price of oil and gas and advances in reservoir engineering (Rice, 1986; Moorehouse, 1997; King, 2008, USGS, 2013). They are also affected by seemingly magic oil-field growth (Cook, 2013), by which total recoverable oil and gas grow through inflow of oil from surrounding rocks and also improved reserve estimates. According to Root and Attanasi (1992), oil-field growth more than doubled the size of reserves in the lower 48 US states between 1960 and 1989. Reserve additions from this effect have averaged 2% per year and some fields have increased enormously (King, 2008: Cook, 2013). Between 2000 and 2009, the reserve growth of existing conventional oil and natural gas fields increased by factors of three and two, respectively, relative to new discoveries (Bui, 2013).

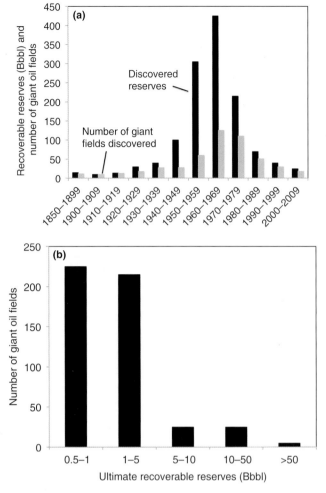

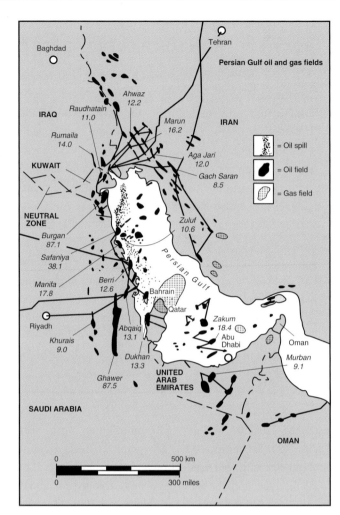

Figure 7.20 (a) The ultimate recoverable reserves of giant oil fields compared with the time of their discovery shows a decrease in new discoveries of these important resources. (b) The number of discovered giant oil fields compared to the total volume of ultimately recoverable oil shows that there that are very few giant oil fields with significant recoverable reserves (modified from Robelius, 2007).

Figure 7.21 Distribution and size (Bbbl of oil and oil equivalent) of oil and gas fields in the Persian Gulf showing the location of the oil spill created during the Gulf War of 1991. Heavy lines connecting oil and gas fields are pipelines (modified from publications of the US Geological Survey and US Energy Information Agency).

A critical feature of any reserve is the rate at which it is replaced by new discoveries. In the United States, the discovery of the Bakken tight-oil field in the Williston Basin changed the landscape for oil exploration. This field was first discovered in 1955 (Langton, 2008), but production was not possible because the oil is tightly held in the shale source rock. In 2007, the use of directional drilling combined with hydraulic fracturing allowed commercial production and reserve estimates were 3.65 Bbbl. As of 2013, the North Dakota Geological Survey and the USGS estimated reserves of up to 500 Bbbl. This would make the Bakken one of the largest oil fields on Earth, but recoverable oil is estimated by state agencies to be only 3–50% of this value. A similar situation exists for the Eagle Ford Shale in Texas, which contains an economically

recoverable reserve of 4.177 Bbbl oil and 8.4 Tcf gas. Production from these reservoirs is highly dependent on high prices and might be impacted by decreased oil prices (Martin, 2013). There is additional concern about the rate at which production from wells will decline through time. Nonetheless, production of greater than 1 Mbbl/day from these fields indicates that the global oil situation will depend increasingly on unconventional oil and gas reservoirs (Maugeri, 2013). In Brazil, most offshore discoveries through the end of the last century were off the southeast coast, including the Libra field with estimated reserves of 8 to 12 Bbbl (Fick, 2013). Globally, reserves in about half of the giant oil fields are declining at an average rate of 6.5% annually (Robelius, 2007). The boom in unconventional reservoirs that made the United States one of the world's top oil producers

Table 7.6 Ten largest conventional oil and gas fields of the world in relation to the largest conventional and unconventional fields in the United States. Fields are ranked in order of declining size with gas reserves converted to barrels of oil equivalent. Numbers show their rank with respect to all fields in the world. Reserves shown here are original and have been depleted by production (data from Carmalt and St. John, 1986; Browning *et al.*, 2013 for Barnett Shale; International Energy Agency World Energy Outlook, various years; United States Geological Survey; US Energy Information Administration; Bbbl = billions of barrels, Tcf = trillions of cubic feet).

Rank	Field name	Country	Year discovered	Oil (Bbbl)	Gas (Tcf)	Oil equivalent (Bbbl)
Largest conventional fields in the world						
1	South Pars*	Iran and Qatar	1971	50.000	1,800.00	360.000
2	Ghawar	Saudia Arabia	1948	140.000	186.00	172.00
3	Burgan	Kuwait	1938	75.000	72.50	87.083
4	Urengoy	Russia	1966	0.000	350.00	60.00
5	Safaniya	Saudia Arabia	1951	50.000	0.00	50.000
6	Upper Zakum	Abu Dhabi	1963	50.000	0.00	50.000
7	Yamburg	Russia	1969	0.000	289.60	49.94
8	Urgenoy	Russia	1975	0.002	285.59	47.602
9	Kashagan	Kazakhstan	2000	38.000	35.00	44.09
10	Bolivar Coastal	Venezuela	1917	44.00	0.00	44.000
Largest producing conventional fields in United States						
	Prudhoe Bay	Alaska	1968	20.00	35.00	26.04
	Panhandle-Hugoton	Kansas	1922	1.5	75.00	14.40
Largest producing unconventional fields in United States						
	Bakken shale	North Dakota	1951	4.844	6.7	5.94
	Eagle Ford shale	Texas	2008	4.177	8.40	5.62
	Barnett shale	Texas	1980	0.118	86.00	11.5
	Marcellus shale	Northeastern USA	1839	0.00	64.90	11.2
	Haynesville shale	Louisiana, Texas	1905	0.00	75.00	12.9

* The South Pars field also contains an estimated 50 Bbbl of natural gas condensates. The Bakken field was originally described in 1951, but production did not occur until about 2000.

is helping keep global reserve to production ratios stable, but in the face of increasing global consumption it is not clear how long this will continue.

Conventional natural gas reserves have also changed greatly. From 1992 to 2012, reserves increased 100% in the Middle East and 66% in Venezuela, but declined 75% in Mexico. In the United States, reserves increased almost 100% between 1998 and 2013. Recent gas-field discoveries off the coasts of Lebanon, Israel, India, Papua New Guinea, and the United States, when added to reserve growth of existing fields, yield reserve to production ratios of about 56 years at current rates of consumption. Increased consumption of natural gas for electricity production, in place of coal, will escalate the global decline.

Oil and gas trade and transport

Oil and gas are among the most important commodities of international trade. Globally, 86.8 Mbbl/day were produced in 2013 and, of this total, 55 Mbbl/day were traded internationally. Exporting countries consume domestically about 20% of their own oil, and importing countries produce about 20% of their domestic consumption. Oil imports resulted in the transfer of $2.3 trillion from importing to exporting countries, accounting for about 24% of aggregate oil exporters' GDP

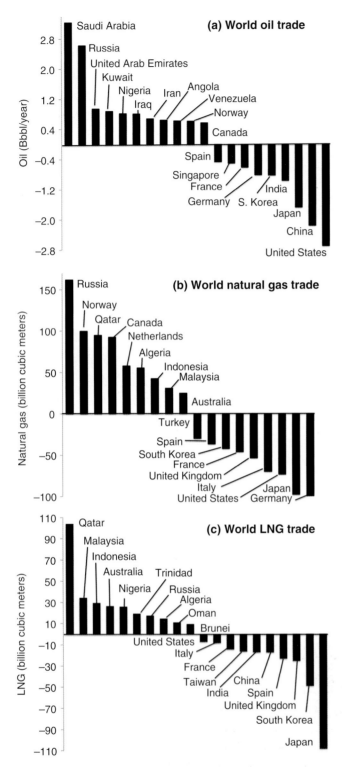

Figure 7.22 World trade in crude oil and natural gas for 2012: (a) crude oil; (b) natural gas, all production; (c) natural gas, LNG. Countries plotting above the line are exporters and those below the line are importers. (Compiled from the CIA World Fact Book, the LNG Report, and the Energy Information Agency.)

(BP, 2012; Andreopoulous, 2012). Globally, two-thirds of natural gas is consumed domestically by producing countries, with another third crossing international borders (Johnson, 2011). These numbers are large enough to have an impact on national budgets, and energy imports account for trade imbalances in many countries. The main oil exporters are Saudi Arabia, Russia, United Arab Emirates, Kuwait, Nigeria, and Iraq, with the main importer being the United States (although the level of imports is decreasing because of increased US production from unconventional sources); other major importers are China, Japan, India, South Korea, Germany, and France (Figure 7.22). World natural-gas exports come largely from Russia, Norway, Qatar, Canada, the Netherlands, and Algeria and go to Germany, Japan, the United States, Italy, the United Kingdom, and France. Historically, natural-gas exporters needed to be connected to major consumers by pipeline, but that has changed with the development of **liquefied natural gas** (LNG) transport (Figure 7.23). Major LNG exporters are Qatar, Malaysia, Indonesia, Australia, Nigeria, Trinidad, Algeria, and Russia going to Japan, South Korea, the United Kingdom, Spain, China, and India (Figure 7.22).

Oil is moved by trains, barges, pipelines, and tankers, with tankers having the advantage on longer hauls. Crude-oil pipelines (Figure 7.23) extend from producing regions to refinery centers. Natural-gas pipelines extend from fields to industrial and residential markets (Kennedy, 1984). The most important oil fields in the Middle East and other producing areas have short pipelines to tanker ports. Long-distance oil and gas pipelines are found largely in Canada, Russia, and the United States (Figure 7.17). Most pipelines operate at capacity all year. Because of the seasonal residential heating market, natural-gas pipelines are used to fill underground chambers near markets during the summer, and this gas is then used to supplement pipeline gas during the winter.

Tankers (Figure 7.23) are more cost effective for really long distances, in terms of both original investment and operating cost, and they can be directed to new destinations as demand changes. The largest tanker in operation in 2014 carried 2 Mbbl oil, enough to make 144 million liters of gasoline, plus 91 million liters of diesel, plus 1 billion plastic toothbrushes. According to the American Petroleum Institute, in 2011 there were 20,000 tanker calls annually into US ports. Shipping oil by tanker accounts for about 1% of the final cost of gasoline at the pump. In practice, world oil tanker shipments are dominated by the Middle East. Tanker transport of natural gas is accomplished by cooling the gas to −162 °C so that it becomes an LNG with a volume 600 times smaller than the original gas.

Figure 7.23 Oil and gas transportation. (a) Alyeska pipeline near Fairbanks, Alaska, where it has been elevated to prevent heated oil from melting permafrost (photo by the authors). (b) The Polar Adventurer oil tanker, shown here (foreground) loading oil at the port of Valdez, Alaska, the southern end of the Trans-Alaska pipeline, transports North Slope oil to refineries along the US west coast. The tanker is approximate 272 m long and has two completely separate hulls. The double hull design, with an inner hull about 3 m inside the outer hull, provides additional protection from punctures, and is required for large tankers operating in United States and most LDC waters (photo courtesy of ConocoPhillips, Inc.) (c) The Arctic Princess LNG tanker, which is 288 m long, carries LNG in four cylindrical tanks at a temperature of –160 °C (photo courtesy of Höegh LNG). See color plate section.

There were 370 operating LNG tankers in 2013, 26 export terminals in 15 countries, and 60 import terminals in 18 countries worldwide (California Energy Commission).

Several decades of operation show that tankers merit more environmental attention than pipelines. Pipelines have closely spaced shut-off valves that limit the size of any spill. In Arctic terranes they have been elevated to prevent blockage of migration routes and to prevent melting of permafrost, which is permanently frozen ground that might become unstable if melted by hot oil moving through the pipeline (Figure 7.23). Early concerns about pipelines were assuaged by the record of the Trans-Alaska pipeline, which measures 122 cm in diameter and extends for 1,265 kilometers from Prudhoe Bay to the tanker port at Valdez (Figures 7.17a and 17.23b). The $7.7 billion line started service in 1977 and had carried a total of 16.5 Bbbl between 1977 and 2012. In 2002, the pipeline experienced no damage during a magnitude 7.9 earthquake that moved the ground 6 m horizontally and almost 1 m vertically (Pemberton, 2012). The two most significant spills from the pipeline were the result of human sabotage. In 1978, someone blew a 2.54-cm hole in the pipeline that released 16,000 bbl oil, and in 2001, someone blew a hole in the pipeline that released 6,100 bbl oil. The largest spill caused by mechanical failure of the pipe released 5,000 bbl oil in 2010, and the line was shut down for 80 hours for repair (Holland, 2010). Tanker oil spills (Table 7.5) range from small losses from all types of ocean-going vessels, which have polluted many shipping lanes and coastal areas, to occasional large spills from oil tankers, which attract a disproportionate share of attention.

Tanker spills release three and four times, respectively, less than is released from tank cleaning operations, and oil refineries and offshore production (Gossen and Velichkina, 2006). There is no question that tanker oil spills are fatal to much marine life, but unlike metals, which are not biodegradable, most oil residues decompose with time to form H_2O and CO_2 if they are sufficiently aerated (although this happens much more slowly if the oil seeps downward into porous beach sand). Experience from the Prince William Sound spill shows that most animal populations rebound rapidly, but that low-level contamination persists for decades in areas where oil is not sufficiently aerated to decompose (Cunningham and Saigo, 1992; Peterson *et al.*, 2003). In general, studies made in areas of spills that occurred 10 to 20 years ago find little or no evidence for the event in the sediment record (Frey *et al*, 1989). We should derive no comfort from these results. Improved marine transport of oil, especially the use of double-hull tankers, discussed below, is essential if we are to continue long-distance trade in oil.

BOX 7.7 | HOW DEPENDENT ARE NATIONAL ECONOMIES ON THE EXPORT OF RESOURCES?

A healthy economy is diversified. In view of this, most governments seek an economy that balances extraction and exports of natural resources with manufacturing and service industries. While this is the goal, it is not a reality for many countries that depend almost completely on export of oil and natural gas to balance their national budgets and fund government infrastructure and social welfare programs. In 2013, oil export earnings accounted for almost 10% of the GDP in Russia, 15% in Iran, 25% in Venezuela, 30% in Qatar, 40% in Iraq, 45% in Saudi Arabia, and 60% in Kuwait. However, these are not secure. In 2014, Iran needed oil prices of about $135/bbl to balance its budget, Venezuela, Nigeria, and Iraq needed $110/bbl–120/bbl, Libya and Russia between $100/bbl and $110/bbl, and even Saudi Arabia needed prices of about $93/bbl (data from Deutsche Bank). The drop in oil prices from $100 in January, 2014 to about $55/bbl in December, means that all of these countries will probably run a budget deficit. For example, Russia's 2015 budget assumed an oil price of $100/bbl. At oil prices below the break-even threshold, countries will have to dip into their currency reserves or sovereign wealth funds, or borrow money and increase national debt. Saudi Arabia has a sovereign fund of about $750 billion, and can afford to borrow from it for at least several years. Many of the other countries do not have such a luxury. The conversation about lowering our dependence on fossil fuels must include long-range plans for transforming these national economies.

Security of oil and gas supplies is of great concern, particularly the high import reliance of many MDCs. The 1970s oil embargo, which crippled these countries, led to formation of oil stockpiles, such as the 691 Mbbl US Strategic Petroleum Reserve, the largest nationally owned stockpile of oil in the world. Although there was widespread interest in development of a national energy policy to decrease future vulnerability, all MDCs, including the United States, are more dependent on oil and gas now than in 1970. Japan, which effectively ceased nuclear power generation in 2011 after the Fukushima nuclear accident, discussed below, is the world's top LNG importer, consuming about 37% of global LNG in 2012, and is the third largest oil importer. Japan imports oil and gas predominantly from the Middle East, but also receives shipments from Indonesia. In spite of the oil and gas boom in the United States, imports of both are required to meet domestic demand. Canada is the top oil supplier to the United States, where more than 2.5 Mbbl of oil are transported by train each day. The proposed construction of the Keystone XL pipeline, while controversial, seeks to increase the security of oil imports to the United States. Much of Europe is supplied by pipelines from the Netherlands, the United Kingdom, and Norway, with important additions from Russia, a connection that was bitterly opposed by the US government during the early 1980s (Jentleson, 1986). Pipelines from Russia continue to allow Russia to exert its political will on Europe. In 2013, Russia used the threat of withholding natural gas to dissuade Ukraine from signing a free trade agreement with the European Union.

Industrial and residential users in most MDC's now depend heavily on natural gas. Slightly more than one-half of US homes heat with natural gas, and the remaining half heat with electricity. Considering that natural gas is predicted to account for 80% of all electricity generation in the United States by 2035, an interruption in natural-gas supply would be just as catastrophic as earlier oil embargoes. Significant attention is paid to security through the Straits of Hormuz, a 34-km sea passage that connects the Persian Gulf to the Indian Ocean, through which is transported one-fifth of all globally traded oil and gas. Such areas are considered chokepoints (EIA, 2012), and are protected by military forces of producer and consumer nations.

The high import dependency for oil and gas among developed nations is cited as the cause for the 1991 Iraq Conflict, and has also been suggested as part of the reason for the US-led invasion of Iraq in 2003 (Roberts, 2003). It is likely that these conflicts will not be the last wars that we fight over oil and gas. In these conflicts, oil and gas installations were major targets of hostile efforts. During the Gulf War of 1991, sabotage of the Sea Island tanker terminal spilled 6 Mbbl of oil into the Persian Gulf creating an oil slick that extended 400 km southward along the western side of the gulf (Figure 7.21). In addition, 752 wells in Kuwait were damaged, spilling out as much as 55 Mbbl of oil into 200 lakes, and burning 5 to 6 Mbbl of oil per day, about the same as pre-war oil production from

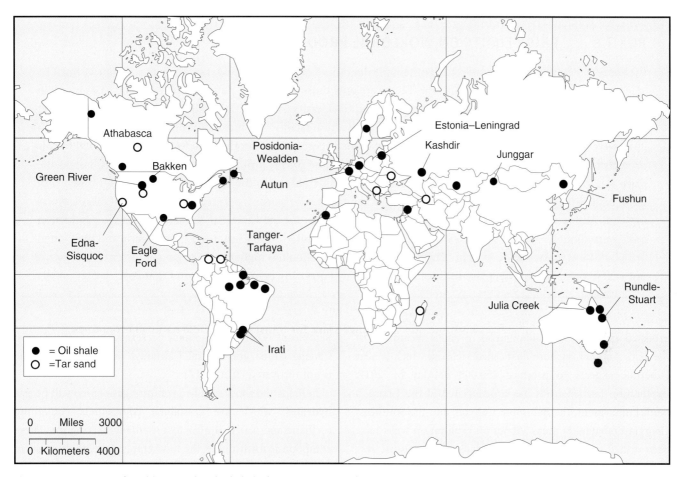

Figure 7.24 Location of world tar-sand and oil-shale deposits. Deposits that are named have had significant production.

these fields (Ibrahim, 1992). Although the fires were extinguished in less than a year, much less than the 5 years originally estimated, and soot from them did not cause local climate changes, as anticipated, oil from surface pools has seeped into the subsurface. Animal populations have recovered at different rates, some birds rapidly but fish stocks slowly, and oil remains along some parts of the coast (Barth, 2002; Poonian, 2003).

Frontier and unconventional sources of oil and gas

Future oil and gas supplies will depend on discoveries of conventional oil and gas in remote, frontier areas, as well as unconventional sources such as the tar sands in Canada, the Bakken tight oil in North Dakota and the Green River oil shale in Utah (Figure 7.24). Frontier exploration areas in North America include the western overthrust belt, the Arctic slope and islands, and the east coast. Interest in the Arctic slope and islands stems from their proximity to the Prudhoe Bay field, the largest field in North America. Four decades of exploration in Arctic Canada have found 1 Bbbl of oil and 10 Tcf of gas in the McKenzie Delta and 90 Bbbl of oil and > 1 Tcf gas

on the northern islands (Government of Canada, 2014). Pipeline transport from fields on the Canadian Arctic islands is complicated by the fact that they are separated from the mainland by channels in which sonar studies show grooves formed by icebergs dragging across the bottom (Lewis and Benedict, 1981; Barnes and Lien, 1988). Even if oil from the island fields reached the mainland, there is no pipeline to move the oil to markets in the south. In 2013, production licenses for the McKenzie delta area were forfeited voluntarily because of the lack of a pipeline to transport oil and gas 1200 km to port (OGJ, 2013; see Figure 7.17). The Arctic National Wildlife Refuge, which is between the Canadian oil fields and the Prudhoe Bay fields, is an important area of frontier exploration in the United States, as discuss in Chapter 5.

The continental shelf of the US east coast is of interest because of its general geologic similarity to the prolific Gulf Coast area. Exploration of the east coast has found producible reserves in Canadian waters, including the large Hibernia field located 315 km east of St. John's, Newfoundland (Figures 7.14, 7.17). Hibernia, which produced 129,000 bbl/day in 2009, is

BOX 7.8 | PRICE LIMITS ON WORLD OIL PRODUCTION

In order for a company, or country, to maintain production of oil without losing money, the market price must be at least equal to their production costs. These costs are only those required to pump the oil to the surface, maintain the facility, and sell the oil, and they are often referred to as cash costs. Cash costs do not take into consideration the cost of finding the oil field, or of constructing a production facility. For most large oil companies, cash costs in 2014 were about $15 to $20/bbl. Finding and development costs added another $30 to $35/bbl for a total break-even cost of a little more than $50/bbl. During periods of low prices, production might continue as long as the price was greater than the cash cost. Cash costs vary considerably with the type of oil production, however. In 2014, cash costs, including royalties, quoted by Morgan Stanley for porous, permeable conventional oil reservoirs in Saudi Arabia were about $4/bbl compared to nearly $40/bbl for the Canadian tar sands. Average production costs were about $5 in Iraq, $8 in Mexico and Iran, $20 in the United Arab Emirates, $25 in Nigeria and the United States, and $30 in Russia. Most tight-sand and oil-shale production has higher costs, although the exact number is not well known. The big decrease in price in late 2014 will probably help us understand that cost.

expected to generate $10 billion in tax revenues over the life of the field. Another offshore field called Hebron, which is near Hibernia, expects to produce 150,000 bbl/day by 2017. Exploration farther south has concentrated in the George's Bank Embayment, the Baltimore Canyon, and the Southeast Georgia Embayment areas. About 50 exploration wells have been drilled into some large traps (detected by seismic surveys) in these areas since 1975, but commercial reserves of oil and gas have not been found, and the United States currently has a moratorium on new drilling in these areas. Another zone of future exploration is likely to center on the area north of the Arctic Circle, where 90 Bbbl oil, 1.7 Tcf gas, and 44 Bbbl natural-gas liquids are estimated to be present in as-yet undiscovered fields (USGS, 2013).

Unconventional resources that might add significantly to our oil and gas supplies include heavy oil, tar sands, tight shale, and oil shale, which are discussed separately below. Another possibility is **coal-bed methane**, with global reserves estimated to exceed 500 Tcf. In the United States, coal-bed methane is present in the Powder River Basin of Wyoming and Montana, the Green River Basin of Colorado and Wyoming, the San Juan Basin of New Mexico, and Uinta–Piceance Basin of Utah and Colorado, where the combined reserves are estimated to be 24 Tcf. Coal-bed methane now accounts for about 7% of US natural-gas consumption, with production stimulated by tax credits provided by the US government in the Fuel Use Act of 1980. Extraction of coal-bed methane involves the co-production of saline water, which must be removed to release the pressure holding gas in the coal seam. One well can release almost 68,000 liters of water per day, and this water requires proper disposal.

Significant coal-bed methane reserves exist in Russia, Canada, China, and Australia, although much of the methane is not recovered (IEA, 2012).

Perhaps the most significant future gas source will be **gas hydrates**, which are crystalline compounds containing methane and water, similar to a methane-bearing ice. These compounds form under the low-temperature, high-pressure conditions that prevail in the shallow parts of near-shore marine sedimentary basins. Widespread reflective horizons seen in seismic surveys of these basins indicate that methane hydrates may be the most important global reservoir of organic carbon and, as such, potentially important both as a fossil fuel and a source of methane contamination to the atmosphere (Kvenvolden, 1988). Seismic exploration for gas hydrates in the Gulf of Mexico indicates 100-m-thick sandstone layers that contain as much as 90% recoverable gas hydrate (USGS, 2015). Japan became the first country to extract gas from methane hydrate, working in the Nakai Trough where reserves of 38 Tcf methane are estimated, enough to satisfy 10 years of Japanese gas consumption (BBC, 2013). China reported 5.2 Tcf gas hydrates below the Pearl River Delta (Qian, 2013). Further technological developments are required to safely extract methane hydrates, owing to the significant overpressure of these reservoirs, and the need to maintain stable flow rates. All of these possibilities pale in comparison to **abiogenic gas** (Gold and Soter, 1980). The abiogenic hypothesis says methane formed in the mantle by inorganic chemical reactions might migrate upward into the crust where it would be trapped by fractured igneous and metamorphic rocks far below sedimentary basins. This possibility was tested by drilling in the Siljan ring meteorite-impact feature in central Sweden, which found only traces of methane, most of which was biogenic (Juhlin, 1991), and the concept has lost support.

7.1.3 Heavy oil and tar sands

Heavy oil and semi-solid bitumen (tar) are unconventional resources that constitute about 15% of global oil with reserves of 100 to 250 Bbbl (Attanasi and Meyer, 2010). They are found in about 20 countries, with the largest reserves in Canada, Kazakhstan, and Russia. Most heavy oils and tar sands appear to be the remains of typical oils that were altered by reaction with groundwater and bacteria at relatively shallow depths. Oxygen-bearing groundwater dissolves and removes lighter molecules, and bacteria in the water use the oil as food, decomposing lighter molecules first. These processes remove hydrogen and increase the fraction of heavy molecules, making the oil viscous. This is an essentially irreversible process in nature; studies show that heavy oil and tar are not easily converted to oil even if they are buried to greater depths at which kerogen would form oil.

Recovery of heavy oil and tar can be done by drilling or mining. For drilling, steam that is injected through a hole softens and dilutes the oil, separating it from the sand grains and allowing oil to flow to production wells. The technique is energy-intensive, with an energy return on energy invested of 5; however, between 25% and 60% of the oil can be recovered using this technique, which is comparable to recovery from conventional reservoirs (CGG, 2015). If the oil is too viscous or is actually tar, the sand must be mined. The tar, which has about 8–14% bitumen, is removed by strip-mining methods that are among the largest in the world (Singhal and Kolada, 1988). Tar is then removed from the sand by washing in very hot water and is converted to synthetic crude oil known as **syncrude** by addition of hydrogen and other chemical processing. About 83% of the bitumen is converted to syncrude, a much better recovery rate than obtained from most oil wells. The tar contains up to 5% sulfur, over 98% of which is recovered during processing.

Heavy oil and tar are less desirable than crude oil because they cannot be converted to gasoline as easily and they yield a larger fraction of heavy oil products including a material known as petroleum coke that is just different enough from coke made from coal to have difficulty finding markets and that has begun to accumulate at plants that treat syncrude (Austen, 2013). Syncrude also contains more sulfur and nitrogen and has locally high metal concentrations, notably nickel and vanadium (Yen and Chilingarian, 1979; Stauffer, 1981).

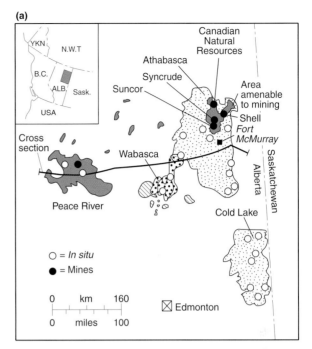

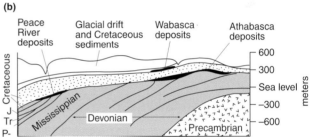

Figure 7.25 (a) Location of Alberta tar sands in the Athabasca, Peace River and Wabasca areas and heavy oil-bearing sands in the Cold Lake area. Syncrude and Great Canadian Oil Sands operations are open pit mines, whereas the other operations produce by EOR from wells (daily production in bbl is also shown). Also shown is the smaller area of tar sands that is amenable to open pit mining. (b) Cross section along the heavy line through the tar sands showing their location in the sedimentary sequence. J, Jurassic; T, Triassic; P-P, Pennsylvanian–Permian.

The largest deposits of tar sand in the world are in the Athabasca area of northeastern Alberta (McConville, 1975), where operations produce over 1.2 Bbbl of oil per year (Figures 7.25, 7.26). These deposits have recently been reclassified as "oil reserves" by most agencies and they now constitute more than 99% of Canadian reserves, making it the third largest global oil reserve, after Saudi Arabia and Venezuela. According to the Canadian Association of Petroleum Producers, in 2013 there were 127 operating heavy-oil and tar-sand projects in Canada, and production is predicted to increase to 2.5 Bbbl/year by 2030. The largest of these produces over 100 Mbbl oil per year and accounts for

Figure 7.26 Tar sands. (a) Overview of processing plant with partly revegetated waste pile on left. (b) Reclaimed mined land with as yet unreclaimed waste dump on right. Processing plant is distant background (photos courtesy of Syncrude Canada). See color plate section.

about 13% of Canadian oil production, as well as important amounts of coke and fuel gas.

The future of tar-sand production in Canada and, indeed, the future of North American energy rests on efforts to export syncrude to the United States via the proposed Keystone XL pipeline. This pipeline has been approved by agencies of the US and Canadian governments but has become victim to environmental and political wrangling. Objections to the tar sands focus on the large scale of the mining, the smaller energy return compared to conventional oil, and possible spills. These arguments fail to recognize that the real alternative to Canadian tar sands is oil from the Middle East where a decade of war has cost numerous foreign and local lives, and where production-related environmental damage is widespread.

7.1.4 Tight (unconventional) oil and gas

Tight or unconventional oil and gas are formed when organic materials in shale, sandstone, and carbonate source rocks are

heated into the oil window, but do not release their hydrocarbons. The permeability of the rock is 100 to 10,000 times lower than conventional oil or gas reservoirs, and the oil and gas remain dispersed in the rock. It is common in the literature for the term tight shale to be used interchangeably with shale oil, shale-bearing oil, gas shale, and gas-bearing shale. Tight oil differs from oil shale, discussed below, which has not entered the oil window.

Production of tight oil and gas in the United States exceeded 1 Mbbl/day oil-equivalent in 2013, and is forecast to increase to 3–4 Mbbl/day by 2017 (US Energy Information Administration, 2014). The most important tight-oil reservoirs are the Bakken shale that underlies western North Dakota and neighboring Montana, the Niobrara Formation and Pierre shale in the Rocky Mountains, and the Eagle Ford Formation in Texas (Figure 7.24). Important US tight-gas reservoirs are located in East Texas, the Piceance Basin of Colorado, and the Green River Basin of Wyoming. These fields contain about 58 Bbbl oil and 665 Tcf gas that are technically recoverable (Dyni, 2010; EIA, 2013a). Tight-shale resources have also been recognized in Argentina, Australia, Mexico, Poland, Russia, and South Africa (Ashraf and Satapathy, 2013). World reserves of tight oil and gas are enormous and include 345 Bbbl oil and 7.3 Tcf gas in 95 basins in 41 countries worldwide (EIA, 2013a). These reserve estimates increased by 1000% and 11% for oil and gas, respectively, from only 2011 to 2013, largely driven by continued reassessment based on engineering advances in directional drilling and hydraulic fracturing. They are heavily dependent, however, on a relatively high price for oil (see Box 7.8; Ratner and Tiemann, 2014).

Production from tight shale is largely done by directional drilling and hydraulic fracturing in relatively flay-lying layers. In the Bakken field two organic-rich shale units, each of about 40 m thickness, and are spread over 520,000 km^2 at a depth of about 3.2 km. Not all tight oil and gas reservoirs are as geologically simple. For example, the Jonah tight-gas reservoir in Wyoming hosts gas in tight sandstone lenses sandwiched between shale sequences, at depths that vary over at least a thousand meters. Seismic imaging is used to identify the gas-containing lenses, and directional drilling allows producers to drill into the lenses where hydraulic fracturing (fracking) is used to enhance permeability and extract the gas.

Horizontal drilling was perfected decades ago to allow a smaller footprint for drills in conventional reservoirs, such as the Endicott Island field discussed in Chapter 4. Fracking has also been done for decades but its use on a larger scale in tight-gas and -oil production has attracted attention to possible

---

BOX 7.9 | **WHAT ARE HORIZONTAL DRILLING AND FRACKING?**

Most wells to extract resources such as water, oil, and natural gas are drilled vertically into the ground. This makes sense for large groundwater, oil, or natural gas reservoirs that have high porosity and permeability. But as the permeability of a rock decreases, vertical wells have to be spaced too close together to be economic. This is particularly true for oil and gas produced from shales and tight sands that have very low porosity. Many of these are relatively flat layers and a horizontal hole drilled along the layer can contact much more rock than a vertical hole, thus lowering the cost of production. Horizontal holes are started as vertical holes and are wedged off toward horizontal as drilling proceeds. Hydraulic fracturing (fracking) involves injection of water with dissolved chemicals and sand grains into the rock under pressure, where it expands fractures in the rock. The sand grains remain in the rock to hold the fractures open and the water is pumped back to the surface (Figure 7.16). Fracking is especially important when attempting to produce from tight reservoirs because they are much less permeable and depend on these fractures to make extraction of oil or gas possible.

---

environmental risks. The main risks involve possible contamination of groundwater by fracking water and increased possibility for earthquakes caused by injected fluid. Individual wells can require up to several million gallons of water, most of which must be trucked to the drilling location. As much as half of the water flows back to the surface in the first days after fracking. This flowback water is not potable and is stored in tanks or pits prior to disposal, which may be deep underground injection or discharge to surface water. The former is regulated by the US EPA Underground Injection Control (UIC) program, and the latter by the National Pollutant Discharge Elimination System (NPDES), which requires that the water is treated prior to disposal. Earthquakes, which have been registered in some fracking areas, could be related directly to the hydraulic fracturing or to lubrication of existing faults. Repeated studies have shown that these risks are very small and that the main concern is proper disposal of fracking fluid that is returned to the surface (Healy, 2012; Holloway and Rudd, 2013).

Oil and gas output from tight-oil and -gas wells appears to decrease more rapidly than it does from conventional wells. For instance, according to the North Dakota Department of Mineral Resources production from wells in the Bakken typically peaked within the first 2 years of operation. Annual rates of decline from individual wells range from 27% to 69% from years 1 to 5 (Hughes, 2013). Similar decline rates have been reported for tight-oil and -gas wells in other areas, including much of the production from the Marcellus shale in Pennsylvania and West Virginia. To sustain production, fracking might be repeated or a replacement well is drilled.

The tight-oil and -gas revolution has had several interesting results. For one thing, it has led to more partnerships between national governments and major oil-exploration companies. These two groups, which have been at odds for decades, are being drawn together by the need on the part of most government groups for technology to deal with these unusual engineering situations. Recent efforts along these lines have happened in Mexico (Montes *et al.*, 2013) and Ukraine (Reed and Kramer, 2013) where legislative changes were enacted to allow foreign exploration companies access to tight-oil and -gas prospects. The other development involves the advantage that lower costs and more dependable supplies of tight oil and gas, as well as electricity generated from them, have given to manufacturing in the United States (McWhitert and McMahon, 2013). Finally, the added cheap energy from these sources gives the world a breather in its search for dependable, cheap sources of renewable energy, and one that we cannot afford to waste.

7.1.5 Oil shale

Oil shale is shale that contains solid organic material known as kerogen, which has not been matured by entrance into the oil window. This resource differs from tight oil-bearing shale, discussed previously, in that it requires open pit mining and *ex situ* processing to recover the oil, whereas tight shale contains liquid oil trapped in pore spaces. The earliest reports of oil-shale mining date to seventeenth-century Sweden where alum-rich oil shale was heated to separate the hydrocarbons from potassium aluminum sulfate, which was used for tanning leather (World Energy Council, 2010).

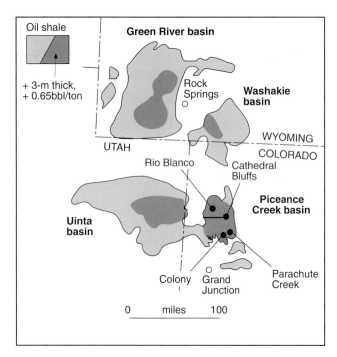

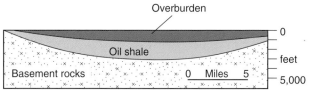

Figure 7.27 Geologic map showing distribution of Green River Formation and areas of high-grade, thick oil shale [>3-m thick and >0.65 bbl/ton (0.6 bbl/tonne)]. The cross section shows the relation of oil and overburden along the line extending west from Cathedral Bluffs. Colony, and Parachute Creek projects used conventional mining and processing and Cathedral Bluffs and Rio Blanco projects used modified *in situ* (MIS) shale combustion (modified from US Geological Survey Scientific Investigations Report 2005–5294).

Oil shale was mined throughout the twentieth century in Germany, China, Brazil, Scotland, Russia, and Estonia, and in the 1970s it was thought to be the answer to US energy self-sufficiency. "Realistic" estimates suggested that oil-shale operations being developed in the Piceance Creek Basin around Grand Junction, Colorado (Figure 7.27), would be producing 400,000 to 500,000 bbl of oil per day by the mid 1990s (Russell, 1981). By the mid 1980s, however, after a total investment of $4–5 billion, these operations had closed without achieving routine production (EMJ, 1982). Oil shale projects in other parts of the world, including the Rundle Oil Shale Project in Queensland, Australia, suffered similar fates (EMJ, 1981).

Most oil shales are relatively typical organic-rich, marine shales that were deposited along the margins of larger basins where they were not buried by later sediment to reach the oil window. Others, however, formed in very different environments. The strangest of all is the Eocene-age Green River Formation, in the Piceance Basin, which is a sequence of lake, or lacustrine, deposits that accumulated in a huge, shallow lake (Lewis, 1985). Water in the lake was alkaline and became saturated with halite and sodium carbonate minerals including trona. The Rundle and Stuart oil shales in Queensland, which are sometimes called **torbanite**, formed in a similar lacustrine environment.

According to the IEA, recovering oil from oil shale is an energy-intensive process, with an energy return on energy investment of around 4 to 5. To recover oil from shale, it must be pulverized and heated to temperatures of 500 to 1,000 °C, where the shale undergoes **pyrolysis**, a process that releases hydrocarbon gases and liquids. Heat for the process can come from combustion of the shale, or injection of hot gases or solids. Most pyrolysis uses containers at the surface in which mined shale is heated, although efforts have also been made to carry out the pyrolysis process in fractured oil-shale layers without actually mining it. In addition to their inherent technical difficulties, oil-shale projects require water in volumes two to five times as large as the volume of oil produced. To produce 500,000 bbl/day, a plant would require most of the precipitation falling on an area of 200–500 km². Water is already a big problem in the drainage of the Colorado River where the best oil shale is located, suggesting that large-scale oil-shale conversion would require water imports into the area (Saulnier and Goddard, 1982).

Although production from oil shale peaked in the late 1970s, sustained high prices during the early 2010s caused renewed interest in this resource. Global reserves are estimated to be as much as 5 Tbbl oil, with up to half of this technically, though not economically, recoverable (IEA, 2013; USGS, 2013). Oil has been produced from shale whenever economics made it an attractive option, beginning with the first patent on a oil-shale process in England in 1694. Jordan, a country with little conventional oil and gas, imports 95% of its energy needs, and in 2013 entered into an agreement with China to help develop oil-shale fields that underlie more than half the country. In late 2013, an Estonian energy company began pursuing oil-shale projects in the Green River Formation with plans to produce up to 50,000 bbl/day by 2020 (Jaber *et al.*, 2008). Cost estimates suggest that a price of $100/bbl is required for economic production. The dramatic and rapid drop in oil prices in late 2014 are not likely to reawaken the sleeping oil-shale giant despite continued global interest.

7.2 Non-fossil mineral energy resources

7.2.1 Nuclear energy

Nuclear energy serves us in the form of heat produced by fission reactions (Häfele, 1990). In present applications, the naturally occurring isotope, ^{235}U, is bombarded by slow neutrons, which increases the rate at which it splits into smaller fission products such as ^{90}Sr and ^{137}Cs. During this reaction, some matter is converted to energy, which we see partly in the form of heat. If the concentration of ^{235}U is high enough and the flux of neutrons is large enough, a self-perpetuating series of ^{235}U fission reactions known as a **chain reaction** will result. If the rate of this reaction is controlled, it becomes a **nuclear reactor**. If not, it is a nuclear weapon (Duderstadt and Kikuchi, 1979).

Use of nuclear energy is complicated because ^{235}U is the only naturally occurring, readily **fissile** isotope and it makes up only 0.711% of natural uranium. Most modern reactors require uranium in which the proportion of ^{235}U has been increased by a process known as **enrichment**. The rest of natural uranium is largely ^{238}U, which is not directly fissile, but which can be converted to fissile ^{239}Pu by absorbing a neutron. The main isotope of thorium, ^{232}Th, is converted to fissile ^{233}U by a more complex series of reactions that begin by absorbing a neutron. These isotopes are the basis for possible future nuclear applications involving breeder reactors, which are discussed below.

Although we think of uranium as the basis for weapons and nuclear reactors, it does have other uses. Uranium oxide was the original yellow pigment in paintings by van Gogh, and small amounts of **depleted uranium**, from nuclear fuel in which the more radioactive ^{235}U has been exhausted, are used in ammunition and ship keels where its very high atomic weight provides compact mass. When uranium was used in the first atomic bombs, demand increased greatly and as this market dwindled, uranium came into demand as a fuel in reactors to generate electric power. Nuclear power grew rapidly through the late 1980s, accounting for 16% of global electricity in 1986 (IEAE, 2013). However, as safety concerns about nuclear power increased following incidents at Three Mile Island and Chernobyl, the growth of nuclear power slowed, causing a second downturn in production (Figure 7.1d), and nuclear's share of global electricity remained relatively flat into the early twenty-first century (Cohen, 1984; IAEA, 2011).

Globally, in 2014, there were 435 nuclear power plants operating in 37 countries, accounting for about 5% of world energy production and slightly more than 11% of world electricity (Nuclear Energy Institute, 2015). World uranium markets are valued as about $7 billion annually, far less than the fossil fuels. The failure of the Fukushima I nuclear power plant in 2010, the largest nuclear power plant disaster since Chernobyl, caused Japan to idle its 50 nuclear reactors and stimulated global awareness of the possible dangers of nuclear power. Germany announced that it will close all nuclear power plants by 2022. The decisions by Japan and Germany to close their nuclear reactors means that alternative sources of energy will be needed to meet growing demand in those countries, as discussed below.

Geology of uranium deposits

Much of what we know about the geology of uranium was learned during the exploration boom of the 1970s. This boom started when people noted that uranium reserves known at that time were only a fraction of the amounts needed to fuel the 1,000,000 MWe (megawatt (electrical)) of nuclear generating capacity estimated to be in place in 1990. This imbalance catalyzed a five-fold rise in uranium prices between 1975 and 1980 (Figure 7.1c) and massive exploration programs, involving expenditures of $6–8 billion. The results were amazing. New deposit types were recognized, new districts were found, and estimates of world uranium reserves more than doubled to levels of about 1.55 million tonnes of contained U_3O_8 as of 1990 (Red Book, 1991). In the late 1980s, increases in nuclear power capacity slowed and the prices dropped, putting many newly discovered mines out of business. Prices increased again in the early twenty-first century and nuclear appeared poised for a resurgence; however, the Fukushima disaster and increased production of cheap, natural gas put renewed pressure on nuclear power. The future of uranium as an electric producer remains uncertain in spite of abundant reserves and the much smaller carbon footprint relative to coal and natural gas.

Uranium forms an unusually complex range of deposit types that reflect its tendency to occur in nature in two oxidation states (Harris, 1979; Nash *et al.*, 1981; Dahlkamp, 1989). The uranic ion (U^{+6}) is stable in oxidized environments, and the uranous ion (U^{+4}) is stable in reduced environments where oxygen is scarce. The uranic ion dissolves readily, especially in the presence of dissolved carbonate ions (CO_3^{-2}), with which it combines to form soluble complex ions; uranic ions precipitate as minerals such as carnotite only if potassium, calcium, phosphorus, or vanadium are relatively abundant. In contrast, the uranous ion is relatively insoluble and forms the common uranium mineral uraninite (UO_2),

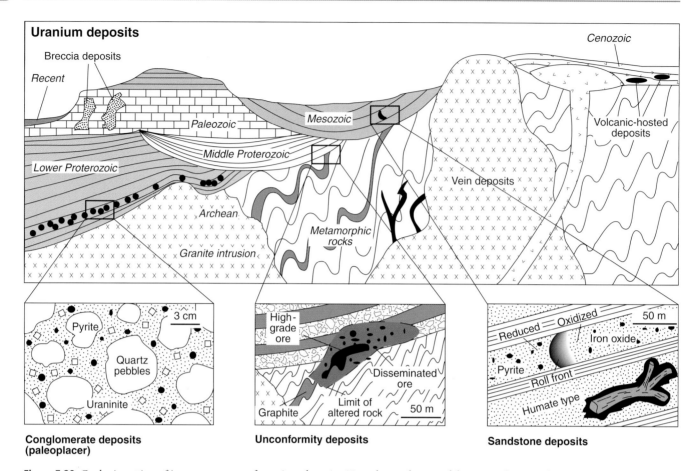

Figure 7.28 Geologic setting of important types of uranium deposits. Note that each type of deposit is shown in host rocks of the most common age.

which is called pitchblende when it is in a microcrystalline form. Obviously, then, the formation of uranium deposits is favored when waters in oxidizing environments scavenge uranium from surrounding rocks, and deposit it where they are reduced.

Three types of uranium deposits contribute most to global uranium production (Figure 7.28). Unconformity-type deposits are found at and near an **unconformity** separating Middle and Upper Proterozoic sediments, particularly in the Canadian Athabasca Basin and Australian Pine Creek Basin (Figure 7.29)(Clark and Burrill, 1981). The ore, which forms veins and disseminations in the sediments, is thought to have been deposited from oxidizing basinal brines that passed through reducing zones rich in carbon or reduced iron. These deposits have unusually high uranium grades, with important by-product gold and nickel. The somewhat similar sandstone-type deposits are most abundant in sedimentary rocks of the Colorado Plateau, parts of Wyoming, the Texas coastal plain, and in west Africa (Fischer, 1970). These deposits formed when oxidized groundwater, that had leached

uranium from surface rocks, flowed downward into aquifers, where it was reduced to precipitate uraninite. In some deposits, reduction took place along curved zones known as **roll fronts**, which represent the transition from oxidized to reduced conditions in the aquifer (Figure 7.30). Elsewhere, reduction took place around accumulations of organic material, including old logs to form the related **humate-uranium deposits**. Both deposit types are also enriched in vanadium, selenium, and molybdenum. In some areas, similar waters came close enough to the surface to deposit uranium by evaporation or reaction with other groundwaters. The largest deposit formed by these processes is Yeelirrie, which is in arid Western Australia.

The final important type of uranium deposit, the **paleoplacer**, is quite different. Paleoplacers (Pretorius, 1981) are quartz-pebble conglomerates containing small grains of uraninite (Figure 7.30). Because uraninite is so heavy and occurs as rounded, clastic grains, these deposits are thought to have formed as placers; because they are preserved in ancient rocks, they are called paleoplacers. The largest deposits of

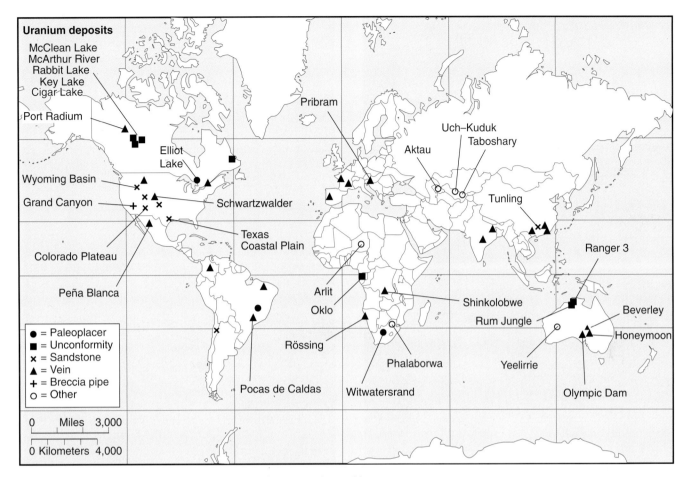

Figure 7.29 Distribution of major types of uranium deposits in the world

this type are at Elliot Lake, Ontario, where several separate conglomerate layers contain mineable ores (Figure 7.29). Similar deposits, in which gold is the dominant mineral and uraninite is a by-product, are found in the Witwatersrand of South Africa and parts of Brazil (Figure 7.29). All of these deposits are pre-middle-Proterozoic in age (pre-2.2 Ga before the Great Oxidation Event) and all formed as stream gravels in thick sedimentary sequences around the early continents. The quartz pebbles and uraninite were eroded from veins and pegmatites in granitic intrusive rocks in the continents. Because uraninite dissolves fairly easily in modern oxygen-rich stream waters, the abundance of these placer deposits in early Proterozoic time is thought to reflect lower oxygen contents in the atmosphere of the early Earth (Holland, 2005).

Uranium is also found in a wide range of veins, stockworks, and breccia pipes (Rich *et al.*, 1977; Dahlkamp, 1989) and it has been produced as a by-product from solutions used to leach low-grade porphyry copper ores and those used in processing phosphate deposits. Several countries have tried, largely in vain, to recover uranium from seawater.

Mining and processing of uranium

Uranium has been produced from open pit and underground mines, and by *in situ* solution mining (*in situ* leach, ISL). The most important hazard unique to uranium mining is the risk of cancer caused by inhalation of radioactive aerosol particles that become lodged in the lungs. This problem is worst in underground mines, where particles of rock dust and diesel fumes are more abundant. The radioactivity of these particles is due largely to **radon**, a gas produced by radioactive decay of uranium and thorium. Radon is liberated during blasting and most of it decays within hours to days to other isotopes that are adsorbed onto the aerosol particles (George and Hinchliffe, 1972). When the particles are inhaled, they are adsorbed onto the lung lining, causing lung cancer and other respiratory diseases. Lung cancer was widespread among workers in the early uranium mines of the western United States during the 1940s and 1950s. Increased mine ventilation, the use of air filters and respirators, and fewer smokers among the miner population has greatly mitigated this problem in modern uranium mines. Direct radiation from ore in the walls of uranium mines is not usually a

Figure 7.30 (a) Sandstone uranium ore containing the bright yellow ore mineral, carnotite, which formed when uranium in the sandstone was oxidized by groundwater. (b) Conglomerate gold ore from the Witwatersrand sedimentary sequence in South Africa consists of large pebbles of quartz in a matrix that contains pyrite, gold, and uranium minerals. Similar conglomerates in the Elliott Lake Group in Ontario, Canada, lack gold but contain uranium minerals. See color plate section.

Figure 7.31 (a) Uranium mining by ISL. The barrels cover well heads that pump oxidizing water into the outer perimeter of the buried uranium deposit. The water flows through the deposit, leaching uranium, and then returns to the surface through wells in the center of the field. The entire field is surrounded by monitoring wells to detect and prevent any leakage of leach fluid beyond the deposit (photo by the authors). (b) The Pickering nuclear power plant on the north shore of Lake Ontario in Canada contains six operating and two inactive CANDU nuclear reactors with a total output capacity of 3,100 MW. The reactors are enclosed in the eight domed buildings, which are constructed of reinforced concrete. The large, light colored cylindrical building is a vacuum enclosure that is connected to the eight reactor enclosures. Any accidental release of radioactive steam from the reactors would be sucked into the vacuum enclosure rather than escaping into the atmosphere (photo courtesy of Ontario Power Generation). See color plate section.

problem, although some very high-grade unconformity deposits require special precautions (Thompkins, 1982). The use of ISL increased in the 1990s and generated about half of all produced uranium as of 2013. The ISL technique essentially reverses the natural process of roll-front deposit formation. An oxidizing solution is pumped underground on the perimeter of a uranium deposit, where insoluble U^{4+} is oxidized to soluble U^{6+}, which is pumped to the surface. ISL production has a small surface footprint, and eliminates the waste associated with open pit and underground mining operations and occupational hazards of active mines (Figure 7.31).

Tailings from open pit and underground uranium mines are also considered a radiation hazard because only about 20%

of the radiation is actually removed from the ore during beneficiation. About 200 million tonnes of tailings are located at more than 15,000 abandoned and active mining sites in the United States (Krauskopf, 1988; http://wwwabandonedmines.gov/). Although the overall radioactivity of mine wastes is lower than natural ore, their powdered form facilitates escape of radon and caused problems with early efforts to use waste rock and tailings in local construction projects. The Uranium Mill Tailings Control Act of 1978 called for clean-up of these wastes at abandoned mines in the United States. The act was updated in 1993 to require all inactive licensed mining sites to define and remediate environmental contamination related to the wastes. Environmental regulations require that the wastes be disposed of underground or in piles on the surface, surrounded and covered by impermeable clay, and that the wastes not contaminate local groundwater aquifers (Gershey et al., 1990; EPA, 2014). In spite of these relatively simple precautions, uranium mining has been banned, at least temporarily in parts of Australia and Canada, as well as on Navajo land in the United States (Yurth, 2012).

Uranium processing involves three main steps (Duderstadt and Kikuchi, 1979). The first step, separation of uranium from the ore, is carried out at the mine by a chemical leaching process that yields a precipitate known as yellowcake, containing about 70 to 90% U_3O_8. Yellowcake is shipped to other locations for the second processing step, enrichment, in which the proportion of ^{235}U in the uranium is increased. Most commercial nuclear reactors require uranium fuel with 4–5% ^{235}U, an enrichment of about five to seven times the natural abundance of 0.7%. Uranium enrichment is achieved commercially by two techniques. Both involve converting uranium to a gas such as UF_6. In one technique, the UF_6 gas is piped through porous membranes with pore sizes that are roughly equal to the size of the UF_6 molecules. The UF_6 molecules with ^{235}U diffuse slightly faster than UF_6 molecules with ^{238}U, allowing separation of the two U isotopes. The second enrichment process feeds the UF_6 gas into a centrifuge, which spins at 90,000 revolutions per minute, about 100 times faster than the spin cycle of a conventional washing machine. Because ^{235}U has a mass that is about 1% lighter than ^{238}U, the lighter ^{235}U is concentrated in the center of the centrifuge and the heavier ^{238}U is displaced toward the walls of the centrifuge. Both techniques produce ^{235}U enriched to a level suitable for commercial nuclear power plants. Enrichment by gas centrifuge consumes about 50 times less energy than gaseous diffusion. A third technique that uses laser energy to enrich ^{235}U from UF_6 gas is not yet commercially viable. Uranium enrichment has been carried out in the United States by government facilities such as the Oak Ridge Laboratories, but all government enrichment ceased in 2013, making the United States dependent on imports for enriched material (Wald, 2013).

Uranium production, resources, and trade

Uranium mine production is about 63 million kg annually, whereas annual consumption is about 90 million kg. After the dissolution of the Soviet Union, excess consumption was satisfied by uranium from reprocessed fuel from nuclear weapons in a program called "Megatons to Megawatts." In 2013, the United States received its last shipment of enriched uranium from Russia, and mine production will likely have to increase to satisfy current nuclear plants and the 23 new reactors that are proposed to come online in the United States between 2020 and 2030 (NRC 2015). The anticipated drawdown of the uranium inventory stimulated increased exploration in the early twenty-first century, with exploration budgets approaching those of the 1970s.

Kazakhstan and Canada are by far the largest producers of uranium, accounting for about one-third and one-fifth of global production, respectively, followed by Australia, Niger, Namibia, Uzbekistan, and the United States. In the United States, uranium is mined in Utah and Wyoming, with a total output of about 680 kg /year. The White Mesa mine in Utah is a conventional underground uranium mine, and Smith Highland Ranch in Wyoming is an ISL mine. Reasonably assured reserves of uranium, available at a price of $130/kg of U_3O_8 or less, are about 5.3 million tonnes, largely in Australia, Khazahkstan, Russia, Canada, Niger, and South Africa. According to the World Nuclear Association, in 2013 reasonably assured reserves at $260/kg are about 7.1 million tonnes uranium (IAEA, 2011). However, some of this uranium is a by-product of other mining; for instance, about half of the Australian reserve is in the Olympic Dam copper deposit. Thus, production of uranium also depends on the price of other mined commodities.

A high import reliance for uranium is common in MDCs. The United Kingdom, Germany, and Japan have been almost completely dependent on imported fuel, and only Canada, among consuming nations, supplies all of its needs. Although this situation is similar to that of the fossil fuels, it does not elicit much comment because of the lower value of uranium production and because imports come, in part, from politically stable countries. Many non-US uranium producers formed a cartel similar to the Organization of the Petroleum Exporting Countries (OPEC), known as the Uranium Producer's Club, which almost caused the bankruptcy of Westinghouse Electric in the 1970s, when prices rose so much that Westinghouse could not supply uranium fuel that they had promised as an incentive for sale of reactors (Taylor and Yokell, 1979). The story is different today, where uranium is traded freely on global markets.

Uranium markets, reactors, and reserves

Reactor fuel is the only market on which uranium can depend if it is to remain an important mineral commodity (World

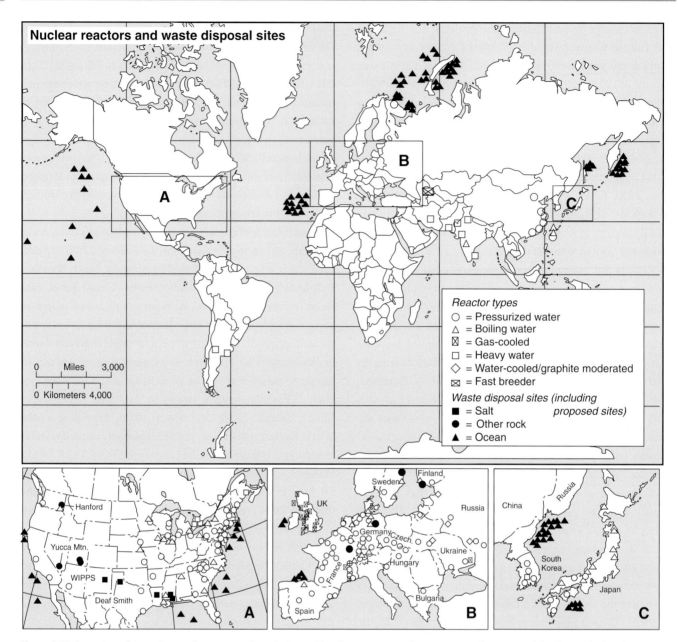

Figure 7.32 Location of operating nuclear power plants in 2014. Also shown are sites that are or were being tested for disposal of nuclear wastes, as well as sites where radioactive materials have been dumped in the ocean. The Asse site (black dot in Germany) has been used so far only for low-level wastes, whereas the Yucca Mountain site has been considered for high-level wastes (from Department of Energy, the International Atomic Energy Agency, and the World Nuclear Association).

Nuclear Association, 2015a). Thirty countries generate electricity using nuclear fuel (Figure 7.32). The United States is the world's largest producer of nuclear electricity measured by total output, with 104 operating reactors in 31 states providing about 20% of US electricity, or about one-third of global nuclear power (World Nuclear Association, 2015b). France and Canada are the second and third largest nuclear power generators by total output, with 58 and 19 reactors supplying 75% and 15% of their electricity, respectively. Nuclear power is also a major component of the electrical energy budget of Ukraine and Slovakia, which each generate about half of their electricity from nuclear fuel (IAEA, 2011). Global nuclear power generation decreased by about 10% between 2010 and 2012, mostly in Japan following the Fukushima disaster. As of 2013, there were about 437 operating reactors, with another 71 under construction, 173 planned, and 314 proposed (Table 7.7). The future of nuclear energy depends on public perception of the safety of nuclear reactors, and an

Table 7.7 Nuclear power plants operating, under construction, planned, and proposed (data from World Nuclear Association as of 2015). The table includes operable reactors in Japan that were temporarily shut down following the Fukushima Incident in 2011.

Country	Operating reactors	Generating capacity (MWe)	Percent of generating capacity	Reactors under construction	Reactors planned	Reactors proposed	Uranium enrichment facilities	Uranium required in 2014 (tonnes U)
United States	99	99,361	19.1	5	5	17	Yes	18,816
France	58	63,130	73.3	1	1	1	Yes	9,927
Japan	48	42,569	1.7	3	9		Yes	2,119
Russia	33	24,253	17.5	10	31	10	Yes	5,456
South Korea	23	20,656	27.6	5	6	0	Yes	5,022
India	21	5,302	3.4	6	22	35	Yes	913
China	20	17,055	2.1	29	59	118	Yes	6,296
Canada	19	13,553	16	0	2	3	No	1,784
United Kingdom	16	10,038	18.3	0	4	7	Yes	1,738
Ukraine	15	13,168	43.6	0	2	10	Yes	2,359
Sweden	10	9,508	42.7	0	0	0	No	1,516
Germany	9	12,003	15.4	0	0	0	Yes	1,889
Spain	7	7,002	19.7	0	0	0	No	1,274
WORLD	435	375,303	11.0	72	174	299		65,908

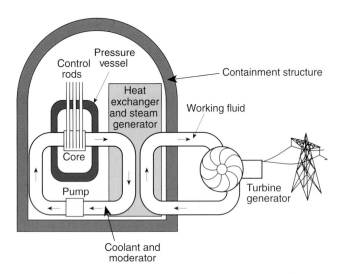

Figure 7.33 Schematic illustration of a nuclear power plant. As discussed in the text, some reactors do not use a heat exchanger and pass coolant fluid directly to the generator turbine. In gas-cooled systems and most breeders, the reactor core is surrounded by graphite rather than water as shown here.

understanding of the controversy requires a brief discussion of reactor design.

In a nuclear power plant (Figure 7.33), the reactor provides heat, which boils a **working fluid**, driving an electric turbine (Duderstadt and Kikuchi, 1979; Kessler, 1987). In some power plants, the working fluid is separated by a steam generator or heat exchanger from another fluid that circulates through the reactor core. The nuclear reactor itself consists of fuel in some form and a **moderator**, which controls the rate of neutron flux through the fuel and often acts as a coolant. All reactors produce some fissile material by neutron bombardment of ^{238}U and ^{232}Th in the fuel or around it. In practice, reactors are divided into **converter reactors**, which produce less fissile material than they consume, and **breeder reactors**, which produce more than they consume. Most commercial reactors in use today are converters, and these are further divided into water-cooled and gas-cooled types. These factors control the life span of world uranium reserves because each reactor type has different uranium requirements per kilowatt of electricity generated.

Water-cooled converter reactors come in four main types. In the United States, the most common types are **boiling water reactors** (BWRs) and **pressurized water reactors** (PWRs), both of which use enriched uranium fuel and a light-water coolant and moderator. Because they have only one fluid cycle, BWRs are more efficient, although they require larger shielding because water expands to steam in a single cycle, going directly from the reactor core to the turbines. PWRs

employ two fluid cycles in which the reactor core is separated from the turbine by a heat exchanger, but they require containers and buildings with thicker walls to protect against possible steam explosions. PWRs account for about 65% of US nuclear power capacity, with the rest in BWRs. **Heavy-water**-cooled reactors (HGRs), such as the Canadian **CANDU reactor**, use heavy water as the moderator and coolant, which permits the use of unenriched uranium fuel, a major advantage for a country without uranium enrichment facilities (Figure 7.31). **Light-water, graphite-moderated reactors** (LWGRs) are used only in the former Soviet Union, where they make up about half of the generating capacity. They are relatively inexpensive to build and operate, but have a higher safety risk because the graphite core can burn in the presence of the water coolant, producing explosive hydrogen gas. The Chernobyl reactor, which exploded and burned in 1986, was of this type. The 12 LWGR reactors operating in Russia as of 2013 were all modified to meet more stringent safety requirements. The LWGRs operating in Ukraine and Lithuania were closed in the early 2000s. Gas-cooled converter reactors, which are used principally in the United Kingdom, include the Magnox reactor, which uses unenriched fuel and a solid graphite moderator, and the advanced gas-cooled reactors (AGRs), which uses enriched fuel. The high-temperature, gas-cooled (HTGR) system, with a helium coolant, has not been used widely in the present generation of commercial reactors.

Most of these reactors were built before about 1990 and are generation II types. Generation III reactors, which began to be built in the late 1990s, include evolutionary changes that emphasize passive safety. One of these, the AP 1000, a two-loop PWR has 50% fewer safety-related valves, 35% fewer pumps, 80% less safety-related piping, and 85% less control cable than earlier versions (Schultz, 2005). Four of these are under construction in China and two have been approved in the United States. Another, the Economic Simplified BWR, has no pumps and is designed to remain stable even in a complete failure of power such as crippled the older reactors at Fukushima, although none have been built. Generation IV reactors are more revolutionary in design and are still being tested (Hylko and Peltier, 2010).

Fast-neutron reactors, also known as breeder reactors, are based on the conversion of ^{238}U to ^{239}Pu, although naturally occurring ^{232}Th can be converted to fissile ^{233}U, and ^{240}Pu can be converted to ^{241}Pu. Breeders use enriched fuel surrounded by a shell of ^{238}U, which allows about 1.5 ^{239}Pu atoms to be produced by each fission event, more than enough to fuel another reactor. These reactors use uranium about 60 times more efficiently than conventional reactors, and could greatly extend the useful life of world uranium reserves. Breeder reactors also offer the ability to burn actinides, a long-lived component of conventional nuclear waste. Unfortunately, commercial development of breeder reactors in the United States effectively ended in the 1970s when the Carter Administration banned the use of ^{239}Pu as a nuclear fuel, instead requiring all ^{239}Pu generated from nuclear reactors to be stored. The outlook for breeders remains poor. They require concentrated fuel and use liquid sodium coolant, which is highly explosive. Breeders have been lightning rods for economic and environmental concern. The Clinch River project in the United States was never built, and the Kalkar and Dounreay reactors in Germany and Scotland, respectively, were shut down in the 1990s. A significant challenge at the Clinch River project involved the cost of operations, which required uranium prices of $450–1,350/kg to recover capital expenses for construction of the breeder reactor. The Phenix and Superphenix reactors in France operated commercially until the late 1990s, but each experienced numerous technical problems, downtime, and significant cost overruns related to fuel reprocessing. They were closed in 2009 and 1998, respectively. Japan's Monju breeder reactor produced power for only the year 1995, and the Japanese breeder program lies dormant owing to significant government research budget cuts (Cochran et al., 2010). The abundant reserves of uranium and proliferation risks make it unlikely that breeder reactors will be able to compete economically or politically with converter reactors in the foreseeable future.

Concerns about reactor safety have limited nuclear power's growth (Figure 7.34). Most attention has focused on emissions during normal reactor operation or failure of some type. It has been demonstrated repeatedly that properly operating reactors are not important sources of radiation, so the problem boils down to what might happen during failure of the system. The principal risk is that of coolant failure, which would cause the fuel to become hot enough to melt. Such an event would create hundreds of radionuclides, some with short half-lives, but many long-lived fission products that could be released to the environment (Burns et al., 2012). Short-lived isotopes such as ^{131}I, ^{134}Cs, ^{137}Cs, and ^{90}Sr are highly radioactive and would have immediate environmental effects, and long-lived isotopes such as ^{129}I and ^{135}I, though less radioactive, would persist for millions of years (Ewing and Murakami, 2012). Failures of this type have been mercifully rare and reassuring. In the 1979 accident at Three Mile Island, Pennsylvania, 27 tonnes of fuel melted and ponded in the base of a PWR containment vessel without any significant radiation leak to the environment (Toth et al., 1986). The tragedy at

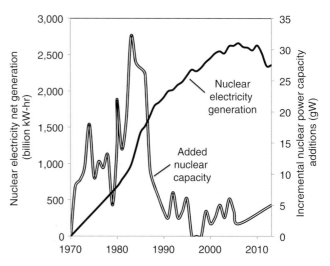

Figure 7.34 Growth of nuclear energy generation for the period 1970 to 2013 showing that the total generation increased; however, the installation of new nuclear capacity decreased because of the lack of construction of new nuclear facilities after the Three Mile Island incident (compiled from data from the EIA and World Nuclear Association)

Chernobyl, Ukraine, which resulted in hundreds of fatalities, was a special, hopefully unique, case. The Chernobyl reactor was an LWGR with a graphite core and no shielding structure, a configuration that was not allowed in the west. It was operating outside its own safety regulations at the time of the accident, when hot fuel caused the graphite to burn, dispersing radioactive wastes throughout northern Europe (Schoenfeld, 1990; Burns et al., 2012).

The Fukushima Daiichi incident resulted from the partial meltdown of the core of three of the six BWRs at that plant. Fortunately, prior to this incident, the fuel had been removed from one of the reactors and two were in temporary shutdown mode. The meltdown of the three operating reactors was caused by ocean water that flooded the power supplies to the reactors, disabling cooling and other pumps. The flooding was caused by an exceptionally large, magnitude 9.0, earthquake off the northeast coast of Japan that triggered a tsunami, which pushed a wave measuring 13 m over a protective 10-m seawall surrounding the power plant. Immediately following the earthquake, each of the three operating reactors went into planned safety mode by inserting control rods to stop the fission reactions. However, the tsunami flooded rooms that housed the electronic controls and coolant systems. A series of events then occurred, resulting in explosions of hydrogen gas and ultimately partial melting of the three cores. The human and environmental impacts of this disaster are not likely to be fully quantified for many years.

BOX 7.10 | ELECTRICITY FROM NUCLEAR VERSUS FOSSIL FUEL PLANTS

The Fukushima incident focused world attention on the safety of nuclear power. Even with these anomalous situations, nuclear power is safer than other types of power, particularly when all aspects of the production process and greenhouse-gas emissions are considered. The relationship between human health and electricity generation, from a fuel-cycle perspective, has been the subject of many studies. Karecha and Hansen (2013 compiled all historical electricity and air-pollution-related mortality data, and concluded that nuclear power prevented upwards of 1.8 million deaths between 1971 and 2009, relative to electricity generated by fossil fuels. The lower mortalities are related to significantly lower emission of greenhouse gases from nuclear plants. For example, nuclear power generation for the period 1971 to 2009 emitted 4 gigatonnes of CO_2-equivalent greenhouse gases. In the absence of nuclear power, the use of coal to generate the same amount of electricity would have resulted in emission of 64 gigatonnes of CO_2-equivalent. Markandya and Wilkinson (2007) assessed the health effects of using lignite, coal, gas, oil, and nuclear to generate electricity and concluded that the use of nuclear fuel to generate electricity is overall much safer for miners and the general public than using fossil fuels.

In spite of reassuring operating statistics, public concern about nuclear power has grown since the late 1970s. Sweden held a national referendum in 1980, which called for the country to phase out nuclear power. However, a new national energy policy adopted in 1997 allowed Sweden to continue operating 10 nuclear power plants that provide 40% of its electricity, and the Swedish Parliament voted in 2010 to repeal the phase out (World Nuclear Association, 2015b). In 1990, Italy closed its nuclear power plants, even though it caused a significant increase in imported electricity. Ironically, 10% of Italy's electricity is imported from neighboring countries that use nuclear fuel to generate the electricity. In 2011, a national referendum rejected new efforts to generate nuclear power in Italy. Although some of this concern resulted from nuclear accidents, over half of US generating capacity had already been canceled before the Three Mile Island accident. These early cancelations reflected escalating reactor construction costs caused by government-mandated, safety-related design changes. Costs for many nuclear projects exceeded original estimates by five to ten times (Cook, 1985). In reality, the Three Mile Island incident could be viewed as validation that its reactor design prevented a core meltdown. Interestingly, the French program, which started later than the US program, did not delay adding nuclear capacity to its electrical grid. France currently generates three-fourths of its electricity from nuclear power, and has a national policy to use nuclear power to ensure energy security in light of the fact that the country relies on imports of fossil fuel to generate about half of its energy needs.

Estimates of global uranium reserves continue to increase and are adequate to provide fuel to conventional nuclear reactors, those that require fuel enriched to about 5% ^{235}U, for about 100 years. Reactors such as the CANDU pressurized heavy water reactor, which uses natural uranium that does not require an enrichment process, and uses slightly less than half of the uranium required for light-water reactors, would allow current reserves to last several hundred years. An additional advantage of the CANDU reactor is its ability to use ^{232}Th as a nuclear fuel. Thorium is a by-product of many other mining operations, including rare-earth metals, and is typically treated as an environmental waste product owing to the lack of industrial uses.

If nuclear power continues to provide the world with electricity, it is likely that future reactor designs will utilize both ^{235}U and ^{232}Th, reprocess fuel that contains ^{239}Pu, and be built with passive safety features to reduce the risk of failure. In 2013, the US Department of Energy funded a new program to design and build small, modular reactors that can be transported easily and assembled in areas without the need for large-scale nuclear power (Wald, 2012). The small size of these reactors, about one-sixth of the power output of conventional reactors, allows them to be cooled by the water used to circulate steam to turbines, which means that they do not require secondary cooling systems in case of an emergency. The reduced size decreases up-front building costs, and the small size and transportability makes these modular reactors ideal candidates to partner with renewable energy sources such as wind and solar, which provide intermittent power and require a partner energy source.

Nuclear waste disposal

Nuclear reactors produce radioactive wastes that give off **alpha particles**, which can be stopped by a sheet of paper or a few inches of air, **beta particles**, which are stopped by thin

Table 7.8 Summary of estimated nuclear waste in the United States (Ewing, 2006; Waste and Materials Disposition Information, Office of Environmental Management, US Department of Energy). The cumulative amount of nuclear fuel produced in the United States is equal to a football field covered to a depth of 6 m, and annual waste from all US reactors contributes an additional 0.3 m (Jogalekar, 2013).

Type of waste	Sources/important isotopes	Amount	Radioactivity (MCi)
High level	Reprocessing of spent fuel; ^{137}Cs, ^{60}Co, ^{235}U, ^{238}U, $^{239-242}Pu$	380,000 m^3	2,400
Spent fuel	Defense; ^{137}Cs, ^{90}Sr, ^{144}Ce, $^{239-242}Pu$	2,500 tHM	
Spent fuel	Power plants; ^{137}Cs, ^{90}Sr, ^{144}Ce, $^{239-242}Pu$	61,800 tHM	39,800
Transuranic, nuclear weapons	Reprocessing waste with Pu; ^{241}Am, ^{244}Cm, $^{239-242}Pu$	71,570 m^3	14
Low level	Power plants, laboratories fission products, ^{235}U, ^{230}Th	6,200,000 m^3	50

MCi = megacuries; tHM = metric tonnes of heavy metal.

metal or a few meters of air, and **gamma rays**, which require centimeters of lead or even more rock shielding. Alpha particles do the most damage because they are the largest, but gamma rays have greater penetration. Wastes are commonly classified according to origin (Table 7.8) and include gases, dilute solutions, and solids (Murray, 1989). Civilian nuclear wastes are classified as low-level (LLW), intermediate- (ILW), and high-level waste (HLW). According to the World Nuclear Association, annually, about 200,000 m^3 of combined LLW and ILW, and 10,000 m^3 of HLW are produced worldwide. LLW constitutes about 90% by volume of all nuclear waste but only about 1% of total radioactivity. The half-lives of LLW are short and do not need shielding or cooling; this waste is disposed of in shallow landfills, typically after compaction or incineration. ILW comprises about 7% by volume and 4% of total radioactivity. It requires shielding, but not cooling and is disposed of in regulated landfills after mixing with concrete or bitumen. HLW comprises the fission products generated in the reactor core. Although HLW is only 1% volumetrically of all nuclear waste, it represents 95% of total radioactivity, and requires shielding and cooling during disposal. Half-lives of many HLW isotopes are so long that they must be isolated for thousands to millions of years before they have decayed to a safe level. It is the 380,000 m^3 of existing HLW in temporary storage worldwide that pose the biggest challenge to long-term nuclear waste storage.

In general, the isolation problem breaks down into two questions: what is the best form for the waste and where should it be put? It was originally thought that spent-fuel reprocessing would be used to extract plutonium for use in breeder reactors, and that all waste would therefore be put into solution to facilitate reprocessing. This did not occur for US electric power reactors, where concerns about the toxicity and security of plutonium led to the previously mentioned ban on reprocessing of plutonium from US reactors. This

questionable decision was largely aimed at reducing the threat of nuclear weapons proliferation rather than maximizing the energy return of the nuclear fuel cycle or minimizing the human health effects of plutonium. Plutonium is less toxic than cadmium, lead, or arsenic, which do not disappear by radioactive decay. It is of greatest concern when it is inhaled where alpha decay particles will damage lung tissue, possibly causing cancer (NEA, 1981). Furthermore, large volumes of plutonium in US defense wastes continue to be recycled, as are wastes from commercial reactor fuel in most European countries. Nevertheless, **spent fuel** from commercial reactors in the US and Canada continues to be stored, largely in water-filled tanks at reactor sites, awaiting final decisions on its ultimate disposal form and location.

The disposal form for HLW will almost certainly be some type of solid because it is more compact and isolates the isotopes from the hydrosphere and biosphere (Tang and Saling, 1990). Most consideration has focused on sealing wastes in a container made of cement, glass, ceramic, or synthetic mineral or rock (known as **synroc**). The French disposal system uses a borosilicate glass. Some Swedish wastes are put in copper drums, because archaeological artifacts made of native copper are preserved for millennia, although concern has been expressed that the copper might be an attractive target for future, resource-starved populations.

Disposal locations for radioactive wastes have been studied for years, with attention shifting from improbable disposal in the ocean or beneath polar ice caps, to the need to put the waste back into the rock from which it came (Chapman and McKinley, 1987; Krauskopf, 1988). Rock disposal locations are favored because they have a better chance of remaining undisturbed until it can decay to an acceptably low level of radioactivity. Rock storage sites must have low porosity and permeability, be free from earthquakes or other natural disturbances and permit reentry to recover the wastes if necessary. Studies of the Oklo uranium deposit in Gabon have been cited as support for rock storage. The proportion of ^{235}U in some uranium from Oklo is much lower than in normal uranium, apparently because it was consumed in natural fission reactions that occurred about 2 billion years ago, shortly after the deposit formed and while it was still deeply buried. This provided a natural laboratory to study the dispersion of fission products in rock. Although the analogy to commercial reactors is not perfect, data from Oklo suggest that non-gaseous fission products are adsorbed onto clays and other minerals in the surrounding rocks, and do not travel far from their source, a result that is encouraging for rock storage (Gauthier-Lafaye, 2002).

Rocks evaluated so far as disposal sites include salt, granite, and volcanic rock. In the United States, an amendment to the Nuclear Waste Policy Act of 1982 assessed $0.001/kW-hr of nuclear power to fund studies that would identify the best HLW disposal site in the country. Of the nine sites chosen for the original study, three locations were investigated as the most geologically plausible long-term storage sites for HLW. These were rhyolite **tuffs** at Yucca Mountain in the Nevada Test Site (Figure 7.35a), the location of early nuclear weapons testing, basalt lava flows at Hanford in Washington, and salt layers in Deaf Smith County Texas. In 1987, Congress stopped the studies as a cost-cutting measure, and Yucca Mountain became the nation's "winning" disposal site by default. Yucca Mountain has some very attractive characteristics. Unlike other disposal sites, where the water table is near the surface and isolation of the waste depends on impermeability of the host rock, the water table is 500 m below the surface at Yucca Mountain and is likely to stay there for thousands of years. In addition, volcanic tuffs that underlie Yucca Mountain contain minerals that react with and adsorb dissolved ions, making migration of waste products less likely. In spite of the geological attractiveness of Yucca Mountain, the site became the locus of two decades of local, state and national political and legal battles. In 2009, the Department of Energy withdrew plans for long-term storage at Yucca Mountain and a second fatal blow was delivered in late 2013 when a federal appeals court ruled that the Department of Environment must stop collecting the nearly $750 million in annual fees paid by nuclear electricity consumers. The 0.001 cent/kW-hr fee generated about $37 billion over its 30-year history, with $7 billion having been spent on Yucca Mountain.

The United States currently has no viable long-term storage facility for HLW now at dozens of sites, and there is no energy fee to generate funds to study a long-term

solution. Unfortunately, the situation is similar in most other MDCs; with the possible exception of Finland and Sweden, no permanent storage solution has been implemented despite the 10,000 m^3 of HLW that continue to accumulate annually.

7.2.2 Geothermal energy

Geothermal energy is the small fraction of Earth's thermal heat that we are able to use, largely in the form of natural hot water and steam in porous, permeable reservoirs (Figure 7.35b). Where the water in these reservoirs has a temperature of at least 150 °C, enough of it will flash to steam when it is pumped upward toward the surface to run turbines that generate electricity. Where the water is not that hot, it can still be used for residential and industrial heating, as is done in China, Italy, Iceland, and towns as widely separated as Klamath Falls, Oregon, and Ebino, Japan (Gupta, 1980; Bowen, 1989). As of 2013, geothermal electric generating facilities in 27 countries throughout the world had a total power output of about 11,765 MWe, and geothermal heating in at least 43 countries accounted for another 48,493 MWt (thermal power) (according to the Geothermal Energy Association website, http://geo-energy.org/currentUse.aspx; see also Lund and Bertani, 2010; IEA, 2011; Boyd *et al.*, 2011). That amount of geothermal energy is roughly equivalent to annual consumption of 1.3 Bbbl of oil, which is about 30 times less than the 32 Bbbl oil consumed worldwide during 2013.

Hot groundwater reservoirs are simply active analogs of the hydrothermal systems that form many mineral deposits, and they have been given the special name geothermal systems (Rybach and Muffler, 1981). Geothermal systems are liquid-dominated if most of the pores in the rock are filled by water, or vapor-dominated if the pores are largely filled by water vapor (steam) (White *et al.*, 1971). Vapor-dominated systems tend to have temperatures of about 240 °C, whereas water-dominated systems can be as hot as 360 °C, although actual temperatures depend on the depth of the reservoir and the resulting pressure, as shown schematically in Figure 7.36. Note in this figure that the boiling curve for water, which separates liquid water from water vapor, is at higher temperatures than most geothermal gradients, the natural rate at which temperature increases downward in the crust. In fact, natural gradients reach typical geothermal temperatures of 250 °C only at depths of 3 to 6 km, considerably deeper than most geothermal fields. It follows that hot geothermal fields must be associated with unusually high geothermal gradients,

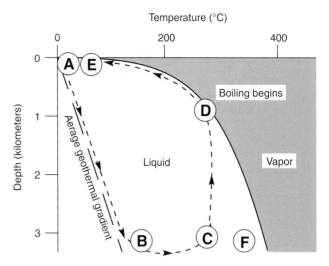

Figure 7.36 Boiling curve for water showing the path of geothermal fluid in a liquid-dominated system. Pressure is represented by the depth of an equivalent column of water (hydrostatic pressure). Recharge water begins at the surface (A) and flows downward in an area with a normal geothermal gradient to reach a temperature of about 100 °C at a depth of 3 km (B) where it is heated and becomes buoyant enough to rise (C). At (D) (whether in the rock or a well), rising water intersects the boiling curve, forming vapor that could be used to drive a turbine. By the time water reaches the surface (E), it has cooled almost back to conditions represented by (A) (modified from Muffler *et al.*, 1977).

Figure 7.35 (a) Yucca Mountain, Nevada has been investigated extensively as a possible repository for high-level nuclear waste. It is in the southern part of the Nevada Security Site, a 3,500 km² area in southern Nevada that was used for testing of nuclear weapons, including 100 atmospheric and 728 underground tests. The drill rig on the left side of the photo was obtaining cores of rock to test properties, including rock type, fracture density, permeability, porosity, and mineral composition, which might affect migration of nuclear waste (courtesy of Eugene I. Smith). (b) Yellowstone Park is one the largest geothermal areas in the world, with numerous boiling springs and steam vents. Geothermal areas make up only about 0.3% of the park, but some of these areas have heat flow of 100 W/km², 2,000 times higher than average heat flow in North America. Geothermal power has not been developed in the park because it would decrease the flow of the hot springs and steam vents (photo by the authors). See color plate section.

usually where magmas have come near the surface (Figure 7.36). Thus, land-based geothermal systems are most common in areas of recent volcanic activity above subduction zones, as in New Zealand, Philippines, Mexico, Japan, Italy, Kamchatka, Russia, and the western United States (Figure 7.37, Table 7.9). Similar systems are present along the mid-ocean ridges, but they are accessible to us only on Iceland where the ridge is above sea level.

Other countries with developing geothermal electricity capacity include Costa Rica (208 MWt), El Salvador (204 MWt), Nicaragua (104 MWt), Russia (97 MWt), Papua New Guinea (56 MWt), Portugal (29 MWt), and Germany (29 MWt).

The simplest geothermal system consists schematically of a heat source and a porous system through which water can reach the heat (Economides and Ungemach, 1987). The dominant type of water is meteoric water, which moves downward along fractures, until it is heated enough to become buoyant and begin to rise (Figure 7.38). Many geothermal reservoirs are overlain by a low-permeability seal that prevents the hot water from leaking out too rapidly, although small amounts will reach the surface as hot springs, geysers, or even steam vents known as **fumaroles**. This seal usually consists of quartz or some other form of silica that has been precipitated in pores in the upper part of the reservoir, usually because of cooling of the geothermal fluid. In some cases, the seal is lacking, and fluid density and temperature control the structure of the system.

Geothermal power comes close to being a "free lunch," but does not make it. After all, a magma should remain hot

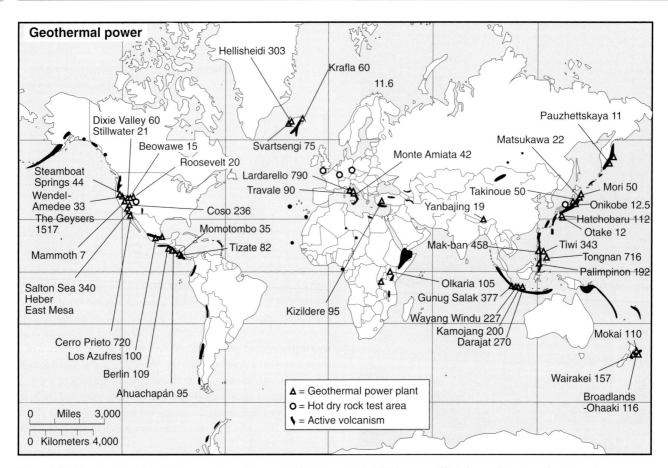

Figure 7.37 Distribution of producing geothermal areas in 2014 (capacity in MW), areas of hot dry rock tests, and active volcanoes (compiled from information supplied by L.J.P. Muffler and Ellen Lougee, from the US Geological Survey, and the International Geothermal Association)

enough to power a geothermal system for hundreds or thousands of years, long after we will have found alternate energy sources. In practice, however, it is hard to balance the rate of steam or water production to the rate at which water is recharged to the system, and the drilling and pumping perturb the chemical balance of the system. These effects can cause precipitation of calcite or silica, known as scale, which clogs piping in the systems and possibly also the natural porosity of the reservoir. This complication is most common for geothermal systems containing large amounts of gas, such as Ohaaki, New Zealand, and Ribeira Grande in the Azores, because gas release triggers calcite precipitation. These complications, along with local over-production of reservoir fluid at fields such as The Geysers, have required that new wells be drilled more frequently to maintain productivity, increasing the cost of geothermal power.

Geothermal systems emit 10–100 times less CO_2 and SO_2 than fossil-fuel power plants per megawatt of electricity generated, but are not without environmental concerns (Axtman, 1975; Bowen, 1989). Some systems contain highly saline water

that corrodes pipes and turbines. Some also deposit scale, metal sulfides, or precious metals in amounts that are too small to be of economic interest but large enough to require constant cleaning of the plumbing system. Most geothermal systems have no significant water emissions because waters are re-injected to recharge the system. This procedure, which is simply prudent operating practice, is also required by law because the waters have locally high dissolved solid contents, including toxic elements such as arsenic, antimony, and boron, which are relatively soluble in low-temperature hydrothermal water and which can enter the vapor phase in small but important amounts (Smith et al., 1987). Gaseous emissions are more difficult to contain and steam loss is common. The main hazard of gaseous emissions is H_2S, a highly toxic gas that is common in some geothermal systems, although minor metal emissions are also observed.

Although geothermal power is very important locally, it is not a major, global source of energy. Even in the United States, which has the largest installations in the world (Table 7.9), geothermal electric power generating capacity is

Table 7.9 World geothermal electric and thermal (heating) capacity (data from Lund and Bertani, 2010; and the Geothermal Energy Association). Number of developing projects includes both electric- and thermal-generating geothermal plants. Thermal applications are considerably more important in than electric power generation at present, despite the strong interest in the latter.

Country	Capacity (MW)	Number of developing projects
Installed electric generating capacity		
United States	3,093	124
Philippines	1,904	29
Indonesia	1,333	62
Mexico	1,005	5
Italy	901	5
New Zealand	895	7
Iceland	664	16
Japan	537	47
Turkey	275	60
Kenya	237	18
Installed thermal generating capacity		
United States	7,820	
Sweden	4200	
China	3690	
Iceland	1,826	
Turkey	1,500	
Hungary	654.6	
Italy	867	

only 3,093 MWt, about 3% of renewable-based electricity consumption, and about 0.2% of total electric generating capacity. California produces about 80% of all geothermal electricity in the United States, followed by Nevada at 15%, with the remainder of commercial geothermal electricity output in Hawaii, Utah, Idaho, Alaska, Oregon, and Wyoming. Large, vapor-dominated geothermal systems, such as The Geysers in California and Lardarello in Italy, produce enough electricity to power a large city, but most geothermal systems are smaller. Existing geothermal plants barely scrape the surface of global geothermal potential, however. Many attractive geothermal areas are known, but development of them has been discouraged by the same factor that challenges the oil-shale industry: low oil and gas prices. Energy-production tax credits have stimulated interest in geothermal energy, but mostly for residential geothermal heat-pump installation. Large-scale commercial geothermal electric plants require significant capital investment and tax credits have not resulted in much new installed capacity relative to total electricity consumption.

Even beyond obvious geothermal prospects, the potential is great. It has been estimated, for instance, that geothermal energy in the outer 10 km of the crust is 2,000 times greater than the energy in the coal resources of the world. We can never harness all of this energy, but some parts of it might be made available. Lower-temperature systems that will not power a turbine directly can generate electricity by use of liquid with a lower boiling point than water. The installed capacity of such binary geothermal systems increased from about 300 to 702 MWt between 1990 and 2014; however, large-scale implementation remains a challenge. Waters in sedimentary basins have attracted attention, and low-level heat might be obtained from brines in unconfined aquifers

BOX 7.11 | WHY IS YELLOWSTONE SO GEOTHERMALLY ACTIVE?

Yellowstone National Park, in northwestern Wyoming, has experienced some of the largest volcanic eruptions on Earth. In fact, some geologists refer to Yellowstone as a supervolcano because it has had major eruptions 2.1 million, 1.3 million, and 640 thousand years ago. Ash from these eruptions covered much of North America and produced a recognizable layer as far away as the Gulf of Mexico. Today, much of the park sits within the 4,000 km^2 volcanic caldera that generated much of these eruptions. Yellowstone is located over a geologic hot spot, where magma from Earth's mantle rises and heats the overlying crust. Geologists have used seismic waves from the nearly 2,000 small earthquakes that occur at Yellowstone each year to determine that a magma chamber measuring approximately 90 km long, 32 km wide, and 10 km deep lies 3.2 to 8 km beneath the surface. This magma chamber is constantly losing heat to the surrounding crust, and part of this heat is transferred to circulating groundwater to generate the 10,000 thermal features, including 500 geysers in the park (Figure 7.35b).

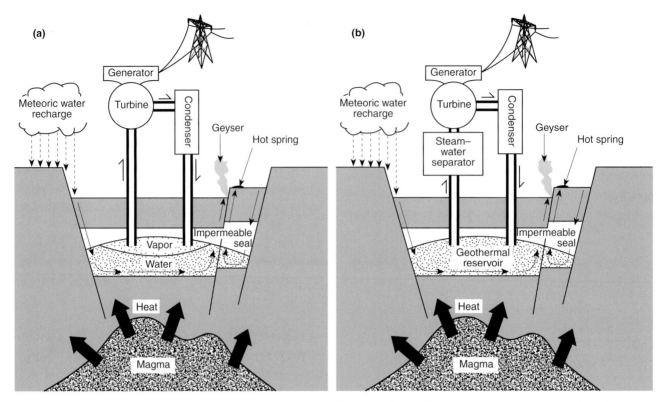

Figure 7.38 (a) Vapor-dominated and (b) liquid-dominated geothermal systems showing how meteoric water recharge is heated until it rises to form hot springs and geysers. Conditions within the reservoir are shown by A to E in Figure 7.36. Note that power generation from vapor-dominated systems does not require a liquid–vapor separator (images courtesy of L. J. P. Muffler).

(Vigrass, 1979). Warm, confined aquifers, which are known as geopressured zones because they exceed hydrostatic pressure, are common in the Gulf of Mexico basin and elsewhere, where they are an unconventional source of natural gas, as noted above (Wallace *et al.*, 1979). Attention has also been given to enhanced geothermal systems, where drilling and hydraulic fracturing of hot, dry rock can be used to enhance permeability allowing a fluid to be circulated through the rock, removing its heat. Systems of this type are most common above shallow magmas, such as in the Rio Grande Rift around Los Alamos, New Mexico, or in felsic intrusions with high uranium contents, such as in the Cornwall area of England (Garnish, 1987; Richards *et al.*, 1992). Most of these possibilities have been evaluated by drilling and pilot plants, and none of them have been shown to compete effectively with cheap fossil fuels, although they will be resurrected if fossil-fuel prices climb and concerns over greenhouse-gas emissions stimulate more government incentives for geothermal infrastructure.

7.3 The future of energy minerals

The future of energy minerals is inextricably linked to international relations and environmental concerns. Figure 7.39

shows the dramatic difference in energy self-sufficiency between major energy-producing and -consuming nations of the world. Russia, Canada, Australia, and Norway are the only MDCs that are major energy exporters. All the rest, including China, are major importers of energy. These relations are controlled by oil, which is the dominant world energy source, and they are likely to prevail for the next few decades. Some of the pressure on oil resources has been removed by the discovery of tight-oil and -gas resources, but these will only delay the much anticipated "peak oil" event, when global oil production stops increasing and actually begins to decline.

There are many among us who will applaud this event because it might slow the use of fossil fuels and their continuing addition of CO_2 to the atmosphere. While the goal of decreased CO_2 emissions is laudable, it is unrealistic to think that this can be done without large amounts of energy from conventional sources. Our job, then, from a geological perspective, is to be sure this energy is available, and that we can perfect carbon capture and storage (CCS) techniques, as others seek ways to migrate to renewable energy.

There are two ways that we might respond to the loss of oil as our major fuel. The first is to shift to natural gas and coal,

Table 7.10 Comparison of energy content of world fossil-fuel resources (data from the International Energy Agency World Energy Outlook in 2010; BP Statistical Review of World Energy in 2013; Hein *et al.*, 2013; US Energy Information Administration, 2014). Oil, gas, heavy-oil, and tar-sand resources represent technically recoverable amounts. Oil shale represents *in situ* resources. Both figures are given for coal, because some might be recovered by *in situ* gasification.

Fossil fuel	Resources	Energy content ($\times 10^{21}$ J)
Coal – recoverable	860 billion tonnes	17.2
– *in situ*	2152 billion tonnes	45.4
Oil	1,688 billion bbl	9.89
Natural gas	6,845 Tcf	7.22
Heavy oil	434 million bbl	2.65
Tar sand	650 million bbl	3.96
Tight oil	345 billion bbl	2.1
Oil shale	1000 billion bbl	6.2
Shale gas	7,299 Tcf	7.7

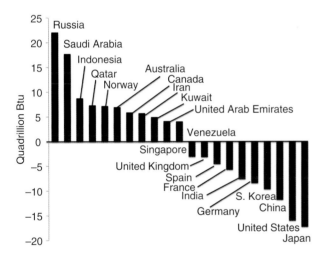

Figure 7.39 Net energy self-sufficiency of major producers and consumers (data from the US Department of Energy, International Energy Review)

which have large reserves (Table 7.10). Coal will have to bear the brunt of our demand, simply because it is more abundant. However, its bad reputation as a primary fuel means that it probably will not be used in its crude form. Instead, more emphasis will be placed on coal gasification and liquefaction. This is not a new idea. Town gas and coal oil, both of which came from coal, were widely used in the 1800s and early 1900s before oil became abundantly available (Perry,

1974). Conversion of coal to gas or liquids requires an increase in the hydrogen to carbon ratio by the removal of carbon or the addition of hydrogen, and offers the opportunity to remove contaminants such as sulfur, nitrogen, and mercury.

Most modern coal conversion plants are based on the Lurgi gasification process and their history has been only partly successful. The Great Plains Coal Gasification Plant in Beulah, North Dakota, is the only commercial-scale operating plant in the United States. It was designed to use 12,700 tonnes of lignite, 12,700 tonnes of steam, and 2,700 tonnes of oxygen to produce 137 million cubic feet of gas each day (along with 82 tonnes of sulfur, 95 tonnes of ammonia, and 818 tonnes of ash). The plant was completed in the mid 1980s at a cost of $2.1 billion, and it currently generates 4.33 million m^3 natural gas/day. A major success of the plant is the capture of 4.3 million m^3 CO$_2$/day (2.5 to 3 million metric tonnes CO$_2$ annually), which is piped 205 miles to a CCS project in Saskatchewan, Canada, where the CO$_2$ is used for enhanced oil recovery.

Another success story is the Sasol operation in South Africa, a system of three plants built to convert domestic coal into synthetic gasoline, diesel fuel, jet fuel, motor oil, and petrochemical products. Sasol was begun in 1955 in response to the worldwide embargo of apartheid South Africa, although most of its capacity was constructed after the 1970s energy crisis. Sasol plants produce about 160,000 bbl/day oil equivalent and provide South Africa with a measure of freedom from international oil suppliers. Sasol has also demonstrated the commercial viability of underground coal gasification, which can be used to convert South Africa's estimated 45 billion tonnes of coal to natural gas. Underground coal gasification is essentially undeveloped in the other parts of the world, although it has great potential. In the UK, gas generated from the large coal resources might even be distributed by parts of the pipeline system that serves declining oil and gas production from the North Sea (McGarrity, 2013). In the United States, the Powder River Basin contains an estimated 510 billion tonnes coal, of which fully 95% is not economically extractable by surface mining techniques and might be exploited by underground gasification (GasTech, 2007) combined with sequestration of CO$_2$, SO$_2$, NO$_x$, and metals.

Our second possible response will be more favorable from the standpoint of global warming. Fossil-fuel combustion currently contributes 57% of anthropogenic CO$_2$ to the atmosphere, as well as 80% of anthropogenic SO$_2$ and almost all of the NO$_x$ and significant **particulates** (IPCC, 2007). Natural gas, which contains no ash and essentially no sulfur and nitrogen offers a solution, albeit not an indefinite one

(Burnett and Ban, 1989). It has the highest hydrogen to carbon ratio of any of the fossil fuels, much higher than coal, and yields about 70% more energy for each unit of CO_2 produced. It can also be converted to liquid fuels. The bad news, however, is that methane itself is a stronger greenhouse gas than CO_2 and small leaks in extensive production and delivery systems contribute to atmospheric pollution. Furthermore, unless tight-gas resources turn out to be larger than expected, reserves are not adequate to support us for more than a few decades (Table 7.10).

Thus, our alternative will be to return to nuclear energy, or to find a substitute among the non-mineral sources such as wind, wave, or solar, all of which lack portability, the dominant advantage of fossil fuels. Although many early studies predicted that nuclear power would dominate our energy future (Figure 7.1; 7.34), the public has lost faith in reactor safety. Current generation II reactors are not fail-safe, and utilize only a small amount of the available energy in the nuclear fuel. If generation IV reactors can be deployed successfully, nuclear energy might make a much greater contribution to our energy needs, although it is highly unlikely that they could take over the entire burden from fossil fuels (Ewing, 2006).

Whatever the outcome, it is clear that we must have a meaningful national energy policy, something that has not been developed by any MDC outside France, Brazil, and South Africa. Programs were started by many countries during the 1970s, but most of them were discontinued in the face of low world oil prices and public concern over nuclear power. The only aspect of energy policy on which action continues is conservation, which is relatively painless, both economically and politically. In fact, most recent energy-related activities in MDCs have been dictated by environmental concerns rather than by the impending shortages. Deterrents such as a carbon tax to be paid for energy produced from fossil fuels or an **energy tax** on all types of energy production regardless of its source have not been successful on a global scale because of different national priorities.

Lack of progress on a national energy policy is, in part, due to confusion about the duration of the problem. In the very short term, a few decades, we must supply cheap fossil-fuel energy to bridge toward our longer-term goal of alternative energy sources. Throughout this process, we must try to decrease CO_2 emissions. To demand the long-term goal before assuring that energy is available to reach it is not a realistic approach. Finally, we must remember that modern society uses an incredible number of products made from fossil fuels. Thus, in our conversation about fossil fuels, it is critical to separate their use for energy from their many other uses.

CHAPTER

8

Iron, steel, and the ferroalloy metals

Iron and steel form the framework for civilization. Steel is a major component of cars, cans, ships, bridges, buildings, appliances, and armament. Even energy minerals are useless without furnaces, pipelines, engines, and reactors that are usually made of steel. This wide range of uses reflects the abundance of iron ore and the relative ease with which it can be converted to steel. The simplest form, carbon steel, contains less than 2% carbon, with minor manganese. Alloy steel, in which carbon is removed and other metals are mixed with iron (Table 8.1), accounts for about 15% of world production. These metals are known as **ferroalloy** metals and include chromium, manganese, nickel, silicon, cobalt, molybdenum, vanadium, tungsten, and niobium and they permit steel to be used in a wide variety of applications.

Annual world iron ore and steel production are worth about $300 billion each (Figure 8.1a). The value of ferroalloy metal production can be quoted in several ways because they are traded as ores, intermediate alloys such as ferromanganese and ferrochromium, and metals. Total world production, worth about $120 billion, is dominated by nickel, manganese, and chromium (Figure 8.1b, c). Prices of ferroalloy metals are higher than that of steel and thus increase its cost (Figure 8.1b, c). Even so, the properties that they impart are so important that consumption for most of them has increased more rapidly than steel, reflecting growth of alloy steel production (Figure 8.1d, e). Major producers and their reserves are summarized in Table 8.1 and important mines of most metals are given in Table 8.2.

8.1 Iron and steel

Steel is produced in 87 countries, making it one of the most widely produced mineral commodities (Fenton, 2011). Only about 42 countries produce iron ore, reflecting the more limited distribution of large iron deposits that can support steel making (Tuck and Virta, 2011). In view of the enormous value of world steel production, one could well ask why the steel industry has little of the glamor of the oil business? The answer lies in profits. Although steel produced some fortunes in the early 1900s, including that of Andrew Carnegie of US Steel, it has been less profitable than oil because it is more closely tied to the overall business cycle. It is much easier to put off buying a car than buying the gasoline to power it.

8.1.1 Iron deposits

Average crust contains about 5% iron, its fourth most abundant element, whereas iron deposits contain 25 to 65% (Rudnick and Gao, 2003). Concentrating iron by 5 to 12 times is an easy job for nature, which makes many different types of iron deposits. The concentration process depends, in part, on the fact that iron exists in three oxidation states in nature. In some meteorites and in Earth's core, where free oxygen is absent, iron is in the **native** (Fe^0) state, whereas in most igneous and metamorphic rocks, where free oxygen is slightly more abundant, iron is largely in the ferrous (Fe^{+2}) state. In some sedimentary rocks and soils that contain abundant free oxygen, iron is largely in the ferric (Fe^{+3}) state.

Common iron ore minerals reflect this differing oxygen availability: siderite ($FeCO_3$) contains ferrous iron and forms in oxygen-poor environments if sufficient carbonate is available, magnetite ($FeO.Fe_2O_3$) contains both ferrous and ferric iron and requires environments with intermediate oxygen availability, and hematite (Fe_2O_3) and goethite ($Fe(OH)_3$) contain ferric iron and form when free oxygen is abundant.

Table 8.1 Major producers and reserves for iron ore, steel, and ferroalloy elements (from US Geological Survey Commodity Summaries)

Commodity	Country	Production (tonnes)	Reserves (tonnes)
Iron ore	China	1,320,000,000	23,000,000,000
	Australia	530,000,000	35,000,000,000
	Brazil	398,000,000	31,000,000,000
Steel (raw)	China	670,000,000	
	Japan	108,000,000	
	Russia	51,000,000	
Manganese	South Africa	3,800,000	150,000,000
	Australia	3,100,000	97,000,000
	China	3,100 000	44,000,000
Chromium	South Africa	11,000,000	200,000,000
	Kazakhstan	4,000,000	230,000,000
	India	3,900,000	200,000,000
Nickel	Indonesia	440,000	3,900,000
	Philippines	440,000	1,100,000
	Russia	250,000	6,100,000
Molybdenum	China	110,000	4,300,000
	United States	61,000	2,700,000
	Chile	36,500	2,300,000
Cobalt	Democratic Republic of Congo	60,000	3,400,000
	China	7,000	80,000
	Canada	6,700	140,000
Vanadium	China	40,000	5,100,000
	South Africa	20,000	3,500,000
	Russia	15,000	5,000,000
Tungsten	China	60,000	1,900,000
	Russia	2,500	250,000
	Canada	2,200	290,000
Columbium (Niobium)	Brazil	45,000	4,100,000
	Canada	5,000	200,000
Tellurium	Japan	45	----
	Russia	40	----
	Canada	10	800

The common iron mineral pyrite (FeS_2), which also contains ferrous iron and requires oxygen-poor environments, is not usually mined for iron because it is more difficult to dispose of the sulfur released during smelting. Modern waters also reflect these oxidation states. Ferric iron is not soluble in surface waters and, instead, forms goethite and other "rusty" iron minerals. Ferrous iron is soluble in surface waters if they do not contain oxygen, although the modern ocean is too oxygenated to contain significant dissolved iron.

By far the most important iron deposits are chemical sediments known as **banded iron formation (BIF)** that contain alternating iron-rich and silica-rich layers (Figure 8.2a) and clastic sediments of similar composition known as **granular iron formation (GIF)** (Clout and Simonson, 2005). These iron-rich sedimentary rocks, which we will refer to here as BIF, are also known as **taconite** in North America, **itabirite** in Brazil, **jaspilite** in Australia, and **banded ironstone** in South Africa.

There are two main types of BIF deposits. Algoma-type deposits, which are relatively small, and Superior-type deposits, which are much larger and are the basis for most of the world iron industry. Superior-type deposits get their name from Lake Superior, where deposits in Minnesota and Michigan have supplied almost all of US production since about 1900. Algoma-type deposits formed in small sedimentary basins with abundant volcanic rocks and Superior-type formed on continental shelves where basins were large and continuous.

Almost all of the world's BIF deposits are restricted to Precambrian rocks (Figure 8.3) and are older than about 1.85 Ga. We know that the ancient ocean and atmosphere lacked free oxygen and therefore that ferrous iron could have been dissolved in the ocean. The source of the iron is thought to have been submarine volcanism and hot springs, possibly driven by rifting caused by mantle plumes. But why was the iron deposited? Originally, it was thought that the iron was deposited during the Great Oxidation Event (GOE) (Holland, 2005), the period around 2.4 Ga when the oxygen content of the atmosphere and oceans increased, and the last time that uranium-bearing conglomerates could have formed, as discussed in Chapter 7. During the GOE, photosynthesis generated oxygen that transitioned the atmosphere and ocean to an oxic state, and this oxygen caused dissolved ferrous iron to deposit as insoluble ferric iron. But, we have since learned that many BIF deposits are hundreds of millions of years older than the GOE and some are as young as 1.85 Ga. So, either the oceans did contain a little oxygen, at least locally, to oxidize and deposit iron even before the GOE, or some other agent oxidized the iron directly. One possible agent would be photosynthetic bacteria that operated in an **anaerobic** (oxygen-poor) environment (Bekker et al., 2010). After the GOE, of course, deposition, driven by dissolved oxygen was more efficient and occurred even in the shallow seas that flooded the shelves of the newly formed continents (Figure 8.4). The fact that iron formations continued to be deposited until

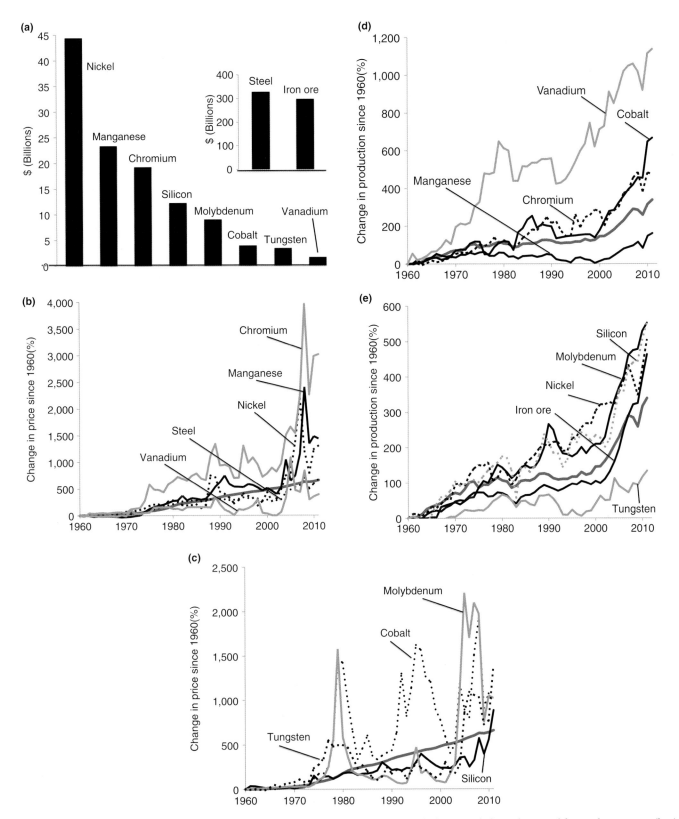

Figure 8.1 (a) Values of 2012 world iron, steel, and ferroalloy metal production, not including recycled metal or metal from other sources. (b, c) Prices of steel and major ferroalloy metals since 1960 relative to the consumer price index (CPI) (broad gray line). (d, e) Growth of world steel and ferroalloy metal production (primary mine production) since 1960. From World Steel Association and minerals statistics of Australia, China, India, Organisation for Economic Cooperation and Development (OECD), the United Kingdom, and the US Geological Survey.

Table 8.2 Large deposits of iron and ferroalloy metals. Most of these deposits are associated with large-scale sedimentary, magmatic or weathering units that do not form individual deposits. Sizes given here represent the entire systems, usually to depths estimated to be economic in the reasonable future, and they provide a comparison of the relative scale of these systems. Data from company and government reports and Laznicka (2006).

Deposit name	Location	Type of deposit	Tonnes of metal
Iron deposits (tonnes Fe)			
Carajas	Brazil	Banded Iron Formation BIF	12,000,000,000
Krivoy Rog	Ukraine	Banded Iron Formation BIF	3,000,000,000
Hamersley	Australia	Banded Iron Formation BIF	2,000,000,000
Manganese deposits (tonnes Mn)			
Kalahari	South Africa	Sedimentary	4,200,000,000
Urucum	Brazil	Sedimentary	600,000,000
Nickel deposits (tonnes Ni)			
Noril'sk-Talnakh	Russia	Magmatic immiscible	24,000,000
Sudbury	Canada	Magmatic immiscible	20,000,000
Goro and Others	New Caledonia	Laterite	65,000,000
Moa	Cuba	Laterite	10,000,000
Chromium deposits (tonnes Cr)			
Bushveld Complex	South Africa	Magmatic layered	2,600,000,000
Great Dike	Zimbabwe	Magmatic layered	325,000,000
Kempirsai	Kazakhstan	Magmatic podiform	100,000,000
Cobalt deposits (tonnes Co)			
Katanga Area	DRC*	Sediment-hosted	10,000,000
Molybdenum deposits (tonnes Mo)			
Climax	United States	Porphyry molybdenum	2,700,000
Henderson	United States	Porphyry molybdenum	1,250,000
Tungsten deposits (tonnes W)			
Shizhuyuan	China	Skarn/Greisen	600,000
Climax	United States	Porphyry molybdenum	280,000

* DRC - Democratic Republic of Congo

about 1.85 Ga shows that removal of iron from the ocean was a slow and complex process.

There are younger BIF deposits, but all of them are smaller and somewhat unusual. A few were deposited in late Proterozoic time (about 700 to 635 Ma) and it has been suggested that they are the result of a "snowball Earth" condition when floating ice from glaciers covered much of the world ocean limiting the input of atmospheric oxygen (Holland, 2005). When the ice melted, dissolved iron was deposited. Even younger Phanerozoic iron formations, such as the Clinton Formation of the Appalachians and the minette ores of Alsace-Lorraine in France, appear to have formed when iron-rich groundwater replaced carbonate sediments (Maynard, 1983).

Iron is also found in hydrothermal and magmatic deposits. Some formed from hydrothermal solutions that were released from intrusions at depths of several kilometers and **skarn** deposits at contacts with limestones (Figure 8.4). Some hydrothermal solutions that formed these deposits were magmatic, but others were simply heated meteoric waters that circulated through iron-rich rock and dissolved iron. At Bethlehem, Pennsylvania, large deposits of this type formed at the contact of mafic intrusions that rose along rifts created when North America and Europe split apart in Triassic time (Eugster and Chou, 1979). A final type of iron deposit, known as iron oxide–copper–gold (IOCG) or iron oxide apatite (IOA), is much disputed. These deposits range from the enormous Kiruna iron deposit in Sweden, which consists of massive magnetite and is the basis for Swedish industrial power, to the similarly gigantic Olympic Dam deposit in South Australia, which consists of copper, uranium, and gold minerals enclosed in hematite-altered rocks. Most geologists feel

Figure 8.2 (a) BIF consisting of layers of hematite (black) and silica (red). The small balls beside the hammer are pellets, which consist largely of hematite grains held together by clay and are the main feed for blast furnaces (photograph by Dale Austin, University of Michigan). (b) Tilden Mine, Marquette County, Michigan, which mines hematite-bearing BIF, has the capacity to produce 8 million tonnes of pellets annually. Note rail loading system with piles of pellets in foreground (courtesy of Dale R. Hemila, Cliffs Natural Resources). See color plate section.

that IOCG deposits formed from magmatic hydrothermal solutions, but opinion on IAP deposits is divided between a simple hydrothermal origin or one involving unusual iron oxide-rich magmas that separated as immiscible melts from parent silicate magmas (Groves *et al.*, 2010).

8.1.2 Iron mining and processing

World iron ore production, amounting to about 3 billion tonnes, is dominated by Australia, Brazil, China, and India. Large Superior-type deposits are present around Lake Superior, in the Hamersley Range of Australia, the Labrador Trough of Canada, the Carajás and Minas Gerais districts of Brazil, Krivoi Rog in Ukraine, and Kursk in Russia (Figure 8.3). Almost all iron ore is mined by open pit methods (Figure 8.2b), although Kiruna and a few other deposits have been mined underground. Mining of fresh BIF is difficult because its high silica content makes it very hard. Enriched ore, which contains as much as 60% iron in contrast to about 25% in fresh BIF, is formed when silica is removed by later hydrothermal or weathering solutions that circulated through the BIF long after it formed (Hagemann *et al.*, 2008; Evans *et al.*, 2013). Enriched ore, which is favored because of its higher grade and softer nature, was largely exhausted in North

America in the 1990s but is widespread in Australia and Brazil.

Iron ores are beneficiated to remove silica and other gangue minerals (Figure 8.5). Iron mineral concentrates are mixed with clay or organic matter to form small balls known as **pellets** (Figure 8.2a). Some coarse-grained ores are simply baked to form a related clinker-like product known as **sinter**. Pellets and sinter are preferred as blast furnace feed because they allow increased flow of gas through the furnace and minimize dust production. Grinding taconite ores to powders fine enough to produce a clean concentrate of iron minerals requires large amounts of energy and has a strong control on the economics of the operation. In unmetamorphosed ores of the Mesabi Range, ore mineral grains are only a few micrometers in diameter, and extensive grinding is required. Where the ores have been recrystallized by metamorphism in parts of the Labrador Trough iron minerals are coarser-grained, permitting lower beneficiation costs that compensate in part for the more remote location.

Beneficiation of taconite ore generates about 2 tonnes of tailings for 1 tonne of concentrate. The large volume of iron ore that is mined generates a large amount of tailings, which are usually disposed of in tailings ponds surrounded by dams. An effort by Reserve Mining to avoid construction of a tailings

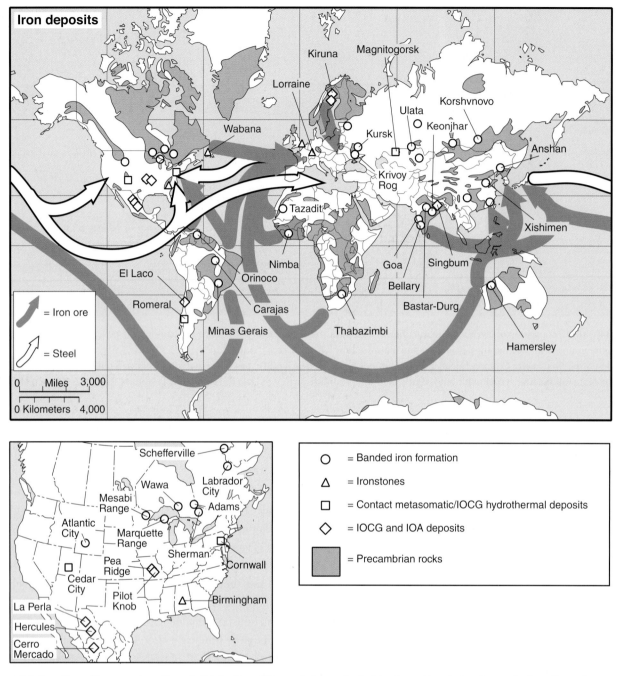

Figure 8.3 Location of iron ore deposits and distribution of Precambrian rocks that host most banded and granular iron formations and Phanerozoic ironstones. Names used here refer largely to mining locations rather than iron formations themselves (see Bekker *et al.* (2010) for names of iron formations) and not all deposits are active mines. Arrows show iron ore and steel trade patterns (compiled from World Steel Association, mineral statistics of Australia, China, India, OECD, United Kingdom and United States; Bekker *et al.*, 2010; Clout and Simonson, 2005).

facility in the Mesabi Range of Minnesota by disposal of tailings directly into Lake Superior led to a series of law suits that resulted in the closing of the operation in 1974. The operation was reopened in 1980 after development of a land-based tailings dam. This litigation is regarded by many as the precedent for government regulation of pollution generated by corporate activities (Huffman, 2000; Schilling, 2013).

8.1.3 Iron ore reserves, transportation, and trade

World iron reserves are estimated to contain 80 billion tonnes of iron metal in about 170 billion tonnes of ore, over 50 times greater than current annual production. Present reserves are slightly larger than those estimated in the early 1990s, but production is currently three times higher. This demonstrates

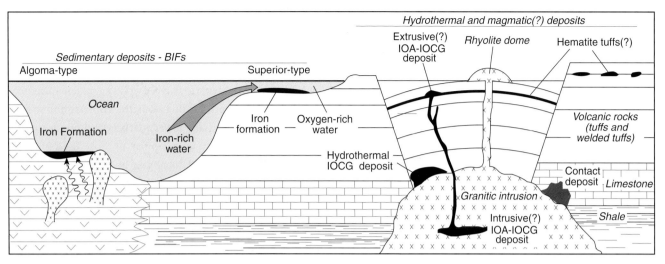

Figure 8.4 Schematic illustration of geologic processes that form iron deposits. Hydrothermal and magmatic deposits are divided into well-understood contact-type deposits and less-well-understood spectrum of deposits of IOCG and iron oxide apatite (IOA) groups discussed further in Chapter 9. Question marks indicate uncertainty about the magmatic origin for these deposits.

BOX 8.1 | **THE IRON AGE – WHY DID IT START AND WHAT DID IT CAUSE?**

Why did the Iron Age start? Bronze is generally harder than iron and would make a superior weapon. In addition, producing iron from iron ore requires higher temperatures than producing bronze. One likely possibility relates to the crustal abundance of iron (~ 4%) compared to copper (28 ppm) and tin (2 ppm) that make up bronze. Because Earth can make deposits from abundant elements much more easily than from scarce ones, iron deposits are much more numerous than copper and tin deposits. To make matters worse for bronze producers, deposits containing both copper and tin are rare, making it necessary to find separate deposits of the two metals. This inconvenience probably drove early metal producers toward iron, which was gradually improved as other elements were added to make steel.

About 140 kg of wood are required to make enough coke to produce 1 kg of iron metal from iron ore, and similarly large amounts of wood were required for production of bronze. Coming on top of the already growing need for wood for construction of homes, ships, and temples, as well as the need for fuel, production of metal was a major contributor to deforestation in the Mediterranean region as well as in China and Japan.

---

not just the successful conversion of iron resources to reserves but also the increasing pressure that is being put on these reserves. The main pressure, of course, comes from China, by far the largest consumer and importer of iron ore, followed by Japan and South Korea (Figure 8.6a). Australia and Brazil dominate exports.

Major iron ore deposits do not coincide with the distribution of major steel making (Figure 8.3). As a result, more than 1 billion tonnes of iron ore products are shipped annually to global markets, largely by sea (Figure 8.3). Ore ships operating on the Great Lakes also haul ore and pellets from the Lake Superior region to steel mills in Detroit, Cleveland, and Pittsburgh. Great Lakes ore ships carry almost 70,000 tonnes

of ore, whereas ocean-going freighters between Brazil and China or Japan carry as much as 400,000 tonnes. Movement of iron ore within individual countries is largely by rail, although slurry pipelines are used in Mexico, Brazil, India, and Australia.

8.1.4 Steel production

Steel making from iron ore consists of two basic steps (Figure 8.5). First, iron ore pellets with about 60% iron are converted to **pig iron** with about 94% iron, 4% carbon and other impurities such as sulfur and phosphorus (Moore and Marshall, 1991). This is usually done in a **blast furnace** (Figure 4.19a), a large, vertical cylinder that is charged with

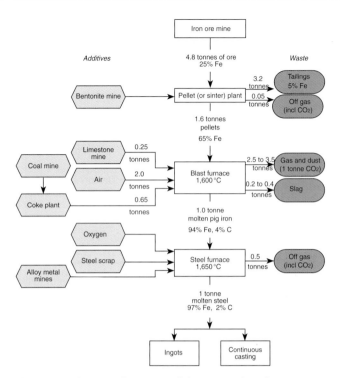

Figure 8.5 Schematic illustration of the iron ore beneficiation–pig iron–steel production process. Additional materials necessary for the process are shown on the left and waste products are shown on the right. In addition to CO_2, other emissions (per tonne of steel produced) include 0.015 tonne SO_x, 0.05 tonne NO_x, 0.02 tonnes CO, 0.05 tonnes particulates and minor volatile organic compounds (VOCs) (compiled from information of US Bureau of Mines and Moore in 1991; Energy and Environmental Profile of the US Iron and Steel Industry in 2000; the IEA Clean Coal Centre (Profile, February 2012).

a mixture of metal ore, coke, and limestone (Peacy and Davenport, 1979). Air is "blasted" in from the bottom to promote combustion of the coke to form carbon monoxide, the iron minerals react with carbon monoxide to form molten pig iron, and the waste minerals react with the limestone to form molten slag. The more-dense pig iron liquid, which accumulates at the bottom of the furnace below the slag liquid, is drawn off for further processing. In the **direct reduction process**, a less common alternative to the blast furnace, iron ore is reduced to metal without actually melting, usually by reacting it with carbon monoxide and hydrogen made from natural gas. Its main appeal has been in countries with abundant natural gas and limited coal, such as Mexico, Saudi Arabia, and Venezuela.

In the second step, pig iron is converted to steel with no more than 2% carbon. About 70% of steel is made in **basic oxygen furnaces** that heat a charge of pig iron and up to 30% scrap iron to 1,400 °C and blow oxygen gas through it. This oxidizes the carbon, silicon, manganese, and phosphorus

impurities, an exothermic process that raises the temperature to 1,650 °C, melting the charge, and removing the impurities as gases.

The most important wastes from iron and steel making are slag and gases. Much of the slag is used as aggregate in cement and other applications, making a positive contribution to cash flow. Dust emissions have fewer markets and require costly particulate collection systems similar to those used on coal-burning electric utilities. Global steel making releases about 2.6 billion tonnes of CO_2, about 6.7% of world emissions (World Steel Associateion, 2014). Most of this comes from blast-furnace reactions and can be reduced only by increasing the energy efficiency of the process. Overall energy use for steel making in MDCs has decreased by about 50% since 1975 and is approaching the theoretical minimum.

Processing of scrap is important in extending the life-cycle of steel. Essentially all of the excess material from steel production and downstream manufacturing (known as pre-consumer scrap) is recycled. Post-consumer scrap is recycled either by melting in **electric arc furnaces** or by remanufacturing, the industry term for reconditioning durable steel items such as motors. Current combined recycling rates in most more-developed countries (MDCs) are close to 90% (World Steel Association, 2014).

8.1.5 Steel markets and trade

World production of crude steel is about 1.5 billion tonnes, up more than 300% since 1960 (Figure 8.1d). Slightly more than half of world steel is used in construction, with another 12 to 15% used in each of automotive, machinery, and metal products. In-use steel stocks in most MDCs have increased along with gross domestic product (GDP) and have begun to level off at about 8 to 12 tonnes per person reflecting more efficient recycling.

The global steel industry employs about 2 million people, although employment in most MDCs has decreased continuously for the last few decades. In the United States, for instance, employment has decreased from about 650,000 in 1960 to about 150,000 in 2013. Other MDCs have also seen decreases, although somewhat less drastic. Much of it reflects increased productivity, which has reached annual levels of about 600 tonnes per worker in the United States, Germany, and Japan and more than 1,000 tonnes per worker in Taipei.

The largest steel producer, by far, is China, with Japan, United States, India, Russia, and South Korea far behind. The growth of Chinese steel production has been one of the most dramatic economic features of the past few decades

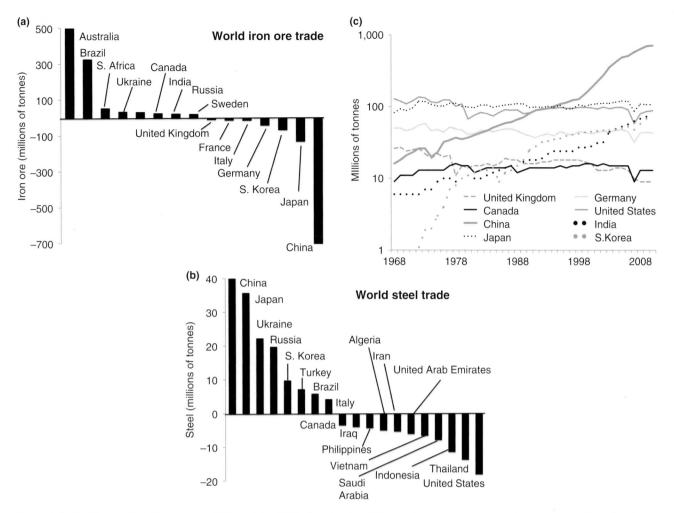

Figure 8.6 World trade in (a) iron ore and (b) steel for 2012, showing the difference between iron ore production and steel production. Countries with exports plot above the line, whereas those with imports plot below (compiled from data of the World Steel Association). Note that China, which produces about half of world steel, does not export as large a fraction of its production as Japan, indicating the heavy use of steel in oil drilling and production. (c) Change in world steel production showing dramatic increase by China and similar but smaller increases by South Korea and India. Note that the scale for this diagram is logarithmic such that each major division is ten times larger than the previous one, the only way that Chinese production can be shown on the same chart as production from other countries. Data from the US Geological Survey.

(Holloway *et al.*, 2010). At one point in the early 2000s, China was increasing steel production annually at a rate equal to the total annual steel production of the United Kingdom. By the mid 1990s, China had surpassed the United States and Japan and now accounts for almost half of global steel (Figure 8.6b). During the early part of this rise, Chinese steel producers could not supply domestic demand and large amounts of steel were imported. By about 2005, however, domestic production exceeded demand and China became the dominant global steel exporter (Figure 8.6). Chinese dominance in steel has come with a heavy price. Steel-making areas such as Hebei Province are heavily polluted and under government orders to increase efficiency and decrease emissions (Price *et al.*, 2011;

Tian *et al.*, 2013c). Steel production in South Korea and India is also growing at a very rapid rate, suggesting that they will soon play a larger role the global steel picture (Figure 8.6b).

As a result of consolidation, major steel-making operations in Canada, France, the United Kingdom, United States, and other MDCs are owned by multinational companies domiciled in other countries. As these multinational steel companies attempt to shut down operations in response to market changes, they are experiencing significant opposition from host governments (Jolly and Clark, 2012). Whether this will limit their flexibility and viability remains to be seen.

8.1.6 Alloy steels

The most common elements that are alloyed with iron include carbon and the ferroalloy elements discussed in this chapter. Although carbon-bearing steels are alloy steels in the strict sense of the term, "alloy steel" usually refers to combinations of iron and other elements.

Iron is amazingly versatile all by itself. When it crystallizes into a solid at 1,538 °C, the iron atoms arrange themselves into a body-centered cube with atoms on each apex and one in the middle for a total of nine atoms. As cooling continues, the body-centered cube changes at 1,395 °C to a face-centered cube with eight atoms at the apices and one in the center of each face of the cube for a total of 14. At 910 °C, the cube reverts to the face-centered form, and then becomes magnetic at 770 °C. If iron is quenched at any point during this cooling sequence, it will retain its high-temperature crystal form and properties. For instance, steels quenched above about 770 °C are not magnetic, something that might surprise most users.

More than 3,500 different steels have been developed and there are numerous classifications for them. One common steel classification has five categories based on compositions and potential applications (ASM, 2002). Carbon steels, which are those with carbon and less than about 1% manganese make up 90% of total steel production. They are divided further into low-carbon (mild) steels with up to 0.2% carbon, medium-carbon steels with 0.2–0.5% carbon, and high-carbon steels with more than 0.5% carbon. These steels are used in the construction of bridges and ships where strength and amenability to machining and welding are important. Steel with more than about 2% carbon is referred to as cast iron. Alloy steels are divided into low-alloy steel with less than 8% alloying elements and high-alloy steels with more than 8%. The main alloying elements include carbon, manganese, silicon, nickel, chromium, molybdenum, and vanadium. Other elements, including boron and lead, are used for special applications. These steels are used in electric motors, pipelines, and other industrial and construction applications. **High-strength, low-alloy (HSLA)** steels are used in automotive and other mobile applications where it is necessary to optimize strength, weldability, fatigue performance, and dent resistance. As attention has shifted to decreasing vehicle weight and increasing crash resistance, various advanced high-strength steels (AHSS) have been developed. The most important type of high-alloy steel is **stainless steel**, which generally contains 10–20% chromium and up to 8% nickel that greatly limit corrosion. They can be either magnetic or non-magnetic, depending on the temperature at which they are quenched. Protective coatings of zinc, chromium, and tin provide cheaper corrosion resistance, and about 15% of world steel is treated in this way. Another type of high-alloy steel known as tool steels contain tungsten, molybdenum, cobalt, and vanadium that make them hard, heat-resistant, and durable for use in cutting equipment.

8.2 Manganese

Manganese is brittle and has few uses as the pure metal. Nevertheless, about 16 million tonnes of manganese, worth about \$23 billion, were consumed in 2011 (Figure 8.1a). About 90% of this manganese goes into the steel industry where it has two uses. The original use was to remove or isolate sulfur and oxygen, both of which degrade steel, and this application still accounts for about 27% of world consumption. As technology has evolved, however, more and more manganese has been used as an alloy with iron in various special steel products, and about 63% of world consumption is used for this purpose today. The remaining 10% of world manganese consumption is alloyed with other metals or used in chemicals. Alloyed with aluminum, manganese increases corrosion resistance of beverage cans, and alloyed with copper, it increases strength. Manganese also forms part of the cathode in 20 billion dry-cell batteries each year, has potential as a component of lithium-ion batteries, and is the main component of chemical compounds such as manganese permanganate that is used to purify water and manganese-organic compounds that are used as a fungicides and octane-boosters in gasoline (Belanger *et al.*, 2008; International Manganese Institute, 2013). Despite its many uses, manganese production has grown more slowly than any of the ferroalloy metals other than tungsten (Figure 8.1b).

8.2.1 Manganese deposits and production

Manganese makes up about 0.1% (1,000 ppm) of average crust, far less than iron and other abundant elements like aluminum, sodium, and potassium, but much more than most other metals such as copper and zinc (Rudnick and Gao, 2003). It is geochemically similar to iron, but has three oxidation states in nature, rather than just two (Krauskopf and Bird, 1994). Under reducing conditions, where it forms Mn^{+2}, manganese substitutes for iron and calcium in some minerals and forms a few of its own, including pyrolusite (MnO_2) and pink rhodochrosite ($MnCO_3$), which is used in jewelry. Under oxidizing conditions, Mn^{+3} and Mn^{+4} form many different oxide minerals including hausmannite (Mn_2O_4) and psilomelane [$(Ba,H_2O)_2Mn_5O_{10}$].

Almost all of our manganese comes from **sedimentary** deposits that precipitated from seawater or **supergene** deposits that resulted from weathering of manganese-rich sediments (Kuleshov, 2011). Sedimentary manganese deposits, which supply most of world production, consist of layers of manganese carbonates and oxides that were deposited as chemical sediments. As was the case with iron, the Great Oxidation Event (GOE) also affected the deposition of manganese sediments. Pre-GOE manganese deposits are dominated by Mn^{+2}-bearing minerals, principally manganese-bearing carbonates including rhodochrosite as well as calcite and dolomite where manganese substitutes for calcium. Post-GOE deposits are dominated by the more oxidized Mn^{+3} and Mn^{+4} in oxides and hydroxides (Beukes and Gutzmer, 2008). Deposits of all ages are associated with iron formation, but the iron and manganese deposits are separated because Mn^{+2} is harder to oxidize (requires a more oxidizing environment) than Fe^{+2} and because Fe^{+2} forms a very insoluble sulfide (FeS_2) whereas Mn^{+2} does not.

By far the largest sedimentary manganese deposit is the Kalahari manganese field (Figure 8.7), which accounts for about 80% of known manganese resources (Gutzmer and Beukes, 2009), making it one of the largest concentrations of any element on Earth. The Kalahari deposits, which formed shortly after the GOE, consist of manganese carbonates and oxides with up to 30% manganese. Layers of typical iron formation are found both below and above the manganese-rich sediments, and reconstructions of the setting indicate that the manganese-rich sediments were deposited closer to shore where oxygen was more abundant (Gutzmer and Buekes, 2009). Much younger, Phanerozoic-age manganese-rich sedimentary deposits, particularly those around the north side of the Black Sea in Bulgaria, Ukraine, and Georgia, are thought to have formed when manganese-rich seawater rose from the reduced, deep ocean into shallow, oxygen-rich water. These deposits are not associated with iron formation because the reduced seawater contained dissolved sulfide that precipitated iron sulfide (Cannon and Force, 1983; Force and Cannon, 1988). Manganese deposits are closely related to global cycles and appear to have formed during periods of high temperature and high sea level (Frakes and Bolton, 1992).

Supergene deposits, which supply the rest of world production, consist of manganese oxides and hydroxides that accumulated on deeply weathered manganese-rich rocks, sometimes in karst terranes. They are, in a way, a special form of laterite soil, similar to bauxite, the aluminum ore. As might be expected, supergene deposits are found largely in humid tropical climates (Figure 8.7), although older, buried deposits are known.

Manganese deposits are usually mined by high-volume, open pit methods similar to strip mines. Despite their enormous size, the Kalahari deposits account for only about 20% of world production because of their remote location and

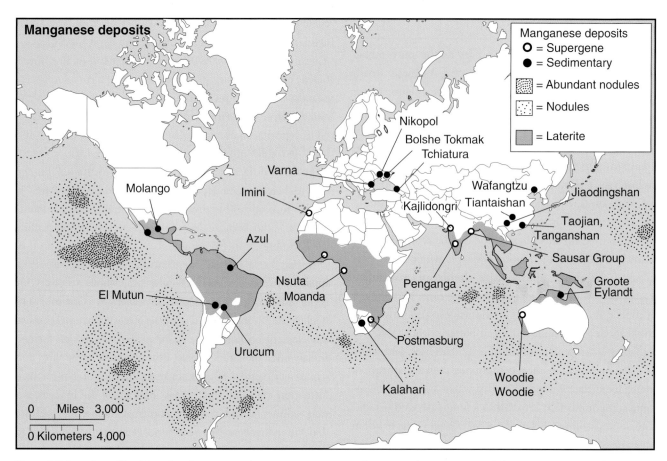

Figure 8.7 Location of land-based sedimentary and supergene manganese deposits and manganese nodule concentrations in the oceans. Variation in base and precious metal content of nodules, which is not shown here, will have a strong effect on the location of any future mining. See Figure 5.3 for the location of offshore areas claimed as an exclusive economic zone by the United States and of manganese nodules claimed by private industry.

limited rail and harbor facilities and because of land-tenure issues that have limited exploration and investment (Gutzmer and Beukes, 2009). Ore is beneficiated by conventional methods to make a concentrate that is fed into a blast or electric furnace with iron (usually in the ore), coke, and limestone to make ferromanganese and/or silicomanganese. Manganese metal and dioxide are made by electrolytic separation from solutions obtained by dissolving manganese ore in sulfuric acid.

8.2.2 Environmental aspects of manganese production and consumption

Manganese is an essential element for life. It plays a role in the formation of tissue and bone, as well as in the development of the inner ear and reproductive functions (WHO, 1981a; EPA, 2003a). Estimated average requirements are about 2–3 mg/day, which comes largely from food (principally nuts, whole grains, and dried fruits) and from inhalation of natural dust.

Exposure to excessive amounts of manganese can cause neurological and respiratory problems. Anthropogenic manganese contributions to the atmosphere are small, only about 10 to 20% of natural contributions, and those to waters are even smaller (Nriagu and Pacyna, 1988). Problems with manganese exposure have been observed in some mines, but more commonly in ferromanganese production facilities where they are caused by inhalation of manganese-rich dust and fumes (Moore, 1991).

About 80% of manganese emissions come from iron and steel production facilities. The remaining 20% comes largely from combustion of coal in power plants and coke ovens (Agency for Toxic Substances, 2012).

8.2.3 Manganese trade and reserves

Manganese is mined in about 33 countries, although only a few of them make a significant contribution to global production (Corathers, 2013a). World manganese ore reserves are

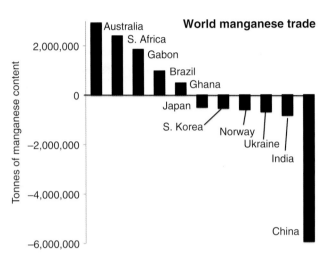

Figure 8.8 World manganese trade for 2011 showing the difference between manganese exports and manganese alloy (ferromanganese and silicomanganese) production for major producers and consumers of manganese (compiled from data of the US Geological Survey). Countries plotting above the line are major exporters of raw ore; those below the line are major importers of raw ore for manganese alloy production. Most countries that plot below the line consume the manganese alloys that they produce in their domestic steel industry; the one exception is Norway, which produces ferroalloy for export.

estimated to contain 630 million tonnes of metal, a figure that is down by about 20% over the last two decades. Unfortunately, most major steel-making nations lack manganese (DeYoung *et al.*, 1984a). North America is in terrible shape, with less than 1% of world reserves even though it has 16% of the planet's land surface. This situation has created an active global trade in manganese ore, ferromanganese, and silicomanganese that are added directly to steel. In addition to South Africa, manganese ore comes largely from the Groote Eylandt deposit in Australia, Moanda in Gabon, and the Urucum area of Brazil, and goes to China, India, Ukraine, Norway, South Korea, and Japan (Figure 8.8). Most of the countries that import manganese ore use it domestically, although Norway exports ferromanganese. In spite of its large reserves, Ukraine imports high-grade manganese ore, apparently to enrich their low-grade production.

By far the largest manganese reserves are in Australia, Brazil, South Africa and Ukraine (Figure 8.7). As noted above, reserves and long-term resources are dominated by the Kalahari deposits with about 4.4 billion tonnes of manganese; other deposits play a role largely because infrastructure is not adequate for expanded export of Kalahari ore. Deposits around the Black Sea are estimated to contain about 500 million tonnes, much of which is either low grade or in the form of carbonate minerals that require additional processing (Cannon and Force, 1983). Grades at Nikopol and Tchiatura, for instance, range from 13 to 33% Mn, considerably lower than most other manganese deposits of the world. Other large manganese deposits in Australia, Gabon, and Brazil (Figure 8.7) have grades of 44 to 50% Mn, but are at least an order of magnitude smaller than the Kalahari deposits. The only large deposit in North America, Molango in Mexico, also has low ore grades and supplies domestic needs. Thus, very little of global manganese reserves that are clearly economic under present conditions are in locations without potential political problems. These facts continue to support interest in deep-sea manganese nodules, which constitute an enormous, untapped resource.

8.2.4 Manganese nodules

Manganese nodules, which litter floors of the ocean and some freshwater lakes (Figure 8.7), are among the most tantalizing of our many ocean resources (Heath, 1981; Baturin, 1987; Abramowski and Stoyanova, 2012). They are grapefruit-sized balls consisting of thin, concentric layers of manganese and iron oxide minerals (Figure 8.9). In addition to manganese and iron, the nodules contain small amounts of nickel, cobalt, and copper, and trace amounts of molybdenum, zinc,

Figure 8.9 Manganese nodules. (a) Cross section of a manganese nodule from the North Pacific Ocean (Ocean Drilling Program Site 886) showing concentric growth layers (photograph courtesy of D. K. Rea and H. Snoeckx, University of Michigan). (b) Manganese nodules on the seafloor. The photo was taken by camera sled in 2003 at the Hawaii-2 Observatory site in the in the central North Pacific at ~ 5000-m depth by K. L. Smith Jr (MBARI) and S. E. Beaulieu (WHOI). See color plate section.

zirconium, lithium, platinum, titanium, germanium, yttrium, and rare-earth elements. Most nodules are on the deep sea-floor at water depths of 5 to 7 km, and others are on elevated submarine plateaus, along the sides of seamounts at depths of 1 and 2.6 km, and even in deep parts of large lakes (Ballard and Bischoff, 1984). The nodules formed where clastic sediments are scarce, and it is thought that they grew by precipitation from surrounding water and pore water, with some addition from hydrothermal fluids (Knoop *et al.*, 1998).

There is a real question about why we see the nodules at all. Isotopic age measurements on growth layers in the nodules indicate that they grew at rates of 1 to 4 mm per million years, which is about 250 to 1,000 times slower than the rates at which clastic sediments are deposited. So, the nodules should be buried by sediment, even in the open ocean where it comes only from wind-blown dust. However, at least 75% of the nodules are at the surface, indicating that some mechanism

works to keep them there. Possibly ocean currents roll the nodules along the surface or sweep finer sediment away, or burrowing organisms push nodules toward the surface. Whatever their origin, the nodules are not homogeneously distributed, and appear to be most abundant in the equatorial North Pacific and smaller areas of the mid-Atlantic and Indian oceans (Figure 8.7). In areas where they are most abundant, nodules form a pavement that covers the ocean floor. The Pacific Ocean, alone, is estimated to contain about 2.5 billion tonnes of nodules with a grade of about 25% Mn, making them similar in abundance to low-grade, land-based deposits, and their content of other elements is even larger.

According to the International Seabed Authority, in 2013, major efforts were made to develop technology for mining nodules. Although one effort billed as a test of nodule recovery methods turned out to be an attempt by the United States to recover a Soviet submarine that sank in the nodule fields south of Hawaii (Polmar and White, 2012), other tests have indeed focused on the nodules. Most equipment employs suction or scraper systems that remove nodules along with the upper few centimeters of the seafloor. Large-scale tests have shown that these methods can produce nodules at sufficiently high rates, although less is known about the feasibility of processing ores onboard ship (Bernier, 1984). Nodule mining would cause major disruption of the immediately surrounding seafloor and studies of the potential effects of these activities are needed (Foell *et al.*, 1992). Commercial production has been stalled by these environmental considerations, along with high costs compared to land-based deposits, uncertainties of ownership, and concern about the Law of the Sea, as discussed in Chapter 5.

8.3 Nickel

About 65% of nickel production is used in stainless steels and another 20% is used in other steel and non-ferrous alloys. The remaining nickel is used in plating steel and other metals, electronics, and in batteries such as the nickel metal hydride (NiMH) that powers the Toyota Prius (Wilburn, 2008). The largest markets for nickel metal products are in turbines and jet engines and in tubing and containers for the chemical industry (Kuck, 2013). Cobalt can substitute for nickel in many of these uses, but it is more expensive. Nickel gave its name to coins in several countries, but most of them have cut nickel contents because of its relatively high price. In the United States, the present nickel coin contains 24% nickel. The Canadian nickel, which contained 99.9% nickel in the late 1940s, now contains only 2% as a plating on a steel–copper

Figure 8.10 (a) Nickel-rich laterite (dark) overlying ultramafic bedrock (light) from which it was derived by weathering, southwestern Puerto Rico. H.L. Bonham provides scale. (b) Shatter cones on the side of a rock outcrop south of Sudbury, Ontario, showing the radiating pattern pointing upward toward the probable impact level, which is now eroded. Key in lower left provides scale (photographs by the authors). See color plate section.

alloy. In the United Kingdom, 5- and 10-pence coins are nickel-plated steel and more valuable coins are copper–nickel alloys. Despite the declining nickel content of coins, nickel markets are increasing relative to steel (Figure 8.1c). Present annual sales of nickel throughout the world, amounting to about 2 million tonnes, are valued at $45 billion (Kuck, 2013).

8.3.1 Geology and distribution of nickel deposits

Nickel is estimated to make up about 1.9% of our planet, but essentially all of it is in Earth's core. In contrast, the upper crust contains 59 ppm nickel and mantle rocks average about 1,960 ppm (Rudnick and Gao, 2003). This is enough, however, to generate two important and very different types of nickel deposits, laterite and magmatic sulfide. As you might expect from the averages noted above, both deposit types form from rocks that originate in the mantle.

Laterite nickel deposits are actually special soils that have undergone extreme weathering that usually takes place in tropical environments (Figures 8.10a, 8.11). Some laterites are sufficiently enriched in aluminum, iron, or nickel to

become ore (Freyssinet *et al.*, 2005). Which of these elements, if any, concentrates in the laterite depends on the composition of the underlying rock that is being weathered. Nickel laterites form on weathered ultramafic rocks that consist almost entirely of magnesium–iron silicate minerals containing olivine [$(Fe,Mg)_2SiO_4$] or their altered counterpart, serpentinite. Nickel, which substitutes for magnesium in olivine and serpentinite, is released into surrounding groundwater during weathering. If the groundwater contains dissolved oxygen, iron oxides precipitate on the spot and adsorb nickel from solution. If the groundwater is reducing, nickel remains in solution and percolates to the base of the weathered zone, where it precipitates as garnierite, a complex nickel silicate mineral. Nickel laterites can be up to 20 m thick with average grades of 3% Ni and 0.1–0.2% Co, and grades are usually higher if the weathered rock is uplifted to allow movement of groundwater (Golightly, 1981; Freyssinet *et al.*, 2005; Berger *et al.*, 2011; Butt and Cluzel, 2013).

Ultramafic rocks do not get many chances to weather because most of them are in Earth's mantle. They reach the surface only in rare convergent margins where the mantle is

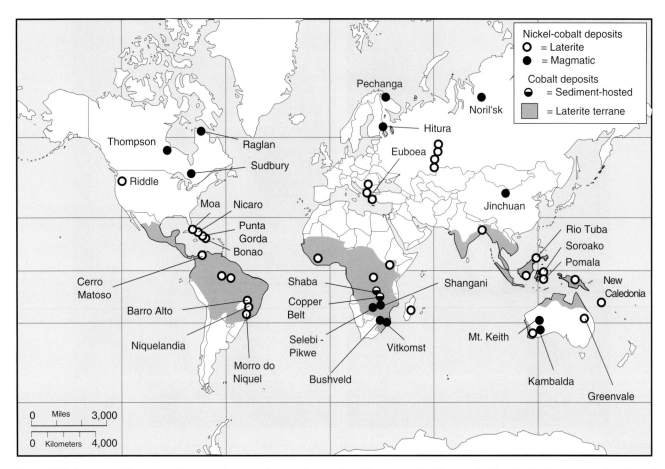

Figure 8.11 Location of laterite and magmatic nickel deposits (most of which contain significant cobalt) and of sedimentary cobalt deposits (based in part on Berger *et al.*, 2011)

pushed up rather than down, a process known as obduction to distinguish it from the more common subduction process. Large **obduction zones** with abundant ultramafic rocks in the present tropics are found in New Caledonia, Cuba, Dominican Republic, and Indonesia (Figure 8.11). The small deposit at Riddle, Oregon, was formed earlier in Earth history and is known as paleolaterite (Cumberlidge and Chace, 1967).

Magmatic nickel sulfide deposits form during crystallization of mafic and ultramafic igneous magmas (Arndt *et al.*, 2005; Barnes and Lightfoot, 2005). As the silicate magmas cool, nickel and iron combine with sulfur to form a new liquid that separates from the silicate magma as an immiscible sulfide magma, a process discussed in Chapter 4. The dense blebs of metal sulfide sink to the bottom of the silicate magma chamber where they coalesce and crystallize to form large bodies of metal sulfide minerals as the intrusion cools (Figure 2.12). The crystallized sulfide magma contains iron–nickel sulfides such as pentlandite [(FeNi)$_9$S$_8$] and pyrrhotite (FeS), which also contain small amounts of cobalt.

Many magmatic nickel deposits form from magnesium-rich ultramafic magma known as **komatiite**, which formed by melting of the mantle. Komatiite magmas have a higher melting temperature than other common silicate rocks (about 1,600 °C) and appear to have formed almost exclusively during Archean time, probably because the mantle was hotter early in Earth's history (Kerrich *et al.*, 2005). Komatiite-related nickel sulfide deposits are found only in Archean parts of Precambrian cratons in western Australia, southern Africa, and Canada (Figure 8.11).

Magmatic nickel deposits also separated from younger, mafic magmas that were not so rich in magnesium, but only under special circumstances. As it turns out, however, two of these intrusions supply about 40% of world nickel production. The most unusual of them is the Sudbury Intrusion in Ontario, where nickel sulfide deposits are found at the base of a large, funnel-shaped mafic intrusion (Figure 8.12). Everything about Sudbury is unusual. The intrusion is not ultramafic and it contains large amounts of felsic rock near its top. Its age is only about 1.8 Ga, much younger than komatiite-related deposits, and the rock surrounding it contains curious features known as Sudbury **breccias** and **shatter cones**, which form only in strongly shocked rock (Figure 8.10b). The strong shock was caused by a **meteorite** impact, which perturbed the crust and

BOX 8.4 | NICKEL DEPOSITS AND THE GOE

Nickel deposits might even have played a role in the GOE. It appears that the Archean Ocean, as indicated by the composition of Archean BIF deposits, contained much more nickel than later oceans. Nickel is a critical element in enzymes of methane-producing organisms, and a drop in nickel in the ocean might have cut off their nutrient supply causing a drop in methane in the atmosphere and a corresponding increase in oxygen (Konhauser *et al.*, 2009).

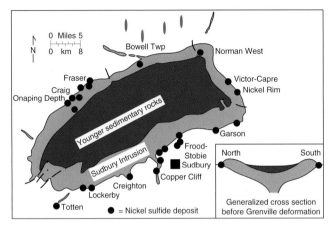

Figure 8.12 Geologic map of the Sudbury area, Ontario, showing the location of the Sudbury layered igneous complex, important nickel mines, and the deep Victor deposit. The generalized cross section shows how the intrusion might have looked after it was emplaced and before it was deformed by later collisional events, especially docking of the Grenville terrane to the east.

underlying mantle, causing formation of magma that rose and mixed with granitic crust depositing large amounts of sulfide ore (Naldrett, 2003). The other major exception is Noril'sk in northern Russia, which is Triassic in age, even younger than Sudbury. Noril'sk formed from basalts that were erupted onto the continent as it began to break apart in Triassic time. Magmas of this type do not usually contain enough sulfur to form magmatic nickel deposits, but the Noril'sk intrusions assimilated sulfur-rich evaporites from surrounding wallrocks.

8.3.2 Nickel production, use, and related environmental concerns

Nickel is mined in 28 countries, including New Caledonia and Cuba that are not commonly thought of as mineral producers. Production of nickel from its ores involves some interesting energy trade-offs between laterite and sulfide deposits. Laterite ores are at the surface and can be mined by

inexpensive open pit methods but processing of the ore requires lots of energy to release nickel (Evans *et al.*, 1979). For iron-rich laterites, this is done by hydrometallurgical methods using acid or ammonia to leach nickel from the ore. Magnesium-rich laterites are heated in the presence of sulfur to form nickel sulfides that are then smelted by pyrometallurgy. Sulfide ores, on the other hand, usually require more expensive underground mining, but can be beneficiated into low-volume sulfide mineral concentrates that are processed by less-expensive pyrometallurgy. Both processes produce nickel metal or ferronickel, which is used directly in steels and other alloys.

Theoretically, these energy considerations should give nickel sulfide ores the economic advantage and, for many years, that was the case. Nickel laterite operations that were started in Australia, Guatemala, and Indonesia during the 1970s were closed because of high processing costs. Recently, however, attention has shifted back to laterites because they are easier to find and their large size gives them a long life. In contrast, nickel sulfide deposits in Sudbury and elsewhere are being exhausted and replacement deposits are hard to find. The new Eagle mine in Michigan is a good example. Eagle required more than a decade of exploration and development and, even so, is so small that it will have an active life of only about a decade. Thus, even though laterite costs are high, the deposits are likely to gain production share (Robinson, 2012).

Sulfide nickel ores contain 8 tonnes of sulfur for each tonne of nickel, causing large smelter emissions. During the 1960s and 1970s, for instance, Sudbury emitted about 2 million tonnes of SO_2 annually that killed vegetation and acidified lakes over a large area (Nriagu and Rao, 1987; Tollinsky, 2013). Emissions from Sudbury have been reduced to only about 45,000 tonnes now, largely by use of flash smelting to make flue-gas SO_2 concentrations high enough to capture and by exclusion of pyrrhotite, which contains much more sulfur per atom of nickel than do other nickel-bearing sulfide minerals. Emissions have also been reduced significantly at Noril'sk, the other major nickel sulfide producer, although the site is heavily polluted from earlier operations (Bellona Foundation, 2010).

Nickel, along with vanadium, is strongly bonded to some high-molecular-weight hydrocarbons and is a common contaminant of heavy crude oils (Lewan, 1984). Combustion of fuel oil and other residual, heavy oils accounts for 62% of anthropogenic nickel emissions, with most of the rest coming from nickel mining and processing facilities and coal combustion (Agency for Toxic Substances, 2005). Power plants are estimated to account for about 30% of the nickel released into the environment annually in the United States, and coal combustion of all types accounts for more than 60% of emissions in China, (Tian *et al.*, 2012). Where scrubbers have been installed in power plants, emissions are lower, but boiler heating probably remains the principal source of environmental nickel. In its inorganic forms, nickel is most readily taken into the human body through the lungs. It is not essential for life, but it also does not appear to be strongly detrimental in small concentrations. Thus, the greatest danger of nickel-related environmental problems are in industrial settings with nickel-bearing dust and fumes such as roasting operations in smelters.

8.3.3 Nickel trade, reserves, and resources

World nickel production comes about equally from laterite and sulfide deposits. Major sulfide nickel miners include Australia, Brazil, Canada, and Russia, and major laterite nickel miners include Indonesia, New Caledonia, and the Philippines. Most mined nickel is exported for the steel industry of other countries (Figure 8.13), and is subject to growing resource nationalism. Indonesia, the world's largest producer of nickel has threatened to embargo export of unprocessed nickel ore in an effort to force producers to build smelters and other treatment plants in the country and negotiation continues over exactly what sort of advanced treatment represents an acceptable investment (Javier and Aquino, 2012).

The demand for primary nickel is dominated by the amount of recycled material. Recycled nickel comes largely from manufacturing waste and stainless steel scrap, although batteries from electric vehicles are a new and potentially growing source of recycled nickel. It is still not clear how depleted batteries can best be recycled. Possibilities include repurposing batteries for other applications or actually separating individual elements for recycling (Kanter, 2011).

World nickel reserves are estimated to contain 75 million tonnes of metal and world resources are estimated to contain

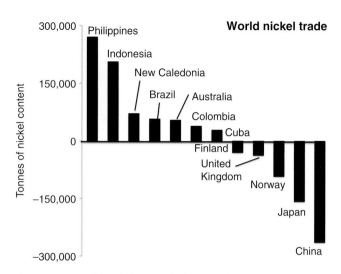

Figure 8.13 World nickel ore trade for 2012 showing major exporters and importers of nickel ore and intermediate products (based on nickel content of products) (compiled from data of the US Geological Survey)

about 130 million tonnes of metal in deposits with an average nickel content of 1% or greater (Kuck, 2013). About 60% of this resource is in laterite deposits with the remaining 40% in sulfide deposits. Both of these resources have their problems. As noted above, the viability of laterite reserves and resources will depend greatly on the price of energy. Furthermore, many of the sulfide resources are at great depth in known mining areas such as Sudbury. The Victor-Capre zone at Sudbury (Figure 8.12), with ore at depths of almost 3,000 m, has been explored for decades but has not been developed (Pickard, 2012). Another strike against nickel sulfide resources is the simple fact that Sudbury, which has been the dominant source of sulfide production, was a unique event in Earth history. Despite continued exploration, no other sulfide resources have been found in impact-related igneous rocks and, even if they were, it is unlikely that their size could match Sudbury, which was one of the largest impact events in Earth history.

These complications have directed attention to several low-grade nickel sulfide resources. Most prominent among them is a zone at the bottom of the Duluth Complex, a mafic intrusion just outside Duluth, Minnesota, where sulfur from wallrocks caused separation of immiscible nickel–copper sulfides. Although this area contains about 6 billion tonnes of nickel-bearing rock, grades are 0.2% Ni, only one-tenth those at Sudbury and Noril'sk. Submarine manganese crusts and nodules are also a possible reserve, of course, but they are not likely to become economic soon. A final type of deposit that is receiving attention contains the strange nickel–iron mineral awaruite (Ni_3Fe). Native iron and nickel require very reducing environments and are rare in Earth's crust. Most awaruite is found in serpentinite and is thought to form when nickel-bearing olivine is altered by H_2-bearing fluids (Britten, 2013), although others have suggested that it might be from the mantle (Bird and Weathers, 1979). Awaruite is getting attention because it is essentially ferronickel, the intermediate material that is manufactured from nickel ore and added to steel. If ferronickel could be mined directly, it would cut processing costs. Recently discovered awaruite deposits in central British Columbia have average nickel grades of 0.1 to 0.2%. If large enough volumes of this material can be found, they might be mined.

8.4 Chromium

Chromium is the essential alloying element for stainless steel, the least expensive corrosion-resistant metal. Stainless steel is used in household utensils, containers for food and chemicals and automobile parts. With a content of 11–30% chromium, stainless steel accounts for about 68% of world chromium consumption. Although other, more expensive metals can be substituted for stainless steel, nothing can do the job at the same price. According to the International Chromium Development Association, another 27% of world chromium is used in specialty steels and in alloys consisting almost entirely of chromium with or without nickel, such as the familiar shiny chrome plating that is highly resistant to corrosion (see also Papp, 2013a). A final 5% of world production is equally divided among chemical products that are used in pigments, tanning leather, and preserving wood. World chromium production of about 7 million tonnes annually is valued at almost $19 billion (Figure 8.1a) and has been growing at a rate considerably faster than steel (Figure 8.1b).

8.4.1 Geology and distribution of chromium deposits

The average content of chromium is about 135 ppm in the crust and 2,600 ppm in the mantle (Rudnick and Gao, 2003). As you might expect, then, most chromium deposits are found in rocks that originated in the mantle. The ore mineral for chromium is chromite, a member of the spinel group with a composition that can be summarized as AB_2O_4, where A can be Mg^{+2} or Fe^{+2} and B can be Cr^{+3}, Al^{+3}, or Fe^{+3}. Iron, aluminum, and chromium substitute widely in spinels and if you put iron in both the A and B sites, you get magnetite (FeO. Fe_2O_3). Chromium-rich spinels (chromite) contain 30% to 60% Cr_2O_3 (Cr contents are commonly reported as the

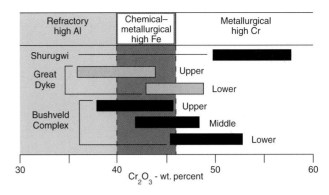

Figure 8.14 Range of chromium contents (expressed as Cr_2O_3) of chromites suitable for refractory, chemical, and metallurgical applications, showing compositions of chromites from major deposits in southern Africa (modified from the US Bureau of Mines, Minerals Facts, and Problems and Stowe, 1987)

compound Cr_2O_3, which contains 68.42% Cr). High-chromium chromites, which are defined as those with at least 46% Cr_2O_3, are the richest source of chromium metal (Figure 8.14). High-iron chromites, with 40–46% Cr_2O_3, can be processed to yield chromium metal and ferrochromium. Chromites with less than 40% Cr_2O_3 melt at very high temperatures and are used in bricks that line blast furnaces and other hot containers (Mikami, 1983).

There are two principal types of chromite deposits, stratiform and podiform (Stowe, 1987; Papp, 2007; Schulte *et al.*, 2012). **Stratiform** chromite deposits are found in **layered igneous complexes** (LICs), particularly the Bushveld Complex of South Africa and the Great Dyke of Zimbabwe (Figure 8.15). LICs are mafic intrusive rocks consisting in part of layers with very distinct mineral compositions, including nearly monomineralic layers that are our main source of chromium. The Bushveld Complex, by far the largest LIC in the world, contains layered mafic rocks over 9 km thick and a thick sequence of genetically related felsic rocks (Figure 8.16). Even though much of the Bushveld has been eroded, it still covers an area of about 66,000 km^2 (Naldrett, 1989).

The Bushveld Complex contains 14 chromite layers, of which six have been mined (Stowe, 1987), especially the Steelpoort and UG-2 layers (Figures 8.16, 8.17). The 1.06–1.8-m-thick Steelpoort layer extends for over 100 km along the surface and probably at least 20 km beneath the surface. Chromite from the UG-2 layer contains 6–7 ppm platinum-group elements, which is the main objective of the mining. Another layer in the Bushveld Complex, the Merensky Reef, also contains chromite, but is mined for platinum-group elements, as discussed later. The Great Dyke is older (2.5 Ga vs. 2.0 Ga for the Bushveld), but its close proximity and

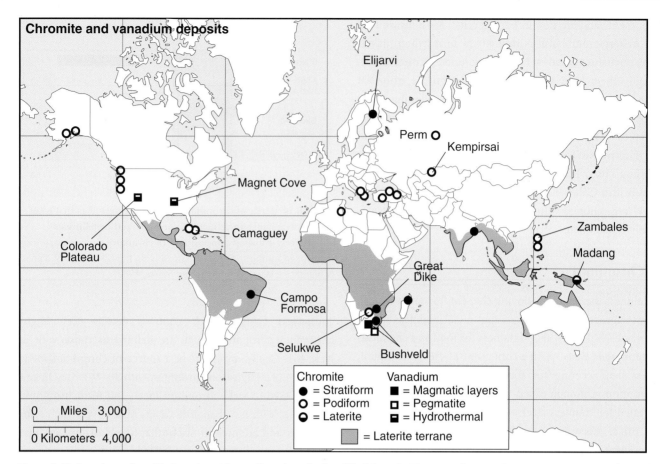

Figure 8.15 Locations of world chromite and vanadium deposits (modified from DeYoung *et al.*, 1984, Stowe, 1987; Mosier *et al.*, 2012). Additional detail on southern Africa is shown in Figure 11.7.

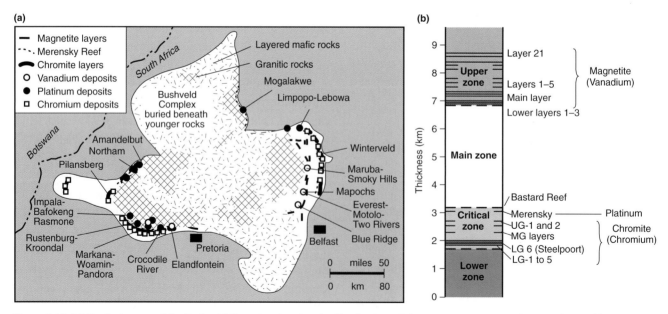

Figure 8.16 (a) Geologic map of the Bushveld Complex showing the distribution of chromite, magnetite, and Merensky Reef layers and producing mines. (b) Schematic stratigraphic column in eastern Bushveld Complex showing position of these layers. This column omits the Marginal Zone and overlying felsic rocks, which contribute to the 9 km total thickness of the Bushveld Complex.

similar chromite-rich nature suggest that there was something unusual about the Precambrian mantle beneath southern Africa during that period in Earth history. Other LICs, including the Stillwater Complex in Montana, also contain chromite

Figure 8.17 Bushveld chromite deposits, eastern Transvaal, South Africa. (a) Chromitite layers (black) at Dwars River. (b) Winterveld chromite mine showing cumulate layers on hillside dipping downward to left. The mine enters at the base of the hill to the right (photographs by the authors). See color plate section.

layers but lack large reserves because the intrusions are smaller, more disrupted by faulting, and are partly tilted and eroded. Least studied of these is the Dufek Complex in Antarctica, the second-largest LIC in the world, which is currently off limits to exploration as noted in Chapter 5.

The other type, **podiform** chromite deposits, are not found in LICs. Instead, they are hosted by mafic and ultramafic igneous rocks that formed near the base of the ocean crust and that have been brought to the surface by obduction. These are the same rocks that generate nickel laterite deposits, and podiform chromite deposits are found in many of the same places, particularly Cuba, Philippines, and Turkey. Chromite in some podiform deposits, such as Selukwe (Shurugwi) in Zimbabwe, is richer in chromium that in stratiform deposits (Figure 8.15). Some podiform chromite bodies are actually faulted and sheared chromite layers that formed in large mafic magma chambers at mid-ocean ridges. Although most podiform deposits are smaller than stratiform deposits, the Donskoy deposit in Kazakhstan is an exception. Because chromite is resistant to erosion, it can accumulate in laterites and placer deposits. Some nickel laterite deposits, which form on ultramafic rocks containing disseminated and podiform chromite, have as much as 1% Cr_2O_3 (Lipin, 1982), although these have not yet been mined economically.

8.4.2 Chromium production and environmental issues

Chromite is mined in 21 countries, but most of it comes from South Africa and Kazakhstan (Figure 8.18). Most chromite is mined by underground methods. Mining is complicated because some layers are only a meter or so thick and their lower contact is locally irregular. Although chromite is

---

BOX 8.5   CHROMITE LAYERS IN THE BUSHVELD COMPLEX

There has been considerable debate about just how chromite layers formed in LICs. The layers are clearly **cumulates**, which means that they formed by settling of chromite crystals from the magma. The problem is that magmas almost always crystallize more than one mineral as they cool and therefore the cumulate layer should contain two or more minerals (experiments show that higher-density chromite would not settle out more quickly if it did crystallize with other minerals). Most people now feel that chromite must have crystallized alone, that periods of "chromite-only" crystallization were long enough to permit 1-m-thick layers to form, and they must have occurred repeatedly to form the many layers. It turns out that this can happen when new batches of magma are injected into the magma chamber; and successive pulses of new magma could account for all of the layers (Irvine, 1977).

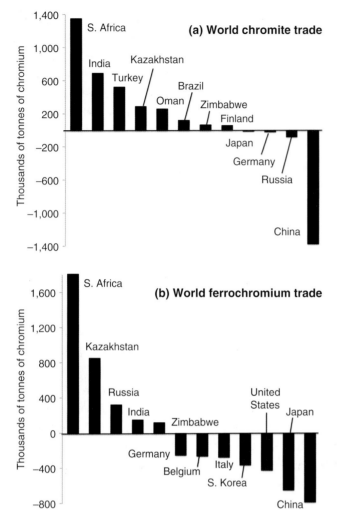

Figure 8.18 World trade in (a) chromite ore and (b) ferrochromium alloy showing exporters above the line and importers below (all amounts based on content of contained chromium) (compiled from data of the US Geological Survey)

oxidation state in which it is found in chromite and other natural minerals, is an essential nutrient. Its principal function is to maintain normal glucose metabolism (WHO, 1981b). Chromium deficiencies can lead to problems in insulin circulation as well as the possible risk of cardiovascular disease. It is estimated that 90% of US adults have a deficiency of chromium in their diets, an interesting statistic in view of experiments suggesting that large amounts of chromium in the diets of animals prolong their lives significantly. There is no evidence that trivalent chromium causes ill effects in workers or the general population in the vicinity of ferrochromium plants.

Despite its good properties, chromium is listed as a toxic substance under Title III of the Superfund Amendments and Reauthorization Act of 1995. This reflects the highly toxic nature of the hexavalent chromium (Cr^{+6}). Depending on how it enters the body Cr^{+6} can cause lung, nasal and sinus cancer, liver or kidney damage, or skin rashes (NIOSH, 2013). Hexavalent chromium does not form naturally and is not produced from trivalent chromium in the human body (Fergusson, 1990). However, it is produced in chemical plants and is a major constituent of chromate compounds that are used as pigments, dyes, wood preservatives, and leather tanning. Chromate compounds used in tanning were among those alleged to have caused environmental damage in Woburn, Massachusetts, although subsequent studies have cast doubt on their role (Rogers *et al.*, 1997; Harr, 1996).

8.4.3 Chromium trade, reserves, and resources

World raw chromium trade involves chromite and ferrochromium, which are treated separately (DeYoung *et al.*, 1984b). The largest ore producers are South Africa, Kazakhstan, India, and Turkey, but much of this ore is converted to ferrochromium before export, particularly in South Africa, Kazakhstan, and India. The largest importers of ore are the large steel makers China, Russia, and Germany and they are also net importers of ferrochromium, along with the United States, Korea, Italy, Belgium, and Germany (Figure 8.18). Reserves to support world chromite trade amount to about 460 million tonnes of ore, enough to supply about 20 years at present annual consumption rates of about 24 million tonnes of ore. The largest reserves are in South Africa and Kazakhstan, with India a distant third. Resources are estimated to be about 12 billion tonnes of chromite ore, almost entirely in Africa and Asia (Papp, 2013a).

The lack of significant reserves outside southern Africa and Kazakhstan is a persistent worry for European and North

relatively insoluble, recent studies indicate that chromium is released from mine wastes that are saturated with oxygenated water (Cherkasova and Ryshenko, 2011). Chromite is converted to ferrochromium in electric-arc furnaces, like those used to make steel. Chromium metal can be produced either by pyrometallurgical reduction of chromic oxide or by electrolytic separation of chromium from a solution of ferrochromium and sulfuric acid. Chromium chemicals are also prepared from sulfuric acid solutions that result from kiln **roasting** of chromite ore. The most important wastes associated with this processing are dust, fumes, and sulfuric acid, which must be caught or neutralized (Papp, 1985). Deposits of both types have locally important platinum-group by-products.

Environmental problems related to chromium ore processing are limited because chromium, in the trivalent (Cr^{+3})

American chromium consumers and has been aggravated by the history of chromite exports. The former USSR embargoed chromite exports to western countries from 1950 until about 1963. Just as this embargo was ending, the United States and some European countries embargoed chromite imports from the racially restrictive government in Rhodesia (now Zimbabwe). The Rhodesian embargo was in effect from 1966 to 1972, when it was lifted by Congressional action designating chromite as a strategic metal. It was reinstated in 1977, however, and not lifted again until the Rhodesian government changed in 1980. Throughout this period, chromite production from Rhodesia/Zimbabwe continued, and the ore was sold to the United States and western Europe through other countries. Furthermore, shortfalls resulting from the embargo were made up largely by sales from the former USSR, with which the United States and western Europe also had significant ideological differences at the time. As a result of this lesson, chromite was not included in economic sanctions imposed by the United States on South Africa during apartheid.

Long-term chromium reserves will depend on the recovery of lower-grade chromite that is disseminated through LICs or obducted ultramafic rocks that host podiform deposits. Obducted ultramafic rocks have the best possibilities because their chromites generally contain more chromium. Laterites on weathered ultramafic rocks are particularly attractive because they are relatively easy to mine and contain nickel and cobalt. During the strategic metal scare of the late 1970s, efforts were made to develop the Gasquet Mountain laterites in northern California, which were reported to contain 0.85% Ni, 0.01% Co, and 2% Cr. Recent discoveries in the Ring of Fire area of northern Quebec offer the first possibility of significant North American chromite production. Ultramafic rocks in this area contain thick chromite layers that would be economically attractive if they were located near transport, although development of road or rail access is mired in ownership disputes (Hill, 2013).

8.5 Silicon

Although most of us think of semiconductors when the word silicon comes up, we should think of iron and steel. Almost 80% of world silicon production is used as an additive in metal alloys, particularly steel making, where silicon ranks second in volume only to manganese. Silicon, largely in the form of ferrosilicon alloys containing 50–90% silicon, is used as a deoxidant in steel and as an alloy to increase tensile strength and corrosion resistance. Much of the silicon that is not used

in the steel industry goes into alloys with aluminum, copper, and nickel, where it increases fluidity and wear resistance. The remaining 20% of world silicon production goes into silicones and other semi-solids that behave like oil, **silicon carbide**, a widely used abrasive material, and silicon metal, which is used in semiconductors. Annual world silicon production amounts to about 7.4 million tonnes worth about $25 billion (Murphy and Brown, 1985; Corathers, 2013b).

Silicon production comes largely from the very abundant mineral quartz (SiO_2), which is found in a wide variety of environments. Most common are quartz-rich clastic sediments that formed in fluvial and coastal environments, as well as their lithified equivalents, sandstone and quartzite. Others include coarse-grained igneous rocks, principally pegmatites that formed during the latest stages of crystallization of granitic magmas, and hydrothermal veins consisting almost entirely of quartz. Although these materials are widespread, not all meet the stringent purity specifications demanded by industry, especially the absence of arsenic, sulphur, and phosphorus.

Ferrosilicon and silicon metal are produced in 32 countries, almost entirely by melting quartz, coke, and scrap iron in an electric furnace. The process requires large amounts of electric power. In fact, silicon production is similar to that of aluminum, in which the raw material is transported long distances to areas of cheap electrical energy, such as Norway and Canada. Iceland, which lacks quartz deposits because of the dominantly basaltic nature of its crust, uses geothermal power to make ferrosilicon from Norwegian quartz. Major importers of ferrosilicon include the steel makers, China, Japan, Germany, and the United Kingdom. Silicon reserves are simply not a problem. Quartz is one of the most abundant minerals in the crust and useable deposits are widely distributed.

8.6 Cobalt

Strictly speaking, cobalt hardly belongs in the ferroalloy group because so little of it is actually used in steel alloys. In fact, **superalloys** containing cobalt with nickel, iron, and chromium are only the second-largest market for cobalt, accounting for about 17% of world demand. They are used in applications requiring strength at high temperature including jet engines and turbines. The largest market for cobalt, about 38%, is in rechargeable batteries including some used in electric vehicles, in magnets in which cobalt is alloyed with nickel and aluminum (Alnico), as the matrix into which tungsten carbide particles are set to make cutting tools, and as a catalyst. It also forms a pigment with the bright blue color that was first used almost 4,000 years ago in Egypt (Cobalt Development Institute, 2013). Cobalt usually falls into the ferroalloy group by default because its alloys are so essential. Annual world production is about 110 thousand tonnes of cobalt, worth about $4 billion (Figure 8.1a).

8.6.1 Cobalt deposits

Although the average upper crust contains only about 27 ppm cobalt (Rudnick and Gao, 2003) there are several very different types of deposits (Hannis et al., 2009). In almost all cases, cobalt is a co-product or by-product from deposits that are mined for nickel or copper (Mishra et al., 1985). By far the most important deposits are in the Central African Copper Belt, which extends from Shaba Province in the Democratic Republic of the Congo (DRC) into Zambia (Cox et al., 2007; Hitzman et al., 2005, 2010) (Figure 8.11). These deposits are copper–cobalt-bearing sedimentary rocks of late Precambrian age (about 1 Ga) that accumulated around older continents. Copper, largely in chalcopyrite, is actually the dominant metal in most of these deposits, but cobalt, mostly in linnaeite (Co_3S_4), is so highly concentrated that it is a major co-product. Early ideas that the deposits were direct chemical precipitates from ocean water have given way to suggestions that the metals were deposited by later hydrothermal solutions moving through the sediments, a concept supported by the occurrence of copper ore in a wide variety of sedimentary rock types including limestone, shale, sandstone, and conglomerate.

Smaller amounts of cobalt are produced by Cuba, Russia, New Caledonia, Brazil, Canada, and Australia, largely as a by-product from nickel sulfide and laterite deposits. Following nickel, cobalt substitutes for nickel and iron in sulfide minerals, particularly pentlandite [$(FeNi)_9S_8$] and pyrrhotite ($Fe_{0.95}S$). It also substitutes for iron and magnesium in silicate minerals; mafic and ultramafic rocks have cobalt contents as high as 300 ppm and laterites formed on them contain cobalt (Rudnick and Gao, 2003; Berger et al., 2011). Cobalt contents of laterites rarely exceed 0.2%, in contrast to nickel grades of several percent. In magmatic deposits, which have similar nickel grades, cobalt grades are somewhat lower, almost never exceeding 0.1%. A special type of magmatic deposit, the platinum-bearing zones of the Bushveld Complex in South Africa, contain even less cobalt, but yield significant production because of the large amounts of ore that are mined for platinum (Wilburn, 2011).

Cobalt is also found in hydrothermal vein deposits of which the most famous are those in Cobalt, Ontario (Andrews et al., 1986). Although these deposits are important sources of silver, as discussed in Section 11.2, they are too small to be an important source of cobalt except at the Bou Azer deposit in Morocco (Ahmed et al., 2009). Finally, a possible new type of cobalt deposit consisting of cobalt, copper, and gold in metamorphosed sedimentary rocks has recently been recognized (Slack et al., 2010).

8.6.2 Cobalt production and environmental issues

Cobalt is produced from only 17 countries of which the DRC is by far the most important (Figure 8.19) (Shedd, 2013a). Cobalt mining and processing vary greatly, depending on the type of deposit. In almost all cases, mining and processing are controlled at least in part by the demand for copper or nickel, which are usually the dominant metal in cobalt deposits. In the Copper Belt, mining is largely in open pits with a few older underground operations. Because of inefficient early operations, tailings piles at Kolwezi and other locations in the Copper Belt contain enough cobalt to be mined again.

Most cobalt-rich ore is processed by making a sulfide mineral concentrate, which is roasted to drive off sulfur and leached of its cobalt with acid solutions. The cobalt is then

---

BOX 8.6 | REAWAKENING OF THE COPPER BELT

The dominant position of the DRC as a cobalt producer is the result of a major reawakening of the country from a steep decline. The first step in the decline was takeover of the original Belgian mining operations of Union Minière du Haut Katanga by the government-owned corporation, La Générale des Carrières et des Mines (Gécamines). Lack of routine maintenance, failure to invest in infrastructure, and misuse of funds caused Gécamines operations to deteriorate by about 95% by 1995. Civil unrest starting in 1996 prevented reconstruction, but things changed in 2002 with the advent of a more stable government and a new mining law. Foreign investment came back into the country leading to an increase in mining, although processing of cobalt ores has not kept pace (Figure 8.19b).

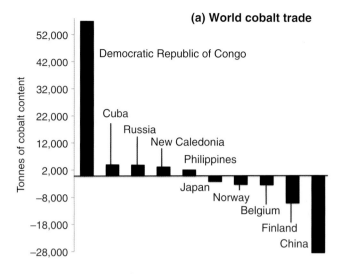

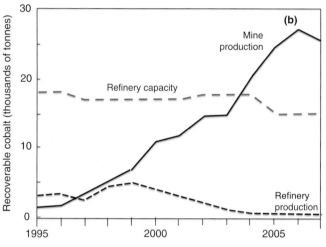

Figure 8.19 (a) World trade in cobalt ore and metal (expressed as cobalt content) for 2012 (compiled from data of the US Geological Survey). Countries extending above the line are major exporters of ore and those extending below the line import ore for domestic use and, in the case of Belgium, Norway, and Finland, for later export of metal or other cobalt products. (b) Change in cobalt production capacity in the DRC showing the common pattern of mine production increasing before refined production capacity (from Wilburn, 2011).

removed from solution by **electrowinning**, a process that is complicated by copper and zinc in leach solutions. Although mining and processing operations in the Copper Belt have local environmental problems related to flooding and rock failures, few of them are uniquely associated with cobalt. One issue that is growing is the desire by the DRC to increase domestic processing of cobalt, a wish that it has underscored recently with efforts to limit or block exports. In 2008, China signed a $9 billion agreement with the DRC in which they agreed to provide $6 billion for construction of infrastructure in exchange for a guaranteed supply of 10 million tonnes of copper and 400,000 tonnes of cobalt (Whewell, 2008).

The main anthropogenic emissions of cobalt are from combustion of fossil fuels and waste materials and exhaust from vehicles and aircraft, both of which reflect the tendency of cobalt to concentrate in fossil fuels. Cobalt and nickel mining and processing are also local sources of emissions (Nriagu, 1989). Cobalt is a component of vitamin B12, which has a role in nucleic acid metabolism and maturation of red blood cells, and has an average abundance of about 0.0015 grams in the human body (Young, 1979; Fergusson, 1990). Excessive cobalt dust, which has been observed in manufacture of cobalt-cemented tungsten carbide tools, can cause chronic bronchitis, and skin contact with soluble forms of cobalt causes dermatitis. Long-term exposure to cobalt, largely by inhalation, causes much more serious problems including respiratory failure, and congestion of the kidneys and liver. It is classified as "possibly carcinogenic" by the International Agency for Research on Cancer, although specific links to cancer have not been recognized (Moore, 1991). Recent research has detected high cobalt levels in patients who have undergone metal-on-metal hip replacements using cobalt alloys (Bozic *et al.*, 2009). Cobalt-60, an isotope of cobalt with a half-life of 5.3 years, is used as a source of radiation in the treatment of cancer, and requires special handling and disposal.

8.6.3 Cobalt trade and resources

World cobalt exports are dominated by the DRC, with comparatively small amounts from Cuba, Russia, New Caledonia, and the Philippines (Figure 8.19). The main importer, by far, is China, although other steel-making countries also import. The concentration of supply in the DRC has created continuing concern about strategic supplies (Blechman, 1985; Wilburn, 2011). Cobalt reserves containing 7.5 million tonnes of metal are quite large relative to annual production of about 110,000 tonnes of metal, but are complicated by two factors. First, cobalt is a by-product in almost all of the mines from which it is recovered. Thus, its production rate is controlled by the demand for copper, nickel, or platinum and cannot be increased substantially in response to changes in cobalt demand. Second, the deposits for which this constraint is weakest, those in the DRC, are the least secure.

Efforts have been made to locate more dependable cobalt resources elsewhere in the world (Wilburn, 2011). Some success has been achieved in the United States, where major cobalt resources are known in Mississippi Valley-type deposits in southeastern Missouri and in sediments of the Proterozoic-age Belt Supergroup in Idaho. Unfortunately, grades of these deposits are not high enough to be economic. A final possibility, cobalt-rich manganese crusts and nodules on the ocean floor, are not ready for production, as noted earlier. Thus, it is likely that cobalt supplies will remain a major concern to industrialized nations and that the Copper Belt deposits will continue to be the focal point.

8.7 Molybdenum

Molybdenum is the quintessential alloy element. In 2009, the latest year for which data are available, 80% of world consumption went into steel, superalloys, and molybdenum metal. In all of these, molybdenum provides hardness and resistance to abrasion, corrosion, and high temperatures. Corrosion-resistant steels with about 5% molybdenum are used in seawater applications, and hard steels with about 10% molybdenum are used in cutting and grinding equipment. The remaining 20% of world molybdenum consumption went into chemicals such as molybdenum-orange pigment, catalysts for oil refining, and lubricants, and into cast iron where it increases strength (Polyak, 2013a). World molybdenum production, amounting to about 260,000 tonnes annually, has a value of about $9 billion.

8.7.1 Molybdenum deposits

Molybdenum is a very scarce element, with a crustal abundance of only about 1 ppm (Rudnick and Gao, 2003). Its most common ore mineral, molybdenite, is also rare and is recovered largely from porphyry-type deposits. Porphyry copper deposits, which are described in the next chapter, contain as much as 0.05% molybdenum and are important sources of molybdenum because they are mined in large volume. Most molybdenum, however, comes from porphyry molybdenum deposits (Figure 9.11), which lack copper, but contain as much as 0.3% molybdenum with smaller amounts of by-product tin and tungsten. The large Climax and Henderson porphyry molybdenum deposits near Denver, Colorado, consist of inverted cup-shaped zones of fractured and veined rock that formed over small porphyritic, granitic stocks (Figure 9.12). Much of the water that formed these deposits is of magmatic origin and separated from the intrusion as it rose into the shallow crust and crystallized. Porphyry molybdenum deposits are not abundant in older rocks and they cluster along **Mesozoic** and **Cenozoic** convergent tectonic boundaries. Although these boundaries extend around the Pacific Ocean, those underlain by continental crust in Canada, the United States, Peru, and Chile contain larger deposits (White *et al.*, 1981).

8.7.2 Molybdenum production and environmental issues

Molybdenum is produced in only 13 countries, far fewer than those producing copper from porphyry copper deposits. In fact, many porphyry copper deposits do not recover their by-product molybdenite because molybdenum grades are too low to pay for the necessary facilities. China and the United States are the largest molybdenum producers, with most production coming from porphyry molybdenum deposits (Zeng *et al.*, 2013). Chile, which is the third largest producer, derives most production from its enormous porphyry copper deposits. Most porphyry molybdenum mines are open pits. The underground Henderson mine in Colorado, which began production in 1976, is an example of highly successful coexistence between mining and other land uses. Although the mine is near major ski areas, it is essentially invisible to passers-by. Beneficiation and tailings disposal, which are located in a valley almost 24 km from the mine, are reached by a conveyor system that includes a 15.4-km tunnel.

Because of their low grades, porphyry molybdenum beneficiation produces large volumes of tailings, which are disposed of in impoundment areas that are revegetated when abandoned, as is the case at Henderson. Processing of molybdenite concentrate involves roasting to convert MoS_2 to MoO_3, the most common form in which it is used in industry. Gases from this process must be recovered to prevent their escape into the atmosphere. Molybdenum is also used as ferromolybdenum and as the metal.

Molybdenum, itself, is an essential element to human life, with a role in the metabolism of copper. Unfortunately, too much molybdenum can limit ingestion of copper, a condition known as molybdenosis (Raisbeck *et al.*, 2006). The condition affects cattle in arid regions where groundwater and surface water is slightly alkaline, a condition that favors solubility of trace amounts of molybdenum leached from common rocks. This molybdenum then finds its way into vegetation and from there into livestock. Alkaline groundwater around molybdenum deposits can also be enriched in the element, although this is not known to have caused environmental problems.

Molybdenum trade consists largely of exports from the United States and Chile, to the rest of the world (Palencia, 1985; Polyak, 2013a). Chinese exports of molybdenum have been affected by taxes with the stated purpose of limiting excess production and loss of reserves. World molybdenum reserves contain about 11 million tonnes of metal and the outlook for further discoveries is relatively good. Because of its large reserves, China is likely to remain the dominant producer and possibly the dominant exporter in the future. Additional molybdenum could be recovered from other porphyry copper deposits where molybdenite recovery circuits have not yet been installed, and from tailings of porphyry copper deposits. Molybdenum is also enriched in some shales that contain relatively large amounts of organic matter and are commonly referred to as black shales. Molybdenum concentrations of several percent have been reported, although methods to mine and separate molybdenum remain untested. The fact that many of the porphyry copper and some of the black shale resources are in North America, Europe, and China suggests that molybdenum will continue to be a metal of choice in important alloys and other applications.

8.8 Vanadium

Although vanadium has an average crustal abundance of 138 ppm (Rudnick and Gao, 2003), more than twice that of copper, it forms very few minerals and and has relatively few uses in society. Annual world vanadium production of about 63,000 tonnes has grown more over the last few decades than any other ferroalloy element (Figure 8.1c) and is valued at less than \$2 billion (Polyak, 2013b). Vanadium minerals are so rare because V^{+3}, the oxidation state in which vanadium is found in much of the crust, is geochemically similar to Fe^{+3}, an abundant ion that is part of many common minerals. The similarity is so great that, rather than forming its own vanadium minerals, V^{+3} substitutes for Fe^{+3} in iron minerals. Magnetite contains as much as 3% V_2O_5 (the chemical form in which vanadium concentrations are quoted). Vanadium did not come to the attention of early chemists and metallurgists until 1830 and, even after that, it languished without significant applications for another 60 years because of the difficulty in separating the element from rocks (Kuck, 1985).

8.8.1 Vanadium deposits and production

The rarity of vanadium minerals means that most vanadium production comes from deposits of other elements (Goldberg et al., 1992). The largest deposits are cumulate magnetite layers in large igneous complexes like the Bushveld Complex (Figure 8.16). The magnetite layers, which are largely mono-mineralic, probably have an origin similar to that of chromite layers lower in the Bushveld. They could not compete with large BIF deposits as iron ores, but their high vanadium content makes them economically attractive. At least 21 magnetite layers are known, many of which are laterally extensive, providing large reserves. Vanadium is also present in cylindrical pegmatite pipes, such as Kennedy's Vale, that cut across layering in the Bushveld and might be original cumulate magnetite layers that funneled downward from the upper part of the complex. Similar deposits are located in China (Panzhihua) and Russia (Kachkanar) (Hao et al., 2012).

Vanadium is also found in **sandstone** vanadium deposits, most of which form from what are essentially groundwaters with dissolved pentavalent (V^{+5}) vanadium that was leached from weathered rocks and minerals. The pentavalent vanadium is deposited if the solution is reduced or evaporated, a situation very similar to the behavior of U^{6+} in groundwaters, as discussed on Chapter 7. In the Colorado Plateau area of the western United States, vanadium precipitated in reduced zones rich in organic matter. This geochemical process is very similar to that which forms uranium ores, and vanadium is a common constituent of many sandstone-type uranium deposits, often in the mineral carnotite [$K_2(UO_2)$ $(VO_4)_2.3H_2O$]. Vanadium is so important in these deposits, that one district is named Uravan. In Western Australia, vanadium was precipitated in the Yeelirrie uranium deposit when groundwater evaporated as it flowed through a near-surface aquifer. Vanadium also concentrates in some alkali igneous rocks and carbonatites, and is produced from vein and disseminated ores in the Wilson Springs area around the Magnet Cove complex in Arkansas.

BOX 8.7 | VANADIUM IN ORGANIC MATTER

Vanadium is also a common constituent of organic material, including coal and oil. In fact, vanadium mining began at the Mina Ragra asphaltite deposit in Peru, although these deposits are not mined at present (Breit, 1992). The largest resource of this type of material is in the tar sands of Alberta. Although the vanadium content of tar sand itself is only 0.02–0.05%, fly ash from the processing plants contains up to 5% V_2O_5. Efforts to recover this vanadium have not been economically successful so far, but they represent a large cushion against future demand.

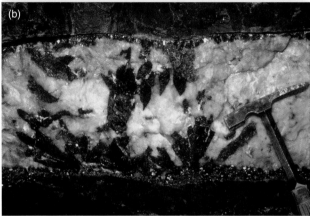

Figure 8.20 (a) Mapochs vanadium mine, eastern Transvaal, South Africa. The vanadium-bearing magnetite layers dip steeply toward the viewer and have been mined by stripping off the side of the hill (photograph by the authors). (b) Vein of quartz (white) with crystals of the tungsten ore mineral wolframite (dark) from the Dajishan mine, China (photograph courtesy of James E. Elliott, US Geological Survey). See color plate section.

Vanadium mining is carried out largely by open pit (Figure 8.20a), although some Colorado Plateau uranium–vanadium mines are underground. The process used for recovery of vanadium from magnetite, which dominates the supply picture, depends on its vanadium content. High-vanadium magnetite (2% V_2O_5) is usually roasted and leached to release vanadium, which is then precipitated to form vanadium chemicals. Low-vanadium magnetite is smelted in a blast furnace to produce pig iron and vanadium-bearing slag that can be roasted and leached to release vanadium. Some South African slags contain more than 20% V_2O_5 and are highly desirable. Some heavy organic residues and ash from petroleum refining have vanadium contents of as much as 34% V_2O_5, and vanadium by-products from oil refining and ash from fossil-fuel combustion account for about 10% of world production.

Problems related to vanadium mining come largely from its close geochemical association with uranium and selenium in Colorado Plateau uranium–vanadium deposits. Both of these elements are enriched in vanadium deposits and both elements require special disposal of beneficiation and treatment wastes. Processing of some vanadium ore has caused dust-inhalation problems including bronchitis and forms of pneumonia, although these can be largely eliminated by use of respirators (WHO, 1988). Point-source anthropogenic emissions of vanadium are highest around vanadium-producing facilities, but are dwarfed on the global scale by emissions from fossil-fuel combustion. Vanadium concentrations in oil are generally higher than in coal, and are highest in heavy fuel oil. Power plants that burn fuel oil are among the largest anthropogenic vanadium emitters (Nriagu, 1989; Visschedijk et al., 2013).

8.8.2 Vanadium markets and trade

About 85% of vanadium is used in steel, with 40% going into HSLA steels in which as little as 0.1% vanadium greatly enhances strength, toughness, and ductility. Vanadium steels are used in high-rise buildings, offshore oil drilling platforms, and pipelines for oil and natural gas. Vanadium is also alloyed directly with titanium and aluminum to improve the strength of these metals. Vanadium chemicals are used in production of nylon and polyester resins, as well as special glass that will not pass ultraviolet radiation and can be used to protect art objects, fabrics, and furniture.

Vanadium production comes from South Africa, China, and Russia, and the main exporter is South Africa (Figure 8.21a). Although North America has domestic production, it imports far more ore and slag than it exports. World reserves contain about 14 million tonnes of vanadium, essentially all of which is in South Africa, China, and Russia. In view of the important uses of vanadium and the uncertain outlook for long-term supplies from these countries, exploration is needed to locate vanadium deposits elsewhere in the world.

8.9 Tungsten

Tungsten is usually classed as a ferroalloy metal, although only about 20% of it is used in steels and related alloys. The main use for tungsten, accounting for about 54%, is the manufacturing of tungsten carbide, one of the hardest synthetic materials used in industry. It is widely used in cutting and wear-resistant materials, particularly at high temperatures. One of its most

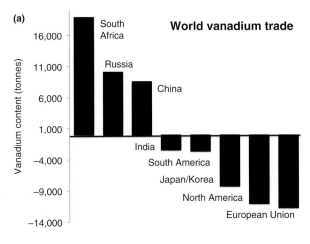

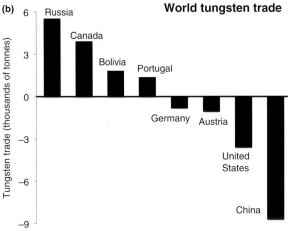

Figure 8.21 (a) World vanadium trade for 2012 based on data of the US Geological Survey and Vanitec. (b) World tungsten trade for 2009 based on data from Pitfield *et al.* (2011). Countries plotting above the line are net exporters; those below the line are net importers.

important markets, drill bits for oil wells, takes advantage of this property. Steel and other alloys, which make up another 27% of tungsten markets, go largely into high-speed and heat-resistant steels. Products made from pure metal are the second major market, mostly for electrical and electronic applications, although the transition away from incandescent light bulbs and cathode-ray television tubes has cut into this market (Pitfield *et al.*, 2011). The metal is also used in superalloys with copper or silver and in the chemical industry. Annual world production of newly mined tungsten is about 70,000 tonnes, worth about $3.5 billion.

8.9.1 Tungsten deposits and production

Tungsten is a scarce element, with an average crustal abundance of about 2 ppm (Rudnick and Gao, 2003). It forms

several minerals, including scheelite [$CaWO_4$) and wolframite ([Fe,Mn]WO_4]. Most tungsten deposits form by late magmatic (**pegmatite**) and hydrothermal processes associated with granitic intrusive rocks and magmatic hydrothermal solutions (Bues, 1986; Pitfield *et al.*, 2009). There are two main types of hydrothermal tungsten deposits. Scheelite skarns are found where granitic magmas intruded limestone. Some of these deposits, including Vostok-2 in Russia, Mactung in Canada and King Island in Tasmania, formed in the limestone wallrock (exoskarn), but others, such as Shizhuyuan in China, formed largely by replacement and veining within the intrusion itself (endoskarn). These deposits are estimated to contain about 40% of world tungsten reserves.

Wolframite deposits, which contain another 40% of world resources, consist of individual veins or stockwork systems of many intersecting veins and veinlets (Figure 8.20b). These are particularly well developed in Hunan and Jiangxi provinces in China (Yaoigangxian, Xihuashan) (Elliott, 1992; Sheng, 2001), but are also found in southwest England (Hermedon), the Erzgebirge of the Czech Republic, and Portugal. At Panasqueira in Portugal, veins surround **cupolas** or protrusions at the top of a granite intrusion. Many tungsten veins are closely associated with tin and molybdenum, which are important co-products. Tungsten is also a by-product of some porphyry molybdenum deposits, as well as some tin deposits in Bolivia.

Most tungsten mining involves small-scale underground operations with the attendant safety concerns. Ores are beneficiated by standard methods, although hand-sorting of ore minerals is still employed widely in China. Tungsten concentrates are processed by dissolving scheelite or wolframite in a concentrated sodium hydroxide solution and precipitating ammonium paratungstate (APT) [$(NH_4)_{10}(W_{12}O_{41}).5H_2O$)], which can be converted into tungsten metal or ferrotungsten (Smith, 1990). Taking advantage of their iron content, wolframite concentrates can be smelted directly to produce ferrotungsten.

Tungsten is an essential nutrient in some bacteria, which makes it the heaviest element used by living organisms (Pitfield *et al.*, 2011). It is not known to play a role in human metabolism and has usually been considered to be of little concern from an environmental standpoint. Occupational exposure to tungsten, usually in facilities that work with tungsten carbide, generates particulate material that can be inhaled or ingested and that does cause illness, but wider dispersal of tungsten is not a known issue. Because of its high atomic weight (184) tungsten has been substituted for the slightly heavier lead (207) or depleted uranium (238) in so-called "green bullets." It was hoped that tungsten would be less **mobile** (see Section 3.3.3) than lead and uranium, and prevent contamination of practice areas and combat zones. Unfortunately, as might have been predicted from the lack of native tungsten in natural environments, tungsten can dissolve in surface waters and is in fact mobile (Dermatas *et al.*, 2006).

8.9.2 Tungsten resources, markets, and trade

Tungsten is mined in 19 countries, but only one of them matters. Chinese tungsten production makes up almost 85% of world tungsten production. Russia, Canada, Bolivia, and Austria supply most of the rest. The Mittersil mine near Salzburg, Austria, is in a protected area and has all operations underground, including disposal of mine waste and tailings. Tungsten markets are under siege on all fronts. New hard materials such as polycrystalline synthetic diamond, boron nitride, and titanium carbide are being substituted for tungsten carbide. Coatings of these materials are also being put on tungsten carbide to make it last longer, thus cutting down on replacements. As a result, tungsten production is growing more slowly than any of the low-volume ferroalloys, although its unique combination of hardness and high melting temperature will probably hold a core of markets.

Tungsten is traded in multiple forms, complicating any simple global summary. Ore and concentrate trade is dominated by exports from Russia, Canada, Bolivia, and Portugal and imports by China and the United States (Figure 8.21b). Two other forms of tungsten also trade globally, both in amounts slightly smaller than tungsten ore and concentrate. These are tungstate chemicals and ferrotungsten, both of which are exported largely by China and imported largely by Germany, Japan, and the United States (Seddon, 2012). World tungsten reserves total 3.2 million tonnes of contained metal, slightly more than half in China. North American reserves are sizeable thanks to large deposits in Yukon, Canada. Large reserves are also found in Russia and Australia, but remaining European reserves are small (Anstett *et al.*, 1985; Shedd, 2013b).

8.10 Niobium (columbium) and coltan

Niobium is an unfamiliar element. Not only is it used in small amounts and in inconspicuous applications, but it also goes by the pseudonym of columbium. Making matters worse, it is often mistaken for the very similar element tantalum.

Columbium was named in 1801 when Charles Hatchett tested a newly discovered mineral named columbite. Hatchett could tell that a new element was present but could not isolate it. Nevertheless, he named the element columbium. In 1809, another English chemist, William Wollaston, compared columbite to another new mineral tantalite that contained tantalum. Wollaston saw no difference between the two minerals, and claimed that columbite was tantalite and therefore that columbium was tantalum. The confusion was cleared up in 1844 when Heinrich Rose isolated the new element in columbite, showed that it was not tantalum, and named it niobium. Unfortunately, industry sells the metal under the name of columbium. Further confusion is caused by the extensive **solid solution** between niobium and tantalum in the columbite–tantalite mineral series [(Fe,Mn,Mg)(Nb,Ta)$_2)_6$]. Finally, a tantalum-rich member of this series is referred to as coltan, which is discussed in Section 10.15.

About 63,000 tonnes of niobium, worth $2 to $4 billion, are sold in various forms each year. The main demand for niobium, accounting for about 90%, is in HSLA and related **microalloy** steels, where addition of as little as 0.1% improves the mechanical properties of carbon steel. Niobium is also used in larger amounts in stainless steel and nickel-based alloys such as Inconel. Most niobium is traded as mineral concentrates or as ferroniobium for direct use in steel making (Crockett and Sutphin, 1993; Papp, 2013B). Annual world production has increased by about 500% since 1990, highlighting its strategic markets.

Niobium has a relatively low crustal abundance of about 8 ppm and forms one main ore mineral, pyrochlore, which has a very complex composition best represented as A$_2$B$_2$C$_7$, in which A is any combination of sodium, calcium, and cerium, B is niobium and/or titanium with minor tantalum, and C is oxygen, fluorine, or hydroxyl (OH). The largest niobium deposits consist of veins and disseminations of pyrochlore in carbonatites and quartz-poor, felsic intrusive rocks that were emplaced as shallow intrusions in rifted areas of continents. Weathered carbonatites have highest grades; at the Barreiro carbonatite complex near Araxá, Brazil, for instance, weathering has almost doubled the grade of niobium ore (Filho *et al.*, 1984). Brazil accounts for about 90% of world production, largely from Barreiro, Catalão, and similar carbonatites, and Canada provides most of the rest from the Niobec unweathered carbonatite deposit near Chicoutimi, Quebec. Smaller niobium deposits take the form of columbite in veins and pegmatites associated with granites, including the not-yet developed Strange Lake and Thor Lake deposits in Canada. Small amounts of columbite are produced from weathered rocks of this type in the Jos Plateau of Nigeria and are a by-product of some lithium pegmatites, mainly in Australia and the DRC (Moller *et al.*, 1989; Shaw *et al.*, 2011). Deposits of this type supply coltan. World niobium reserves contain about 4 million tonnes of metal, divided between Brazil with about 90% and Canada with almost all of the rest.

8.11 The future of ferrous metals

Our strong global interdependence for essential steel-making minerals is undeniable. The United States, Russia, and Canada have good supplies of iron ore, but most other major steel-making countries need to import it. Of the major steel makers, Russia, has good supplies of ferroalloy elements, but with a few exceptions such as tungsten in China and Canada, none of other the major steel-making countries have an adequate supply. Regional groups can overcome some of these issues, but by no means all. For instance, in North America, the United States and Canada, together, have excellent iron ore supplies and important supplies of four ferroalloy metals. However, they must depend on outside sources for manganese (and for chromium if the Ring of Fire deposits are not developed). Western Europe, China, and Japan depend on imports for all of the ferroalloy metals, and China and Japan even supply their iron ore needs through imports.

South Africa and the Commonwealth of Independent States (CIS) countries are major producers of all six of the important ferroalloy metals, not counting silicon, for which reserves are not a problem anywhere. This fact has been of continuing concern to strategic planners in China, Japan, North America, and western Europe. Changing political alignments in the CIS countries have made some of these reserves more available to outside markets, but geographic proximity is the ultimate factor in determining strategic availability of mineral resources. South Africa has developed over the last few decades as the best alternative to CIS mineral supplies. It is a major source of manganese, nickel, chromium, and vanadium, the most essential of the ferroalloy elements. The highly efficient and dependable operation of South African industry has earned it an impressive list of customers.

Although steel and ferroalloy metals will probably not become growth industries, their future is a solid one, and all industrial countries must take the necessary steps to assure access to adequate supplies. The main structural change that will affect these steps is **forward integration** in which supplier countries produce ferroalloys rather than simply shipping mineral concentrates to their customers. This appears to be an inevitable consequence of growing **resource nationalism** in producer countries. The response to this in customer countries will probably be to dismantle domestic plants for production of ferroalloy products from imported ore and concentrates. This will produce a level of dependence on imports that is considerably more serious than one based on imports of ore concentrates. In times of supply emergency, it is often possible to find small supplies of domestic ore. It will be much more difficult to locate emergency supplies of ferroalloy products.

All of this is written with the traditional "strategic mineral" mindset that has prevailed in the western world since the end of the World War II. Since the end of the Cold War, countries have become more heavily interdependent with mineral-related trade crisscrossing the globe. It is just possible that we have reached a stage where embargoes and supply disruptions would be equally damaging to all sides. If this attitude can dominate, and demands for a greater degree of local processing by ore producers can be met, all partners in the ferrous metal business should be able to find suitable supplies.

CHAPTER

9 Light and base metals

Like it or not, the dreary term **non-ferrous metals** is widely used for the many metals that are not closely related to steel making. They are divided further into the **light metals**, aluminum, magnesium, and titanium and the **base metals**, copper, lead, zinc, and tin, which are discussed in this chapter. Other non-ferrous metals, including antimony, arsenic, bismuth, cadmium, germanium, hafnium, indium, mercury, rare earths, rhenium, selenium, tantalum, thallium, and zirconium are referred to here as the technology metals, which are discussed in Chapter 10. A final group of non-ferrous metals, the **precious metals**, gold, silver, and platinum-group elements, is discussed in Chapter 11. The total value of world light and base metal production, about $300 billion annually, is almost three times that of the ferroalloy metals, but well behind the $660 billion plus of iron ore and steel. The dominant non-ferrous metals are aluminum and copper, with global production worth $100–150 billion each, followed by zinc at about $30 billion, lead and bauxite (aluminum ore) at $10 billion, and then tin and magnesium in the $3–9 billion range (Figure 9.1a).

Production of most non-ferrous metals has increased following two different trends. Figure 9.1b shows that prices of all metals lagged the consumer price index (CPI) until about 1980. After about 2000 prices of copper, lead, zinc, and magnesium rose spectacularly, reflecting demand from China. Prices of most other metals, especially aluminum and titanium, have not participated significantly in the China-driven price increase. Production of the light metals, especially aluminum and magnesium, has increased at a rate almost double that of steel (Figure 9.1d). Production of the base metals, has not matched this pace; copper and zinc production have essentially followed steel, and lead and tin have lagged

considerably. Table 9.1 summarizes major producers and reserves and Table 9.2 provides information on sizes of important deposits.

9.1 Light metals

The light metals have unusually low densities compared to most metals (Polmear, 1981). In contrast to densities of 7.87 g/cm^3 for iron and 8.96 g/cm^3 for copper, densities of the light metals are only 1.74 for magnesium, 2.7 for aluminum, and 4.51 g/cm^3 for titanium. All three light metals are relatively strong, but aluminum and magnesium markets are restricted by their low melting temperatures of 650 and 660 °C, respectively, in comparison to 1,535 °C for iron. Titanium, which melts at 1,678 °C, competes in high-temperature applications, and aluminum alloys containing iron, silicon, and vanadium are also being developed. Their combination of light weight and strength has generated strong demand for the light metals, with annual production valued at almost $120 billion.

Light metals have high crustal abundances. Aluminum is the third most abundant element in the crust, magnesium is seventh, and titanium is ninth, putting them in the same abundant metal group with iron, the fourth most abundant element. That the light metals came into widespread use so much later than iron and the scarcer base metals reflects the difficulties involved in liberating them from their ores.

9.1.1 Aluminum

Aluminum was first isolated as a metal in 1825, five millennia after iron began to be used. For several decades after that, it

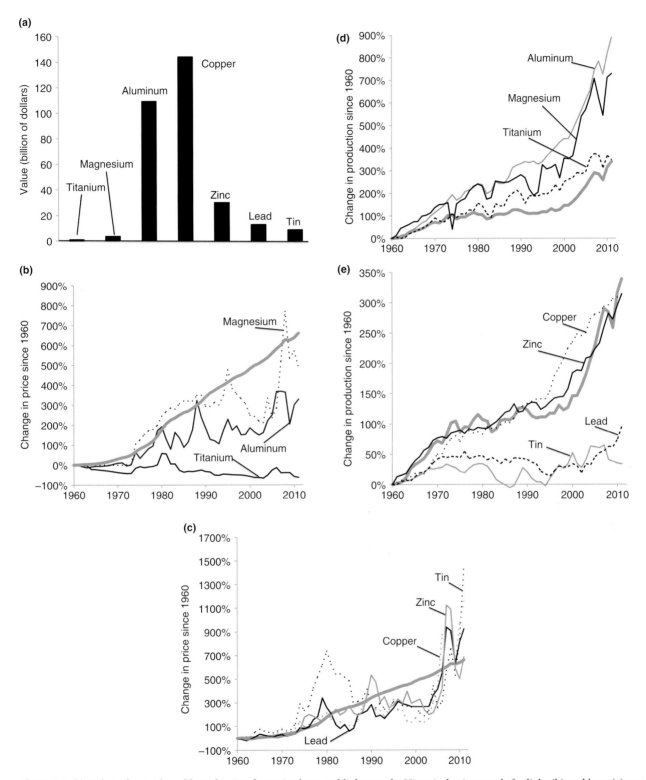

Figure 9.1 (a) Value of annual world production for major base and light metals. Historical price trends for light (b) and base (c) metals relative to CPI (thick gray line). Growth of world light- (d) and base-metal (e) mine (primary) production compared to steel (thick gray line) (all data from the US Geological Survey).

Table 9.1 Major producers and reserves for aluminum, bauxite and the light and base metals (from US Geological Survey Commodity Summaries)

Commodity	Country	Production (tonnes)	Reserves (tonnes)
Light metals			
Aluminum	China	21,500,000	
	Russia	3,950,000	
	United States	1,950,000	
Bauxite	Australia	77,000,000	6,000,000,000
	China	47,000,000	830,000,000
	Brazil	34,200,000	2,600,000,000
Titanium*	South Africa	1,100,000	63,000,000
	China	950,000	200,000,000
	Australia	940,000	160,000,000
Magnesium**	China	698,000	
	Russia	30,000	
	Israel	28,000	
Base metals			
Copper	Chile	5,700,000	190,000,000
	China	1,650,000	30,000,000
	Peru	1,300,000	70,000,000
Zinc	China	5,000,000	43,000,000
	Australia	1,400,000	64,000,000
	Peru	1,290,000	24,000,000
Lead	China	3,000,000	14,000,000
	Australia	690,000	36,000,000
	United States	340,000	5,000,000
Tin	China	100,000	1,500,000
	Indonesia	40,000	800,000
	Peru	26,100	91,000

* Data for titanium are for ilmenite mineral concentrates.

** Magnesium reserves are in seawater

was so difficult to produce that it was essentially a precious metal, used only in jewelry and other ornaments. Aluminum did not become available in large amounts until the Hall–Heroult electrolytic process was developed in 1886 (Burkin, 1987), starting it on a path of steady growth to the present annual production of 44 million tonnes worth about $110 billion in 2012.

Geology of bauxite and other aluminum ore deposits

Aluminum makes up over 8% of Earth's crust and is a major constituent of many common minerals. Despite its abundance in a lot of minerals, almost all global aluminum production comes from **bauxite**, which is made up largely of diaspore and böehmite, which have the same composition [AlO(OH)] but different crystal structures, as well as gibbsite [(Al(OH)₃]. Bauxite contains 40–60% Al_2O_3 (the chemical form in which aluminum abundances are reported), with the rest made up of clay minerals, iron and titanium oxides, and quartz and other forms of silica. It is the best ore of aluminum because it lacks elements like magnesium, calcium, and potassium that would complicate the aluminum-extraction process.

Bauxite is actually a special, aluminum-rich **laterite**. It forms by intense weathering that dissolves most other elements in the rock, leaving a residuum enriched in aluminum (Patterson *et al.*, 1986; Bardossy and Aleva, 1990). Formation of aluminous laterites and bauxite deposits requires abundant rainfall to dissolve unwanted rock constituents and high temperatures to speed the dissolution reactions, conditions found in the broad equatorial band of tropical and subtropical climates (Figure 9.2). The setting must allow precipitation to move downward through the rock, and erosion must be limited so that weathering affects the same rock for a long time. This sort of lateritic weathering affects all types of rocks, but it makes bauxite only where it affects rocks with a high aluminum content, such as felsic igneous rocks and shales (Figure 9.3). Nickel and cobalt are mined from laterites that form on ultramafic, mantle rocks, as discussed in Chapter 8.

The absence of iron in bauxites is curious. It is present in many rocks that are weathered to produce bauxite and we know that iron forms numerous red to brown oxides in many soils. Iron even forms laterites and "hard-pan" soils in many tropical areas. Its absence in bauxites requires special weathering conditions that dissolve iron but retain aluminum. This happens where decay of abundant vegetation consumes oxygen in groundwater, thus keeping iron in the more soluble ferrous state. In areas with less vegetation, ferric iron dominates and iron oxides are deposited in the laterite (Petersen, 1971).

Other less-abundant elements also concentrate in bauxite. Much of the world's gallium, an element with geochemical properties similar to aluminum, is a by-product of the processing of bauxite to form alumina. A serendipitous example of bedrock contributions to bauxite deposits is Boddington in Western Australia (Figure 9.2), which developed over an underlying, previously unknown gold deposit (Davy and El-Ansary, 1986). Bauxite is also enriched in titanium, although this metal has not yet been recovered economically.

Most bauxite deposits formed on aluminum-rich bedrock of various kinds (Figure 9.3). The large deposits around

Table 9.2 Large deposits of light and base metals. Deposit sizes shown here include both original and remaining reserves and are intended to provide an order-of-magnitude measure of the size of mineral deposits of each commodity. Data are based on annual reports of companies and government

Deposit name	Location	Types	Metal content (tonnes)	
Aluminum deposits (tonnes)			Aluminum	
Weipa	Australia	Laterite	406,510,000	
Sangaredi–Bidikoum–Silidara	Guinea	Laterite	81,000,000	
Worsley (Boddington)	Australia	Laterite	57,494,400	
Gove	Australia	Laterite	41,075,000	
Titanium deposits (tonnes)			Titanium	
Namalope	Mozambique	Beach placer	56,000,000	
Richards Bay	South Africa	Beach placer	20,000,000	
Copper deposits (tonnes)			Copper	
El Teniente	Chile	Porphyry	106,624,000	
Pebble	United States	Porphyry	81,000,000	
Olympic Dam	Australia	IOCG	78,952,500	
Chuquicamata	Chile	Porphyry	60,481,200	
Lubin District	Poland	Sediment-hosted	52,000,000	
Tenke-Fungurume	Democratic Republic of Congo	Sediment-hosted	36,716,000	
Lead–zinc deposits (tonnes)			Lead	Zinc
Howards Pass	Canada	SEDEX	11,000,000	27,500,000
Red Dog	United States	SEDEX	7,590,000	27,390,000
Broken Hill	Australia	SEDEX	20,500,000	22,550,000
McArthur River	Australia	SEDEX	9,717,000	21,804,000
Brunswick No. 12	Canada	VMS	9,721,920	4,018,820
Tin deposits (tonnes)			Tin	
Potosi (Cerro Rico)	Bolivia	Tin–silver vein	2,898,000	
Kinta Valley	Malaysia	Placer	2,000,000	
Longtaoshan (Dachang)	China	Skarn	1,500,000	
Tongkeng-Changpo (Dachang)	China	Skarn	800,000	
Laochang (Gejiu)	China	Skarn	720,000	

Weipa, Australia, and in Suriname and Guyana (Figure 9.2) formed on sediments rich in kaolinite, an aluminum-rich clay mineral. Other important deposits in Western Australia and Guinea formed on felsic igneous and metamorphic rocks. However, some very important deposits, sometimes referred to as **karst** bauxite deposits, do not obey the rules. In Jamaica, bauxite is found on limestone, a rock consisting almost entirely of calcite and dolomite with almost no aluminum. The limestone has been extensively dissolved, forming an irregular surface and underlying caves, known as a karst terrane. Bauxite forms a layer on the top of the karst terrane; it is thought to have come from clay minerals that mixed into the limestone or volcanic tuff that was deposited on top of it, and that concentrated into a surficial layer as the limestone was dissolved (Comer, 1974). Smaller, generally similar deposits are found in Hungary and central Kazakhstan.

Although most bauxite deposits are eventually destroyed by erosion, some ancient ones, known as **paleobauxite**, are preserved in older sedimentary rocks (Bogatyrev and Zhukov, 2009). Most of the bauxite in southern Europe, Kazakhstan, and Russia is probably of this type, as are deposits in the Magnet Cove area of Arkansas. In some areas, these deposits were eroded, transported, and deposited as bauxite-rich sediment that can also be mined.

As might be expected, bauxite deposits are scarce or absent in cooler climates and in areas where Pleistocene glaciation has scraped the regolith away (Retallack, 2008). Countries like Canada and Russia, which have this problem, have considered recovery of aluminum from aluminium-rich igneous rocks like syenite, a quartz-poor intrusive rock that also forms important laterite deposits in Arkansas (Bliss, 1976; Patterson, 1977; Hosterman et al., 1990).

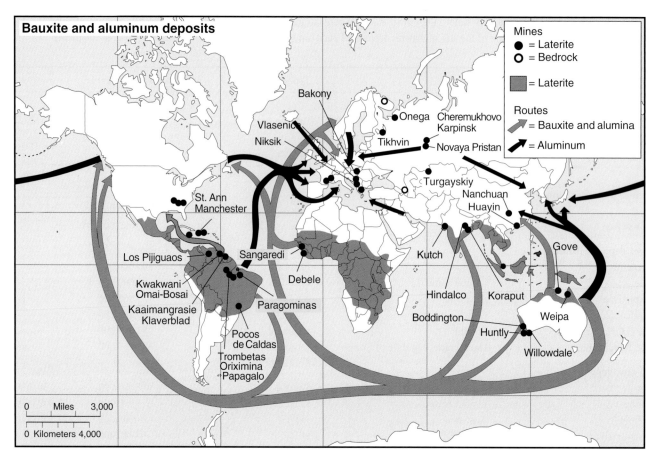

Figure 9.2 Distribution of bauxite and bedrock aluminum deposits showing worldwide distribution of lateritic soils and general trade routes for both bauxite and aluminum metal (compiled from Patterson *et al.*, 1986; Hosterman *et al.*, 1990; Menzie *et al.*, 2010)

Production of bauxite and aluminum

Aluminum ore comes from 29 countries (Bray, 2013), with Australia, China, Brazil, India, Guinea, and Jamaica accounting for 90% of production. Because most bauxite forms flat layers at or near the surface, it is extracted largely by strip-mining type methods (Figure 9.4). Bauxite is converted to aluminum in a two-step process (Figure 9.5). In the first step, known as the Bayer process, a solution of caustic soda (NaOH) is used to leach aluminum from the bauxite. The dissolved aluminum is then precipitated as $Al_2O_3.3H_2O$, which is heated, or calcined, to drive off the water, leaving pure **alumina** (Al_2O_3). Waste from the process, known as red mud, is toxic and must be disposed of properly. The main producers of alumina are China, Australia, Brazil, India, United States, and Russia, which account for 80% of production. The appearance of the United States and Russia on this list reflects efforts by both countries to have a greater control of their aluminum supplies.

In the second step of aluminum production, known as the Hall–Heroult process, alumina is dissolved in a 950 °C molten "pot" of cryolite and other fluorides of calcium and aluminum. Electric current is passed through this bath between a carbon electrode and the carbon walls of the container, reducing aluminum ions to molten aluminum metal that is cast into ingots or other forms (Figure 9.5). Conversion of alumina to aluminum metal yields gas emissions consisting largely of CO_2 and perfluorocarbon (PFC) gases, CF_4 and C_2F_6. Fluorocarbons are among the most potent greenhouse gases being emitted by industry today, about 6,500 to 9,200 times more powerful than CO_2, and they account for as much as 1% of global anthropogenic greenhouse effects (Abrahamson, 1992). Engineering improvements have decreased emissions from aluminum production by about 90% since 1990, and they currently average about 0.6 tonnes CO_2e (CO_2-eqiuvalent, the form in which carbon-bearing gases are measured) (World Aluminum Institute, 2013). Little is known about natural sources of these gases.

Aluminum production requires about 542 GJ/tonne compared to only about 15 GJ/tonne for steel (Price *et al.*, 2011). Production of alumina uses largely heat and can be carried out

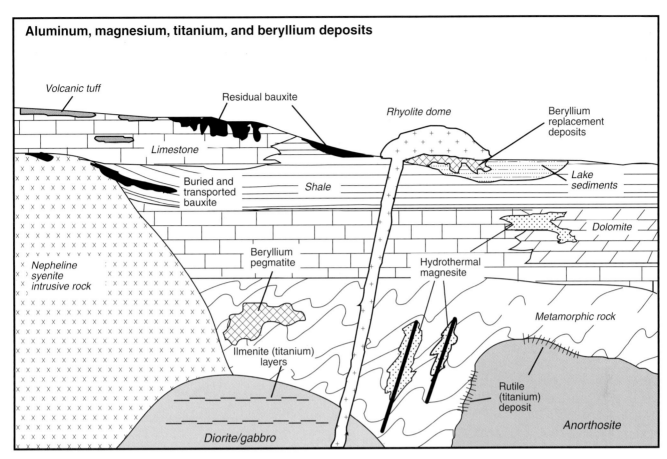

Aluminum, magnesium, titanium, and beryllium deposits

Figure 9.3 Schematic illustration of bauxite, titanium, magnesium, and beryllium deposit types. As shown here, bauxite is residual if it is still in the location where it formed by weathering, or transported if it has been eroded and deposited as a sediment in another location.

in areas with abundant fuel of any sort, but aluminum reduction, which accounts for about 80% of total energy consumption, depends on electricity (Boercker, 1979). The need for cheap electricity caused early aluminum reduction facilities to be located in areas of abundant hydroelectric power, including the Tennessee River valley in the United States, although most newer plants are in less-populated areas with lower overall power demand. Abundant hydroelectric power allowed countries with no bauxite ore, like Canada and Norway, to become aluminum producers (Figure 9.4). Hydroelectric power has not grown significantly during the last few decades and the increased level of aluminum production has been powered largely by coal, which accounts for more than half of the present power mix. Elsewhere, geothermal power is used in Iceland, and natural gas is used in Abu Dhabi, Dubai, and Bahrain. Countries with limited energy supplies, such as Japan, have curtailed aluminum production and increased imports.

Aluminum is often referred to as "packaged power" and "congealed electricity" to emphasize the importance of power in its manufacture. The "embodied energy" of aluminum, which refers to the energy required for its entire life-cycle from manufacture to disposal, is 155 MJ/kg, much higher than other materials such as steel (20.1 MJ/kg) or common brick (3 MJ/kg) (Hammond and Jones, 2008; Gutowski et al., 2013). The overriding importance of energy in aluminum production creates considerable debate about the real cost of power, and the nature of energy subsidies provided by competing countries (OECD, 1983; Peck, 1988). Recycling is a particularly important part of global aluminum production because aluminum metal can be remelted and reformed for only about 5% of the energy needed to produce metal from bauxite, and with 95% lower emissions. About 30% of current global aluminum consumption comes from recycled material, compared to less than 20% in 1980. Recycling of containers has been most successful, with rates reaching almost 70% globally, especially where regulations require that cans be returned (Gatti et al., 2008). Use of alternative energy sources, especially geothermal, has been shown in life-cycle analyses to minimize overall CO_2 emissions, even with long-distance transport of alumina or bauxite (Institute of Economic Studies, University of Iceland, 2009).

Figure 9.4 Aluminum production. (a) Strip mining of paleobauxite at MacKenzie, Guyana, which is buried beneath younger sediments. (b) Red mud waste from processing of bauxite, Aughinish, Erie (photograph courtesy of International Aluminum Institute). (c) Potroom where alumina is converted to aluminum, Dubai Smelter (photograph courtesy of International Aluminum Institute). (d) The largest aluminum refinery in Europe at Sunndalsora, Norway, uses hydroelectric power to process imported alumina (photograph courtesy of Oivind Leren/ Hydro Sunndal). See color plate section.

Aluminum does not dissolve significantly in neutral waters, and therefore does not commonly enter organisms through water or food. It is a known neurotoxin in some vertebrates, especially in lakes with unusually low or high pH, where it might have greater solubility. Accumulation of aluminum in the human body is not common and is usually associated with health irregularities. Suggestions that it plays a role in Alzheimer's disease have been a subject of continuing investigation (Romero, 1991; Kolata, 1992; Kawahar and Kato-Negishi, 2011).

Aluminum markets, reserves, and trade

About 90% of world bauxite production is converted to aluminum metal, with the remainder used to produce aluminum oxide abrasives, refractory materials, and chemical compounds. Aluminum markets are growing largely at the expense of steel. The main market for aluminum is the transportation sector, which accounts for 27% of global demand, followed by construction with 24%. Packaging, including aluminum cans, makes up only 13% of the global market. In the transportation sector, automobile manufacturing accounts for about 50% of consumption and is growing (Figure 9.6). For instance, use of aluminum rather than steel for the body and other parts in the 2014 Ford F-150 truck allowed it to shed 318 kg . The use of aluminum in US autos is expected to increase from today's average of 168 kg to about 250 kg by 2025 (Kelly, 2012). Airframe manufacture, which accounts for only about 20% of the transportation market, is also a growing market, although it faces competition from titanium and carbon. In addition to the low ratio of weight to strength, which makes aluminum so useful in the construction and transport industries, it is also valued for its high electrical and thermal conductivities. It has replaced copper in high-voltage electric transmission lines where there is a

Figure 1.4 (a) The German ultra-deep borehole project was undertaken to provide information on geologic conditions at depth in Earth's crust. Two separate holes, which were drilled from the station shown here, reached a total depth of 4 km. Temperatures at the bottom of the holes were 120 °C and pressures were 40 megapascals, conditions that are extremely challenging for drilling equipment (photograph courtesy of KTB-Archive, GFZ Potsdam). (b) The Finiston Pit (also known as the Super Pit) at Kalgoorlie, Western Australia is one of the largest open pit mines in the world, measuring 3.5 km long, 1.5 km wide, and 0.6 km deep. The pit moves about 15 million tonnes of rock annually containing about 20,000 kg of gold. Waste rock removed to reach the gold ore is placed on the gray waste-rock dumps to the right of the mine and pulverized ore after processing to remove gold is placed in the white tailings ponds in the upper right.

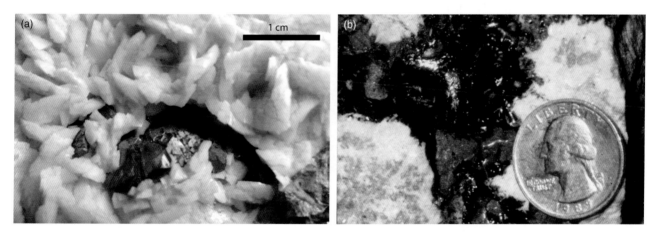

Figure 2.2 (a) The dark crystals of galena grew on top of smaller crystals of white dolomite. Whereas the dolomite contains almost no zinc, sphalerite contains as much as 64% by weight and is our principal ore mineral for zinc. (b) The dark material is solidified crude oil known as **bitumen**, which is lining a cavity in the white limestone.

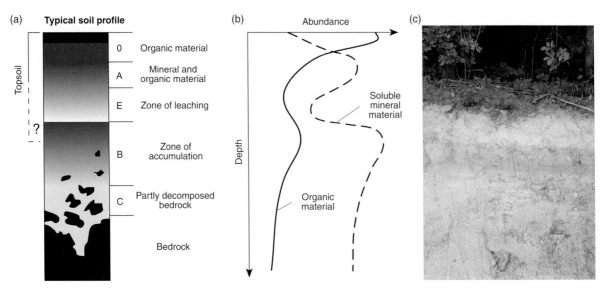

Figure Box 2.4.1 Typical soil profile (a) with the related abundances of soluble mineral and organic material (b) of a well-drained soil in a humid temperate region showing zones in which soluble mineral and organic matter are leached and accumulated. (c) Photo of soil profile showing organic matter at the top (gray), leached A zone (white) and zone of accumulation (reddish-brown).

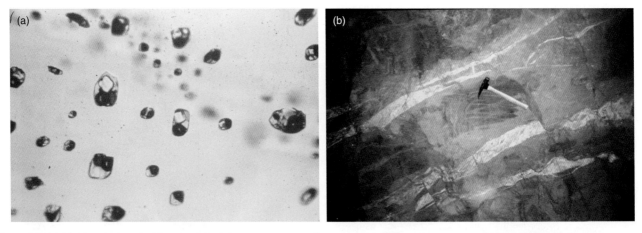

Figure 2.8 (a) Fluid inclusions inside a crystal of transparent quartz from the Granisle porphyry copper deposit, British Columbia, as seen through a microscope. Each inclusion is about 30 micrometers in diameter and they contain transparent saline water, a vapor bubble (dark) and several daughter minerals including hematite (red) and halite (clear cubes). The inclusion was originally trapped as a liquid, but during cooling the liquid shrank to form the vapor bubble and deposit the daughter minerals. If the quartz crystal is reheated under the microscope the vapor bubble will shrink and the daughter mineral will dissolve, providing information on the temperature at which the liquid was trapped in the inclusion (photograph by the authors). (b) Hydrothermal alteration (light zones) in wallrock around quartz veins (white) at the Mt. Charlotte gold deposit, Kalgoorlie, Australia (photograph by the authors).

Figure 2.10 Hydrothermal systems. (a) The famous Pamukkale travertine (calcite) deposits in Turkey were produced by warm, meteoric (~ 35 ° C) water flowing from springs that have been active for at least the last 55,400 years. The Greco-Roman city of Hierapolis, which was built at the springs, has been partly enveloped by younger deposits. Epithermal precious-metal deposits in the same area of Turkey are probably deeper extensions of systems such as this one (photograph courtesy of David Stenger). (b) Submarine hydrothermal "black smoker" vent surrounded by shrimp (white) at the Mid-Cayman rise in the Caribbean Sea. Water expelled from this vent at a temperature of 400 °C mixes with cold seawater to precipitate fine-grained black particles of metal sulfide that form the "smoke" (photo by Chris German, WHOI/NASA/ROV Jason © 2012 Woods Hole Oceanographic Institution).

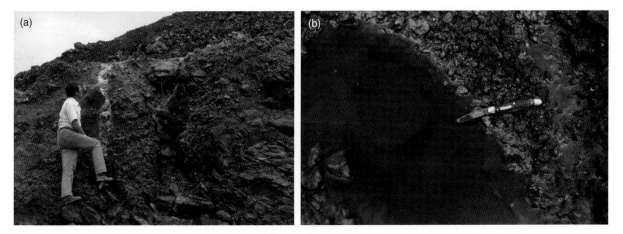

Figure 4.6 (a) Weathered veins in the Pueblo Viejo gold–copper deposit have released copper, some of which has precipitated as blue-green carbonates and oxides. (b) This pool of water near the weathered veins contains so much dissolved copper that the knife blade was coated by copper that was deposited from the water.

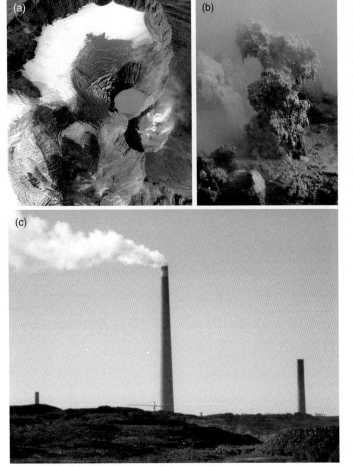

Figure 3.7 Natural and anthropogenic sulfur sources. (a) Mutnovsky volcano (seen from above) consists of a large crater and lake with fumaroles releasing 620 °C gases consisting of H_2O, CO_2, SO_2, H_2S, HCl, HF, and H_2. The fumaroles release about 4×10^5 tonnes of SO_2 annually. Some of the sulfur is deposited in vents and fumeroles on the volcano (such as the 2-m-high chimney in (b)) but most of it goes into the atmosphere. The eruption of Mount Pinatubo, Philippines, in 1991, released about 2,000 times more SO_2 in only a few hours. (c) Treatment of nickel–copper ores at Sudbury, Ontario, have been significant sources of sulfur to the atmosphere. Since mining began, roasting and smelting have released about 10^8 tonnes of sulfur (up to 2×10^6 tonnes annually), contributing to the lack of vegetation as seen in the foreground of the photo. Improvements began in the early 1970s with construction of the 380-m superstack, which disperses emissions over a much broader area. The later addition of sulfur capture systems has cut annual emissions by at least 90% (photos by the authors).

Figure 4.8 (a) Helicopter in Alaska towing a large wire loop that measures electromagnetic fields in rock below (photograph courtesy of Kirsten Ulsig, SkyTEM Surveys APS). (b) Magnetic anomaly over a possible diamond-hosting kimberlite pipe. The anomaly consists of a zone of high magnetic intensity (red-purple) and a zone of low magnetic intensity (blue). A line of smaller anomalies that trends directly north and south from the kimberlite is caused by a related kimberlite dike (image courtesy of Brooke Clements and Jennifer Pell, Peregrine Diamonds Ltd.).

Figure 4.10 (a) Gravels carried into the area by glaciers cover many parts of North America making mineral exploration difficult. In this photo, from the Soudan mine in Minnesota, a layer of gravels about 10 m thick covers iron ore. Geophysical methods are useful in exploring country of this type because they can "see through" the gravels to determine the character of underlying rocks. (b) The final stage of mineral exploration involves drilling to obtain samples of buried rock, either chips or cylinders known as core. The core samples shown here are cut by dark veins of sphalerite.

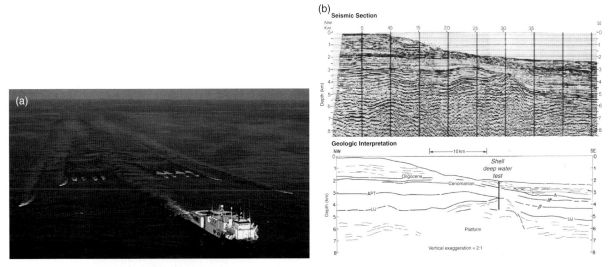

Figure 4.11 (a) Marine seismic survey ship, which uses collapsing air bubbles to send shock waves (that are harmless to marine life) through the water and into the underlying rocks. Returning waves are detected by measuring devices known as hydrophones, which are towed behind the ship (courtesy of Society of Exploration Geophysics). (b) Seismic survey across the Baltimore Canyon offshore from the eastern United States showing the edge of the carbonate platform that has been the target of exploration drilling for oil (from Klitgord and Watkins, 1984).

Figure 4.12 Drill rigs. (a) Small "back-pack" drill rig in use on a Mexican hillside. Drill rods are rotated down the hole cutting a core of rock, while drilling mud is pumped in from the top to wash away rock cuttings (photograph by the authors). (b) The *D/V Chikyu* holds the world marine record of 7,740 m for depth of hole drilled below sea level. The ship, which is operated by the Japan Agency for Marine–Earth Science and Technology (JAMSTEC), is designed to drill through the ocean crust into the underlying mantle (photograph courtesy of JAMSTEC).

Figure 4.14 Mineral production in LDCs. (a) Overview of the KingKing porphyry copper deposit on Mindanao Island, Philippines. The upper part of this deposit has been excavated by hand by thousands of unemployed people to recover gold from the weathered ore. Small houses scattered around the deposit provide scale. (b) Young miner panning gold from weathered KingKing ore (photographs by the authors).

Figure 4.15 (a) The Escondida porphyry copper mine in Chile, with annual production of more than 1 million tonnes of metal, is the world's largest copper producer. The mine reaches a depth of almost 700 m and measures 4 km by 3 km at the surface. About 1.4 million tonnes of ore and waste are removed from the mine each day using more than 150 large trucks that move up the inclined ramp at the back of the pit (photograph courtesy of BHP Billiton Ltd). (b) Side of the main pit at Grasberg, Irian Jaya in Indonesia, showing a landslide that formed where the pit wall collapsed. Pit-wall failures of this type are not uncommon and are caused by faults, water-saturated ground, and oversteep pit walls. The truck at the bottom center of the pit provides scale (photograph courtesy of Karr McCurdy).

Figure 5.11 (a) White arrows show the location of four man-made islands immediately offshore from Long Beach, California, that were constructed with gravel and are used for production of oil from the Wilmington field. The islands, which are named after US astronauts, Grissom, Freeman, White, and Chaffee, have produced 930 Mbbl of oil through 2014, with a revenue sharing agreement signed in 2012 that gives 49% of revenue to the oil producers, 49% to the state of California, and 2% to the city of Long Beach (Belk, 2012). (b) Since the upper photo was taken, development of Long Beach harbor has put pleasure-boat docks very close to the islands, and the islands themselves have been modified with drill rigs enclosed in structures that look like high-rise residential complexes as seen in this photo of Island Grissom (photos from US Department of Energy and "Donielle from Los Angeles, USA" (oil island uploaded by PDTillman) [CC-BY-SA-2.0 (http://creativecommons.org/licenses/by-sa/2.0)], via Wikimedia Commons).

Figure 4.17 Offshore oil production in increasingly deep water. (a) The artificial ice-proof island that produces from the Endicott oil field near Prudhoe Bay on the north coast of Alaska was constructed in the mid 1980s for over $1 billion. The 445-acre (18-hectare) island has close spacing of wells designed to minimize its area and environmental impact. Initial oil production at a rate of about 100,000 bbl/day has declined to about 10,000 bbl/day today, and total production so far amounts to about 460 Mbbl of oil (photo courtesy of BP America). (b) The Monopod platform, in 20 m of water in Cook Inlet, Alaska, began operation in 1967. Its single leg, which is less vulnerable to floating ice, has 32 wells that reach the entire oil field. With enhanced recovery methods that were applied in 2014, it produces about 2,900 bbl/day (photo courtesy of Chevron Corporation). (c) The Jasmine platform, at a depth of 81 m in the central North Sea, consists of a production unit with 24 slots for holes into the field and a living-quarters unit. The platform, with gross capacity of 140,000 bbl/day of oil equivalent (oil and gas), began production in 2013 (photo courtesy of ConocoPhillips). (d) In deep water where platforms do not work well, floating platform storage and offloading (FPSO) systems are used. The FPSO Aseng, shown here, began production from the Aseng field offshore Equatorial Guinea in November, 2011, at a water depth of about 1,000 m. The unit produces 80,000 bbl/day using injection of 150,000 bbl of water and has a storage capacity of 1.695 Mbbl (photo courtesy of Glencore Ltd.).

Metal reduction facilities

(a) Blast furnace

(b) Smelter - Noranda continuous system

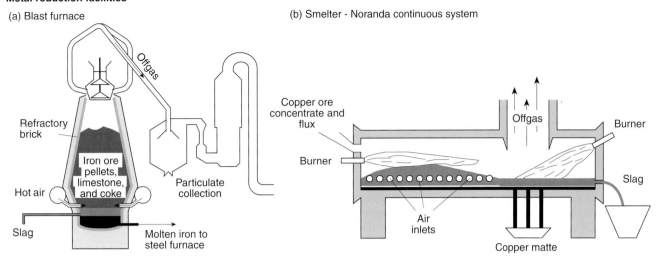

Figure 4.19 Metal ore reduction facilities. (a) Blast furnace used to convert iron ore concentrates (usually pellets) to pig iron, which is then transferred to steel furnaces. (b) Smelter used to convert metal sulfide concentrates to molten sulfide (matte), which is then transferred to **converters** to remove remaining sulfur as SO_2. (c) Blast furnace at Bremen, Germany (courtesy of ArcelorMittal). (d) Molten matte at the Alto Norte smelter, Chile (courtesy of Glencore plc.).

Figure 5.1 Living near mineral production. (a) The Terry Peak ski area in the Black Hills of South Dakota looked down on three gold mines that produced more than 60 million ounces of gold and were a major factor in the local economy. At the base of the hill is the Golden Reward open pit mine, in the far right distance is the Gilt Edge open pit mine, and barely visible in the far left distance is the headframe for the giant Homestake underground mine (photograph by Paul A. Bailly). (b) The city of Butte, Montana, is directly adjacent to the Berkeley open pit and overlies the extensive underground operations that mined veins (photograph by the authors).

Figure 6.1 (a) The Kimberley mine in South Africa, began as small diggings that gradually grew to form the deep pit shown here, which continues to depth as the underground mine shown in Figure 11.18. The dark, high-rise building in the background is Harry Oppenheimer House, where diamonds from South Africa and Botswana are sorted and graded (photograph by the authors). (b) The Chuquicamata mine in northern Chile is one of several mines owned by US companies that were nationalized by the Chilean government in 1970 to form the state-owned corporation, Codelco. This mine, one of the largest in the world, measures 4.3 km long, 3 km wide and is almost 1 km deep. (c) This shovel is one of the tiny dots at the bottom of the Chuquicamata pit in the middle picture. The person standing below the bucket of the shovel, and another just beside it, show just how large this machine is (photographs by Antonio Arribas).

Figure Box 6.4.1 Highland Valley Copper Operation in British Columbia, showing the Lornex pit, beneficiation plant, and waste rock piles. Note the revegetated waste rock pile on the left behind the red domes. (Photograph courtesy of Teck Resources Ltd.)

Figure 7.6 Surface coal mining methods. The Freedom lignite mine near Beulah, North Dakota, supplies about 16 million tons of coal each year to the basin Electric Power Cooperative. (a) Mining is carried out with a dragline (background) that removes the light-colored sediments that cover the horizontal coal layer piling it on the left side of the photo. The black coal layer in the bottom of the pit is removed by shovels and loaders (small equipment in center of photo). (b) After mining, the dragline replaces the waste rock and the area is graded to near-original contours (including wetlands), covered with topsoil and revegetated, as shown here. Additional information on the operation can be seen at http://www.basinelectric.com/About-Us/Organization/Subsidiaries/Dakota-Coal-Company/ (photos courtesy of Bill Suter, Coteau Properties).

Figure 7.14 (a) Production on land. The anticlinal structure at Elk Basin, Wyoming, is a typical simple oil trap. Sedimentary layers dipping away in both directions from the center of the photo show the crest of the anticline. Oil is produced from the buried part of the crust by pumping units. (b) Offshore production. The Hibernia field, 315 km east–southeast of St. John's. Newfoundland, is the largest oil deposit along the east coast of North America. The field contains an estimated 2.1 billion bbl (Bbbl) of oil in place, of which about 700 million are recoverable. It was discovered in 1979 and put into production in 1997 with the platform shown here, which sits on the seafloor (80 m deep). A fleet of tugs protects the platform from passing icebergs calving off Greenland (photo courtesy of Suncor Energy).

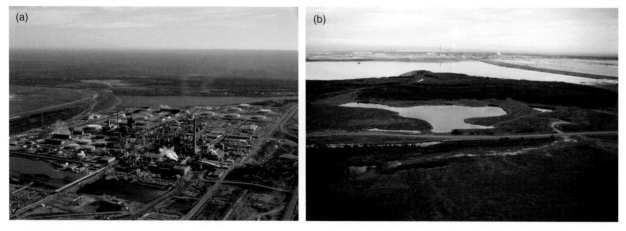

Figure 7.26 Tar sands. (a) Overview of processing plant with partly revegetated waste pile on left. (b) Reclaimed mined land with as yet unreclaimed waste dump on right. Processing plant is distant background (photos courtesy of Syncrude Canada).

Figure 7.23 Oil and gas transportation. (a) Alyeska pipeline near Fairbanks, Alaska, where it has been elevated to prevent heated oil from melting permafrost (photo by the authors). (b) The Polar Adventurer oil tanker, shown here (foreground) loading oil at the port of Valdez, Alaska, the southern end of the Trans-Alaska pipeline, transports North Slope oil to refineries along the US west coast. The tanker is approximate 272 m long and has two completely separate hulls. The double hull design, with an inner hull about 3 m inside the outer hull, provides additional protection from punctures, and is required for large tankers operating in United States and most LDC waters (photo courtesy of ConocoPhillips, Inc.) (c) The Arctic Princess LNG tanker, which is 288 m long, carries LNG in four cylindrical tanks at a temperature of –160 °C (photo courtesy of Höegh LNG).

Figure 7.30 (a) Sandstone uranium ore containing the bright yellow ore mineral, carnotite, which formed when uranium in the sandstone was oxidized by groundwater. (b) Conglomerate gold ore from the Witwatersrand sedimentary sequence in South Africa consists of large pebbles of quartz in a matrix that contains pyrite, gold, and uranium minerals. Similar conglomerates in the Elliott Lake Group in Ontario, Canada, lack gold but contain uranium minerals.

Figure 7.31 (a) Uranium mining by ISL. The barrels cover well heads that pump oxidizing water into the outer perimeter of the buried uranium deposit. The water flows through the deposit, leaching uranium, and then returns to the surface through wells in the center of the field. The entire field is surrounded by monitoring wells to detect and prevent any leakage of leach fluid beyond the deposit (photo by the authors). (b) The Pickering nuclear power plant on the north shore of Lake Ontario in Canada contains six operating and two inactive CANDU nuclear reactors with a total output capacity of 3,100 MW. The reactors are enclosed in the eight domed buildings, which are constructed of reinforced concrete. The large, light colored cylindrical building is a vacuum enclosure that is connected to the eight reactor enclosures. Any accidental release of radioactive steam from the reactors would be sucked into the vacuum enclosure rather than escaping into the atmosphere (photo courtesy of Ontario Power Generation).

Figure 7.35 (a) Yucca Mountain, Nevada has been investigated extensively as a possible repository for high-level nuclear waste. It is in the southern part of the Nevada Security Site, a 3,500 km^2 area in southern Nevada that was used for testing of nuclear weapons, including 100 atmospheric and 728 underground tests. The drill rig on the left side of the photo was obtaining cores of rock to test properties, including rock type, fracture density, permeability, porosity, and mineral composition, which might affect migration of nuclear waste (courtesy of Eugene I. Smith). (b) Yellowstone Park is one the largest geothermal areas in the world, with numerous boiling springs and steam vents. Geothermal areas make up only about 0.3% of the park, but some of these areas have heat flow of 100 W/km^2, 2,000 times higher than average heat flow in North America. Geothermal power has not been developed in the park because it would decrease the flow of the hot springs and steam vents (photo by the authors).

Figure 8.2 (a) BIF consisting of layers of hematite (black) and silica (red). The small balls beside the hammer are pellets, which consist largely of hematite grains held together by clay and are the main feed for blast furnaces (photograph by Dale Austin, University of Michigan). (b) Tilden Mine, Marquette County, Michigan, which mines hematite-bearing BIF, has the capacity to produce 8 million tonnes of pellets annually. Note rail loading system with piles of pellets in foreground (courtesy of Dale R. Hemila, Cliffs Natural Resources).

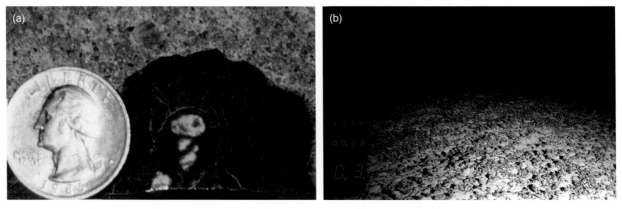

Figure 8.9 Manganese nodules. (a) Cross section of a manganese nodule from the North Pacific Ocean (Ocean Drilling Program Site 886) showing concentric growth layers (photograph courtesy of D. K. Rea and H. Snoeckx, University of Michigan). (b) Manganese nodules on the seafloor. The photo was taken by camera sled in 2003 at the Hawaii-2 Observatory site in the in the central North Pacific at ~ 5000-m depth by K. L. Smith Jr (MBARI) and S. E. Beaulieu (WHOI).

Figure 8.10 (a) Nickel-rich laterite (dark) overlying ultramafic bedrock (light) from which it was derived by weathering, southwestern Puerto Rico. H.L. Bonham provides scale. (b) Shatter cones on the side of a rock outcrop south of Sudbury, Ontario, showing the radiating pattern pointing upward toward the probable impact level, which is now eroded. Key in lower left provides scale (photographs by the authors).

Figure 8.17 Bushveld chromite deposits, eastern Transvaal, South Africa. (a) Chromitite layers (black) at Dwars River. (b) Winterveld chromite mine showing cumulate layers on hillside dipping downward to left. The mine enters at the base of the hill to the right (photographs by the authors).

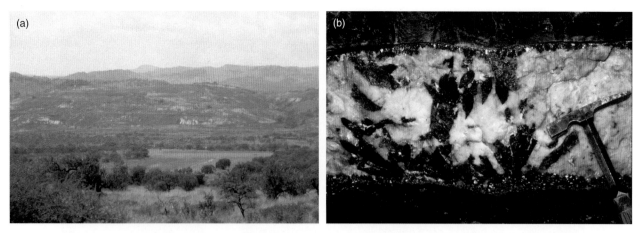

Figure 8.20 (a) Mapochs vanadium mine, eastern Transvaal, South Africa. The vanadium-bearing magnetite layers dip steeply toward the viewer and have been mined by stripping off the side of the hill (photograph by the authors). (b) Vein of quartz (white) with crystals of the tungsten ore mineral wolframite (dark) from the Dajishan mine, China (photograph courtesy of James E. Elliott, US Geological Survey).

Figure 9.9 (a) The Moma titanium operation mines ilmenite, rutile, and zircon from beach sands along the northeast coast of Mozambique. The dredge shown here scrapes sand from the bottom of the water-filled pit and sends it through a pipeline to a concentrating plant. (b) Concentrating plant at the Moma operation showing spiral sluices down which the sand slurry flows. Heavy minerals move to the outside of the spiral and are captured. Waste sand is used to fill the mining pond, which advances into unmined sand (photographs courtesy of Kenmare Resources plc).

Figure 9.4 Aluminum production. (a) Strip mining of paleobauxite at MacKenzie, Guyana, which is buried beneath younger sediments. (b) Red mud waste from processing of bauxite, Aughinish, Erie (photograph courtesy of International Aluminum Institute). (c) Potroom where alumina is converted to aluminum, Dubai Smelter (photograph courtesy of International Aluminum Institute). (d) The largest aluminum refinery in Europe at Sunndalsora, Norway, uses hydroelectric power to process imported alumina (photograph courtesy of Oivind Leren/ Hydro Sunndal).

Figure 9.12 (a) Intersecting veins of quartz, pyrite, and chalcopyrite in porphyry copper ore from the Granisle deposit, British Columbia. (b) Veinlets of chalcopyrite that make up the feeder zone beneath the Millenbach VMS deposit, Noranda area, Quebec (photographs by the authors).

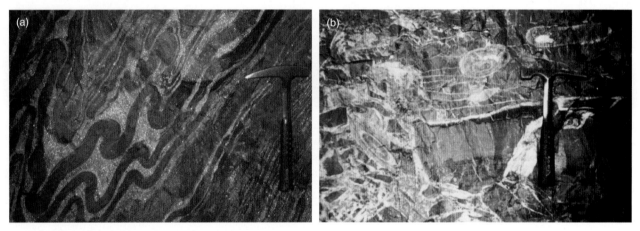

Figure 9.17 (a) Folded layers of sphalerite (red) and galena (gray) in the Sullivan SEDEX deposit, Trail, British Columbia. (b) Breccia zone in which fragments of limestone and dolomite (dark) and elliptical chert nodules have been cemented by hydrothermal dolomite crystals (white) that were deposited by basinal brines in the Coy MVT zinc deposit, Tennessee (photographs by the authors).

Figure 9.22 (a) Tin-bearing veins in granite at Cligga Head, Cornwall, England. Narrow vein systems of this type are referred to as sheeted veins. (b) Larger veins were mined individually. This photo looks outward from a stope that mined a 0.7-m-wide vein of this type at Bottalack in Cornwall (photographs by the authors).

Figure 10.6 Lithium deposits. (a) Pegmatite from the Kings Mountain mine showing large light green spodumene crystals (photo by authors). (b) Salt encrustations around a transient pond at the southern end of the Salar de Atacama, the world's largest brine-evaporite deposit of lithium (photo courtesy of Martin Reich, University of Chile).

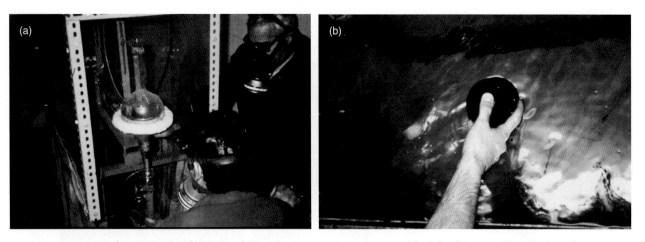

Figure 10.10 Mercury production at Almadén, Spain. (a) Pouring mercury into a 76-pound flask for shipment. (b) Ball of iron floating in a tank of liquid mercury.

Figure 11.4 Close-up view of two types of epithermal gold veins. (a) A vein from the rich Hishikari LS (adularia-sericite) gold mine in Japan. The vein consists of alternating layers of quartz, adularia, clays, and gold. (b) A close-up view of the San Albino vein at the Parra mine in Mexico. The sharp white crystals are feldspar and the darker crystals are mostly sphalerite and galena. This type of vein produces largely silver, lead, and zinc rather than gold and is part of the IS group of deposits with characteristics between those of LS and HS deposits.

Figure 11.6 Two extremes in gold mining. (a) Miners at work in the Driefontein Mine, South Africa, which extends to depths of 3,300 meters in the Witwatersrand Basin and produces about 450,000 ounces of gold per year from about 1.9 million tonnes of rock. Note the steel rock-support columns. Specks of white and black in the wall in front of the minerals are pebbles in the Witwatersrand conglomerate (photograph courtesy of Gold Fields Limited). (b) View of the Post-Betze open pit at Carlin, Nevada, which produces about 900,000 ounces of gold per year from more than 1 million tonnes of rock. At the bottom of the pit is an entrance to an underground mine that exploits the deeper continuation of this deposit (photograph by the authors).

Figure 11.8 (a) Heap leach pads showing heaps of ore in left background. Cyanide solution is sprayed on the heaps, flows through the rock dissolving gold, and collects in the basins to the right foreground. Note thick plastic layer beneath the ponds and heaps to prevent recharge of solution into underlying groundwater. (b) Molten gold pouring from a furnace into molds. (c) Gold bars removed from molds. Each bar weighs about 27 kg and contains about 87% gold, 8% silver and small amounts of lead and zinc. The hand (lower right) provides scale (photos by the authors).

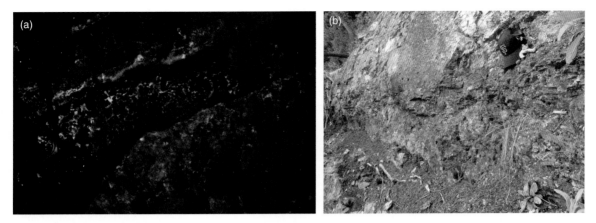

Figure 11.15 Platinum deposits. (a) Bushveld Complex – a close-up view of the Merensky Reef in the Rustenburg Platinum Mine showing its coarse-grained nature with abundant chromite (dark grains). The reef increases in thickness abruptly to the left as it enters a pothole, a feature discussed in the text. (b) Outcrop of the J–M Reef (the 10-cm-thick dark layer extending across the photo just below the baseball cap) at the Stillwater Mine, Montana (photographs by the authors).

Figure 12.7 (a) Underground salt mine at Avery Island showing large openings (stopes) in salt, which is relatively rigid at shallow depths typical of Gulf Coast salt mines (image courtesy of Corporate Archives, Cargill, Incorporated). (b) Continuous-mining machine in the Laningan potash mine in Saskatchewan, which creates a relatively small opening because the potash is not strong. The potash layer occupies the entire height of the photo and the continuous mining machine has cut curved grooves into it as it advanced toward the right side of the photo (photo courtesy of Potash Corporation, Saskatoon, Saskatchewan).

Figure 11.18 Diamond deposits and mines. (a) Kimberlite breccia in the upper part of the Kimberley pipe, South Africa showing a large boulder (upper left) that was transported from the mantle. (b) Beach placer diamond mine in southern Namibia showing pumps to remove seawater (lower left). The mining area and pumps are several meters below sea level but are separated from the ocean by a dike made of sand, which is seen in the lower right corner of the photo. Partly flooded mining areas down the coast can be seen in the background (photographs by the authors). (c) Diavik diamond mine, Northwest Territories, Canada, which produces about 7.5 million carats per year from three kimberlite pipes, is surrounded by a 3.9 km dike to protect it from waters of Lac de Gras. (d) Uncut diamonds from the Diavik Mine (copyright © 2014 Rio Tinto).

Figure 12.13 (a) Dragline moving overburden from phosphate mine in central Florida. (b) Overview of phosphate mine in central Florida showing active mining in left corner and two settling ponds where land is being reclaimed (lakes in center). Most other land has already been mined and reclaimed, usually as wetlands, the form in which it began. (c) Close-up view of reclaimed wetland on the site of a former phosphate mine in central Florida (photographs courtesy of Mosaic Corporation).

Figure 12.17 Iodine mining in the Atacama Desert of Chile. (a) Overview of mining operation showing their location on hillsides rather than in playa lake beds. (b) Iodine minerals (light colored) filling holes in soil and rock debris that cover the hillsides (photographs courtesy of Martin Reich, University of Chile).

Figure 13.5 (a) Aggregate quarries on the outskirts of Monterrey, the third largest city in Mexico. Although the quarries were located perfectly from an economic standpoint, they created too much particulate pollution (note plumes of dust coming from the quarry at left) and most were closed during the late 1980s. (b) Aggregate quarry on the south side of metropolitan Chicago. This quarry is divided into two parts by a rib of rock left that supports Interstate 80, which can be seen in the background. A tunnel through the ridge, seen in the center, background, allows access to both parts of the mine (photos by the authors).

(a)

(b)

Figure 13.7 (a) The Pantheon showing the 12-m granite columns supporting its portico. (b) Marble quarries occupy most of the hillsides around Carrara, Italy, and extend underground. The best stone comes from rock that is below the zone of weathering and iron staining.

Figure 13.10 (a) Borax ore in 20-mule team wagons leaving an early mine. (b) Modern borax mine and processing plant, Boron, California (courtesy of Florence Yaeger, Rio Tinto Corporation).

Figure 13.13 (a) Kaolin deposit in the coastal plain of Georgia, United States. Four layers can be seen in the photo. The upper layer, which is behind and to the left of the trees at the top of the mine, is a pile of overburden. Immediately below the trees is the darker, Tertiary-age Twiggs clay, which overlies lighter-colored Cretaceous sands. The whitest layer at the bottom of the wall and making up the floor of the pit is kaolin (photo by Thomas Kesler). (b) Goonbarrow kaolin mine near St. Austell in Cornwall, England. Kaolinite here formed by hydrothermal alteration of granite and is mined by spraying water from large cannons called monitors, as seen in the left-center part of the picture (photo by the authors).

Figure Box 13.6.2 Round Loch of Glenhead, in Galloway, Scotland, provides a good example of a relatively fragile lacustrine system. The lake is underlain and surrounded by granitic rocks that are covered by peat and peaty soils, all of which do not react readily with acid water (photo courtesy of Ewan Shilland, University College, London).

BOX 9.1 | BAUXITE AND RED MUD

Red mud, the main waste product from conversion of bauxite to alumina, is the residue that remains after alumina has been leached from the bauxite. It consists of silica and iron and titanium oxides, and remnants of the highly alkaline leach solution in a slurry with a pH of about 13, which is highly caustic. Some red mud tailings can be used to neutralize acid waste from other industrial processes, but this is not a very large market and the tailings are usually not near the area of need. Mixing with seawater, which has a pH of about 7, can bring the pH of red mud slurry down to about 9, but this can be done only if the facility is on the ocean. Attempts have also been made to use red mud as the basis for bricks and other building material (Johnston *et al.*, 2010; Liu *et al.*, 2009) In the absence of these alternatives, red mud slurry must be put into a tailings reservoir, usually a pond that is held in place by dikes or a dam (Figure 9.4b). These enclosures are well engineered, although one did rupture in Ajka, Hungary, in 2010, releasing about a million cubic meters of slurry over about 40 km² of low-lying ground (Gura, 2010). The sludge eventually reached the Danube River, where dilution lowered its pH and toxicity.

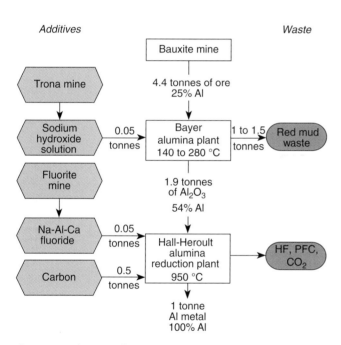

Figure 9.5 Schematic illustration of typical aluminum production process. Additional materials necessary for the process are shown on the left and waste products are shown on the right (compiled from information of the US Bureau of Mines; Polmear, 1981; Burkin, 1987; Menzie *et al.*, 2010; Grandfield, 2014).

premium on light weight. Aluminum consumption patterns vary from country to country, however, especially for packaging with only 3% of total consumption used for this in Russia compared to 29% in Brazil (OECD, 2010).

World bauxite reserves of about 28 billion tonnes are largely in Guinea, Australia, Brazil, Vietnam, and Jamaica and are estimated to be sufficient for almost two centuries at present rates of production (Meyer, 2004). US resources are

small, and only France and Hungary, among European countries, have reserves. Other possible aluminum raw materials include fly ash produced by coal-fired power plants, which contains enough aluminum to meet the present world demand.

Although only 26 countries produce bauxite and alumina, 46 produce aluminum, confirming its importance in world trade (Figure 9.2). The trade pattern is complex because all three commodities, bauxite, alumina, and aluminum metal are exported and imported. By far the largest exporter of bauxite is Guinea, which has no alumina production. Most other bauxite producers convert at least some of their product to alumina; the largest exporters of alumina are Australia and Brazil. Norway, Canada, Iceland, and the United Arab Emirates import bauxite and alumina that are used to produce aluminum metal from their cheap energy sources. The major exporters of aluminum metal are Russia, Canada, Australia, Norway, Bahrain, and the United Arab Emirates, and the major importers are the United States, Japan, Germany, and Italy (Figure 9.6c).

Changes are likely for the world aluminum market. You probably noticed that this is one of the few commodities in which China is not a major player. Even though China's share of global aluminum production grew from 3% in 1980 to about 35% now, they have been able to supply most of their needs from domestic bauxite. Unfortunately, these deposits are small and low grade. As they are exhausted, China will probably become an important factor in global markets. At the same time, some more-developed countries (MDCs) might drop out of the market because, as suggested by materials flow analyses, their large stocks of in-use aluminum will probably supply some demand (Hatayama *et al.*, 2009).

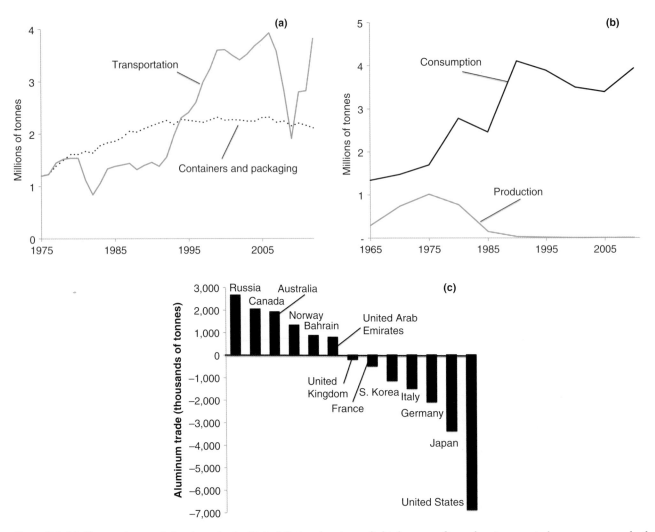

Figure 9.6 (a) Changes in use of aluminum in the United States, showing a slight decrease after a slow increase in beverage cans and other containers compared to a stronger, but more volatile increase in transportation, including aircraft. (b) Change in the primary aluminum production and the total aluminum consumption in Japan. The decreased production reflects the lack of extra energy for this purpose. (c) World trade in aluminum metal for 2006. Countries plotting above the line produce more metal than they use and those below the line import metal (based on data of the US Geological Survey and the Organization for Economic Cooperation and Development (OECD)).

9.1.2 Magnesium

Although magnesium is an important metal, its non-metallic forms account for almost 90% of global consumption. Non-metallic magnesium comes in many forms, but the simplest is in magnesium minerals such as magnesite and olivine. These are the main forms in which non-metallic magnesium is sold, and they are used mainly as refractory linings in furnaces and molds in the iron and steel industry. Magnesium carbonates, hydroxides, and chlorides, both natural and made from other magnesium compounds, are used in the manufacture of rubber, textiles, and chemicals. Non-metallic magnesium is produced and traded in so many different forms that it is difficult to place a simple value on them, although they are worth at least a few billion dollars.

The 770,000 tonnes of magnesium metal produced annually throughout the world are worth slightly more than $3 to $4 billion. Magnesium metal is used as an alloy to harden aluminum and increase its resistance to corrosion; it makes up 2.5% of the aluminum in cans. It is also used in aircraft and the automotive industry, as an agent to remove sulfur from steel, and as a reducing agent in the production of other metals such as titanium and zirconium from their ores. The real hope for magnesium, however, is as components of motors, seats, and other automobile parts, a growing market (Kramer, 1985; Kramer and Plunkert, 1991; Couturier, 1992).

Magnesium has been produced from a wide range of deposits (Wicken and Duncan, 1983). First to be used were olivine-rich rocks in obducted mantle along some convergent

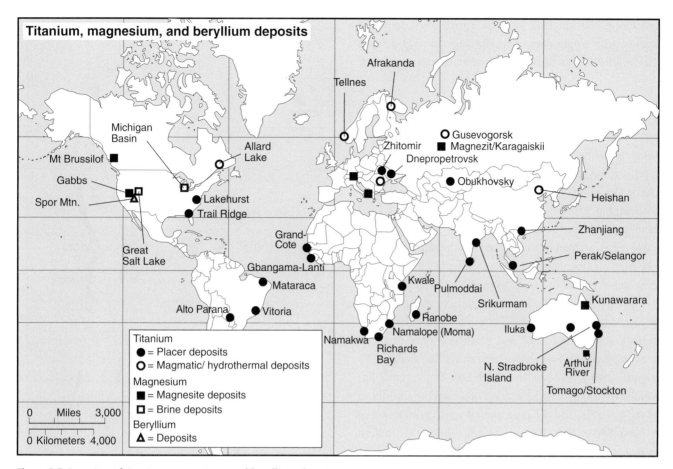

Figure 9.7 Location of titanium, magnesium, and beryllium deposits

margins, and magnesite, a magnesium carbonate, that formed where olivine was altered by CO_2-rich hydrothermal solutions. Magnesite is also mined where magnesium-rich hydrothermal solutions replaced limestone or dolomite in sedimentary basins. Olivine deposits are mined in North Carolina, and deposits of magnesite are found in Nevada, Austria, and Greece (Figure 9.7). As demand for magnesium and its compounds grew, attention shifted to **dolomite**, a calcium–magnesium carbonate mineral, even though it contains less magnesium than magnesite. Small amounts of magnesium were also produced from evaporite minerals, and this drew attention to dissolved sources of magnesium. The first such source was brine in the central part of the Michigan basin, which contained about 3% magnesium. As technology improved, it became possible to process solutions with lower concentrations of magnesium, such as the waters of Great Salt Lake and, ultimately, seawater, which contains only 0.13% magnesium. At present, seawater and brines account for 60% of US production, making it the only metal to be recovered largely from seawater.

Olivine is used directly as a refractory mineral, but most other forms of magnesium must be processed before they can be used. Magnesite and dolomite are converted to MgO by calcining in a kiln to drive off CO_2. Magnesium sulfate (Epsom salt) is produced by dissolving MgO in sulfuric acid, and magnesium carbonate is precipitated from this solution by addition of sodium carbonate. Seawater processing can yield either magnesium chloride or magnesium hydroxide [$Mg(OH)_2$]. Magnesium metal is produced by an electrolytic process in which molten magnesium chloride is put into a 1,292 °C bath with carbon electrodes, or by reacting calcined dolomite with silicon. Neither of these processes is as energy-intensive as aluminum or titanium production. Magnesium metal is very easy to shape and machine, although it produces dust that will react with water to form potentially explosive hydrogen.

Magnesium is essential for life and is used by humans largely to activate enzymes and facilitate protein synthesis reactions (Itokawa and Durlach, 1988). Deficiencies of magnesium have been linked to growth failure and behavioral problems. It is also important to plant growth as the basic constituent of chlorophyll.

Magnesium reserves and resources are essentially unlimited. World reserves of magnesite are estimated to be several billion tonnes. Backing this up is the entire ocean, which is resupplied with magnesium by weathering and river flow, as well as by alteration of basalt at mid-ocean spreading centers.

9.1.3 Titanium

Commercial use of titanium metal began only in 1906, although it has been used longer in its mineral form (Lynd, 1985). Currently, about 95% of world titanium production is consumed as an oxide (TiO_2), which forms a white pigment. The remaining 5% is used as titanium metal, which has had unusually rapid market growth. Titanium metal is valued because it has a significantly lower weight than steel, but a similar high melting temperature and strength. Thus, it can be used in high-temperature applications where weight is important, especially in engines and other parts of air and space-craft. In commercial aircraft, titanium usage has been increasing since about 1980 and reached a peak of about 14% of the weight of an empty aircraft for the Boeing 787 (Sankaran, 2006). The relatively high cost of the metal has prevented it from making inroads on other steel or aluminum markets, however, and it has an annual world production of only about 100,000 tonnes, worth about $1 billion (Figure 9.1a).

Titanium deposits

Titanium forms two very common minerals, rutile and ilmenite, as well as leucoxene, a form of ilmenite from which iron has been removed by weathering or alteration, thus increasing the titanium content (Force, 1991). Rutile and ilmenite are heavy and resistant to weathering and erosion. Not surprisingly, more than half of world titanium mineral production comes from **placer** deposits in which rutile and ilmenite are accompanied by other heavy minerals such as monazite, garnet, and zircon (Towner *et al.*, 1988). Most placer deposits containing titanium minerals are beach dunes and sands that have been uplifted and preserved from erosion (Figure 9.8). Deposits of this type are widespread along the southeastern coast of the United States, eastern Brazil, the southern and western coast of Australia, and the coast of southern Africa (Figure 9.7). Many of these deposits formed as the seas retreated from a major incursion onto the continents in Pliocene time, about 3.5 to 3.0 Ma ago (Carpenter and Carpenter, 1991). A smaller number of titanium placer deposits, including those in Sierra Leone, formed in a stream, or fluvial, environment.

Much small production of titanium minerals comes from **bedrock** deposits, most commonly mafic intrusive rocks.

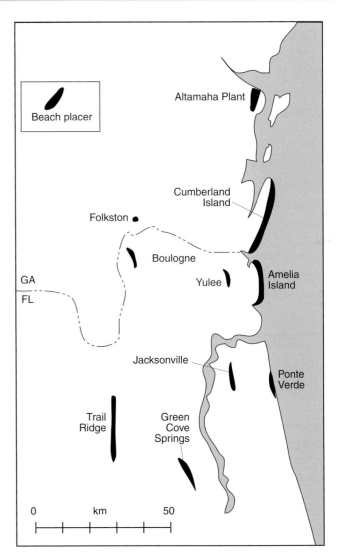

Figure 9.8 Map of titanium mineral placer deposits along uplifted beaches in the Trail Ridge district near Jacksonville district, Florida (after Force, 1991)

During cooling of these magmas, ilmenite crystals form cumulate layers near the bottom of the magma chamber, probably by the same processes that form the chromite and magnetite layers discussed in the previous chapter. Rutile is also found along the margins of **anorthosite**, a rock rich in plagioclase feldspar, where it appears to have been concentrated when other elements were melted and removed during intrusion of the anorthosite.

Titanium production and reserves

Mining of bedrock titanium deposits requires considerably more energy than placer mining. Bedrock deposits make up for this disadvantage because they have ore grades of as much as 35% TiO_2, much higher than grades commonly found in placer deposits. Mining of placer deposits is done largely by

Figure 9.9 (a) The Moma titanium operation mines ilmenite, rutile, and zircon from beach sands along the northeast coast of Mozambique. The dredge shown here scrapes sand from the bottom of the water-filled pit and sends it through a pipeline to a concentrating plant. (b) Concentrating plant at the Moma operation showing spiral sluices down which the sand slurry flows. Heavy minerals move to the outside of the spiral and are captured. Waste sand is used to fill the mining pond, which advances into unmined sand (photographs courtesy of Kenmare Resources plc). See color plate section.

dredges that dig their own lakes, which they fill with tailings from ore processing that is carried out on-board (Figure 9.9). Although dredged land is reclaimed, the location of dredging operations in and near beach areas has led to disputes over land use. Controversy has been most severe along the coast of Queensland, Australia, where important rutile placer deposits will probably never be mined.

Processing of rutile is usually done by reacting it with chlorine gas at 800 °C to form $TiCl_4$, which is purified by fractional crystallization and then reduced to titanium metal by reaction with magnesium or sodium. Ilmenite concentrates can be converted to a titanium-rich slag by reaction with coal, which is treated with sulfuric acid or chlorine gas, depending on its titanium content (Guéguin and Cardarelli, 2007). Neither process is strongly polluting, although the sulfuric acid leachate solution is caustic and must be neutralized. Titanium is not toxic, as shown by its use in medical implant devices (WHO, 1982).

In spite of its high crustal abundance, titanium is mined in only 17 countries. Most production comes from beach sands in Australia, South Africa, Mozambique, Ukraine, India, the United States, and Sierra Leone, and from bedrock deposits in Canada and Norway. Titanium pigments are produced in at least 18 countries, led by China, the United States, Germany, Japan, and the United Kingdom. Titanium metal is produced in just six countries, China, Japan, Russia, Kazakhstan, the United States, and Ukraine. Because they have limited domestic reserves of titanium minerals, the United States and most large European manufacturing nations consider titanium to be a **strategic metal**.

From a purely numerical standpoint, titanium reserves do not look so bad. In terms of contained TiO_2, rutile and ilmenite reserves amount to about 700 million tonnes each, almost a hundred times larger than current consumption of both types of ore. Estimates made a few decades ago predicted imminent reserve shortfalls, and this improvement reflects both an increase in exploration and the capability to mine lower grade deposits. The deposits are also helped along by their content of other heavy minerals, especially zircon, the main mineral containing zirconium. In the future, submerged beach placer deposits on continental shelves might be mineable. By-product rutile is also likely to be important in the future, especially from the tar sands of Alberta.

9.2 Base metals

The base metals were named by medieval alchemists who could not convert them to gold or other precious metals. Since that time, their reputation has been improved through many important applications in modern industry, which support annual world production worth about $200 billion, somewhat more than the value of light-metal production.

It could well be asked why base metals came to the attention of alchemists at all. Their concentration in average crust is very small, usually only a few tens of parts per million (ppm). The answer is that they commonly form sulfide and oxide minerals rather than substituting in silicate minerals. Even where rocks contain only traces of a base metal, that metal is usually present in sulfide or oxide minerals that are bright and shiny and that would have attracted attention. Because most

BOX 9.2 | THE HISTORY OF COPPER

Copper was probably the first metal to be used widely by humans. Iron from meteorites and gold panned from streams were almost certainly in use at about the same time or even earlier, but neither was abundant enough to have a real impact on every day life. In contrast, elemental, or **native** copper as it is known, is relatively widespread around the world. It was used for ornaments as early as 10,000 years ago and by about 6,000 years ago, it was a major trade commodity. The huge native copper deposits of northern Michigan supplied the Old Copper Complex trade that persisted until Europeans came to North America. Early copper implements that were hammered into the desired shape were brittle. This was solved by **annealing**, in which the copper was heated to change the shape and size of individual copper grains. By about 5,600 years ago, copper was being smelted from its ores in the Middle East and copper artifacts, including weapons, were in wide use. The next big step was to alloy copper with other elements to make bronze. Although we usually think of bronze as a copper–tin alloy, earliest bronzes were actually copper–arsenic alloys, probably because these two elements are commonly associated in some deposits. It was only later that the much less common tin was found in enough quantity to make copper–tin bronze. Bronze was a far superior metal for weapons and implements and its use ushered in the Bronze Age, which started about 5,000 years ago in Egypt, Mesopotamia, India, and China. Copper mines in Cyprus, the Timna Valley of Israel, and Huelva, Spain, were early sources of copper, and the Cornwall area of England became an important tin mining center about 4,000 years ago. Shortly afterwards, copper was combined with zinc to make brass, which was also used widely. Bronze gave way to iron in the Middle East and Europe by about 3,000 years ago, and the market for copper was restricted to cooking implements, pipes, and related uses that depended on its resistance to corrosion and its germicidal properties. With the start of the industrial age, copper found a large number of new markets based on its high electrical and thermal conductivity.

sulfide minerals melt at low temperatures, metallic liquids oozing out of stones around early camp fires probably started society on the metal-dependency path that we travel now.

9.2.1 Copper

Copper is a remarkable metal; it is easily melted and can be combined with other metals to make very useful alloys, especially with tin to make **bronze** and with zinc to make **brass**. Even in the solid state, it is very malleable and can be worked into shape by hammering. It has an unusually high electrical conductivity, conducts heat very well, and is also a low-level bactericide. It is no surprise, then, that copper was one of the first metals to be used by early civilizations and that it continues to be in demand with annual production worth almost $150 billion annually.

Geology of copper deposits

Copper forms a wide variety of minerals, of which the sulfide, chalcopyrite ($CuFeS_2$), is most common. It is found largely in hydrothermal deposits, of which there is a truly impressive variety, as well as magmatic and supergene deposits.

The most important hydrothermal deposits are porphyry (**porphyritic**) copper deposits, which supply about 60% of world copper production, as well as important amounts of molybdenum and gold (Sillitoe, 2010). The largest deposits are found at convergent tectonic margins (Figure 9.10) in the United States, Canada, the western Pacific, and especially Chile, which has an unusually large number of high-grade and giant deposits including El Teniente and Chuquicamata that form the basis for the state-owned copper company Codelco. These deposits consist of closely spaced, intersecting veinlets or stockworks containing quartz, chalcopyrite, and other minerals, which surround felsic intrusions that protrude upward from underlying batholiths (Figure 9.11). They formed when **magmatic water** was expelled from the intrusion as it cooled and crystalized, explosively shattering the surrounding rocks and forming a highly fractured zone (Figure 9.12) through which hydrothermal fluids flowed, depositing chalcopyrite and other copper minerals. The term "porphyry" refers to the fact that the intrusive rock contains large crystals surrounded by smaller ones, a texture suggesting that its crystallization was interrupted, possibly by separation of the magmatic water. Fluid inclusions show that quartz in the veinlets was precipitated from very hot (> 500 °C), highly saline hydrothermal solutions of probably magmatic origin, but that meteoric water invaded the hydrothermal systems as they cooled. Although individual veinlets are

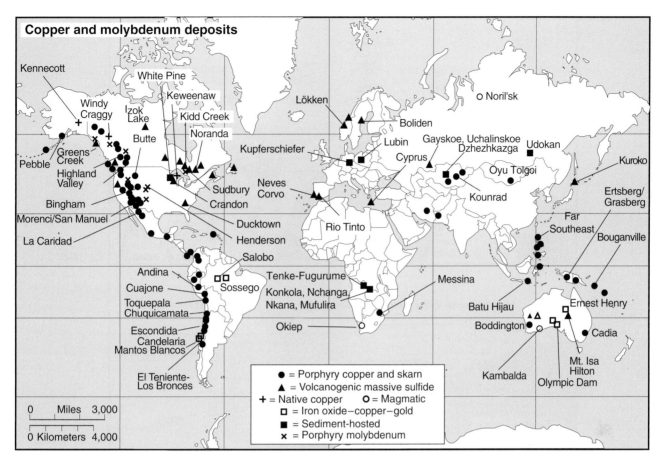

Figure 9.10 Location of major copper and molybdenum deposits. Not all deposits shown in this map have been mined.

small, they are so numerous and closely spaced that the bulk rock averages 0.5–2% copper and can be mined by inexpensive, large-scale methods. A related type of deposit known as iron oxide–copper–gold (IOCG) deposits supplies only about 5% of world copper production, most of which comes from the giant Olympic Dam deposit in South Australia (Figure 9.10). IOCG deposits formed from high-salinity fluids that appear to be of magmatic origin, but that lack the sulfur that characterizes porphyry copper deposits (Richards and Mumin, 2013).

Basinal hydrothermal systems formed two different types of sediment-hosted copper deposits that account for another 20% of world production (Hitzman et al., 2005). The best known are shale-hosted sulfide deposits like the Kupferschiefer shale, which underlies an area of 20,000 km^2 extending from northern England to Poland and has been mined over 140 km^2 in Germany and Poland. By far the most important of these deposits is the Copper Belt, which extends for several hundreds of kilometers from the Democratic Republic of Congo (DRC) into Zambia (these deposits are the most important global source of cobalt and were discussed

briefly in the previous chapter). Although early theories suggested that these deposits formed as chemical sediments, it is now recognized that they formed later when basinal brines flowed upward into the sediments replacing pyrite and depositing copper sulfides (Figure 9.11c). Native copper deposits formed by similar processes in basins that contained volcanic rocks. One such area in the Keweenaw Peninsula of Michigan (Figure 9.10) contains native copper mixed with small amounts of silver in **vesicles** in basalt flows and porosity in interbedded conglomerates of the Mid-Continent rift zone that formed when North America almost broke apart about 1 billion years ago.

Another 10% of world copper production comes from volcanogenic massive sulfide (VMS) deposits, which formed from seawater hydrothermal systems. They consist of massive lenses of iron, copper, zinc, and lead sulfides that were deposited where very hot fluids vented onto the seafloor (Galley et al., 2007). The massive ore lenses are commonly underlain by a system of veinlets that formed the feeder zone through which fluids reached the surface (Figure 9.12a). Two main types of ancient VMS deposits are named for areas in which

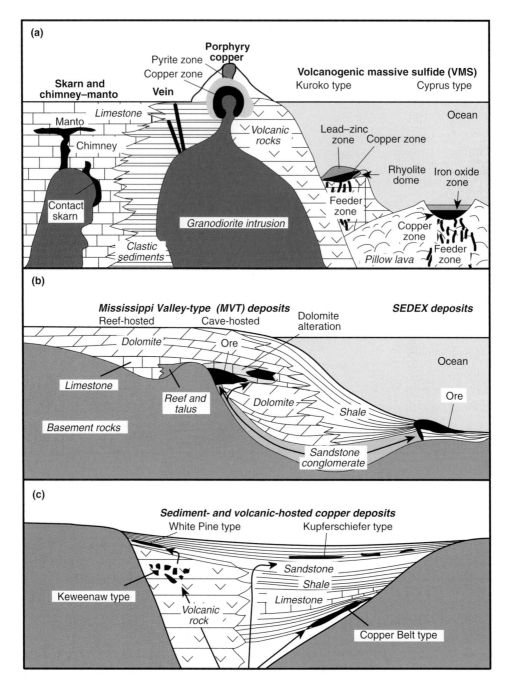

Figure 9.11 Geologic environments and characteristics of important types of base-metal deposits. (a) Deposits in volcanic-intrusive environments; (b) deposits in basinal environments with minor volcanic rocks; and (c) deposits in basinal environments with locally important volcanic rocks.

they were first recognized. Cyprus-type deposits get their name from Cyprus, which was itself named for copper. These deposits formed in ocean crust at divergent tectonic margins; the TAG copper–zinc deposit at 26° N along the mid-Atlantic ridge is an actively forming deposit of this type. Kuroko-type deposits, which are named for an area in northern Japan, formed at convergent tectonic margins, where **island arcs** were cut by rift zones. Deposits similar to the

Kuroko-type are also common in Precambrian rocks, and include the giant Kidd Creek deposit in Ontario (Figure 9.10). The shape of VMS deposits is controlled in part by the relative densities of the venting fluid and surrounding seawater. Where the venting fluid is denser than seawater, it collects in depressions and forms saucer-shaped deposits. Where it is less dense, it rises and forms the black smokers discussed in Chapter 2, which grow together and deposits, and have formed separate ore bodies in a few places, such as the Exotica deposit near Chuquicamata, Chile.

Copper production and environmental concerns

Most porphyry copper deposits, as well as **secondary deposits** and **supergene deposits**, are mined in open pits. The largest of these, Escondida in Chile, removes almost 500 million tonnes of ore and waste rock each year. A few porphyry copper deposits, such as El Teniente, Chile, and Oyu Tolgoi, Mongolia, are mined by large-scale, block- and panel-caving underground methods. Higher-grade magmatic, VMS, and sedimentary deposits are also mined by open pits where they are near surface, but have been followed to depths of at least 2,000 m by underground mining (Jolly, 1985; OTA, 1988).

Copper sulfide ores are beneficiated to produce a sulfide mineral concentrate for smelting (Figure 9.13). Because of the low copper content of porphyry copper deposits, as much as 98% of the volume of the ore must be discarded as tailings. These are usually placed in piles or valley-fills, which occupy a large fraction of the total footprint of the mining operation. **Reclamation** of abandoned mines involves both the pit and waste piles. Open pits are rarely filled after mining, because the ore bodies are not flay-lying or near-surface as in strip mines and to put the ore back would require "mining" adjacent waste and tailings piles to fill the pit, long after exhaustion of the ore body had ended cash flow to pay for the operation. However, abandoned pits are regraded and, if possible, revegetated or left as lakes; waste piles are given a protective covering and revegetated (Dutta *et al.*, 2005). The Flambeau deposit in Wisconsin, an exception to these rules, was refilled (Applied Ecological Services, 2015). Reclamation was not practiced at many older copper mines, such as Butte, and legacy pollution from them remains a major problem, as discussed in Chapter 4.

At many mines, low-grade ore and waste are leached by acid waters to dissolve the copper. Much of the acid for this process comes from natural dissolution of pyrite already present in the rock. Copper is precipitated from these solutions by **solvent-extraction**–electro-winning (SX–EW) methods in which the dissolved copper is plated out onto the cathode of an electrolytic cell, or by reacting it with steel scrap to produce cement copper, which is then smelted (Kordosky, 1992). In the western United States, cans from municipal waste as far away as the San Francisco Bay area have been used to recover copper from deposits. This same process is used as the primary production method in secondary copper deposits because copper oxides and carbonates are highly soluble.

Figure 9.12 (a) Intersecting veins of quartz, pyrite, and chalcopyrite in porphyry copper ore from the Granisle deposit, British Columbia. (b) Veinlets of chalcopyrite that make up the feeder zone beneath the Millenbach VMS deposit, Noranda area, Quebec (photographs by the authors). See color plate section.

collapse to make a mound. Well-preserved ancient black smokers have been found in the large VMS deposits in the Ural Mountains (Herrington *et al.*, 2005).

Magmatic copper deposits, which supply about 5% of world copper production, consist of immiscible metal sulfide magmas that separated from mafic and ultramafic silicate magmas. Copper is almost always associated with nickel and minor amounts of platinum-group elements in these deposits, particularly at Kambalda, Australia, Sudbury, Ontario, and Noril'sk, Russia, which are discussed in Section 8.3 on nickel deposits (Naldrett, 1989).

Most copper sulfide minerals are not stable at Earth's surface and dissolve during weathering. Where weathering acts on hydrothermal or magmatic copper deposits exposed at the surface, it can form two types of deposits (Sillitoe, 2005). If the dissolved copper precipitates as copper oxides and carbonates, secondary copper deposits are formed. If it percolates downward into underlying unweathered rock and precipitates new sulfide minerals, supergene copper deposits are formed. Both processes have contributed to the grade of some copper

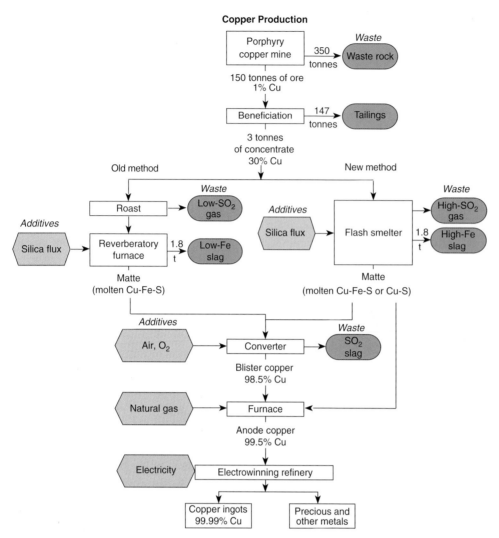

Figure 9.13 Schematic illustration of copper production process showing the old method that produced low-SO₂ off-gas and the new flash smelting method that produces off-gas with higher SO₂ content, which makes recovery more efficient (modified from Goonan, 2006)

About 80% of world copper smelting is carried out by **pyrometallurgy** (Davenport *et al.*, 2002). In this process, concentrate containing copper minerals is dried or roasted and then melted in a furnace to produce **slag**, an iron–silicon melt that captures the impurities in the concentrate, and **matte**, a copper–sulfur melt that captures the copper. Slag floats on the matte and can be removed leaving matte to be treated by a process known as converting that blows through oxygen to remove sulfur as SO₂, leaving **blister copper** that is refined electrolytically. Early smelters were bad SO₂ emitters because the melting and converting steps were done in separate furnaces. **Reverberatory furnaces**, where melting was done, gave off a gas with an SO₂ concentration too low for efficient capture in acid plants or scrubbers. SO₂ was also lost in transferring matte to the converter and in the converter operation. In response to environmental regulations, numerous new smelting and converting methods were developed,

with the main innovations being continuous operation in which smelting and converting were connected in the same container, and the use of flash smelting in which small grains of concentrate or matte are mixed with hot air and melted very rapidly, using energy released by oxidation of the sulfur and iron to heat the charge. Attention was also given to methods to keep the ratio of air to released sulfur as low as possible so that the concentration of SO₂ in the off-gas was very high, as discussed in Chapter 4. Hydrometallurgical methods were also improved, although they remain a distant second in terms of total copper produced.

The result has been an enormous improvement in smelting capability in most MDCs and a dramatic decrease in SO₂ and other metal emissions (Figure 9.14). The large copper operation at Bingham, Utah, is a good example. In 1976, it installed a Noranda continuous smelter system that captured 93% of process SO₂. By 1995, however, it had installed a continuous

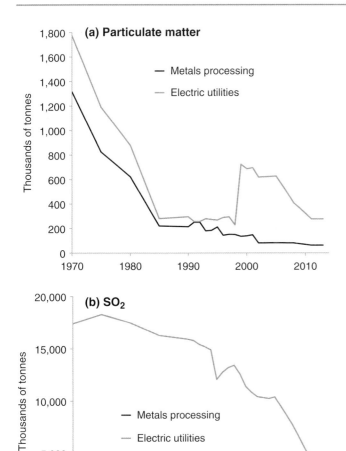

Figure 9.14 Comparison of particulate matter (a) and sulfur dioxide (b) emission histories for smelters and electric power plants in the United States (from the EPA National Emissions Inventory for 2012)

anemia (Howell and Cawthorne, 1987; Nriagu, 1979). Copper in water tubing is actually beneficial because is discourages the growth of bacteria. Exposure to elevated copper levels can cause symptoms ranging from green hair to death, although even the least severe symptoms are rare, apparently reflecting the ability of the body to cope with elevated exposure to essential elements, a behavior not observed for nonessential elements such as lead and mercury.

Copper markets, reserves, and trade

The largest market for copper is in construction and electrical motors, generators, and wiring (Figure 9.15). Aluminum substitutes for copper in long-distance power transmission lines where lighter cables permit construction of less-expensive towers, and silver and even gold substitute for copper where high electrical conductivity is needed (West, 1982). Most of these battles ended in the 1970s, however, and the new assault on copper comes from glass fibers for telephone transmission lines. Glass fibers have taken over the trunk lines; although copper remains the cheaper option for local connections fiber optic cables now reach individual homes. Copper is also used in construction, for roofing, decorative trim, and plumbing, although plastic pipe is already substituted in wastewater lines. Another major copper market, brass for ammunition casings, depends largely on the status of world conflict (Mikesell, 1988). With all of this competition, it should not be a surprise that the production of copper is growing at a rate about the same as steel (Figure 9.1b). One of the few bright spots is the use of copper in automobiles. Today's cars contain about 25 kg of copper, including about 1 km of wire compared to the 46 m of wire in a 1948 car, and the amount should rise as computer controls increase. Coinage, another potentially large market, was helped by the use of copper in Euro coins, which range from 75 to 89% copper.

The leading copper producer is Chile, with about one-third of world production, followed by China, Peru, the United States, Australia, and at least 45 other countries. Six of the 10 largest copper mines in the world are in Chile, led by Escondida. By far the largest smelter of copper is China, which produces about one-third of world copper metal, followed by Japan, Chile, Russia, and India in that order. Japan, with no copper mines, has the fifth and sixth largest smelters in the world (International Copper Study Group, 2013). This should make clear that copper moves around the world in both concentrate and metal form. The major exporters of concentrate are Peru, Australia, Canada, Indonesia, and Chile and the major importers of concentrate are China,

flash smelter system at an additional cost of $880 million that recovers 99.9% of SO_2 emissions (Bon *et al.*, 1995). In Peru and other less-developed countries (LDCs), older copper smelters have not yet attained such high SO_2 recovery levels and might be shut down rather than improved. That has already happened in remote areas of MDCs, such as Flin Flon, Manitoba, where the old smelter was closed after polluting an area more than 100 km around the town (Zoltai, 1988). Recent reports show that satellite methods can be used to monitor emissions in LDCs, providing a new metric that has been used by environmentalists to encourage improvements (Carn *et al.*, 2007).

Environmental concerns with copper production and use rarely focus on the metal itself. In fact, copper is an essential element for life. Copper deficiency in humans, which is rare, is manifested in the failure to use iron effectively, causing

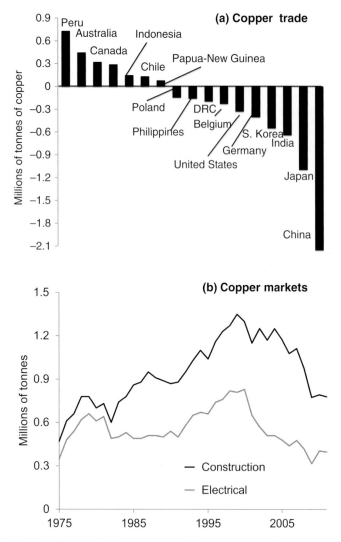

Figure 9.15 (a) World copper ore trade for 2012 expressed as copper content of mined vs. refined copper. Countries that plot above the line produce more ore than they smelt and refine; those that extend below the line import ore concentrates for their smelters and/or refineries (based on data from the US Geological Survey and International Copper Study Group, 2013). Note that Chile, which is the world's leading miner of copper, does not show up as a major exporter of copper concentrate because it smelts and refines most of its ore. Peru, the third largest copper miner after Chile and China, does not smelt and refine all ore and shows up as a major exporter, as does Canada. Japan, India, Korea, and Belgium lack significant mines and are big importers of refined copper. (b) Change in major copper markets in the United States, showing the growth of the construction market compared to the electrical and electronic market (based on data of the US Geological Survey).

Japan, India, Korea, and Germany (Figure 9.15). Japan and Germany turn right around and export copper metal, along with Chile, Australia, and Peru. The United States comes in tenth on the list of concentrate exporters and Canada is fourteenth. Both produce less copper than they mine because the closure of old smelters in response to clean-air legislation has limited capacity (Jolly and Edelstein, 1992; Crozier, 1992). Only three smelters remain in the United States and one in Canada.

The change in the US and Canadian copper trade status coincided with a gradual increase in the proportion of blister copper exported by many LDCs. This reflected the LDCs' desire to reap the economic benefits of a more refined, higher margin product rather than the concentrates that they were exporting. Regression of the United States to the role of concentrate exporter is due to its failure to build new smelters to replace obsolete ones. In response to delays and public concern, many new smelter projects have been located in LDCs, further increasing their capacity to deliver finished copper and decreasing jobs in the United States. It is ironic that smelters should become so difficult to locate at the same time that the technology has become available to make them cleaner.

Copper ore reserves are estimated to contain 680 million tonnes of copper metal, of which Chile has almost 30% and Australia and Peru have about 12% each (Edelstein, 2013a). This is enough to supply about 40 years at the present annual consumption rate of about 17 million tonnes. Although this might seem like an optimistic time cushion, it is not. Large copper deposits will be exhausted within 10 to 30 years, and must be replaced by new discoveries. Even more importantly, these discoveries must be developed into new mines, a task that is becoming more and more difficult as discussed at the end of this chapter.

9.2.2 Lead and zinc

Lead and zinc are usually found together in ore deposits (Figure 9.16), but have very different biological effects. Zinc is an essential element for life, whereas lead has no known biological function and is an important environmental hazard. This might make you think that lead would not be produced in significant amounts. In fact, annual world lead production is worth almost $13 billion, not peanuts even when compared to zinc production worth $30 billion. Production of both commodities has grown at about the same rate as copper and prices of both commodities have followed a similar pattern. This might be surprising to you, in view of lead's bad environmental reputation. The reason lies in lead's pivotal role in electric storage batteries. Ironically, lead consumption may even have to increase in order to minimize other environmental problems, as discussed below.

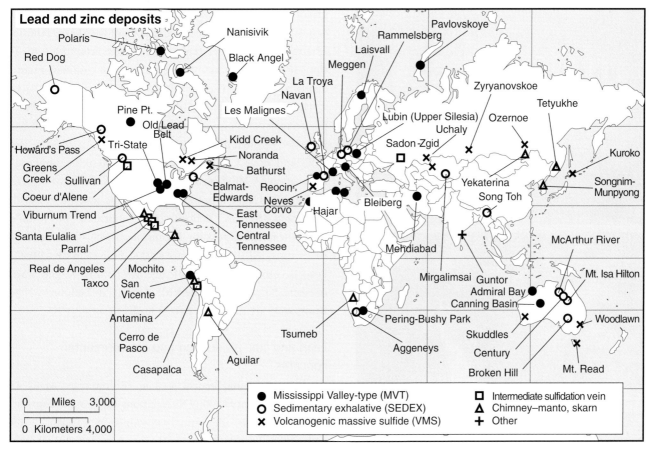

Figure 9.16 Location of major lead and zinc districts. Many of the VMS and skarn deposits shown in Figure 9.10 are also major lead and zinc producers

Geology of lead and zinc deposits

Lead is usually obtained from the common ore mineral galena (PbS) and zinc comes from sphalerite (ZnS), which are common in two types of hydrothermal deposits, sedimentary exhalative (SEDEX) and Mississippi Valley-type (MVT), that form largely in basinal environments (Leach *et al.*, 2005).

SEDEX deposits, which have very high metal contents between 10 and 20% Pb+Zn, consist of sulfides that were deposited as layers of sulfide sediment (Figure 9.17) or by replacing sediment just below the seawater interface. They are similar to VMS deposits, but form from brines that flow onto the seafloor in sedimentary basins rather than at the mid-ocean ridges. Most SEDEX deposits formed when the hot springs emerged into reduced water with abundant H_2S that precipitated the metal sulfides. The H_2S was generated by abundant organic activity in the seawater and it managed to persist only in rift-bounded basins where the water was isolated from the open ocean. The largest SEDEX deposits, including Broken Hill, Mt. Isa, and McArthur River in Australia and Sullivan in British Columbia (Figure 9.16), are

in a system of 1.8- to 1.3-Ga-old rifts that cut the early continents (Young, 1992). The more recently developed Red Dog deposit in Alaska is of similar size, but is much younger, at about 330 Ma. Several large deposits in Europe, especially Meggen and Rammelsberg in Germany, were important historical sources. Howard's Pass, an especially large deposit in the Canadian Yukon, has been known for decades but is in too remote a location to justify mining. A 50% interest in the deposit was purchased in 2014 for about $50 million (Selwyn Chihong Mining, Ltd).

MVT deposits, which account for another 20% of global production, are so-named because they were first found in the valley of the Mississippi River in the United States (Figure 9.16). Huge amounts of lead and zinc were produced from now-exhausted MVT deposits in the Tri-State district near Joplin and the Old Lead Belt near Bonne Terre, both in Missouri (Figure 5.10). MVT deposits generally have lower metal contents than SEDEX deposits, usually between 3 and 10% Pb+Zn, which makes them less attractive. Production continues from deposits in the Viburnum Trend of Missouri

Figure 9.17 (a) Folded layers of sphalerite (red) and galena (gray) in the Sullivan SEDEX deposit, Trail, British Columbia. (b) Breccia zone in which fragments of limestone and dolomite (dark) and elliptical chert nodules have been cemented by hydrothermal dolomite crystals (white) that were deposited by basinal brines in the Coy MVT zinc deposit, Tennessee (photographs by the authors). See color plate section.

and East Tennessee in the United States, as well as the Lubin area of Poland and the Canning Basin of Australia. These deposits consist of galena, sphalerite, barite, fluorite, and other minerals that fill secondary porosity in limestone and dolomite in much the same way that oil and gas fill other porous sediments (Figure 9.17b). The deposits are on the margins of ancient sedimentary basins and formed from basinal brines that migrated out of the basin until they reached areas where they were deposited by H_2S that was present or that formed by reduction of sulfate in the brines.

Deposits with characteristics intermediate between MVT and SEDEX are found in Ireland. These Irish-type deposits, including Navan and Lisheen, have supported a strong mining industry for several decades. As discussed in Chapter 6, they were discovered when the Irish government established tax regulations that encouraged mineral

exploration. Lead and zinc are found in numerous other deposit types; among the most important are **skarn** deposits where magmatic water replaced limestone around an intrusion, such as at Antamina in Peru and the chimney–manto deposits of Mexico (Figure 9.16).

The distribution of lead deposits in Earth history reflects the average concentration of lead in crustal rocks, which has increased through time by the decay of uranium. Archean-age MVT and SEDEX deposits are very rare and of the numerous Archean-age VMS deposits, only the huge Kidd Creek deposit in Ontario, Canada, produces remarkable amounts of lead. Important SEDEX deposits first appeared in Middle Proterozoic time, possibly because lead-rich, felsic igneous rocks intruded the continents at that time (Sawkins, 1989). Lead is common in Paleozoic-age MVT, SEDEX, and VMS deposits. Some, such as the VMS deposits in Bathhurst, New Brunswick, are unusually rich in lead.

Lead and zinc production and environmental concerns

Lead is mined and beneficiated in 38 countries versus 48 countries for zinc, making them among the most widespread metals in terms of primary production (Tolcin, 2013; Guberman, 2013a). Lead and zinc are mined largely by highly mechanized open pit and underground methods. With a few rare exceptions, such as the extremely high-grade ore that was shipped from the Pine Point MVT deposits in Canada early in their history (and that paid for building the railroad to Pine Point), lead–zinc ores must be beneficiated to produce concentrates for smelting. SEDEX and VMS deposits, which were deposited by rapid cooling of submarine hot-spring fluids entering cold ocean water, can be too fine-grained for ore and gangue minerals to be separated by conventional beneficiation methods. The large McArthur River SEDEX deposit in Australia was so fine-grained that it took almost 30 years after discovery before new methods allowed beneficiation of its ore. Where these deposits have been metamorphosed, recrystallization has produced larger crystals that are easily separated. The outstanding examples of this process are at Aggeneys, South Africa and Broken Hill, Australia, which contain galena crystals measuring tens of centimeters on a side.

Primary smelting of lead and zinc concentrates is carried out in 24 countries for lead and 31 for zinc. Most present primary zinc smelting begins with roasting, which drives off sulfur as SO_2 and converts the sphalerite to zinc oxide. Zinc oxide from the roasting step is recovered by SX-EW methods using a sulfuric acid solution. Conventional lead smelting is

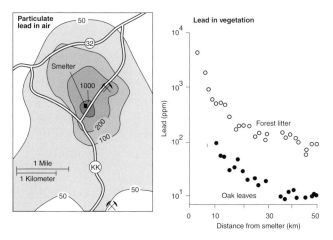

Figure 9.18 Distribution of settleable particulate lead and lead in oak leaves and forest litter around lead smelters in Missouri (from Jennett *et al.*, 1977). Most metal smelting created similar haloes until advanced gas and particulate recovery systems such as those discussed in Chapter 4 were installed.

done in a blast furnace, producing lead bullion, which must be refined to remove any zinc and silver that were also in the concentrate. When a mixed galena–sphalerite concentrate must be smelted, it is roasted and then put through the imperial smelting process, in which zinc is driven off as a vapor (to be condensed elsewhere), leaving less volatile lead in the residue. Secondary lead smelters, which remelt lead metal products, are located in 53 countries and treat largely lead from electric storage batteries.

Lead and zinc mining and processing are responsible for a considerable legacy of pollution, in part because lead vaporizes at such low temperatures. Older lead smelters emitted galena particles from handling of ore concentrates, as well as an aerosol of lead oxides and sulfates that condensed in smelter exhaust gas (Jennett *et al.*, 1977). Smelters built as recently as the 1970s produced measurable emissions from these sources (Figure 9.18). Metal-rich dispersion zones consisting of galena particles, which are usually near the smelter or mine, are less damaging because the lead is relatively immobile and not easily taken up by plants and animals. Lead oxides and sulfates from flue gases are much more soluble, however, and are dissolved and moved to deeper levels of the soil by precipitation. They are also dispersed farther from the source because of their small size.

Lead and zinc differ greatly in their effects on humans (Boggess, 1978; Nriagu, 1978, 1987; WHO, 1989a). Zinc is an essential constituent of enzymes, DNA, RNA and protein synthesis, carbohydrate metabolism, and cell growth. Lead

has no biological use and is toxic even in low amounts, causing both physiological and neurological damage in humans. The physiological effects, which are best understood, include nausea, anemia, coordination loss, coma, and death, with increasing exposure. The neurological effects of lead range from reduced nerve function to encephalitis. Lead causes many of these problems because it takes the place of other metals that are essential for life in important biochemical reactions. For instance, it causes anemia by taking the place of iron in compounds used in the biosynthesis of hemoglobin; and calcium deficiency by accumulating in the bones, the principal reservoir of calcium; it also interferes with zinc enzymes. In spite of these negative effects, lead is not known to be carcinogenic. It can be removed from the body by use of organic chelating agents such as ethylenediaminetetraacetic acid (EDTA), which form stable dissolved complexes with lead that are excreted. As discussed below, although the general population has some risk from lead-production activities, their risk is much greater from lead in widely used products such as gasoline and paint.

The legacy of lead persists throughout the world. It is well established that even low levels of lead can cause cognitive impairment in young children. Although elimination of lead in gasoline and paints allowed an 80% decrease in lead poisoning in young children by the early 2000s, about 2% of the young population remains at risk. This group is overwhelmingly in areas of high-density population near roads where soils remain strongly enriched in lead, and where evidence is accumulating that this lead is being remobilized (Filippelli *et al.*, 2005).

Lead and zinc markets, trade, and reserves

Lead is believed to have been used in glazes on pottery by 7000 to 5000 BCE and in metal objects and medicines in ancient Egypt and China (Blaskett, 1990). By the time of the Roman Empire, it was used to make pipes, eating and cooking utensils, coins, weaponry, and writing equipment (whence the term lead for the graphite in pencils), and to sweeten wine and other foods. Per capita lead consumption during Roman times is estimated to have been equal to that today, but for distinctly different uses. In fact, it has been suggested that lead caused the decline of the Roman Empire, whose later rulers displayed symptoms of lead poisoning (Nriagu, 1983; 1992). The hypothesis is supported by elevated concentrations of lead found in bones of Roman nobility (Aufderheide *et al.*, 1992). Despite millennia of damage to humans, lead is still in wide use simply because it has continued to find new markets.

BOX 9.3 | TETRAETHYLLEAD

At one time, it was thought that phasing out lead plumbing and the use of lead as a pigment in paints would help decrease the lead content of humans. Unfortunately, tetraethyllead came into wide use just as lead plumbing was phased out. During the tetraethyllead days, annual anthropogenic lead emissions of about 300,000 tonnes far outweighed natural sources of less than 20,000 tonnes (Nriagu and Pacyna, 1988; Nriagu, 1990a). Some of this lead came from coal combustion, and some still does, but the high concentration of lead around roads confirms tetraethyllead's role. Lead has moved around the world and is also enriched in recent ice layers in Greenland and in the ocean (Rossman *et al.*, 1997). In hindsight, use of tetraethyllead was a terrible mistake, and possibly one that might have been predicted at the time in view of the well-known toxicity of lead; but the pressure for an anti-knock compound overcame common sense (Kitman, 2000). Discontinuing tetraethyllead use in gasoline and installation of scrubbers in power plants have caused lead fallout to decrease dramatically on a global scale, although it continues to accumulate in coral and other marine organisms in areas of the ocean adjacent to countries that were late to remove lead from gasoline (Figure 9.19).

The most important modern market for lead is in electric storage batteries for fossil-fuel-powered cars and trucks, battery-powered electric vehicles, and emergency power supplies for large computer systems, hospitals, and similar fail-safe consumers (Figure 9.19). Over 80% of world lead production goes into batteries. Additional markets include ammunition, pigments, roofing materials, and items such as lead weights. The only one of these markets that might endure is batteries. As we learned in the last chapter, lead is being replaced by tungsten and uranium in ammunition, and titanium has taken its place as the basic pigment in paint. It has also been replaced by tin in solder, especially for use on copper pipes used for water. The biggest market that lead lost was the use of **tetraethyllead** [$Pb(C_2H5)_4$] and related compounds as anti-knock additives in gasoline, which has been phased out worldwide over the last three decades. Even earlier than that, lead was gradually replaced by copper for plumbing.

Zinc was not heavily used until well into the 1800s, but it has penetrated a large number of markets since that time. Zinc's pervasive role in world industry is not widely appreciated by the casual observer because it is used in forms other than the pure metal. For instance, about half of zinc consumption goes to make **galvanized steel**, in which thin coatings of zinc are used to prevent corrosion. Galvanized steel is the major component of power transmission towers, culverts, tanks, and nails, where it protects the steel both by insulating it from atmospheric corrosion and through galvanic decomposition, in which two metals form a natural electrolytic cell. Galfan, an alloy with 90% zinc, 5% aluminum, and rare earths, outperforms galvanized steel, but is more expensive. Sacrificial anodes of zinc are also used to prevent corrosion on ships, oil drilling platforms, and submerged pipelines. A third of zinc production goes into brass, bronze, and other zinc alloys that are used as ammunition shell casings and in motors, refrigeration equipment, and cars, and zinc metal is used in roofing in Europe. The use of zinc in die castings has declined as the size of cars has decreased.

World trade in lead and zinc is similar to that for copper with considerable movement of concentrates from ore-rich countries to those with smelting capacity. Australia, and Peru dominate the export scene for both metals, and China and South Korea are the major importers (Figure 9.20). Many of the European countries that import lead and zinc concentrates turn around and export metallic zinc and lead. Reserves are not particularly good, amounting to about 70 million tonnes of contained lead and 150 million tonnes of contained zinc. The largest reserves are in Canada, Australia, Peru, Mexico, the Commonwealth of Independent States (CIS) countries. With the exception of the Red Dog SEDEX deposits in Alaska, reserves in the United States are declining rapidly.

9.2.3 Tin

Tin is an element with lots of friends but dwindling markets. It was the first commodity to have a global group of producers and consumers to encourage its consumption, the International Tin Council. As can be seen in Figure 9.1e, tin production has remained just about constant for the last half century, while production of almost all other metals has increased. Tin prices, on the other hand, have been more volatile than most metals, including a recent price increase

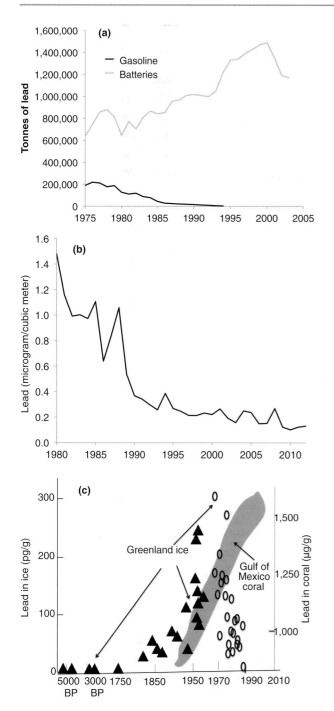

Figure 9.19 (a) Change in lead markets in the United States. (b) Decrease in atmospheric lead emissions in the United States since 1980 showing a dramatic decrease caused by a drop in the consumption of leaded gasoline. (c) Change in lead content of Greenland ice and Gulf of Mexico coral (compiled from Boutron *et al.*, 1994; Horta-Puga and Carriquiry, 2014).

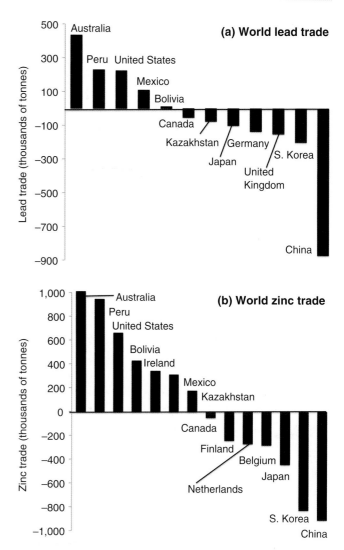

Figure 9.20 World lead (a) and zinc (b) ore trade for 2012 expressed as metal content of products. Countries above the line produce more ore than they smelt; those below the line import concentrate for smelting. Note that the United States is a major exporter of both lead and zinc concentrates; a situation that reflects an absence of smelting capacity. Canada imports zinc concentrates from the large Red Dog mine in Alaska, which is owned by Teck, a Canadian company (all data from the US Geological Survey).

Geology and mining of tin and related environmental concerns

The most important tin ore mineral is cassiterite (SnO_2), which forms ore deposits in only a few parts of the world (Figure 9.21). Almost three-quarters of world production comes from only six countries and three of these, Malaysia, Thailand, and Indonesia, actually host a single tin-bearing region. One of the first areas to be mined was around Cornwall, England, which supplied lead for bronze in the Roman empire (Figure 9.22).

that almost matches that of copper (Figure 9.1c). At present, the value of world tin production, amounting to about 250,000 tonnes annually, is a respectable $8 billion (Figure 9.1a).

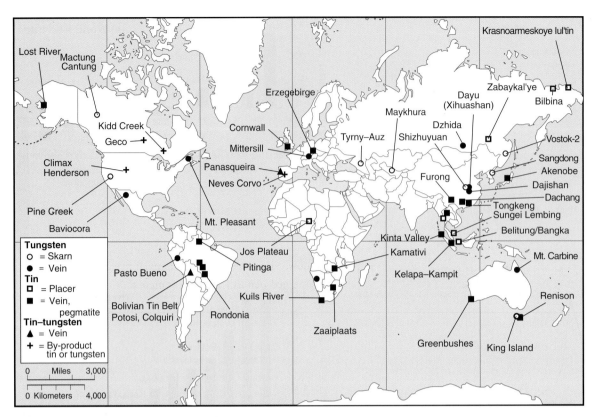

Figure 9.21 Location of world tin and tungsten deposits. Tin pegmatite deposits in particular are also sources of elements such as lithium and cesium.

Figure 9.22 (a) Tin-bearing veins in granite at Cligga Head, Cornwall, England. Narrow vein systems of this type are referred to as sheeted veins. (b) Larger veins were mined individually. This photo looks outward from a stope that mined a 0.7-m-wide vein of this type at Bottalack in Cornwall (photographs by the authors). See color plate section.

Geologists are fascinated by tin because of its highly specialized distribution in Earth. The reason that tin is found in so few places, known as metallogenic provinces, has been debated for years, with two major competing hypotheses (Lehmann, 1990). One holds that tin deposits are found only in tin-rich parts of the crust, possibly inherited from the early history of Earth. The other holds that tin can be found anywhere that special tin-concentrating processes were active.

Both appear to be correct because tin deposits are associated with granites and their rhyolitic volcanic equivalents that probably formed in special settings by partial melting of sedimentary rocks (Kesler and Wilkinson, 2014). The most common deposits, known as lode deposits, are in pegmatites, quartz veins, stockworks, and disseminations clustered around cupolas at the top of these intrusions. In the Cornwall area, tin ore consists of coarse-grained muscovite

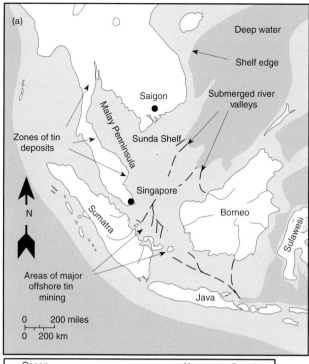

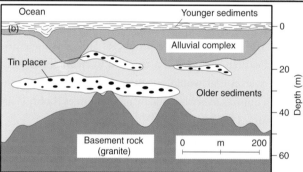

Figure 9.23 Tin placer deposits on the Sunda Shelf, southeast Asia. (a) Location of tin-producing areas on and offshore (from Sainsbury, 1969 and Batchelor, 1979). (b) Cross section showing submerged tin placer deposits on the Sunda Shelf (after Batchelor, 1979).

mica, quartz, and other minerals known as greisen, which formed by high-temperature (500 °C) alteration of granite (Figure 9.22). Where the wallrocks consist of limestone, such as at Renison, Tasmania and Dachang, China, sulfide-rich, tin-bearing replacement deposits are found. Tin is also found in veinlets and stockworks associated with volcanic domes such as the huge Cerro Potosi (Cerro Rico) tin–silver deposit in Bolivia that is discussed in the silver section.

Tin is an important by-product from tungsten vein and stockwork deposits and porphyry molybdenum deposits, where it probably has the same magmatic origin. It is also found as a trace constituent in some VMS and SEDEX deposits. Interestingly, tin is most common in the largest of these deposits, such as the huge Kidd Creek and Geco VMS deposits and the Sullivan SEDEX deposit, all in Canada, and the lore of

mineral exploration says that you have found a big VMS or SEDEX deposit if it contains tin.

The tin ore mineral, cassiterite, is heavy and highly resistant to weathering, and forms placer deposits during erosion of bedrock tin deposits. Placer tin deposits have been the dominant source of world tin for many years. The most important deposits of this type are in northern Brazil around Pitinga, and in the southeast Asian province, which extends through Thailand, Malaysia, and Indonesia as well as adjacent parts of the seafloor in the Sunda Shelf (Figure 9.23). The Sunda Shelf was exposed to erosion during the last global glacial advances because water was trapped in the ice caps, lowering sea level, and streams flowed much farther out onto the shelf than they do now. Thus, the shelf hosts alluvial placer deposits that are submerged beneath only a few meters of water and, in some cases, a few tens of meters of sediment.

Tin mining takes different forms, depending on whether it is from lode or placer deposits (Batchelor, 1979). Lode mining uses small-scale open pit and, more commonly, underground methods. Because of the close association of tin and uranium in some granitic rocks, radon can be a hazard in underground mines. Placer mining is done by gravel pumps and dredges, which creates large volumes of sediment-laden water that is captured and impounded on land but not in marine operations. Many tin mines, especially in Indonesia and China, are small operations that violate common labor and safety regulations; closing of many of these in China led to shortages and increased prices during the late 2000s (Hodai, 2012; Carlin, 1985b, 2013). Production of tin from cassiterite concentrate is done by heating it to about 1,200 °C with carbon in a blast, flash, or electric furnace to drive off the oxygen as CO_2, leaving molten tin metal. Unless the tin concentrates contain sulfide minerals, this process does not cause significant environmental problems.

Tin is not a problematical element from a health perspective and it has no essential function in the body. Inhalation or ingestion of relatively high levels are required before negative symptoms are observed. Dust from tin oxide, such as in cassiterite, produces a benign **pneumoconiosis** known as **stannosis**, which is a potential problem only in tin processing and can be prevented by proper air handling and use of respirators. Organotin compounds are used widely in fungicides and similar chemicals and are toxic, but they do not occur in nature (WHO, 1980).

Tin reserves, markets, and trade

Tin reserves are estimated to be about 4.9 million tonnes of metal, most of which is in China, Indonesia, and Brazil. In

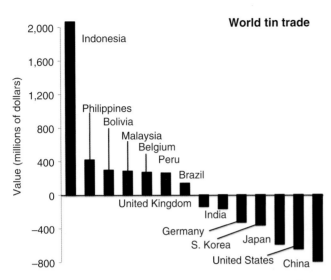

Figure 9.24 Value of world tin ore trade for 2012. Countries that extend above the line export tin or tin concentrate and those that extend below the line import tin (from data of UN Comtrade). The presence of Belgium as a tin exporter reflects the sale of previously imported stocks.

view of declining tin consumption, this is a relatively comfortable margin, especially when compared to other metals in this base-metal group.

About half of world tin consumption is in solder. Another 30% goes into **tinplate**, a thin coating of tin on steel cans and other metal products, and tin chemicals that go into wood preservatives, antifouling paints, and plastics. Most of the rest of tin consumption goes into bronze, the copper–tin alloy that started the world on its metal-using trajectory. The market for tin-bearing solder has grown enormously during the last two decades because of the increase in consumer electronics. Traditional solder, which consists of 60% tin and 40% lead, has an unusually low melting temperature (188 °C) and makes a strong electrical connection. Disposal of consumer electronics, which makes this lead available for leaching into soil and groundwater, led the European Union to prohibit the addition of lead to consumer electronics in 2006 and spurred the search for a lead-free solder. The best candidate contains 96% tin and 4% copper and silver and melts at 217 °C. Whether the change was beneficial from the standpoint of overall life-cycle has been debated widely (Titus, 2011). Similar problems affect the use of lead in solder for joining copper pipes, where it can be dissolved into drinking water. In the United States, use of lead in pipe-fitting solders was stopped in 2014. Although these bans expand the amount of tin in solders, this is balanced by efforts to use less solder, thus limiting the overall growth in tin consumption.

Tinplate faces similar challenges. Food containers must be sterilized at high temperature and welded shut, processes for which tinplate is particularly well suited. Although glass can also be treated at high temperature, it is not a threat to tinplate because glass containers are heavier and expose food to light, which can change its color. In spite of manufacturing improvements that have decreased the weight of steel cans, however, they are still much heavier than aluminum cans and this translates into higher transportation costs. Thus, tinplate has lost markets to plastic and aluminum containers, as well as to chromium-coated steel, particularly in the beverage sector.

Tin concentrates are exported by most ore-producing countries except Malaysia, which imports tin concentrates and is the world's most important producer of tin metal (Figure 9.24). Other important tin smelting operations are in the United Kingdom and the Netherlands (Thoburn, 1981; Baldwin, 1983). Although tin is not critical to defense needs, production of tin is so restricted geographically that it is sometimes regarded as a strategic metal, particularly in North America, which lacks important tin deposits. The relatively small number of important tin-producing countries and the fact that none of them are major consumers led to formation of a unique cartel, known as the International Tin Council, which included both producers and consumers. As discussed earlier, the council attempted to stabilize the price of tin at a level that would encourage production and consumption but was overwhelmed by market forces (Lohr, 1985; Kestenbaum, 1991).

9.3 The future of light and base metals

The future of light, base, and chemical–industrial metals depends greatly on environmental issues related to the use of batteries in electric cars. If batteries are used widely and other trends in consumption continue, supply shortages could be experienced within the next few decades for the base metals. These shortages can be averted only if we succeed in recycling metals more efficiently and in finding more deposits, neither of which are simple tasks. Even with recycling, many deposits of these metals are small in relation to global demand, and they have to be replaced by new ones at a regular rate. Thus, exploration for these metals is particularly active and development of new mines is a common focus of public concern.

Whether you are a producer or a consumer, you should be deeply concerned about the difficulty that we have experienced in finding new metal deposits and, particularly, in bringing newly discovered deposits into production. Continued base- and light-metal supplies require that new deposits be found and put into production to replace the ones that are exhausted. We are not doing this. Between 1998 and

---

BOX 9.4	WHERE WILL TOMORROW'S COPPER COME FROM?

The recent history of efforts to bring copper deposits into production has been dismal. **Resource nationalism**, which is on the rise around the world, has led to steep increases in the share of profits that are demanded by host countries. Mongolia is a case in point. The large Oyu Tolgoi deposit, which contains about 16 million tonnes of copper, was discovered by deep drilling in 2001 and brought into production in 2013 at a cost of about $6.6 billion. Ownership of the mine is divided between the Mongolian government and Turquoise Hill, the original discoverer, which is controlled by Rio Tinto, one of the world's largest mining companies. When the mine is in full production, it is expected to account for 30% of the GDP of Mongolia, making it an obvious focus of attention to the government. Disputes about government ownership (now 34%), the level of taxation, the timing of tax payments, and even whether payments are taxes or loans, have bedeviled the operation for years (Donville, 2013). These and other financial uncertainties have led to major delays in the development of the deep ore body, which has limited production and caused the loss of an important long-term contract. All projects in LDCs have teething problems, but these are especially troublesome and it is not hard to imagine that they will lead to stronger demands by the government, possibly even *de facto* nationalization.

A logical response to this risk is to expand operations in MDCs where ownership and tax regulations are clearer. This brings up the Pebble deposit in Alaska. With about 36 million tonnes of copper, Pebble is the second-largest porphyry copper deposit in the world, and the best hope that the United States has to continue as an important copper producer. However, Pebble is mired in environmental controversy because of its location in a major salmon-bearing drainage, and it received a poor draft assessment by the Environmental Protection Agency (Harden, 2012). Two major mining companies have written off their interest in the project and the final company is considering it, which would leave the deposit in the hands of its original discoverer, a small Canadian company without funds to develop the deposit. Thus, after an exploration expenditure of at least $500 million, the likelihood that the project will ever reach production is declining rapidly.

2012, about 100 copper deposits were discovered, but only about 10 have been put into production (Komnenic, 2013). Reasons for this poor success range from low-grade ore to environmental opposition, and they show the challenge that we must face if we are to supply ourselves with metals. Not only will we have to accept lower-grade ores, but we will also have to produce metals from deposits in unfavorable locations, whether near population concentrations or in land that we treasure. In all of this, we must keep in mind that, as technology develops, we will be better able to use land sequentially and to reclaim mined land in a way that shows proper respect to the land and to its next user.

CHAPTER

10 Technology elements

Technology elements have changed the world, and not just the modern world. Mercury was a magical metal to the ancient Egyptians, and thousands of years later its remarkable properties were the basis for gold mining that drew Europeans to the New World. Then came arsenic, which played a role in speeding dynastic successions around the world. And now, we have gallium, germanium, indium, and tantalum that are essential to computers and mobile communications. Whereas we have referred to the ferroalloy and the base and light elements as metals, this group is best referred to as elements because it includes both **metals** and **metalloids**.

Although many of the elements included in this group are in the news frequently, it might surprise you to learn that their total global production is valued at only about $14 billion annually (Figure 10.1), far behind the metals we have discussed so far. Rare earths, by far the value leader of the group at about $6.5 billion, are followed by antimony and lithium with totals of about $2.5 billion each. The rest bring up the rear with individual production values of $500 million or less (Figure 10.1).

Also surprising is the fact that only rare earths, gallium, indium, lithium, and rhenium have grown in production at a greater rate than steel since 1960; all the rest of have lagged steel production (Figure 10.1). The greatest laggards are mercury, beryllium, and tellurium, for which global production has actually decreased, and arsenic and tantalum production have remained just about constant. Of the remaining elements, antimony, bismuth, germanium, and selenium have increased production most, although they still lag behind steel.

This pattern of relatively low consumption reflects the fact that technology elements go into products that are used largely in MDCs. As these products become more widely used in LDCs, consumption will rise, a pattern that has begun to show up during the last decade. Thus, it is time to look into these elements and learn what they are used for, where they come from, and whether resources are sufficient to supply the expected increasing demand. Because of their tremendous variety of properties and markets, there is no obvious system to our discussion. In order to help readers locate their element of interest, we discuss them here in alphabetical order. This order bears no resemblance to relative values of their global production or to their geological, economic, and environmental significance. Consequently, some valuable elements, such as indium for which the story is relatively simple but demand is growing, receive less attention than mercury that is losing markets but has important environmental legacies. Production and reserve data for the technology elements are summarized in Table 10.1 and major uses are listed in Table 10.2

10.1 Antimony

The biggest market for antimony, by far, is in fire-retardant chemicals that are added along with bromine to plastics and textiles in everything from clothing to airline seats (Figure 10.2). Earlier, antimony was alloyed with lead in electric batteries for cars, but this market has dwindled as substitutes have been found; present automotive batteries contain only about 0.6% antimony. Lead–antimony alloys have other uses, however, including ammunition, antifriction bearings, cable sheaths, and solder (Carlin, 2013). Annual antimony production of 180,000 tonnes is worth about $2.5 billion (Figure 10.1a).

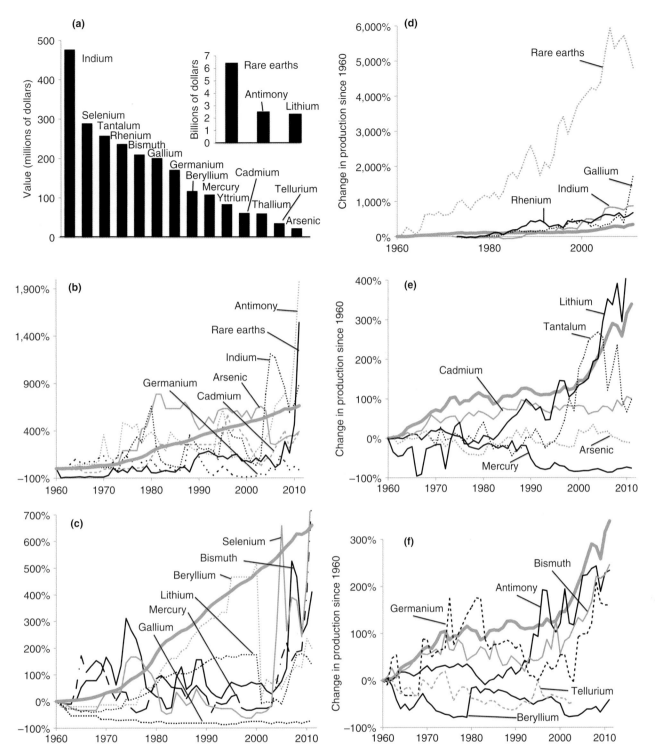

Figure 10.1 (a) Value of global production for the technology elements in US dollars for 2012. (b, c) Change in price since 1960 for technology elements relative to the consumer price index (CPI) (broad gray line). Note that prices for most elements have grown at rates lower than the CPI. (d, e, f) Change in global production since 1960 for technology elements relative to steel (broad gray line). The vertical scale differs on the three plots to accommodate production changes. Note the increase in production for rare earths, gallium, indium, and rhenium. Production of most other elements has grown at a rate less than that of steel and some, including arsenic, tellurium, and mercury have remained constant or fallen. Data for all plots are from publications of the US Geological Survey.

Table 10.1 Major producers and reserves for the technological elements (from the US Geological Survey Commodity Summaries)

Commodity	Country	Production (tonnes)	Reserves (tonnes)
Antimony	Bolivia	30,000	950,000
	China	6,500	350,000
Arsenic*	China	26,000	-----
	Chile	10,000	-----
Beryllium	United States	220	-----
	China	20	-----
Bismuth	China	6,500	240,000
	Mexico	940	10,000
Cadmium	China	7,400	92,000
	South Korea	3,900	-----
	Mexico	1,630	47,000
Cesium	Canada	----	99,000,000
	Zimbabwe	----	64,000,000
Germanium	China	0.110	-----
	Russia	0.005	-----
Indium	China	410	-----
	South Korea	150	-----
Lithium	Chile	13,500	7,500,000
	Australia	13,000	1,000,000
	China	4,000	3,500,000
Mercury	China	1,350	21,000
	Kyrgyzstan	250	7,500
Rare earths	China	7,000	220,000
	India	56	72,000
Rhenium	Chile	27	130,000
	United States	8.1	390
Selenium	Japan	780	-----
	Germany	700	-----
Tantalum	Brazil	140	36,000
	Democratic Republic of Congo	----- 110	
Zirconium (Hafnium)	Australia	600	40,000,000
	South Africa	360	14,000,000

* Production of arsenic trioxide.

Table 10.2 Uses for technology elements (from the US Geological Survey; Hedrick, 2014)

Antimony	Flame retardants, batteries, chemicals, and specialty glass
Arsenic	CCA preservatives, batteries, ammunition, bearings, weights
Bismuth	Pharmaceuticals, metal alloys, solder, glass, ammunition
Cadmium	Nickel–cadmium batteries, pigments, coatings
Gallium	Semiconductors, optoelectronic devices
Germanium	Fiber-optics, infrared optics, polymerization catalysts, electronics, and solar electric
Indium	Flat-panel devices (liquid crystal devices (LCDs)), solder, alloys
Mercury	Chloralkali process, compact fluorescent lights (CFLs) and neon lights, dental amalgam

Rare earths

Light rare earths

Lanthanum	Special glass, batteries, catalyst
Cerium	Glass and glass production, catalytic converters, CFL bulbs, pharmaceuticals
Praseodynium	Special glass, catalyst in plastic manufacturing
Neodynium	Lasers, magnets
Promethium	Phosphors, CFL bulbs
Gadolinium	Medical, optoelectronics, nuclear reactors

Heavy rare earths

Terbium	Lasers, phosphors, data storage
Dysprosium	Lasers, magnets, sensors
Holmium	Lasers, radar
Erbium	Lasers, special glass, jewelry
Thulium	Phosphors, metal halide lamps
Ytterbium	Lasers, medical chemicals
Lutetium	Lasers, optoelectronics
Rubidium	Superalloys, catalyst for petroleum refining
Scandium	Biomedical research, electronics, glass, pyrotechnics
Selenium	Glass, electronics, metal halide lamps, lasers
Tellurium	Cadmium–tellurium solar cells, alloy in steel
Thallium	Medical tracer, scintillometers, superconductors
Yttrium	Phosphors, temperature sensors, trichromatic fluorescent lights

Antimony is found in a wide range of hydrothermal deposits and forms several important minerals, especially stibnite (Plunkert, 1985a). It is the dominant element in a few deposits (Figure 10.3), but most antimony production comes from lead–zinc–silver and tin–tungsten deposits. It is also common in epithermal and Carlin-type gold deposits but is not produced from them. Antimony is usually recovered during smelting of the primary metal ores and is traded in metal ingots or as the trioxide (Sb$_2$O$_3$), the major chemical form in which it is traded. Although antimony contamination has been observed in groundwater around some lead–zinc and precious-metal deposits, most global antimony pollution is atmospheric fallout from fossil-fuel combustion, (Forstner and Wittmann, 1981; Hutchinson and Meema, 1987). There is growing concern about the use of flame retardants in fabrics. Although most of this concern focuses on bromine, it is likely to spill over to antimony (Cooper and Harrison, 2009; Grossman, 2011).

China accounts for about 85% of global mine production of antimony, followed distantly by South Africa and Bolivia.

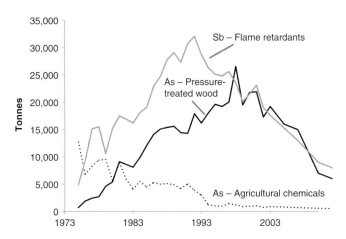

Figure 10.2 Changing uses for antimony (Sb) and arsenic (As) in the United States based on data of the US Geological Survey. The decrease in antimony used in flame retardants in the United States reflects the decline in textile manufacturing rather than a global decline in its use in this application.

Antimony reserves of about 1.8 million tonnes are very large relative to consumption, with most in China, Russia, and Bolivia.

10.2 Arsenic

Arsenic is a remarkable example of the compromises that we face in modern society. It is a well-known poison, but it has been used widely in many applications that we come into contact with on a regular basis. The most obvious of these is chromated copper arsenide (CCA), which is injected into wood to prevent decay. CCA extends the life of wood by a factor about ten, thereby limiting the number of trees that must be harvested and the number of structures that must be rebuilt (Henke, 2009). It is toxic to humans, however, and its use in residential construction has been discontinued in the United States and many other countries (Figure 10.2). Arsenic is also added as a hardener in lead batteries, ammunition, wheel weights, and anti-friction bearings, and it is used in fertilizers, fireworks, herbicides, and insecticides (Edelstein, 1985, 2013b). Arsenic's more modern markets are in semiconductors for optoelectronic and related devices. Annual world arsenic production of about 35,000 tonnes has a value of only about $22 million, making it the least valuable of all the elements discussed here.

Arsenic is common in many hydrothermal deposits where it forms sulfide minerals including arsenopyrite, realgar, and orpiment. Large arsenic deposits are rare (Figure 10.3), however, and it is much more common as a minor element in precious-metal and base-metal deposits. It is remarkably common in most hydrothermal gold deposits, suggesting that it is dissolved and precipitated by the same factors that

---

BOX 10.1 | **SEMICONDUCTORS**

Semiconductors are an important market for many of the technology elements, and arsenic was one of the first to be used in this market. Semiconductors have electrical conductivity properties between those of a metal and an insulator. In metals like copper and silver, which are good conductors, electrons move freely from atom to atom, making it easy for an electrical current to flow. In insulators like glass and silicate minerals, atoms are firmly held and current does not flow. In semiconductors, a few electrons move, leaving vacancies known as "holes," and this movement of electrons in one direction and apparent movement of holes in the other direction carries an electrical current. Silicon, germanium, gallium nitride, and gallium arsenide are natural semiconductors, but their properties can be modified by addition of other elements, a process known as "doping." Addition of phosphorus or arsenic increases the number of electrons available to conduct electricity, making N-type semiconductors, and addition of boron or aluminum increases the number of holes, making P-type semiconductors. These two types of semiconductors can be combined in various ways to make diodes and transistors, as discussed in Box 10.3.

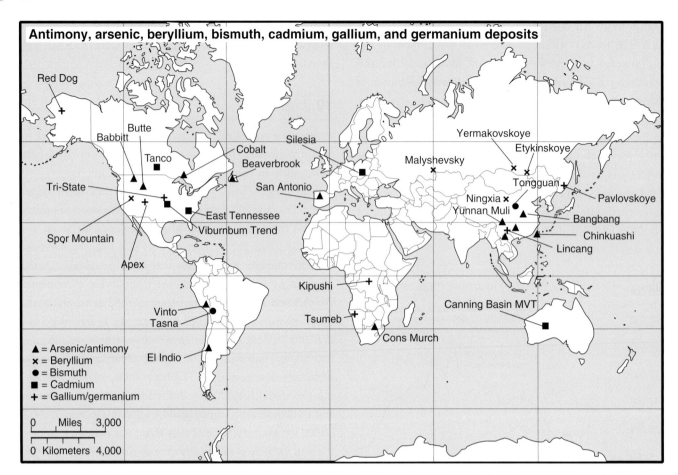

Figure 10.3 Location of antimony, arsenic, beryllium, bismuth, cadmium, gallium, and germanium deposits. In most cases, the deposits shown here are not mined for these specific elements. Instead, the elements are unusually enriched and more easily recovered as a by-product of major metal production.

control gold migration. It is also found in late-formed veins in the upper part, or lithocap, of porphyry copper deposits; the large arsenic-bearing veins at Butte, Montana (Figure 10.3), are of this type. Smaller amounts of arsenic are found in the Cobalt-type silver ores and in some tin deposits.

Because of its widespread association with other metals, arsenic is recovered largely as a by-product. It vaporizes at only 615 °C and so is released during roasting of gold or base-metal ores. Although modern smelters capture most of this vaporized arsenic in scrubbers, older smelters that did not are surrounded by zones of arsenic pollution, such as the arsenic-rich lake sediments around the Sudbury nickel smelters in Ontario (Nriagu, 1983). Because of its toxic characteristics, there are now stringent limitations on arsenic emissions and few smelters will take ore that contains arsenic.

Arsenic is a well-known environmental bad actor. Large amounts (more than 100 mg) induce acute arsenic poisoning resulting in death. Smaller exposure causes symptoms such as white lines on fingernails, and might cause cancer of the lungs and skin. It can be converted to toxic methylated forms by **bacterial action**. Even before anthropogenic arsenic emissions became a problem, it was known that high natural levels of arsenic were dangerous. Elevated arsenic concentrations in water at Three Forks, Montana, have been attributed to emissions from hot springs in nearby Yellowstone Park. Around the Chinkuashi gold deposits in Taiwan, natural arsenic dispersion in groundwater entered water wells causing severe arsenic poisoning (Tseng, 1977; WHO, 1981c; EPA, 2013).

Early problems with anthropogenic arsenic emissions were caused by pesticides used before the 1940s, which contained large concentrations of soluble arsenic. More recently, concern has focused on ash from CCA-treated wood, which has very high concentrations of arsenic and is a major source of arsenic to soil and groundwater if it is used or disposed of improperly. Disposal of unburned wood in landfills also opens the possibility for release into groundwater (Bleiwas, 2000). Tailings from arsenic-bearing mining operations, such as the Summitville, Colorado gold deposits, have also been point sources of pollution in downstream areas (Bigelow and Plumlee, 1995).

From a global perspective, however, two other sources of arsenic are of much greater concern. One of these is combustion of coal, which commonly contains 1 to 1,500 ppm arsenic, only some of which is removed with pyrite during coal cleaning (NRC, 1978). This arsenic is usually vaporized and condenses on fly ash, which is commonly trapped in the outgoing gas stream and put into impoundments. Improper impoundment provides an opportunity for leaching of arsenic into surrounding groundwater (Yudovich and Ketris, 2004). More widespread is the contamination of groundwater by arsenic that is dissolved naturally from arsenic-bearing pyrite and iron oxides. As many as 137 million people are thought to live in areas with groundwater containing more than 10 ppm arsenic, far above permissible limits (Ravenscroft, 2007; Jing and Laurent, 2013). Changes in aquifer characteristics, commonly caused by large numbers of wells, can increase arsenic concentrations, exacerbating this problem.

Because it comes largely from specially equipped smelters, arsenic is produced in only 10 countries, led by China, Chile, and Morocco. Reserves are estimated to be about 20 times greater than annual production, although actual supplies depend on base- and precious-metal production rates (Bedinger, 2013).

10.3 Beryllium

About 250 tonnes of beryllium worth about $110 million are used annually in alloys with aluminum, copper, and other metals that are valued for their extreme light weight (Petkof, 1985; Jaskula, 2013a). These are very specialized, however, and demand has actually decreased over the last 50 years (Figure 10.1e).

Beryllium is produced largely from the minerals beryl and bertrandite, which form in **pegmatites** and hydrothermal veins around felsic intrusions and volcanic rocks (Figure 9.3). Beryl is most common in pegmatites, coarse-grained igneous bodies discussed in Sections 10.10 and 10.15 on lithium and tantalum. Particularly good beryl crystals with specific green and light blue colors qualify as the precious gems, emerald and aquamarine, respectively. Bertrandite is more common in veins and disseminations in shallow rhyolite intrusions and volcanic tuffs, such as those at Spor Mountain, Utah (Figure 10.3), which is by far the largest active reporting source of beryllium in the world. China is thought to be a large producer, but does not report production figures.

Production of beryllium is usually carried out by open pit mining, melting the ore mineral, and then leaching the residue to liberate beryllium. Inhalation of beryllium dust in and around processing facilities causes **berylliosis**, a serious chronic lung disease, although significant problems have been limited by preventive measures taken since it was recognized in the 1940s (WHO, 1990a; Rossman *et al.*, 1991). Global reserve estimates are not available for beryllium, although reserves at Spor Mountain are estimated to be 15,200 tones of contained beryllium, which is very large in relation to present consumption.

10.4 Bismuth

Many of us are familiar with bismuth as the active agent, bismuth subsalicylate, in remedies for mild digestive ailments, one of its most important markets. Bismuth is toxic to organisms that cause the digestive problems and it combines with H_2S in your system to form the typical black tongue associated with bismuth remedies. Bismuth is one of the few metals that expands when it crystalizes from a melt, which makes it very useful in applications that require a tight fit at high temperatures (Norman, 1998). It is used in alloys with antimony, cadmium, gallium, indium, lead, and tin in fuel-tank safety plugs, holders for precision machining and grinding, and fire sprinkler mechanisms (Carlin, 1985). The chemical properties and the atomic weights of bismuth (208) and lead (207) are very similar, and this has given bismuth an edge in substitution for lead in a number of applications, including ammunition, "leaded" glass, and potentially in lead-free solder.

Annual bismuth production of about 8,000 tonnes, worth about $200 million, puts it near the middle of the value list for technology elements (Figure 10.1). The price of bismuth has held constant for decades but has increased significantly since about 2000, along with most other technology elements (Figure 10.1).

Bismuth is common in low-temperature hydrothermal deposits but rarely forms deposits of its own; the only actual bismuth deposit is in the Tasna–Rosario district in Bolivia, which is not active (Figure 10.3). Instead, it is usually produced as a by-product of lead, molybdenum, and tungsten mining where it is recovered during smelting. Bismuth is produced by 12 countries, led by China with about 80% of world production, and followed distantly by Mexico and then Belgium and Japan, which use imported concentrates. Although bismuth has not been implicated in any specific environmental or health problem, it was banned in France in 1978 due to a suspected relation to brain disease and its use in cosmetics was stopped in Austria in 1986. Bismuth reserves are estimated to contain a comfortable 320,000 tonnes of metal.

10.5 Cadmium

Cadmium presents a environmental conundrum. Its main use is in rechargeable batteries with nickel–cadmium batteries that provide convenient, portable electric power for a wide range of tools, machines, and instruments, but its toxicity makes it difficult to produce, use, and recycle (Plunkert, 1985b; Sigel *et al.*, 2013). Batteries account for about 85% of global cadmium consumption; the remaining 15% includes anti-corrosion coatings on steel, stabilizers in special engineering plastics and, somewhat ironically in view of its toxicity, cadmium telluride, a semiconductor used in some solar panels (Morrow, 2011). World cadmium production of about 22,000 tonnes is valued at only $62 million (Figure 10.1).

Although cadmium forms the mineral greenockite (CdS), most cadmium substitutes for zinc in sphalerite (ZnS), which can have cadmium concentrations as high as 1.3%. As a result, primary cadmium production comes largely from hydrothermal zinc deposits, especially Mississippi Valley-type (MVT) and sedimentary exhalative (SEDEX) deposits. It is recovered during smelting of sphalerite concentrates, and significant secondary production comes from recycling of batteries. Primary cadmium is produced in about 20 countries, led by China, Korea, and Japan. Reserves, which are essentially based on reserves of zinc, are a relatively large 500,000 tonnes of metal (Tolcin, 2013b).

In an interesting twist of nature, cadmium affects health by interfering with the body's natural uptake and utilization of zinc, the metal to which it is most similar chemically (Hutchinson and Meema, 1987; Sigel *et al.*, 2013). In natural soils and waters, the zinc:cadmium ratio is high enough for zinc to dominate, but if extra cadmium is introduced to the body, it can take the place of zinc. Cadmium also interferes with metabolism of calcium, copper, and iron, all of which are essential. It accumulates mainly in the kidneys and liver and can lead to irreversible kidney failure. The most famous incident of cadmium poisoning is Itai-Itai (ouch-ouch) disease, which involves vitamin D deficiency and osteopenia (weakening) of the bones.

Before the problems of cadmium poisoning were understood, mine wastes were major local sources of contamination. The worst problems occurred in 1947 downstream from a lead–zinc mine in Toyama Prefecture, Japan, where tailings that had not been properly impounded caused as many as 100 deaths (Kobayashi, 1971; Forstner and Wittman, 1981). At present, over half of the cadmium that is released into the environment comes from phosphate fertilizers and combustion of fossil fuels. Some Chinese coals are strongly enriched in cadmium, which is adsorbed onto kaolinite (Song *et al.*, 2008). Soil, water, and other natural sources of cadmium account for about 20% of emissions (Nriagu and Pacyna, 1988; Nriagu, 1989).

10.6 Cesium

Cesium is used in very small amounts but has some big applications. The main market, based on the high atomic weight and solubility of cesium compounds, is to make very high-density brines for used in oil-well drilling (to prevent blowouts when the well intersects high-pressure zones). It is also used in atomic clocks, cell phones, and global positioning systems (GPS). Radioactive cesium isotopes are used in radiation therapy for cancer, as well as for the sterilization of food and surgical equipment.

Cesium is one of the four elements that are recovered from pegmatites (Bradley and McCauley, 2013), which are discussed in the lithium and tantalum sections below. It is contained in the mineral pollucite and the largest and best-known deposit is at Tanco, Manitoba (London, 2008). Only a few tonnes of cesium are consumed in the United States each year and data for world consumption are not available (Tuck, 2013), making it impossible to estimate the value of global production. Global resources are estimated to be 163 tonnes, based largely on the amount of pollucite at Tanco.

10.7 Gallium

Gallium was discovered in 1875 in France and is named for Gaul, the Roman name for France. The existence of gallium was predicted by Mendeleev, who noted an opening in his newly developed periodic table and predicted that it would have properties similar to its neighbor, aluminum (Figure 3.1). Gallium melts at an unusually low temperature (about 29 °C) and was originally used in high-temperature thermometers and metal alloys with low melting temperatures (Aldridge *et al.*, 2011; Foley and Jaskula, 2013). It is also used by practical jokers to make spoons that melt when put into hot coffee (Kean, 2010).

The breakthrough for gallium came in the 1960s with development of gallium-based semiconductors, which now dominate gallium markets. These are used in integrated circuits for cell phones and optoelectronic components

BOX 10.2 | METALS IN RECHARGEABLE BATTERIES

Demand for rechargeable batteries in consumer electronic and automotive applications has increased enormously during the last two decades. Nickel–cadmium batteries were the first to be used, followed by nickel–metal hydride, and more recently lithium-ion and lithium-polymer batteries. All of these batteries contain cobalt, which prevents corrosion and extends the life of the units. Changing battery design has caused significant changes in the market for individual metals, with cadmium consumption declining, nickel growing and then gradually declining, and lithium and cobalt increasing (Figure Box 10.2.1). Significant amounts of cadmium and nickel remain for recycling in spent batteries (Wilburn, 2008).

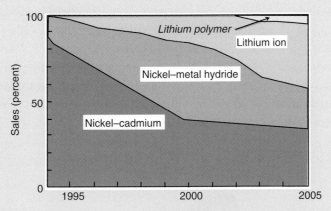

Figure Box 10.2.1 Change in metal content of batteries showing decreasing nickel and cadmium and increasing lithium (from Wilburn, 2008)

including lasers and light-emitting diodes (LEDs). Gallium semiconductors are also a component of photovoltaic cells, including ones that power the Mars Exploration Rovers (Moskalyk, 2003). Finally, it is the main component of gallistan alloys with indium and tin, which have ultra-low melting temperatures with a myriad of actual and potential uses from replacing mercury in dental amalgam to manufacture of brilliant mirrors and cooling of computer chips and nuclear reactors. Latest-generation cell phones use up to six times more gallium than earlier versions, almost assuring continuing demand for the metal.

Gallium production has increased significantly over the last decade or so to about 290 tonnes annually (Figure 10.1d). Its price started in the 1960s at very high levels but declined rapidly, a rarity for technology elements, and has increased only slightly over the last decade (Figure 10.1c). Current world mine production is valued at $200 million, an amazingly low value when you consider the value of the electronic and other devices for which it is the critical component.

Gallium has a crustal abundance of about 20 ppm (Rudnick and Gao, 2003), almost the same as lead, but it does not commonly form its own minerals. Instead, it substitutes for aluminum in common rock-forming minerals like feldspar and for zinc in sphalerite from hydrothermal deposits. Even in these minerals concentrations are low, and it takes a second step to raise gallium concentrations to economically interesting levels. The best second step is weathering that creates aluminous laterites (bauxite), which has gallium contents ranging as high as 160 ppm and averaging about 50 ppm (Dittrich *et al.*, 2011). Gallium is also concentrated in coals and in zones of highly acidic hydrothermal alteration around some high-sulfidation gold deposits, but is not recovered from either source at present (Zhao *et al.*, 2009; Rytuba *et al.*, 2004).

Gallium is a captive by-product of aluminum production and zinc smelting. Most newly mined gallium comes from China, Germany, Kazakhstan, and Ukraine. Recycled gallium from electronic products almost equals primary production and comes largely from China, Japan, the United Kingdom, and the United States. Gallium resources in bauxite deposits are estimated to total about 1 million tonnes, far above present consumption rates. However, because of its captive by-product status, gallium can only be produced at the rate of its primary metal host, either aluminum or zinc.

BOX 10.3 | DIODES, TRANSISTORS, AND INTEGRATED CIRCUITS

Most modern electronic equipment needs devices that control the flow of electric current. This is where semiconductors come in. When a P-type semiconductor is joined with an N-type semiconductor it forms a PN junction, which behaves as a diode (Figure Box 10.3.1) in which resistance to current flowing in one direction is low and in the other direction relatively high. Depending on the properties of the semiconductors used and the fabrication method, diodes can switch current on and off, reverse the direction of current flow, or even convert electricity to light as in photoelectric devices such as LEDs and solar cells. Electronic equipment also needs to change the strength of the current, a process known as amplification. This is done with a triode, which has one more part than a diode. Early triodes were vacuum tubes, which purists often refer to as the source of the best sound in old audio systems. Vacuum tubes were far too cumbersome for portable operation, however, and it took the invention of triodes made of semiconducting materials to start the modern portable electronic age. Semiconductor triodes are transistors, and they are made by combining two PN junctions to form either an NPN transistor or a PNP transistor. The layers in a transistor operate somewhat like the plate, grid, and cathode of a vacuum tube, controlling a large current by varying a small base current. Because of their small size and the increasing complexity of modern electronics, many transistors are now combined into integrated circuits along with diodes and other devices, for use in telephones, radios, televisions, cell phones, and photoelectric electric devices.

Figure Box 10.3.1 Metal oxide field effect (MOSFET) transistor consisting of two PN junctions on either side of a metal oxide diode (courtesy of Infineon Technologies North America)

10.8 Germanium

Immediately after the French gallium, comes germanium, which is named for Germany. It was predicted to exist in 1871 to fill the gap between silicon and tin, and was separated as an element in 1886 from silver sulfide minerals. It was used to make early solid-state, signal-processing diodes and then triodes. Early germanium semiconductor diodes and triodes were used in telephones, televisions, and computers and production grew until the early 1970s (Plunkert, 1985c). At that time, it became possible to produce ultra-high-purity silicon, which performed better in diodes and transistors. Since then, germanium's main markets have been in glasses, with the two main uses being infrared optics for night vision and fiber optics consisting of silicon fibers with small amounts of germanium dioxide. Together, these account for a little more than half of global consumption, with most of the rest being used in germanium-based solar cells to power satellites, high-brightness LEDs for cars and televisions, and as a catalyst in the manufacture of plastics (Butterman and Jorgenson, 2004; Guberman, 2013b). About 120 tonnes of germanium are produced each year, worth just under $200 million, placing germanium just below gallium in the value hierarchy (Figure 10.1a).

Germanium is found largely in hydrothermal zinc deposits where it substitutes for zinc in sphalerite, and it is recovered as a by-product from smelting of zinc concentrates. Germanium is present in relatively low concentrations in most deposits but is unusually enriched in distal skarn deposits in Namibia, including Tsumeb and Khusib Springs (Melcher et al., 2006). The smaller, similar Apex germanium–gallium deposit in Utah (Bowling, 1988) has been mined occasionally (Figure 10.3). Germanium is also enriched in coal ash and in veins formed by later hydrothermal alteration of the coal layers (Hu et al., 2009; Du et al., 2008) and is recovered from coal ash in China and Russia. Production of germanium is not associated with specific environmental issues, and the element has no known role in human health, although it is thought by some natural health advocates to enhance immunity.

China is the leading world germanium producer, with about 75%, followed by Canada. World reserves are simply not known and estimated US reserves have been quoted at 450 tonnes for the last several decades without declining. This is a small number when compared to present consumption and the fact that most germanium is a by-product of zinc production. Future production might well depend on germanium from coal-ash and ash-related deposits in view of the larger demand for coal.

10.9 Indium

Indium is one of the newest technology elements (Aldridge et al., 2011). A few decades ago, its main market was special solders that depend on its unusually low melting temperature of 156.6 °C. Indium metal was also distinguished for making a crying noise when bent, although this property did not generate extra sales. Now, indium is a key component of indium–tin oxide (ITO), which consists of 90% indium oxide and 10% tin oxide. ITO is a semi-transparent semiconductor that can be deposited in very thin films or coatings for flat-panel displays in computers, televisions, cell phones, and other devices. It is also used in organic LEDs and special semiconductors, and continues to supply the now-growing market for lead-free solders. Indium production and consumption has increased dramatically during the last two decades, far more than steel (Figure 10.1d) and is now at a level of about 660 tonnes with a value of almost $500 million, mostly in flat-panel displays (Figure 10.4).

Production of only 660 tonnes is not very large, possibly raising hopes that you could find an indium deposit and corner the market. Unfortunately, indium won't cooperate. It rarely forms its own minerals and instead it substitutes for zinc, tin, and tungsten in the **crystal lattice** of their minerals (Schwarz-Schampera and Herzig, 2002). Most indium is recovered from sphalerite, which contains hundreds of parts per million of indium. Sphalerites with the highest indium concentrations are found in tin–tungsten vein and skarn deposits in Bolivia, China, and India, volcanogenic massive sulfide (VMS) deposits of Russia, Australia, Canada, and

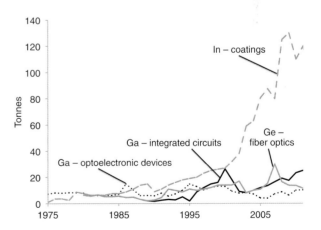

Figure 10.4 Changes in use of gallium (Ga), germanium (Ge), and indium (In) in the United States, showing the increase in the integrated circuit market for gallium and fiber-optic market for germanium and the much greater increase in indium for flat-panel displays (based on data of the US Geological Survey)

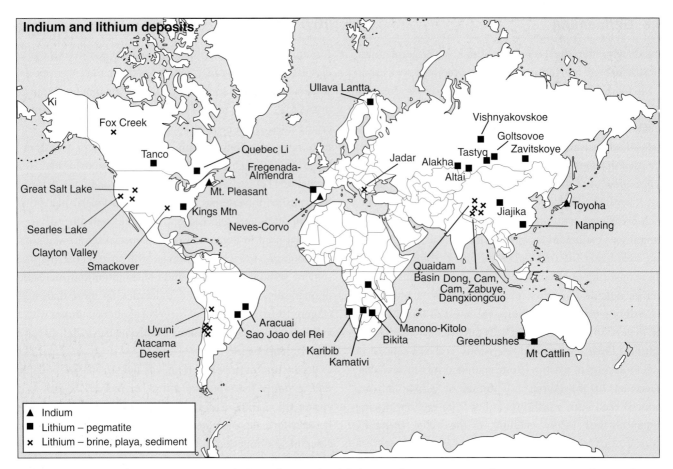

Figure 10.5 Location of important indium and lithium deposits (modified in part from Jorgenson and George, 2004; Murakami and Ishihara, 2013; Kesler *et al.*, 2012)

particularly Neves Corvo in Portugal (Figure 10.5). Two hydrothermal deposits with particularly high indium contents are Toyoha in Japan and Mt. Pleasant in Canada.

Recovery of indium from sphalerite concentrates requires **vaporization** or dissolution followed by separation of other trace elements including antimony, arsenic, and germanium (Jorgenson and George, 2004). Indium has no known role in human metabolism and is relatively inert, although it is toxic in large concentrations. Indium is produced by 14 countries, including the major zinc refiners China, Korea, Japan, Canada, Belgium, and the United States. Murakami and Ishihara (2013) identified about 75,000 tonnes of indium resources in major tin–tungsten and VMS deposits, which is large in relation to current consumption. However, the indium can become available only at the rate at which the zinc, tin, and tungsten are produced. This has generated concern and confusion in the electronics industry, and even questions such as what to do when indium "runs out" (Worstall, 2012).

10.10 Lithium

Lithium has emerged as the leading element in the rechargeable battery race. Batteries are only the latest in a long line of applications for lithium. Even now, they are second to lithium-bearing ceramics and glass, which do not expand and contract as much as other ceramics and are not damaged by rapid heating or cooling. Lithium is also used to expand the useable temperature range of lubricating greases, as a dehumidifier in commercial air conditioners, as an alloying metal with aluminum, and in pharmaceuticals (Jaskula, 2013b).

In response to the battery market, world production of lithium has increased significantly since about 2000 to more than 600,000 tonnes (Figure 10.1). Prices, however, have just begun to regain the high levels that they had before the change from pegmatite-related to brine-related production, as discussed below. Even at low prices, the total value of world production is about $2.5 billion, second only to rare earths in the technology group (Figure 10.1a).

BOX 10.4 **LITHIUM BATTERIES**

Batteries are commonly compared on the basis of the energy and power that they will deliver relative to their density, and on the speed with which they can be recharged. Battery weight is a critical factor in hybrid elective vehicles, where a large amount of energy needs to be stored, and lithium-ion batteries have a considerable weight advantage over competing lead–acid, nickel–cadmium or nickel–metal hydride batteries (Gruber *et al.*, 2011; Kawamoto and Tamaki, 2011; Sullivan and Gaines, 2012). The optimum configuration that will deliver maximum energy density, thermal stability, and load characteristics are all under investigation (Etacheri *et al.*, 2011). Research is also under way on the best elements that will be combined with lithium in the cathode of batteries. Possibilities include cobalt, manganese, iron, phosphorus, aluminum, and titanium, all of which are more abundant then lithium. Thus, lithium is likely to be the element that might limit production of lithium-ion batteries.

10.10.1 Geology of lithium deposits

Two very different types of lithium deposits dominate present production (Kesler *et al.*, 2012). The first is pegmatites, which are igneous rocks of generally granitic composition with crystals that can be greater than 1 m in length (London, 2008). Pegmatites form at the margins of granitic intrusions and appear to be the last phases of the magma to crystallize. One group of pegmatites, the lithium–cesium–tantalum (LCT) group, which is enriched in various combinations of lithium, cesium, tantalum, and tin (Figure 10.6a). One of the largest lithium pegmatite deposits in the world, at Kings Mountain, North Carolina (Figure 10.5), has unusually uniform, high lithium and spotty high tin values but does not contain much cesium or tantalum. Kings Mountain was the first deposit to be mined on a large scale and the basis for the development of an industrial demand for lithium, a lead that it has since ceded to the Greenbushes pegmatite in Australia. At other LCT pegmatites in Canada and especially the Democratic Republic of Congo, tantalum and cesium are more important, as discussed in the Section 10.15 on tantalum below. Additional LCT pegmatites in Brazil, Portugal, and Zimbabwe have also been mined for lithium.

The second type of lithium deposit is found in dry inland lake beds known as **playas** or **salars**, which contain **brines** and salts formed by evaporation of freshwater (Figure 10.6b). Brines formed by evaporation of lake waters differ from those formed from seawater, because surrounding rocks contribute unusual elements to the water, including lithium. Brine deposits with lithium were first recognized in Clayton Valley, Nevada, but much richer deposits have since been discovered in the Atacama Desert of Chile and adjacent Argentina and Bolivia, and even more recently in China. If

Figure 10.6 Lithium deposits. (a) Pegmatite from the Kings Mountain mine showing large light green spodumene crystals (photo by authors). (b) Salt encrustations around a transient pond at the southern end of the Salar de Atacama, the world's largest brine-evaporite deposit of lithium (photo courtesy of Martin Reich, University of Chile). See color plate section.

there is enough lithium in the brine, it is obviously easier to pump some brine from a well and precipitate lithium from it than to mine spodumene and recover its lithium, and this has allowed the brine deposits to dominate the lithium market.

10.10.2 Lithium production and reserves

Despite their beautifully large crystals, pegmatite deposits present several problems to miners. Individual pegmatite bodies are relatively small, have irregular shapes, and many have zoned compositions that require selective mining to recover only lithium-rich material. On top of all that, lithium in pegmatites is hosted by silicate and phosphate minerals of which spodumene is by far the most important. Processing of spodumene and other minerals to release lithium involves dissolution and precipitation, which requires a lot of energy.

Production of lithium from brines also has its challenges (Houston *et al.*, 2011). The playa must be underlain by an aquifer with a large volume of brine and enough porosity and permeability to release it when pumped. Even if the brine can be recovered, it must be evaporated to separate lithium and this can only be done in areas with an arid climate. Finally, separation of lithium can be greatly complicated by the presence of magnesium and other dissolved elements that interfere with precipitation of lithium. This complication might prevent economic recovery of lithium from the large Salar de Uyuni in Bolivia (Figure 10.5). At present, brine deposits in Argentina, Chile, and China dominate world production, although the large and high-grade Greenbushes pegmatite continues to be mined and allows Australia to share top spot with Chile as a global producer.

World lithium trade, amounting to about 37,000 tonnes of contained lithium, is based on lithium mineral concentrates from pegmatites as well as lithium compounds, mainly lithium carbonate produced from both spodumene and brines (Jaskula, 2013b). Argentina and Chile, the dominant brine producers, have no domestic market for lithium chemicals and are major exporters, largely to the United States, Japan, and Germany (Figure 10.7). Australia, the dominant pegmatite producer, sends mineral concentrates to China and other countries, as do smaller pegmatite operations in Zimbabwe, Portugal, and Brazil.

Global reserves of lithium are estimated by the US Geological Survey to be about 13 billion tonnes, which is only a two-decade cushion for present consumption. During the early phases of interest in batteries, this small cushion generated considerable debate about the capacity of lithium to meet the anticipated increased demand. This encouraged

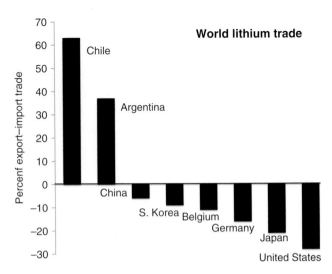

Figure 10.7 Global trade in lithium carbonate for 2009. Lithium chloride and lithium hydroxide trade in significantly smaller amounts. Very large amounts of mineral concentrates from Australia, Canada, Portugal, and Zimbabwe go largely to China. Figure based on data from the US Geological Survey and Polinares (2012).

exploration, which discovered additional lithium brines and pegmatites, as well as new lithium deposits containing lithium-bearing clays. It also generated several studies that compared long-term demand based on anticipated automobile production, battery technology, and recycling rates to the resource potential for lithium. The general conclusion was positive, suggesting that lithium resources are adequate for many decades (Gruber *et al.*, 2011).

10.11 Mercury

It might be surprising that we would include mercury among the technological metals. Unlike many other elements in this group, except perhaps arsenic, mercury has been known and used for millennia. It began as **quicksilver**, prized because it is the only common metal that is a liquid at surface temperatures. Its main ore mineral, cinnabar, was also valued for its bright red color, which provided an early pigment for the early Egyptians, Greeks, and Romans. During the discovery of the New World, mercury found two new uses. One was as a treatment for syphilis, which is thought to have spread from the Americas at that time (Rothschild, 2005; Harper *et al.*, 2012). The other was in the **patio process**, a variety of the amalgamation method for recovering gold from ore. In the patio process, gold ores were scattered onto a flat patio, covered with mercury, and then crushed, often by having mules walk over them. Mercury dissolved the gold from the

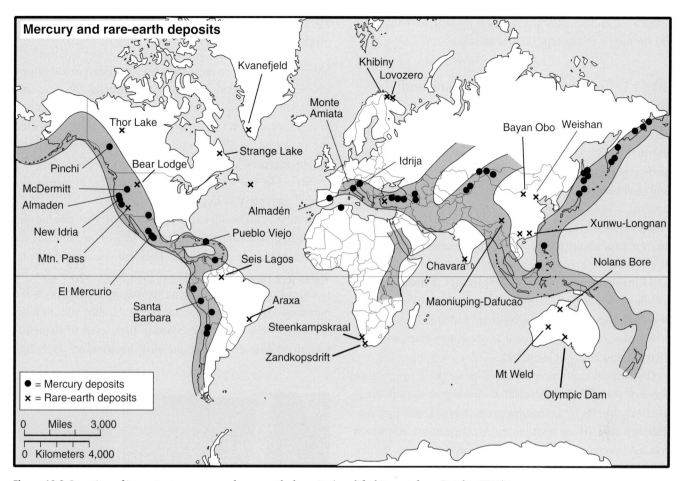

Figure 10.8 Location of important mercury and rare-earth deposits (modified in part from Rytuba, 2003)

ore and the resulting liquid was collected and heated to drive away mercury vapor, leaving a residue of gold. The patio process was the dominant form of gold ore processing during European colonization of the Americas (Prieto, 1973). Spain, which controlled much of these lands, required that a fifth of any gold that was produced be paid to the Crown. Policing of this requirement was not difficult because most mercury for the process came from the Almadén mine in Spain (Figure 10.8). Thus, mercury was a key factor in Spain's accumulation of riches from the New World (Flawn, 1966). Variants of the amalgamation process are still used in many developing countries to recover gold from ores, as discussed in the Sections 11.1.2 and 11.1.3 on gold deposits.

By the mid-eighteenth century, mercury was being used widely as a medicine, in coatings on mirrors, and in solutions to prepare felt hats, for which its detrimental effects were immortalized by the term "mad as a hatter." As the industrial age got under way, mercury was used in electrical switches, barometers, and thermometers that take advantage of its

liquid properties; special exterior paints and coatings in which its toxicity discouraged mildew; herbicides to kill unwanted plants; and dental amalgam, the magical alloy that can be inserted into a cavity before it hardens. Somewhat later, it found additional uses in high-performance electrical batteries, lamps for outside lighting, and as a catalyst in the manufacture of chlorine and sodium hydroxide.

Because of its strongly negative impact on health, major efforts have been made to limit mercury losses from these processes and, in some cases, to replace mercury altogether. Its use in batteries has been banned in most countries and more efficient recovery of mercury from chlorine–caustic soda plants, or the use of alternative processing methods, have cut mercury use significantly. These efforts, combined with the ban on the use of mercury in indoor paints and thermometers, and limits on its concentration in most electronics, has caused world mercury consumption to decline from a peak of about 10,000 tonnes in 1971 to about 2,000 tonnes in 2013 (Figure 10.1e). The value of this production, about $100 million, puts mercury near the bottom of the value list for metals of any type (Figure 10.1a).

10.11.1 Geology of mercury deposits and mercury reserves

Mercury is readily soluble in very low-temperature hydrothermal solutions and usually accumulates in the near-surface parts of hydrothermal systems, such as hot springs and geothermal areas. It is even vaporized and carried upward to condense in overlying groundwater (Figure 10.9). Most mercury deposits formed at temperatures of only about 100 °C from dilute, meteoric waters and some deposits even contain petroleum residues (Peabody and Einaudi, 1992; Rytuba, 2003). Mercury is also recovered as a by-product from many gold–silver deposits, especially the Carlin-type deposits of Nevada, and from smelting of many base-metal ores where it is present in trace amounts (Rytuba and Heropoulos, 1992).

On a global scale, mercury deposits are most common in areas of young felsic volcanism at convergent tectonic margins (Figure 10.8). In northwestern Nevada, for instance, mercury deposits are found in the McDermitt caldera, which formed by explosive rhyolitic volcanism only a few million years ago, and the Monte Amiata mercury deposits in Italy are on the slopes of a young volcano. Curiously, the world's largest mercury deposit at Almadén, Spain, does not match these generalizations. Instead, mercury at Almadén fills pores in an Silurian-age sandstone and fractures in overlying Devonian-age basalts (Fernández-Martinez and Rucandio, 2014).

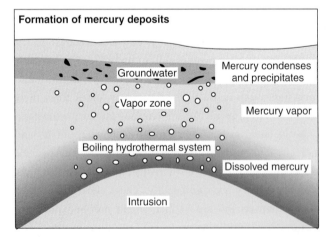

Figure 10.9 Schematic illustration of boiling hydrothermal system with mercury transported in the vapor phase and deposits where the vapor condenses in overlying groundwater

10.11.2 Mercury production, uses, and related environmental concerns

Mercury has been produced largely from underground mines, where proper ventilation is a primary concern. After conventional beneficiation, mercury is liberated by heating in retorts or furnaces to form mercury vapor, which is then condensed and collected in 76-pound (35-kg) flasks (Figure 10.10). Global mercury trade is limited by export bans in some countries. In the United States, for instance, export of elemental mercury was stopped in 2012.

Pollution and health effects related to mercury mining and its use in gold production have been important problems for centuries (Nriagu *et al.*, 1992). They arise because elemental mercury is more soluble in blood than water, which allows it to enter the body easily, where its principal effect is on the nervous system (US Department of Health and Human Services, 1999). Workers exposed to high levels of elemental mercury vapor experience mercury intoxication, including

Figure 10.10 Mercury production at Almadén, Spain. (a) Pouring mercury into a 76-pound flask for shipment. (b) Ball of iron floating in a tank of liquid mercury. See color plate section.

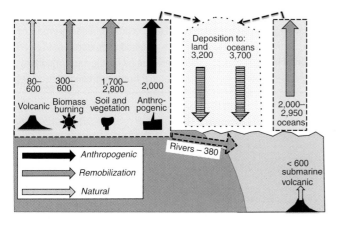

Figure 10.11 Global atmospheric mercury cycle showing amounts (in tonnes) released from natural and anthropogenic sources and amounts deposited on land and the ocean. Anthropogenic sources are much greater than natural sources, which are largely from volcanic activity on land and in the ocean (modified from UNEP, 2013).

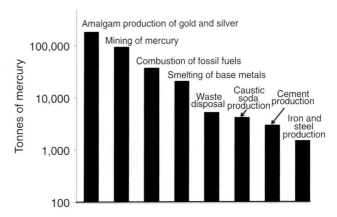

Figure 10.12 Cumulative emissions of mercury to the atmosphere since 2000 BCE (compiled from data of Street *et al.*, 2011)

psychotic behavior, extreme irritability, tremors, and even death. At Almadén it was a routine procedure to detoxify workers by placing them in a hot room where they would sweat out mercury. Industrial or other exposure to **methylmercury** (CH_3Hg^+) is much more dangerous and prenatal exposure carries particularly high risks of neurological disorders (WHO, 1990b).

Mercury is released to the environment from both natural and anthropogenic sources (Figure 10.11), but the distinction between these two is confused by mercury released from earlier-legacy-mercury pollution from anthropogenic activity. Legacy mercury is estimated to total about 350,000 tonnes, almost two-thirds of which has been released since 1840 (Figure 10.12). This legacy is so large that the present atmosphere contains about 7.5 times more mercury than it did in 2000 BCE. For this reason, much of the mercury now being emitted naturally from the ocean and other sources was probably originally of anthropogenic origin. In fact, legacy mercury is estimated to account for as much as 60% of present atmospheric deposition, with only 27% coming from primary anthropogenic activity and 13% from natural sources (Amos *et al.*, 2013). The two big anthropogenic sources of mercury through time have been the processing of mercury and gold–silver ores, and combustion of fossil fuels, largely coal. During the late-nineteenth century when newly discovered gold–silver deposits were attracting settlers to western North America, mercury releases reached levels of about 2,500 tonnes per year. Releases from coal combustion are only about 25% of this amount presently.

Mercury is released from most natural and anthropogenic sources largely as elemental mercury, which is dispersed into the atmosphere in gas form and adsorbed on aerosol particles (Table 10.3). Precipitation washes this mercury into the soil and eventually into lakes and the ocean. Sediment cores in lakes throughout the world show significant increases in mercury content through the late twentieth century, as do ice cores in Greenland and elsewhere (Krabbenhoft and Schuster, 2002). Once in the water, mercury is taken up by plants and animals. During **bioaccumulation**, some elemental mercury is converted to methylmercury (CH_3Hg) and other organomercury compounds through metabolic processes and bacterial action (Merritt and Amirbahman, 2009; Hsu-Kim *et al.*, 2013; Gilmour *et al.*, 2013).

The general population's exposure to the increasing global mercury load comes from several sources. The most important one appears to be biomagnification of mercury in aquatic organisms, which reflects the fact that fish do not excrete mercury as rapidly as they take it in, partly because some elemental mercury is converted to methylmercury. High mercury concentrations are widespread in modern freshwater and marine fish, with highest biomagnification rates in cold, low-productivity environments (Lavoie *et al.*, 2013). Some mercury pollution has been traced to industrial activity, including methylmercury disposal at Minimata Bay, Japan, and mercury used in pulp and paper production in Ontario, but most appears to be of more distant anthropogenic origin, almost certainly from coal combustion, with additional spikes in mercury emissions from volcanic eruptions (Porcella, 1990; Krabbenhoft and Schuster, 2002).

The other major concern about mercury stems from its use in **dental amalgam**, a mixture of about 50% mercury, 35% silver, 12% tin, and smaller amounts of copper and

Table 10.3 Natural and global anthropogenic emissions of mercury. The first column shows annual emissions for all sources; second and third columns show cumulative emissions since 2000 BCE for each major anthropogenic source, along with the amount of each source-commodity produced or used (Pirrone *et al.*, 2010; Streets *et al.*, 2011). The final column shows the ratio of emissions to amount of each commodity produced or used (in tonnes of mercury produced per million tonnes of commodity). Note that production of mercury and of gold–silver were by far the most important sources of total global emissions. As discussed in the text, much of the so-called natural emissions are recycled material from anthropogenic sources

	Annual emissions (tonnes/yr)	Total emissions (tonnes)	Commodity produced/ used (million tonnes)	Emission/production ratio
Anthropogenic sources (emissions to atmosphere and hydrosphere)				
Gold–Silver	400	185,000	1.636	1.13×10^5
Mercury	50	95,000	0.949	1.00×10^5
Fossil fuel	810	37,420	634,000	5.98×10^{-2}
Base metals	310	20,940	1,368	1.53×10^1
Waste	187	5,310	2,310	2.30
Caustic soda	163	4,240	1,710	2.48
Cement	236	3,000	61,200	4.90×10^{-2}
Iron and steel	43	1,518	77,500	1.96×10^{-2}
Coal-bed fires	32			
Vinyl chloride production	24			
Natural sources (emissions to the atmosphere)				
Oceans	2,682			
Lakes	96			
Forest–grass–desert	1,336			
Agricultural land	128			
Biomass burning	675			
Volcano–geothermal	90			

zinc. This alloy is used in dental work because it acts like a liquid for a few minutes but crystalizes into a solid within a few hours. It is well known that amalgam releases metallic mercury during chewing and that some of this is ingested or inhaled, and there is also a positive correlation between the amount of dental amalgam and the concentration of mercury in urine. No statistically significant health effect has been found to correlate with the amount of dental amalgam, however, possibly because levels of mercury exposure are low (Kolata, 1992; Brownawell *et al.*, 2005). In an ironic twist, cremation of bodies containing dental amalgam results in complete release of the mercury unless entrapment systems are used, a point that has raised concerns in several countries including the United Kingdom (Mari and Domingo, 2010).

10.12 Rare earths, scandium, and yttrium

The rare earths (Table 10.2) are a group of closely related metals, from lanthanum to lutetium in the periodic table. Yttrium and scandium are commonly included with the rare earths because of their chemical similarities and tendency to be found in the same deposits (Hedrick, 1985; Hedrick and Templeton, 1990). The value of annual global production of the rare earths and related elements totals almost $7 billion, far above all other technology elements (Figure 10.1). Rare-earth production and prices have risen dramatically over the last decade, reflecting their wide range of markets (Figure 10.1).

Rare earth markets are amazingly varied, complex, and constantly changing, both in the product that is sold and the

BOX 10.5 | **MERCURY'S SECOND CHANCE**

Despite all of its bad press, mercury might make a comeback in the form of compact fluorescent light (CFL) bulbs. Along with LED lights, CFL bulbs are being substituted for incandescent light bulbs. The substitution trend began in Brazil and Venezuela in 2005, moved into Europe and Australia by 2009, Russia in 2012, and reached the United States, Canada, and Mexico in 2012. CFL bulbs consist of phosphor-coated glass with a small amount of mercury. When voltage is applied to the bulbs, the mercury emits ultraviolet light, which causes the phosphor to fluoresce and emit visible light. CFL bulbs require three to four times less energy than incandescent bulbs for the same amount of light (Goonan, 2006). Energy Star, the US government website, estimates that use of only one CFL light in each home in the United States would prevent greenhouse-gas emissions equivalent to that of 800,000 cars annually.

Just how important might this be to the ailing mercury industry? Well, each light contains about 4 mg of mercury, considerably less than the approximately 500 mg that filled an old-fashioned thermometer. An estimated two billion light bulbs are sold in the United States each year. If all of these were CFL bulbs, they would require only about 8 tonnes of mercury. Compared to current production of about 2,000 tonnes, this seems very small, even if we increase it by 10 times as a wild guess of global light-bulb demand. Furthermore, because CFL bulbs last eight to fifteen times longer than incandescent bulbs, the market would soon decrease to a smaller number of bulbs needed for replacements.

This low number is important from another perspective, however, because it gives an indication of the degree to which CFL bulbs might pollute the environment. If CFL bulbs are not disposed of properly, the mercury escapes. If all the bulbs in the world were broken at the same time, however, they would release only about 80 tonnes of mercury, roughly equivalent to annual global emissions from volcanoes, although the emissions would be more concentrated in populated areas (Lim *et al.*, 2013).

use to which it is put. Specific uses for individual rare earths are summarized in Table 10.2. In terms of value, the main markets for rare earths include high-intensity permanent magnets for use in everything from ipods to wind turbines, phosphors for television and computer screens and CFL bulbs, alloys in steel and other metals including nickel–cadmium batteries, catalysts for chemical reactions ranging from automotive catalytic converters to petroleum refining, special glasses and ceramics, and polishing compounds. In terms of volume of material, magnets, catalysts, and metal alloys are about equal with 60% of the total market, and phosphors account for only about 7% (Kingsnorth, 2009)

10.12.1 Geology and deposits of rare-earth elements

The term rare earths for these elements is somewhat misleading because they have surprisingly high crustal abundances, led by cerium at about 63 ppm and neodymium at 27 ppm (Rudnick and Gao, 2003). Although these crustal abundances are similar to those of copper, lead, and tin, the rare earths substitute in common rock-forming minerals rather than

forming minerals of their own. The rare-earth elements are also divided into light and heavy groups on the basis of the electron structure. The most important rare-earth minerals are bastnasite and monazite, which concentrate the light rare earths (especially cerium, lanthanum, neodymium, and praseodymium), and xenotime, which is enriched in the heavy rare earths.

Monazite and xenotime commonly form trace minerals in common rocks. Both minerals are heavy and not easily decomposed during weathering, and they are common by-products in gold, ilmenite, rutile, cassiterite, or zircon placer mines. Monazite has been recovered from titanium placers in Florida and Georgia (Figure 9.7) in the United States and in Australia, Brazil, and India (Figure 9.21) and from tin placers in Malaysia, but its content of radioactive thorium, for which there is a lack of markets, has discouraged commercial production. Monazite is also present in the Witwatersrand gold and Elliot Lake uranium paleoplacer deposits, but is not recovered (Moller *et al.*, 1989; Castor, 2008).

Almost all rare-earth deposits in bedrock rather than placers are associated with carbonatites or alkaline igneous rocks. **Carbonatites** are strange intrusive rocks composed of more

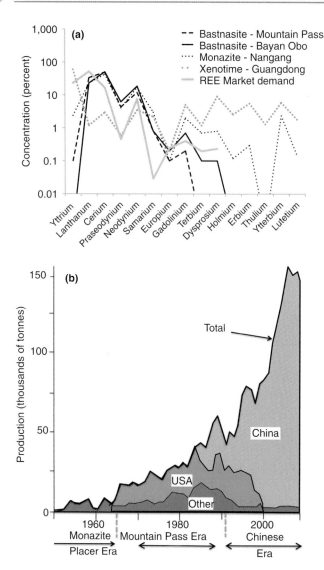

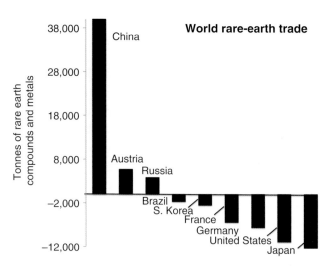

Figure 10.14 Exports and imports for rare-earth compounds and metals for 2009 based on data from Walters and Lusty (2011)

Figure 10.13 (a) Content of rare-earth elements in bastnasite from Mountain Pass and Bayan Obo, monazite from Nangang and especially xenotime from Guangdong relative to market demand for rare earths in US manufactured and imported products (based on data in Bleiwas and Gambogi, 2013). Note that market demand concentrates on light rare earths and roughly follows their abundance in rare-earth minerals, but that bastnasite (especially that from Mountain Pass) is depleted in heavy rare earths relative to monazite and xenotime. (b) Change in major mining sources of rare earths since 1950 (from Bleiwas and Gambogi, 2013).

than 50% carbonate minerals that form in areas where thick continental crust has been cut by rifts. They commonly form pipe-like bodies several kilometers in diameter and are surrounded by rocks that are strongly altered by hydrothermal solutions. The two most important rare-earth deposits in the world, Mountain Pass in the United States and Bayan Obo in China (Figure 10.8), are both of this type. Other rare-earth deposits associated with carbonatites are known in Brazil,

India, and Africa. Deposits associated with alkaline igneous rocks are found in New Mexico, Greenland, Russia, and Canada, but are generally of too low grade to compete with carbonatites (Castor, 2008; Verplanck and van Gosen, 2011).

The economic viability of rare-earth deposits depends in part on their relative abundances of light and heavy rare earths. Figure 10.13a shows how relative abundances of rare-earth elements vary in the common ore minerals and how these relate to the market demand for rare earths. Fortunately, market demand roughly follows the relative abundances of the rare earths in bastnasite, but you can see that Bayan Obo has an edge because it contains just a little more of the intermediate rare earths. Along with its higher overall grade and enormous size, this makes it the dominant deposit in the world.

10.12.2 Rare-earth production, reserves, and trade

Rare earths are produced in only six countries, led by China with about 90% of world production and the United States with about 5% (Gambogi, 2014). Production was dominated by the Mountain Pass mine in the United States from about 1965 to 1985, but has since been overtaken by production from Bayan Obo and other deposits in China (Figure 10.13b). China is by far the largest exporter of rare earths, followed by Austria and Russia, which produce rare-earth metals and compounds from imported ores (Figure 10.14). The major importers, including Japan, the United States, Germany, France, and South Korea, are the principal manufacturers of products that contain rare earths.

> ## BOX 10.6 | TECHNOLOGY ELEMENTS AND RESOURCE NATIONALISM
>
> The presence of large deposits of technology elements has inspired **resource nationalism** around the globe. China has dominated world rare-earth markets because production costs from the Bayan Obo deposit are below those of other possible suppliers. This position has been used to withhold rare earths from possible competing markets, leaving China as the only country with assured rare-earth supplies. Bolivia has aspired to dominate world lithium markets in the same way, using the large deposits at Salar de Uyuni. Here, however, geologic and processing complications have prevented development of the deposits, and the widespread presence of other lithium-rich salars in Argentina, Chile, and China makes it unlikely that they will succeed.

Essentially all rare-earth mining is by open pit or dredge operations (Shannon, 1983; Walters and Lusty, 2011). *In situ* mining using ammonium sulfate solutions to leach rare earths from clay minerals has also been used in China. Where concentrates are made, the elements are recovered by hydrometallurgy involving acid or base leaching and solvent extraction, and then separated into the individual rare-earth elements by complex solution chemistry (Gupta and Krishnamurthy, 2005). Because consumption of individual rare earths does not exactly match their relative abundances in ores, some elements are stockpiled in order to obtain enough of others. The radioactive thorium in monazite requires special storing and handling. Although mining of rare earths is not associated with significant health hazards, large doses of the pure metals or their compounds cause severe toxic effects that require precautions during production and use (Hedrick, 1985; Viljayan *et al.*, 1989).

Current annual production of rare earths, about 110,000 tonnes, is comfortably less than world reserves of about 110 million tonnes (Gambogi, 2014). Although China hosts about 50% of world reserves, significant reserves are found in the United States and exploration is underway in many other countries. Although these statistics are superficially reassuring, they mask the fact that the large size and high grade nature of the Bayan Obo deposit is likely to allow it to dominate world rare-earth markets for the foreseeable future.

10.13 Rhenium

Rhenium was predicted to exist by Mendeleev in 1860 and was discovered in 1925 and named for the Rhine River (Blossom, 1985). It has two main markets based partly on its extremely high melting temperature of 3,180 °C. About 70% of world consumption goes into superalloys with nickel for high-temperature turbine engines, and most of the rest is used as catalysts in production of high-octane, lead-free gasoline.

Smaller amounts go into tungsten and molybdenum alloys that are used in heating elements, thermocouples, and related thermal devices. Annual world rhenium production grew more rapidly than steel until about 1990 after which it declined and then followed steel to present production of about 50 tonnes. Rhenium prices have fluctuated more than many other commodities, but have been rising since about 2000. Even at today's higher prices, annual production is worth only about $250 million, placing it in the middle of the technology element value chain (Figure 10.1).

Getting rhenium out of the Earth is no small job. It has a crustal abundance of only about 1 ppb, making it one of the least abundant elements in Earth's crust. To make matters worse, it does not form a common mineral and instead usually substitutes for molybdenum in molybdenite. Not just any molybdenite will do. Molybdenite in porphyry molybdenum deposits contains only 10 to 100 ppm rhenium, but molybdenite from porphyry copper deposits contains up to 2,000 ppm (Sinclair *et al.*, 2009). Most rhenium production comes from porphyry copper deposits in Chile, the United States, and Kazakhstan, and some comes from sediment-hosted copper deposits in Poland. In all cases, it is a by-product of other metals and its availability is subject to their production rates. World reserves are estimated to contain about 2,500 tonnes of metal, a value that has not changed for decades. Although this reserve is comfortably larger than current production, it is poorly documented and highly dependent on copper and molybdenum production.

10.14 Selenium

Selenium's claim to fame comes from its photoelectric properties, which offer hope for solar energy and have been used widely in plain-paper copying machines. It is also used in electrolytic cells for production of manganese, in glass to prevent the green tint caused by trace amounts of iron, as the active ingredient in dandruff-preventing shampoos, as a curing agent in rubber, and as a dietary supplement. Most of these uses are not amenable to recycling (Woollins and Laitinen, 2011; George, 2013). Annual production of about 2,000 tonnes is worth about $300 million (Figure 10.1).

For such a scarce element, selenium is produced in a surprisingly large number of countries because it comes entirely from residues from electrolytic refining of blister copper. Although selenium-rich ore deposits exist, none are mined exclusively for the element (Sindeeva, 1964). Coal contains up to 12 ppm selenium, which is considerably more than is in copper deposits, but there is no simple system to recover it (Yudovich and Ketris, 2004). Similarly, some gold ores are enriched in selenium, but it is not recovered.

Selenium is necessary for cellular function but toxic in large amounts, causing selenosis, a disease characterized by breath with a garlic odor, gastrointestinal problems, and eventual liver damage and death (Institute of Medicine, 2000). Its possible link to cancer and heart disease has also been investigated (Wei et al., 2004). Most selenium taken into the body comes from food grown on selenium-enriched soil, particularly areas with selenium-rich coal in China (Yudovich and Ketris, 2006). The only industrial process in which selenium emissions might be of importance is oil refining, where some wastewaters are enriched in the element. World selenium reserves are estimated to contain 98,000 tonnes of metal, largely in Chile, Russia, and the Philippines. In view of the widespread distribution of selenium in coal, there is little likelihood of significant shortages.

10.15 Tantalum

Tantalum is geochemically similar to the ferroalloy metal columbium (niobium), which we discussed in Chapter 8, but it has very different uses. The main market is in capacitors for electronic applications in cars, cell phones, and computers, with much smaller amounts used in ferroalloys (Schwela, 2011; Aldridge et al., 2011). The value of annual production of tantalum is difficult to specify because it trades in so many forms and because prices are not widely quoted. In keeping with our effort to compare prices of semi-finished metals rather than raw ore, however, world tantalum production of about 765 tonnes would be worth about $250 million if it were sold as metal (Papp, 2013c).

Tantalum is found in the minerals microlite, which is the tantalum-rich equivalent of the columbium mineral pyrochlore, and tantalite, the tantalum-rich equivalent of columbite. Tantalum deposits are found largely in LCT pegmatites,

which were described in Section 10.10. Tantalite, in particular, resists weathering and accumulates in residual and stream placer deposits. Mined tantalum comes largely from pegmatites, mostly Mozambique and Brazil (Papp, 2013c) with smaller amounts from lithium-bearing deposits in Zimbabwe and the Democratic Republic of Congo. Deposits in the Democratic Republic of Congo have attracted attention because of their role as conflict minerals. Here, the mineral concentrate containing tantalum and niobium minerals is known as coltan. (The "col" part of the term refers to niobium, which is also known in the industry as columbium.) World reserves contain about 150,000 tonnes of metal, which is large in comparison to annual production.

10.16 Tellurium

Tellurium is used largely in solar cells, usually with cadmium, as discussed earlier. It is also combined with bismuth in thermoelectric semiconductors, in which a temperature difference creates an electric potential; these are used largely in cooling infrared detectors, integrated circuits, laser diodes, and medical instruments, but have the potential to recover thermal energy that would otherwise be lost and use it for power generation and refrigeration. Smaller amounts of tellurium are used in high-strength, low-alloy (HSLA) steel where only 0.04% tellurium makes carbon steel easier to machine. It is also used in alloys with copper, tin, aluminum, and lead (Woollins and Laitinen, 2011; George, 2013; Kavlak and Graedel, 2013). Total world tellurium production of possibly 100 tonnes has a value of only about $35 million, placing it just above arsenic at the bottom of the technology element value list (Figure 10.1a).

Tellurium forms the large telluride family of minerals, much like the sulfides, in which tellurium combines with metals, especially precious metals. The best known of the tellurides is calaverite, which takes its name from Calaveras County, California, the site of Mark Twain's story of the famous jumping frog and an area of gold telluride-bearing quartz veins. Although calaverite and other telluride minerals are found in many precious-metal deposits, particularly the famous Emperor Mine in Fiji, they are rarely a source of tellurium because the ores are not processed in a way that permits its recovery. Instead, almost all of world tellurium production comes from refining of copper metal. Tellurium is present in most copper ores both in solid solution in copper sulfides and as minor amounts of copper tellurides. When the ores are beneficiated and smelted, tellurium follows copper and is not removed until the electrolytic refining step (Sindeeva, 1964). Tellurium is not known to be essential to

BOX 10.7 | CONFLICT MINERALS

Tantalum has been in the news because of its identification as a conflict mineral. Conflict minerals are defined as minerals that are mined in areas with armed conflict and human-rights abuses. Conflict minerals are usually used to provide funds to support armed efforts. They are mined by primitive methods and transported in small amounts in order to reach markets without being detected. Materials best suited to this are diamonds, gold, and some of the technology elements. All of these are resistant minerals that concentrate in weathered zones above bedrock deposits and that are amenable to primitive placer-type mining.

These commodities are widespread in central Africa, including Angola, Burundi, Congo Republic, Central African Republic, Democratic Republic of Congo, Rwanda, South Sudan, Tanzania, Uganda, and Zambia, many of which are or have been the scene of armed conflict. Production of diamonds from this area first raised the issue of conflict minerals and provided a stimulant for the "ice diamond" marketing initiative used by Canadians to distinguish their diamonds. Diamonds can be identified by trace-element and trace-mineral contents and they can be marked around their girdle to identify their source (although this is not done at the mine), and this has done much to minimize trade in conflict diamonds. Metals, on the other hand, are harder to trace, and have become the next-generation conflict-mineral product. Coltan mineral concentrate from the Democratic Republic of Congo accounts for as much as 20% of world supplies. Recent rulings by several MDCs, such as the Dodd-Frank Section 1502 ruling in the United States (Courts, 2012), attempt to identify these minerals and require that manufacturers do not include them in their products. Initial compliance with these rulings, which will require supply chains with rigorously identified transfer points, is estimated to cost up to $4 billion. A recent report on these efforts by Apple Computer (Chen, 2014) was one of the first to appear in the public press.

the human body. In soluble forms it is toxic in moderate to large concentrations, producing first a strong garlic breath, followed by more serious symptoms involving nausea and unconsciousness.

Tellurium is produced in only five countries, including Canada, Japan, Peru, Russia, and the United States, all from copper smelting. Reserves of tellurium depend on world copper reserves and the way in which they are smelted and refined. Tellurium reserves of about 24,000 tonnes, largely in the United States and Peru, are based entirely on world copper resources and their availability is subject to global copper production.

10.17 Thallium

Thallium is hidden from view in a lot of technological corners. It was originally used as a rodenticide until people realized that it is retained in the body and is toxic to humans (Kazantzis, 2000; Aldridge et al., 2011). The toxicity is thought to result from interference in metabolism of potassium, an element with which thallium has many chemical similarities (Moore, 1991). Now, thallium is used as an activator in sodium iodide crystals that detect gamma radiation, is alloyed with barium, calcium, and copper in high-temperature super-conductors, goes into lenses, prisms, and windows for

infrared equipment, combines with arsenic and selenium to make crystals for diffraction of light, and alloys with mercury for low-temperature measurements. It is also used as radio-active thallium-201 for medical imaging (Guberman, 2014). These uses add up to global thallium production of about 10 tonnes, worth only about $60 million (Figure 10.1a).

Thallium is sufficiently concentrated to be of economic interest in some sediment-hosted Carlin-type deposits in China (Zhou et al., 2005). It is also enriched in coal, but is not recovered from it (Anton et al., 2012). The main source of thallium is sphalerite, where it substitutes for zinc, and it is recovered largely from zinc smelter flue dust, although not all smelters make the effort (Plunkert, 1985d). World reserves are thought to contain 380 tonnes of the metal, a number that has remained about the same for several decades. Larger amounts should be available from coal and other sources if needed.

10.18 Zirconium and hafnium

These two metals are recovered from zircon, a heavy mineral that accumulates in placer deposits with the titanium miner-als, ilmenite, and rutile (Garnar, 1983). Zircon is used in mineral form in ceramics and refractory products, where it is valued for its high melting temperature of 2,550 °C. A much smaller amount of zircon is processed by chemical leaching to yield elemental zirconium, as well as by-product hafnium, which has similar geochemical characteristics and substitutes for zirconium in zircon (Adams, 1985). The best known use for zirconium and hafnium metals is in nuclear reactors, where zirconium contains the fuel and hafnium forms control rods (Kessler, 1987). The two contrasting applications stem from the fact that neutrons can pass through zirconium easily, whereas hafnium has a high ability to stop neutrons. Global production of zircon mineral amounts to only about 1,500 tonnes, worth about $3 billion. Very little of this is converted to zirconium and hafnium metal, which would have higher values. Reserves of 48,000 tonnes of contained ZrO_2 are large compared to present annual production.

10.19 The future (and present) of technology elements

What have we learned here? One obvious point is that a lot of relatively rare elements have found important uses in modern technology. Nevertheless, the value of global production for almost all of these elements is amazingly low when compared to the value of the products for which they are essential. No one is going to make a fortune mining deposits of these elements. Fortune hunting in technology elements is further discouraged by the fact that only a few of the elements, such as lithium, mercury, and rare earths, actually form their own deposits. Almost all of the others must be recovered as by-products from deposits of other, more abundant and more commercially important elements. In that sense, then, supplies of many of these elements are hostage to production rates for the major element with which they are associated. This by-product status makes resource estimates for many of these elements fuzzy, although major shortages in the next few decades seem unlikely. Finally, we can get additional assurance about resources from the fact that many of these elements are enriched in materials that have not yet been used as a source. The most obvious ones are coal and coal ash, which are enriched in many of the technology elements and these elements might be recovered from them.

11 Precious metals and gems

Precious metals and gems were among the first minerals to be valued by civilization and they have kept their attraction for millennia. Even in their raw state, gems and precious metals are easily recognizable as truly special and beautiful. Although other minerals, such as pyrite (**fool's gold**), might be confused for gold, it is rare to confuse gold for another mineral. Its rich color and luster are simply unique. The same is true for the brilliant play of light and color from gemstones. On top of it all, precious metals and gems are rare. We even use special nomenclature to discuss their abundances. Gold and silver abundances are commonly reported in terms of the troy ounce, which equals about 31 g; prices of precious metals are commonly quoted in troy ounces, and we have used them (referred to here as ounces) to describe the metal endowment of individual deposits. Weights of diamonds and other precious gems are measured in an even smaller unit, the carat, which amounts to 0.2 g. Because these non-metric units are widely understood, we use them in parts of this discussion.

In spite of their rarity, precious metals and gems are a big business, with world production worth almost $200 billion (Figure 11.1). Gold is the most important of the group, with silver, diamonds, platinum-group elements (PGEs), and the colored gems emerald, ruby, and sapphire following in that order. Gold production increased in the 1990s, but has declined annually since 2000 in all countries except China. Global production of PGEs increased by more than 1,000% from 1960 to 2008, and plateaued thereafter. Production of gem diamonds increased by almost 1,400% at the start of the twenty-first century, dropped slightly during the economic crisis of 2008–2010, and leveled off after that. Mine production of industrial diamonds, which are discussed further in

Chapter 13, increased from 1960 to 2008, although with cyclicity, but fell dramatically after 2008 owing to the proliferation of synthetic diamonds for industrial applications.

Prices of the PGEs, gold, and silver have increased dramatically since 1960 (Figure 11.1). In the early 1980s, prices rose to unprecedented levels largely on the basis of speculative buying fueled by increasing global inflation. Prices declined in the 1990s, but then increased, most dramatically for gold and silver, to near record levels in the mid 2000s. Much of this price increase was fueled by hedges against poor currencies and low interest rates, and can be tied to the 2008 financial crisis that started in the United States and reverberated globally. Despite all the press about control of diamond prices, they have not kept up with the metals. Even with more price decreases for gold and silver (Figure 11.1), the market for precious metals and gems remains strong. Production and reserve data for precious metals and gems is given in Table 11.1 and important deposits of each commodity are shown in Table 11.2.

11.1 Gold

Few people lose track of gold. Only about 18% of the approximately 171,300 tonnes that have been mined through history are lost (Thompson Reuters GMFS, 2013; Grabowski *et al.*, 1991). Of the gold that can be accounted for, 18% is held by central banks (Table 11.3), and the rest is in private stocks of bullion, coins, jewelry, or art objects. As of 2013, more than 52% of world gold consumption went into jewelry, 16% was held in investment funds, and 12% was used in industrial applications ranging from catalytic converters to cell phones. Gold is also used as a dental material, as a guide for X-rays

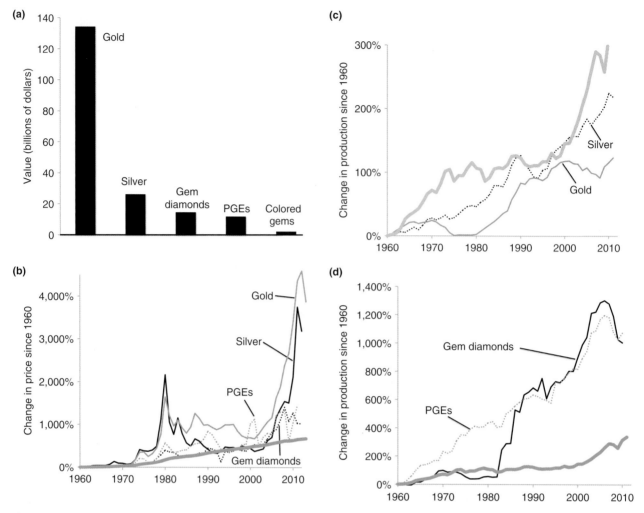

Figure 11.1 (a) Value of 2013 world production of gold, silver, gem diamonds, colored gems, and platinum-group elements (PGEs). Values shown here are for primary production from mines and do not include recycled or other metals. (b) Change in price of gold, silver, PGEs, and diamond since 1960. Heavy gray line shows change in CPI. (c, d) Change in production of gold, silver, PGEs, and diamonds since 1960. The heavy gray line shows the change in steel production (all data from the US Geological Survey; diamond prices represent cut stones of all sizes imported into the United States).

used to perform laser surgery, and as a medication for facial paralysis and rheumatoid arthritis. These markets support annual gold production of about 2,500 tonnes (80 million ounces) worth almost $135 billion, ranking it along with copper and aluminum as one of the most valuable metals produced.

11.1.1 Geology of gold deposits

Gold is found largely as **native** gold or in combination with silver in electrum. If there is enough tellurium present, it can also form telluride minerals, including calaverite, which was discussed in Chapter 10. It also substitutes for copper in some copper minerals and enters pyrite if arsenic is also present. Because so little gold is needed to form economically

interesting concentrations, this sort of **solid solution** gold is an important source of production.

There are two main types of gold deposits. **Placer deposits** have supplied about 43% of total world gold production and were the first to be mined (Frimmel, 2008). **Hydrothermal deposits**, which began to be mined when people followed the placers upstream to their "mother lode," have supplied the rest.

Hydrothermal gold deposits

Epithermal and Carlin-type deposits Epithermal gold deposits consist of gold-bearing veins, veinlets, and disseminations that formed from low-temperature hydrothermal solutions (200 to 300 °C) at depths of about 1 km

Table 11.1 Major producers and reserves for precious metals (tonnes) and diamonds (million carats) (from US Geological Survey Commodity Summaries)

Commodity	Country	Production	Reserves
Gold	China	420	1,900
	Australia	255	9,900
	United States	227	3,000
	Russia	220	5,000
Silver	Mexico	5,400	37,000
	China	4,000	43,000
	Peru	3,500	87,000
	Australia	1,700	88,000
Platinum	South Africa	140	63,000
	Russia	25	1,100
	Zimbabwe	12	
Palladium*	South Africa	82	
	Russia	82	
	Canada	13	310
Diamond (gem)	Republic of Congo (Brazzaville)	21.5	
	Russia	20.7	
	Botswana	14	
	Zimbabwe	11	
	Canada	10.5	
Diamond (industrial)	Botswana	22	130
	Democratic Republic of Congo	17	150
	Russia	15	40
	Australia	8	270

* Reserve figures for platinum and palladium refer to all platinum-group elements.

(Figure 11.2). They are Earth's most common type of hydrothermal deposit, with over 1,000 examples known (Kesler and Wilkinson, 2009). Epithermal deposits have been divided into three groups with the non-euphonious names of high-, intermediate-, and low-sulfidation, based on their mineralogy and the acidity of their parent hydrothermal solutions (White and Hedenquist, 1995; Simmons *et al.*, 2005; Cline *et al.*, 2005).

High-sulfidation (HS) epithermal deposits contain gold in strongly altered rock containing alunite, kaolinite, and pyrophyllite, all of which form in very acid solutions. These solutions were dominantly magmatic and they were made acid by magmatic gases including CO_2, SO_2, and HCl. HS deposits are associated with intermediate and felsic intrusions, many of which formed porphyry copper deposits at deeper levels.

Many HS deposits are relatively small and occupy fracture systems on the perimeters of volcanic calderas where magmatic gases were channeled upward. A few deposits are very large, however, including Goldfield in Nevada, El Indio in Chile, Pueblo Viejo in the Dominican Republic, and Yanacocha in Peru (Figure 11.3). Ore in HS deposits probably formed as the fluids cooled, boiled, or reacted with surrounding rocks. In general, these deposits have higher gold:silver ratios and much more arsenic and copper than their low- and intermediate-sulfidation cousins.

Low (LS)- and intermediate (IS)-sulfidation epithermal deposits are hosted by rocks that have been altered to feldspar and sericite, a fine-grained form of muscovite, both of which are deposited by neutral hydrothermal solutions. Many LS deposits are associated with bimodal basalt–rhyolite (mafic–felsic) volcanism that formed where continents underwent rifting (Sillitoe and Hedenquist, 2003). IS deposits are hosted by large volcanic piles in convergent margins, but are more distal to obvious intrusions. They contain gold and silver with abundant pyrite and variable amounts of galena and sphalerite (Figure 11.4). Salinities of the ore-forming solutions are usually less than about 1% but are higher if lead and zinc are present. Isotopic data suggest that the hydrothermal solutions are dominantly meteoric with minor magmatic additions, possibly containing the metals. The deposits usually form veins up to several meters wide and more than a kilometer long containing high-grade zones called ore shoots that formed where gold or base metals were deposited when the hydrothermal fluid boiled or mixed with shallow groundwater (Figure 2.9). Important LS deposits include Waihi in New Zealand, Hishikari in Japan, and Round Mountain in Nevada. Important IS deposits include the large silver–lead–zinc deposits of Parral, Zacatecas, and Fresnillo in Mexico and the Comstock in Nevada, which are discussed further in Section 11.2 on silver deposits.

Carlin-type gold deposits are known for their invisible gold. Whereas most gold in the epithermal deposits is in a visible part of the veins, gold in Carlin-type deposits is in solid solution in arsenic-rich pyrite known as arsenian pyrite. The pyrite is disseminated through carbonaceous, silty limestones from which most of the carbonate was leached by hydrothermal solutions. By far the largest and most numerous Carlin-type deposits are around Carlin, Nevada, from which they get their name, although smaller deposits have been found in the Yukon province, China, and Iran (Figure 11.3). Deposits in the Carlin district have produced about 200 million ounces of gold, and have reserves of over 100 million ounces, making the area the most important source of mined gold in the

Table 11.2 Large deposits of precious metal and diamonds. The list does not contain deposits that produce precious metals as by-products. Data from corporate and government reports.

Deposit name	Location	Deposit type	Content	
Gold deposits			*Gold (oz)*	
Muruntau	Uzbekistan	Orogenic	170,000,000	
Ashanti	Ghana	Orogenic	67,000,000	
Golden Mile	Australia	Orogenic	64,000,000	
Telfer	Australia	Orogenic	50,000,000	
Yanacocha	Peru	Epithermal	41,000,000	
Homestake	United States	Orogenic	40,000,000	
Pueblo Viejo	Dominican Republic	Epithermal	39,000,000	
Silver deposits			*Silver (oz)*	
Potosi (Cerro Rico)	Bolivia	Tin–silver vein	2,600,000,000	
Fresnillo	Mexico	Epithermal IS	1,900,000,000	
Penasquito	Mexico	Epithermal LS	1,500,000,000	
Pachuca-Real del Monte	Mexico	Epithermal IS	1,400,000,000	
Guanajuato	Mexico	Epithermal IS	1,100,000,000	
Zacatecas	Mexico	Epithermal HS	750,000,000	
Platinum-group-element deposits			*Platinum (oz)*	*Palladium (oz)*
Mogalakwena (Bushveld)	South Africa	Magmatic	141,000,000	
Zimplats (Great Dike)	Zimbabwe	Magmatic	95,000,000	77,000,000
Impala (Bushveld)	South Africa	Magmatic	59,000,000	28,000,000
Diamond deposits			*Diamonds (carats)*	
Jubilee (Yubileyny)	Russia	Kimberlite	153,000,000	
Udachny	Russia	Kimberlite	152,000,000	
Mir	Russia	Kimberlite	141,000,000	
Argyle	Australia	Lamproite	140,000,000	
Catoca	Angola	Kimberlite	130,000,000	
Venetia	South Africa	Kimberlite	102,000,000	

United States. The deposits are located along linear zones that trend north–northwest for tens to hundreds of kilometers. The gold is thought to have been deposited from either basinal fluids heated by intrusive rocks, or from magmatic–hydrothermal fluids released from upper-crustal magma chambers (Cline *et al.*, 2005; Muntean *et al.*, 2011). Gold deposition was caused by sulfidation, where gold- and sulfur-bearing aqueous fluid reacted with iron in the host rock minerals or pore waters.

Orogenic gold deposits

Orogenic gold deposits, which have produced about 32% of all mined gold (Frimmel, 2008), consist of gold-bearing quartz veins that formed dominantly in metamorphic rocks at crustal depths of 5–15 km (Figure 11.2). The term orogenic is used to indicate that the deposits formed during collisional mountain-building events. Most orogenic gold deposits formed in three discrete periods

of time, between 2700 and 2400 Ma, 2100 and 1800 Ma, and 650 Ma to the present, which were periods of continental amalgamation and crustal thickening (Goldfarb *et al.*, 2001, 2005; Groves *et al.*, 2003). Most of the Precambrian deposits are hosted by metamorphosed volcanic–sedimentary belts and they form large vein systems that are hosted by second- and third-order structures associated with continent-scale transcurrent faults (Figure 11.2). Individual vein systems have vertical extents of up to 3 km and can host millions of ounces of gold, making these the largest, on average, of hydrothermal gold deposits.

Most orogenic gold deposits consist of quartz and gold, with very minor sulfides or tellurides. The veins are surrounded by rock in which calcium, magnesium, and even iron silicate have been replaced by carbonates and iron oxides have been replaced by iron sulfide. The fluids had temperatures of about 300–550 °C and contained abundant CO_2.

Table 11.3 Monetary gold reserves (not geological gold reserves) of the top 40 reported official gold holdings at the end of 2013 (data from the World Gold Council). One tonne contains 32,150.7466 troy ounces

Country	Gold reserves (tonnes)
United States	8,133.5
Germany	3,390.6
International Monetary Fund	2,814.0
Italy	2,451.8
France	2,435.4
China	1,054.1
Switzerland	1,040.1
Russia	1,015.1
Japan	765.2
Netherlands	612.5
India	557.7
European Central Bank	502.1
Turkey	490.2
Taiwan	423.6
Portugal	382.5
Venezuela	367.6
Saudi Arabia	322.9
United Kingdom	310.3
Lebanon	286.8
Spain	281.6
Austria	280.0
Belgium	227.4
Philippines	193.0
Algeria	173.6
Thailand	152.4
Kazakhstan	137.0
Singapore	127.4
Sweden	125.7
South Africa	125.1
Mexico	123.5
Libya	116.6
Bank for International Settlements	115.0
Greece	112.1

Table 11.3 (cont.)

Country	Gold reserves (tonnes)
South Korea	104.4
Romania	103.7
Poland	102.9
Australia	79.9
Kuwait	75.0
Indonesia	75.9
Egypt	75.6
World total	**30,263.8**

Prevailing opinion favors a metamorphic source for the waters that formed orogenic gold deposits, based on the observation that most deposits are underlain by large thicknesses of metamorphic rocks and there was very little contemporaneous igneous activity. Release of the metamorphic fluid apparently occurs during episodic earthquake events, which allow pulses of fluid to ascend rapidly through focused conduits where gold is deposited by changes in pressure and temperature, mixing with other fluids, and reaction with host rocks (Sibson *et al.*, 1975; Cox, 2005; Weatherly and Henley, 2013).

Most of the large orogenic gold deposits are Precambrian in age and include the Porcupine (Timmins) and Red Lake districts in Canada, Golden Mile (Kalgoorlie) in Western Australia, Kolar in India, and Ashanti in Ghana. Younger examples include the Mother Lode and Alleghany district in California and Linglong in China, as well as several very large deposits hosted by fine-grained sediments, including Ballarat and Bendigo in Victoria, Australia and the largest known hydrothermal gold deposit, Muruntau in Uzbekistan.

Porphyry copper–gold deposits Because porphyry copper deposits are mined in such large volumes, their trace levels of gold are economically important (Cox and Singer, 1992). The gold is easy to produce because most of it is in solid solution in copper minerals and it is recovered during smelting of the copper concentrate. The gold content of porphyry copper deposits varies over a wide range from large deposits such as El Teniente in Chile with less than 0.1 g/tonne gold to Far Southeast (see Figure 9.10), a deeply buried porphyry copper deposit in the Philippines with about 1.3 g/tonne gold (Kesler *et al.*, 2002). Because of their large size, the gold content of large porphyry copper deposits rivals that of large epithermal and orogenic gold deposits. Gold, however, is a by-product of

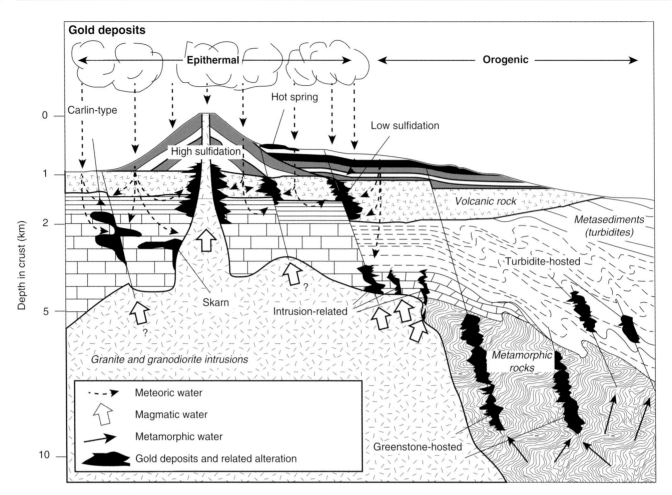

Figure 11.2 Schematic illustration of geologic environments in which hydrothermal gold deposits form, with epithermal deposits on the left and deeper orogenic gold deposits on the right. Arrows show sources of hydrothermal water thought to have formed these deposits.

copper production. For instance, in 2011, the Bingham deposit, which was the source of the gold medals for the 2002 Winter Olympics in nearby Salt Lake City in Utah, produced 237,000 tonnes of copper and 13 tonnes of by-product gold. Grasberg in Irian Jaya produces even more by-product gold, about 60 tonnes, and is one of the largest gold mines in the world.

Placer and paleoplacer gold deposits

Placers have been the main source of gold for thousands of years, reflecting the extremely high specific gravity of gold and its resistance to weathering. Most placers are approaching exhaustion, however, with the possible exception of those in Siberia, some of which were prison camps during the Stalin years (Kizny, 2004). Large-scale placer mining in North America started during the late 1800s, when gold seekers in California followed them upstream into the veins of the Mother Lode (Figure 11.5). From there, they continued northward through British Columbia into the

Klondike goldfields of the Yukon and then to Fairbanks and Nome. At the same time, diggers in Australia moved into Victoria to mine the placers at Bendigo and Ballarat. Most of these were typical stream placers, except at Nome in Alaska, where much gold was in beach placers. Many deposits exhibited the curious feature known as **bedrock** placer values in which gold was concentrated at the bottom of a thick layer of stream gravels. This probably happened when floods disrupted the entire layer, shaking down gold that was originally deposited throughout the sediment (Cheney and Patton, 1967). Another curious feature of placers, giant **nuggets** that are larger than gold in veins from which they were derived. These large nuggets are thought to form by dissolution and re-precipitation of smaller ones (Youngson and Craw, 1993).

As placer deposits began to be exhausted, the incredible Witwatersrand paleoplacers of South Africa were discovered in 1886 (Figure 11.3). The Witwatersrand is the remains of a late Archean-age sedimentary basin that contains several

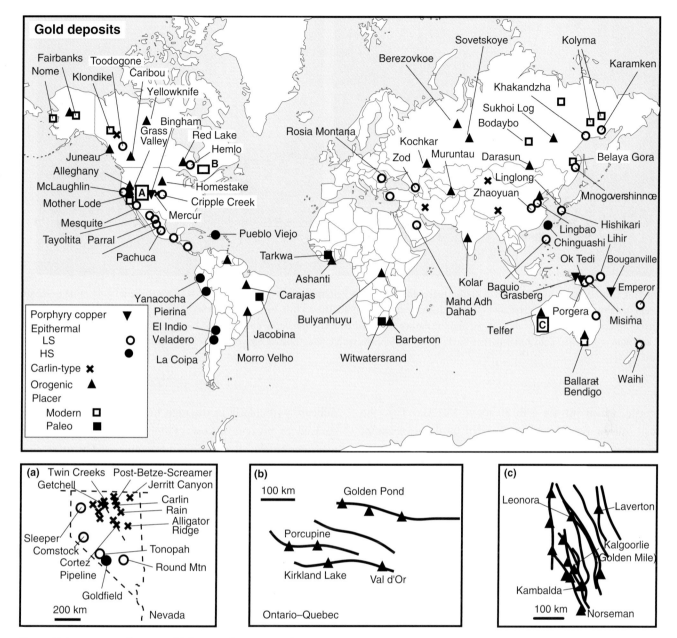

Figure 11.3 Location of major gold deposits discussed in the text. Insets show location of: (a) epithermal and Carlin-type deposits in Nevada and (b) orogenic gold deposits along major faults in Ontario and Quebec and (c) Western Australia.

gold-bearing conglomerate layers, as well as the Carbon Leader, a layer of carbonaceous material, possibly originally an algal mat, that also contains gold. The basin formed over approximately 180 million years, as sediments accumulated at a rate of about 14 m per million years to cover an area of 2000 km² (Minter, 2006). Where these gold-bearing sedimentary layers have been mined along the 300-km northern margin of the basin, they have yielded about 41,000 tonnes of gold and they probably contain another 40,000 to 60,000 tonnes (Robb and Meyer, 1991; Conradie, 2008). Regardless of the exact reserve of gold, it is at least 20 times larger than Muruntau, the largest hydrothermal deposit.

The huge size of the Witwatersrand caused doubt that it is really a paleoplacer, a skepticism fueled by the very fine-grained nature of most of the gold. This encouraged an alternative hypothesis, in which the gold was deposited by basinal and low-grade metamorphic fluids that were released from the Witwatersrand sediments as they were buried and heated (Phillips and Law, 2000). Although debate persists, most observers have converged on a modified placer origin in which the gold was largely of placer origin but was remobilized over short distances by later basin-related hydrothermal fluids (Hayward *et al.*, 2005). In addition to burial, fluid migration might have been triggered by the Vredefort

Figure 11.4 Close-up view of two types of epithermal gold veins. (a) A vein from the rich Hishikari LS (adularia-sericite) gold mine in Japan. The vein consists of alternating layers of quartz, adularia, clays, and gold. (b) A close-up view of the San Albino vein at the Parra mine in Mexico. The sharp white crystals are feldspar and the darker crystals are mostly sphalerite and galena. This type of vein produces largely silver, lead, and zinc rather than gold and is part of the IS group of deposits with characteristics between those of LS and HS deposits. See color plate section.

meteorite, which hit the area at about 2.02 Ga. That the deposit started as a placer is supported by the presence of uraninite, another heavy mineral found in paleoplacer deposits at Elliot Lake, Ontario, as discussed in Section 7.2.1 on uranium deposits, and by rhenium–osmium age measurements on gold grains showing that they are much older than the host conglomerate and therefore must have been derived from erosion of pre-existing rocks (Kirk, 2002).

11.1.2 Gold mining and Witwatersrand deep mining

Placer gold mining requires only that gold-bearing stream gravels be diverted over a riffle system made of wood, rough fabric, or skin to catch the gold. In ancient times, sheepskins were used, whence the name "golden fleece"; now, it's Astroturf. Large-scale dredges use more sophisticated gravity-based separation methods such as discussed Section 9.1.3 on titanium deposits.

Many modern vein and Carlin-type deposits are mined by open pit methods and followed deeper by underground mines (Figure 11.6). Underground mining operations in the Witwatersrand offer several distinct challenges. The gold-bearing layers have been followed to extremely deep levels approaching 4 km where surrounding rock is very hot (air temperature in the TauTona mine, the world's deepest mine at 3.9 km, is 55 °C). This makes working conditions very

difficult, particularly in the high humidity created by water sprayed to prevent silicosis. Pumping chilled air into the workings barely helps because it is compressed by increasing pressure that almost doubles its temperature before it reaches miners at depth. Chilled water undergoes less compression and is widely used, but too much water causes broken rock to flow out of stopes in highly dangerous mud rushes. Air-conditioning equipment, essentially heat exchangers that use a combination of salt and water to produce slushy ice, reduces the air temperature to 28 °C, but the relative humidity can exceed 90%. Cooling costs, including training and screening of workers, were about 15% of mining costs in 2013. Finally, the deep mines experience rock bursts or bumps, which occur when rock under high confining pressures is exposed in mine openings and fails by explosions. Some bumps are large enough to be felt as earthquakes at the surface; they are monitored by micro-earthquake, seismic systems to help predict future bumps.

South African gold mining is highly labor intensive, with about 150,000 unskilled and semi-skilled employees from throughout southern Africa (Figure 11.7), a significant drop a peak of about 500,000 in 1988 (Madongo, 2014). The mines have been widely criticized for paying too low a wage to their unskilled and semi-skilled laborers, a charge that began in the early twentieth century and continues today (Wilson, 1985; Smith, 2013). Wages that can be paid to any gold miner are limited by the amount of gold produced. Gold production

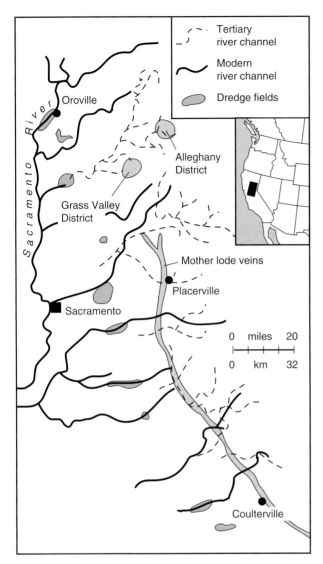

Figure 11.5 Distribution of placer gold associated with the Alleghany, Grass Valley, and Mother Lode orogenic gold veins. Gold is found in river systems of two different ages, an older Tertiary system and a modern system that redistributed some older gravels and eroded new gold from some veins. In wider parts of rivers, placer deposits were mined by dredges (compiled from US Geological Survey Professional Paper 73 and publications of the California Division of Mines).

Figure 11.6 Two extremes in gold mining. (a) Miners at work in the Driefontein Mine, South Africa, which extends to depths of 3,300 meters in the Witwatersrand Basin and produces about 450,000 ounces of gold per year from about 1.9 million tonnes of rock. Note the steel rock-support columns. Specks of white and black in the wall in front of the minerals are pebbles in the Witwatersrand conglomerate (photograph courtesy of Gold Fields Limited). (b) View of the Post-Betze open pit at Carlin, Nevada, which produces about 900,000 ounces of gold per year from more than 1 million tonnes of rock. At the bottom of the pit is an entrance to an underground mine that exploits the deeper continuation of this deposit (photograph by the authors). See color plate section.

from South African mines decreased by 80% from 1990, when South Africa was the world's top gold producer, to 2014, when it ranked fifth in global production, producing annually only about 250 million tonnes. Table 11.4, which compares major gold mines in North America and South Africa, shows that average annual gold production per employee is over 300 ounces outside the Witwatersrand versus only 30 to 40 ounces in South Africa. This enormous, essentially 10-to-1 ratio reflects the highly mechanized nature of North American gold mining and is not a comment on the skill of the South

African miner; no one could do better under the same geologic and mining conditions. It does mean, however, that wages should be about 10 times lower for South African miners, assuming that the companies have similar operating margins. According to the South African Chamber of Mines, the average wage for unskilled and semi-skilled workers on the mines in 2012 was about 156,000 Rand (~$15,600), which is 10% or more of the average North America mine wage. This comparison underscores the enormous challenges faced by South

Table 11.4 Major South African gold producers compared to those in other parts of the world where deep mining is not needed (based on information from corporate annual reports)

Company	Gold production (ounces)*	Fraction of gold produced in South Africa# (%)	Employees	Ounces per employee
Agnico–Eagle	1,100,000	0	3,250	338.5
Newmont	5,100,000	0	15,100	337.7
Kinross	2,600,000	0	8,230	315.9
Barrick	7,160,000	0	23,000	311.3
Goldcorp	2,670,000	0	11,500	232.2
AngloGold Ashanti	4,100,000	37	61,000	67.2
Sibanye	1,440,000	100	35,227	40.9
Harmony	1,300,000	100	39,440	33.0

* 32.105 ounces of gold in 1 kg.

No other African contry is a major producer of gold for these countries.

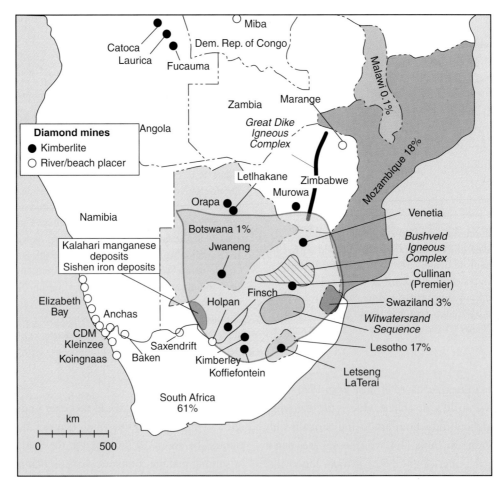

Figure 11.7 Major mineral deposits in southern Africa showing the location of placer and bedrock (kimberlite pipe) diamond, Bushveld and Great Dike layered igneous complexes (LICs), Witwatersrand sedimentary basin (and gold deposits), and the Kalahari manganese–Sishen iron ore deposits. Most of these mineral deposits are part of the Kaapvaal craton, a 3.6–2.5-billion-year-old Precambrian terrane that underlies much of South Africa. Detail of Bushveld Complex is shown in Figure 8.17. Witwatersrand gold mines are along the northern edge of the basin. Also shown are countries supplying labor to Witwatersrand gold mines (showing percentage of total work force from each location, from publications of the South African Department of Labour).

African mines as increasing depth causes lower productivity and even higher wages. Mechanization of mining operations might allow a wage increase, but would result in decreased employment, which has negative societal consequences.

11.1.3 Environmental issues in gold production

Placer gold mines were early environmental offenders. In California during the late 1800s, hydraulic mining used high-powered blasts of water from hoses to break up gravels and move them into gold-concentrating systems. Sediment stirred up by these operations in the Sierra Nevada reached as far as San Francisco Bay. In 1883, Congress passed the California Debris Commission Act requiring that wastewaters be impounded until sediment had settled out, and many placer mines were closed because they could not support the added costs (Koshmann and Bergendahl, 1968).

Gold is recovered from ores by two main methods, both of which involve environmental concerns (Salter, 1987; Stanley, 1988). Early recovery used **amalgamation**, in which ore is mixed with mercury that selectively dissolves the gold and is then removed by heating to vaporize the mercury, leaving behind pure gold. This approach, known as the **patio process**, was widely used during Spanish settlement of the New World, as discussed in the Section 10.11.2 on mercury. Mercury from most amalgamation operations was rarely recovered and remains as a legacy pollutant in many old mining areas. The cyanide process, which replaced amalgamation in the early

1900s, is based on the fact that precious metals form soluble complex ions with cyanide (CN^-) anion (Popkin, 1985; Constantine, 1985).

About 90% of the gold produced worldwide is extracted by the cyanide process, although only 6% of the 1.1 million tonnes of cyanide produced annually is used for gold processing. Cyanide will dissolve gold, but will not dissolve quartz, iron oxides, and other common gangue minerals in gold ore; thus, it can be mixed with ore and yield a simple gold-bearing solution. In some gold mines, gold is dissolved from the ore by crushing and grinding it and then mixing it with cyanide solution in large vats. In mines in arid areas, ore is put in piles onto which a sodium cyanide solution is sprayed and then collected as it leaks out the bottom of the pile (Figure 11.8). This process, known as heap leaching, requires less investment than conventional processing using crushing and vats, and has made it possible to mine low-grade gold ores that would otherwise be uneconomic (Wargo, 1989). The gold-bearing leachate, which is known as a pregnant solution, is processed by adding zinc to form soluble zinc cyanide and a precipitate of gold and silver. The pregnant solution can also be passed over activated carbon, which adsorbs up to 99.5% of the dissolved gold in less than 24 hours. Gold from either process is cast into bars of **bullion**, known as **doré** when it contains silver, which must be further refined to remove impurity metals such as mercury, arsenic, and copper (Figure 11.8). Most unoxidized Carlin-type ores cannot be treated by the cyanide process because their gold is in small

---

BOX 11.1 | GOLD, AMALGAMATION, AND MERCURY

Amalgamation is still being used by the 10–15 million artisanal miners in Brazil, Ghana, Thailand, Latin America, and the Pacific rim (Hilson *et al.*, 2007; Swain *et al.*, 2007; Styles *et al.*, 2010). It is estimated that amalgamation releases annually over 100 tonnes of mercury, of which about half is discharged to local surface waters and half to the atmosphere. Much of this mercury is converted to **methylmercury**, which undergoes **bioaccumulation** via consumption of fish from local waterways. Artisanal miners in Ghana had urinary mercury levels of about 170 micrograms/liter, more than three times the guideline of 50 microgram/liter established by the World Health Organization (Paruchuri *et al.*, 2010). Mercury contents of fish from the Balbina Reservoir, Brazil, which are a major part of the local diet, range from 0.03 to 0.9 microgram/gram, often exceeding the limit (0.5 µg/g) set by the World Health Organization (Kehrig *et al.*, 2001). Efforts to stop amalgamation have been thwarted by popular resistance because gold mining is one of the few means of local employment. Miners in Ghana indicate that they are aware of the health effects of mercury, but continue to use it because there is no good alternative (Amankwah *et al.*, 2010). Adoption of simple mercury retorts, in which vaporized mercury is condensed and recycled, would be a major improvement. However, miners complain that the retorts are too time-intensive, requiring more than 2 hours of work to process amalgam, relative to 5 minutes to process by using a blow torch (Hilson *et al.*, 2007; Tschakert and Singha, 2007).

Figure 11.8 (a) Heap leach pads showing heaps of ore in left background. Cyanide solution is sprayed on the heaps, flows through the rock dissolving gold, and collects in the basins to the right foreground. Note thick plastic layer beneath the ponds and heaps to prevent recharge of solution into underlying groundwater. (b) Molten gold pouring from a furnace into molds. (c) Gold bars removed from molds. Each bar weighs about 27 kg and contains about 87% gold, 8% silver and small amounts of lead and zinc. The hand (lower right) provides scale (photos by the authors). See color plate section.

inclusions or even in solid solution in minerals such as pyrite. This gold is treated by roasting, which converts pyrite into porous iron oxides containing small grains of gold that can then be dissolved by cyanide. Roasting is more expensive because it requires a special oven and releases SO_2 and arsenic gases that must be recovered by scrubbers. Biological decomposition of ores might take the place of roasting at some point in the future.

Cyanide is a highly toxic compound. Although it is found in common plants such as almond and cassava, concentrations in ore solutions are higher and require special handling. During ore treatment, the pH of cyanide solutions must be kept above about 11 to prevent cyanide from reacting with hydrogen ion to produce HCN, a deadly gas. Cyanide solutions contain between 0.1 and 0.5 micrograms per gram of dissolved sodium cyanide and at least 1 liter of solution would have to be consumed to produce a fatal dose for most humans (Moore, 1991). It is not surprising, then, that accidental human deaths resulting from the use of cyanide in gold recovery plants are extremely rare. Spills can be highly toxic to smaller aquatic organisms, however. A dam failure at a gold mine near Baia Mare, Romania, in 2000, released 100,000 m^3 of water containing 100 tonnes of cyanide into the Tisza River and devastated the aquatic ecosystem. This event stimulated the creation of the International Cyanide Management Code, which established a set of industry best practices for cyanide use. Montana is the only US state that bans the use of cyanide for gold production, and worldwide, Costa Rica, Germany, the Czech Republic, Hungary, Japan, and Turkey prohibit cyanide use for gold leaching, and Argentina has banned the use of cyanide in some mining operations. Although less toxic substitutes for cyanide are known (Table 11.5), they are not cost-effective or environmentally benign.

Cyanide oxidizes to produce H_2O, CO_2, and nitrogen compounds, which are not toxic. Originally, cyanide-bearing wastes were oxidized by placing them in holding ponds, but more modern operations speed up the process and remove other metals from solution by agitating waste solutions in the presence of SO_2 or microbes (Devuyst et al., 1991). This is particularly helpful in arid areas, where migrating birds use ponds as resting spots, in spite of aggressive prevention efforts including screens, plastic flags, remote-control chase boats, and noisy cannons. Although this problem has been highly publicized, the number of birds killed by cyanide is less than 0.1% of that killed by hunters each year (Kennedy, 1990; Conger, 1992). Runoff or groundwater recharge from abandoned heap **leach pads** could be an environmental hazard, particularly in areas where oxidation is slow. In the early

Table 11.5 Solvents that might be used in place of cyanide in processing of gold ores, showing their toxicities. Toxicity figure given is LD_{50} in mg/kg measured on rats by ingestion of compound, as discussed in Chapter 3. Smaller numbers indicate greater toxicity (from Hiskey and DeVries, 1992).

Compound	Composition	Toxicity
Thiosulfate	$S_2O_3^{-2}$ ion	7,500
Sodium iodide	NaI	4,340
Sodium bromide	NaBr	>3,500
Bromochlorodimethylhydantoin (BCDMH)	Bromo-organic molecule	600
Thiourea	$CS(NH_2)_2$	125
Malononitrile	$CH_2(CH)_2$	61
Sodium cyanide	NaCN	6.4

1990s, concern centered on the Summitville mine in Colorado, where snow melt creates a heavy spring runoff over abandoned leach pads, some of which were not positioned properly with respect to existing drainages. Because the original operator went bankrupt in 1992, supervision of the property reverted to the Environmental Protection Agency, which partnered with other government agencies to remediate the environmental contamination at a cost of $155 million (Bigelow and Plumlee, 1995).

11.1.4 Gold production and reserves

Although gold is produced in 67 countries, China, the United States, Australia, Russia, and South Africa account for about 40% of production. Production from South Africa is declining and the other countries have not been able to completely replace this lost production (Figure 11.9). This is partly due to declining ore grades. For the world's top ten mines ranked by gold grade, grade decreased from 23 g per tonne in 1998 to about 16 g per tonne in 2011, whereas ore production for the mines increased by 20% over the same period. The decline in gold grade in South African mines is striking, where grade decreased from 4.3 g per tonne in 1998 to about 2.8 g per tonne in 2011. Despite having more than half of global reserves, South Africa's ability to increase production will be a complex function of geologic factors, mining costs, labor issues, and politics. South Africa had only about 18% of global gold reserves as of 2014, Canada has 36% of global reserves, the United States 12%, Russia 10%, and Australia 8%.

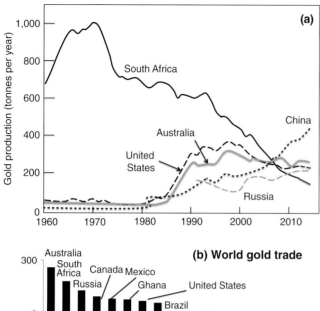

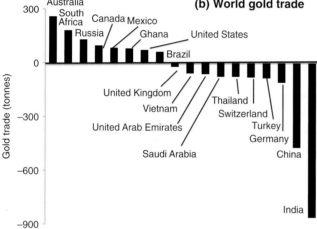

Figure 11.9 (a) World production of gold. Note the decline in production from South Africa and the increase from China and Russia (compiled from data of US Bureau of Mines and South African Department of Mines). (b) World gold trade in 2013 showing difference between gold mined and gold consumed for all markets, including investments. Countries plotting above the line are exporters of gold and those below the line are importers. Countries that consume all the gold that they produce do not appear, regardless of how much they produce. Compiled from data of the US Geological Survey and World Gold Council.

China has seen the fastest rise of gold production, which increased from 50 tonnes in 1987 to 425 tonnes in 2014 (Figure 11.9). This rate of increase has not been enough to prevent China from being a major importer, and it cannot continue; China is now dealing with increasing labor and environmental concerns, and mineable reserves are estimated to be only about 2,000 tonnes. Many other gold producers have a similar challenge. At present annual production, the United States will exhaust its gold reserves in about 15 years, Peru in 13 years, and Canada in 9 years. The rapidly decreasing reserves are stimulating exploration budgets, with $5.3 billion spent globally in 2010, and gold exploration now dominates global mineral exploration budgets, outpacing base metals and uranium.

All in all, the gold reserve picture is not comforting. In spite of intense exploration for a century, no province similar in size to the Witwatersrand has been found. Other discoveries have all been much smaller. For instance, the largest discoveries of recent years, the Post-Betze Carlin-type gold deposit and the Lihir epithermal vein deposit in Papua New Guinea, each contain less than 1,000 tonnes of gold. And remember that much gold is a by-product of mining other metals, such as copper from porphyry-type ore deposits. The economics of co-producing gold from these types of deposits are strongly influenced by the price of copper and therefore are less certain. Since 1950, 871 gold deposits with a total of 32 tonnes (165 million ounces) have been discovered, of which 706 are gold-only deposits, and 165 produce gold as a by-product metal. From 2000 to 2010, when gold exploration budgets grew rapidly, new discoveries averaged about 170 million ounces of gold annually. Over that same period of time, there were 4.5 deposits with more than 1 million ounces of gold, and 0.45 deposits with 10 million ounces of gold discovered per $1 billion of exploration expenditures. And not all of these will be economical. Thus, unless demand for gold changes dramatically in the near future, the price must rise significantly to make lower-grade reserves economic (McKeith *et al.*, 2010).

11.1.5 Gold trade and related monetary issues

Gold differs from all other elements in its historical relation to money. It was among the first metals used in coins, and it gradually became dominant because of its rarity, beauty, resistance to corrosion, and extreme malleability, the property that permits it to be shaped easily. Gold bars were used in trade as much as 6,000 years ago in Egypt, and gold coins came into use by about 600 BCE in the Lydian civilization of Turkey (Butterman and Amey, 2005). Gold was gradually adopted as the standard for settling international debts, and by 1816 the United Kingdom established the **gold standard**. In its purest form, the gold standard set the price for gold in the currency of the country and allowed free import and export of the metal. Thus, the amount of money that a country could put into circulation was controlled by the amount of gold in its central bank, a powerful deterrent to inflation. Other countries, including the United States, adopted the gold standard, making it the basis for international payments. This caused the demonetization of silver, which fell to a gold:silver value ratio between 40:1 and 100:1.

The gold standard was stable during global conflicts and economic crises, but was discontinued during the two major world wars and much of the Great Depression. The International Monetary Fund, which was established during World War II, returned the world to a modified gold standard in which global currencies were valued against the US dollar, which was valued at $35 per ounce of gold (Yaacob and Ahmad, 2014). This system failed because the United States ran large balance-of-trade deficits, putting so many dollars in foreign hands that it could not honor its pledge to exchange gold for dollars at a rate of $35 dollars an ounce. The link between gold and currencies was discontinued in 1974 and non-monetary gold was allowed to be valued on the open market. At that time, Americans were freed from a number of restrictions that had made it a criminal offence to own gold. This enlargement of the market for gold and the growing inflation that gripped the United States and parts of Europe, started the price of gold on an upward trend that peaked at about $800/ounce ($24,918/kg) in 1980 ($2,508/ounce in 2012 dollars; $80,903/kg) (Figure 11.1b). The price declined after that, but reached almost $2,000 an ounce in a second resurgence during the financial crisis of 2007–2010. Although there is no formal link between gold and money at present, government gold stocks (Table 11.3) are still widely regarded as an indication of the health of their economies and the value of their currencies (Hawtrey, 1980; Drummond, 1987; Crabbe, 1989; Hart, 2013). The US gold treasury stock is valued officially at $11 billion.

Gold has been regarded historically as a secure haven for wealth in troubled times. In the 2000s the creation of exchange trade funds (ETFs) that held gold made it easier for average citizens to own gold. The rapid rise and fall of the gold price during and following the 2007–2010 economic crisis revealed how sensitive the gold market is to the whims of speculators and hoarders. The gold price increased from ~ $700/ounce to ~ $1,200/ounce in only 7 months from August, 2007 to March, 2008 and dropped almost 17% during the collapse of Bear Stearns in March, 2008. In a reversal, prices increased about 25% following the bankruptcy of Lehman Brothers in September, 2008, but then fell 25% when the Federal Reserve Bank of New York was allowed to loan money to economically troubled American Insurance Group. Gold prices continued to climb following the economic crisis, driven up both by demand from ETFs and hoarding in less-developed countries (LDCs). In 2013, for instance, China imported an estimated 1,100 tonnes of gold, almost three times its domestic production, Hong Kong imported 597 tonnes, and India 975 tonnes (Macleod, 2014).

China is now the world's largest consumer of gold, even though it was illegal for citizens to purchase gold prior to 2002. Because hoarded gold does not earn interest or pay dividends, it rewards holders only if its price increases more rapidly than other investments and inflation. Viewed in this way, gold has not been a bad investment. Comparing the price of gold to the Dow Jones Industrial Average (DJIA) from 1900 to 2013, the value of each increased, but gold outpaced the Dow from 1928 to 1940, 1972 to 1980, and 2000 to 2012. From 1970 to 2011, the price of gold increased by almost 3,000%, whereas the US consumer price index (CPI) increased by about 500% (Figure 11.1)

Investment demand for gold will probably continue, particularly in countries with unstable currencies and rapid inflation. Jewelry is also a major consumer of gold. Jewelry makers in India and China dominate world gold trade, consuming 552 and 518 tonnes, respectively, in 2012, compared with 108 tonnes in the United States. Although most jewelry enters world markets, some is hoarded in China and India. In 2013, India raised gold import duties to 10% and passed a law that requires re-export of 20% of all legally imported gold, owing to concerns about devaluation of the rupee. In Dubai, a major trading center for Indian gold, trade decreased by 60% in 2013 (Reuters, 2013), although imports of gold increased in neighboring Pakistan and Thailand as it flowed illegally across their borders with India.

11.2 Silver

Silver is evolving from a precious metal to an industrial metal, a path also being followed by platinum. In LDCs, where consumption rates are low, silver is still a precious metal and is used largely in jewelry and tableware. In more-developed countries (MDCs), it is also used in industrial applications and coins, and consumption rates are higher. Historically, the main industrial market for silver was photographic film, in which silver halide compounds embedded in film are reduced to native silver in proportion to the light exposure. The proliferation of digital cameras caused this market to decline by about 75% between 2003 and 2012. Now, about 55% of world silver is used in a wide variety of electrical and electronics applications that depend on its very high electrical and thermal conductivity. It is also used in pharmaceuticals, in highly reflective mirrors, in batteries with zinc, and as a major component in dental amalgam with mercury. These markets support annual world production of 33,000 tonnes worth almost $32 billion (Figure 11.1a), not a bad value for a metal that lost its major market recently.

11.2.1 Geology of silver deposits

Silver forms the gold–silver solid solution electrum, the silver sulfide argentite, as well as numerous complex sulfide minerals containing lead, copper, antimony, and arsenic, such as tennantite–tetrahedrite (Boyle, 1968). It is found in a wide variety of hydrothermal deposits as well as in a few placer deposits. It is the dominant metal in three deposit types, but is more commonly a by-product.

The most familiar silver-dominant deposits are the epithermal vein deposits, which were discussed above as sources of gold. Many of these deposits in the United States and Mexico (Figure 11.10) have silver:gold ratios greater than 100, making silver the economically dominant metal (Simon, 1991). Two other types of silver-dominant deposits merit mention.

Tin–silver deposits, mostly in Bolivia, consist of quartz–silver–tin veins that cut silica-rich volcanic domes (Figure 11.11, 11.12). By far the largest of these is Cerro Rico (Turneaure, 1971). Whereas typical epithermal silver deposits formed at temperatures below about 300 °C from dilute fluids, veins at Cerro Rico began to form at temperatures of 500 °C from very saline fluids, and continued to precipitate ore down to temperatures of less than 100 °C as the solutions became less saline (Sillitoe *et al.*, 1975; Suttill, 1988; Bartos, 2000). This pattern suggests that the hydrothermal system began with hot magmatic waters and that cooler meteoric waters invaded and finally dominated the hydrothermal system as time went on. The second type of silver-dominant deposit is cobalt–nickel–arsenide veins. The largest such deposit, at Cobalt, Ontario

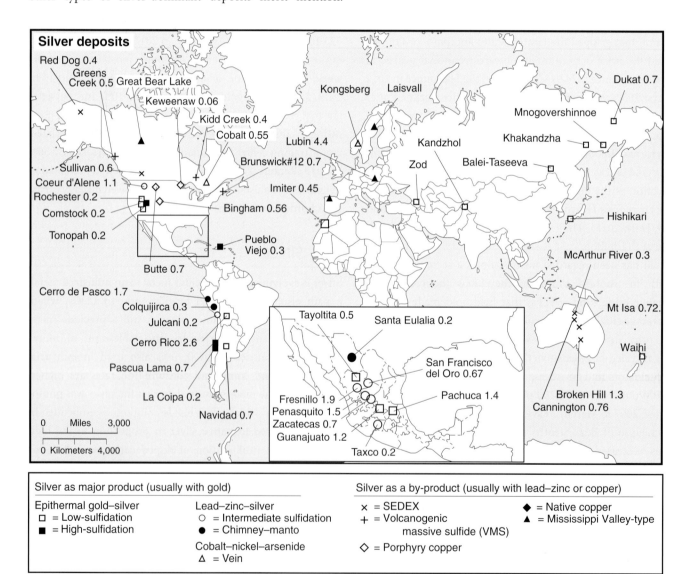

Figure 11.10 Location of important silver deposits showing historic production plus reserves in billions of troy ounces (compiled from data of US Geological Survey, Simmons *et al.*, 2005; Kesler and Wilkinson, 2009; Graybeal and Vikre, 2010 and reports on individual deposits)

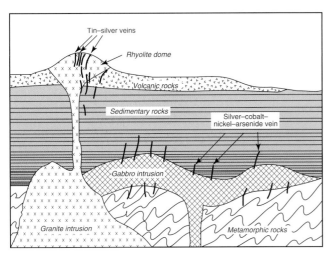

Figure 11.11 Schematic illustration of processes that form typical silver-rich hydrothermal deposits including Cerro Rico-type deposits around a volcanic dome and Cobalt-type deposits around tabular gabbro intrusions. Large amounts of silver are also produced from epithermal deposits shown in Figure 11.2 and as a by-product of base-metal deposits discussed in Chapter 9.

(Andrews *et al.*, 1986), consists of veins with native silver and cobalt, nickel, and iron arsenides in calcite and quartz. Sulfur is scarce in these deposits and its place is taken by arsenic. The veins are found in sediments above and below a large, tabular **gabbro** intrusion and they appear to have formed from basinal brines that were heated by the intrusion (Figure 11.11), rather than from meteoric water.

By-product silver comes from many different types of gold and base-metal deposits. At the Homestake orogenic gold deposit, in South Dakota, which operated for more than a century until it closed in 2002, the silver:gold ratio of production was about 1:5 (Caddey *et al.*, 1991). The huge Proterozoic-age sedimentary exhalative (SEDEX) deposits at Broken Hill and Mt. Isa in Australia are also major silver producers (Both and Stumpfl, 1987), as are the Bingham porphyry copper deposit in Utah and the Kidd Creek volcanogenic massive sulfide (VMS) deposit in Ontario. According to the Silver Institute, the Greens Creek VMS deposit in Alaska was the biggest silver producer in the United States in 2012, producing 6,390 tonnes. Silver is also important in chimney–manto lead–zinc–silver deposits, such as Santa Eulalia and Naica in Mexico and Cerro de Pasco in Peru, where it dominates mining economics (Megaw *et al.*, 1988), and it is found in almost all types of lead–zinc vein deposits, including the large Coeur d'Alene district in Idaho.

Sediment-hosted copper deposits at White Pine in Michigan and the Kupferschiefer in Germany and Poland (see Figure 9.10) have by-product silver, as did the Keweenaw native

Figure 11.12 Cerro Rico, seen from the city of Potosi, Bolivia, is a dome of volcanic rock cut by quartz veins containing silver and tin. The top of the dome (dark) was altered by the addition of silica. During the sixteenth and seventeenth centuries, this was the second largest city in the New World, contributing greatly to the wealth of Spain (from Cunningham *et al.*, 2005).

copper deposits in Michigan (Olson, 1986). Because silver has an even higher electrical and thermal conductivity than copper, copper metal from these deposits was particularly desirable. In fact, the term **Lake Copper**, named for the Keweenaw area on Lake Superior, refers to a premium brand of copper that contains up to 400 g of silver per tonne. Silver is not found in all copper deposits of this type; it is not an important constituent of the Copper Belt ores of the Democratic Republic of Congo and Zambia, for instance.

11.2.2 Silver production, markets, trade, and reserves

Production of silver from its ores depends largely on whether it is present with gold or with base-metal sulfides. Where silver is in electrum and other gold–silver minerals, usually

BOX 11.2 | SILVER AND THE GALLEON TRADE

During early European exploration of Central and South America silver-rich epithermal deposits were controlled by the Spanish, who used the silver not only to finance operations in Europe, but also as currency for trade. The most famous of these trade patterns was the Galleon Trade, which began in 1565 and continued until 1815. During this period, large, multi-decked ships called galleons left Mexico once or twice a year carrying silver mined from Cerro Rico in Bolivia and Guanajuato and Pachuca in Mexico. In the Philippines and other Asian ports, the silver was traded for ivory, porcelain, silk, spices, and other items from China. On the return trip, the galleons carried the Chinese goods to Acapulco, where they were taken overland to the east coast of Mexico for trans-shipment to Spain. As much as a third of New World silver production went to China in this exchange, and silver became the basis for its trade and commerce (Mann, 2011).

in epithermal deposits, it is recovered by cyanide leaching, as discussed in Section 11.1 on gold. Where it is present in base-metal sulfides, as in lead–zinc and most copper deposits, it must be recovered by a special step in the smelting process in which a separate, silver-rich mineral concentrate is formed or at the final stage of refining of the main production metal. Silver recoveries in all of these processes are relatively low and it is very rare for all silver in a deposit to actually make it to market.

Silver is produced in 56 countries, led by Mexico, China, Peru, Australia, and Russia. The United States ranks ninth globally and is the only major silver producer that is a net importer of the metal, importing almost five times more silver than domestic production (Figure 11.13). The silver market is dominated by industry, which uses almost half of all produced silver. Jewelry, coinage, photography, and silverware account for 18, 9, 6, and 4% of demand, respectively (Thompson Reuters GFMS, 2014). The remaining demand comes from government purchases and ETFs. ETF demand increased 200% in the decade ending 2014, with a 21% increase in 2012 alone.

The use of gold and silver in coinage creates a temptation for governments that maintain precious metal **stockpiles**, such as that held by the United States in the National Defense Stockpile until 2003 when it was transferred to the US Mint for use in coins. By using this material in coins, governments can show a "profit" on the coins that would not be possible if they purchased the metal on the market in the same year. The allure of this approach is strongest in years of high budget deficits. For instance, coin minting reduced the US silver stockpile by 50% between 1986 and 1992, including 187 tonnes of silver used in the US American Eagle coin. By 2002, the stockpile was depleted, and Congress introduced the Support of American Eagle Bullion Program Act to

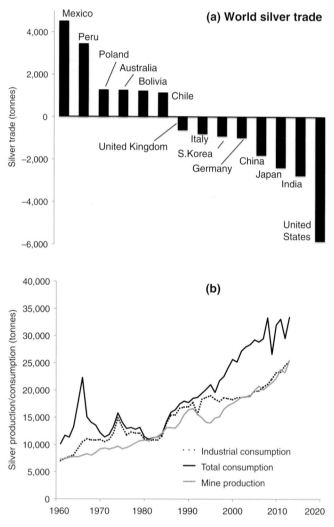

Figure 11.13 (a) World silver trade in 2012. Countries plotting above the line produced more silver than they consumed in coins and fabrication; those plotting below the line imported silver for these markets. Recycled silver is not included in this diagram. (b) Historic trends in silver mine production and consumption showing the large shortfall supplied by recycled silver. Based on information from the US Geological Survey and the Silver Institute.

"authorize the Secretary of the Treasury to purchase silver on the open market whenever the stockpile is depleted."

Silver production has fallen short of consumption for many years, frequently by a wide margin (Figure 11.13). Most of the shortfall is made up by recycled silver from old scrap and fabrication waste, which accounts for about 25% of consumption in the United States. During periods of peak demand, however, silver comes from other sources. The largest of these are silver jewelry in India and bullion, coins, and silverware in MDCs, particularly the United States where government sales of silver accounted for 1% of total silver supply in 2012. Silver holds a special allure in India, where it is used in everything from jewelry to edible silver foil, called vark, consumed as an aphrodisiac and health aid. Silver consumption by non-industrial sources, largely investors and hoarders, was particularly high during the 1960s. Much of this silver was purchased as a hedge against inflation, but some of it was accumulated as a gamble that the shortfall in silver mine production would lead to a dramatic price increase. In an especially unfortunate example of such speculation, the price of silver ran to $48/ounce ($1495/kg) as the Hunt brothers tried and failed to control the global silver market (Fay, 1982).

World silver reserves of 540,000 tonnes are adequate for only about 17 years of production at present rates and are highly dependent on by-product silver from lead–zinc and copper deposits. About 30% of produced silver comes from silver-dominant mines, whereas the remainder is by-product silver. Most silver-dominant deposits form at shallow levels of the crust and are relatively easy to discover compared to more deeply buried deposits. Thus, it is likely that a larger proportion of them have already been found. This means that an increasing proportion of future silver is likely be by-products of lead–zinc and copper mining. Here, too, the outlook is poor. Important by-product deposits, such as the silver-rich SEDEX deposits at Mt. Isa and Broken Hill, Australia, are being exhausted, and new deposits such as McArthur River contain much less silver. It unlikely that by-product silver from porphyry copper and VMS deposits will be able to supply world demand.

11.3 Platinum-group elements

The platinum-group elements (PGEs) include platinum, palladium, rhodium, ruthenium, iridium, and osmium, which have similar chemical properties and occur together in nature. Platinum and palladium, with crustal concentrations of about 5 ppb each, are about as scarce as gold and the other PGEs are even scarcer. Compared to gold and silver,

the PGEs are late-comers. Platinum was recognized as an element only in 1750, with osmium the last to be discovered in 1844. PGEs were first observed in some placer deposits, where they formed curious steel-gray nuggets that looked like silver but behaved like gold. In fact, the name platinum comes from platino, the diminutive term in Spanish for silver, which was used to describe nuggets found in the San Juan and Atrata rivers in Colombia (Loebenstein, 1985).

The main market for PGEs, as a catalyst in chemical reactions, is definitely industrial. Catalytic converters in automobile exhaust systems account for 38% of global platinum demand, 72% of palladium demand, and 79% of rhodium demand. These converters speed oxidation reactions that convert hydrocarbons, nitrous oxides, and carbon monoxide in exhaust gases to less-harmful carbon dioxide, nitrogen, and water. Use of PGEs in catalytic converters depends on the cost and type of engine. Palladium is cheaper and is used widely in gasoline engine exhaust systems. In diesel exhaust systems, which are cooler and more oxidizing, palladium reacts with oxygen to form palladium oxide and with sulfur to form sulfides, both of which decrease the catalytic activity. Platinum and rhodium remain in their metal forms in a highly oxidizing environment and are preferred for diesel engines. This may change owing to legislative mandates that have reduced the sulfur content of diesel fuel from about 300 ppm in 2000, to less than 50 ppm in 2010 in Europe, the United States, China, and Japan. PGEs are used also as a catalyst in oil refining, in the production of nitric acid, and in fuel cells.

Other markets for PGEs include the electrical and electronic industry, which accounts for 11% of demand for palladium and 3% for platinum. Here, PGEs are used in electrical contacts, high-resistance wires, memory devices, and special solders. They are also used as sensors to measure the oxygen content of automotive fuel, and adjust the mixture for more efficient combustion. Dental and medical applications consume about 8% PGE production, largely to strengthen crowns. Platinum is also fundamental to a wide range of pharmaceuticals, including those that treat cancer and arthritis. For example, the chemotherapy drug cisplatin is synthesized by using platinum as the inorganic chemical substrate. Other markets for the PGEs include jewelry, which constitutes about 40% of platinum demand and 4% of palladium demand, and the chemical industry, which consumes about 12% of PGEs. Future applications may include use in fuel cells for hydrogen-powered vehicles. Worldwide production of newly mined PGEs is almost 460 tonnes worth about $19 billion. An additional 150 tonnes of PGEs worth $5 billion are derived from recycling, although much is lost in catalytic converters in used cars that are exported to LDCs.

11.3.1 Geology of platinum deposits

Most PGE production comes from the same LIC deposits that are our main source of chromium and vanadium, particularly the Bushveld Complex (Figure 11.7, 11.14). Present PGE mining in the Bushveld focuses on the Merensky Reef, which is found between the chromite- and vanadium-bearing magnetite layers (Figure 8.16), the UG-2 chromite layer, and the Platreef layer in the Potgietersrus area (as noted in Section 8.4.1 on chromite, the term "reef" refers to a layer in the intrusive rock). The Merensky Reef is a complex layer consisting of coarse-grained, mafic silicate minerals and one or more chromitite layers a few centimeters thick, which contains PGE-bearing sulfide minerals and alloys (Figure 11.15). The entire reef is about 1 m thick and extends

laterally for about 250 km. Although the Merensky Reef forms a well-defined layer, 10–20% of it consists of depressions known as **potholes** where the reef did not accumulate or was removed (Ballhaus, 1988). The entire reef usually contains 3–20 g of PGEs per tonne, similar to the grade of gold in the Witwatersrand ores. Two processes have been proposed for the origin of Merensky Reef-type PGE deposits. The most widely accepted theory is that they are the result of magmatic immiscibility similar to the process that formed magmatic nickel deposits but with the immiscible sulfide reacting with a larger amount of magma to account for its high PGE content; the alternative theory involves extremely high-temperature fluid scavenging of PGEs (Boudreau *et al.*, 1986; Hanley *et al.*, 2008; Simon and Pettke, 2009; Naldrett, 2010;). The UG-2 chromitite in the Bushveld Complex consists largely of chromite, but contains over three times as much rhodium per tonne as the Merensky Reef. The Platreef, which is at the base of the Bushveld on its northern

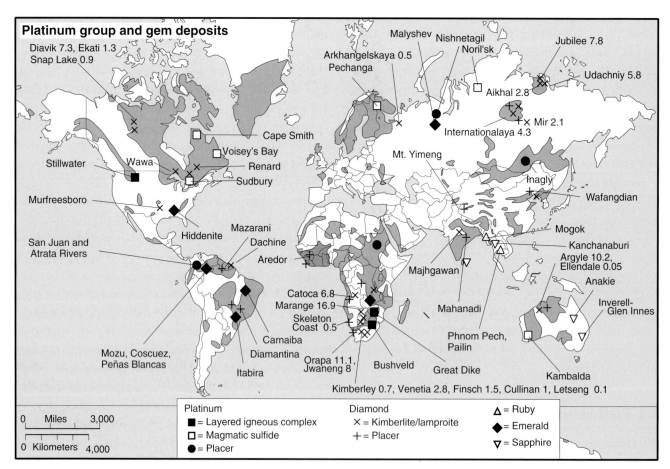

Figure 11.14 Locations of important deposits of platinum-group elements and gems (based on Barnes and Lightfoot, 2005; Gurney *et al.*, 2005; Shirey and Shigley, 2013). Numbers following diamond deposits show production (billions of carats) for 2013 (from http://www.kitco.com/ind/Zimnisky/2013-08-20-Ranking-Of-The-World-s-Diamond-Mines-By-Estimated-2013-Production.html). Shaded areas show distribution of Precambrian cratons that host most diamond deposits.

Figure 11.15 Platinum deposits. (a) Bushveld Complex – a close-up view of the Merensky Reef in the Rustenburg Platinum Mine showing its coarse-grained nature with abundant chromite (dark grains). The reef increases in thickness abruptly to the left as it enters a pothole, a feature discussed in the text. (b) Outcrop of the J-M Reef (the 10-cm-thick dark layer extending across the photo just below the baseball cap) at the Stillwater Mine, Montana (photographs by the authors). See color plate section.

side (Figure 8.16), consists of blebs and veinlets of PGE-bearing sulfide minerals in ultramafic rocks that contain large, partly digested fragments of wallrocks. This ore is of lower grade, but larger tonnage than that in the UG-2 or Merensky Reef and was the last to be mined by modern efforts.

PGEs are also found in other LICs, but in smaller amounts. Best known of these is the Stillwater Complex in Montana, where platinum production concentrates on a layer known as the J-M Reef. Although the J-M and Merensky reefs have many geological similarities, the J-M averages about 21 g of PGE per tonne, as much as three times higher than the Merensky Reef. However, platinum:palladium ratios in the J-M are 1:3 versus about 3:1 in the Merensky, and its

platinum:rhodium ratios are about 32 compared to only 15 in the Merensky. Because of the larger demand for platinum and rhodium, and their higher prices, this gives the Merensky Reef an economic edge. Furthermore, reserves in the J-M Reef are considerably smaller than in the Bushveld reefs because the Stillwater Complex has been extensively folded, faulted, and eroded, and its limits are not as well defined, with some PGE values extending above and below the main layer (Maier, 2005; Cawthorne *et al.*, 2005).

Smaller amounts of by-product PGEs come from nickel–copper sulfide deposits that formed by separation of immiscible sulfide magmas at Sudbury and Noril'sk, as discussed in Section 8.3.1 on nickel (Figure 11.14). PGE concentrations in these deposits are considerably lower than in the LICs, and production is significant only where there is large-scale nickel production.

11.3.2 Production of platinum-group elements

Although some PGEs are produced from open pit mines in the Bushveld Complex, most are underground. All mines face the problem of extracting enough ore from thin layers to meet production requirements, and the consequent need to mine over large areas. The effort is further complicated in the Merensky Reef by the potholes, which require expensive diversions of working levels, causing lower overall ore recoveries than would be attained from a truly planar layer. To meet production quotas, mines have been opened throughout the Bushveld Complex wherever the Merensky Reef is of adequate thickness and grade. Although these mines are only about half as deep as the nearby Witwatersrand gold mines, the geothermal gradient in the Bushveld Complex is twice as great, making operations at depths of only 1,500 m roughly equivalent to the most hostile conditions seen in the gold mines.

Production of PGE metals differs from operation to operation, but always involves separation of a concentrate containing sulfide minerals and PGEs. In some cases, a special PGE-rich concentrate is formed before the other concentrate and is sent directly to the smelter. Once a PGE-rich matte is formed during smelting, it is dissolved and different metals are separated by ion exchange. PGE smelters and refineries are subject to the same environmental problems as those affecting base-metal smelters. Although some forms of the PGEs are toxic (for instance only a few ppm of palladium sulfate or nitrate can cause blood to coagulate), the elements do not cause environmental problems in mines because ore grades are so low and the elements are not mobile enough to enter the biosphere. In refineries and chemical plants, where PGE concentrations are higher, problems are observed. The main issue is platinosis, which involves respiratory and dermatological symptoms from exposure to PGE's.

11.3.3 Platinum-group element markets, trade, and reserves

PGEs are produced in only 10 countries, one of the smallest totals for any mineral commodity. South Africa accounts for about 70% of the global platinum supply, and South Africa and Russia produce about 40% of global palladium. Smaller contributions come from LICs in the United States and Zimbabwe, magmatic nickel sulfide deposits in Australia, Canada, and Finland, and placer deposits in Colombia and Ethiopia. Production also comes from smelters in Japan that are supplied by imported nickel–copper ore. Most of this production goes to the large car manufacturers in Germany, Japan, the United States, and Korea.

There is also demand for PGEs as jewelry and a hedge against inflation. Investment demand was very important in the United States during the late 1980s, and accounts for about 10% of consumption. The jewelry market is very strong in Asia but not in the western world. In 2013, jewelry and investments accounted for 33% and 10% of world PGE demand, respectively (Johnson Matthey, 2014).

About 97% of world PGE reserves of 66,000 tonnes of contained metal is in South Africa and Russia, where geological, political, and labor uncertainty cloud the supply picture. In South Africa, much of the reserve is in deeper levels of the Merensky Reef or in the UG-2 or Platreef, where mining is more complex. Although the UG-2 contains chromite, it is discarded by most miners, and PGE production has to bear the complete cost of mining. Perhaps the biggest challenge for South African mining operations is labor relations tied to working conditions and wages. Periodic strikes crippled production from 2012 to 2014, when unskilled and skilled workers demanded wage increases. The supply situation in Russia is similarly clouded, with major uncertainties surrounding the future of the outdated and environmentally troubled Noril'sk operation. PGE production at Noril'sk is also a by-product of nickel and copper mining, which means that the market for these base metals ultimately dictates the ability for Russia to maintain a strong output of PGEs (Loferski, 2012).

11.4 Gems

Although diamonds are the best known of the world's gem stones, over 150 natural compounds have been used as gems (Jahns, 1983; Pressler, 1985; Austin, 1991; Olson, 2013). Of these, diamond, emerald, and ruby consistently sell for the highest prices, with alexandrite and sapphire only slightly behind (Table 11.6). Amber, aquamarine, jade, opal, pink topaz, spinel, and tourmaline have intermediate values. Many other minerals, from agate and amethyst to zircon and zoisite, make up a third tier of gems that are usually inexpensive, although special examples can be quite expensive, a fact that has discouraged use of the term "semi-precious stone."

As can be seen in Table 11.6, gems do not have distinctive chemical compositions. Instead, it is their appearance and rarity that confer value on them (Hurlbut and Kammerling, 1991; Schumann, 2006). The crystal structure of most gems

---

BOX 11.3 | PGES AND RECYCLING

PGE production has increased dramatically since 1974, when the automotive catalyst market opened. From 1983, when the United States required catalytic converters, to 2014, the world PGE market doubled and was one of the fastest-growing of metal markets (Figure 11.1). Today, more than 90% of vehicles sold worldwide are equipped with catalytic converters, and this proportion should grow. Recycling generated about 150 tonnes of PGEs in 2013, although many products that contain PGEs are not recycled. Recycling rates for PGEs are 50–60% worldwide in automotive applications, and 5–10% in electronic applications (Hagelüken, 2012). This is ironic because the average content of PGEs in many technology products is actually greater than the grade in primary ores. For example, a computer motherboard contains 80 g/tonne palladium and a mobile phone contains 130 g/tonne palladium (Hagelüken, 2012), whereas the average grade of palladium in ores is about 10 g/tonne. Recycling was historically hindered by a lack of economically feasible technology. However, this is certain to change, and efforts to recycle PGEs and other metals from high-grade electronic products must improve in order to satisfy increasing consumption of the products that rely on these metals.

Table 11.6 Major gem stones and their mineralogical classification. Many other stones, including agate, apatite, coral, epidote, feldspar, garnet, hematite, kyanite, lazurite, malachite, olivine, pectolite, prehnite, sphene, and zircon, are used widely in jewelry and especially striking specimens can have high values (compiled from Jahns, 1983; Sauer, 1988; Sinkankas, 1989; Austin, 1991; Schumann, 2006).

Gem stone	Mineral	Hardness (Mohs scale)	Major elements	Important sources
Highly valuable				
Alexandrite	Chrysoberyl	8.5	Be,Al,O	Russia, Brazil
Diamond	Diamond	10	C	South Africa, Namibia
Emerald	Beryl	7.5–8	Be,Al, Si, O	Colombia, Brazil, Russia
Ruby	Corundum	9	Al,O	Cambodia, Myanmar
Sapphire	Corundum	9	Al,O	Sri Lanka, Thailand, Brazil
Less valuable				
Amber	Not a mineral	low	C,H	Russia, Dominican Republic
Amethyst	Quartz	7	Si,O	Many locations
Aquamarine	Beryl	7.5–8	Be,Al,Si,O	Brazil
Jade	Jadeite	6.5–7	Na,Al,Si,O	Myanmar, China
	Nephrite	6.5–7	Ca,Mg,Fe,Si,O,H	Russia, China
Opal	Opal	5.5–6.5	Si,O,H	Australia, Brazil, India
Pearl	Calcite	2.5–4	Ca,C,O	Many locations
Topaz	Topaz	8	Al,Si,O,H,F	Brazil, Russia, India
Tourmaline	Tourmaline	7–7.5	Na,Mg,Fe,Al,B,Si,O,H	Namibia, Brazil, USA
Turquoise	Turquoise	5–6	Cu,Al,P,O,H	USA, Egypt, Australia

imparts clarity, color, and brilliance, including fire, which is produced by refraction of white light into different wavelengths and colors. Other gems, such as amber and pearls, are valued for their luster. The highest-priced gems are very hard, measuring 8 to 10 on the 10-point Mohs scale for mineral hardness (Table 11.6). In fact, diamond, is the hardest natural or synthetic substance known. Because they are so hard, precious stones resist scratching and preserve their polish, brilliance, and value (although they will shatter if struck by a hard object).

Most of the common gems of the world are retailed in cut and polished forms that differ from the natural shape of the gem crystal. For instance, the most common natural crystal form for diamond is an octahedron, but it is commonly cut into other shapes that enhance light refraction and color, and that change with fashion. Thus, the world gem industry consists of two parts, mining and cutting–polishing, which are usually followed by jewelry manufacture (Green, 1981). The majority of the world's rough gem diamonds, about 80%, are processed in Antwerp, Belgium, where 30,000 people are employed. Gem diamond cutting is also carried out in New York, and newer centers in Mumbai and Toronto are increasing their production of small cut stones. The most important cutting–polishing center for small stones and chips is India, where more than 1 million people cut and polish slightly more than 85% of all diamonds by volume.

Important colored-stone preparation is centered in India, Hong Kong, Thailand, and Brazil.

Gems are formed by a wide variety of geological processes, many of which have not been adequately investigated because of the rarity and small size of their deposits. Even diamonds, which have been mined on a large scale in several locations, are still only partly understood. The following summary focuses on the most important of the gems but, for lack of space, cannot cover many other interesting ones.

11.4.1 Diamonds

Gems are the ultimate in value among mineral commodities. Even though a carat weighs only 0.2 g, good quality one-carat diamonds wholesale for about $7,500. In fact, all of 2012 annual world production of about 128 million carats worth about $14.8 billion (Figure 11.1) would occupy a cube only about 3 m on a side. In contrast, the same fortune in gold would occupy a volume almost eight times larger. Diamonds are clearly the best form in which to concentrate wealth, if you can find them.

Geology of diamond deposits

Diamonds are **metastable** at Earth's surface. As can be seen in Figure 11.16, the range of temperatures and pressures at which diamond is stable does not include surface conditions

(Kirkley *et al.*, 1992; Boyd and Meyer, 1979; Harlow and Davies, 2005). As the diagram shows, carbon takes the crystal form of graphite at surface conditions, and transforms to the

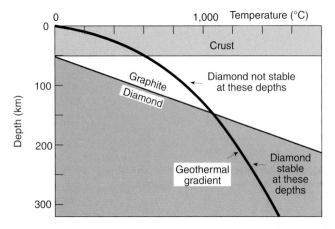

Figure 11.16 Schematic cross section through continental crust and upper mantle showing the downward increase in temperature (geothermal gradient) beneath thick continental crust (cratons) and stability fields for graphite and diamond, both crystalline forms of carbon (C). Note that graphite is stable at the surface of Earth, and that the geothermal gradient reaches the diamond stability field at a depth of about 150 km in the mantle. Thus, diamond is metastable at Earth's surface.

crystal form of diamond only at very high temperatures and pressures (Meyer, 1985). Fortunately for anyone with diamond jewelry, the rate at which this transition takes place is vanishingly small at normal surface temperatures, where diamond persists as a metastable compound.

Most diamonds are found as xenocrysts, or foreign fragments, in magnesium-rich ultramafic rocks known as **kimberlites** (Gurney *et al.*, 2005; Shirey and Shigley, 2013). Diamond-bearing kimberlites are found largely in thick continental crust or **cratons** of Archean age, where they form cylindrical pipes or tabular, dike-like bodies (Figure 11.17). They are hybrid rocks that consist of fragments of many rocks and minerals, including diamonds, held together by small amounts of magma (Figure 11.18a). Combined with the diamond stability relations shown in Figure 11.16, this suggests that kimberlite magmas originated in the mantle and that they brought up pieces of it, including diamonds, when they were intruded into the crust.

Most diamond-hosting magmas were extruded onto the surface. They had to come from depths of about 150 km or more where diamonds are stable (Figure 11.16) all the way to the Earth's surface. They must have come up very rapidly to prevent the diamonds from reverting to graphite. The rapid

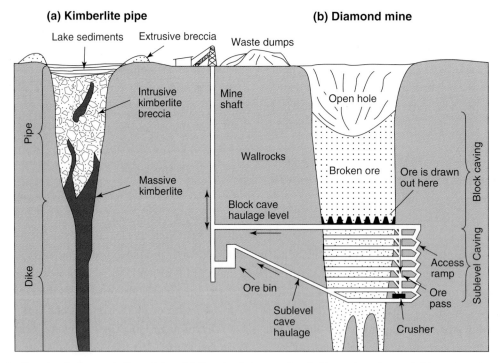

Figure 11.17 Kimberlite pipes before and after mining. (a) Unmined pipe surrounded by a ring of extruded breccia at the surface. The top of the pipe is filled with lake sediments, which are underlain by kimberlite breccia and then massive kimberlite, which grades into a dike at depth. (b) Kimberlite is being mined by block caving (upper part) and sublevel caving (lower part). Block-caved ore is drawn out at the main haulage level below. Sublevel ore is drawn out on each level, dropped through the ore pass to the crusher, and hauled to the ore bin. Ore from both parts of the mine is taken to the surface through the shaft (based on diagrams for the Kimberley mine, Kimberley, South Africa).

Figure 11.18 Diamond deposits and mines. (a) Kimberlite breccia in the upper part of the Kimberley pipe, South Africa showing a large boulder (upper left) that was transported from the mantle. (b) Beach placer diamond mine in southern Namibia showing pumps to remove seawater (lower left). The mining area and pumps are several meters below sea level but are separated from the ocean by a dike made of sand, which is seen in the lower right corner of the photo. Partly flooded mining areas down the coast can be seen in the background (photographs by the authors). (c) Diavik diamond mine, Northwest Territories, Canada, which produces about 7.5 million carats per year from three kimberlite pipes, is surrounded by a 3.9 km dike to protect it from waters of Lac de Gras. (d) Uncut diamonds from the Diavik Mine (copyright © 2014 Rio Tinto). See color plate section.

ascent is thought to have been aided by high water and carbon dioxide contents in the intruding magma, which lowered its viscosity and provided explosive gas pressure to help drill upward through the crust. That all diamonds did not rise so rapidly is shown by the presence of graphite with the external crystal shape of diamond (**pseudomorphs**) in mantle rocks that are exposed at the surface in Spain and Morocco (Pearson *et al.*, 1989). Unlike emplacement of kimberlite intrusions, these diamonds moved toward the surface by slower tectonic processes (Gurney *et al.*, 2005). In what might be regarded as the reverse of this process, aggregates of small, black diamonds known as carbonado, are found where some think that meteorite impacts converted organic material to diamond (Smith and Dawson, 1985; Heaney *et al.*, 2005). Age measurements show that kimberlites that host diamonds are mostly less than 300 million years old whereas the diamonds are much older, indicating that the mantle under young cratons is enriched in volatile constituents.

Diamonds are not found in all kimberlite intrusions. Of the 6,400 or so kimberlite bodies that are known, only about 15% are estimated to contain diamonds and only about 0.5% of them have been mined. Until the 1950s, diamonds had been mined successfully from kimberlite pipes only in the Kaapvaal craton of southern Africa (Figure 11.7). Since then, however, they have been found in kimberlites in other parts of the world and even in other rock types. One surprising discovery was the Argyle deposit in northern Australia (Figure 11.14), where diamonds are hosted by **lamproite**, a rock that was not known previously to contain them (Lang, 1986; Mitchell, 1991). They are also found in lamprophyre at Wawa, Ontario, one of the oldest occurrences of diamond in the world. Exploration, spurred in part by these developments, found rich diamond-bearing kimberlites and lamproites in the Northwest Territories of Canada, allowing Canada to become a major diamond producer (see Figure 11.19) (Erlich and Hausel, 2002).

Diamond placer deposits have a wider geographical distribution than diamondiferous kimberlite pipes, and are not necessarily confined to thick Archean crust because they can be transported long distances by rivers. Some diamond placers have been traced to pipes that are of too low grade to mine, such as those near Murfreesboro, Arkansas (Bolivar, 1986), and the sources of others are not known. Erosion of known diamond pipes has been surprisingly deep. Around Kimberley, where it all began, about half of the original vertical extent of the pipes has been eroded. Because the grade and quality of diamonds is greatest in the upper part of most pipes, a sizeable fraction of their diamonds is now in placers, along the Orange River and the Atlantic coast (Figure 11.14). Elsewhere in Africa, placer deposits may be derived from deeply eroded kimberlites in Angola, Ghana, Guinea, Sierra Leone, Tanzania, and the Democratic Republic of Congo (Ellis, 1987).

Diamond mining and production

Diamonds are mined by both open pit and underground methods (Reckling *et al.*, 1983). Along the western coast of southern Africa, beach placers are mined by building large dikes to hold out the surf (Figure 11.18c). Submerged beach placers, which formed farther offshore from Namibia and South Africa when sea level was lower during Pleistocene glaciation, are also mined by vacuuming the sediments onto large ocean-going vessels. Diamond placer mining is among the most thorough of mining operations. Even though large equipment is used to remove the sand and gravel, every last grain has to be swept from irregularities in the bedrock surface to find the diamonds. Diamond grades reported for 2013 for Diamond Fields' Marshall Fork placer deposit off the coast of Namibia were 1.07 carats per tonne of sediment processed. The volume of diamonds in a tonne of ore with

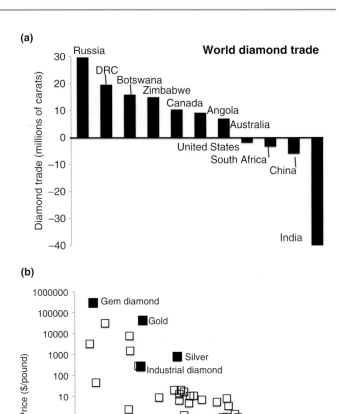

Figure 11.19 (a) World raw diamond trade showing the difference between diamonds exported (above the line) and imported (below the line) based on data of Kimberley Process. These data represent raw diamonds, most of which are exported for cutting to the major centers, especially India. Trade in cut gem diamonds is not available. (b) Relation between US consumption and price per unit for mineral commodities showing that prices for gem diamonds, gold, and silver are generally near the high edge of the price range to be expected based on the amount of material traded.

these grades would be 0.01 cm^3, a truly small volume to recover.

Kimberlite mining usually begins with an open pit but changes to an underground operation as it gets deeper (Figure 11.18). Most underground diamond mining is carried out by block caving or **sub-level caving**, which are needed to supply the large volumes of rock needed to meet diamond-production goals. Diamond grades reported for 2013 for kimberlite mines in southern Africa ranged from 0.37 carats per tonne at Finsch to 0.05 carats per tonne at Koffiefontein.

Diamonds were recovered from ore in the early days by concentration in rotary pans followed by hand-sorting, an arduous task that produced security problems, limited the scale of the operation, and yielded poor recoveries (Joyce and Scannell, 1988). Discovery, in 1896, that diamonds adhere to grease-coated cloth opened the way for large-scale mining and processing. Modern kimberlite diamond mines begin by crushing ore to fragments small enough to liberate most diamonds. The heavy minerals are then concentrated by pans, similar to those used to pan gold, and cyclones, similar to those used to separate particles from combustion flue gas. Fine-grained heavy-mineral concentrates from these processes go to grease tables and coarse-grained material goes to X-ray sorters, which scan each grain and use air jets to remove any diamonds that are detected. Large stones can show up at any point in this process. The largest diamond of all time, the 11-by-6-cm, 3107-carat Cullinan diamond, was spotted by a mine worker in 1906 on the side of the Premier mine pit, whereas the 426.5-carat Ice Queen from the Premier mine was found on the grease table during the final stages of concentration (Erlich and Hausel, 2002).

Diamond classification and gem-diamond trade and reserves

Natural diamonds are hand-sorted and graded according to size, quality, color, and shape at central locations such as Oppenheimer House in Kimberley (Figure 6.1). The processes is guided by a set of representative diamonds kept at major grading centers and which is, itself, compared against a master set of diamonds kept in London. Diamonds are first divided into 11 groups on the basis of size, and then sorted according to type or shape, quality, and color. The main divisions of the classification are between gems and industrial diamonds, which are widely used in abrasives and cutting tools. An intermediate class of near-gem diamonds depends on markets for the other classes. In South Africa and most other areas, diamonds that have been graded are examined by government inspectors to assure that they meet specifications (Optima, 1985).

All stones, including the occasional enormous gem (Table 11.7), are cut into a wide variety of shapes to enhance their appearance. The size and number of gems that can be cut from a single stone are controlled by its shape and the distribution of imperfections. For instance, when the 995-carat Excelsior was cut into 21 gems of which the largest was only 69 carats, about 63 percent of the stone was wastage caused largely by black spots. The great Cullinan diamond was cut into 96 small brilliants, nine polished fragments and nine

Table 11.7 Important large diamonds of the world (after Joyce and Scannel, 1988; Deakin, 2014)

Name	Size (carats)	Source deposit
Sergio	3,167	Brazil
Cullinan	3,107	Premier, South Africa
Excelsior	995	Jagersfontein, South Africa
Star of Sierra Leone	969	Diminco, Sierra Leone
Incomparable	890	MIBA, Democratic Republic of Congo
Millenium Star	777	Mbuji-Mayi, Democratic Republic of Congo
Great Mogul	787	Kollur, India
Vargas	727	San Antonio River, Brazil
Woyie River	770	Woyie River, Sierra Leone
Jonker	726	Elandsfonetein, South Africa
Jubilee	634	Jagersfontein, South Africa
Sedafu	620	Diminco, Sierra Leone
Lesotho Brown	603	Letsent, Lesotho
Lesotho Promise	601	Letseng, Lesotho
Goyaz	600	Brazil
Centenary	599	Premier

major stones, including the 317-carat Lesser Star of Africa and the 530-carat Greater Star of Africa, which belongs to the British Crown and are the largest cut diamonds in the world.

Diamonds are produced in 21 countries and are designated as gem-quality, near-gem-quality or industrial. In terms of total value, the most important producers are Botswana, Russia, Canada, Angola, and South Africa, which account for about 63% of the worldwide market of about $10 billion. In terms of number of carats mined, these countries also dominate, along with Zimbabwe, the Democratic Republic of Congo, Angola, and Australia, which mine large numbers of smaller, industrial diamonds. In terms of value per carat, however, the most important producers are Lesotho, Namibia, Sierra Leone and Liberia, which are mining large diamonds. In terms of world trade, the major exporters are Russia, the Democratic Republic of Congo, Botswana, Zimbabwe and Canada (Figure 11.19). Australia is the world's largest producer of industrial diamonds, largely from the

Argyle and related deposits in northwest Australia. Profit margins for rough diamond producers are 16–20%, although it is less than the profit from jewelry manufacturing and sales, where the price per carat increases by a factor of about eight during the transition from rough to gem diamond (Bain, 2014). The relative value of mining gem quality versus industrial diamonds is evident when comparing the market value of diamonds in Canada, the world's third largest producer of gem-quality diamonds (Figure 11.18c, d), and Australia, a major producer of industrial diamonds. In 2012, Canada produced 10.45 billion carats at a value of $2.01 billion, whereas Australia produced 9.18 billion carats at a value of $269 million. Production of industrial diamonds in Australia has decreased significantly, from 30–40 million carats per year in the 1990s to the current level of about 10 billion carats.

In addition to mined diamonds, there are synthetic diamonds, which can be made by either high-temperature, high-pressure, or vapor-deposition processes. Although gem diamonds up to 2 carats in size can be synthesized, most of the demand for synthetic diamonds is in industrial applications such as

abrasives, computer chips, drilling, and transportation systems. World production of synthetic industrial diamonds was about 100 million carats in 2012. In the United States, synthetic diamonds comprise about 97% of all industrial diamond use, where they are preferred because they can be synthesized for specific applications (Kane, 2009).

International scrutiny of the diamond trade has focused on blood or conflict diamonds, which are mined by artisanal methods and used to finance guerrilla activity. In an effort to slow this trade, global diamond production is monitored by the Kimberly Process Certification Scheme (KPCS), which tracks all diamond sales by volume and value (http://www.kimberleyprocess.com/). The 74 Kimberley Process countries certify that their diamond exports are from conflict-free sources. This system is not perfect because some diamonds lack inclusions or other diagnostic features (see Box 10.7).

Diamond consumption is tied directly to disposable income. The United States is the world's largest consumer of gem diamonds, and consumption in China and India continues to increase in parallel with the rapid expansion of their

BOX 11.4 | WORLD DIAMOND PRODUCTION AND THE CENTRAL SELLING ORGANIZATION

The world's diamond trade was historically controlled by the Central Selling Organization (CSO), which was founded in 1888 and operated by De Beers Centenary (van Zyl, 1988; Bergenstock, 2006). This company perpetuated the early philosophy of Cecil Rhodes that a stable diamond market benefits everyone from producers to consumers. Before Rhodes imposed his vision on the industry, it was plagued by price fluctuations spurred by miners who competed for market share by producing more diamonds, often far exceeding demand, which fluctuated considerably with economic conditions. Although several groups competed with the CSO over the years, most eventually joined it. The CSO even handled sales for the old USSR and after the breakup, it contracted to represent the Russian province of Yakutia, the major source of Russian diamonds. Production from the Democratic Republic of Congo, Australia, and other locations was also represented by the CSO, which was estimated to control about 90% of world diamond sales through the 1990s. The CSO was widely regarded as the most effective cartel in the mineral industry (Austin, 1991). Between 1949 and 1990, it increased the asking price for rough diamonds sold to cutters and polishers by 1,800% and since then prices of large diamonds have doubled. In contrast, the price of industrial diamonds decreased during this period because of competition from synthetic stones.

The CSO rigidly controlled the gem-diamond market, selling packets or groups of uncut diamonds for its clients at "Sights" in London, Lucerne, and Kimberley that were attended only by approved bidders. It expanded the market for its diamonds through very effective advertising campaigns and maintained a $2 to $5 billion stockpile of surplus gem diamonds. Even with this support, the price of diamonds falls barely on the high side of the consumption-price correlation followed by most mineral commodities, including gold and the colored gems (Figure 11.19b). By the end of the twentieth century, production began from new diamond deposits that were not affiliated with the CSO, and DeBeers' control began to decline from about 90% market share in 1987 to about 35% today. At present, much of the production from Australia, Canada, and Russia is sold outside the DeBeers group.

middle class. One interesting sign of the value that is placed on diamonds is the commercial development of a process to make a synthetic diamond from the ashes of deceased humans and animals, a service now offered in MDCs.

World gem and near-gem diamond reserves and resources are estimated to be about 2.3 billion carats, only about six times higher than annual world production. As discussed further in the section on industrial diamonds, it is not likely that synthetic gem diamonds will help make up this impending shortage. Diamond demand worldwide is predicted to grow 5% annually for the next decade. This has not gone unnoticed by the world mining community and exploration is underway to capitalize on it. Eighteen newly discovered deposits currently in the evaluation stage could add up to 18 million carats to the supply chain.

11.4.2 Colored gems

Colored gems, especially emeralds, aquamarines, rubies, and sapphires, which are discussed here, are an even smaller business. Natural deposits of these and other colored gem stones amounts to about $2.5 billion (Olson, 2014), a small number compared to the other commodities in this group. The rarity of these minerals has encouraged efforts to make synthetic gems (Kane, 2009).

The beryl group – emeralds and aquamarines

Beryl, a relatively common beryllium–aluminum silicate, forms several important gem stones where it develops crystals with good color and few imperfections. These include green emeralds, pale blue or bluish-green aquamarines, yellow heliodor, and pink morganite. The different colors of these beryls are commonly ascribed to differing contents of trace elements, known as **chromophores**, that substitute in their crystal structures. The most common chromophores are chromium and vanadium for emerald, iron for aquamarine, manganese and iron for morganite, and manganese, iron, and titanium for heliodor (Sinkankas, 1989; Rossman, 2009). Prices of these stones vary greatly, with some high-quality emeralds commanding up to $100,000 per carat, and average prices for cut emeralds of $2,400–4,400 in 2011 (Groat et al., 2005; Olson, 2014).

Gem beryls are found in beryllium deposits, which have already been discussed. The best gems come from pegmatites and hydrothermal veins that formed at deep crustal levels, where slow cooling probably permitted growth of the near-perfect crystals typical of gems (Groat et al., 2008). The famous emeralds of Colombia are found in narrow calcite

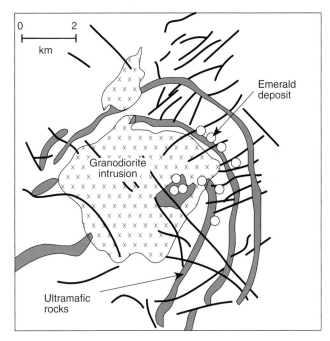

Figure 11.20 Geologic map of the Carnaiba emerald district in Brazil showing the concentration of deposits in a chromium-rich ultramafic rock that surrounds part of a granodiorite intrusion (compiled from Guiliani et al., 1990; 1995; Branquet et al., 1999)

veins that cut carbonaceous shales. In some areas, availability of the chromophore appears to be an important factor. For instance, emeralds are found where veins and pegmatites around granitic intrusions cut ultramafic rocks that are relatively rich in chromium (Figure 11.20). In other deposits, such as Santa Terezinha de Goias in Brazil, the beryls are not found in well-developed veins and appear, instead, to have been deposited by hydrothermal solutions that pervaded the entire rock (Guiliani et al., 1990). For reasons that are not entirely clear, emeralds tend to be smaller and to contain more inclusions than aquamarines, which have been found in very large stones.

Emeralds are mined largely from Brazil, Colombia, Russia, Zambia, and Zimbabwe (Figure 11.14) and aquamarines come largely from Brazil. The only emerald mine in the United States produces red emeralds, which demand prices higher than diamonds. Most production is from bedrock deposits or related regolith, although some aquamarines come from placers. In all of these deposits, the gems are distributed very irregularly, making it essentially impossible to determine the size and grade of an ore body before mining. In most cases, the veins and pegmatites are too small for use of mechanized mining and most work is done by hand, an arrangement that encourages theft of stones and requires constant supervision by management personnel. As a result, most emerald mines tend to be small (Mowatt, 1992). Colombia produces 90–95% of emeralds worldwide; the government owns two of the three largest mines, and handles emerald sales. Pilferage is a major problem, even forcing intermittent closing of mines.

Emerald reserves are simply not known, even in the deposits that are being mined. In the absence of the discovery of a large, new deposit that can be mined by mechanized methods, emerald and aquamarine production will remain a small business. Most small operations cannot deal with the expense and complexities of deeper mining, suggesting that production will decrease as near-surface deposits are exhausted. Thus, the outlook is not good for future emerald supplies.

The corundum group – rubies and sapphires

Corundum, the oxide of aluminum, forms gems when it occurs in well-developed transparent crystals with good color (VanLandingham, 1985; Schumann, 2006). Red corundum is known as ruby, and blue corundum is known as sapphire. Star sapphire and the less common star ruby are varieties in which a six-rayed cross or star in the mineral is caused by the pattern of light refraction. Chromophores, the elements that add color to the minerals, are thought to be chromium for ruby, and iron and titanium for sapphire. Many other colors of corundum are known, some of which form gems. Prices for fine-quality, 1-carat stones ranged from $4,000–10,000 for rubies to $4,500–7,500 for sapphires in 2013. In addition to its use as a gem, corundum is widely used in industry because of its extreme hardness.

Under most conditions in Earth's crust, corundum (Al_2O_3) is not stable in the presence of quartz (SiO_2), with which it will react to form the minerals kyanite, sillimanite, or andalusite all of which have the same composition (Al_2SiO_5). If quartz and corundum react in the presence of water, they can form kaolinite or pyrophyllite. Because quartz is so common in crustal rocks, these reactions limit the geologic environments in which corundum can form to very special, quartz-free rocks where aluminum is abundant. The most common such rock is bauxite, the ore of aluminum, although the weathering conditions under which bauxite forms are not hot enough to form corundum. Corundum can form if bauxite is metamorphosed, which is a truly rare occurrence. More commonly, corundum forms in other low-silica, high-aluminum rocks such as **peridotite** and limestones that have been altered by hydrothermal solutions of magmatic or metamorphic origin (Garnier *et al.*, 2008).

Ruby and sapphire production is largely confined to Australia, Cambodia, Myanmar, Madagascar, Nigeria, Sri Lanka, Thailand, and Vietnam (Figure 11.14). In most of these areas, production is from placer deposits, which take advantage of the relatively high specific gravity (about 4) of these gems. With the exception of Australia, production is dominated by poorly capitalized, small operations and is highly dependent on local politics. During fighting for control of Cambodia in the 1980s the Khmer Rouge, which occupied the western part of the country, opened the Kanchanaburi ruby fields to Thai miners in an effort to finance their activities. Geological and other uncertainties afflict most ruby-producing areas, making its supply the least secure of the precious gems, and reserves are not known (Shigley *et al.*, 1990).

11.5 The future of precious metals and gems

The real question about precious metals and gems is whether we have reached a major crossroads in their history. For centuries, they have been valued for their beauty and as a safe haven for monetary Cassandras. During this time, we have grown to accept the miner's version of the Golden Rule, "She who has the gold, makes the rules." At the same time, industrial applications for these commodities have increased until they dominate markets for the PGEs and silver, and impact those for gold, diamonds, and even colored gems. Every new renewable-energy project or battery design increases consumption of one of these commodities.

There is talk that the world has outgrown the need for mineral commodities as investment vehicles, and possibly even as ornaments and art objects. Virtual currencies, or cryptocurrencies, such as E-Gold, Bitcoin, Litecoin, Peercoin, and Namecoin have sprung up to take the place of traditional currencies and, in so doing, to push gold, silver, and gems even farther into the monetary background. Some suggest that the availability of so many other financial instruments may siphon money from potential gold investors and synthetic gems are supposed to satisfy the world's gem buyers.

Predictions of this type have been made periodically for centuries, however. If we do turn our backs on these commodities now, sociologists and economists will probably look back on the event as a fundamental transition in human psychology. As egocentric as all generations are about the significance of their own era, it is hard to believe that we have reached that point. So, we will probably continue consuming these commodities as we have before, with increasing industrial uses adding pressure to demand from the arts and investments. To satisfy this demand, exploration is badly needed. No group of mineral commodities has a dimmer reserve outlook than the precious metals and gems, particularly in the face of declining grades and increasing labor challenges. Making matters worse is the fact that relatively few unconventional resources are known for these mineral commodities.

CHAPTER

12 Agricultural and chemical minerals

Agricultural and chemical minerals are the most important of the numerous industrial minerals, a term that includes almost any commercial Earth material that is not a metal or fuel (Harben and Bates, 1984). The spectrum of industrial minerals extends from stone used for driveways to fluorine for your teeth and includes all of the minerals used for fertilizers, glass, and most chemical products. In the following two chapters, we have divided the industrial minerals into those used in agricultural and chemical markets, which are discussed in this chapter, and those used in the construction and manufacturing markets, which are discussed in the next chapter. Although some minerals, such as gypsum, have uses related to both groups, almost all of them have dominant markets on which this grouping is based.

The value of world agricultural and chemical minerals, almost $200 billion, is dominated by nitrogen, lime, and the fertilizer minerals potash and phosphate (Figure 12.1). In order of declining value of world production, the other minerals include salt, sulfur, fluorite, sodium sulfate, iodine, and bromine. The production history since 1960 for these commodities can be divided between nitrogen, iodine, and bromine, which have grown very rapidly, and all the rest that have not (Figure 12.1d, e). Within the laggard group, phosphate, potash, sulfur, salt, and sodium carbonate have not even kept pace with our bellwether, steel. Prices of phosphate, potash, and iodine have greatly exceeded the CPI, but all of the others have fallen behind significantly (Figure 12.1b, c). Production and reserve data for these commodities are given in Table 12.1.

With the exception of nitrogen and fluorite, deposits of the major agricultural and chemical minerals are formed during the normal evolution of a marine sedimentary basin, that is, a basin filled with sediment and seawater or freshwater. The distribution of these mineral resources reflects the location of ancient oceans, particularly shallow areas that extended onto the continents (Reading, 1986). Some of the less important minerals in this group formed by evaporation of non-marine lakes.

The agricultural and chemical minerals are associated with an unusually large number of environmental concerns, reflecting the dual facts that the minerals are highly soluble and are consumed in large amounts. As can be seen in Figure 12.2, consumption of fertilizers and de-icing salt for roads, both highly beneficial but important environmental culprits, has increased tremendously, even though the area of farms and length of roads has not changed significantly. The resulting dispersion of these constituents into soil, surface water, and groundwater is a growing problem, forcing us into a more careful comparison of their relative benefits and disadvantages.

12.1 Chemical minerals

12.1.1 Limestone, dolomite, and lime

The common sedimentary rocks limestone and dolomite account for almost 70% of US crushed stone production, as discussed further in the next chapter. They are also our main source of calcium and magnesium for chemical applications, the most important of which is **lime** (CaO) (Pressler and Pelham, 1985; Kogel *et al.*, 2006). Lime (CaO) production in the United States, which amounts to about 20 million tonnes, is valued at about $37 million (Beach *et al.*, 2000).

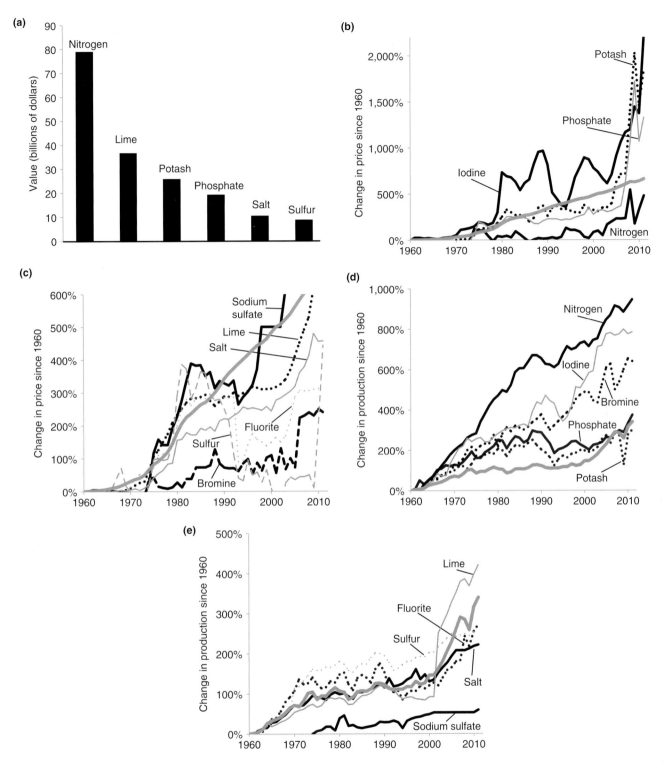

Figure 12.1 Agricultural and chemical minerals. (a) Value of 2012 world production (including that used outside agricultural and chemical markets). (b, c) Change in prices since 1960. (d, e) Growth in production since 1960. Note that only potash, phosphate, and iodine have increased in price at a rate significantly greater than the consumer price index (CPI) and that only nitrogen, bromine, and iodine have increased production at a rate significantly greater than steel (from data of US Bureau of Mines). Heavy gray lines in (b) and (c) show CPI and those in (d) and (e) show steel production.

Table 12.1 Major producers and reserves (millions of tonnes) for fertilizer and chemical minerals (from the US Geological Survey Commodity Summaries)

Commodity	Country	Production (tonnes)	Reserves (tonnes)
Lime	China	220,000,000	
	United States	18,800,000	
	India	15,000,000	
Salt	China	71,000,000	
	United States	40,100,000	
	India	18,000,000	
Sulfur	China	10,000,000	
	United States	9,100,000	
	Russia	7,300,000	
	Canada	6,000,000	
Potash	Canada	10,500,000	1,000,000
	Russia	5,300,000	600,000
	Belarus	4,900,000	3,300,000
	China	4,300,000	210,000
Phosphate	China	97,000,000	3,700,000
	United States	32,300,000	1,100,000
	Morocco (West Sahara)	28,000,000	50,000,000
Nitrogen	China	46,000,000	
	India	12,000,000	
	Russia	10,000,000	
Fluorite	China	4,300,000	24,000,000
	Mexico	1,240,000	32,000,000
	Mongolia	350,000	22,000,000
Iodine	Chile	18,000	1,800,000
	Japan	9,400	5,000,000
Sodium sulfate	United States	860,000,000	
	Spain	180,000,000	
	Mexico	170,000,000	
Bromine	Jordan	200,000	
	Israel	180,000	
	China	100,000	

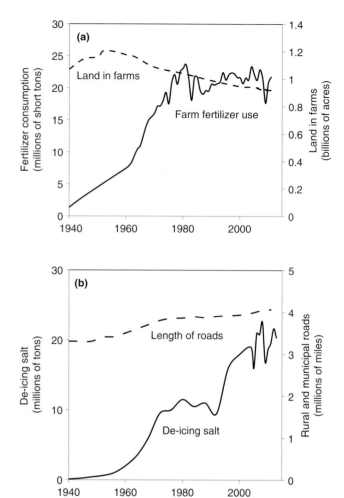

Figure 12.2 (a) Fertilizer consumption versus farmland in the United States. (b) De-icing salt use versus length of highways and rural roads. (Data from the Salt Institute http://www.saltinstitute.org/.)

Limestone and dolomite deposits and production

Limestone is a rock made up of calcite and aragonite, which have the same composition ($CaCO_3$) but slightly different crystal structures, and **dolomite** [$CaMg(CO_3)_2$], with lesser amounts of chert, apatite, pyrite, hematite and **sand**, **silt**, or **clay**. Pure limestone ($CaCO_3 > 95\%$) is known as high-calcium limestone and that with a high proportion of magnesium ($MgCO_3 > 86\%$) is known as high-magnesium dolomite. If the limestone contains silicate fragments, the rock is **marl**

(Carr and Rooney, 1983; Reading, 1986). Limestone and dolomite are widespread on the continents, which were flooded by shallow seas during Paleozoic and Mesozoic time.

Although small amounts of calcite will precipitate from seawater as it evaporates (Table 12.2), most limestones are the direct or indirect product of organic activity (Figure 12.3). Since most forms of life require light, this means that limestones usually form in relatively shallow water. Some limestones consist of the skeletal remains of coral, molluscs, and algae, which form **reefs** and similar structures. Others are composed dominantly of fine-grained material known as **micrite**, which consists of fecal matter, shells of small organisms such as algae, and calcite that was precipitated from seawater, possibly by organic activity. Aragonite and calcite dissolve at ocean depths between 500 and 3,000 m and 750 and 4,300 meters, respectively, varying among ocean basins and latitude, preventing accumulation of carbonate mud in extremely deep seawater (Blatt *et al.*, 1980; Reading, 1986).

Dolomite is not usually deposited as a primary sediment. Instead, it is an alteration product of limestone (Figure 12.3), which undergoes a wide range of chemical changes, collectively known as **diagenesis**, after it is deposited. The most important of these is dolomitization in which magnesium-bearing water replaces calcium with magnesium to form dolomite (Machel, 2004). Solutions that can form dolomite include evaporated seawater, such as might form in a saline lagoon, and freshwater that reacts with seawater-saturated limestones, such as might form where meteoric water collects on an island (Figure 12.3). Dolomite can also form where magnesium-rich hydrothermal brines invade limestones as happens in some Mississippi Valley-type (MVT) lead–zinc deposits. Other minerals that form in limestones include

chert, a form of silica, and pyrite, both of which are undesirable impurities from a commercial standpoint (Blatt *et al.*, 1980).

Even with these quality restrictions, there is plenty of suitable material available. Limestone is mined in 33 US states. Perhaps surprisingly, about 6% of US mines, mostly in Illinois, Missouri, and Kentucky, are underground to minimize environmental disturbance (Freas *et al.*, 2006). Lime is made by **calcining**, the process in which a gas is driven off from a solid by heating. In this case, limestone is heated to 700–1,100 °C to drive off CO_2 (Boynton, 1980; Boynton *et al.*, 1983). Depending on how much magnesium was present in the original limestone, this leaves either quicklime (CaO) or dolomitic quicklime ($CaMgO_2$). Water can be added to these to produce hydrated lime [$Ca(OH)_2$] or its dolomitic equivalent. The manufacture of lime produces particulates, SO_2, and CO_2 emissions, that total considerably less than those from smelting, cement production, and refining, not to mention electric power generation.

The main concerns of the lime industry are energy consumption, pollution control, and the gradual loss of mineable land to population growth, as discussed in the next chapter in connection with aggregate mining. The high cost of lime, relative to other industrial minerals, reflects the energy involved in heating limestone to drive off CO_2 to make lime. Estimates of energy requirements are 3.6–7.5 GJ/tonne in the EU (IPCC, 2001), 7.2 GJ/tonne in Canada (Miner and Upton, 2002), and up to 13.2 GJ/tonne for small kilns in Thailand

Table 12.2 Percentage (by weight) of various minerals deposited from seawater

Mineral or group	Percentage
Calcite	0.3
Gypsum	4.5
Halite	77.2
Potash	1.4
Other salts	16.5

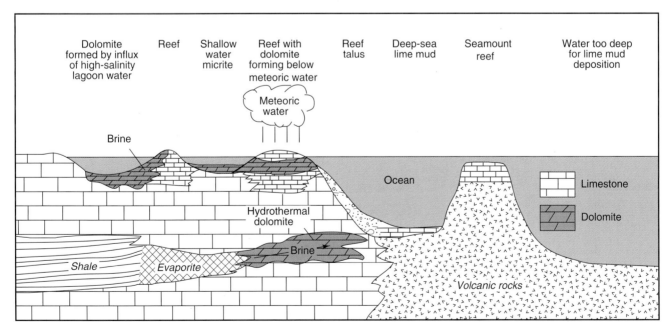

Figure 12.3 Schematic illustration of geologic environments in which limestone and dolomite form

(Dankers, 1995). Lime producers must limit filterable particulate matter (PM) from kilns to 0.12 pounds PM/tonne of limestone for dry systems and 0.60 pounds PM/tonne of limestone for kilns equipped with wet scrubbers (EPA, 2003b).

Limestone and lime markets and trade

Lime's largest market is in steel making, where it is used as the **flux** that lowers melting temperatures and scavenges impurity elements such as phosphorus, silicon, and aluminum into the slag. The use of lime for steel making declined rapidly in the 1980s (Figure 12.4) owing to the declining capacity in the US steel industry and to changes in furnace technology from open-hearth furnaces to basic oxygen furnaces, but has recovered in part (Miller, 2011). Lime is also widely used to control the pH of natural and industrial chemical processes because it consumes H^+ (Table 3.4). It has been used for this purpose in processes ranging from beneficiation of copper ores to purification of drinking water. Lime is also essential in paper and aluminum manufacturing and is the main source of calcium for manufacture of calcium hypochlorite bleaches and other chemicals. It has recently been used to make precipitated calcium carbonate (PCC), a very fine-grained form of calcite. Although it might seem self-defeating to mine limestone, calcine it, and then convert it back to calcite, the shape and size of the newly precipitated grains can be controlled so well that they make an excellent filler and extender, a market discussed in the next chapter (Pressler and Pelham, 1985). The use of lime to mitigate air pollution, as regulated by the US Clean Air Act, is anticipated to have continued strong market growth. Lime is used to remove >95% of SO_2 and >99% of HCl from flue gases of power plants and smelters,

and can also be mixed with activated carbon to control mercury emissions (National Lime Association, 2014). Current consumption of lime for this purpose, almost 5 million tonnes, has exceeded that used in metallurgical markets (Figure 12.4).

The future market for lime appears strong. Globally, annual demand for lime increased by almost 20% per year between 2005 and 2011, and demand is predicted to reach nearly 80 billion tonnes in 2016 (Market Wired, 2012). Steel is likely to remain the dominant use of lime, with consumption for environmental markets continuing to increase, especially for flue-gas desulfurization (Reuters, 2011). Lime must still compete with its own raw material, limestone, which is used for many of the same applications even though it reacts more slowly and weighs more. High energy costs favor limestone because of the heat needed to produce lime. Limestone consumption is also expected to increase to meet the demands of the construction industry and cement manufacturing. It must compete with **fly ash**, dust recovered from cement manufacturing, and other industrial wastes that have many of the same properties and that are themselves searching for a market now that they cannot be spewed into the environment. As a result, lime manufacture and marketing will remain highly competitive and will be especially constrained by energy costs.

12.1.2 Salt

Salt protects your car at the same time that it destroys it. About 45% of US salt production, almost 17 million tonnes in 2014, was used to remove ice from roads in winter (Kostick, 2013a). The second leading use of salt, at about 43% is in the preparation of sodium and chlorine chemical products, largely chlorine gas and caustic soda (NaOH). Remaining salt production is used as a food additive, in agricultural products, water purification, and other applications (Kostick, 2013a). About 300 million tonnes of salt worth about $10 billion are produced in the world each year.

Salt deposits and production

Salt, which is the mineral halite (NaCl), is extremely abundant. It makes up about 2.9% of seawater and is widespread as rock salt deposits that formed where the ocean evaporated. Deposits formed by evaporation of water of any type are known as **evaporites**. Marine evaporites, which are by far the most common, have two main forms, bedded salt and salt domes (Lefond and Jacoby, 1983; Warren, 1989; 2006).

Bedded salt consists of layers of salt that preserve the original depositional form of the sediment (Figure 12.5). Bedded salt deposits can be hundreds of meters thick and

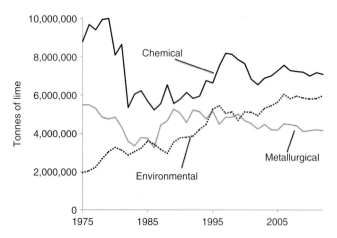

Figure 12.4 Lime use in the United States since 1975 showing growing environmental markets and relatively static chemical and metallurgical (steel-making) markets (from data of the US Geological Survey)

BOX 12.1 | LAKE LIMING

A small but growing amount of limestone is used for liming to mitigate acidification of surface waters, and to reduce the toxic effects of metals such as aluminum, copper, cadmium, lead, and zinc that can threaten aquatic species and human health (Helfrich *et al.*, 2009). Although limestone is favored because of its low cost, lime, quicklime, soda ash, and sodium carbonate have also been used for this purpose. Material is usually added in slurry form, with dosages ranging up to 50 tonnes per hectare of lake surface. Liming causes an abrupt increase in pH along with an increase in dissolved calcium as well as a decrease in aluminum, an agent of fish mortality in acid lakes (Figure Box 12.1.1). A rule of thumb is that liming lasts for about three to five times the residence time, which is less than a year for most shallow northern lakes. Despite the cost and inconvenience, liming has been carried out in thousands of lakes in Canada, Norway, Sweden, and the United States with results that are widely regarded as favorable (Hasselrot and Hultberg, 1984; Lessmark and Thornelof, 1986; Wentzler and Aplan, 1992; Clair and Hindar, 2005). One case study focused on White Deer Lake and Bruce Lake, both about 50 acres and located in Pennsylvania (Stevens, 1989). Through the 1980s, aquatic life in each lake declined owing to acid rain. Over several years in the mid 1980s, White Deer Lake was limed, while Bruce Lake was not. Within the first 3 years of liming, acid-sensitive organisms such as clams, snails, zooplankton, and diatoms completely re-established themselves in White Deer Lake, whereas the same species continued to decline in Bruce Lake. Diatoms, which are discussed in the next chapter, provide a good record of long-term pH changes in lakes.

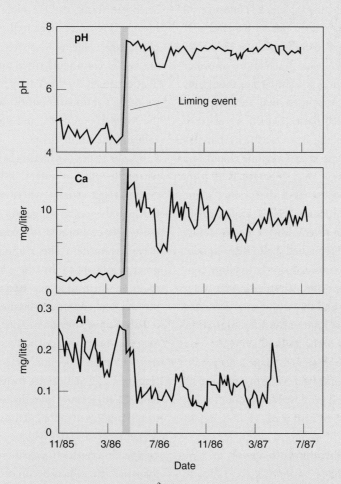

Figure Box 12.1.1 Changes in chemical characteristics of 1 km^2 Loch Fleet, Scotland after addition of 300 tonnes of limestone to streams and bogs draining into the lake. Note the abrupt decrease in aluminum and increase in calcium and pH (based on data of Longhurst, 1989).

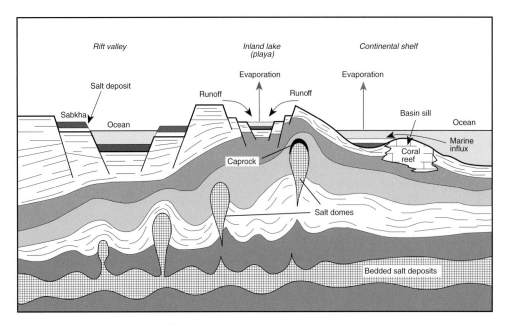

Figure 12.5 Schematic illustration of processes that form evaporite deposits

their genesis presents a real geologic problem. For instance, all of the seawater in a basin 10,000 m deep, deeper than any part of the modern ocean, would have to evaporate to form a layer of salt only 150 m thick. Obviously, much more seawater needs to evaporate to form salt deposits that are thousands of meters thick. It has been proposed that this happens because evaporites form in silled basins, which are arms of the sea that are isolated from the open ocean by barriers known as sills. Basins of this type in arid regions would lose more water through evaporation than they received from incoming seawater or freshwater, thus becoming saline enough to deposit evaporite minerals. Silled basins might form thick evaporite deposits in several ways. One possibility is that water depths in the basin remained shallow throughout its life but that the basin floor subsided slowly to allow continuous deposition of new evaporites. Another is that water depths in the basin were deep and that evaporites deposited from dense brines formed by continuous evaporation at the surface and sank to the bottom of the basin. Finally, the deep basin might have evaporated to dryness to make a deep, salt-covered valley fed by a cascade of seawater flowing over the sill from the open ocean, as has been proposed to account for salt deposits in the Mediterranean Sea (Hsü *et al.*, 1977; Schmalz, 1991; Schreiber *et al.*, 2007).

Several different geologic environments appear to have created evaporite-depositing silled basins (Figure 12.5). In Paleozoic time, the ocean flooded much of the continents and arms of these seas were isolated by reefs, as in the Williston basin of Saskatchewan and North Dakota, the Michigan Basin, the Permian Basin of west Texas and New Mexico (Figure 12.6), as well as the Zechstein Basin of northern Europe. During Mesozoic time, evaporite basins formed where the continents rifted apart, forming valleys that were flooded slowly by the sea, such as the Danakil depression, the southern continuation of the Red Sea rift between Egypt and the Arabian peninsula, and the Rhine **graben** that forms the valley of the Rhine River.

Evaporite deposits have also formed from freshwater lakes where the influx of water fell short of the amount removed by evaporation, usually where rifts in the continents never reached the ocean (Figure 12.5). In desert areas, lakes in these rifts evaporated to form salt accumulations known as playa evaporites. The composition of playa evaporites is greatly influenced by the chemistry of precipitation and rocks in their drainage basins and can be rich in boron, lithium, and sodium carbonate.

Salt domes, the other major type of salt deposit, are tabular and tear-drop shaped bodies of salt that intruded upward from buried bedded salt into overlying sediments (Figure 12.5). Salt domes form because the specific gravity of halite (2.16) is more than that of unconsolidated sediment (1.2 to 1.8) but less than that of most consolidated sediment (> 2.4) As salt layers and newly deposited clastic sediment are buried, compaction causes the clastic sediment to become more dense than the salt, which becomes buoyant. Salt domes are most common where the original bedded salt has been relatively deeply buried, as around the North American Gulf Coast and the Persian Gulf, and they can rise to the surface (Figure 12.6).

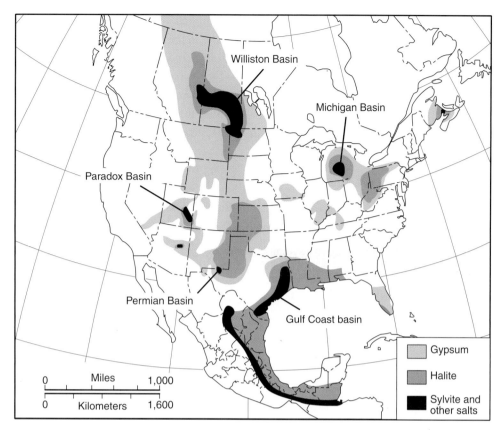

Figure 12.6 Distribution of evaporite basins in North America showing location of major salt and potash (sylvite) deposits.

In addition to these rock-salt deposits, salt can be obtained directly from the ocean, saline lakes, basinal brines, and salt springs, which form where meteoric water passes through salt layers on its way to the surface.

Salt is extracted from conventional open pit and from underground mines with unusually large stopes (Figure 12.7a). It also comes from solution mines, in which it is dissolved *in situ* by water circulating through a system of wells (Morse, 1985). Sea salt and some spring and playa salt are also harvested from evaporation ponds. Abandoned salt mines have been used for storage of documents, processing and industrial wastes, and summer storage of oil and natural gas (which allows more efficient use of pipelines and other transportation facilities during off-peak periods). The US Strategic Petroleum Reserve is in salt domes along the Gulf of Mexico coast in Texas and Louisiana. Salt deposits have also received attention as disposal sites for nuclear wastes. For example, plutonium waste from weapons production is stored in Permian salt beds in the Waste Isolation Pilot Plant (WIPP) in New Mexico, and these salt deposits are being studied as candidates for long-term storage of commercial radioactive waste (Wald, 2014).

Salt processing begins with crushing, grinding, and sizing. Impurities such as gypsum, anhydrite, and shale, are removed by scanning a falling curtain of small grains and ejecting non-translucent grains with an air jet, or by heating all of the grains, causing darker waste grains to become hotter and stick to a resin-coated belt (Bleimeister and Brison, 1960). Salt for human consumption is dissolved and recrystallized in a multi-step, vacuum process that uses steam from earlier steps to heat later ones. Wastewater from processing creates important disposal problems and is sometimes injected back into unused parts of salt mines.

Salt markets, trade, and environmental issues

The advantages and problems of salt center on its use in removing ice from roads (Terry, 1974; Carlson, 1986), an application that has grown by almost 20,000% in the United States since 1940 (Figure 12.2). The use of de-icing salt is based on its ability to depress the freezing temperature of water. Grains of salt put on roadways react with small amounts of water, gradually melting surrounding ice. De-icing is considered a necessity for the United States, where a 1-day shutdown owing to icy roads results in an economic loss ranging from a low of $62 million for Iowa to $600 million for New York (Feldman, 2006).

De-icing salt has made an immense contribution to the safety and convenience of winter driving. Studies in Germany

consumption, (Bubeck *et al.*, 1971). Salt increased the density of water in the bay so much that it stabilized the **hypolimnion**, impeding convective overturn. The US EPA recommends that the NaCl concentration of drinking water be limited to between 30 and 60 mg/liter, although some organizations urge limits as low as 25 mg/liter for people on low-sodium diets. These moves have spurred the search for a salt substitute.

Salt is produced by 110 countries, with major production coming from the United States, China, India, and Germany (Kostick, 2013a). Domestic production satisfies about four-fifths of US consumption, and imports the remaining one-fifth from Chile, Canada, Mexico, and the Bahamas. Australia exports 90% of its salt, mostly to Asia, and Europe obtains additional supplies from China and India. Salt reserves are essentially limitless, and production is available from solar evaporation even if energy costs become prohibitive.

12.1.3 Sulfur

About 90% of US sulfur consumption is used to make sulfuric acid, the most important industrial acid and the basis for the sulfur chemical industry (Apodaca, 2013b). Most sulfuric acid is used to make superphosphate fertilizers, with about 70% of total US sulfur consumption going for this purpose. Sulfur used in catalysts for production of high-octane gasoline during petroleum refining accounts for about one-fifth of total sulfur consumption. Smaller amounts are used in acid solutions for base-metal refineries and in sulfur chemicals. World sulfur production amounts to about 70 million tonnes worth about $9 billion. Sulfur has been produced from a wide array of sources, including mined sulfur, which comes from conventional, solid mineral deposits, and recovered sulfur, which is a by-product of other mineral products and processes.

Mined sulfur comes from three sources. The most important are deposits of **native** sulfur in sedimentary rocks (Figure 12.8). These deposits consist of sulfur in the elemental state that forms when **sulfate-reducing bacteria** convert sulfate in gypsum (hydrous calcium sulfate) to H_2S, which then reacts with oxygen dissolved in groundwater to precipitate native sulfur (Barker, 1983; Wessel and Wimberly, 1992). Native sulfur deposits are found where gypsum is abundant. One surprising place is in **caprocks** of salt domes where dissolution of the rising salt by groundwater has created a residue, or cap, of gypsum and limestone. Sulfur-bearing salt domes are not common. Less than 1% of the domes in the large Gulf of Mexico province (Figure 12.9) contained sulfur that could be produced economically. Generally, similar sulfur deposits

Figure 12.7 (a) Underground salt mine at Avery Island showing large openings (stopes) in salt, which is relatively rigid at shallow depths typical of Gulf Coast salt mines (image courtesy of Corporate Archives, Cargill, Incorporated). (b) Continuous-mining machine in the Laningan potash mine in Saskatchewan, which creates a relatively small opening because the potash is not strong. The potash layer occupies the entire height of the photo and the continuous mining machine has cut curved grooves into it as it advanced toward the right side of the photo (photo courtesy of Potash Corporation, Saskatoon, Saskatchewan). See color plate section.

and the United States indicate the accident rate decreased by an order of magnitude within the first 2 hours after salt was applied to winter roads. Unfortunately, this has come at an environmental price. Spreading rates range from 36 to 635 kg salt per mile for a two-lane road (Kelting and Laxson, 2010; Salt Institute, 2013). Saline runoff from the salt corrodes cars, bridges, and other metal structures, kills plants along the roadside, and flows into streams and groundwater where it has become a major contaminant (Schell, 1985; Rowe, 1988; Kelting and Laxson, 2010). Chloride contents of the lower Great Lakes, for instance, have more than doubled since the beginning of the century. The concentration of chloride ion in Irondequoit Bay, which borders Rochester, NY, on the east and is crossed by an interstate highway increased from 30 mg/liter in 1950 to 120 mg/liter in 1970, paralleling de-icing salt

BOX 12.2 | SUBSTITUTES FOR ROAD SALT

Several substitutes for de-icing salt are available, but they also have disadvantages. Calcium chloride is more effective because it melts ice at lower temperatures and produces heat during the reaction. It suffers from the same corrosion and contamination problems, however, and is about four times more expensive than salt because it is manufactured from limestone and salt. Of greater interest is calcium–magnesium acetate (CMA), another manufactured compound, which has the added advantage of being biodegradable into CO_2, water, and calcium–magnesium oxides such as lime (Boice, 1986). Unfortunately, the median cost of CMA is about $1,500/tonne compared with $40/tonne for sodium chloride. Proponents of CMA argue that this price difference is balanced by the cost of salt damage to cars and highway structures, although this argument has had little effect on municipal governments, which can postpone repair expenditures but must remove snow and ice each winter. Kelting and Laxson (2010) assessed the environmental impact of using sodium chloride, CMA, calcium chloride, and magnesium chloride salts to de-ice roads in the Adirondacks region of New York, and found that the environmental cost to surface waters and forests from application of salt was $2,320 per lane mile per year. Adding this to the $924 per lane mile per year spent for the salt brought the cost of salt closer to that of alternatives, although environmental impacts are essentially hidden external costs and 95% of municipalities continue to use sodium chloride (Transportation Research Board, 2007).

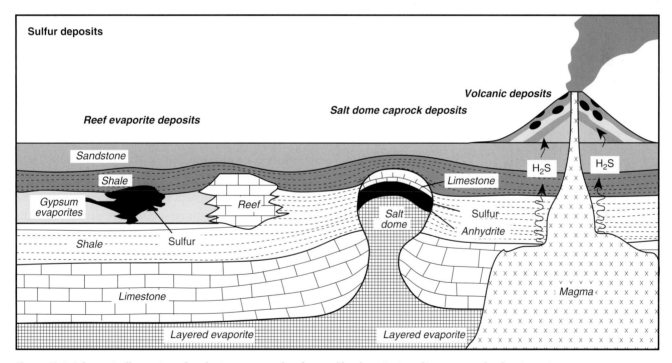

Figure 12.8 Schematic illustration of geologic processes that form sulfur deposits in sedimentary and volcanic environments

are found around bedded gypsum-rich evaporite deposits in the Delaware Basin of west Texas, and in Poland, Ukraine, and Iraq. Native sulfur in volcanic rocks is considerably less common, but has been mined in Japan, New Zealand, and a few other areas of recent volcanism (White, 1968). This sulfur also forms by oxidation of H_2S, although the gas comes from underlying magmas rather than from bacterial reduction of

sulfate. The third source of mined sulfur is metal sulfide minerals such as pyrite, marcasite, and pyrrhotite (Russel, 1989). Deposits of this type have been mined at Ducktown, Tennessee, Rio Tinto in Spain, and in Russia, Japan, and Cyprus. Production of sulfur from sulfide deposits is declining, making up only 7% of global production compared to 17% in 1990 (Ober, 2006b). China, however, remains heavily

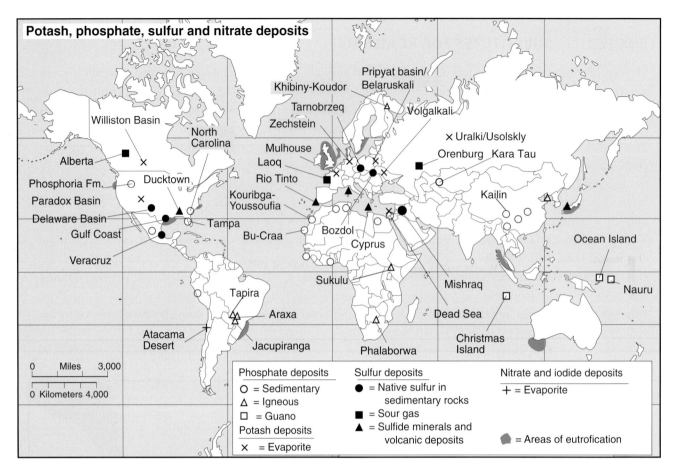

Figure 12.9 Location of important potash, phosphate, sulfur, and nitrate deposits, showing areas of ocean eutrophication

dependent on sulfide sulfur, with half of its production coming from that source as recently at 2011 (Apodaca, 2013b).

Recovered sulfur from oil, natural gas, and smelters now accounts for essentially all of new sulfur production (Figure 12.10). The first source of recovered sulfur was sour natural gas, which was discovered in large amounts in Alberta during the 1950s. As discussed in Chapter 7, even small amounts of H_2S must be removed before gas can be transported and sold. Because sulfur from sour gas is a by-product, it can be sold at lower prices than sulfur from mined sources. The advent of sulfur production from the Alberta gas fields had a major impact on world sulfur prices, which was mitigated only partly by the long rail transport needed to get the sulfur to shipping points in Vancouver. As environmental regulations grew, sulfur was recovered in a marketable form from crude oil, which is now the dominant source of recovered sulfur (Ober, 2002).

Mining of native sulfur has been done by a solution mining technique known as the **Frasch process** (Niec, 1986). This process depends on the fact that native sulfur melts at 112 °C and becomes a less viscous liquid as the temperature rises to about 160 °C (Lee *et al.*, 1960). Thus, superheated water or steam pumped into porous rock containing native sulfur will melt it to form a liquid that can be lifted to the surface by pumping compressed air into the deposit, making a froth of liquid sulfur. The Frasch method works best on ore that is surrounded by impermeable rock, conditions met by most salt-dome and other sediment-hosted sulfur deposits. The method made it possible to mine offshore deposits from drilling platforms in the Gulf of Mexico, although production ceased in 2000 because of competition from recovered sulfur (Ober, 2002). The only operating Frasch mines as of 2011 were located in Poland and Mexico (Sulphur Institute, 2014).

The only important sulfur emissions that are not recovered in these countries come from electric power plants, which seem unlikely to contribute recovered sulfur to the market soon. Most presently used flue-gas systems at power plants recover sulfur as some form of sulfate mineral such as synthetic gypsum, which can be used for wallboard and cement but is not usually used as a source of elemental sulfur. Further pressures could come from less-developed countries (LDCs),

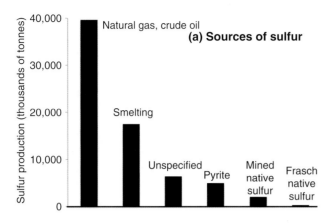

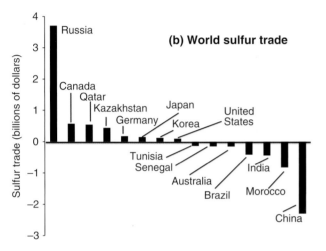

Figure 12.10 (a) Sources of world sulfur production (from Ober, 2006b). The first category includes all sulfur from oil and natural gas processing and refining. The smelting category includes sulfur from production of coke and from the recovery of SO_2 from smelters. (b) World sulfur trade (compiled from UN Comtrade Statistics).

however, if they enforce environmental regulations similar to those in the more-developed countries (MDCs).

Slightly more than one-half of elemental sulfur is traded internationally. Current sulfur trade consists largely of exports by Russia, Canada, Kazakhstan, and Qatar, with each shipping more than 1 million tonnes annually (Figure 12.10). China, the world's leading importer of sulfur, along with Morocco, India, and Brazil each import more than 1 million tonnes annually. The import market is driven by the use of sulfur to produce phosphate fertilizers, a market that is predicted to increase significantly owing to world population growth and food demand (Heffer and Prud'homme, 2014). In 2013, China entered into a partnership with Saudi Arabia to build a refinery at Yanbu, Saudi Arabia, that will supply 438,000 tonnes/year sulfur to China in addition to producing

400,000 barrels of oil per day (Apodaca, 2013b). Although sulfur mines will continue to be of strategic interest, future sulfur supplies will be controlled by recovered sulfur. The quantity of sulfur recovered from oil, gas, and mining operations fluctuates depending on the global demand for these commodities. However, production of sulfur exceeded demand for most of the last two decades (Ober, 2006b), and should be sufficient to meet future demand. Production of oil from the Alberta oil sands, for instance, yielded 9 million tonnes of recovered sulfur in 2012. Proposed reduction of sulfur in gasoline will also result in increased recovery of sulfur. To reduce environmental concerns surrounding long-term storage of sulfur, companies are now advocating new uses of sulfur, including sulfur-enhanced asphalt modifier to improve roadway durability. World resources of sulfur in oil, gas, tar sands, sulfides, coal, evaporite, and volcanic deposits are estimated at about 5 billion tonnes (Apodaca, 2013b).

12.2 Agricultural minerals

12.2.1 Potash

Potash is an industry term that refers to a group of water-soluble salts containing the element potassium, as well as to ores containing these salts. This ambiguity developed because potassium salts were originally recovered by leaching ashes left from burning trees in large pots, a logical source for an element that is essential for plants. The most common mineral in potash is sylvite, potassium chloride, which chemists call muriate of potash. Potash abundances are quoted in terms of yet another compound, potassium oxide (K_2O).

The vast bulk of potash consumption, about 85% in the United States and 95% worldwide, is used in fertilizers, with the rest going into potassium compounds (Jasinski, 2013b). World potash production, which comes from only 13 countries and amounts to about 36 million tonnes, has a value of almost $25 billion (Figure 12.1).

Potash deposits, production, and environmental concerns

Modern potash production comes from evaporite deposits known as sylvinite, which consists of a mixture of halite, sylvite, carnallite, and other potassium, magnesium and bromine minerals known as **bittern salts** because of their bitter taste in comparison to halite (Adams and Hite, 1983; Warren, 2006). The deposits are thought to have formed during evaporation of seawater in restricted marine basins (Schmalz,

1969). Sylvite, carnallite, and the bittern salts are among the last to crystallize from evaporating seawater and they make up a relatively small fraction of total seawater evaporites (Table 12.2). Thus, potash-bearing evaporites are not as common as halite or gypsum, and they usually occupy a more central position in evaporite basins (Figure 12.6).

Sylvinite deposits are so soluble that they are dissolved at the surface and must be mined from deeply buried deposits that were protected from groundwater (Figure 12.5). Unfortunately, sylvinite is not strong enough to support mine openings at great depth (Coolbaugh, 1967; Haryett, 1982). Rock stability has been a particular problem in the kilometre-deep deposits of the Williston Basin in Saskatchewan (Figure 12.9), where openings created by mining close by flow and rock failure within a month or so. Most mines in the area use **continuous miners** (Figure 12.7b), but recover only 35–45% of the ore, leaving the rest in place to prevent collapse and destruction of the mine (Prud'homme and Krukowski, 2006). Additional problems are presented by groundwater, which can dissolve sylvinite and destroy the mine. The water comes largely from formations directly above and below the potash levels, and it enters the mine through faults and collapsed areas created by mining. To make matters worse, shafts that reach the sylvinite must pass through an overlying unit, the Blairmore Formation, which contains sand and water under high pressure. Shafts can be excavated through the Blairmore only by freezing the rock and lining it with concrete and steel shields, which must remain sealed to prevent influx of water (Prugger and Prugger, 1991; Property Saskatchewan, 2014). At the Scissors Creek mine, construction of a 6-m diameter shaft that extends more than 1,100 m deep required drilling 32 holes around the perimeter of the shaft and pumping calcium chloride brine at –35 °C to a depth of 600 m to create a frozen ring around the shaft (Godkin, 2012). The shaft was lined with concrete and steel to prevent influx of water. This technology is now being exported and used to develop new potash mines in places such as the Ust-Jaiwa mine in Russia (Moore, 2012).

The expense of building deep mines and their relatively low recovery of potash have given mines at shallower depths some room to compete. For instance, mines at depths of 200–450 m in the Permian Basin of New Mexico can extract up to 90% of their thinner ore layer without collapse and water problems. Solution mining has been used on sylvinite layers in Saskatchewan and Utah, where water leakage, collapse, or tilting of the layers prevent conventional mining, and it will be the main method for mining deep reserves (Adams and Hite, 1983; Prud'homme and Krukowski, 2006).

Potash can also be recovered from brines that fill pores in marine evaporites, as well as non-marine evaporite such as Searles Lake, California, and the Bonneville Salt Flats west of Great Salt Lake, Utah, which are discussed in Sections 13.2.3 and 13.2.2 on boron and soda ash. In a few areas, including Great Salt Lake and the Dead Sea, saline lake waters are processed for potash and other elements.

Processing of solid potash ores usually involves crushing and grinding, followed by flotation to separate waste sediment and halite from sylvite. More complex ores are dissolved and sylvite or other potassium salts are precipitated from the resulting solutions. Potassium sulfate is a dual nutrient fertilizer providing both potassium and sulfur for plants such as grapes, citrus, tobacco, and vegetables that cannot tolerate the chloride released when sylvite dissolves (Baird, 1991). Potassium lost from fertilizer applications is not as big an environmental contaminant as phosphorus or nitrates. Excess potassium is usually adsorbed onto the surfaces of clay minerals in fine-grained soils, although sandy soils have less ability to retain the element. Increases in dissolved potassium have been observed in surface and groundwater in agricultural areas, sometimes above permissible potassium contents of about 10 mg/liter for drinking water. However, this limit was set because sewage and other municipal wastes are relatively high in potassium. The element itself is not a significant problem (even milk contains about 1400 mg/liter of potassium) and it is not usually a major cause of eutrophication (Bockman et al., 1990; Manahan, 1990).

Potash trade and reserves

Potash is produced in 12 countries, and the most important potash producers, Canada, Russia, and Belarus, together account for almost 60% of world production and 90% of estimated reserves (Figure 12.11a). Germany, where the world's first potash mine was built in 1856, produces from mines in the Zechstein Basin around Göttingen, Hannover, and Stassfurt. The United States imports almost 90% of the potash that is used as fertilizer, largely from Canada and also from Russia, Israel, and Chile (Jasinski, 2013b). China produces only about 1% of global potash output and imports about one-fifth of global potash shipments. To meet China's huge demand for potash, in 2014 the China Investment Corporation, the national sovereign wealth fund, bought a 12.5% stake in Russian producer Uralkai, the world's second-largest potash company (Fedorinova and Kolesnikova, 2012). World trade in potash is highly dependent on transportation costs because so many of the important sources are not near port facilities. Canadian deposits in Saskatchewan have an

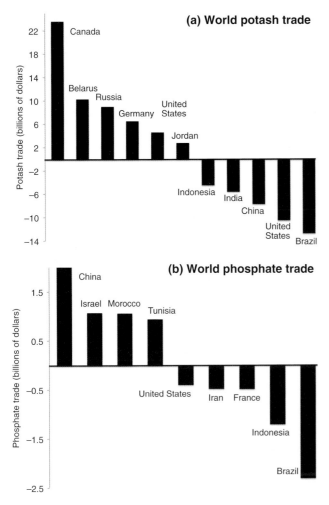

Figure 12.11 World (a) potash and (b) phosphate trade (compiled from UN Comtrade Statistics)

unfortunate location in this respect, but compensate by being high grade and close to the North America prairie agricultural belt. Uralkali in Perm Krai, Russia, is worse, with low ore grades and long distances to ports, especially Vanino and Vostochnyy on the Pacific rim.

Potash reserves of 6 billion tonnes of contained K_2O are enormous in relation to present annual production. Resources are even larger and include extensive deposits below 2-km depths, where information is limited and comes largely from oil and gas drilling. Deeper parts of potash-hosting basins with little hydrocarbon potential remain poorly explored. In terms of immediately mineable potash, Belarus rules the roost, with slightly more than 50% of world reserves. Potash consumption is expected to increase significantly, especially in developing markets such as Asia and Latin America, as world population increases, and active exploration and mining activities are being pursued in places such as Australia, Brazil, and Ethiopia.

12.2.2 Phosphate

Phosphate is the chemical form in which the element phosphorus is mined and traded. Phosphorous is a critical nutrient for all life, where it is a critical component of the genetic material ATP, DNA, RNA, and phospholipid cell membranes (Miller and Urey, 1959; Emsley, 1980; Madigan *et al.*, 2014). Its role is so important, in fact, that Isaac Asimov once wrote, "We may be able to substitute nuclear power for coal power, and plastic for wood, and yeast for meat, and friendship for isolation – but for phosphorus there is neither substitute nor replacement" (Asimov, 1974). As much as 90% of world phosphate production of about 224 million tonnes is used as a fertilizer, about 5% is used for animal feed supplements, and 5% for products ranging from detergents, fire retardants, toothpaste, and other products (FIPR, 2010; Jasinski, 2013a). At current prices of $100/tonne, world phosphate production is worth almost $20 billion and the value of fertilizers prepared from them is many times that (Figure 12.1).

Geology of phosphate deposits

Naturally occurring phosphorus is found largely in the mineral apatite $[Ca_5(PO_4)_3(OH,F)]$, which is a common constituent of most rocks, as well as skeletal material in many organisms, including humans. Phosphate deposits, which are concentrations of various forms of apatite, are found in three very different geological settings.

About 85% of world phosphate production comes from sedimentary deposits, which consist of accumulations of apatite that formed by biologic activity, and the remaining 15% is from igneous alkalic rocks (Emigh, 1983; Slansky, 1986; Notholt *et al.*, 1989; Zhang *et al.*, 2007; Danielou, 2012). Sedimentary phosphate deposits form on the continental shelf or slope where organic productivity is high and there is limited influx of clastic sediment (Figure 12.12). In some cases, productivity is stimulated by upwelling currents carrying phosphorus-rich cold waters from deeper ocean levels. Ancient deposits of this type are found where deep-water shales grade into shallow-water limestones along the margin of an old continental shelf. Some deposits also form where rivers with abundant phosphate flow into estuaries. Most sedimentary phosphate deposits consist of layers of organic debris such as fishbones, teeth, and plankton shells that are interbedded with shale and other chemical deposits, such as chert and diatom-rich silica sediment. As the organic debris is buried, phosphorus is dissolved and re-precipitated to form pellets of fine-grained apatite. Wave action winnows out non-phosphate sediment and moves the pellets into layers of mineable grade and thickness. Where deposits are exposed to later weathering, phosphate will be further concentrated because it is less soluble than the limestone in which it is usually enclosed.

The formation of sedimentary phosphate deposits is highly dependent on ocean currents, which are controlled by the distribution of continents and land. Plate-tectonic processes have had a strong effect on the location of ancient phosphate deposits. Many ancient deposits occur in long zones that were once along continental margins (Figure 12.9). Deposits in Idaho, Montana, and adjacent states are in the Permian-age Phosphoria Formation, a 330,000 km^2 area that formed the shelf along the western margin of North America at that time. In Florida, Georgia, and the Carolinas, phosphate deposits are found in younger, Miocene-age sediments that accumulated in small basins along the margin of the continent (Cathcart *et al.*, 1984). Along the northern and western margins of Africa,

a major phosphate province occupies the shelf that formed as North and South America separated and as Africa separated from Europe. The largest of these deposits is in western Sahara (Figure 12.9), a former Spanish colony administered largely by neighboring Morocco, which was attracted to the area by the potential revenues from phosphate production. Mining from this area is a source of considerable controversy (Wearden, 2010).

The final type of phosphate deposits are igneous phosphate deposits that are rich in apatite (Krauss *et al.*, 1984; Stowasser, 1985). Apatite is present in trace amounts in most igneous rocks, but it is abundant only in rocks that are poor in silica, such as syenite and the carbonate-rich magmas known as **carbonatite**. These magmas usually form along rift zones that cut older continental crust and they are mined at Jacupiranga and Araxa in Brazil, in the Kola peninsula of Russia, and at Phalaborwa in South Africa (Figure 12.9).

Phosphate production, use, and environmental concerns

Phosphate is extracted largely by surface mining (Figure 12.13), although limited underground mining has been used in Russia, Tunisia, and Morocco. Standard beneficiation techniques separate phosphate minerals from sand and clay to form concentrates. Because plants cannot release phosphorus easily from apatite, concentrates destined for fertilizer markets are dissolved in sulfuric acid to make phosphoric acid and other soluble forms of phosphate known as **superphosphate** (16–21% P_2O_5) and **triple superphosphate**

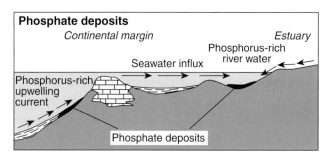

Figure 12.12 Schematic illustration of processes that form phosphate deposits in marine and marginal environments

---

BOX 12.3   GUANO

Guano, bird and bat excrement, used to be the dominant phosphate source but has largely been exhausted. Before European settlement of the Americas and Australasia, it covered the floors of caves and the entire surface of many small ocean islands in tropical areas. Modern guano deposits are concentrated along zones of upwelling associated with El Niño events (where abundant marine life provides food for birds), and are therefore directly related to the upwelling currents to form modern and ancient sedimentary phosphate deposits. Guano deposits were widely mined during the 1800s and formed the main source of fertilizer to the sugar-cane growers of northern Australia and the West Indies. The largest deposits were on Christmas Island in the Indian Ocean, where production ceased in 1991, and on Ocean and Nauru islands in the Pacific Ocean (Figure 12.9). Phosphate mining on Nauru removed material from approximately four-fifths of the 21 km^2 island, and was responsible for up to 80% of local income throughout the twentieth century (Gowdy and McDaniel, 1999). In 1990, per capita income for the 9,000 citizens of Nauru amounted to about $20,000, largely from phosphate mining. However, by the early twentieth century, phosphate reserves on Nauru were exhausted, and in 2006 per capita income had dropped to $2,564 and unemployment had reached 90% (Asian Development Bank Nauru 2015; Das, 2013), a testament to the dangers of **Dutch disease** when the economy is almost completely dependent on one natural resource.

Figure 12.13 (a) Dragline moving overburden from phosphate mine in central Florida. (b) Overview of phosphate mine in central Florida showing active mining in left corner and two settling ponds where land is being reclaimed (lakes in center). Most other land has already been mined and reclaimed, usually as wetlands, the form in which it began. (c) Close-up view of reclaimed wetland on the site of a former phosphate mine in central Florida (photographs courtesy of Mosaic Corporation). See color plate section.

(43–48% P_2O_5). Concentrates intended for chemical markets are heated in enclosed electric furnaces in which phosphorus vaporizes and is then condensed to make phosphoric acid (Emigh, 1983; Stowasser, 1985).

Florida accounts for almost 80% of US phosphate mining and is one of the major sources of world phosphate (Figure 12.13). Mining in Florida takes place in densely settled land and deals with a wide range of environmental challenges (McFarlin, 1992; Beavers, 2013). The phosphate-bearing layers are 4–5 m thick and are covered by 3–6 m of overburden. About 1,190 km^2, slightly less than 1% of the land area of Florida, had been mined for phosphate by 2000 (FIPR, 2010). Ore is beneficiated by flotation of apatite and tailings, which contain only about 5% solids, and must remain in settling basins for about a year to reach a level of 20% solids that is sufficient for disposal. This long delay reflects the very fine-grained nature of the phosphate ore, which contains clay minerals that form a colloidal suspension in the wastewaters. Efforts to speed the dewatering are complicated by the fact that any reagent added to the tailings to flocculate the colloids could also become a contaminant. Mining and processing wastes also contain uranium, thorium, and fluorine. Some fluorine is recovered during processing as H_2SiF_6, which is used largely in fluoridation of municipal water. Uranium is associated with its entire suite of radioactive decay products, including radon gas. Even if by-product uranium is recovered, as it was during the 1970s and 1980s when uranium prices were high, daughter products remain in the wastes. Anomalous radon levels were observed in homes built on older reclaimed land, but this problem has been alleviated by installation of adequate ventilation. Reclamation in modern mines largely eliminates the problem because most of the natural radioactivity is concentrated in a stratum found just above the ore level, and this radioactive rock is put in the bottom of the reclaimed pit, buried at a level even deeper than it is in unmined ground.

The final concern of phosphate producers is **phosphogypsum**, a waste material formed during production of superphosphate and other fertilizer materials. Production of 1 tonne of P_2O_5 yields about 5 tonnes of phosphogypsum. About 32 million tonnes of phosphogypsum are generated annually by US phosphate mining operations, and Florida alone has accumulated about 1 billion tonnes (Zhang *et al.*, 2006; Beavers, 2013). Phosphogypsum is similar to gypsum, except that it contains about 1% P_2O_5, 1% F, and 10 to 30 times more radon, none of which is desirable. These properties, and the radon scare of the 1980s, resulted in a 1989 Environmental Protection Agency (EPA) ruling that phosphogypsum is unsuitable for sale as common gypsum. Research indicates that phosphogypsum is suitable for road construction and agriculture; however, it is currently treated as waste. Phosphogypsum waste is stored in large piles, known as gyp stacks, that must be underlain by an impermeable clay layer and a plastic liner, and they have to be covered with a plastic membrane to retard escape of radon and leaching by groundwater. Water used to transport gypsum slurries to the gyp stacks is acidic, with high dissolved solids, and must be recirculated to the plant or treated with lime before being released.

The use of phosphate has also come under scrutiny. Most attention has been given to its role in stimulating growth of algae and other organisms in surface water, a process known as **eutrophication** (Bowker, 1990; Manahan, 1990; Correl, 1998). Eutrophication can be a deleterious process if it causes blooms of algae, which consume dissolved oxygen when they die. The process is widespread in lakes near populated areas and is even observed in shallow, isolated arms of the ocean (Figure 12.9). Although all nutrients contribute to organic productivity and eutrophication, phosphorus is usually the limiting one because its natural concentration is the lowest, making additions from pollution that much more important.

As it turns out, phosphate fertilizers are probably not the only or even the major cause of phosphate-induced eutrophication. Fertilizer phosphate does not leach readily from the soil, partly because it is adsorbed onto clay minerals or combines with calcium in soil waters to form apatite (Bockman *et al.*, 1990; Borda *et al.*, 2014). In fact, after 100 years of fertilization, phosphate in some soils has migrated downward only about 20 cm. In addition to internal eutrophication, which does not require addition of new material (Smolders *et al.*, 2006), a major culprit appears to be municipal wastewater. Some wastewater contains 25 mg/liter of dissolved phosphates coming largely from detergent, in contrast to permissible concentrations of less than 0.5 mg/liter for surface water. One of the best ways to remove this phosphate is the addition of lime, which causes precipitation of apatite, although the procedure is relatively costly and has not been widely applied. New York State passed laws in 2010 and 2013 that regulate the use of phosphate fertilizer on lawns and in detergents, respectively. Fertilizers that contain phosphorous, nitrogen, or potassium cannot be applied between December 1 and April 1, and can only be applied to new lawns or to existing lawns if a soil test indicates a deficiency in natural phosphorus. The law does not apply to farmland or gardens, however. The Dishwasher Detergent and Nutrient Runoff Law, amended in 2010, bans the sale of phosphorus-containing dishwater

detergent and other household cleaning products. Seventeen states have followed suit to ban the use of phosphate in detergents (Shogren, 2010).

Phosphate trade and reserves

Phosphate production comes from about 37 countries, although about 90% comes from China, the United States, Morocco, and Russia (Figure 12.1b; Danielou, 2012). China consumes almost all domestic production, which grew rapidly from 16% of the world total in 1992 to 32% in 2011. In fact, much of the growth of the phosphate market from 1992 to 2011 was driven by local consumption, whereas global trade remained relatively constant at about 30 million tonnes per year (Danielou, 2012). Domestic production meets about 90% of US consumption, and US reserves are adequate to supply about 40 years at present consumption levels (Jasinski, 2013a). Morocco leads the global export market and ships most of its phosphate to Europe. Although world reserves of about 67 billion tonnes of mineable phosphate rock are large, Morocco and adjacent territory holds almost 75% of these reserves, and it is not clear whether the unprocessed ore will remain easily available to world commerce. The Moroccan government, which owns 94% of the domestic reserves, plans to diversify away from exports of phosphate rock and begin beneficiating the ore to make fertilizer and other products for export (Wellstead, 2012). As of 2014, Morocco entered into trade agreements with European, Middle Eastern, and North American countries to ensure market access for beneficiated products such as fertilizer and phosphoric acid.

12.2.3 Nitrogen compounds and nitrate

Nitrogen compounds are the most valuable and fastest growing of the agricultural and chemical commodities (Figure 12.1), with about 87% of world production used in fertilizers (Apodaca, 2013a). It is an essential component for the formation of **amino acids**, which are fundamental building blocks for proteins. In plants, nitrogen is an important component of chloryphyll pigment, which allows plants to photosynthesize carbohydrates from solar energy. Although nitrogen is very abundant, making up 78.8% of the atmosphere (Table 3.1), plants and animals cannot absorb this nitrogen. Rather, the unreactive triple-bonded N_2 in the atmosphere must be converted to reactive ammonia (NH_3) or nitrate (NO_3^-) before plants can use it efficiently and transfer it to the food chain. In nature, this process is known as **nitrogen fixation** and is carried out by bacteria. In modern agriculture, we assist the process by addition of nitrogen

BOX 12.4 | SYNTHETIC FERTILIZERS AND THE NITROGEN CYCLE

The use of synthetic fertilizers has changed the global nitrogen cycle greatly. Natural nitrogen fixation is estimated to be about 85 million tonnes annually compared with about 150 million tonnes for anthropogenic fixation (Oldfield, 2005; Miyamoto *et al.*, 2008). This dramatic change to the nitrogen cycle is reflected in the finding that almost 80% of nitrogen in human tissues originated from the Haber–Bosch process rather than natural nitrogen fixation (Howarth, 2008). Fertilizer applications are estimated to provide up to 80% of total fixed nitrogen to agricultural soils, with the rest derived from the atmospheric aerosol and natural biological fixation. Most of the nitrogen fertilizer that is added to soil is oxidized to form soluble nitrate, the main form in which nitrogen nourishes plants. Unfortunately, soils retain little of this nitrogen and any that is not taken up by plants is usually released into waters where it can have a deleterious effect on aquatic and terrestrial ecosystems (Goolsby *et al.*, 1997; Bernhard, 2010). Even nitrogen taken up by plants and subsequently eaten by animals is excreted as ammonia and related compounds and adds to the load of nitrates in surface waters. The addition of nitrogen and nitrates to terrestrial systems can lead to acidification and eutrophication, with attendant declines in biodiversity, forest health, and nutrient imbalance, as well as changes in carbon storage capacity of soils and their abundance of infectious and non-infectious pathogens (Johnson *et al.*, 2010; Bernhard, 2010).

fertilizers, usually ammonium nitrate (NH_4NO_3) or urea [CO($NH_2)_2$], to the soil (Foth, 1984; Brady, 1990). The 13% of world nitrogen production that is not used for fertilizers goes into nitric acid (HNO_3), the important industrial acid ammonium sulfate [$(NH_4)_2SO_4$], and other compounds that are used in explosives, plastics, resins, and synthetic fibers. Nitrogen is also used to produce sodium azide (NaN_3), the propellant in automotive airbags (Davis, 1985a), and it is present in almost all pharmaceutical drugs, the medical anesthetic nitrous oxide, nylon, plastic, and coolants for computer central processing units (CPUs) and X-ray detectors.

World nitrogen production, which comes from about 66 countries, amounts to about 140 million tonnes annually worth an estimated $79 billion, although very little of this comes from mineral deposits (Figure 12.14).

Nitrogen deposits and production

Almost all of world nitrogen production comes from the Haber–Bosch process in which air reacts with hydrogen in natural gas to make ammonia. Before this process became commercially feasible in the 1920s, nitrogen was obtained from natural deposits, including guano, the bird and bat excrement that was a major source of phosphate (Box 12.3). Production also came from non-marine nitrate evaporite deposits in the Atacama Desert (Figure 12.9, 12.17), one of the driest parts of Earth with an average annual rainfall of less than 0.1 cm. Extreme dryness is essential for the formation and preservation of nitrate minerals, which are highly soluble.

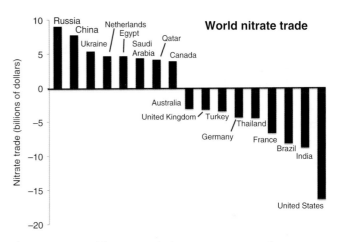

Figure 12.14 World nitrate trade for 2010–2013. Note that most production comes from countries with large natural gas reserves and most importers are large agricultural countries (from United Nations Comtrade Statistics).

The Atacama nitrate deposits, which are also referred to as Chilean saltpeter, occupy a 15–80-km-wide belt that extends along the eastern side of the coastal mountain ranges for about 700 km. Unlike boron, iodine, and lithium, which form in playas, nitrate deposits form crusts and cements up to 2 m thick in soil and rock debris along the sides of valleys (see Figure 12.17). The crusts consist of about 25% nitrates, with halite, sodium sulfate, and other evaporite minerals containing calcium, magnesium, bromine, iodine, and chromium. Bromine, iodine, and sodium sulfate are all by-products of these deposits, which are also the only natural setting for abundant chromate (Ericksen, 1981, 1983). The isotopic composition of oxygen in the Chilean nitrate deposits is unequivocally atmospheric (Michalski *et al.*, 1994), whereas cosmogenic iodine (^{129}I) is similar to deep sedimentary marine pore waters and shales, and stable chromium isotope data indicate chromium cycling owing to groundwater transport (Perez-Fodich *et al.*, 2014). These data are consistent with dry atmospheric deposition of photochemically produced nitrates, sea-spray inputs for sulfates and chlorides, and precipitation of iodates and chromates driven by groundwater. Tectonic uplift and hyperarid conditions that persisted since Miocene time allowed preservation of these water-soluble compounds at the surface.

The Atacama nitrate deposits were so important, and lucrative, to the global market in the middle to late nineteenth century that Chile, Peru, and Bolivia fought to gain control of them in the 1879 War of the Pacific, also known as the Saltpeter War (Bonilla, 1978). In 1912, the Chilean industry employed 46,500 workers and was a major contributor to the national economy (Harben and Theune, 2006). Naval blockades during World War I prevented Germany from importing Chilean saltpeter for the production of gunpowder, which led German scientists Fritz Haber and Carl Bosch to invent the Haber–Bosch method of converting nitrogen and hydrogen gases into ammonia (McConnell, 1935; Erisman *et al.*, 2008). This invention effectively ended large-scale mining of nitrates in the Atacama. Today, about 1 million tonnes per year of Chilean nitrate, less than 1% of global nitrogen production, is mined for the niche organic fertilizer market and industrial applications such as glass batch refining, steel tempering, and as a food additive to inhibit botulism. Production from the nitrate deposits involves direct open pit mining followed by a complex solution–re-precipitation process that concentrates nitrates (Harben and Theune, 2006). Reserves are enormous in relation to production, but the future of the deposits seems grim owing to the inability to withstand competition from synthetic sources.

Nitrogen markets and environmental issues

The US EPA established as part of the Safe Drinking Water Act a maximum contaminant level goal of 10 mg/liter for nitrate in drinking water. Nitrogen in drinking water is of concern because nitrate is reduced by bacteria in saliva and the stomach to nitrite (NO_2^-), which has been linked to **methemoglobinemia**, a potentially fatal condition that limits the oxygen-carrying capacity of hemoglobin in the blood. The problem, also known as **blue-baby syndrome**, is most common in babies fed with bottled milk made from nitrate-rich water. Carcinogenic compounds formed by reactions between nitrite and food have also been suggested as a cause of gastric cancer, although a World Health Organization study found no such

link and noted that gastric cancers decreased during the same period that nitrate concentrations in drinking water increased (Bockman *et al.*, 1990). In general, MDCs have done a good job at reducing nitrogen in water supplies. Municipal water suppliers remove nitrate from drinking water. The concern lies in the consumption of water from wells that supply drinking water to about 15% of the US population, with shallow wells in areas of heavy fertilizer use presenting the greatest concern.

Nitrogen fertilizers are produced in about 60 countries. Asia contributes about half of global output, and India, Russia, and the United States each produce about 6% of world supply (Apodaca, 2013a). World trade in nitrate products is dominated by Russia and China as exporters and the United States, India, and Brazil as importers (Figure 12.14). Imports comprise about one-third of US consumption, and the main exporters to the US market are Canada, Trinidad (relying on their large oil and gas production), Russia, Egypt, and Venezuela. Low prices for US natural gas, the primary raw material for ammonia used in fertilizers, stimulated companies to begin construction of more than a dozen new fertilizer plants in the United States, the first since 1990 (LeCompte, 2013). Consumption of nitrogen fertilizer in the US increased 455% between 1960 and 2005, and future demand is predicted to increase owing to population growth and expansion of agriculture to meet regulatory goals for commodities such as corn ethanol.

12.3 Other agricultural and chemical minerals

12.3.1 Fluorite

Fluorite is our main source of the element fluorine, which has gone from one environmental storm to another. Just as concern about fluorine in drinking water waned in the latter twentieth century, it was implicated in the disappearance of Earth's ozone layer.

Fluorite, or fluorspar as it is called in the industry, has two main uses. In MDCs, about 70% of fluorspar consumption goes into hydrofluoric acid, which is the basis for the world fluorine chemical industry, whereas about the same amount is used in LDCs for steel production (Simandl, 2009; Miller, 2013). Among the most important products prepared from hydrofluoric acid are elemental fluorine that goes into UF_6 for enrichment of uranium, synthetic cryolite for use in aluminum production, hydrofluorocarbons (HFCs) and hydrochlorfluorcarbons (HCFCs) for refrigeration and propellant

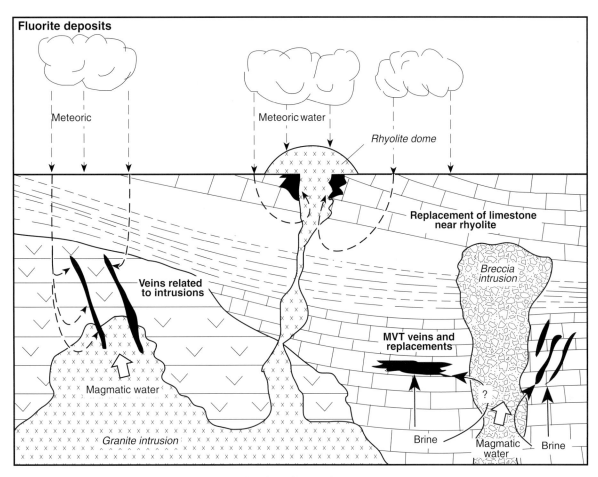

Figure 12.15 Schematic illustration of geologic processes that form fluorite deposits

fluids, and fluoropolymers for the aerospace industry. Fluorspar itself is used as a flux in steel making and in glass and ceramics (Pelham, 1985). World fluorspar production to supply these markets amounts to about 7.5 million tonnes annually. Prices vary depending on purity of the product, with the acid-grade material containing 97% CaF_2 being the most expensive and the metallurgical-grade material with at least 60% CaF_2 the least expensive. Averages for all grades are in the range of $200/tonne, giving annual world production a value of almost $1.5 billion (Figure 12.1).

Most fluorspar deposits consist of fluorite that was deposited by low- to medium-temperature hydrothermal solutions (Figure 12.15). The source of these solutions and the nature of the deposits vary greatly. Deposits in the Cave-in-Rock area of southern Illinois (Figure 12.16), historically the most important area of fluorite production in the United States, are Mississippi Valley-type (MVT) deposits that were contaminated with fluorine from a nearby magmatic system (Plumlee *et al.*, 1995). Veins in the Burin peninsula of Newfoundland were deposited from magmatic hydrothermal solutions from a granitic intrusion that mixed with meteoric solutions

(Strong *et al.*, 1984; Fulton and Miller, 2006). In both of these deposits, as well as the Pennine district of England, fluorite is found with barite, and lead and zinc sulfides. In contrast, large deposits in the San Luis Potosi area of Mexico are at the contact between limestone and shallow rhyolite intrusions and appear to have formed when meteoric water circulated through the hot rhyolite shortly after it was emplaced, leaching out fluorine and depositing it as fluorite where the solutions encountered calcium-rich limestone (Ruiz *et al.*, 1985). The most unusual fluorite deposits in the world are found at Vergenoeg in the upper part of the Bushveld Complex in South Africa, where a large body of magnetite and fluorite was deposited by saline magmatic fluids associated with granites of the Bushveld Complex layered igneous complex (Borrok *et al.*, 1998; Goff *et al.*, 2004).

China and Mexico are responsible for 62% and 17%, respectively, of world fluorite production, with smaller contributions from Mongolia, South Africa, Spain, and Kenya totaling about 15% of the global market. The United States does not mine fluorspar (fluorite), which is now classified as a critical mineral by the US and European governments.

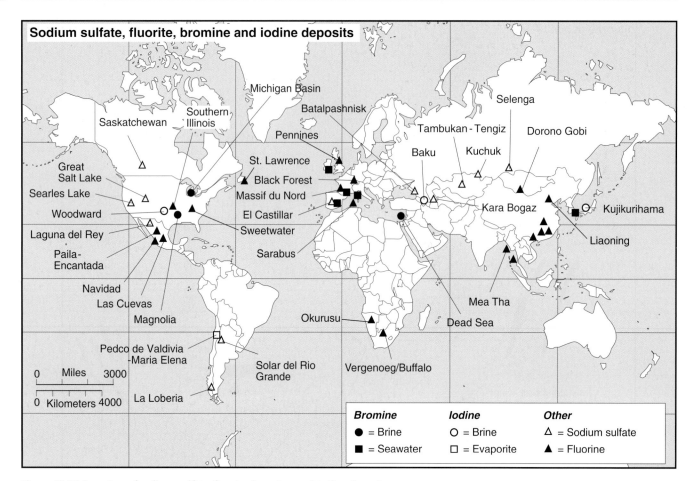

Figure 12.16 Location of sodium sulfate, fluorite, bromine, and iodine deposits

Fluorine is toxic in high concentrations but beneficial in smaller ones. At concentrations of 0.7–1.2 mg/liter, the amount used in most fluoridated drinking water in the United States, it substitutes for OH⁻ in apatite, the phosphate mineral in teeth and bone, making them less soluble and decreasing the incidence of dental caries. In 1945, Grand Rapids, Michigan, became the first US municipality to fluoridate its drinking water, and after a longitudinal study of children indicated a 50–70% decrease in tooth decay, fluoridation spread to the point that today nearly three-fourths of the US population consumes fluoridated drinking water (Zernike, 2012). Although fluorine has been under attack ever since it began to be used in water, the only significant health problem with which it has been linked is **fluorosis**, a disease that involves dental defects and bone lesions caused by concentrations of fluoride much higher than those in municipal water supplies (Osterman, 1990; Mahoney *et al.*, 1991; Hamilton, 1992). Fluorosis has been observed most frequently in grazing animals around old industrial facilities that use

fluorite or in people and livestock that use farm wells with high natural fluorine concentrations, usually from phosphate-rich rock or fluorine-rich volcanic glass (Felsenfeld, 1991). One economic evaluation of fluoridation for human health reported that each $1 spent to fluoridate water supplies results in a $38 reduction in dental costs (Griffin *et al.*, 2001; Kullgren, 2014). Fluorite particle emissions can come from mining operations, brick works, and aluminum processing facilities and gaseous HF or SiF₄ emissions can be produced by phosphate processing plants and other industrial operations (Thompson *et al.*, 1979; Polamski *et al.*, 1982; WHO, 1984).

Legislative requirements to reduce emissions of CFCs and HFCs resulted in stagnant fluorspar markets in MDCs, and replacement of these compounds with non-fluorocarbons in MDCs will reduce this part of the fluorspar market by up to 50% (Will, 2004; Fulton and Miller, 2006). Global reserves of 240 million tonnes are certainly adequate for the foreseeable future.

BOX 12.5 | FLUORITE AND THE OZONE LAYER

Fluorine became the focus of renewed controversy about chlorofluorocarbons (CFCs), which are known to destroy the atmospheric ozone layer that protects us from ultraviolet radiation, a major cause of skin cancer (Glas, 1988). CFC emissions climbed rapidly in the 1970s and 1980s, at the same time that decreasing concentrations of stratospheric ozone were observed globally (Figure Box 12.5.1). Even though the United States banned the use of CFCs in most aerosol propellants in 1978, other countries did not and widespread global restrictions on CFC emissions took force only in 1988 with the Montreal Protocol. The CFC problem is not simple, for fluorite or the environment. Hydrofluorocarbon (HFC) and hydrochlorofluorocarbon (HCFC) compounds that have been developed as alternatives to CFCs do not deplete stratospheric ozone. However, they are climatically important, with atmospheric residence times of up to 270 years and demonstrated greenhouse warming potential (Velders *et al.*, 2012). This resulted in industrial use of HFCs and HCFCs being phased out as part of the 1992 Kyoto protocols. All production and importation of these gases should have been reduced by 90% as of 2015 and completely ceased by 2030.

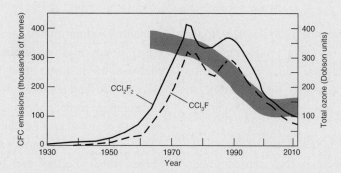

Figure Box 12.5.1 US fluorocarbon emissions between 1930 and 1990 showing coeval change in ozone concentration in the atmosphere at the South Pole beginning in 1960. Emissions peaked in 1991 and declined to very low levels by 2010. The ozone layer of the stratosphere is recovering; however, the years to decades atmospheric residence time of HFCs and HCFCs means that ozone levels may not fully recover for several decades (World Meteorological Organization, 2010).

12.3.2 Iodine

The world iodine industry was boosted with the discovery in the 1820s that iodine is used by the thyroid to produce the hormone thyroxin, which prevents goiter (Warren, 1989; Lyday, 1999; Obregon *et al.*, 2005). Food additives, such as iodized salt, directed at this problem for both humans and livestock, remain a major market, accounting for about one-fourth of iodine use in the United States (Krukowski and Johnson, 2006). Other markets include silver iodide, the light-sensitive compound on X-ray and graphics film, and various compounds for use as disinfectants and biocides, wood stabilizers, catalysts, and pharmaceuticals (Lyday, 1985; Krukowski and Johnson, 2006; Polyak, 2015). Growing uses for iodine include the production of optical polarizing film for liquid crystal displays, and the production of radiopaque contrast media. The latter are ingested by, or

injected into, humans and concentrated in a targeted area of soft tissue to allow X-ray imaging of medical conditions such as heart disease, brain tumors, and kidney disease (Novelline, 1997). World iodine production, amounting to about 27,000 tonnes, has a value of about $1 billion (Figure 12.1).

Iodine, like bromine, is produced from brines and evaporites, often as a by-product of bromine. The world's largest iodine suppliers are Chile and Japan, comprising about 63% and 33%, respectively, of global supply. Chile obtains most of its iodine from tailings and waste from nitrate mining (Figure 12.17) (Ericksen, 1981, 1983; Jan and Roe, 1983; Velasco and Gurmendi, 1988; Krukowski and Johnson, 2006). Japan obtains most of its production from brines associated with the Kanto natural gas fields in the Chiba peninsula (Figure 12.16). These brines, which contain up to 160 mg/liter iodine, actually produce twice as much bromine as iodine, and the high price of domestically produced natural

Figure 12.17 Iodine mining in the Atacama Desert of Chile. (a) Overview of mining operation showing their location on hillsides rather than in playa lake beds. (b) Iodine minerals (light colored) filling holes in soil and rock debris that cover the hillsides (photographs courtesy of Martin Reich, University of Chile). See color plate section.

gas makes iodine production economic (Polyak, 2015). In the United States, the only important iodine production comes from brines of the Pennsylvanian-age Morrowan Formation in Oklahoma, which contain 150 to 1,200 mg/liter iodine. At present, these brines do not contain enough bromine to support economic extraction.

In spite of its therapeutic effects on goiter, elemental iodine is even more toxic than bromine in high concentrations. Iodine vapor in working environments is limited by the US EPA to concentrations of 0.1 ppm, although its major applications have not been implicated in important environmental problems and its markets remain relatively undisturbed. The only environmental problem of significance associated with iodine production is subsidence around brine wells in Japan. Iodine reserves are large and are augmented by both the ocean, with 0.05 ppm iodine, and some forms of seaweed, which burn to produce ash with 1.4–1.8% iodine that was the main source of iodine prior to discovery of brine and evaporite sources.

12.3.3 Sodium sulfate

Sodium sulfate is recovered in about equal amounts as a mined material and as a by-product or waste product from the manufacture of rayon, cellulose, lithium carbonate, boric acid, and paper. Natural sodium sulfate is used largely in the production of soaps and detergents (35% of the US market), glass (18%), pulp and paper (15%), and carpet fresheners and textiles (4%) (Kostick, 2013b). World production of natural sodium sulfate amounts to about 8 million tonnes valued at about $1.2 billion (Figure 12.1).

Natural sodium sulfate comes largely from lake, or lacustrine, brines and evaporites in playas (Weisman and McIlveen, 1983; Austin and Humphrey, 2006). Many inland lakes are far enough away from the ocean to avoid major contributions from airborne sea spray, allowing their composition to be dominated by elements weathered from rocks in the surrounding watershed rather than NaCl from the sea (Jones, 1966). Eugster and Hardie (1978) divided inland lakes into four main types on the basis of their water compositions. Lakes containing $Na–CO_3–Cl$ and/or $Na–SO_4–Cl$ waters are most common and form soda ash and sodium sulfate evaporite deposits, respectively. Less common lakes contain $Na–Mg–Ca–Cl$ or $Na–Mg–SO_4–Cl$ brines and evaporites.

Playa evaporites almost always form in enclosed valleys with no easy outlet for incoming water (Jones, 1966). Most enclosed valleys form by faulting such as in the Basin and Range province in Nevada and adjacent states. The floors of many of these valleys, including the Imperial Valley and Death Valley in southern California, actually reach below sea level. These valleys contained large lakes during Pleistocene time when continental glaciation created humid climates over much of North America (Figure 12.18). When the area became arid, the lakes evaporated to form playas. Enclosed valleys also form in northern areas where sediment freed by melting of glaciers forms a hummocky surface with an irregular drainage pattern.

Deposition of sodium sulfate in playa lakes is controlled by its main mineral mirabilite. Mirabilite becomes much more soluble as temperature increases, and this causes it to precipitate from brines as they are cooled and redissolve as they are

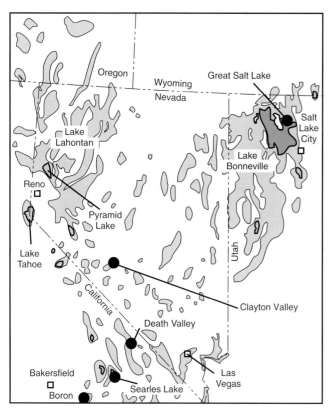

Figure 12.18 Distribution of Pleistocene lakes formed by increased rainfall during continental glaciation in the western United States, showing location of evaporite and brine deposits, including the Clayton Valley lithium brine deposit

warmed, either on a daily or seasonal cycle. It can also precipitate from water that cools as it flows into the ground below the playa, forming layers of crystals that fill pores in underlying sediment. Resulting sodium sulfate deposits, some of which come and go with the seasons, can be tens of meters thick and cover large areas. The main area of sodium sulfate mining in North America, in Saskatchewan, is part of a belt of glacial lakes that extends from Alberta to North Dakota (Figure 12.16). Sulfate in these lakes could be from springs and groundwaters draining gypsum-rich parts of the underlying Williston Basin potash deposits that were discussed above. In the United States, sodium sulfate is produced from brines underlying Searles Lake, California, and Cedar Lake, Texas, both of which are the remains of much larger Pleistocene glacial lakes. The concentration of sodium sulfate in brines at these lakes, 12% and 10.5% respectively, makes it a major product from both operations (Austin and Humphrey, 2006). Salt formed by evaporation of water from the Great Salt Lake, Utah, also contains 12% sodium sulfate, although production is not currently economic.

Sodium sulfate mining is split among open pit mining of near-surface layers, solution mining of layers that are too

deeply buried to merit open pit mining, and recovery from brines pumped from lakes or wells drilled below playa evaporites. The only problems specific to sodium sulfate production are the need for large areas of flat land in solar evaporation ponds and the difficulty of disposing of waste solutions.

The major sodium sulfate producers are China, with about two-thirds of the world market, followed by Mexico, Spain, Russia, Kazakhstan, Canada, and the United States. About one-half of US consumption is from natural sources, with the remainder obtained as a by-product from the waste stream of other chemical processes. The sodium sulfate industry in MDCs is declining owing to recycling and substitutes that have decreased the paper market from about 60% of US consumption to only 15%. Some of the lost market was recovered by an increase in the soap and detergent market, where sodium sulfate acts as a filler and makes up 50% of the volume of the product. Globally, the sodium sulfate market in LDCs, especially Asia and South America, is expected to increase by several percent per year. Reported world reserves of 3.3 billion tonnes can last for several hundred years at present consumption.

12.3.4 Bromine

Bromine is beset by environmental complications that are having a strong impact on its markets. Its first major market was in methylene bromide (CH_3Br) and ethylene dibromide ($C_2H_4Br_2$) that were added to **leaded gasoline** to prevent formation of lead coatings on cylinders and spark plugs (DeLong, 1926). An average of 1 g of bromine was required for every 3.86 g of lead in these fuels, creating a substantial demand. Bromine was dispersed into the atmosphere from combustion of leaded gasoline and is a common constituent of the upper part of soils, especially in urban areas (Sturges and Harrison, 1986a, b). This market disappeared with the abolition of leaded gasoline, but bromine found new markets in fire-retardant compounds in plastics and epoxy in electrical and electronic equipment, which now comprise almost half of the global bromine market. Bromine fire retardants are present in about 90% of all electrical components. Bromine is used in soil fumigants and other agricultural chemicals to prevent pests from damaging produce and grain, oil-well drilling mud, and compounds for treatment of water and sewage, all of which are under environmental attack (Ober, 2013). The largest market for bromine is as organobromines that are used to manufacture flame retardants, currently the largest application with annual growth of almost 10%, biocides, and dyes. World annual consumption of organobromines is predicted to reach nearly 11 million tonnes by 2018. World bromine production is estimated to be about 683,000 tonnes, worth about $700 million.

Brines in the Devonian Detroit River Group in the Michigan Basin were an important early source of bromine (Figure 12.16). In the late nineteenth century, Herbert Dow, who later founded the Dow Chemical Company, invented a method to produce bromine electrolytically from brines sourced below Midland, Michigan, and production continued until 2005 (ACS, 2007). Dow's production method was much less expensive than the competing methods used by Germany at the time, which attempted to drive him out of business by exporting bromine to the United States at a price much below even Dow's cost. However, Dow bought the **dumped** German bromine and resold it to Europeans at a higher price, thus breaking the monopoly (Whitehead, 1968). Today, US bromine production comes from basinal brines of the Jurassic Smackover Formation in the northern part of the Gulf Coast Basin in Arkansas, which contain up to 6,000 mg/liter bromine (Jensen et al., 1983; Ober, 2013). The bromine:chlorine ratio of these brines is also higher than that in seawater. This ratio does not change during evaporation of seawater until halite begins to precipitate, taking out chlorine. From then on, the brine is enriched in bromine. Thus, brines with bromine:chlorine ratios greater than seawater could be residual solutions from late stages of evaporation of seawater. Bromine-rich, saline lakes such as the Dead Sea probably formed in this way.

Most bromine extraction facilities re-inject spent brine into the formation from which it was pumped originally. Injection of this type is strictly monitored and can be curtailed if brine quality or formation fluid movement are not judged to be acceptable. Separation of elemental bromine from brines involves the use of stone or cement towers through which the brine falls as it reacts with chlorine or sulfuric acid. Iodine is a common by-product of most bromine brine operations.

Elemental bromine and its compounds are toxic endocrine disruptors and must be treated with great care; overexposure to bromine affects the ability of human tissue to retain iodine. Concentrations of bromine-bearing vapors above 1 ppm are a health hazard and concentrations of 500 to 1,000 ppm can produce death. The highest risk from these compounds is confined to bromine producing and manufacturing facilities, although there is concern about possible effects of bromine products on the general population. The use of brominated Tris as a flame retardant in children's pajamas was banned by the Consumer Product Safety Commission in 1977 after

studies showed that the compound caused mutations in DNA and could be leached from pajamas (Blum, 2006). In 1983, the EPA suspended the use of ethylene dibromide in soil fumigants, and the use of methyl bromide as a soil fumigant and agricultural pesticide ceased in 2005 to comply with the Montreal Protocol. Recent studies have implicated the use of polybrominated diphenyl ethers (PBDEs) in thyroid disruption, memory and learning impairment, advanced puberty, fertility, among other health issues (McDonald, 2002; Eskenazi et al., 2013). The only major application in which bromine lacks satisfactory competitors is in additives to so-called "clear drilling muds," which dissolve high-molecular-weight, bromine–metal compounds in brines to achieve a high specific gravity without the use of barite or other conventional solid materials. These fluids are used in completion operations on oil and gas wells.

The value of US bromine production, which accounts for about one-third of global capacity, is estimated to be about $900 million (Cottingham, 2012). Other major producers are Jordan, Israel, and China, which together with the United States account for about 95% of world supply, with smaller amounts from Ukraine, France, and Japan (Ober, 2013). Demand for bromine in the United States increased with gas and oil production in the early 2000s (Frim and Ukeles, 2007; Ober, 2013), and the market for bromine as brominated, powdered, activated carbon used to sorb mercury from flue gases is expected to grow owing to regulatory controls. Globally, consumption in China increased by nearly 100% between 2000 and 2012, and the Asia market for bromine is predicted to grow by several percent per year (Persistence Market Research, 2014). Reserves for future production are essentially limitless, with the entire ocean as a back-up to the many bromine-rich brines that have yet to be tapped.

12.4 The future of chemical and agricultural minerals

Demand for agricultural and chemical minerals is expected to increase by almost 100% by 2050. Reserves vary from small for sulfur to enormous for potash, and exploration for many of them must continue. Substitutes for most of these commodities are not widely available, although attention has been given to the use of sewage sludge. This material, which is the purified solid residue from municipal sewage treatment, is an unusually large-volume waste product and disposal of it from large coastal cities has become more difficult since ocean dumping in the United States was banned in 1992. Biosolids, as this material is called, are now used in every US state on about 1% of total farmland (Artiola, 2011). However, some sewage sludge has been found to contain brominated flame retardants, heavy metals, pharmaceuticals, and other hazardous materials, which limits its use. Switzerland, Sweden, and Austria have banned the use of sewage sludge, and controversy and legal action surround its use in other areas.

Aggravating the reserve situation for most of the chemical and fertilizer minerals is the fact that few of them can be recycled in a meaningful way. Many of them are used in highly soluble forms that dissolve and disperse in ionic form into surrounding soils, rocks, and waters. If we do not find a way to limit this dispersion, we may well have to limit their use. This problem is even more serious because of the small amounts of metals and other trace elements that are also released during applications of salt and fertilizers. Because of their greater capacity to be adsorbed by clays and other soil minerals, many of these elements are retained in the soil. In fact, soil metal levels appear to be climbing gradually and could become sufficiently enriched in the future to create problems for plants growing on them. In view of the extremely beneficial nature of fertilizer minerals, it is unlikely that this will cause their use to decline greatly. Instead, we must find ways to apply them that minimize their dispersion beyond the desired area and that limit the addition of undesirable elements.

CHAPTER

13 Construction and industrial minerals

Construction and industrial minerals are the least known of the world's mineral commodities, even though they are essential for housing and infrastructure (Harben and Bates, 1984). The average US home contains 60 tonnes of concrete products, 25 tonnes of **sand**, gravel, and stone, 7 tonnes of gypsum products, and 0.1 tonne of glass. The anonymity of these mineral resources is curious because mines and processing facilities for most of them are much closer to most of us than are gold mines and oil wells. This is because their unit values are low and transportation to market can increase costs by as much as 100% (Juszli, 1989). The good news about this is that production of many essential construction minerals is high in most more-developed countries (MDCs) and is an important source of employment. The bad news is that their proximity to populations makes these mines a common focus of land-use and environmental concerns, which directly affect our cost of living.

The mineral resources discussed here can be divided into four groups based on their main uses. Ranked in order of decreasing value and tonnage, these are the construction minerals, the glass raw materials, the fillers, pigments and filters, and the abrasives. World production data for crushed stone, sand and gravel, and common clay are not tabulated and have been estimated here by assuming that US consumption has the same relation to world consumption that US gross domestic product (GDP) has to world GDP. On that basis, the total value of global construction and industrial minerals is about $460 billion.

13.1 Construction minerals

The main mineral commodities used directly in construction include cement, aggregate and lightweight aggregate, dimension stone, gypsum, and common clay (Tepordei, 1999). Cement is by far the most valuable member of the group (Figure 13.1), largely because it is actually a processed mineral material like steel rather than a raw mineral commodity. Production growth for these minerals since 1960 has diverged significantly, with cement far outpacing the growth of steel, gypsum matching steel, and clays and aggregate falling far behind. Prices of all of the commodities have also diverged significantly, with clays and sand and gravel keeping pace with the CPI, cement falling slightly behind, and gypsum and crushed stone showing essentially no increase (Figure 13.1).

13.1.1 Cement

Cement is a world-class, high-tech business camouflaged as a stodgy old material (Struble *et al.*, 2011). The 3.6 billion tonnes of cement produced annually have a value of about $315 billion, among the highest of all mineral commodities (Figure 13.1). Cement is produced in 154 countries, an amazingly large number that reflects widespread availability of limestone and other commodities from which it is manufactured. Almost all of this cement is **hydraulic cement**, which will harden under water. Hydraulic cement has been used for more than 3,000 years and is the principal material in many famous ancient structures.

Cement raw materials and production

In simplest terms, cement is produced from a mixture of calcite and silica. Rocks with the right mixture of calcite and silica, known as cement rock, form where limestone reefs are breached by rivers carrying silicate sediment (Ames and

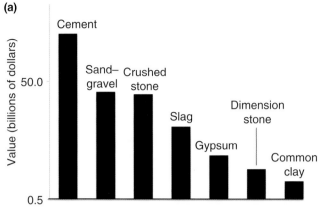

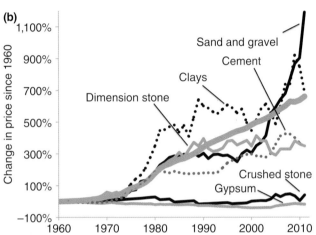

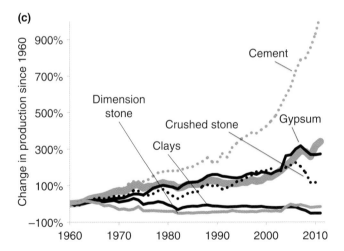

Figure 13.1 Comparative values (a) and price (b) and production (c) histories for construction minerals. Prices are compared to the consumer price index (CPI) and production is compared to the price of steel, both of which are shown as broad gray lines (compiled from data of the US Geological Survey). Note that the scale for (a) is logarithmic in order to show all values clearly.

Cutcliffe, 1983). Where cement rock is not available, limestone is mined separately and mixed with silica in the proper proportions. The silica can be mined from shaley sediment or fine-grained volcanic ash deposits, or collected as fly ash from coal-fired power plants, and it is often referred to as pozzolan, a term that refers to its ability to react with the calcium in limestone. The most common type of hydraulic cement, **Portland cement**, is named after a limestone mined at Portland on the south coast of England and used widely as a building stone.

Certain relatively common elements and minerals can render potential cement raw materials useless. Most problematical is magnesium, usually in the form of dolomite, which prevents formation of calcium silicates and is not acceptable in concentrations above about 5% magnesium oxide. Iron can also be a problem if it is present as pyrite, which will create SO_2 during heating. In the Great Lakes region of the United States, Paleozoic-age limestones grade westward into dolomite that cannot be used, placing constraints on the location of cement resources (Kendall *et al.*, 2008). Where limestone is not available, shells dredged from offshore areas can be used. Silica and alumina can be obtained from fly ash, red mud waste from alumina production, and from slag. In rare cases, gypsum and calcium feldspar have been used as the primary source of calcium (Johnson, 1985).

Most mines associated with cement production are large, open pits that are situated as close as possible to the processing plant. Many mines and plants are located along waterways to minimize transportation costs, and cement, along with aggregate, production is moving toward the use of megapits that serve large areas (Kendall *et al.*, 2009). Cement production requires crushing and grinding the raw material to make a fine-grained powder followed by heating the mixture in a rotating cylinder known as a **rotary kiln** (Figure 13.2). Temperatures in the kiln range from about 500 to 1,500 °C; during the heating calcite first **calcines** or loses CO_2 to become CaO, which then reacts with silica to form calcium silicates. Calcium silicate comes out of the kiln in chunks known as **clinker** that are ground into a powder. When this powder is mixed with water, the calcium and silicates take on new crystal forms containing OH⁻ or H_2O, and it is this network of new crystals that makes cement so strong.

Emissions from cement production depend in part on the type of fuel used to heat the kiln. On average, about 1 tonne of CO_2 is emitted per tonne of cement produced. SO_2 emissions from cement manufacture rank behind petroleum refining and metal smelting, but can be important locally. Particulate emissions from cement plants in MDCs have declined considerably and are improving in many less-developed countries (LDCs). Cement plants also emit trace elements that are vaporized by high temperatures in the kiln. Although most cement raw materials are not sufficiently enriched in trace elements for this to be an important problem, high mercury emissions have been observed (Haynes and Kramer, 1983; Fukuzaki *et al.*, 1986; Arslan and Boybay, 1990; Zemba *et al.*, 2011).

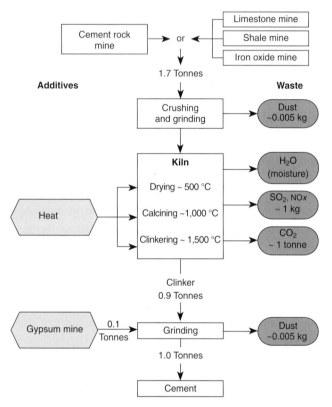

Figure 13.2 Schematic illustration of the cement-producing process (compiled from Ames and Cutcliffe, 1983 and Johnson, 1985; EPA Cement Factsheet, 2010)

Cement markets and trade

Although cement can be used alone, it is expensive, and it is more common to combine with sand, gravel, or other mineral or synthetic materials, known as **aggregate**, to make the synthetic rock called concrete. For most uses, concrete contains about 60–75% aggregate, 15–20% water, 10–15% cement, and 5–8% entrained air. Depending on the material and proportions that are added, concrete can have a wide range of properties and useful lifetimes.

About three-quarters of US consumption goes into ready-mix concrete, which is used to make foundations, sidewalks, and driveways as well as large buildings and highways. Another 11% goes into concrete block, pipe, and other products and the remainder is used in special larger construction projects (Van Oss, 2014). About 60% of the 76,000-km US interstate highway system is paved with cement (Sullivan, 2006). Concrete is not used widely in single-family homes, but it is a major component of apartment buildings, and new high-strength versions compete with steel framing for office buildings.

The leading cement producers include China, with over a third of world production, followed distantly by the United States, India, Iran, and Russia. The large amounts of energy required for clinker production give cement producers in areas of cheap energy an edge. At a price of about $100/tonne for cement, a further advantage goes to producers located near consumers or with access to cheap barge or ship transport. Only about 3% of world cement production is traded across borders, but that is still valued at $4 billion. The main exporters are China, Italy, Turkey, Spain, and India and the main importers are the United States, Japan, Korea, Netherlands, and France. With the exception of the United States, most importers are relatively close to their source of cement.

BOX 13.1   CEMENT AND ENERGY

Cement production is a major energy consumer, accounting for 12–15% of industrial energy use (Madlool *et al.*, 2011). Even so, it has improved; in 1972, production of a tonne of cement in the United States required an average of about 8.43 GJ and by the late 2000s that had dropped to about 2.5 GJ (Ullman, 1991; NRMCA, 2012). Even with these improvements, cement is the most energy-intensive manufacturing industry in the United States, with a share of energy that is 10 times more than its contribution to the GDP (EIA, 2013b). About 80% of cement's energy consumption goes into heating the kiln, with the remaining 20% going largely into grinding the clinker. Coal is the main fuel used for cement production, although use of alternative fuels is increasing (Johnson, 1985; Capone and Elzinga, 1987). Cement rock can be pulverized either wet or dry before it is put into the kiln; the dry process requires less energy because it does not need to vaporize water, and is becoming the dominant method, although it requires more effort to minimize particulate emissions.

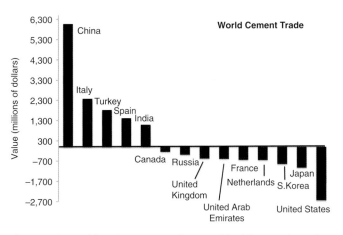

Figure 13.3 World trade in cement, lime, and building products for 2012 showing the difference in value of exports and imports (from United Nations Trade Statistics Yearbook (Comtrade))

Environmental damage to concrete

Concrete's reputation has been damaged by the alarming decay of bridges, highways, and other structures around the world. Many of these, such as the interstate highway system in the United States, were built with the expectation that they would have a long life. Instead, the concrete is crumbling and governments are faced with large repair costs. This has made many people question whether concrete is the miracle rock that it was originally thought to be. The answer lies in a better understanding of what concrete actually is and how it reacts to its environment such as climate change (Stewart *et al.*, 2011).

Hydraulic cement hardens when water is added because elongated crystals of hydrated calcium silicate grow into an interlocking matrix that has very high strength (Kerkhoff, 2007). When these crystals form, they leave small pores that were originally occupied by water, and it is in these pores that the damage begins. Water seeping into them freezes and expands, breaking the concrete. Minerals such as gypsum, which precipitate as water evaporates, have the same effect. De-icing salt on highways reacts with steel reinforcing bars in the concrete to form rust that breaks up the concrete even more. Acidic water also dissolves the concrete, particularly in concrete sewer pipe where bacteria acidify the water. Interestingly, this acidification has become more problematical as heavy metals that poisoned the bacteria were banned from sewers (Sedgwick, 1991; Mays, 1992).

New approaches are being tested to deal with the environmental challenges to concrete. Corrosion of reinforcing steel rods is being limited by epoxy coatings and use of low voltage direct currents to reverse oxidation reactions, and fiberglass rods are being tested. One experimental concrete for use in bridges uses recycled synthetic carpet fibers, which give the concrete added flexibility and greater resistance to cracking (Li *et al.*, 2004). Organic and silicon compounds are being added to fill pores left during consolidation of concrete. Although many efforts are still experimental, it is widely felt that concrete will continue to improve and find wider uses. With these developments and essentially inexhaustible raw material reserves, cement should have a good future.

13.1.2 Construction aggregate – crushed stone and sand and gravel

Construction aggregate consists of fragments of rocks and mineral that are used alone or combined with concrete, asphalt, and plaster. Between 80 and 90% of most paved roads are made of aggregate, either as a base for the road or mixed with the asphalt or cement. The US interstate highway system contains about 1.5 billion tonnes of aggregate along with 35 million tonnes of asphalt, 48 million tonnes of cement and 6 million tonnes of steel (Sullivan, 2006). The most common constituents of aggregate are crushed stone and sand and gravel. The price of crushed stone has remained essentially unchanged since 1960 (Figure 13.1b), far behind the rise in the CPI. Prices for sand and gravel for construction are not available but probably followed a similar trend. So, you cannot blame the increased cost of a home on the 25 tonnes of aggregate that it contains.

The US production of construction aggregate depends on economic activity, but averages about 2 billion tonnes annually from at least 10,000 mines. It includes almost all of the sand and gravel and about half of the crushed stone produced (the other half of crushed stone is largely limestone used to make cement, lime, and other chemicals). Although statistics on world construction aggregate production are not available, it would amount to about 2 billion tonnes, worth about $60 billion, if production in other countries has the same relation to GDP as it does in the United States. The extremely low unit value of aggregate means that it cannot be transported more than a few kilometers from its source without becoming prohibitively expensive, an important factor in land-use decisions around cities.

Geology and production of aggregate

The best type of aggregate is sand and gravel from streams and rivers (Figure 13.4). Where rivers reach the ocean, finer sand and gravel are deposited in deltas and in beaches, and ocean currents can move beach sand to form bars and spits (Robinson, 2004; Langer, 2006). During Pleistocene time, sand and gravel were deposited in **moraines** at the ends of

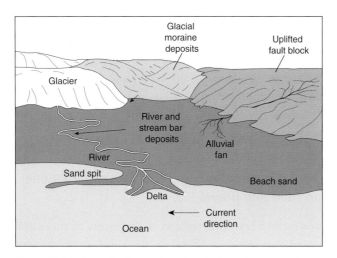

Figure 13.4 Schematic illustration of geologic environments in which sand and gravel deposits form

widespread melting glaciers. Sand and gravel also accumulate in **alluvial fans** where rivers cross faults into valleys. Although sand and gravel can be mined from active streams and beaches, this interferes with normal sediment transport and can cause undesirable erosion; it is much better to mine ancient deposits that are not in active stream beds (Kondolf, 1997).

Where sand and gravel are not available, crushed stone is produced from rock (Schenck and Torries, 1985). The best rock for this purpose is limestone, which makes up about 70% of production largely because it is relatively soft and easily mined. Harder granite and basalt supply another 20%, and sandstone, quartzite, and other materials make up the rest. Even though crushed stone requires more processing and is more expensive, it supplies slightly more than half of US aggregate production because sand and gravel deposits are scarce in so many areas (Langer, 1988).

Not all sand and gravel or massive rock deposits meet construction-aggregate specifications (Stewart *et al.*, 2007). Some of the more important specifications include resistance to abrasion, chemical attack, and splitting due to freezing water. Even small amounts of pyrite, a common trace constituent in many rocks, can render the material completely unacceptable because it oxidizes to produce a rusty color. Thinly layered shales are also useless because they are easily broken apart by the freeze–thaw cycle. As a result, many large areas of rock or gravel are not useable as aggregate.

About 95% of US aggregate comes from open pit mines, but interest in underground aggregate mines is rising as a measure to minimize environmental impact. This trend actually began long ago in Europe where cities found it easier to mine downward than to transport aggregate from distant locations outside town. Much of central Paris was

undermined by operations of this type, which collapse occasionally. Some modern underground aggregate mines are used as commercial storage facilities. Processing of sand and gravel consists essentially of washing and screening the material to produce different size fractions. In an interesting twist, sand and gravel operations in areas downstream from gold-mining areas, such as those around Sacramento and the Mother Lode in California, recover small amounts of placer gold as a by-product (Sander, 2007).

Environmental challenge of aggregate production

Aggregate deposits are being squeezed out by other land uses in MDCs, presenting a critical problem to regional planners and construction projects. No one wants to have aggregate mines nearby, but everyone wants cheap aggregate. With each 5–10-km distance adding at 10–20% in transportation costs, it is hard to replace nearby deposits with more distant ones. At the same time, the mines do not enhance the scenery and they can have particulate emissions (Figure 13.5).

The situation is particularly critical around large cities such as Los Angeles, where aggregate supplies come largely from alluvial fans along the south side of the San Gabriel Mountains. In the suburb of Irwindale, which is on a 13-km-long fan that extends into the Los Angeles valley, mining became increasingly difficult as land prices and reclamation requirements increased. Mining also interfered with groundwater in the alluvial fan, an important local aquifer. Alternative aggregate deposits farther away from Los Angeles added considerably to transportation costs (Henderson and Katzman, 1978; Goldman and Reining, 1983). By 2004, Irwindale had produced over 1 billion tonnes of aggregate and was reclaiming pits for use as home sites and industrial parks (Sauerwine, 2004). In some parts of Denmark, the Netherlands, Japan, and the United Kingdom, adequate land sources are no longer available, and aggregate is dredged from the continental shelf, which contains a huge resource for future use (Williams, 1992). Although aggregate is not a major factor in international trade, China, Singapore, and the Netherlands have significant imports (Figure 13.6). Long-distance, ocean transport is still limited, although crushed stone was exported from Scotland to Houston, Texas, in 1985, largely because of the lack of material in the area and the availability of cargo for the return trip (Zdunczyk, 1991).

Before concluding that society will simply have to make the change to more distant aggregate sources, give some thought to its implications. In a study of this matter, it was concluded that an 8-km increase in the distance between aggregate mine

Figure 13.5 (a) Aggregate quarries on the outskirts of Monterrey, the third largest city in Mexico. Although the quarries were located perfectly from an economic standpoint, they created too much particulate pollution (note plumes of dust coming from the quarry at left) and most were closed during the late 1980s. (b) Aggregate quarry on the south side of metropolitan Chicago. This quarry is divided into two parts by a rib of rock left that supports Interstate 80, which can be seen in the background. A tunnel through the ridge, seen in the center, background, allows access to both parts of the mine (photos by the authors). See color plate section.

and consumer in the United States would add $450 million to total transportation costs, require 2 million additional barrels of oil for fuel, and greatly increase truck traffic and highway accidents (Goldman and Reining, 1983). Between 1978 and 2005, the average distance traveled by crushed stone operators increased by 17% (Robinson, 2004), so the situation is only getting worse, and it goes directly into the cost of construction.

Aggregate resources are large, of course, because so many sand and gravel or stone sites are suitable for mining. Forward-looking governments recognize this and include aggregate mining in land-use plans. This limits mining to areas that are deemed best for operations of that type. In the United Kingdom, a levy is placed on aggregate production (£2 per tonne in 2014). The effect of this is to limit mining only to

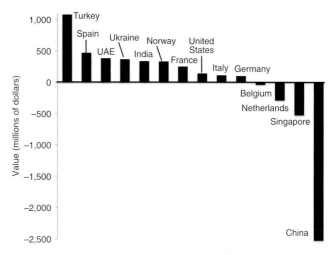

World stone and sand and gravel trade

Figure 13.6 World trade in stone and sand and gravel for 2012. Note that the main importing nations (other than China) are countries with small areas and limited rock resources (from United Nations Trade Statistics Yearbook (Comtrade) 2014).

aggregate that is essential to construction in the area and, especially, to limit export of aggregate to other areas that for one reason or another are not producing aggregate locally.

13.1.3 Lightweight aggregate and slag

Lightweight aggregate and slag, which are special forms of aggregate, are at opposite ends of the density spectrum. **Slag**, the waste product from iron and steel making, has a density much larger than average rock. This makes it valuable for applications that require heavy concrete or a heavy foundation material beneath pavement or buildings. About 500 million tonnes of slag worth about $8.5 billion are sold each year (van Oss, 2014).

Lightweight aggregate is lighter than normal aggregate but retains the strength of natural rock. It is especially useful in construction of building and bridges where weight is an important factor, but is held back in part because of its more brittle nature (Grotheer, 2008; Hassanpour *et al.*, 2012). It is also used in insulation and as an additive to soils and other materials where increased porosity is desired (McCarl, 1985). Volcanic rocks make the best natural lightweight aggregate. The lightest aggregate consists of **pumice**, in which abundant vesicles lower the bulk density of the rock so much that it will float in water. Vesicular **andesite** and *basalt*, as well as **scoria**, which can also be used, rarely float but have greater strength. Some forms of volcanic tuff with tiny angular glass fragments also have high enough porosity to be considered lightweight aggregate. World pumice production, which amounts to

BOX 13.2 | LIGHTWEIGHT AGGREGATE AND THE PANTHEON

The Pantheon, built in Rome in 126, was one of the first projects to make large-scale use of lightweight aggregate (see Figure 13.7a). This remarkable building has a circular base that measures 43.3 m in diameter and is topped by a dome with the same height, which is among the largest unreinforced domes in the world (Wilson-Jones, 2003). To prevent the dome from collapsing, Roman engineers used several imaginative methods to lessen its weight, including the large hole or oculus at its top, embayments or coffers inside the dome, decreasing the thickness of the dome from about 6 m at its base to about 1.5 m at the oculus, and a ring of weights around its base. They also used lightweight aggregate in the upper part of the dome, one of the first times that concrete had been lightened intentionally. The Pantheon is also remarkable for the granite columns that support its portico. These columns are single pieces of granite almost 12 m in length and 1.5 m in diameter at the base, and that weigh about 60 tonnes each. They were mined at Mons Claudianus in the eastern desert of Egypt, dragged 100 km to the Nile River and transported by barge and ship to Rome.

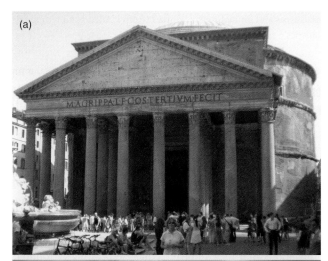

Figure 13.7 (a) The Pantheon showing the 12-m granite columns supporting its portico. (b) Marble quarries occupy most of the hillsides around Carrara, Italy, and extend underground. The best stone comes from rock that is below the zone of weathering and iron staining. See color plate section.

about 17 million tonnes worth $400 million, is used largely as lightweight aggregate, although some is used as an abrasive, including in machines that give a worn look to "stone-washed" denim pants.

Manufactured lightweight aggregate, which is used more widely than natural material, is produced from minerals and rocks that contain water, which expands on heating to create pore space. These include clay-rich shales, the mica mineral, vermiculite, and volcanic glass known as **perlite** (Arioz et al., 2008). Water is held in the crystal structure of vermiculite and is in the structure of clay minerals in shale. In perlite, it is trapped as molecules that did not have time to escape from the cooling magma. Whereas shale is relatively widespread, pumice and perlite are found only in areas of rhyolitic volcanic rocks and vermiculite forms deposits only where magnesium-rich igneous rocks have been altered by hydrothermal solutions. Annual world production of pumice, perlite, and vermiculite is worth about $100 million each, far less than the value of other construction materials and too small to show up on Figure 13.1. The lightweight aggregate market is small, but will probably grow because of the advantages of transporting and using strong, light construction materials.

13.1.4 Dimension stone

Dimension stone, which includes decorative slabs, large blocks, monuments, and tombstones, is the quintessential shrinking business in the MDCs. In the United States, for instance, dimension stone accounted for about half of the stone produced in the early 1900s. Since that time, crushed stone, concrete, and other materials have taken much of the dimension-stone market and the 2 million or so tonnes of dimension stone consumed annually in the United States amount to only about 0.2% of crushed-stone consumption. Over the last few decades, dimension-stone production in the United States has declined slightly to a value of about $300 million. Using the GDP relations discussed above, world production might amount to about 6 million tonnes worth about $1.5 billion, far less than aggregate or cement (Figure 13.1). Although the situation has been generally similar in most MDCs, dimension-stone production has increased in Brazil, China, and India, LDCs where wages are low enough to be competitive. Italy also is a major dimension-stone exporter, based largely on its Carrara marble quarries (Figure 13.7b).

The most common types of rock used for dimension stone are granite and related intrusive igneous rocks, limestone and its metamorphosed equivalent marble, sandstone, and **slate**, a form of metamorphosed shale that splits into large, flat slabs. Their most common use is in cut stone that forms a decorative veneer on buildings, but rarely part of the supportive framework of the building. Smaller blocks are used as cabinet tops, fireplace surrounds, and wall and floor tiles. Larger blocks of rough stone are used as foundations for bridges and other structures, and walls are built of rectangular pieces known as ashlar. Slabs of slate and other stone are also used for paving and roof tiles (Power, 1983; Taylor, 1985).

Dimension stone must resist abrasion, as well as corrosion during weathering. Slate, granite, brick, and clay products are highly resistant to corrosion by acid, whereas steel, limestone, and marble have poor acid-resistance. Copper, aluminium, and painted wood fall between these extremes. Urban areas in humid environments have particular problems with corrosion caused by acid rain and discolored gypsum coatings caused by dry deposition from the atmospheric aerosol. Even granite is damaged under these conditions, as shown by the extensive deterioration of Egyptian obelisks that were moved to London and New York by early explorers. Although SO_2 emissions have declined since that time in most cities, problems are still being encountered, particularly with limestone and marble, which are more easily corroded. For instance, marble that covered the Amoco building in Chicago weakened due to corrosion and had to be removed in the early 1990s. It was replaced by more resistant granite from Mt. Airy, North Carolina, at a cost of $60–80 million (Taylor, 1985, 1991). Special coatings for stone that have been developed recently might solve some corrosion problems.

Mining and processing of dimension stone require special deposits and techniques. Most mines are open pit, although underground quarries are common. With the exception of slate, dimension-stone deposits must lack closely spaced joints or fractures, which are actually very common in most rocks. The rock must be removable in large blocks that can be cut and shaped without breaking apart. For stone that will be used inside, an additional consideration is the possibility of radon emissions related to uranium content, which is usually greatest in felsic igneous rocks (Amaral *et al.*, 2012). Blocks of suitable rock were originally removed by drilling a line of holes and driving in wedges to split the rock, or by use of wire saws and chipping machines. More recently, granitic rock has been cut by jet channeling units, which use heat from a fuel oil–oxygen torch to cut a groove by spalling off chips of rock. This method will not work on limestone and marble, which calcines under the intense heat, and it has come under environmental attack because it is loud and creates dust. High-speed water jets might take over the rock-cutting process. Once the stone has been removed, it must be cut and polished into slabs and other shapes, which is usually done with the abrasive materials discussed later in this chapter.

In spite of its relatively low unit value, Carrara marble from northern Italy is exported throughout the world. This marble, which has an amazingly uniform texture, has been used in some of the most famous sculptures in the world, including Michelangelo's Pieta. Although bathrooms lined with "Italian marble" are often assumed to be a sign of luxury, Carrara marble is almost always the cheapest form of cut stone available anywhere in the world and it even competes in cost with tile. Another widely travelled stone is the black, iridescent larvikite, a type of mafic rock named for the small town in Norway from which it is quarried. In the United States, limestone from Indiana is the most widely used stone and quarries that dot the landscape around Bloomington supplied much of the stone used for the Federal Triangle in Washington, DC.

The outlook for dimension stone depends more on public taste than other factors. By its very nature, dimension stone varies from piece to piece. Architects are understandably nervous about this, not wanting to construct a building with abrupt differences in color or texture. This imposes high quality control on producers, causing increased waste and higher prices that make way for competitors such as synthetic polished stone made of rock fragments in special cements. Regardless of what happens on the demand side, dimension-stone reserves should be adequate.

13.1.5 Gypsum

Gypsum is the main raw material from which we manufacture **plaster**, an essential construction material (Henkels, 2006; Crangle, 2012). Smaller amounts are made from anhydrite, the anhydrous equivalent of gypsum. Gypsum is also used in cement and as a soil additive. About 150 million tonnes of gypsum and anhydrite are produced annually from at least 80 countries. The value of this production, based on US prices, would be about $600 million (Figure 13.1).

Gypsum deposits and mining

Gypsum deposits are almost entirely marine evaporites that formed by one of two processes (Schreiber *et al.*, 2007). Thick deposits of gypsum associated with halite and potash are thought to have formed during the early stages of evaporation of seawater, as discussed in the previous chapter. Other thick gypsum deposits, which are not associated with salt, are thought to have formed in **salt flats**, or sabkhas, such as those along the margin of the Persian Gulf. Sabkhas are tidal flats. During high tide, seawater flows onto them and is trapped in pools. During low tide, water in the pools is concentrated by evaporation and sinks into the ground where it deposits gypsum. Some salt is also deposited, but it is usually dissolved during the next tidal influx. Through time, this process can deposit a large thickness of gypsum-rich evaporite. As discussed in Section 12.1.3 on sulfur, gypsum also accumulates at the top of salt domes where they rise into and react with bacteria in groundwater.

Gypsum deposits are widespread, but are rare in Archean rocks because the lack of atmospheric oxygen at that time limited the sulfate content of its oceans and because old evaporites are too soluble to be preserved (Babel and Schreiber, 2014). Gypsum also loses water to become anhydrite when it reaches temperatures of about 50 °C. Thus, as gypsum is buried beneath younger sediments, where it becomes hotter, it converts to anhydrite. As it is exposed again during erosion, it can revert back to gypsum.

Gypsum is usually mined by open pit methods, although a few underground mines are in operation. Some of the caverns under Paris resulted from gypsum mining, from which the term Plaster of Paris was derived. About 75% of mined gypsum is calcined at temperatures of 100 °C or so to produce the hemihydrate, $CaSO_4.0.5H_2O$, which is the true form of Plaster of Paris. When mixed with water, this material forms a network of gypsum crystals that makes a strong, soft solid. Plaster was originally used as a coating to finish other building surfaces, but is now used largely in wallboard, a sandwich of gypsum between two sheets of heavy paper (Appleyard, 1983).

Gypsum trade, resources, and problems

China is the leading gypsum producer, followed distantly by Iran, Spain, and Thailand. In spite of its low cost (Figure 13.1), gypsum is relatively widely traveled, largely because domestic production in the United States, Japan, and many European countries is not adequate to satisfy demand. Metropolitan areas in the northeastern United States are supplied in part by gypsum brought in on ships from Canada, Mexico, Spain, and Thailand, which is cheaper than gypsum moved by train from deposits in the Great Lakes area. Many cities in northern Europe are supplied in a similar way by Spain, and Japan's extra needs are supplied largely by Thailand. Resources to supply the gypsum trade have not been quantified but are very large.

Additional gypsum comes from two other sources. New wallboard contains about 10–15% recycled material from sulfur-gas recovery systems (Crangle, 2012). Synthetic gypsum, mostly from scrubbers that remove SO_2 from flue gas in coal-fired power plants, supplies about half of the market in many MDCs today. Flue-gas gypsum, as this material is called, varies greatly in purity depending on the recovery method that is used. Impure material of this type is thought to be the source of sulfur gases that were emitted from wallboard imported from China during the building boom of the mid 2000s (Wayne, 2009; Halford, 2009).

13.1.6 Clays

Clay is a term with many meanings and the potential confusion is greatest when talking about clay as an industrial mineral (Ampian, 1985b; Harvey and Murray, 2006). To the lexicographer, clay is any fine-grained material that becomes plastic when mixed with water. To the geologist and environmentalist, it is a layered silicate mineral and important constituent of soils, as well as the finest grain size for clastic sediments, regardless of their composition. To the miner, it is any saleable clay-size mineral material that can be used in the construction or filler–extender markets. In this section, we are dealing with materials that are known as common clay and shale and ball clay, which should be distinguished from kaolin and bentonite, two more valuable clay products that are part of the filler–extender market discussed below. About 13 million tonnes of common clay and shale and ball clay, worth about $200 million, were mined in the United States in 2011. Although global production is not available, it would be worth about $1 billion if US production is proportional to US versus global GDP.

Common clay is used largely for brick, drain, roof-tile, sewer-pipe, and other construction materials. Much of it is also heated to produce lightweight aggregate, as discussed above. The most common clay deposits are fine-grained sediments that are low in iron and calcium. Deposits of this type are widespread and are usually mined relatively near large construction markets. Reserves of the materials are enormous, although deposits close to markets are often the focus of disputes about land use.

13.2 Glass raw materials

Glass is an amorphous solid without a well-defined crystal structure (Zarzycki, 1991; Whitehouse, 2012). Most glass is made by melting quartz and other minerals and rocks, and then cooling the melt in a way that prevents it from crystallizing (Mills, 1983). This makes it analogous to steel and cement in that it is a processed product of several mineral materials. The main constituents of glass are industrial sand, boron, feldspar, soda ash (sodium carbonate), and strontium. The total value of annual production for these commodities, based in part on estimates from US production, is about $75 billion (Figure 13.8). Production of boron, feldspar, and strontium have grown enormously since 1960, whereas soda ash has kept pace with steel. Strontium is the only commodity to have seen a dramatic increase in price.

13.2.1 Industrial sand and gravel

About 25% of US industrial sand and gravel production goes into glass, which used to be the dominant use. In the last few years, however, proppant sand to hold fractures open during fracking of oil and gas wells has grown to account for almost half of the market (Dolley, 2013). The remainder goes into foundry sand, abrasives, playground sandboxes, and sand traps in golf courses. The total value of world industrial sand and gravel production is only about $6 billion, a surprisingly small number until you stop to think that it must be inexpensive if it is used largely to make glass, which is not a particularly expensive material itself.

Quartz sand deposits suitable for glass markets are scarcer than common sand and gravel deposits used for construction (Heinrich, 1980; Zdunczyk, 1991). Minerals containing iron, which would impart color to glass, must be avoided, as must refractory minerals such as cassiterite, corundum, kyanite, chromite, or zircon, which would remain unmelted, creating imperfections in the glass. The best North American glass and proppant sands are found in blanket sand deposits of Paleozoic age that cover parts of the eastern United States (Figure 13.9). These sands were deposited in beaches and dunes along marine coasts and are especially pure because they were derived from erosion of sandstones that were deposited during Precambrian and earlier Paleozoic time. Thus, they have been purified by two and sometimes even three cycles of erosion and sand concentration. The largest of

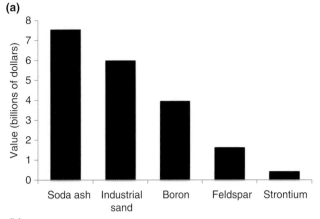

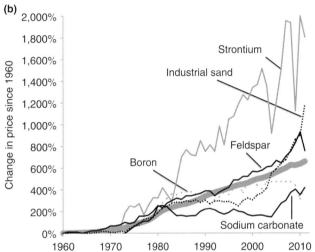

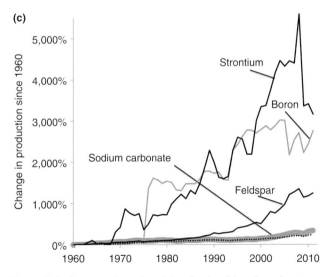

Figure 13.8 Comparative values (a) and price (b) and production (c) histories for glass raw materials. Prices are compared to the CPI and production is compared to the price of steel, both of which are shown as broad gray lines. Compiled from data of the US Geological Survey. World production data for industrial sand are not available.

BOX 13.3 | SAND MINING IN WISCONSIN

Mining of clean sands in the United States for use as proppants in fracking has grown along with production of gas and oil from shales. One area of significant new mining is in central Wisconsin where the 1.7-billion-year-old Baraboo Quartzite covers low-lying terrane. The Baraboo sands were deposited in a system of braided streams that gradually subsided to be covered by a shallow sea where sands formed beaches that were modified by tides. The sands were cemented to form quartzite, which is now exposed in nearly flat layers that form buttes and other curious landforms. Mining is from open pits that can extend to depths of only a few meters before they are flooded by the abundant groundwater in the area, much of which is part of the floodplain of the Wisconsin River.

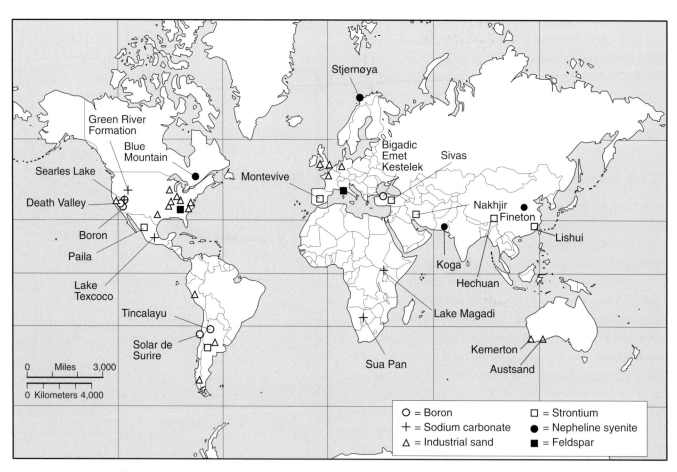

Figure 13.9 Location of important deposits of minerals used in the glass and ceramic industries

these are the Ordovician St. Peter, Silurian Clinch, and Devonian Oriskany and Sylvania sandstones, which are mined from Oklahoma to Pennsylvania. Smaller glass-sand deposits are found in beaches and dunes that formed along the Atlantic, Pacific and Great Lakes coasts during Pleistocene glaciation. Quartz suitable for some glass making is also found as large crystals in veins and pegmatites.

The largest glass-sand producers are the United States, Italy, Spain, and Germany. Most mines are on land, but some is dredged from offshore bars. Processing involves some crushing and grinding and, where necessary, gravity separation and flotation to remove impurity minerals. The main environmental concerns during all of this is the escape of silica dust, which can cause the serious lung disease silicosis, as discussed in the next section. A quick search of recent publications will show that sand mining is probably the hottest topic in the mineral sector throughout the world (Younger, 2013). This reflects the fact that many land-based mines are in dunes or other sand accumulations that are favorites as recreational land. Although reserves of sand for future glass making are enormous, the conflict between silica mines and other land uses will almost certainly restrict production in many areas.

13.2.2 Soda ash

Soda ash is the industry term for sodium carbonate (Garrett, 1992). About half of soda ash production is used to make glass; the rest goes into chemicals, soaps and detergents, paper manufacturing, flue-gas desulfurization systems, and water treatment. World production amounts to about 52 million tonnes annually; only about one-quarter comes from natural sources, largely in the United States (Kostick, 2013c). The rest is synthesized from salt and limestone in chemical plants. The total value of world soda ash production, including both synthetic and natural, is about $7.5 billion, with natural deposits accounting for only about a quarter of this. Although soda ash is much more expensive than silica sand, it is used in considerably smaller amounts in glass and therefore does not contribute more to its cost.

Natural soda ash comes from extensive deposits of complex sodium carbonate minerals such as trona, which are found in evaporite deposits that formed from lake water. These lacustrine evaporites form far from the ocean where sea spray has little effect on water compositions, and where addition of volcanic CO_2 might have helped the process (Mannion, 1983; Earman *et al.*, 2005). As discussed in Section 12.3.3, inland lakes can be divided into four compositional groups, of which the sodium carbonate type is one of the most important. The largest trona deposits in North America are in the Eocene-age Green River Formation of Wyoming, which hosts the oil-shale deposits discussed earlier. In the Green River Basin in Sweetwater County, Wyoming, these sediments contain at least 25 layers or beds of sodium carbonate that are more than 1 m thick, and only five of these have been mined. In the Piceance Creek area in Utah, these sediments contain thick layers of nahcolite and other sodium carbonate minerals that have not yet been mined. Sodium carbonate is also produced from playa brines at Searles Lake, California (Figure 13.9) and from playas and springs along the East African rift system and in Botswana. US soda ash reserves in the Green River deposits make up 96% of the world reserve of about 23 billion tonnes (Kostick, 2013c).

Soda ash mining in Wyoming is carried out by underground methods very similar to those used on coal layers, using continuous mining machines that are reinforced to cut the abrasive trona and shale. Collapse of overlying rock must be prevented in order to keep methane in nearby oil-shale layers from entering the mine and causing an explosion. During treatment, ore is heated to drive off CO_2, and then dissolved and re-precipitated in a purer form. The main environmental problems associated with trona mining and processing involve tailings settling ponds containing highly alkaline (pH = 10.5) water, which leaches oil from feathers of migratory birds that land on the ponds. Water birds have also been covered by salt that crystallized during cold evenings, preventing them from flying or even floating. Although extensive efforts are made to prevent birds from using the ponds, bird rescue and rehabilitation is a common practice.

World soda ash trade is basically a competition between natural material from the United States and synthetic material largely from Europe and Mexico. Long transport distances from the mines in Wyoming leave room for the synthetic material to compete. The main problem for the synthetic soda ash plants is disposal of large volumes of calcium chloride waste. Substitution of waste glass, known as **cullet**, is a real threat to both types of soda ash as recycling programs gain momentum in MDCs. Nevertheless, soda ash is so widely used and world reserves are so large that production is not likely to decline.

13.2.3 Boron

Almost 80% of US boron production is used in the manufacture of glass and ceramics (Crangle, 2014). The huge volume of glass manufactured in the United States makes this the dominant world market. Present world boron production amounts to about 4.9 million tonnes of ore with a value of about $4 billion.

Boron comes almost entirely from lacustrine evaporites, also known as **playas**, that remain from evaporation of ancient lakes (Figure 13.10). Many such lakes that formed in western North America during wet periods associated with Pleistocene glaciation dried up after the glaciers retreated (Figure 12.18). Boron is also enriched in brines that fill pores deep in some playa lakes including Searles Lake brines in California. Boron production from brines avoids the crushing, grinding, and dissolving stages associated with evaporite mining, but has to deal with a more complex, dilute solution and separation of boron from other elements, including sodium, potassium, and lithium.

Boron in low concentrations is essential for plants, but can be toxic in large concentrations (Nable *et al.*, 1997). Synthetic boron–hydrogen compounds known as boranes, which do

---

BOX 13.4 | DEATH VALLEY BORAX

Lakes filled Death Valley in southern California during Pleistocene glaciation, and their evaporation left lacustrine evaporites containing abundant hydrous sodium borate minerals including kernite and borax that constitute the world's largest source of boron. The boron in these deposits is thought to have come from volcanic rocks and been released into the lakes by hot springs (Carpenter and Kistler, 2006). Deposits in the area were discovered in the early 1900s, but were at first mistaken for gypsum, which is also white and easily dissolved. Boron production started quickly, however, and was made famous by the 20-mule team wagons that were used to haul borates out of Death Valley during the early part of the 1900s (see Figure 13.10). Early mining was by underground methods although most present mining is in open pits. Mined material is crushed and ground to make a powder that is dissolved to recover borax and boric acid. Boron is also present in many of the playa deposits of South America where it is largely a by-product of lithium.

---

not occur in nature, are used in semi-conductor manufacturing and are toxic (Tsan *et al.*, 2005). The most important suppliers of boron are the United States and Turkey and world reserves amount to 210 million tonnes of boron ore, a relatively large number in relation to present production (Crangle, 2014).

13.2.4 Feldspar and related mineral commodities

Feldspar, one of the most common silicate minerals, is used almost exclusively in glass and ceramics, where they act as a **flux** to lower melting temperatures, and as a source of aluminum. Annual world feldspar production is just less than 20 million tonnes, worth about $1.6 billion. Production of nepheline syenite and aplite is smaller, but probably increases the annual world value to about $2 billion (Tanner, 2014).

Although feldspars are found in almost all types of igneous rocks, they are mined largely from special rock types that are depleted in mafic and other minerals, which would not melt easily during glass making or which might add iron or other undesirable elements to the mix (Potter, 2006). The most common feldspar-rich rocks of this type are aplite, alaskite and pegmatite, all of which form during the latest stages of crystallization of granitic magmas, after most of the mafic minerals have already formed. Deposits of these rocks are mined at Spruce Pine, North Carolina, and the Black Hills of South Dakota. Feldspar-rich beach sand deposits have also been mined in Spain. Nepheline syenite, a feldspar-rich intrusive rock with little or no quartz, is also used in glass making and as a source of feldspar. The largest deposits of nepheline syenite are at Blue Mountain, Ontario, and on the island of Stjernøya in Norway (Figure 13.9).

13.2.5 Strontium

Strontium has probably prevented an entire generation of children from being damaged by radiation from color television sets. The discovery in the late 1950s that glass containing strontium blocked radiation from cathode-ray, color television tubes without damaging the quality of the image, produced a new market for strontium, causing a rise in exploration and global production. In the early 1990s, glass for television accounted for almost 80% of global production. However, this market disappeared when flat-panel displays were developed, and only about 10% of global production goes into glass of this type now (Ober, 2006b). At present, the two dominant markets for strontium are pyrotechnics and signals, where it gives the familiar red color, and ferrite ceramic magnets, which are used in a wide variety of generators and motors, especially those that need to be lightweight. World strontium production of about 380 thousand tonnes with a value of almost $500 million comes largely from China, Spain, Mexico, and Argentina (Figure 13.9).

Deposits of strontium consist of layers and disseminations of the strontium sulfate mineral, celestite, almost always in limestone. In the best deposits, celestite forms extensive, nearly pure layers, known as mantos, that have replaced layers of limestone. The replacement took place where groundwater containing dissolved strontium came into contact with sulfate-rich waters. The groundwater became enriched in strontium when it reacted with aragonite, the crystal form of calcium carbonate that was originally deposited in the limestone, converting it to calcite, the crystal form in which the calcium carbonate is found today. Aragonite accommodates larger amounts of strontium substituting for calcium than does calcite, and releases it when it changes to calcite. The sulfate-rich waters with which the groundwater reacted to

Figure 13.10 (a) Borax ore in 20-mule team wagons leaving an early mine. (b) Modern borax mine and processing plant, Boron, California (courtesy of Florence Yaeger, Rio Tinto Corporation). See color plate section.

precipitate celestite probably came from salt-flats or sabkhas. Most deposits of this type are found in and around Cretaceous-age reefs (Kesler and Jones, 1981; Bearden, 1988).

Celestite is mined by both open pit and underground methods. Where inexpensive labor is available, the more selective underground mining is widely employed. The main forms in which strontium is used are strontium carbonate and strontium nitrate, which are produced by heating celestite in

kilns with coke. World celestite ore reserves of at least 6.8 million tonnes are very large relative to current production (Ober, 1992, 2014).

13.3 Fillers, filters, and pigments

Fillers and **extenders** are used to provide special characteristics to materials or to cut their cost by substituting a less expensive

(a)

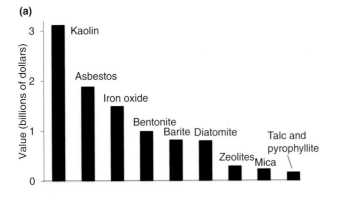

(b)

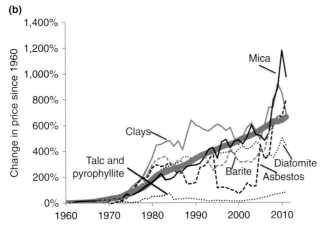

(c)

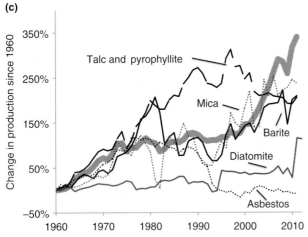

Figure 13.11 Comparative values and price and production histories for filler, filter and pigment minerals (based on data from the US Geological Survey). Thick gray lines in the price and production graphs show data for CPI and steel, respectively. Separate data for kaolinite and bentonite clays are not available.

material (Severinghaus, 1983). The list of mineral fillers is very long and their list of applications is even longer, with many curious examples. Paint containing the heavy mineral barite has been sprayed inside car doors to make them close with the proper sound, and small amounts of finely ground mica provide the sheen in lipstick. Materials that provide color

are known as pigments. Others are used largely as filters through which liquids and gases can be strained. Distinctions among these uses can be blurred, and some commodities discussed elsewhere in this book are also used as fillers, extenders, pigments, or filters. Lime, for instance, competes with kaolinite as a filler in paper. Finally, rutile and ilmenite, which are our sources of titanium metal, are also important white pigments.

Annual world production of these commodities is valued at about $10 billion, and is dominated by kaolinite clays (Figure 13.11). None of these commodities have grown production more than steel since 1960; only barite, mica, and talc–pyrophyllite have kept pace with steel, whereas asbestos, clays, diatomite, and iron oxide pigments have fallen behind. Prices of the commodities have generally followed the CPI, with the exception of diatomite and asbestos.

13.3.1 Kaolinite and bentonite clay

Kaolin and bentonite are special types of clay mineral products that are used largely as fillers and extenders. The trade name for kaolinite is kaolin, the main mineral in the kaolinite clay group discussed in Chapter 3, and bentonite is one of the minerals included in the smectite clay group. Kaolinite clays do not change composition when immersed in water, but smectite clays undergo cation exchange and adsorb water, which causes them to swell. The commercial smectite clays, bentonite and fuller's earth, are valued for these properties.

The main kaolinite-group products are kaolinite and halloysite, a generally similar mineral with a tubular structure in its dehydrated state. Kaolinite is used as a filler and coater in paper, rubber, and paint, as the main component in refractories and fine china, and as a catalyst in petroleum refining and other manufacturing processes. The most recent market is in ceramics that are used as ceramic proppants to hold fractures open during oil and gas well fracking (Virta, 2013a). Factors that affect kaolinite quality and value are its crystal size and brightness, with iron-staining particularly undesirable for many applications. **Coating clays**, which are used to make paper smooth and glossy, are mostly less than 2 micrometers in diameter, whereas filler clays are coarser. Kaolin is mined from open pits and processed to remove impurities and separate it into products of different crystal size. World production of about 22 million tonnes of processed kaolin comes largely from Georgia in the United States and Cornwall in England. Although the total value of this production is not tabulated, it would be about $3 billion based on an average value of US kaolinite production.

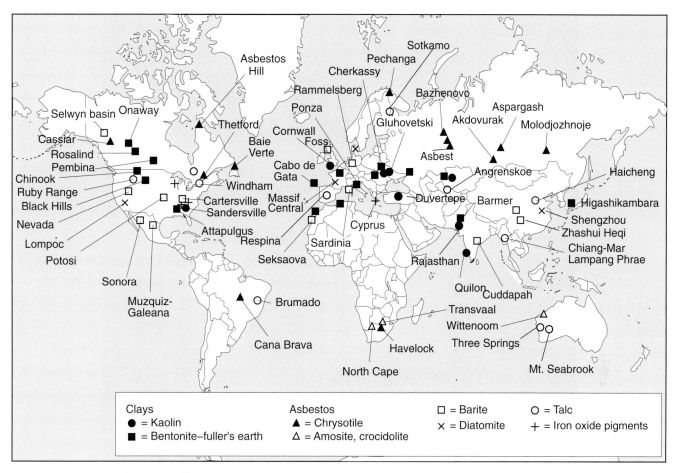

Figure 13.12 Location of important filler, extender, filter, and pigment deposits

The main smectite-group products are bentonite and **fuller's earth**. Bentonite is used largely in drilling mud where it makes the clay–water slurry act like a solid when it is not agitated. This property of a suspension, known as **thixotropy**, is very important. Without it, rock chips cut during drilling would sink to the bottom of the hole when pumping stopped, trapping the drill bit and pipe, and causing the hole to be abandoned at great cost. Bentonite is also used as a binder in foundry sand and iron ore pellets. Fuller's earth is a similar swelling clay that is valued for its ability to absorb other liquids. It is used as a cleaning compound and as a carrier for liquid pesticides and other liquids that must be distributed in solid form. Attapulgite, a special form of these minerals, comes from deposits around Attapulgus, Georgia (Figure 13.12). World bentonite and fuller's earth production is about 12 million tonnes, 80% of which is bentonite. At average US prices for these products, this would have a maximum value of about $1 billion.

Both types of clay minerals form where water leaches cations from feldspars and other minerals, leaving aluminum and silicon, the two main elements in clay minerals. This occurs in two environments, the weathering zone and hydrothermal systems (Patterson and Murray, 1983). Clay formation in the weathering zone is part of the soil-forming process, particularly in humid, subtropical, and tropical climates where waters are acid (Figure Box 2.4.1). Clay-rich soils were formed in large amounts during the last major global warming period from Cretaceous to Eocene time. Soils themselves rarely form good clay deposits, but they can be eroded to form fine-grained, clay-rich shale, which is mined widely for common clay, as discussed earlier. Even richer deposits form where special sedimentary processes occur, such as the kaolinite deposits of the coastal plain of Georgia (Figure 13.13). These deposits are lenses of pure kaolinite that were deposited in meanders of rivers that drained the Appalachians (Kesler, 1956). Bentonite deposits, on the other hand, are found in thin, very extensive layers that are the weathered remains of fine-grained volcanic glass that was deposited as ash (Spencer *et al.*, 1990).

Hydrothermal alteration also forms clays where acid solutions leach cations and replace them with hydrogen ions, H^+. Most clay deposits of this type are associated with meteoric hydrothermal systems that derived heat from intrusive or

Figure 13.13 (a) Kaolin deposit in the coastal plain of Georgia, United States. Four layers can be seen in the photo. The upper layer, which is behind and to the left of the trees at the top of the mine, is a pile of overburden. Immediately below the trees is the darker, Tertiary-age Twiggs clay, which overlies lighter-colored Cretaceous sands. The whitest layer at the bottom of the wall and making up the floor of the pit is kaolin (photo by Thomas Kesler). (b) Goonbarrow kaolin mine near St. Austell in Cornwall, England. Kaolinite here formed by hydrothermal alteration of granite and is mined by spraying water from large cannons called monitors, as seen in the left-center part of the picture (photo by the authors). See color plate section.

volcanic rocks (Figure 2.12). Large kaolinite deposits in Cornwall in southern England are in the outer parts of hydrothermal systems associated with granitic batholiths and formed at depths of several kilometers (Kesler, 1970; Bray and Spooner, 1983). Other kaolinite deposits in young volcanic rocks along convergent margins are like those that host epithermal gold–silver veins, and probably formed at shallower depths.

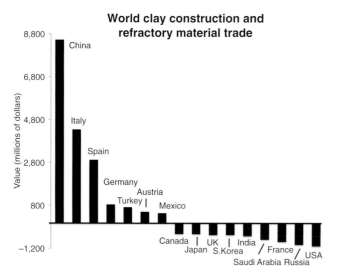

Figure 13.14 Value of world clay trade showing major exporters (above the line) and importers (below the line). This trade includes all clay-based material. In terms of value, the most important ones are floor and wall tile and kaolinite for ceramics and fine paper.

Almost all clay minerals are produced from open pit mines (Figure 13.13). Reclamation of mined land is a major factor in clay mining, with costs often exceeding the original cost of the land. Processing of clays for market ranges from simple drying and washing to complex beneficiation involving flotation for kaolinite clays. World trade in clays and clay products is dominated by exports from China, Italy, and Spain. Imports are not well documented and, at least as far as international statistics go, are far less than exports (Figure 13.14).

13.3.2 Asbestos

Asbestos is a general term for any fibrous mineral with a thread-like or acicular shape (Lowers and Meeker, 2002; Gunter, 2009). In commercial usage, it refers to flexible fiber with a length greater than three times its width, that is resistant to acid and has a high tensile strength (Schreier, 1989; Skinner *et al.*, 1988). Long fibers of commercial asbestos (Figure 13.15) can be woven into fabrics that resist heat and corrosion, and short fibers have been used as a filler to impart strength to cement pipe, asphalt shingles, and automotive brake parts. The importance of many of these applications and the lack of ready substitutes made asbestos a rapidly growing mineral commodity until the mid 1980s when production dropped by half and never recovered (Figure 13.11). Current world asbestos production, which amounts to about 2 million tonnes, has a value of almost $2 billion and is not likely to grow.

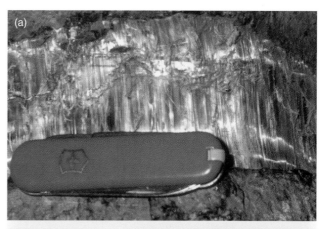

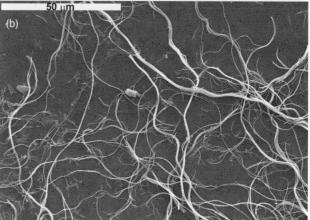

Figure 13.15 (a) Chrysotile asbestos fibers filling vein from Thetford, Quebec (photo by the authors). (b) Electron microscope image of chrysotile fibers (http://usgsprobe.cr.usgs.gov/images/Chrysotile_UICCA.jpg).

Geology and production of asbestos

Commercial asbestos comprises six distinct mineral groups, including the serpentine asbestos mineral, chrysotile (or white asbestos), which makes up about all of present world production, and five amphibole asbestos minerals of which only crocidolite (blue asbestos) and amosite (brown asbestos) have been produced in significant amounts, although production of both has been discontinued (Table 13.1). All of these are essentially hydrous silicates of magnesium and/or iron (Table 13.2), and their fibrous form is a direct reflection of their crystal structure (Figure 3.2). Chrysotile consists of separate layers of silicate tetrahedra and brucite [$Mg(OH)_2$] that form hollow tubes. Amphibole asbestos consists of two ribbon-like chains of silicate tetrahedra linked by Mg^{+2}, Ca^{+2}, Fe^{+2}, Na^+, and OH^- ions. Exactly why these minerals form elongated fibers is not clear, although it is thought that aluminum substitution increases this tendency (Virta, 2002, 2006).

Table 13.1 Major producers and reserves for construction and industrial minerals (from US Geological Survey Commodity Summaries). Reserves are not shown for most of these commodities either because they are large or have not been quantified.

Commodity	Country	Production	Reserves
Construction minerals			
Cement	China	2,300,000,000	
	India	270,000,000	
	United States	74,000,000	
Gypsum	China	50,000,000	
	United States	16,300,000	700,000,000
	Iran	14,000,000	
Glass raw materials			
Boron	United States	40,000,000	
	Turkey	3,000,000	60,000,000
	Argentina	700,000	2,000,000
Feldspar	Turkey	7,000,000	240,000,000
	Italy	4,700,000	
	China	2,100,000	
Strontium	Spain	97,000	
	China	95,000	6,800,000
	Mexico	45,000	
Fillers, filters, and pigments			
Kaolinite	Uzbekistan	7,000,000	
	United States	5,950,000	
	Germany	4,500,000	
Bentonite	United States	4,950,000	
	Greece	1,200,000	
	Brazil	570,000	
Asbestos	Russia	1,000,000	
	China	400,000	
	Brazil	300,000	
Barite	China	3,800,000	100,000,000
	India	1,500,000	34,000,000
	Morocco	850,000	10,000,000
Diatomite	United States	770,000	250,000,000
	China	420,000	110,000,000
	Denmark	324,000	
Talc and pyrophyllite	China	2,200,000	
	India	650,000	75,000,000
	United States	531,000	140,000,000
Abrasive, lubricant, and refractory minerals			
Graphite	China	810,000	55,000,000
	India	160,000	11,000,000
	Brazil	105,000	58,000,000
Kyanite	South Africa	220,000	
	United States	95,000	

Table 13.2 Classes of asbestos minerals (from Ross, 1987; Schreier, 1989)

Mineral name	Chemical composition
Serpentine asbestos	
Chrysotile	$Mg_6[SiO_5(OH)_4]$
Amphibole asbestos	
Crocidolite	$Na_2Fe_5[Si_4O_{11}(OH)]_2$
Amosite	$Fe_7Si_8O_{22}(OH)_2$
Anthophyllite	$(Mg,Fe)_7[Si4O_{11}(OH)]_2$
Actinolite	$Ca_2Fe_5[Si_4O_{11}(OH)]_2$
Tremolite	$Ca_2Mg_5[Si_4O_{11}(OH)]_2$

part of the ocean crust or upper mantle. These rocks are found along ancient convergent tectonic margins, including a suture formed by collision of North America and Europe in Paleozoic time, which hosts the largest asbestos deposits in North America near Thetford, Quebec (Figure 13.12). Asbestos occurs in these deposits as veins or mats of disoriented short fibers that replaced the massive rock during hydrothermal alteration. Asbestos-forming hydrothermal solutions probably came from seawater that circulated deep into the ocean crust or magmatic water from granites that intruded the deposits. A few chrysotile deposits are found in other hydrothermal alteration environments. The large Asbestos Hill deposit in the Ungava peninsula of Quebec is in a sequence of magnesium-rich, komatiite volcanic rocks that host magmatic nickel deposits. Deposits at Msauli, South Africa, and Havelock, Swaziland, are in similar rocks. Asbestos can also form in dolomite, the magnesium-rich carbonate rock, where it is altered by solutions containing silica, although few deposits of this type are large enough to mine. The only important amphibole asbestos deposits, which are in South Africa, are in magnesium-rich layers of banded iron formations, where they overlie dolomite. The hydrothermal solutions that formed them were basinal brines and meteoric waters heated by the Bushveld layered igneous complex.

Asbestos is mined largely from open pits, and is concentrated from the ore by dry methods in which ore is shattered by a rapid, sharp impact and the liberated fibers are then blown or sucked away. Large amounts of air are used in this process and all of it must be thoroughly filtered before release. Wet processing would make it easier to suppress fiber release, but it has been used only in deposits consisting largely of short, matted fibers. Asbestos fibers are usually classified according to length, which can vary from rare fibers more than about 2 cm long to those only a few millimeters in length (Mann, 1983).

Almost all asbestos deposits form by hydrothermal alteration of rock that is rich in magnesium and/or iron (Riordan, 1981). Chrysotile asbestos deposits are usually found in altered ultramafic igneous rock that came from the lower

Asbestos markets, trade, and environmental issues

The United States and most other MDCs banned the manufacture, importation, processing, and distribution of most asbestos-containing products in the 1980s. The results of this ban were dramatic, with US consumption falling from about 500,000 tonnes in the late 1970s to only 1,000 tonnes in 2010. Asbestos was originally used in insulating materials, cement pipe, vinyl floor tile, roofing shingles, and brake pads. The only markets that remain are cement pipe, rubber gaskets, and brake pads, where asbestos gives the products strength and heat resistance and where it is bound into a matrix from which it cannot escape easily.

Concern about asbestos centers on its role as a cause of lung diseases, of which there are several. **Asbestosis** is a chronic affliction resulting from inhalation of asbestos fibers and is commonly seen in workers who have been associated with high levels of asbestos dust. It involves an increase in the amount of fibrous protein in the lung, which decreases its flexibility and oxygen-absorbing capacity. The incidence of severe asbestosis has been declining for years because of dust control in the workplace. Lung cancer, which is associated with increased inhalation of fibrous minerals of all types, is not necessarily associated with asbestosis. Although the incidence of lung cancer among non-smoking asbestos workers is normal to slightly high, it is much higher among miners who are smokers as well as in some textile plants. **Mesothelioma**, which can be either benign or malignant, is a tumor in tissue that encases the lung. Malignant mesothelioma has been shown to be a statistically significant cause of death in mining districts in Northern Cape Province, South Africa, and Wittenoom, Australia, (Figure 13.12) both of which produced crocidolite, the amphibole asbestos known as blue asbestos. Mortality rates were high among these workers, their families, and even people with short exposure periods to crocidolite. For the more common chrysotile asbestos, studies show a less clear correlation with mesothelioma (Skinner *et al.*, 1988; Ross, 1987; Donovan *et al.*, 2012; Goswami *et al.*, 2013). At present, however, all asbestos is considered to be a group I carcinogen (International Agency for Research on Cancer, 2012, p. 294)

Research into the association between asbestos and lung disease has been hampered by the long period between asbestos exposure and symptoms, known as the **latency period**, which can exceed 30 years. Further complications came from confusion about the type of fibrous material to which people were exposed and the nature and duration of exposure. The asbestos problem is almost certainly a wider issue involving all fibrous materials, both natural and man-made. Strong evidence for this came from observations that residents of two small towns in Turkey exhibited high mesothelioma mortality with no significant asbestos exposure. This mortality was thought to be caused by erionite, a fibrous zeolite mineral discussed below, which was present in the air in concentrations of 0.01 fiber/cm^3 (Ross, 1987; Dogan *et al.*, 2008). Fiber size and shape appear to be important characteristics that determine the danger of fibrous substances, and problems appear to be associated with both natural and synthetic fibers (Mossman *et al.*, 1989; Magon, 1990; Cohen, 1991).

Currently, asbestos is produced and consumed largely in Russia, China, Brazil, and Kazakhstan. World reserves are no longer quantified, but are large in relation to present world production of about 2 million tonnes.

13.3.3 Mineral pigments – iron oxides

Mineral pigments, in the form of aboriginal paintings and ornamental coatings, were among the first uses for minerals. Although many minerals are used as pigments locally, the only natural pigments that are present in large enough amounts to be produced commercially are the iron oxides (Severinghaus, 1983; Tanner, 2014). Present annual world production of natural iron oxides is not publicly available, but is likely to be more than a million tonnes worth about $1.5 billion. This number has increased significantly over the last few decades as consumers have become more interested in natural materials and colors.

Natural iron oxides have a wide range of colors that reflect the main mineral present. Limonite has a yellow color, hematite has a red color, and magnetite has a brown to black color. These iron oxides are rarely found in the pure state in nature and the minerals with which they are mixed can also influence their color. Of particular importance are manganese oxides and organic material, both of which are black. The most common trade names used for iron oxide pigments are **ocher**, the yellow pigment rich in limonite, *sienna*, a reddish pigment containing hematite, and *umber*, a purplish pigment that also contains manganese oxide (Table 13.3). Ocher

Table 13.3 Composition (in weight percent) of common iron oxide pigments (after Siegel, 1960; Podolsky and Keller, 1994)

Pigment	Fe_2O_3	SiO_2	Al_2O_3	MnO_2
Ocher	17–60	35–50	10–40	--
Umber	37–60	16–35	3–13	11–23
Sienna	25–75	10–35	10–20	Small

BOX 13.5 | MINERAL PIGMENTS AND EARLY ART

Mineral pigments, especially ocher, umber, sienna, goethite, and hematite, were the mainstays of early artists. Fine-grained hematite was reported among flint and bone artifacts at the 250,000 year old Maastricht–Belvédère Neanderthal site in the Netherlands. The hematite is thought to have been used in a water slurry and the nearest place from which it could have been obtained is about 40 km distant (Roebroeks *et al.*, 2012). Evidence for the use of mineral pigments by early humans stretches as far back as about 200,000 years and it reached its apex in the amazing cave paintings of Lascaux, which are about 17,500 years old (Bahn, 2008).

deposits in Brixham, England, were used to coat sails of fishing boats, given rise to sayings about red sails in the sunset.

Concentrated iron oxide deposits form largely by weathering of iron minerals in rocks and ore deposits, and by direct sedimentation at submarine hydrothermal vents. The famous deposits of umber and ocher associated with the copper deposits of Cyprus (Figure 13.12) appear to have formed as fine-grained sediment during the waning stages of hydrothermal vent activity after deposition of metal sulfides (Robertson and Hudson, 1973). Ocher and umber are also mined at Cartersville, Georgia (Figure 13.12), where locally manganiferous iron formation has been weathered (Kesler, 1939). The large ocher and associated white clay deposits in Rajahstan, India are derived from weathered igneous rocks (Cavallo and Pandit, 2008). The deposits are almost always mined by open pits, and the material is prepared for sale by washing and separation into different grain sizes. The main producers are India, the main source of ocher, followed by Germany, a distant second.

The largest market for natural iron oxide pigments is in construction material, where it is used to color brick, tile, and cement products. Paints and other coatings are the second largest market. The disadvantage of natural pigments, and it is a big one from the standpoint of modern manufacturing, is their variability. Just a glance at the side of a hill will show how the color of weathered rock varies from place to place. In large deposits, blending can be used to maintain a specific color, but most deposits will not support this type of production for long. The suitability of natural pigment is also affected by its grain size, which affects color. As a result, there is a relatively large market for synthetic iron oxide pigments. In the United States, synthetic pigment production is almost double natural pigment production and it is valued at almost ten times as much because of its better, more consistent quality. In spite of the technical advantages of synthetic pigments, and their increasing availability as a result of environmentally induced iron oxide recovery in steel making and other processes, there

will always be a place for the cheaper natural pigments. Reserves for future production are not quantified, but are thought to be large relative to production (Tanner, 2014).

13.3.4 Barite

Barite is almost a one-use commodity. More than 90% of world consumption is used as the main filler in **drilling mud**, the water-rich slurry of finely ground barite, bentonite clay and other compounds that are used in oil and mineral drilling (Mills, 2006; Miller, 2013). Barite's high specific gravity (about 4.5) makes it an ideal material for drilling mud because it significantly increases the bulk density of the mud, causing rock chips to float and increasing pressures on pore fluids in the rock that is being drilled, thus preventing blow-outs. World barite production of about 8.4 million tonnes comes largely from China, India, Algeria, the United States, and Morocco, and is valued at about $800 million (Figure 13.11).

Barite deposits are almost entirely hydrothermal in origin. Many smaller deposits were precipitated from hydrothermal solutions that circulated through veins and other open spaces. They are often found in Mississippi Valley-type (MVT) lead–zinc–fluorite deposits or in veins around intrusions. The really large deposits, however, consist of lenses and layers of barite in shale, with or without associated lead and zinc sulfides. They are similar to the previously discussed lead–zinc sedimentary exhalative (SEDEX) deposits and take that name. SEDEX barite deposits are associated with lead and zinc in Paleozoic-age shales around Meggen and Rammelsberg, Germany, and in the Selwyn Basin in the Canadian Yukon (Figure 13.12), where they appear to have formed when barium-rich hydrothermal solutions vented into sulfate-rich seawater. Other SEDEX barite deposits consist entirely of barite, with no lead, zinc, or other metals, such as those in the Devonian-age Slaven chert in Nevada and in the Mississippian-age Stanley shale in Arkansas (Papke, 1984; Mitchell, 1984). These deposits probably formed when cool, reducing pore waters, possibly created by high organic productivity, dissolved barium from the sediment and then vented into seawater containing sulfate. They are similar in origin to the extensive sedimentary manganese deposits discussed previously (Brobst, 1983; Ampian, 1985a; Maynard and Okita, 1991; Orris, 1992). Similar barite deposits are forming today around active cold springs in the Gulf of Mexico (Feng and Roberts, 2010).

Because barite is almost inert during normal weathering and soil formation, weathering of barite-bearing rock forms residual deposits in which the regolith is enriched in barite. Deposits of this type, which developed on MVT deposits that cropped out at the surface, are common in the Potosi area of Missouri and near Cartersville, Georgia (Kesler, 1950).

Barite is mined largely by open pit methods and processing involves gravity separation methods and grinding to make a fine, pure powder. Environmental regulations in most areas require that barite used for drilling mud contains very low amounts of galena, sphalerite, and other sulfide minerals that might release metals. Barite itself is chemically very inert in oxidizing seawater and effects from released drilling mud are largely related to burial of any local organisms, although organic fluids mixed with the barite can create greater mobility of barium (Neff, 2000; Contreras-León, 2013). Barite reserves of 240 million tonnes are adequate for many decades.

13.3.5 Diatomite

Diatomite is a rock consisting of tiny siliceous fossils known as **diatoms**, which are single-celled aquatic plants similar to algae (Kadey, 1983). Only 3 million tonnes of diatomite are produced each year worth about $800 million (Figure 13.11). Most of this is used as a filtering agent, taking advantage of the unique cellular form of diatomite, with its numerous very small holes (Figure 13.16). Almost every liquid that you can imagine, including beer, fruit juice, organic chemicals, swimming pool water, and varnish, is filtered through diatomite to remove suspended solids. Smaller amounts of diatomite are used as fillers and extenders in cement, paint, paper, rubber and plastic and as the abrasive ingredient in some polishing compounds. It is also used as a filler in dynamite, where Alfred Nobel found that nitroglycerin adsorbed onto diatomite was stable (Crangle, 2013).

Diatomite deposits form as sediments in marine to freshwater. The best environment is a relatively shallow body of water with a good supply of dissolved silica to make diatoms, and limited contamination from clastic sediments. The most common source of silica appears to be glassy volcanic rock

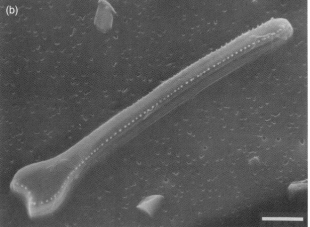

Figure 13.16 Examples of diatom morphologies. (a) *Stephanodiscus sp.* is a centric diatom that grows in the plankton of large, basic, hardwater lakes. (b) *Actinella punctata* is a pennate diatom that grows attached to underwater surfaces in very soft water, acid lakes, and bogs (the 10 micrometer bar in bottom photo provides scale for both images; courtesy of Mark Edlund, University of Minnesota).

that is easily dissolved during weathering. The presence of other elements such as boron in trace amounts in diatoms suggests that they, too, are needed for growth (Ibrahim, 2012).

Deposits of diatomite are mined largely in the United States, Denmark, China, and Japan (Figure 13.12). Mining is by open pit methods in the United States, although underground methods have been used in Spain and dredging is used in Iceland. Processing consists largely of heating to dry the material, followed by grinding and sizing. Additional heating can be used to fuse smaller particles together, changing filtration characteristics, if necessary. Because of the risk of silicosis, processing areas must be kept free of dust. Diatomite reserves are estimated to total about 1 billion tonnes, a very ample supply at present consumption rates, although they are not located close to their markets (Crangle, 2013).

BOX 13.6 | DIATOMS AND ACID LAKES

Diatoms have become an important source of information in the war against acid rain. It turns out that diatoms and another species of algae known as chrysophytes have very specific tolerances to water acidity. Because these organisms form in most lakes and their shells are well preserved after they die, they can be used to determine the pH of lake and stream water prior to the time when direct measurements began (Meriläinen, 1986; Dixit *et al.*, 1992). This sort of historical information is critical to assessments of the importance of anthropogenic sources of acidity. Figure Box 13.6.1 shows how the pH declined starting in the late 1800s in two well-studied lakes. One of these lakes, Round Loch (Figure Box 13.6.2), is typical of upland lakes in Scotland that have been impacted by airborne pollutants. Monitoring of the lake since 1988 has shown that it is recovering slowly but will require much more time to reach the pre-industrial levels that are indicated by the diatom record (Battarbee *et al.*, 2011).

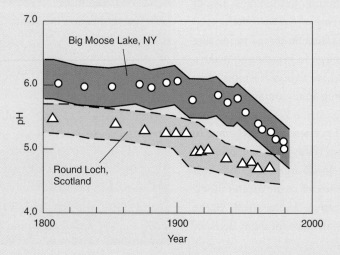

Figure Box 13.6.1 Historic record of pH changes in Round Loch, Scotland and Big Moose Lake, New York, as recorded by diatoms in lake sediment. The band surrounding data points shows measurement uncertainty (drawn from data of the US National Acid Precipitation Assessment Report, 1990 Integrated Report).

Figure Box 13.6.2 Round Loch of Glenhead, in Galloway, Scotland, provides a good example of a relatively fragile lacustrine system. The lake is underlain and surrounded by granitic rocks that are covered by peat and peaty soils, all of which do not react readily with acid water (photo courtesy of Ewan Shilland, University College, London). See color plate section.

13.3.6 Zeolites

Zeolites are a family of hydrous silicate minerals with peculiar crystal structures that allow them to adsorb or trap other atoms or molecules (Auerbach *et al.*, 2003). Industrial and environmental applications for zeolites are growing rapidly and world production of natural zeolites amounts to almost 3 million tonnes annually (Virta, 2014). Individual zeolites have very different prices, making an estimate of the value of global production difficult, although it would almost certainly be more than $300 million (Figure 13.11). Production comes largely from China, with smaller amounts from Korea, Japan, Jordan, and Turkey.

Zeolites are made of silicate tetrahedra, many of which contain aluminum rather than silicon. Substitution of Al^{+3} for Si^{+4} in the tetrahedron, produces a charge imbalance that is satisfied by other cations that are held loosely to the structure. Substitution is facilitated by the crystal structure of zeolites, which have passages that allow atoms and ions to enter.

Zeolites are used as **exchange** media and as molecular sieves. One of the first applications of zeolites involved water softeners, where sodium ions on the zeolite crystal exchanged places with calcium ions in the **hard water** to produce a sodium-bearing **soft water**. When all of the sodium had been exchanged by the zeolite, salty water was passed over it to replace calcium ions with sodium ions and it could be used again. More recently, the same principle has been used to remove lead and other undesirable elements from surface waters and process waters in chemical plants and smelters (Groffman et al., 1992). Mol sieves, as molecular sieves are known in industry, are widely used in separation of gases from one another and in petroleum refining (Bekkum et al., 1991). Some natural zeolites are used in less sophisticated applications, ranging from kitty litter to fillers in cement. As it turns out, synthetic zeolites can be made with a much wider range of crystal structures than natural zeolites (Auerbach et al., 2003). Many of these have larger holes in their structure that are especially useful for separating large organic molecules, and they are used widely in crude-oil refining and petrochemical manufacturing.

Natural zeolite minerals form in low-temperature, hydrothermal, and sedimentary environments (Olson, 1983). The main ions present in them other than silicon and aluminum are sodium, calcium, and potassium (Table 13.4). Although there are many natural zeolites, only analcime, chabazite, clinoptilite, and mordenite form large enough deposits to support mining. Zeolites are among the first new minerals to form as igneous and sedimentary rocks are buried and begin to undergo diagenesis and metamorphism, but these are usually dispersed through the rock and form mineable deposits only where the original rock was very rich in feldspar or volcanic glass and was extensively altered. Most deposits are found in volcanic tuff layers, especially where they were deposited in old lake beds. Most deposits contain more than one type of zeolite and are not as pure as synthetic zeolites.

Table 13.4 Commercially important natural zeolite minerals ranked in approximate order of declining production (compiled from Eyde and Holmes, 2006)

Mineral	Composition
Clinoptilite	$(Na_2,K_2,Ca)_3Al_6Si_{30}O_{72}.24H_2O$
Chabazite	$CaAl_2Si_4O_{12}.6H_2O$
Analcime	$NaAlSi_2O_6.H_2O$
Mordenite	$(Na_2,K_2,Ca)Al_2Si_{10}O_{24}.7H_2O$

Concern about the widespread use of zeolites was raised with recognition of the high incidence of mesothelioma associated with erionite, as mentioned in Section 13.3.2 on asbestos. Mordenite has also been implicated in increased lung fibrosis, but not lung cancer (Guthrie, 1992). Erionite differs from other zeolites in having a fibrous form, suggesting that crystal morphology is the important factor in determining acceptability of zeolite products (Ilgren et al., 2008).

Zeolite reserves are not known and are difficult to quantify because different applications have very different requirements. Even without new markets, demand is likely to increase substantially. The market for natural zeolites would increase if ways could be found to separate mixed zeolites.

13.3.7 Mica

Mica was valued by aboriginal people because of its sheen, which comes from light reflected off its hexagonal, platy crystals. The fact that large crystals could be cleaved into very thin sheets that were nearly transparent led to its early use in small windows in dwellings. Its main use now is as fine-grained fillers and extenders in paint, rubber, wallpaper, cement, and drilling mud, where it provides body and sheen. Smaller amounts are used as sequins and even as the gloss in lipstick. Larger crystals of mica are used as electrical insulators and as windows in special applications that require electrical and thermal resistance (Chapman, 1983; Davis, 1985b; Willett, 2013). World production of all types comes largely from China, Russia, the United States, and South Korea and amounts to about 400,000 tonnes annually worth around $400 million (Figure 13.11).

Most commercial mica is muscovite, the colorless, potassium-rich mica that is named for Moscow. Fine-grained muscovite mica comes from aluminum-rich metamorphic rocks that were derived from shales. Coarser-grained mica comes largely from granitic pegmatites. Significant fine-grained mica is produced as a by-product of kaolin and feldspar mining. Most mining focuses on deposits where the surrounding rock has been partly decomposed by weathering, leaving mica grains that are easier to free and concentrate by washing the mined material. Mica reserves have not been measured, but are large.

13.3.8 Talc and pyrophyllite

Talc may have been the first mineral that most of us came into close contact with. It is the main ingredient in talcum powder, which is widely used to prevent diaper rash. Most of talc's markets result from its sheet-like crystal form, which is similar to that of mica. Both can be cleaved into thin laminae, those of mica firm and resistant and those of talc soft and easily pulverized. The main use for talc powders is in ceramics and as a filler in paints, paper, cosmetics, plastics, roofing materials, rubber, and textiles. Rocks with large amounts of talc, known as **soapstone**, are used widely for aboriginal sculptures. Pyrophyllite, an aluminum silicate that is physically similar to talc, has many of the same uses, with a greater proportion of sales to the ceramics industry (Virta, 2013b). World production of these commodities is about 8 million tonnes worth about $200 million (Figure 13.11).

Talc is a hydrous magnesium silicate that forms by hydrothermal alteration of magnesium-rich rock, an origin similar to that of asbestos (Chidester, 1964). Talc deposits are most common where fault zones cut mafic and ultramafic igneous and metamorphic rocks or where silica-bearing hydrothermal solutions flowed through magnesium-rich dolomite. Some talc deposits contain the non-economic asbestos minerals tremolite and actinolite, which has led to concern about the environmental effects of talc use. Although no major problems have been identified for pure talc, products with asbestos-type minerals are considered dangerous (Honda *et al.*, 2002; Huncharek *et al.*, 2003).

Pyrophyllite deposits are found in similar hydrothermal settings where the original rock was rich in aluminum, or where hydrothermal solutions leached all cations from the rock. As noted earlier, pyrophyllite is a common alteration mineral around many high-sulfidation epithermal gold–silver deposits. These deposits do not normally contain asbestos-like minerals. Large amounts of talc and pyrophyllite are produced in China, Brazil, the United States, and India; reserves are not quantified but estimated to be very large in relation to present annual production.

13.4 Abrasive, lubricant, and refractory minerals

You are familiar with the use of abrasives in sandpapers, whetstones, grinding wheels and in powders for clearing paint and corrosion from metals and discoloration from brick and stone surfaces. But did you know that they are also used in soaps, detergents, toothpastes, cleansers, car polishes and silver polishes (Hight, 1983; Crookston and Fitzpatrick, 1983)? The most valuable natural abrasive material is *industrial diamond*, the hardest substance known, with silica sand a distant second and other products such as silica stone, garnet, tripoli, emery, feldspar, diatomite (Round *et al.*, 1990), and pumice bringing up the rear. The natural abrasive market is complemented by synthetic abrasives including diamonds, aluminum oxide prepared from bauxite, and silicon carbide prepared from quartz and coke. At the other end of the abrasive spectrum are lubricants, of which *graphite* is the only really important natural material.

Refractory materials are a similar motley lot. They provide heat-resistant bricks, blocks and other forms that are used in a wide range of industrial applications. Many of the materials used in glass are also used in refractories, including industrial sand itself, about 20% of which is used to make foundry sand. In addition to the new materials that are discussed here, other important refractory raw materials that have are discussed elsewhere in this book include magnesite, magnesium-rich chromite, zircon and bauxite.

The value of world production of these minerals is about $4 billion, led by industrial diamonds, which make up more than half of the total (Figure 13.17). The price of graphite has far outpaced the CPI since 1960, reflecting its unique properties as a non-flammable lubricant, although its production has not increased greatly. Production of garnet, feldspar and to a lesser extent kyanite have outpaced steel significantly, but all others have lagged.

13.4.1 Industrial and synthetic diamond

Industrial diamonds are those that cannot be used as gems. They range from imperfect stones of several carats to very fine-grained diamond powder (Reckling *et al.*, 1983). Although many industrial diamonds come from natural deposits, over 70 times as many are synthetic diamonds; in 2011, natural industrial-diamond production amounted to about 60 million carats worth several hundred million dollars compared to 4.4 billion carats worth about $2 billion for synthetic diamonds (Olson, 2013). The main producers of

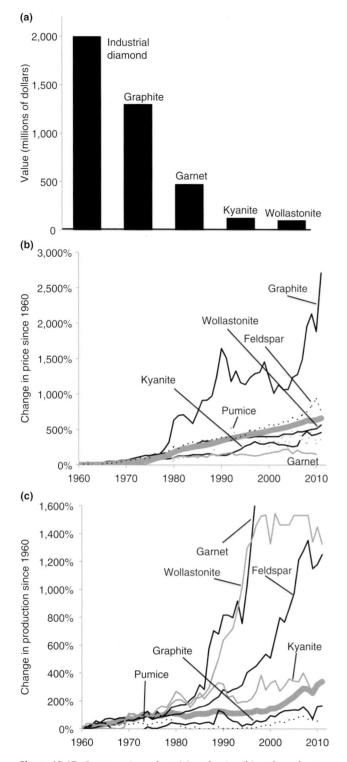

Figure 13.17 Comparative values (a) and price (b) and production (c) history for abrasive, lubricant, and refractory minerals (compiled from data of the US Geological Survey). Heavy gray lines in (b) and (c) are the price of steel and the consumer price index (CPI), respectively.

natural industrial diamonds are the Democratic Republic of Congo, Russia, Zimbabwe, and Australia and the main producers of synthetic diamonds are China, Ireland, Russia, South Africa, and the United States.

The main industrial markets for diamond depend on its hardness, which is greater than that of any other substance. Large stones are used in tools and bits for cutting rocks, and small stones, known as dust, grit, and powder, are used in grinding wheels, saws, and other devices for cutting and polishing rocks, minerals, metals, ceramics, and other hard substances. Diamond-tipped knives are used in everything from surgical procedures to cutting pasta. In addition to its use as an abrasive, diamond's hardness makes it an excellent form or die through which metal can be drawn to make special wires. Although 98% of natural diamonds and almost all synthetic diamonds contain small amounts of nitrogen substituting for carbon atoms, the small fraction that is free of nitrogen has very special thermal, light, and electrical properties. One type of nitrogen-free diamond is used as a heat sink in electronic devices, as windows and lenses in analytical equipment, and as semiconductors (Pressler, 1985).

Diamond synthesis, which began in 1955, is carried out by a high-pressure–high-temperature (HPHT) process that recreates the extreme pressures and temperatures of Earth's mantle where diamond is stable (Figure 11.16), or by chemical vapor deposition (CVD) (May, 2000; Meng *et al.*, 2008). Other methods to make synthetic diamonds include detonating explosions in carbon, or cavitating a suspension of graphite in organic liquid, but these are not widely used (Mackay and Terrones, 1991; Werner and Lochner, 1998; Ferro, 2002). The first of these methods produces most industrial diamonds, especially those smaller than about 0.01 carats. Larger synthetic crystals, up to about two carats in size, are available for specific applications but are not sold as gems, and synthetic polycrystalline aggregates can also be manufactured (Austin, 1992). CVD has the ability to place thin films of diamond onto surfaces of other materials at considerably lower temperatures and pressures than those used for traditional diamond synthesis. These diamond coatings could make glass and other materials resistant to abrasion, but their potential for making surfaces smooth is even more attractive. Thicker diamond films might also be used as semiconductors or sound conductors.

Natural industrial diamonds come from the mines that produce gem diamonds, with largest production from the placer deposits of Democratic Republic of Congo and the Argyle lamproite pipe in Australia (Figure 11.14). Of the natural-diamond producers, only South Africa and Russia also produce synthetic stones, accounting for about 40% of world synthetic-diamond production. Currently, natural industrial stones larger than 20 carats constitute less than 10% of the US market, with the rest dominated by synthetics. In fact, natural and synthetic industrial diamonds are in strong competition, and prices for both types of industrial diamonds have declined steadily since 1980, even though world consumption has increased. Very few opportunities exist to compare the environmental impact of a mined commodity to its equivalent synthetic commodity. This has been done for diamond, but with unclear results in view of the wide variety of mine settings from which they are recovered (Ali, 2011).

Natural industrial diamonds are likely to remain in heavy demand because of the scarcity of competitively priced synthetics in the large size ranges. World reserves of industrial diamonds, which are not tabulated separately, are probably similar to the 300 million carats estimated for gem diamonds and are concentrated in Australia, Botswana, Democratic Republic of Congo, and South Africa. The increasing rate of production in recent years suggests that they will be exhausted relatively rapidly, putting further pressure on the search for new diamond reserves, as discussed in the Chapter 11.

13.4.2 Other natural abrasives

Silica sand, the second most important abrasive material, comes largely from industrial sand deposits, which were discussed in Section 13.2.1 on glass raw materials. Because of corrosion of metals by salt, sand used for abrasive purposes is mined only from deposits that formed in freshwater.

Other forms of silica include silica stone and Tripoli, which are forms of **chert**, flint, and other amorphous to microcrystalline forms of silica that are found largely in sedimentary rocks. Some of these deposits formed as accumulations of siliceous organisms such as diatoms in deep marine environments where clastic sediments were scarce. Others formed during diagenesis shortly after deposition of the sedimentary rock. Tripoli, an industrial term derived from a deposit of this type near Tripoli, Libya, is a porous rock that forms by removal during weathering of carbonate and other minerals that were intergrown with the silica. The most famous of the many natural silica abrasive rocks is Arkansas novaculite, a Mississippian-age sediment that makes excellent whetstones.

Garnet is a common metamorphic mineral that is abundant enough to mine in a few rocks. In some locations, such as Gore Mountain, New York, it forms crystals as much as a foot (0.3 m) across that must have formed by hydrothermal recrystallization of original metamorphic minerals (McLelland and Selleck, 2011). Emery is an impure form of corundum, the aluminum oxide that forms ruby and sapphire, which is also found in metamorphic rocks. The high aluminum content of these rocks makes them relatively rare; they are likely to have been derived from metamorphism of a soil, laterite, or shale. Wollastonite is a calcium silicate mineral that forms during metamorphic reactions between quartz and calcite. It is used in special refractory ceramics such as the white tops of spark plugs.

The main environmental concern associated with the use of silica-rich abrasive products is the respiratory disease silicosis (Table 4.1). Silicosis is caused by inhalation of crystalline silica or quartz. It causes nodular areas in lung tissue and is commonly associated with tuberculosis, which it accelerates (Cowie, 1994; Rosner and Markowitz, 2006; Thomas and Kelley, 2010). Although it has been a serious problem in many mines that contain silica, including chert, its greatest effects have been seen in manufacturing settings that use silica in powder form, usually as an abrasive or ceramic ingredient. Silica sand remains an important contaminant in the workplace in many areas, partly because fine-grained sand remains suspended in air long after its use, where it can be inhaled by persons without respiratory protection. Broader recognition of these problems is causing substitution of garnet and other materials for quartz sand in abrasive applications. This is readily apparent from the enormous growth of world garnet production during the last two decades (Figure 13.17c).

13.4.3 Graphite

Graphite is the crystal form of carbon in which the atoms form hexagonal plates. It is used as the "lead" in pencils because of its color, softness, and crystal form, which causes it to cleave or smear off into black lines when drawn across paper. Other uses for graphite include facings for metal foundries, batteries that supply low current levels, bearings in high heat environments, brake linings, and lubricants (Kraus *et al.*, 1988). These uses depend in part on graphite's high temperature resistance and its excellent cleavage (Olson, 2013). Graphite is also used to make graphene, which consists of a one-atom thick sheet of carbon with the graphite crystal structure. World graphite production of about 1.1 billion tonnes is worth about $1.3 billion (Figure 13.17).

Most world graphite is synthetic and is produced from petroleum coke. A smaller amount is mined, largely from China, India, and Brazil, with some high-quality material also coming from Sri Lanka and Madagascar. Graphite deposits form layers, disseminations, or veins in metamorphic rock or contact-localized skarn deposits near granitic intrusions. Some of this graphite is the highly metamorphosed remains of original organic sediment such as coal, and layers found in ancient Precambrian rocks could be accumulations of algae and other early life forms. Other graphite deposits probably formed by direct deposition from hydrothermal solutions containing dissolved carbon. The crystallinity of material sold as graphite can range from excellent to almost non-existent, or amorphous. Synthetic graphite production will probably increase as more efforts are made to capture CO_2 emissions.

13.4.4 Kyanite and related minerals

Kyanite, sillimanite, and andalusite are different crystal structures of the aluminum silicate, Al_2SiO_5. These minerals are valued largely for their refractory characteristics and over 90% of US consumption goes into furnaces for steel, other metals, and glass. A smaller amount is used in high-temperature glasses and ceramics (Tanner, 2014). The value of world production, which amounts to about 400,000 tonnes, is worth about $140 million (Figure 13.17)

These minerals form by metamorphism of aluminum-rich rocks that are poor in other cations such as sodium, calcium, or iron. The three minerals form under different temperature and pressure conditions, which prevail in different parts of the crust. Andalusite is found in low-pressure environments such as intrusive rock contacts and regional metamorphic rocks that have not been deeply buried, whereas kyanite is found in more deeply buried, regionally metamorphosed rocks. Most metamorphic rocks that contain these minerals started as shale or some other aluminum-rich rock. Because they are resistant to erosion, these minerals can also be mined from weathered overburden or from beach sands derived from erosion of the overburden.

Andalusite is produced in the largest amount, coming from South Africa and France. Kyanite comes from the United States and sillimanite from India. Most is mined by open pit methods and processing includes crushing and grinding. Kyanite expands during heating and all three of the aluminum silicates react to form mullite and a silica-rich glass when heated to about 1,300 °C. Reserves have not been quantified, but are large.

13.5 The future of construction and industrial minerals

Construction and manufacturing minerals will remain very important for the foreseeable future. As composite materials are developed for specific applications, we will use less and less of the raw materials directly, but they will remain the basis for the more complex forms of cement, plaster, and glass. With the exception of glass, none of these materials are amenable to recycling. Thus, like the chemical and fertilizer minerals, we will have to rely on new deposits for these minerals for the foreseeable future. Fortunately, reserves for most of them, except diamonds, are relatively large, as long as we can resolve land-use and environmental disputes related to them.

CHAPTER

14 Global mineral reserves and resources

If history can be our guide, global mineral supplies will be a major factor in future world relations, as well as the focus of important environmental and economic controversies. At the heart of these disputes will be the fundamental questions of how much is left and how long it will last. Throughout this book, we have quoted reserve and resource figures, which are half of this equation. It is time now to review the ways that mineral reserves and resources are estimated, and then turn attention to the other side of the equation, mineral consumption.

14.1 Reserve and resource estimation methods

Recall from our first chapter that Earth's mineral endowment is referred to as a **resource** and the part that can be extracted at a profit is known as a **reserve** (Figure 1.3). Estimates of reserves and resources rely on geology and statistics (Singer and Menzie, 2010). Geological approaches range from simple enumeration of known deposits and their probable extensions to more complex assessments based on specific ore deposit types or favorable geological environments. Statistical methods range from those based on assumptions regarding characteristics of the population of mineral deposits to others that extrapolate exploration, production, or consumption trends (Wellmer and Becker-Platen, 2002; Sinclair and Blackwell, 2006; Singer, 2013; DeYoung, 2014).

14.1.1 Geological estimates

Geological estimates are the most traditional and probably the most dependable because they are based on actual observations about rocks. In the simplest method, reserve estimates for individual deposits are collected, evaluated, and compiled to determine the total reserve. Such estimates are useful only for measured and perhaps indicated reserves, and can therefore tell us only how much material has actually been observed and quantified. Geological estimates of resources are much more complex. In most cases these estimates involve the search for favorable geological environments likely to host deposits of interest. If more than one type of mineral deposit is involved, then each environment must be evaluated separately.

The search for conventional oil deposits provides a good example of geological methods. As you know, oil is released by thermal maturation of organic matter in source rocks and some of it is trapped in porous, permeable reservoir rocks from which it can be extracted. All of this happens in sedimentary basins and use of these geological insights to find oil is referred to as **basin analysis**. This involves a search for kerogen-rich source rocks in a basin that was buried deeply enough to enter the oil window and that contains good reservoir rocks and traps. The search can be assisted by any number of measurements on the rock. For instance, organic material can be extracted from shales and analyzed to assess its favorability for generation of oil, and whether or not it has passed through the oil window. We can study possible traps and processes that might modify them. We know, for example, that sandstone with large quartz grains makes a very porous reservoir rock. Thus, basins containing sediment eroded from coarse-grained, quartz-rich igneous rock such as granite would be likely to contain good sandstone reservoirs. Because granite makes up much of the continental crust, but is absent from the ocean crust and many oceanic island arcs, we would expect the best reservoirs of this type to be in basins adjacent to or on continents. Reefs that form on the margins of some basins, even in oceanic areas, can also be good reservoirs, and we could also direct our trap search toward rocks of the right age and geographic location to have formed large reefs.

Metal resources can be estimated in the same way. Consider nickel, for instance. As we have learned, nickel deposits are associated with ultramafic igneous rocks, either as laterites that formed when these rocks weathered or as nickel sulfide deposits that formed when immiscible metal sulfide melt separated as the ultramafic silicate magma cooled. Thus, the search for nickel deposits involves a search for ultramafic rocks that are (or were) in the right climatic and topographic setting to generate thick laterite soil, or that would have generated an immiscible sulfide melt during crystallization.

Once the processes that form a mineral deposit have been identified, geologic information can be used to rank areas in terms of the probability that they contain one or more deposits of a given type. In order to make rankings less subjective, or at least less dependent on a single person, it is common to incorporate criteria developed by several experts in these assessments. This is the approach that was used to evaluate mineral potential of wilderness areas in the United States, as discussed in Chapter 5 (Figure 5.9).

Although this approach provides an estimate of resources in deposits of a known type, it is almost useless for estimating resources in deposits that have not yet been recognized. For instance, Carlin-type gold deposits were not discovered until the early 1960s, and unconformity-type uranium deposits were not known as a distinct class when the uranium exploration boom began in the 1970s. Thus, early estimates of gold and uranium resources failed to include these deposits. Unknown deposit types can be included in the estimates by imaginative geological hypotheses, but are more commonly evaluated by methods that emphasize a statistical or model-based approach, as discussed below. Although these methods do not identify the type of unknown deposit, they can suggest that large resources of some type remain to be found.

14.1.2 Statistical estimates

One of the most interesting applications of a statistical approach to resource estimates is the crustal abundance relation (McKelvey, 1960), which holds that reserves for some elements exhibit a constant ratio to their average **crustal abundance** (Figure 14.1a). If this relation holds true for all mineral commodities, reserves for less-explored commodities can be estimated from reserves for well-explored ones. Furthermore, if the relation is true for reserves, it might also be true for resources, thus permitting estimation of resources for a wide range of minerals from resource estimates for a few commodities.

Another approach, known as the **Lasky relation** (DeYoung, 1981), is based on the fact that the volume of ore for most commodities increases logarithmically with arithmetic decreases in grade (Figure 14.1b). Although the exact form of this relation differs for different deposit types and areas, it provides a comforting indication that the volume of material available will increase greatly for even a small decrease in grade. By extrapolating to an extremely low grade, we might even postulate the existence of an enormous, almost infinite resource, but such a conclusion runs afoul of common sense and is not consistent with the original formulation of the Lasky relation (DeYoung, 1981; Gerst, 2008). It also assumes that the mineral form of the element or commodity of interest remains the same in all types of rocks. As we have seen, most elements form their own ore minerals, such as metal sulfides,

---

| BOX 14.1 | **THE OUTLOOK FOR WORLD POPULATION** |

Mineral consumption will be a function of world population, which reached 7.3 billion in early 2015 according to the United Nations Population Fund. World population is thought to have grown continuously from about 370 million since the plague and famines that ended in about 1350. Much of this increase has been very recent. There were only 1 billion people on Earth in about 1800, 2 billion by 1927, and 3 billion by 1959. Since then, we have added another billion people every 15 years or so. How population will grow in the next century will determine the future of Earth's mineral supplies. For 2100, UN population estimates range from about 6 billion to 16 billion, depending heavily on birth rates and assuming no global pandemics. The medium growth rate projection of about 10 billion people will put enormous pressures on world mineral resources, particularly if incomes increase for a larger proportion of the population.

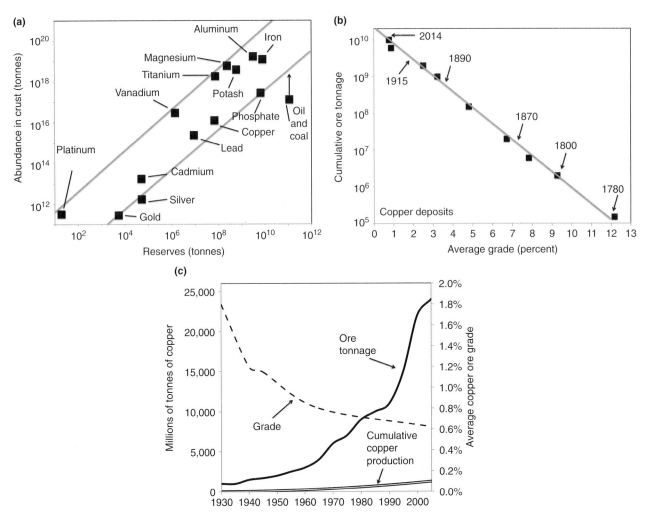

Figure 14.1 Geological relations used to estimate mineral reserves. (a) Relation of world mineral reserves to abundance of various elements in the crust. Oil and coal fall below this line because they come only from sedimentary rocks in which the abundance of carbon is higher than the 200 ppm estimated for the entire crust (data from Rudnick and Gao, 2003). (b) Lasky relation for copper deposits showing linear relation between arithmetic decrease in grade and logarithmic increase in tonnage of ore; arrows indicate calendar year corresponding to average worldwide copper ore grade; note the significant decline in grade (data from Crowson, 2012; USGS Mineral Commodity Summaries). (c) Relation among ore tonnage, copper grade, and cumulative copper production for worldwide copper mines, showing that production is increasingly dependent on lower-grade ore bodies (Crowson, 2012; USGS Mineral Commodity Summaries).

when they are present in high concentrations, but substitute in the crystal structure of common rock-forming silicate and oxide minerals when they are present in trace abundances. Skinner (1975) suggested that this difference could greatly limit our ability to exploit low-grade ores because it takes so much more energy to liberate elements that substitute in silicate and oxides, and that this will place an important upper limit on world mineral resources.

Statistical methods can also make use of exploration drilling data. Cargill *et al.* (1981) suggested that there is a logarithmic relation between the amount of exploration carried out (as measured by cumulative length of exploration drill holes) and the amount of ore or oil discovered, and that we

should find larger deposits with early exploration holes than with later ones. For instance, the average size of oil discoveries in the lower 48 states decreased from 23 million barrels of oil (Mbbl) in 1941–1942 to 1 Mbbl in 2010, confirming that early exploration finds most large, easily discovered deposits. This relation permits us to estimate the ultimate amount of ore or oil that might be found by extrapolating to some intense level of exploration drilling. The tendency to extrapolate to unrealistically large amounts of drilling, thereby arriving at unrealistically large reserve estimates similar to those mentioned above for the Lasky relation, can be constrained by studies of target geometry. For instance, statistical studies of the size and shape of the targeted mineral deposits can provide an estimate

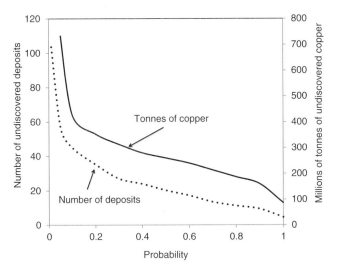

Figure 14.2 Probability distribution of the number of undiscovered porphyry copper deposits and tonnage of copper in the Tibet region of China determined by using a lognormal deposit density model (modified from Singer, 2013)

of the area of influence of a single drill hole (Singer and Drew, 1976; Singer and Kouda, 1988). This information can be used, in turn, to constrain the spacing and therefore the number of holes that can be drilled in an area to fully assess its mineral potential.

Recent attempts to estimate undiscovered resource availability involve Monte Carlo simulations that use the spatial density of known mineral deposits and reported metals tonnages and grades. For example, Singer (2013) used reported data for 3,021 known mineral deposits, from 28 ore deposit types, to determine that a lognormal distribution of metal abundances reproduces existing data, but only if applied to single deposit types and not across deposit types. An example of this model approach is shown in Figure 14.2, where the statistically predicted number of undiscovered porphyry copper deposits and total copper tonnage of undiscovered deposits are shown for the Tibet Plateau of China.

14.1.3 Combined estimates

The most effective reserve and resource estimates use all available geologic data as well as other information on production or other aspects of the resource. It was an estimate of this type by M. K. Hubbert that focused public attention on our dwindling oil reserves (Hubbert, 1985). The estimate was based on the prediction that mineral production will increase smoothly, reach a peak, and then decline at roughly the same rate that it increased, as shown in the theoretic model in

Figure 1.8a. Application of this concept to estimate ultimately recoverable reserves of a mineral commodity requires that the peak of the curve be identified. For a commodity that is consumed as rapidly as it becomes available, such as oil, the consumption rate should parallel the discovery rate (as indicated by the amount of reserve available each year), with the two curves separated by the time it takes to put the discovery into production. Hubbert found that this was indeed the case for US oil, where production followed discovery by about 10.5 years. On the basis of this relation, he estimated that oil production from the lower 48 states would peak in 1971 and decline steadily after that. A curve of this form indicated that oil reserves from the lower 48 states were about 170 billion barrels (Bbbl).

As can be seen in Figure 14.3, Hubbert's prediction was surprisingly accurate. The actual curve for the entire United States turned out to be larger because of the huge Prudhoe Bay discoveries, which caused a second peak in production in the 1980s, but the general shape of the curve followed the predicted form. What did change the shape of the curve, however, was the discovery of a new type of oil deposit, shale-hosted oil, which accounts for the large upswing in production that began in the late 2000s. Thus, the curve for US oil production now represents two different types of deposits, one consisting of conventional deposits with production on the decline and another consisting of new shale and tight-sand deposits with production still increasing. Just how far upward the new curve will go and when it will peak are important questions for US and ultimately for world oil production. It is important to note that while US oil production did peak in the 1970s, world production increased from 23 Bbbl in 1979 to 32 Bbbl in 2014. Much of this increase was driven by frontier exploration, especially deep-water production, and speaks to the difficulty of distinguishing a regional peak from a simple fluctuation in a rising trend until after the fact, particularly because so many other factors affect world mineral supplies.

Recent estimates for many commodities have been based on model calculations that combine information on resources of the mineral of interest with estimates of future price and trade patterns, as well as waste and recycling. In two studies of this type focusing on world copper supplies, it was estimated that peak copper production would take place in the twenty-first century and that conventional deposits would be exhausted sometime between 2100 and 2400 (Sverdrup *et al.*, 2014a; Northey *et al.*, 2014). A similar study suggested that conventional silver deposits would be exhausted by about 2240 (Sverdrup *et al.*, 2014b).

BOX 14.2 | ULTIMATE GLOBAL RESOURCES AND TECTONIC DIFFUSION

What is the ultimate resource that civilization will be able to wring from Earth's crust? How much time do we have left on this planet? A recent model-based approach sheds new light on this issue. As we know, many hydrothermal and magmatic deposits form at a specific depth in the crust. After they form, they move up and down through the crust with some reaching the surface and being eroded and others remaining buried at various depths. It turns out that this process can be modeled mathematically by assuming that movement of the deposits is a random process. This tectonic-diffusion model approach indicates that resources of copper in conventional deposits at all depths in the entire crust are about 3×10^{11} tonnes, almost 17,000 times greater than current annual copper consumption (Kesler and Wilkinson, 2008). However, most of the deposits are too deep to be discovered or produced. Limiting the estimate to crustal depths of about 3 km, which are more likely for exploration and production, and assuming that only about half of the deposits at these depths are discoverable and mineable, yields a resource only about 2,500 times larger than current consumption, suggesting exhaustion of conventional deposits sometime during the fifth millennium. This time period probably applies to many other mineral resources, and it would be even shorter if we could estimate future consumption rates. Regardless of exact consumption rates, however, the life of resources in conventional mineral deposits is probably much shorter than the history of civilization, a fact that is disturbing in its implications for future residence on the planet.

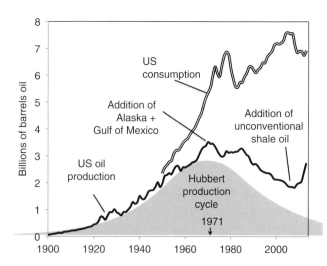

Figure 14.3 Predicted reserves and production cycle for crude oil in the United States (originally made by Hubbert in the early 1970s on the basis of data available through 1971 for oil production only in the lower 48 states) compared to actual production (including the lower 48 states, Alaska, and offshore production in the Gulf of Mexico) and US consumption through 2013 (from Hubbert, 1985; oil data from US Energy Information Administration).

14.2 Factors that affect the adequacy of world reserves

The adequacy of world mineral reserves and resources will be determined by our rate of consumption (Singer, 1977; Rogich and Matos, 2008). Consumption can change rapidly, as shown by the drop in uranium demand caused by negative public opinion about the safety of nuclear power, reinforced by the Fukushima Daiichi incident (Ewing and Ritsema, 2011). Market action, including all forces working to change demand for a product, has been discussed throughout this book, and new products are increasing demand for a wide range of mineral commodities. Of most importance is substitution, by which one commodity takes the place of another. As pointed out earlier, energy mineral markets have evolved through successive substitution of coal for wood, then oil for coal, then natural gas for oil. Metals have experienced similar substitution, with copper losing ground to glass in fiber optics, and steel losing markets to cement and aluminum. Government actions also have a strong effect on consumption, largely through their approach to stockpiling and recycling.

14.2.1 Stockpiles

There are many types of mineral stockpiles. Stockpiles maintained by the London Metal Exchange and other commodity exchanges are used to supply immediate industrial demand. Some private investment firms stockpile metals such as aluminum and copper in order to regulate market supply, hence artificially influencing commodities prices (Berthelsen and Tracy, 2014). Others, such as the large silver holdings in India, are held by individuals for personal or speculative reasons. The most important stockpiles, which are those held by governments, started as accumulations of strategic minerals essential for national defense that would be in short supply if world trade patterns were interrupted by conflict (van Rensburg, 1986; Paone, 1992). The present US National Defense Stockpile, under the supervision of the US Secretary of Defense, was authorized by the Strategic and Critical Stockpiling Act of 1939 and augmented by a related act of 1946. During World War II, mineral supplies to the United States were greatly disrupted by German submarines, which sank 96 ships containing Australian bauxite.

The nature of the stockpile has changed over the years, gradually losing importance. It was intended to last for 5 years in 1942, but that was cut to 3 years by 1958. By 1970, amounts required for the stockpile were lowered for many commodities, thus permitting further savings. In 1973, President Nixon asked to reduce the stockpile to a 1-year limit, and to dispose of 90% of its existing material. In 1985, President Reagan proposed to reduce the goals of the stockpile by over 50% (Bullis and Mielke, 1985; Mikesell, 1986). Part of the motivation for these changes was economic, because the stockpile represented a major investment that paid no dividends or interest. The present US stockpile contains about 25 mineral commodities, and has a market value of about $1.3 billion (Department of Defense, 2014).

Oil became part of the stockpile when the Strategic Petroleum Reserve (SPR) was authorized as part of the Energy Policy and Conservation Act of 1975. Early plans envisioned a stockpile of 500 Mbbl, at a cost of about $8 billion, to prevent a recurrence of the oil embargo of 1973. By 2014, the stockpile contained 691 Mbbl in five salt domes in the Gulf of Mexico (Figure 14.4), representing a supply adequate for 37 days, worth almost $35 billion at $50/bbl and $70 billion at $100/bbl.

Stockpiles can affect global exploration. Buying for a stockpile, as was done for uranium in the 1960s and 1970s, can stimulate exploration. Sale from a stockpile, as was done for gold from monetary funds in the 1990s and 2000s, can depress exploration. The stockpile is also a tempting rainy-day fund for governments when they are running budget deficits, and annual sales from the US stockpile have ranged from $462 million in 1998 to $116 million in 2013 (Department of Defense, 2014). Although there is pressure to completely dispose of the stockpiles, current tensions in the oil-rich Middle East and northern Africa and the concentration of cobalt, tantalum, and other commodities in the Democratic Republic of Congo make it unlikely that less-developed countries (LDCs) will abandon stockpiles completely.

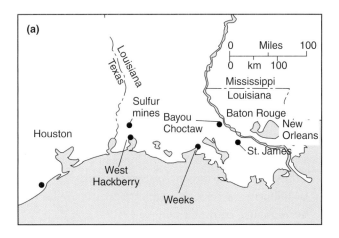

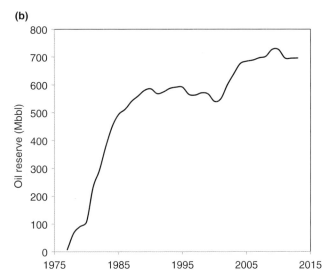

Figure 14.4 US Strategic Petroleum Reserve showing the location (a) and change in oil content of the reserve since 1977 (b) (data from the US Energy Information Administration)

14.2.2 Recycling

Recycling is the most perfect form of mineral use, a state in which we use the same material over and over again. The impact of this on mineral consumption would be overwhelming. We would only have to extract new material to provide for population increases and improvements in standard of living. However, some commodities simply cannot be recycled (Figure 14.5). They are consumed when they are used. Foremost among these are the fossil fuels, oil, natural gas, and coal, which are converted to water and carbon dioxide when they burn. Uranium is also depleted in ^{235}U during use in reactors and cannot be used further (although breeder reactors can create new nuclear fuels). The fertilizer minerals and salt are commonly dissolved and dispersed during use, preventing recycling. Thus, recycling can actually affect world

BOX 14.3 | HOW WILL POPULATION GROWTH AFFECT FERTILZER DEMAND?

In the United States and the European Union, daily per capita food consumption exceeds 3,000 kilocalories, whereas 0.5 billion people in LDCs survive on less than 2,000 kilocalories. To feed the world's growing population, demand for food is expected to double by 2050. Currently, worldwide, a land area equivalent to Africa is used to grow grains, and an area equivalent to South America is used to grow animals for human consumption. Good agricultural land is limited, so yield from land will have to be increased. The consumption of fertilizers is expected to increase from 166 million tonnes in 2007 to 263 million tonnes in 2050 (Alexandratos and Bruinsma, 2012). Consumption in developing regions such as sub-Saharan Africa is predicted to increase by 600% by 2050, including a 250% increase in fertilizer use per hectare of land. The potential impact to the environment is huge. To minimize eutrophication and saline runoff, as well as energy required for mining and transportation, we will have to improve our fertilizer application patterns.

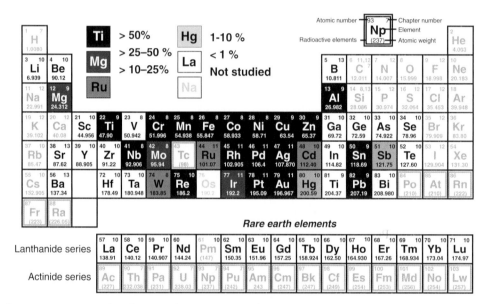

Figure 14.5 Rates of post-consumer recycling for natural resources; note that some resources such as the rare-earth elements used in all technology products are recycled at rates of less than 1% (modified from United National World Environment Program)

mineral supplies only for the metals and some industrial minerals and products such as glass and some plastics. Even the metals are subject to corrosion, the process by which they oxidize or undergo other reactions that convert the metals into forms, such as rust, that are more stable at Earth's surface; thus, in the long term (centuries and beyond) most materials are not recyclable (Waldman, 2015).

There are two major settings in which recycling can be undertaken, industrial and municipal. Industrial recycling is simply good sense and it is practiced widely (EPS, 2012). Many industrial facilities produce concentrated wastes that can be processed easily to recover metals and other compounds. Among the metals, steel, aluminum, copper, and lead are most widely recycled and trade in them constitutes a major industry (Figure 14.5). In the United States, 8,000 companies recycle scrap metal worth about $80 billion annually, with iron and steel for the auto, construction, and steel industries comprising half of all recycled metal. In 2013, the United States and European Union exported a combined 35 million tonnes of scrap metal, reducing the demand for primary mine production and landfill space. Today, in North America, more steel (65 million tonnes) is annually recycled

BOX 14.4 | WHAT IS SUSTAINABLE DESIGN?

Sustainable design is a philosophy that applies to both manufacturing and construction with the goal of minimizing material requirements and environmental impact (Vallero and Brasier, 2008). In the United States, buildings account for 40% of energy use, 70% of electricity use, 12% of water use, and are responsible for 40% of all CO_2 emissions to the atmosphere. And this does not count the energy and water used to extract and transport the resources necessary to construct the buildings. Almost two-thirds of building emissions are caused by cooling and heating the space. To reduce energy and water consumption, architects practice sustainable building design including site selection, energy efficient windows and insulation, sensors that monitor activity levels and adjust energy consumption accordingly, low-flow or no-flow toilets, and onsite collection and re-use of water for landscaping. The Leadership in Energy and Environmental Design (LEED) program certifies buildings built to reduce their environmental footprint. Sustainable design of products aims to incorporate recycling into the design and manufacture by making them easier to recycle. This, in turn, reduces the long-term need for new resources to replace products at the end of life. In simplest terms, sustainable design aims to mine a resource once, and use it forever.

than aluminum (4 million tonnes), glass (2.7 million tonnes), paper (45 million tonnes), and plastic (2 million tonnes) combined. The growing importance of industrial recycling has led to greater attention to the entire life-cycle of elements from production to end of life with an emphasis on minimizing waste (Izatt *et al.*, 2014).

Complications can arise. For instance, early use of steel scrap ran afoul of environmental regulations because it contained polychlorinated biphenyls (PCBs), lead, and other elements that are considered hazardous (Coppa, 1991; Porter, 2014). The US Environmental Protection Agency (EPA) banned production of PCBs in 1977. However, an estimated 1 billion pounds of PCBs were produced between 1929 and 1976, and an estimated 300 million pounds remain in building materials. Combustion of plastics can also produce PCBs, making recycling of plastics challenging. These problems have led to efforts to make products that are more amenable to recycling, a process that is part of sustainable design (McDonough and Braungart, 2002).

Annual production of solid municipal waste is almost 300 million tonnes in the European Union, 250 million tonnes in the United States and 80 million tonnes in the United Kingdom. According to the US EPA, the average composition of this wastes in the United States is 27% paper and paperboard, 14.5% food waste, 13.5% yard wastes, 12.7% plastics, 8.9% metals, 8.7% rubber, leather, and textiles, 6.3% wood, 4.6% glass, and 3.4% classified as other. Although it should be possible to separate these components by the same beneficiation processes used on ores, this does not work. The simplest system would involve a crusher/shredder to break up the

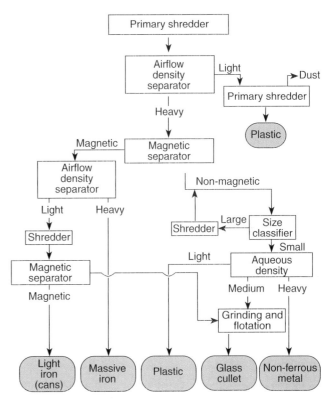

Figure 14.6 Schematic processing plant for recycling of municipal wastes (modified from Philipps, 1977)

material, a magnetic circuit to remove iron and steel, and some sort of gravity-based circuit to separate glass and other metals from paper and plastics (Figure 14.6). Further separation could be undertaken by flotation or other gravity methods. However, there are some very real differences between mines and municipal recycling facilities. The nature of the

BOX 14.5 | RESOURCES FOR A SUSTAINABLE FUTURE

Solar and wind power, along with electric cars, are often thought of as the sustainable future, but there are complications. In the United States, if everyone switched to an all-electric Tesla, annual electricity consumption would double. In most areas, electricity is generated by combustion of fossil fuels. In the absence of CO_2 recovery systems in these power plants, electric cars have a larger carbon footprint than those powered by gasoline. Can we change this by transitioning to renewable energy? The United States, Germany, and many other more-developed countries (MDCs) have enacted goals for a gradual transition from fossil-fuel-based electricity to renewable forms of energy. In the United States, tax credits were designed to stimulate addition of wind and solar energy to the electrical grid, and new installed wind-energy capacity increased annually by 30% from 2005 to 2015. Progress has been much better in Europe. In Germany, wind and solar combine to produce almost one-third of the country's electricity, and Denmark aims to produce 100% of its electricity from wind and solar. However, wind turbines require copper for generators, and the rare-earth elements dysprosium and neodymium for high-performance permanent magnets. Solar panels require copper, silver, tellurium, indium, and rare-earth metals (see Table Box 14.5.1). If new wind capacity increases annually by 15%, the demand for rare-earth metals will exceed current mine production by the year 2020. This will require increased production at existing mines and also development of new mines.

Table Box 14.5.1 Mineral resources required to manufacture hybrid cars, solar panels, and wind turbines

Hybrid car	Solar panel	Wind turbine
Bauxite (aluminum)	Arsenic (semiconductor)	Aggregate (for concrete)
Cadmium (batteries)	Bauxite (aluminum)	Bauxite (aluminum)
Chromite (chromium)	Boron (semiconductor)	Clay and shale (cement)
Coal (coke for steel)	Cadmium (thin-film solar cells)	Coal (coke for steel)
Cobalt (alloy; batteries)	Coal (coke for steel)	Cobalt (magnets)
Copper (wiring)	Copper (wiring; thin-film cells)	Copper (wiring)
Gold (circuitry)	Gallium (solar cells)	Gypsum (cement)
Iron ore (steel)	Indium (solar cells)	Iron ore (steel)
Lead (batteries)	Iron ore (steel)	Limestone (cement)
Lithium (batteries)	Molybdenum (photovoltaic cell)	Molybdenum (alloy in steel)
Manganese (steel alloy)	Lead (batteries)	Dysprosium (magnet)
Nickel (batteries; alloy)	Phosphate rock (phosphorous)	Neodymium (magnet)
Platinum (circuitry)	Selenium (solar cells)	Zinc (galvanizing)
Lanthanum (batteries)	Tellurium (solar cells)	Silica sand (cement)
Neodymium (electric motor)	Titanium dioxide (solar panels)	
Silica (silicon)		
Sulfur (chemical solution)		
Tungsten (wiring)		
Vanadium (alloy)		
Zinc (galvanizing)		

feed (i.e. the waste) can shift rapidly as a result of seasonal changes, holidays, or market forces. This causes inefficient separations because most plants are designed for a feed of roughly constant composition. The volume of waste also varies greatly, far outside the limits that can be accommodated by most plants, which always operate near capacity to be economic. But organic material in municipal wastes spoils too rapidly to be stored until the volume of feed drops enough to accommodate extra material.

As a result of these complications, current practice in most areas is to collect and treat easily recyclable material including paper, plastics, metals, and glass separately and then to separate them by beneficiation. The volume of material is controlled by collecting waste in special bins. While total US waste volume has increased, per capita waste generation has actually decreased since 1990, and recycling rates have increased from 5.6% in 1960 to 34.5% in 2014. Per capita waste generation in the United States peaked in 2000, and if we apply the Hubbert treatment to waste-generation data, this predicts that per capita waste generation will continue to decline as recycling becomes more efficient and effective. In some states and municipalities, addition of a mandatory front-end deposit fee on glass bottles has proved successful, which speaks to the power of financial incentives to stimulate consumer recycling. According to the US EPA, current rates of overall recycling in the United States are about 65% for paper and paperboard, 28% for glass, 20% for aluminum, 34% for steel and other metals, 8.8% for plastics, 18% for rubber and leather, 16% for textiles, 15% for wood, and 58% for yard waste.

One clear success story has been the municipal waste recovery of natural gas, stimulated by programs such as the US EPA Landfill Methane Recovery Outreach Program. It has been known for many years that decomposition of organic matter in municipal wastes generates large volumes of methane, accounting for 18% of total methane emissions in the United States. By installing perforated pipe systems in landfills, this gas is collected and used to generate electricity at about 400 landfills in the United States. However, the number of landfills in which to put unrecyclable waste in the United States has declined from 18,500 in 1979 to 3,091 in 2014. The alternative for some waste is incineration, which has come under scrutiny because of its potential to release toxic metals to the atmosphere (Seltenrich, 2013; Antonopoulos *et al.*, 2014). Incinerators can be fitted with scrubbers and particulate removal systems to limit volatile metal and organic emissions, however. The future probably lies in sustainable designs and recycling, likely followed by incineration. The need for this approach is underscored by the growing shortage of space for landfills and the recognition that many plastic and paper items do not degrade as rapidly as had been thought.

14.3 World reserves and the challenge for the future

The best estimates of world mineral production and reserves are tabulated in Table 14.1. A quick overview of our mineral supply situation can be obtained by dividing reserve figures by present consumption for each commodity of interest, giving us the period of years for which current reserves are adequate. This approach shows that reserves of 19 of the 48 mineral commodities for which estimates are available are adequate for more than a century (Figure 14.7). Coal, bauxite, vanadium, potash, and the rare-earth and platinum-group elements are among this group. Unfortunately, another eight minerals, including silver, lead, zinc, gold, and tin, have reserves that will last for only 10 to 25 years. Just above them in the 25-to-50-year category are copper, manganese, and nickel. Natural gas is a bit better at 55 years. With world population and standards of living increasing rapidly, the time periods obtained using current consumption rates are almost certainly overly optimistic, however, possibly by a factor of two or three.

The real significance of these numbers to society depends on just how long it takes to find reserves and put them into production. Reserves are of no use whatsoever if they remain in the ground. As we have seen, exploration requires long time periods and access to large areas because we do not know exactly where some of these deposits are. In fact, we do not even know if the resources exist at all, although geologic estimates suggest that they do (Figure 14.2). Even after a deposit is discovered, it can take a decade to get it into production. By making environmental aspects of mineral exploration and production matters for public scrutiny, we have involved more people in the process and lengthened the time required to find and develop mineral deposits (Wellmer, 1992; Matthews, 2010). In most cases, these delays have led to better engineering and operations that are more acceptable environmentally. In some cases, lack of agreement among negotiators has led to excessive delays and cancellation, such as at the Pebble porphyry copper deposit in Alaska.

As noted earlier, any decision not to produce a mineral deposit necessarily requires that another be put into production. To be sure, some deposits should never be produced. But for most deposits we will probably have to accept that careful

Table 14.1 World reserves and annual production for mineral resources (compiled from the United States Geological Survey Annual Commodity Summaries 2014)

Resource	World reserves	World production	Unit of measure
Antimony	1,800,000	163,000	Tonnes
Barite	350,000	8,500	Thousand tonnes
Bauxite	28,000,000	259,000	Thousand dry tonnes
Bismuth	320,000	7,600	Tonnes
Boron	210,000	4,900	Thousand tonnes
Cadmium	500,000	21,800	Tonnes
Chromium	480,000	26,000	Thousand tonnes
Coal	861	7.9	Million tonnes
Cobalt	7,200,000	120,000	Tonnes
Copper	690,000	17,900	Thousand tonnes
Diatomite	800,000	2,150	Thousand tonnes
Fluorspar	240,000	6,700	Thousand tonnes
Gold	54,000	2,770	Tonnes
Graphite	130,000	1,190	Thousand tonnes
Ilmenite	700,000	6,790	Thousand tonnes
Iodine	7,600,000	28,500	Tonnes
Iron ore	170,000	2,950	Million tonnes
Lead	89,000	5,400	Thousand tonnes
Lithium	13,000,000	35,000	Tonnes
Magnesium	2,400,000	5,960	Thousand tonnes
Manganese	570,000	17,000	Thousand tonnes
Mercury	94,000	1,810	Tonnes
Molybdenum	11,000,000	270,000	Tonnes
Natural gas	185.7	3.3,699	Trillion cubic meters
Nickel	74,000,000	2490,000	Tonnes
Niobium	4,300,000	51,000	Tonnes
Oil	1,687.9	31.66521	Billion barrels
Peat	12,000,000	25,000	Thousand tonnes
Phosphate	67,000,000	224,000	Thousand tonnes
Platinum-group metals	66,000,000	403,000	Kilograms

Table 14.1 (cont.)

Resource	World reserves	World production	Unit of measure
Potash	6,000,000	34,600	Thousand tonnes
Rare earths	140,000,000	110,000	Tonnes of rare-earth oxide
Rhenium	2,500,000	53,000	Kilograms
Rutile	750,000	7,550	Thousand tonnes
Selenium	120,000	2,000	Tonnes
Silver	520,000	26,000	Tonnes
Soda ash	24,000,000	14,200	Thousand tonnes
Sodium sulfate	30,000,000,000	6,000,000	Thousand tonnes
Strontium	6,800,000	245,000	Tonnes
Tantalum	100,000	590	Tonnes
Tin	4,700,000	230,000	Tonnes
Tungsten	3,500,000	71,000	Tonnes
Uranium	14,000	140	Million pounds
Vanadium	14,000,000	76,000	Tonnes
Vermiculite	60,000	420	Thousand tonnes
Yttrium	140,000,000	110,000	Tonnes of rare-earth oxide
Zinc	250,000	13,500	Thousand tonnes
Zirconium	67,000	1,440	Tonnes

disruption of the land for a few decades is a necessary price for the benefits that accrue to society from the minerals that are extracted. This puts us in a difficult position. On the one hand, we need to find minerals and produce them, and on the other, we want to keep the world as clean and undisturbed as possible. Our messy history is a warning that past ways are not adequate for the job. Mineral wastes are among our worst legacies from earlier times and we have spent many decades cleaning them up. It is important to keep in mind that we all made these messes. Individual prospectors, big companies, and even the government at all levels created pollution as we dug for Earth's riches. However, production only occurs when someone consumes. Thus, we consumers are ultimately responsible for much of the mess and must join in the clean-up.

We must not let this clean-up effort cause us to forget the need to continue producing minerals. Modern society is based

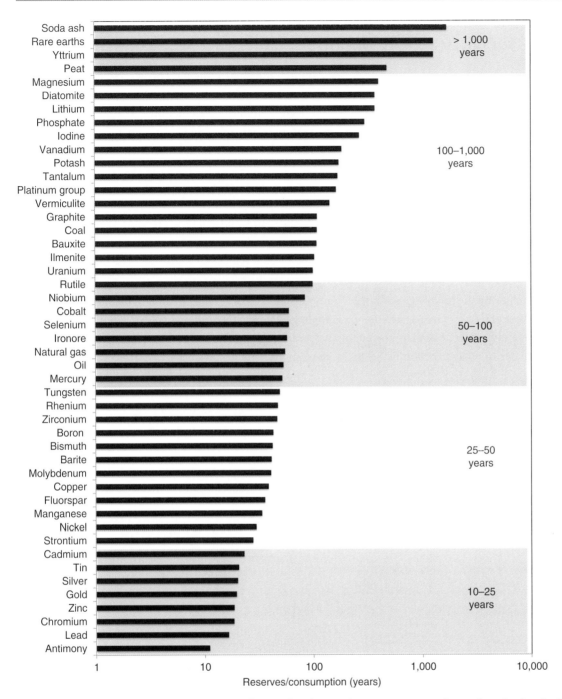

Figure 14.7 Adequacy of world mineral reserves illustrated as the ratio between reserves and annual production for 2013; resources such as indium and gallium, among others, that are co-products of mining other resources (indium and gallium are produced with zinc from sphalerite) are not shown (data from United States Geological Survey Mineral Commodities Summary)

on mineral consumption and no amount of wishful thinking will change that. In the United States, population grew by a factor of 2.4 from 1960 to 2012, whereas consumption of steel, cement, aluminum, and plastic increased by factors of 1.5, 1.4, 2.6, and 43. Worldwide, over the same time period, population grew by a factor of only 2.3, compared to consumption increases by factors of 4.3, 11.1, 9.5, and 48.

This big difference reflects the enormous increase in demand from China, and we can expect more of the world to do the same. As consumption increases in other LDCs and as population increases, we will require even more minerals. Although recycling will help, much of the minerals will have to be newly mined or pumped. Some of us respond to this challenge by trying to stop everything. Others forge ahead as

---

BOX 14.6 | THE SOCIAL LICENSE TO OPERATE

Mineral deposits have always had to overcome geological, economic, and environmental hurdles. However, a new hurdle has appeared, the Social License to Operate (SLO). SLOs are required for exploitation of a mineral deposit in democratic countries, even if governmental economic and environmental permission has been provided. Mineral companies usually seek SLOs by disseminating information on the proposed operation and conducting public question-and-answer sessions. This process puts a new stakeholder, with a completely different perspective, into the approval process. Whereas the discovering company and the host government stand to benefit economically from the operation, the general-public SLO stakeholder rarely sees this benefit, and instead focuses mostly on environmental risk. It should be apparent by now that few human activities, including mineral operations, are without risk. Thus, we require an informed public to help understand and balance the risks of a potential operation with its rewards, especially since we cannot stop all mineral projects without jeopardizing our supplies.

---

if there were no problem. Achieving the middle ground will require informed debate. Those of us who have learned more about mineral matters owe a large debt to society for our good fortune. We might be able to repay part of it by being responsible sources of information in the impending debate about global mineral resources, keeping in mind always that we are simply stewards of the planet for forthcoming generations.

APPENDIX 1
Minerals, rocks, and geologic time

Table A1.1 Classification of selected minerals based on anions (oxidation state of sulfur in sulfide minerals can be more complex than shown here). Some mineral compositions, elements, or subscripts are indicated by W, X, Y, x, and y as defined in table. More complete mineral compositions for some ore minerals are given in Appendix 2.

Name	Anion	Common example
Native elements	None	Gold – Au
Sulfides	S^{-2}	Galena – PbS
Oxides/hydroxides	O^{-2}, OH^-	Magnetite – Fe_3O_4
Halides	F^-, Cl^-, Br^-, I^-	Halite – NaCl Sylvite – KCl
Carbonates, nitrates	CO_3^{-2}, NO_3^-,	Calcite – $CaCO_3$
Borates	BO_2^{-4}, BO_4^{-5}	Dolomite – $CaMg(CO_3)_2$
Sulfates, chromates	XO_4^{-2}	Gypsum – $Ca_2SO_4.2H_2O$
Molybdates, tungstates	Barite – $BaSO_4$	
Phosphates, arsenates	PO_4^{-3}, AsO_4^{-3}, AsO_4^{-5}	Apatite – $Ca_5(PO_4)_3(F,Cl,OH)$
Silicates	$Si_xO_y^{-z}$	Quartz – SiO_2
(W = Ca, Na; X = Mg, Fe;		Kaolinite – $Al_4Si_4O_{10}(OH)_8$
Y = Al, Fe; Z = Si, Al)		Montmorillonite- $Al_2Si_4O_{10}(OH)_2.xH_2O$
		Mafic silicates
		Biotite – $KX_3(AlSi_3O_{10})(OH)_2$
		Pyroxene – $(W,X,Y)_2Z_2O_6$
		Amphibole – $(W,X,Y)_{7-8}(Z_4O_{11})_2(OH)_2$
		Olivine – $XSiO_4$
		Felsic silicates
		Potassium feldspar – $KAlSi_3O_8$
		Plagioclase feldspar – $WAl_{1-2}Si_{2-3}O_8$
		Muscovite – $KAl_2(AlSi_3O_{10})(OH)_2$

Table A1.2 Classification and compositions of major rock types (parentheses indicate whether rock is felsic or mafic, as defined in text)

Rock Type	Main Minerals, Grains or Characteristics
*Igneous – (volcanic/plutonic)**	
Rhyolite/granite* (felsic)	Quartz, potassium feldspar, plagioclase feldspar, biotite, muscovite
Latite/syenite (felsic)	Potassium feldspar, plagioclase feldspar, biotite, hornblende
Andesite/tonalite (intermediate)	Plagioclase feldspar, pyroxene
Basalt/gabbro (mafic)	Olivine, pyroxene, plagioclase feldspar
Komatiite/peridotite/kimberlite (ultramafic)	Olivine, pyroxene
Sedimentary	
Clastic	
Conglomerate	Gravel-sized grains > 2 mm
Sandstone	Sand-sized grains (0.0625 to 2 mm)
Quartz sandstone	Largely quartz grains
Graywacke	Largely rock grains
Arkose	Largely feldspar grains
Shale	Silt–clay-sized grains < 0.0625 mm
Chemical/organic	
Limestone	Largely calcite, often with fossils
Dolomite	Largely dolomite
Chert	Quartz and amorphous silica
Rock salt	Halite
Gypsum	Gypsum
Coal	Plant fragments, humic material
Metamorphic	
Greenstone, amphibolite	Metamorphosed basalt or gabbro
Marble	Metamorphosed limestone or dolomite
Slate	Metamorphosed shale
Quartzite	Metamorphosed quartz sandstone

* Pumice and obsidian are igneous rocks of granitic composition that consist entirely of glass; they did not cool slowly enough to form crystals. Pegmatite is an igneous rock with unusually large crystals. Carbonatite is an igneous rock (usually intrusive) made up largely of calcium and magnesium carbonate. Serpentinite is altered ultramafic rock, commonly seen in obduction zones and slivers along faults. Schist and gneiss are metamorphic rocks named for their textures, rather than their compositions. Schist consists of many flat or rod-shaped minerals that have a planar or linear arrangement; gneiss consists of layers of schist separated by layers of more equant minerals such as quartz and feldspar. Schist is commonly of metasedimentary derivation; gneiss is commonly of metaigneous derivation.

Table A1.3 Divisions of geologic time (from Walker *et al.*, 2012)

Eon	Era	Period	Epoch	Age range (Ma)
Phanerozoic	Cenozoic	Quaternary	Holocene	0–0.01
			Pleistocene	0.01–2.6
		Neogene	Pliocene	2.6–5.3
			Miocene	5.3–23.0
		Paleogene	Oligocene	23.0–33.9
			Eocene	33.9–56.0
			Paleocene	56.0–66.0
	Mesozoic	Cretaceous		66.0–145
		Jurassic		145–201
		Triassic		201–252
	Paleozoic	Permian		252–299
		Pennsylvanian		299–323
		Mississippian		323–359
		Devonian		359–419
		Silurian		419–444
		Ordovician		444–485
		Cambrian		485–541
Proterozoic	Neoproterozoic			541–1000
	Mesoproterozoic			1000–1600
	Paleoproterozoic			1600–2500
Archean				2500–4000
Hadean				4000–4650

APPENDIX 2
Ore minerals and materials

Table A2.1 Elements commonly recovered from natural ores showing the main ore minerals and their content of the desired element (in weight percent). Some of these minerals are also used in mineral form without processing, as noted in Table A2.2 below.

Element	Symbol	Ore minerals	Composition	Maximum content of desired element
Aluminum	Al	Bauxite		
		Boehmite	$Al_2O_3.H_2O$	43
		Diaspore	$Al_2O_3.H_2O$	43
		Gibbsite	$KAl_3(SO_4)_2(OH)_6$	43
		Alunite	$Al_2O_3.H_2O$	26
		Nepheline	$NaAlSiO_4$	19
Antimony	Sb	Stibnite	Sb_2S_3	72
		Tetrahedrite	$Cu_8Sb_2S_7$	25
Arsenic	As	Arsenopyrite	FeAsS	46
		Realgar	Smaltite	70
		Orpiment	AsS	61
		Löllingite	As_2S_3	73
		Enargite	$FeAs_2$	72
			$CoAs_2$	19
			Cu_3AsS_4	
Barium	Ba	Barite	$BaSO_4$	59
Beryllium	Be	Beryl	$Be_3Al_2Si_6O_8$	5
		Bertrandite	$Be_4Si_2O_7(OH)_2$	15
Bismuth	Bi	Native bismuth	Bi	100
		Bismuthinite	Bi_2S_3	81
Boron	B	Borax	$Na_2B_4O_7.10H_2O$	12
		Kernite	$Na_2B_4O_7.4H_2O$	16
		Colemanite	$Ca_2B_6O_{11}.5H_2O$	16
Cadmium	Cd	Greenockite	CdS	78
		Sphalerite	(Zn,Cd)S	< 1
Cesium	Cs	Pollucite	$(Cs,Na)AlSi_2O_6$	43
Chromium	Cr	Chromite	$(Mg,Fe)(Cr,Al,Fe)_2O_4$	44
Cobalt	Co	Linneaite	Co_3S_4	58
		Cobaltite	(Co, Fe)AsS	35
		Pyrrhotite	$(Fe, Ni, Co)_{1-x}S_x$	< 1

Table A2.1 (cont.)

Element	Symbol	Ore minerals	Composition	Maximum content of desired element
Copper	Cu	Native copper	Cu	100
		Chalcopyrite	$CuFeS_2$	35
		Chalcocite	Cu_2S	80
		Bornite	Cu_5FeS_4	63
		Enargite	Cu_3AsS_4	48
Fluorine	F	Fluorite	CaF_2	49
		Cryolite	Na_3AlF_6	54
Gold	Au	Native gold	Au	100
		Electrum	(Au, Ag)	Variable
		Calaverite	$AuTe_2$	44
Hafnium	Hf	Zircon	$(Zr,Hf)SiO_4$	<2
Iron	Fe	Magnetite	Fe_3O_4	72
		Hematite	Fe_2O_3	70
		Geothite	$Fe_2O_3.H_2O$	60
		Siderite	$FeCO_3$	48
		Pyrite	FeS_2	47
Lead	Pb	Galena	PbS	87
Lithium	Li	Spodumene	$LiAlSi_2O_6$	8
		Lepidolite	$KLi_2AlSi_4O_{10}F_2$	8
		Petalite	$LiAlSi_4O_{10}$	5
Magnesium	Mg	Olivine	$(Fe, Mg)_2SiO_4$	34
		Magnesite	$MgCO_3$	28
		Dolomite	$CaMg(CO_3)_2$	13
Manganese	Mn	Braunite	$(Mn, Si)_2O_3$	70
		Pyrolusite	MnO_2	63
		Psilomelane	$BaMn_9O_{18}.2H_2O$	52
Mercury	Hg	Native mercury	Hg	100
		Cinnabar	HgS	86
Molybdenum	Mo	Molybdenite	MoS_2	60
Nickel	Ni	Garnierite	$(Mg, Ni)_6Si_4O_{10}(OH)_8$	47
		Pentlandite	$(Fe, Ni)_9S_8$	36
		Pyrrhotite	$(Fe, Ni, Co)_{1-x}S_x$	<2
Niobium (Columbium)	Nb	Columbite	$(Fe, Mn)Nb_2O_6$	66
Phosphorus	P	Apatite	$Ca_5(PO_4)_3(F,Cl,OH)$	17
Platinum group				
Platinum	Pt	Sperrylite	$PtAs_2$	57
Palladium	Pd	Froodite	$PdBi_2$	20
Iridium	Ir	Laurite	$(Ru, Ir, Os)S_2$	61, 75, 75
Osmium	Os			
Rhodium	Rh			
Ruthenium	Ru			

Table A2.1 (cont.)

Element	Symbol	Ore minerals	Composition	Maximum content of desired element
Potassium	K	Sylvite	KCl	52
		Carnallite	$KCl.MgCl_2.6H_2O$	14
Rare-earth metals and related elements				
Yttrium	Y	Monazite	$(Ce,La,Th,Y)PO_4$	–
Important lanthanide elements		Bastnasite	$CeFCO_3$	64
		Xenotime	YPO_4	48
Cerium	Ce			
Europium	Eu			
Lanthanum	La			
Rhenium	Re	Molybdenite	$(Mo, Re)S_2$	< 1
Selenium	Se	Copper ores	–	–
Silicon	Si	Quartz	SiO_2	
Silver	Ag	Native silver	Ag	100
		Argentite	Ag_2S	87
Sodium	Na	Halite	NaCl	
Strontium	Sr	Celestite	$SrSO_4$	48
Sulfur	S	Native sulfur	S	100
		Gypsum	$CaSO_4$	24
		Pyrite	FeS_2	47
Tantalum	Ta	Tantalite	$(Fe, Mn)Ta_2O_6$	79
		Microlite	$(Na, Ca)_2Ta_2O_6(OH)$	69
Tellurium	Te	Calaverite	$AuTe_2$	56
		Copper ores	–	Trace levels
Thorium	Th	Monazite	$(Ce,La,Th,Y)PO_4$	71
Tin	Sn	Cassiterite	SnO_2	79
Titanium	Ti	Rutile	TiO_2	61
		Ilmenite	$FeTiO_3$	32
Tungsten	W	Wolframite	$(Fe,Mn)WO_4$	76
		Scheelite	$CaWO_4$	80
Uranium	U	Uraninite	UO_2	88
		Coffinite	$U(SiO_4)_{1-x}(OH)_{4x}$	60
		Carnotite	$K_2(UO_2)_2$ $(VO_4)_{2.3}H_2O$	57
Vanadium	V	Carnotite	$K_2(UO_2)_2$ $(VO_4)_{2.3}H_2O$	12

Table A2.1 (cont.)

Element	Symbol	Ore minerals	Composition	Maximum content of desired element
Zinc	Zn	Sphalerite	ZnS	67
		Willemite	Zn_2SiO_4	56
Zirconium	Zr	Zircon	$(Zr, Hf)SiO_4$	49($<$ 2)

Ore minerals and other solid, natural materials commonly used in mineral or natural form, with major uses for each. Some of these minerals are also processed for specific elements, as noted above and others, such as trona, sylvite, and halite, undergo some chemical processing before actual use.

Table A2.2 Ore minerals and other solid, natural materials commonly used in mineral or natural form with major uses for each. Some of these minerals are also processed for specific elements as noted previously and others, such as trona, sylvite, and halite, undergo some chemical processing before actual use.

Ore mineral	Composition	Major uses
Anhydrite (see gypsum)		
Asbestos		
Chrysotile	$Mg_6Si_4O_{10}(OH)_8$	Filler in cement and plaster
Barite	$BaSO_4$	Used in oil-well drilling to "float" rock cuttings and prevent blowouts; also in glass
Boron minerals		Glass, B chemicals
Borax	$Na_2B_4O_7.10H_2O$	
Kernite	$Na_2B_4O_7.4H_2O$	
Colemanite	$Ca_2B_6O_{11}.5H_2O$	
Calcite (limestone, marble)	$CaCO_3$	Fillers, lime (CaO), Ca chemicals
Chromite	$(Mg,Fe)(Cr,Al,Fe)_2O_4$	Refractory compounds
Clays		Construction materials, fillers,
Kaolinite	$Al_4Si_4O(OH)_8$	Fillers and extenders
Montmorillonite	$Al_2Si_4O(OH)_2.xH_2O$	
Diamond	C	Abrasive
Diatomite	$SiO_2.nH_2O$	Filler, filter
Feldspar		Ceramics, abrasives
Orthoclase	$KAlSi_3O_8$	
Plagioclase	$(Na,Ca)Al_2Si_2O_8$	
Fluorite	CaF_2	Blast furnace flux, F chemicals
Graphite	C	Refractories, lubricants, electrical equipment
Gypsum (anhydrite)	$CaSO_4.2H_2O$; $(CaSO_4)$	Plaster, cement
Kyanite	Al_2SiO_5	Refractory compounds
Limestone (see calcite)	–	–
Magnesite	$MgCO_3$	Refractory compounds

Table A2.2 (cont.)

Ore mineral	Composition	Major uses
Mica		Filler, electrical insulation
Muscovite	$KAl_2(AlSi_3O_{10})(OH)_2$	
Phlogopite	$KMg_3(AlSi_3O_{10})(OH)_2$	
Vermiculite	$Mg_3Si_4O_{10}(OH).xH_2O$	Thermal insulation, packaging
Ocher, umber, sienna	Fe-oxide/hydroxide	Pigment
Phosphate minerals		Fertilizer, phosphorus chemicals
Apatite	$Ca5(PO4)3(F,Cl,OH)$	
Potassium minerals		Fertilizer, potassium chemicals
Sylvite	KCl	
Carnallite	$KCl.MgCl_2.6H_2O$	
Pyrophyllite	$Al_2Si_4O_{10}(OH)_2$	Filler
Silica sand		
Quartz	SiO_2	Glass
Sodium minerals		Sodium chemicals
Halite (salt)	$NaCl$	
Trona	$Na_2CO_3.NaHCO_3.2H_2O$	
Nahcolite	$NaHCO_3$	
Talc	$Mg_3Si_4O_{10}(OH)_2$	Filler
Titanium dioxide minerals		Pigment
Rutile	TiO_2	
Ilmenite	$FeTiO_3$	
Wollastonite	$CaSiO_3$	Filler
Zeolites	Hydrous Na, Al silicates	Cation-exchange medium for water purification, etc.

APPENDIX 3
Units and conversion factors

Weight

1 tonne (t)	= 1 metric ton (mt)
	= 1000 kilograms (kg)
	= 1,000,000 grams (g)
	= 32150.7 troy ounces (t oz)
	= 1.102 short tons (st)
	= 0.984 long tons (lt)
	= 2204.6 pounds (lb)
kilogram	= 1000 grams (g)
	= 1,000,000 milligrams (mg)
	= 2.2046 pounds (lb)
1 troy ounce	= 31.1 grams (g)
	= 1.09714 avoirdupois ounces (oz)
1 short ton	= 2000 pounds (lb)
1 long ton	= 2240 pounds (lb)
1 pound	= 453.6 grams (g)
1 flask (mercury)	= 76 pounds (lb)
1 carat (diamond)	= 200 milligrams (mg)
1 karat (gold)	= 1 twenty-fourth part

Distance

1 kilometer	= 1000 meters (m)
	= 100,000 centimeters (cm)
	= 1,000,000 millimeters (mm)
	= 0.62 miles (mi)
1 meter	= 100 centimeters (cm)
	= 3.281 feet (ft)
1 mile	= 5280 feet (ft)
	= 1.609 kilometers (km)

1 foot	= 12 inches (in)
	= 30.48 centimeters (cm)
1 nautical mile	= 1.152 miles (statute miles) (mi)

Area

1 hectare	= 10,000 square meters (m^2)
	= 0.1 square kilometers (km^2)
	= 2.471 acres (ac)
1 square mile	= 640 acres (ac)
	= 259 hectares (hc)
1 acre-foot	= 43,560 square feet (ft^2)
	= 325,851 gallons (gal)

Volume

1 liter	= 1,000 cubic centimeters (cc)
	= 0.264 US gallons (US gal)
	= 1.0566 quarts (qt)
1 cubic meter	= 35.32 cubic feet (ft^3)
1 trillion cubic feet (Tcf)	= 28.3 billion cubic meters
	= 264.2 US gallons (US gal)
1 quart	= 0.9464 liters (l)
1 US gallon	= 4 quarts (qt)
	= 3.785 liters (l)
1 barrel	= 31.5 US gallons (US gal)
1 Imperial gallon	= 4 Imperial quarts (Imp qt)
	= 4.546 liters (l)
1 Imperial barrel	= 36 Imperial gallons (Imp gal)
1 barrel (oil)	= 42 US gallons (US gal)
	= 0.15899 cubic meters (m^3)
1 barrel (cement)	= 170.5 kilograms (kg)
	= 345 pounds (lb)

Energy

1 kilojoule	= 1,000 joules (J)
	= 239 calories (cal)
1 calorie	= 3968.5 British thermal units (Btu)
1 kilowatt-hour	= 3600 joules (J)
	= 3412.9 Btu

1 British thermal unit	= 1.055056 kilojoules (kJ)
1 kilowatt	= 3412.9 Btu per hour (Btu/hr)
	= 1 joule/second (J/s)
	= 14.34 calories/minute (cal/min)

Energy equivalents

1 tonne of coal	= 27,200,000 kilojoules (kJ)
1 bbl oil	= 6,000,000 kilojoules (kJ)
1,000 ft^3 (natural gas)	= 1,278,000 kilojoules (kJ)

GLOSSARY

abiogenic gas – natural gas formed by deep earth processes not involving organic matter

accuracy – degree to which an analysis approaches the correct number

acid mine drainage – acidic water, usually from oxidation of pyrite, which drains from areas disturbed by mining

acid rain – acidic rain, usually be caused by dissolution of CO_2, SO_2, and other gases

activated charcoal – charcoal or similar forms of carbon with a surface that has been treated to enhance its capacity to adsorb dissolved ions or molecules

ad valorem tax – an assessment of tax imposed on the value of property or goods

adit – a roughly horizontal tunnel with one open end, used as an entrance to a mine

adsorption – the process by which liquids, gases, or dissolved substances attach to the surface of solids; compare to *ion exchange*

advanced high-strength steels (AHSS) – steels with higher strength and, usually, lighter weight than HSLA steels; their properties can be due to special alloy elements or production methods, or both

aerobic – referring to an environment in which oxygen is present

aggregate – natural and synthetic rock and mineral material used as a filler in cement, asphalt, plaster, and other materials

alluvium – unconsolidated sediment deposited by streams – contrast with *colluvium*

alluvial fan – a fan-shaped deposit of alluvium

alpha particle – the nucleus of a helium atom consisting of two protons and two neutrons; one of the products of radioactive decay

alumina – Al_2O_3, the intermediate product in aluminum production

amalgamation – (1) the process by which adjacent mining operations are joined to make a larger one; (2) the process by which gold is dissolved from its ore by mercury, which is then separated from the dissolved gold by heating to drive off mercury vapor

amino acid – organic acids with carboxyl (COHOH−) and amino (NH_2−) groups

amorphous – without crystal structure, as in a glass

amortization – payment or deductions related to repayment of debt

andesite – an intermediate igneous (volcanic) rock

anaerobic – referring to a lack of oxygen

anion – a negatively charged ion

annealing – commonly used to refer to changes in metals or metal alloys during cooling

anomalous – samples that differ significantly from all others in a group or population

anorthosite – an intrusive igneous rock consisting largely of plagioclase feldspar

anthracite – high-grade coal with a very high C:H ratio

anthropogenic – formed or contributed by human activity (in contrast to those from natural Earth processes)

API gravity – measure of density of liquid petroleum; oils with greater density have lower API values reflecting the fact that lighter oils float on them

appropriation rights – doctrine that allows first (or sometime most important) user of water to acquire rights to the water

aqueous – containing water or pertaining to water

aquifer – rock or regolith through with groundwater moves easily (e.g. with high porosity and permeability)

Archean – eon of geologic time between about 4.0 and 2.5 billion years (Ga)

artesian well – a well that flows naturally, without pumping

asbestos – fibrous, acicular or thread-like minerals with minimum length:diameter ratio of three, usually serpentine asbestos (chrysotile) and amphibole asbestos (crocidolite, amosite)

asbestosis – chronic lung ailment resulting from inhalation of asbestos

assay – analysis of the grade of ore or concentrate

asthenosphere – a zone of highly viscous, mechanically weak mantle below the lithosphere, beginning at a depth of ~ 80 km and extending downward for at least 200 km

atmosphere – the shell of gas and atmospheric aerosol that surrounds Earth

atmospheric aerosol – particles of liquids and solids suspended in the atmosphere

atom – the smallest particle of an element that can enter into a chemical combination

atomic number – number of protons in the nucleus of an atom

atomic weight – weight of an atom in relation to oxygen as 16, determined largely by number of protons and neutrons

background – a term used to refer to values characteristic of the average or most common sample in a population (see *anomalous*)

backwardation – a situation where the future price is less expensive than the cash (spot, present) price

bacterial action – changes in organic matter caused by bacterial activity, including formation of kerogen

balance statement – statement of income and expenses provided by a company or other financial entity

banded iron formation (BIF) – chemical sediment with alternating layers of iron and silica or silicate minerals; also called taconite, itabirite, jaspilite, or ironstone

banded ironstone – see banded iron formation

basalt – mafic, extrusive igneous rock, commonly formed at mid-ocean ridge spreading centers; formed by partial melting of mantle

base metals – copper, zinc, lead, tin; so-named by ancient alchemists who could not convert them to gold

basic oxygen furnace – steel-making furnace with a jet of pure oxygen

basin analysis – evaluation of the likelihood that a sedimentary basin contains crude oil, natural gas, or other mineral deposits

batholith – large body of intrusive igneous rock, commonly measuring tens of kilometers in length

bauxite – ore from which aluminum is obtained; consists of hydrous aluminum oxide, clay, and silica minerals

bedrock – solid rock that underlies softer rocks or unconsolidated material near the surface

bench – step-like zones around walls of open pit to prevent rockfalls and collapse

beneficiation – physical separation of an ore mineral from its ore by crushing, grinding, froth flotation, and other methods

berylliosis – chronic lung disease caused by inhalation of beryllium compounds

beta particle – negatively charged particle identical to an electron or its positively charged equivalent; radioactive decay product

bioaccumulation – accumulation of an element or compound in living organisms; extreme bioaccumulation is biomagnification

biosphere – all living and dead organic matter on Earth

bittern salts – salts that crystallize from highly evaporated seawater, rich in potassium and magnesium

bitumen – semi-solid organic material similar in composition to heavy crude oil

bituminous coal – intermediate-grade coal; the most commonly used type

black lung disease – see coal worker pneumoconiosis

black smoker – submarine hot spring in which water contains suspended black particles of sulfide minerals precipitated by rapid cooling of water

blast furnace – furnace used to convert iron ore to pig iron

blister copper – smelter copper containing some gold, silver, and other metals that must be removed in a refinery

block caving – underground mining method involving collapse of large blocks of ore that is withdrawn through tunnels below the block

blowout preventer – large valve on oil, gas, and other deep drilling rigs to prevent escape of pressurized fluids

blue-baby syndrome (see *methemoglobinemia*)

body fluid – water, blood, and other natural fluids

boil – to change state from liquid to gas at the boiling point

boiling-water reactor – a type of nuclear reactor that uses boiling water as the working fluid

bonus bid – non-refundable (money) bid for the right to explore a tract of land

bottom ash – ash from coal combustion that falls to the bottom of the combustion chamber

brass – an alloy of copper and zinc

breccia – a rock consisting of angular fragments in a matrix of finer-grained rock or chemically precipitated minerals

breeder reactor – a nuclear reactor that produces more fissionable fuel than it consumes (see *converter reactor*)

brine – an aqueous solution with a salinity higher than that of seawater (about 3.5% by weight)

bronze – an alloy of copper and tin

bullion – strictly speaking, metal with a high content of gold or silver; in practice, impure gold and silver; sometimes also refers to lead metal

by-product – in mineral resources, any product that results from production of the main one; examples are molybdenum from some copper production and natural gas from some oil production

calcine – to heat and drive off a gas that is part of a compound; often used for process by which CO_2 is driven off from limestone ($CaCO_3$) to form lime (CaO)

CANDU reactor – a type of nuclear reactor that uses unenriched fuel

cap-and-trade system – a system of trade in which organizations (usually companies) with emissions below a set value can sell (trade) the unused part of their allowed emissions to other organizations with emissions that exceed the limit

caprock – impermeable rock that seals the top of some oil and gas traps; gypsum-rich insoluble residue at the top of a rising salt dome

carbon leakage – the process by which organizations (usually companies) with emissions that exceed a limit in one country transfer the operation to another country with less stringent emission limits

carbon sequestration – any process, natural or anthropogenic, used to remove carbon dioxide from the atmosphere or a waste gas stream and put it into a liquid or solid form that will be retained in Earth

carbon tax – a tax on CO_2 generated during consumption of fossil fuels

carbonate – pertaining to rocks or minerals containing CO_{3-2}, such as calcite, dolomite, and limestone

carbonatite – an igneous rock composed largely of calcite and dolomite

cartel – organization designed to control availability and price of a commodity

cash flow – total amount of money that moves into or out of a business or project, usually measured over a specific time period

casing – pipe used to line a drill hole to prevent collapse of the walls and escape or entry of unwanted fluids

catagenesis – geologic process intermediate between diagenesis and metamorphism, by which oil and gas are formed

catalyst – a substance that enhances the rate of a chemical reaction without being changed itself

cation – a positively charged ion

cellulose – a carbohydrate that is the main constituent of the cell walls of woody plants

cement – any new mineral that is precipitated from aqueous solution in pores in a rock; a mixture of powdered lime, clay, and other minerals that crystallizes to form a hard solid when water is added (hydraulic cement); the binding material in concrete

Cenozoic – the Era of geologic time from 66 million years (Ma) to the present

chain reaction – a self-sustaining chemical or nuclear reaction in which the products of the reaction cause new reactions

chelate – a compound with a ring-like structure, in which a central cation is connected to each surrounding anion by at least two bonds; commonly seen in organometallic molecules including chlorophyll and hemoglobin

chemical bond – force that causes atoms or molecules to link into more complex crystals or compounds; can involve

sharing (covalent bond) or capture (ionic bond) of electrons

chert – sedimentary rock consisting of fine-grained silica

chimney – a tube-shaped ore body, usually lead-zinc skarn

chromophores – elements or compounds that create color in minerals

claim – a parcel of land obtained under a law that permits public land to be obtained for mineral exploration and extraction; in the United States claims can be for bedrock deposits (lode) or alluvial deposits (placer)

clastic sediment – sediment consisting of fragments of minerals, rocks, and other material; when lithified, forms clastic sedimentary rocks such as conglomerate, sandstone, and shale

clay – (1) any of several hydrous aluminum silicate minerals, including kaolinite and smectite; (2) clastic material smaller than 1/256 mm

clinker – agglomeration of calcium silicate and other compounds formed by calcining limestone, clay, and other minerals in a cement kiln

coal – combustible rock consisting largely of the partly decomposed remains of land plants

coal-bed methane – natural gas recovered from coal deposits

coal combustion residuals (CCRs) – solid material remaining from combustion; coal ash

coal workers' pneumoconiosis (CWB) – lung disease caused by inhalation of coal dust (also known as black lung disease)

coalification – the process by which dead land plants are converted to coal

coating clay – kaolinite and other clays used to coat paper and other products

coking coal – coal used to make coke for steel plants and other industrial uses

colloid – a suspension of particles with a diameter between 10^{-9} and 10^{-6} meters in size, so small that surface characteristics control their solubility

colluvium – regolith that has not been moved (eg. as contrasted with alluvium)

coltan – informal term for a columbite–tantalite (mineral) solid solution

commodity future – a contract for sale or purchase of a commodity at a set date for a set price (commonly sold on margin)

complex ion – an ion that contains more than one atom

compound – a substance containing more than one element and having properties different from its constituent elements

concentrate – a product in which the proportion of ore mineral has been increased significantly above its concentration in ore, usually by crushing, grinding, and mineral separation by flotation or some other method

concession – the right to explore for minerals or produce them from a parcel of land; commonly granted by governments

concrete – an artificial rock consisting of a combination of cement and aggregate

condensation – the process by which a gas changes to a liquid

conglomerate – a coarse grained, clastic sedimentary rock containing rounded to semi-rounded pebbles and cobbles

continental crust (see *crust*)

continuous miner – machine that advances as it digs rock in an open pit or underground mine and transports it away from the mining face

continuous smelting – smelting process that includes roasting, matte production, and conversion to metal in a single continuous process

contour mining – a form of open pit mining in which a flat-lying layer in hilly terrane is mined from a bench cut around the side of a hill

convergent margin – a boundary where two lithospheric plates of Earth move together; to accommodate this movement, one plate will subduct or obduct

converter – the last step in conventional smelting of copper and other metals involving production of metal by blowing air through molten metal sulfide (matte)

converter reactor – a nuclear reactor that produces some fissionable material, but not enough to fuel another similar reactor

core – central part of Earth, consisting largely of iron; a cylinder of rock obtained by drilling

cost depletion – see *depletion allowance*

contango – a situation where future prices are more expensive than the present (spot, cash) price

cracking – second step in conventional refining of crude oil involving breaking large organic molecules into smaller ones

craton – unusually thick, stable continental crust

critical zone – zone of Earth extending downward from the top of the vegetation canopy to and including underlying aquifers; includes soil formation, plant growth, and groundwater

crude oil – liquid consisting largely of organic molecules containing 5 to 30 carbon atoms

crust – outermost shell of Earth; includes continental crust (granitic rocks and sediments) with thickness of 40 to 50 km, and ocean crust (basalt) with thickness of 5 to 10 km

crustal abundance – average concentration of an element in the continental crust

crystal – solid with a regular arrangement of atoms or ions

crystal fractionation – process by which crystals sink or float in their parent magmas, commonly changing composition of residual magma

crystal lattice – the orderly arrangement of atoms in a crystal

cullet – waste glass, commonly recycled

cumulate – a textural term referring to minerals that appear to have accumulated in layers within layered igneous complexes

cupola – protrusion on the top of an igneous intrusion; magmatic fluids and mineral deposits are thought to concentrate around cupolas

cut-and-fill mining – underground mining method that removes overlying rock in stages, with access to successive stages provided by waste rock that fills the opening

cut-off grade – minimum grade of ore that can be extracted profitably

cuttings – rock fragments cut by bit when rock is drilled

cyclone – a device that separates particles from fluids by centrifugal force

delta – deposit of clastic sediment where a stream enters a lake, river or ocean

dental amalgam – an alloy of mercury, silver, and tin used to fill dental caries

depleted uranium – uranium from which most daughter products of radioactive decay have been removed

depletion allowance – deduction from taxable income derived from mineral production, reflecting future exhaustion of deposit; variants include cost depletion (fraction of cost of property), percentage depletion (percentage of gross income), and earned depletion (fraction of funds expended in exploration)

desertification – process by which fertilel and becomes desert, usually by decreased precipitation

detection limit – the level below which an analytical method cannot detect the element or compound of interest

deuterium – isotope of hydrogen with one neutron and one proton

development – the process of preparing a mineral deposit for production

diagenesis – the chemical and physical processes that occur in sediments shortly after deposition

diatom – single-celled, siliceous aquatic plant related to algae

dike – tabular intrusive rock that cuts across enclosing strata, usually vertical

diorite – mafic to intermediate intrusive igneous rock

direct reduction process – production of iron without complete melting; produces sponge iron

discounted present value – the present value of a sum of money that will be received in the future based on a specific rate of return (discount rate)

discount rate – interest or inflation rate used to calculate present value of future income

dispersion – outward migration of ions, elements, compounds, minerals, rocks or other substances from a relatively concentrated source

distillation – the process of boiling a liquid and then condensing the resulting gas, usually to separate it from other liquids or solids

dividend – that part of after-tax income paid to owners of an operation as a return on their investment

divergent margin – a boundary where two lithospheric plates move apart, commonly a mid-ocean ridge

dolomite – carbonate mineral of that name [$CaMg(CO_3)_2$] and sedimentary rocks consisting dominantly of that mineral

doré – gold–silver bullion that must be purified in a refinery

dredging – open pit mining done from a barge that floats in a flooded pit digging its way along through regolith

drill – machine used to bore holes into rock for exploration or production; it uses a drill bit to cut through the rock

drill stem test – a test method in which a layer of rock intersected in a drill hole is isolated, allowing measurement of its capacity to release an economic flow of gas or oil

drilling mud – high-density slurry of pulverized barite, bentonite clay, and other minerals and compounds used to wash rock drill cuttings to the surface

drive – natural pressure of rocks, water, and gas forcing oil from a reservoir toward the surface

dry deposition – deposition of minerals or compounds from the atmosphere

dump – sell a commodity at a price below its real cost

Dutch disease – in mineral business, a negative impact on a national economy caused by a large inflow of foreign currency (usually paying for material sold by the country, such as oil); usually happens after a large discovery

earned depletion – see *depletion allowance*

economic rent – difference (positive) between the actual return earned on land, labor, or capital and the return that might be expected for reasons such as scarcity

electric arc furnace – steel-making furnace that is used largely with scrap iron or sponge iron

electron – very small, negatively charged particles that surround the nucleus of an atom, balancing the charge of the protons

electrowinning – concentration of one or more dissolved ions onto electrodes placed in a solution; the basis for hydrometallurgy

element – any substance that cannot be broken into smaller parts by chemical or physical reactions

eminent domain – the power of government to appropriate privately owned land for public use

energy resources – fuel minerals, including fossil fuels and nuclear fuels

energy tax – see *carbon tax*

enhanced oil recovery (EOR) – methods used to increase yield of oil from a reservoir, including CO_2 flooding, gas injection, thermal effects, and water flooding

enrichment – process by which the proportion of ^{235}U in uranium is increased sufficiently to permit fission to take place

epilimnion – the upper layer of water in a stratified freshwater lake

Equator Principles – a risk management framework adopted by financial institutions for determining, assessing and managing environmental and social risk in projects

equilibrium – in chemistry, the conditions under which a chemical reaction proceeds both forward and backward at the same rate; in economics, the conditions under which supply and demand are equal

ethanol – ethyl alcohol; grain alcohol (C_2H_5OH)

eutrophication – process by which increased organic productivity in surface waters consumes dissolved oxygen

evaporation – process by which a liquid becomes a gas

evaporite – rock consisting of minerals that precipitated from evaporated water

US exclusive economic zone – 200-mile zone of ocean water and floor around the United States and its territories claimed by the US Federal government

exhalative – refers to fluids that flow out onto the ocean floor from submarine springs, not necessarily hot fluids

exinite – see *liptinite*

expropriation – most commonly used when governments take assets from private owners with or without compensation

extender – anything added to dilute a more costly material; a filler

extralateral rights – in US mineral law, the right to mine minerals outside vertical boundaries of a claim because they are a continuation of ore within the claim

extrusive rocks – igneous rocks formed where magma reaches Earth's surface as flows or fragments

fast-neutron reactor – nuclear reactor that has no moderator and therefore uses fast neutrons to sustain the fission reaction; requires highly enriched fuel

feasibility analysis – test made to determine whether a deposit can be produced economically

felsic – refers to rocks rich in feldspar and its common elements, silicon, sodium, potassium, calcium, and aluminum

ferroalloy – any alloy made with iron

ferrous metals – iron and metals such as manganese, nickel, chromium, cobalt, vanadium, molybdenum, and tungsten, which are used to form ferroalloys

filler – see *extender*

fissile – (1) describes an isotope that can undergo nuclear fission; (2) describes rock that can be split into sheets or slabs

fission – nuclear reaction in which a radioactive isotope splits apart to form two daughter isotopes of approximately equal size, subatomic particles, and energy

flash smelting – process in which small grains of mineral or other material that are suspended in a stream of hot air or oxygen melt rapidly and sink to the bottom of the container to be collected

flocculation – electrostatic attraction of colloidal particles to one another to form larger grains

flue-gas desulfurization (FGD) – removal of sulfur from gases emitted by power plants, smelters, refineries, and other industrial installations

fluid inclusion – imperfection in a crystal that trapped part of the fluid from which the crystal grew

fluorosis – disease involving dental defects, bone lesions, and other effects of excess fluorine

flux – (1) substance added to lower the melting temperature of a material; (2) the amount of a substance that moves between reservoirs in a specific period of time

fly ash – small ash particles that go up the stack during combustion of coal and other materials

fool's gold – pyrite (FeS_2)

forward integration – business strategy in which activities are expanded to include control of the direct distribution and sale of the product

fossil fuel – coal, crude oil, natural gas, tar sand, oil shale, and related materials consisting of organic matter that has been preserved in rocks

fracking – see hydraulic fracturing

Frasch process – method used to extract sulfur from deeply buried deposits by melting it with superheated steam

freshwater – water with little or no dissolved solids

froth flotation – method used to make a concentrate from a slurry of pulverized ore by coating one or more minerals with an organic liquid that causes them to attach to bubbles and float to the surface

fuller's earth – any clay material that has sufficient absorbent capacity to clean and decolorize textiles

fulvic acid – humic material that remains in solution when alkaline extract (humic acid) is acidified

fumarole – volcanic vent through which steam and other gases reach the surface

fusion – melting of a substance; process by which two isotopes combine to form a new isotope and release energy

gabbro – mafic, intrusive igneous rock consisting of feldspar and pyroxene

galvanized steel – steel coated with a thin film of zinc to prevent corrosion

gamma rays – electromagnetic energy, with a shorter wavelength than X-rays, emitted during radioactive decay

gangue mineral – waste minerals in an ore deposit, commonly quartz and calcite

gas – a state of matter in which the atoms or molecules are essentially unrestricted by cohesive forces

gas hydrate – solid clathrate compound containing eight CH_4 molecules in a cage of 46 water molecules ($CH_4.5.75H_2O$)

gas injection – pumping natural gas into an oil field to maintain drive on the oil or to store oil or natural gas

coal gasification – the process by which coal is converted to natural gas

Geiger counter – a device used to measure alpha particles emitted by radioactive decay

geochemical cycle – flow of elements or compounds through reservoirs on Earth; can be at any scale from local to global

geochemical exploration – mineral exploration based on analysis of the chemical composition of rocks, soils, water, gas and living organisms

geochronology – determination of the age of a rock or mineral based on its content of radioactive and daughter isotopes

geomedicine – study of the effects of rocks and minerals on health

geophysical exploration – mineral exploration by measurement of magnetic, gravity, electrical, radioactive or other physical features of rocks and minerals

geostatistics – statistical treatment of spatially arranged data such as assay values from a mineral deposit

geothermal gradient – natural increase in Earth temperature downward in the crust and mantle (usually 20–40 °C/km in the upper crust)

geyser – hot spring that boils intermittently

ghost town – town that has been deserted, often after exhaustion of a nearby mineral deposit

glass – amorphous solid

global change – geologically rapid changes in climate caused by natural and anthropogenic changes in composition of the atmosphere and hydrosphere

goiter – swelling of the neck resulting from enlargement of the thyroid gland, a common result of iodine deficiency

gold standard – use of gold to determine the value of a currency

graben – valley formed when the crust is pulled apart (by extension)

granite – a common felsic, intrusive, igneous rock consisting of quartz, feldspar and mica

granular iron formation (GIF) – iron-rich sedimentary rock consisting of reworked grains of iron-rich sediment

greenhouse gas – a gas (especially H_2O, CO_2, CH_4) that prevents incoming solar energy from being radiated back into space, thus warming Earth

groundwater – water stored in the ground; commonly refers to water that can be extracted by wells for domestic, industrial, and agricultural use

guano – accumulations of bat and bird excrement

gusher – oil well that flows to the surface uncontrollably

Hadean – eon of geologic time prior to about 4 billion years (Ga) ago

half-life – period of time necessary for half of the amount an isotope to undergo radioactive decay

hard water – water that does not lather easily with soap; rich in dissolved calcium and magnesium

heap leaching – the process by which a fluid is sprayed onto a pile of crushed ore and allowed to percolate through, it leaching the desired element, which is recovered from the fluid; usually applied to gold ore using cyanide solutions

heavy oil – liquid petroleum with a high viscosity and API (American Petroleum Institute) gravity of less than 20 degrees

heavy water – water that is enriched in deuterium

hemoglobin – iron-containing protein in red blood cells that reversibly binds oxygen

heterogeneous mixture – mixture containing more than one phase, mineral or substance

high-strength, low-alloy (HSLA) steel – alloy steel with a wide range of compositions designed to provide better mechanical properties or greater resistance to corrosion than carbon steel

humate-uranium deposit – concentration of uranium in and around decayed plant material

humic acid – humic material that is soluble in dilute alkali solutions but precipitates when the solution is acidified

hydraulic cement – see *cement*

hydraulic conductivity – rate at which water can move through porous rocks (equivalent to permeability for water)

hydraulic fracturing (fracking) – injection of pressurized water-rich fluid into rock to create fractures and enhance porosity and permeability

hydraulic mining – use of a high-pressure stream of water to disaggregate gravels for mining and processing

hydrocarbon trap – hydrocarbon-bearing reservoir rock that is isolated by caprock from overlying porous, permeable rocks and the surface

hydrometallurgy – separation of desired metal from an ore or concentrate by dissolution and later precipitation or electrowinning

hydrosphere – that part of Earth consisting of water at and near the surface, including the ocean

hydrothermal alteration – mineral changes caused by the reaction between a rock and a hydrothermal solution

hydrothermal deposit – a valuable mineral accumulation that was precipitated from hydrothermal solutions

hydrothermal solution – natural hot water that circulates through Earth

hypogene deposit – mineral deposit formed by ascending water, usually hydrothermal (hot) water of magmatic, meteoric, basinal, metamorphic, or seawater origin

hypolimnion – deeper water level (below epilimnion) in a stratified freshwater lake

igneous rock – rock that formed by crystallization or cooling of magma; includes volcanic (extrusive) rocks that form when magma reaches surface and plutonic (intrusive) rocks that form below surface

immiscibility – process by which a homogeneous fluid separates into two fluids

impermeable – term used to refer to rocks through which fluids pass very slowly or not at all

import quota – limitation placed on the amount of a product that can be imported

industrial mineral – geological material of commercial value, not including fossil fuels or metals

inertinite – complex mixture of fungal remains, unoxidized wood or bark, and other altered plant material in coal

intermediate – a term used for igneous rocks that have compositions between mafic and felsic

intrusive rock – igneous rock that formed below Earth's surface

ion – atom or group of atoms that has lost (cation) or gained (anion) one or more electrons

ion exchange – exchange of ions between a solution and a solid

ionizing radiation – radiation that ionizes atoms by removing electrons

island arc – arc of volcanic islands above a convergent margin subduction zone

isotope – atoms with the same number of protons but different numbers of neutrons

itabirite – see *banded iron formation*

jaspilite – see *banded iron formation*

karst – landscape underlain by limestone that has been dissolved by weathering to produce sink holes and other collapse features

kerogen – complex organic molecules resulting from modification of organic matter preserved in sediments

kiln – long, horizontal cylindrical furnace that is tilted slightly to allow material to progress through it; widely used in the production of cement, iron ore pellets, and other mineral products

kimberlite – porphyritic peridotite with high sodium or potassium content, containing olivine, mica, and garnet; source of diamonds

kinetics – refers to rate at which chemical reactions take place

komatiite – ultramafic volcanic rock consisting largely of olivine and pyroxene; commonly associated with nickel deposits

lacustrine – term used to refer to water, sediments, and other features of lakes

Lake Copper – copper produced from mines in the Keweenaw district of northern Michigan, valued for its small amounts of silver

lamproite – porphyritic mafic to ultramafic rock commonly containing biotite and amphibole, similar to kimberlite

land withdrawal – restriction of the uses to which public land can be put

Lasky relation – relation suggesting that arithmetic decreases in grade will be accompanied by an exponential increase in amount of ore

latency period – period between exposure (to the cause of a health problem) and appearance of first symptoms

laterite – type of soil formed by extremely intense weathering that removes all but the most immobile elements

layered igneous complex (LIC) – mafic, intrusive igneous rocks containing (cumulate) mineral layers

lava – magma that extrudes onto Earth's surface

leach pad – pile of ore onto which reactive solution is sprayed to leach element of interest; commonly used to recover gold from low-grade ores using cyanide solutions

lead time – the time between the decision to invest in a project (or the earliest stages of evaluation of the project) and the time that it begins production

leaded gasoline – gasoline to which lead has been added, usually in the form of tetraethyllead $[Pb(C_2H_5)_4]$

legacy pollution – pollution that remains in the environment long after it was introduced

life-cycle assessment – the process of quantifying the environmental impact of a product, process, or service; done by inventorying the energy and materials consumed and the emissions released

light metals – aluminum, magnesium, titanium, beryllium, and other metals that have relatively low densities

light oil – liquid petroleum with a low density and low viscosity

light-water, graphite-moderated reactor (LWGR) – nuclear reactor that uses regular water as the coolant and graphite as the moderator; Chernobyl-type reactor

lightweight aggregate – aggregate with a relatively low density, including natural materials such as perlite and pumice, as well as shale and other rocks that have been expanded by heating

lignin – hard material embedded in the vascular matrix of plant cells

lignite – a form of low rank coal

lime – CaO; produced by calcining limestone or calcite

limestone – a sedimentary rock consisting largely of the minerals calcite and aragonite, which have the same composition ($CaCO_3$)

liptinite – remains of spores, algae, resins, needles, and leaf cuticles in coal

liquid – state of matter in which molecules can change position freely but maintain a relatively fixed volume by mutual attraction

liquefied natural gas (LNG) – natural gas that has been liquefied by cooling to temperatures of -162 °C

liquefied petroleum gas (LPG) – ethane, ethylene, propane, propylene, normal butane, butylene and isobutane produced at refineries or natural gas processing plants

lithosphere – entire solid Earth (environmental geochemistry); the rigid, outer 80 to 100 km of Earth including crust and upper mantle (geology)

lode – a vein or lenticular ore body

lode claim – see *claim*

longwall mining – underground mining in which flat-lying, tabular ore bodies (usually coal) are removed, usually by continuous miners, along long fronts and transported to a shaft by conveyor

maceral – grains or subdivisions of coal, including vitrinite, liptinite, and inertinite

mafic – rocks rich in magnesium and iron and minerals containing these elements, such as olivine and pyroxene

magma – molten igneous rock with minor amounts of suspended crystals and dissolved gases

magmatic water – water that was dissolved in a magma and later expelled (exsolved) when the magma crystallized to form water-poor silicate minerals

manganese nodule – ball of manganese oxide, usually consisting of concentric shells, found on the floor of the ocean and some lakes

mantle – middle shell of Earth below the crust and above the core; consists of ultramafic rocks

margin – term referring to the use of credit in commodity and stock purchases

marginal demand – excess demand for a commodity; usually related to short-term market changes

marine – related to seawater or the ocean

marl – limestone containing a significant proportion of clay or silt

matte – molten metal sulfide, formed by melting sulfide minerals in the early stages of smelting, usually involving loss of some S as SO_2

maturation – progressive alteration of natural organic material by burial and reaction with microbes and water to form kerogen

mean – sum of all numbers in a group divided by their number; average

median – central value in a group, with same number of values above and below

mesothelioma – tumor in the mesotheliomum, the tissues that encase the lung; can be benign or malignant

Mesozoic – Era of geologic time between 252 and 66 million years (Ma)

metal – hard solid material with good electrical and thermal conductivity; includes elements such as iron, copper, and gold, as well as alloys such as brass and steel

metalloid – element that looks like a metal but has different properties, including being a semiconductor rather than a conductor of electricity

metamorphic rock – rocks formed by recrystallization of preexisting igneous, sedimentary or metamorphic rocks, without significant melting

metamorphic water – water driven off from micas and other hydrous minerals during metamorphism, when they are recrystallized to form minerals that contain less water

metastable – refers to minerals or other substances that persist at temperature, pressure, and other conditions outside their range of stability because their breakdown reactions are slow

meteorite – fragment of extraterrestrial material that reaches Earth's surface

meteoric water – atmospheric precipitation; in ore deposits, precipitation that seeps into the ground and descends to levels at which it is heated and incorporated into a hydrothermal system

methemoglobinemia (blue-baby syndrome) – condition caused when nitrite (NO_2^-) oxidizes Fe^{+2} in hemoglobin, forming methemoglobin that cannot carry oxygen to tissue; causes oxygen deficiency

methanol – CH_3OH; wood alcohol

methylmercury – CH_3Hg^+, an organometallic cation containing mercury

micrite – a very fine-grained limestone

mineral – a naturally occurring, inorganic, crystalline solid with a regular chemical composition

mineral deposit – any unusual mineral concentration, regardless of whether it can be extracted at a profit (see *ore deposit*)

mineral right – ownership or control over real or potential mineral deposits on a parcel of land

microalloy – alloy, usually with iron, in which only very small concentrations of the additive are used

mobile (mobility) – in geochemistry, the term refers to the degree to which an element or compound is dispersed, usually by dissolution in water, from its rock or mineral source

mode – the most common number in a population

moderator – substance used to slow the flux of neutrons in a nuclear reactor, thus controlling the rate of fission

molecule – the smallest particle of a substance that retains the chemical and physical properties of the substance and is composed of two or more atoms; a group of like or different atoms held together by chemical forces.

moraine – ridge of gravel and sand deposited at the front of a stationary glacier by conveyor-like action of the ice, or along the sides of the glacier as it moves

native – adjective used to refer to elements that occur alone as solids or liquids in nature without being combined with other elements

natural gas – methane (CH_4) with small, but locally important amounts of ethane (C_2H_6) and heavier hydrocarbons, H_2S, He, CO_2, and N_2

natural gas liquid – hydrocarbons in natural gas that are separated as liquids (includes ethane, propane, butanes, and pentane) at surface conditions

neutron – a subatomic particle with a mass of one and no electrical charge; part of the nucleus of an atom

NIMBY – "not in my backyard" – a widely used defensive attitude manifested toward land use

nitrogen fixation – the processes by which microbes convert nitrogen in air or fertilizers to forms that can be taken up by plants

non-ferrous metals – metals that are not commonly used in or with steel; commonly refers to the base and light metals

non-renewable resources – natural resource that is not replaced by normal Earth processes at a rate equal to the rate at which it is consumed

normal distribution – term used to describe the arrangement of individual values in a data set in which the mean, mode, and median are the same

normal population – a group of samples or population in which the mean, median, and mode are the same

nuclear reactor – apparatus in which a nuclear chain reaction can be initiated, maintained and controlled, releasing energy at a specific rate

nuclide – a species of atom distinguished by its number of protons and neutrons; a specific isotope

nugget – unusually large piece of native, usually found in a placer deposit

obduction zone – zones where there is movement of one lithospheric plate onto another at a collisional tectonic margin

ocean crust – see *crust*

ocher – iron oxide pigment with a yellowish-orange color

oil – see *crude oil*

oil shale – kerogen-bearing shale that yields oil or gas when distilled

oil window – the range of depths (and related temperatures and pressures) to which sediments must be buried to generate oil from organic material

open cast mine – see *open pit mine*

open pit mine – mine that consists of an open hole or pit (as opposed to an underground mine consisting of tunnels and stopes)

operating profit – profit before tax but after deduction of operating and financial expenses

ore deposit – mineral deposit that can be extracted at a profit

ore mineral – mineral of value in an ore deposit; it is either useful in mineral form or contains an element or elements that are useful

organic material – material consisting largely of carbon and hydrogen

organic sedimentation – deposition of sediment containing abundant organic matter

organometallic compound – compounds in which metals are connected to complex molecules consisting largely of carbon and hydrogen

overburden – rock or regolith covering a mineral or ore deposit

overthrust belt – belt of rocks in which one block of rocks has been pushed or thrust over another, concealing those beneath

oxidation – loss of electrons from elements; commonly characterized by availability of free oxygen

ozone – a colorless, light blue gas containing three oxygen molecules, which is formed in the stratosphere by ultraviolet radiation, a process that shields Earth's surface from ultraviolet radiation

paleobauxite – old bauxite preserved by burial beneath younger sediments

paleoplacer – geologiocally ancient placer deposit

particulates – governmentese for "particles"

patenting – transfer of ownership in a mineral claim from government to claim holder

patio process – gold recovery process in which ore is spread onto a patio, crushed, and covered with mercury (see *amalgamation*)

payback period – the amount of time needed to recover the initial investment in a project plus relevant interest

peat – a sedimentary layer of unconsolidated plant material, commonly deposited in a swamp, from which coal is derived

pegmatite – a very coarse-grained, intrusive igneous rock thought to form from water-rich magma

pellets – (1) spheres of iron oxide and clay made during beneficiation to facilitate steel making; (2) natural spheres of phosphate in sediments, probably formed during diagenesis

percentage depletion – see *depletion allowance*

peridotite – ultramafic igneous rock consisting largely of olivine and pyroxene

perlite – rhyolitic volcanic glass with a concentric crack pattern

permeability – the capacity of a rock to transmit fluid (see *hydraulic conductivity*)

permanent normal trade relations status – the successor status to "most-favored-nation" (changed in 1998) allowing for free trade among members of a group

Phanerozoic – eon of geologic time between 541 million years (Ma) and the present

phenocryst – large crystal in a matrix of smaller ones in an igneous rock

phosphogypsum – calcium phosphate waste product produced during processing of phosphate ore

pig iron – iron with several percent carbon as well as other impurities such as phosphorus, silicon, and sulfur

placer – sand or gravel in a stream bed or beach containing grains of a heavy mineral such as gold or rutile

placer claim – see *claim*

placer deposit – a valuable mineral accumulation that formed by gravity accumulation during sedimentary processes; usually a heavy mineral although can apply to light minerals as well

plaster – usually refers to Plaster of Paris ($CaSO_4.0.5H_2O$), which is prepared by partial calcining of gypsum, and which crystallizes as gypsum when water is added

plate tectonics – movement of lithospheric plates that make up Earth's outer shell causing earthquakes, volcanoes, and other geologic processes

playa – flat central area of an enclosed desert basin with no river outlet

pneumoconiosis – any of several diseases of the lungs characterized by fibrous hardening caused by inhalation of irritating particles

podiform – lenticular, commonly refers to cumulate layers of chromite that have been sheared during deformation

population – in statistics, a group of samples

porosity – volume percent pore space in a rock

porphyritic – textural term referring to rock with large crystals (phenocrysts) in a matrix or groundmass of smaller crystals; rocks with this texture are porphyries

Portland cement – see *cement*

pothole – (1) rounded depression in bedrock over which a stream flows, formed by abrasion by rocks in the stream; (2) depression in platinum-bearing Merensky Reef of the Bushveld Complex, probably formed by chemical corrosion

polymer – a large molecule consisting of connected simple molecules of the same type

Precambrian – period of Earth history before about 541 million years ago

precious metals – gold, silver, and platinum; sometimes includes other platinum-group elements, palladium, rhodium, ruthenium, iridium, and osmium.

precipitate – the process by which dissolved ions or compounds combine to form a solid substance, or that substance

precision – agreement between replicate analyses of the same sample by the same method

pressurized water reactor – nuclear reactor in which coolant water is pressurized and separated from turbines by a heat exchanger

property tax – see *ad valorem tax*

prospector – an individual who searches for mineral deposits largely by examination and analysis of rocks at the surface, using little scientific theory to select areas for further work

protein – a biological polymer based on amino acids

Proterozoic – eon of geologic time between 2.5 and 0.54 billion years (Ga)

protons – elementary particle with a positive charge equal to that of the electron, but with a mass about 1,837 times greater

pseudomorph – mineral that has the external form of another mineral

pumice – felsic volcanic rock containing numerous holes known as vesicles that formed by expansion of gas from the cooling magma; often sufficiently buoyant to float

pyroclastic – fragments of magma, crystals, or other rocks ejected during a volcanic eruption

pyrolysis – chemical decomposition by heat

pyrometallurgy – smelting process based on thermal decomposition of ore minerals

quicksilver – mercury

radioactive – capable of spontaneously giving off alpha and beta particles, and gamma and other radiation

radionuclide – a nuclide that is unstable and undergoes radioactive decay with corpuscular and/or electromagnetic emission

radon – a radioactive chemical element formed from radioactive decay of radium that is, itself, a radioactive decay product of uranium and thorium

random drilling – mineral exploration conducted by drilling holes at locations that have been selected randomly

rank – degree of maturation of coal

recharge – flow of precipitation into an aquifer

reclamation – remedial action taken to restore areas of mineral production to conditions that prevailed before production

reduction – chemical process in which valence electrons are added to elements; commonly characterized by scarcity of free oxygen

reef – (1) accumulation of marine skeletal material; (2) layer of rock consisting of this material, (3) planar ore layer in some clastic sedimentary rocks and layered mafic complexes

refining – conversion of crude oil to gasoline, heating oil, asphalt, and other products; removal of trace amounts of impurity elements from a smelted metal

reforming – the third step in conventional oil refining, involving reconstitution of organic molecules

refractory – refers to substances that are difficult to melt

regalian – pertaining to royalty or royal perogatives

regolith – unconsolidated rock and soil material overlying bedrock

reserve – that part of the resource that can be extracted at a profit at the time of determination, whether or not facilities are available on the property

renewable resource – resource that is replenished by natural processes at a rate similar to the rate at which it is extracted

reserve tax – ad valorem tax on mineral reserves

reservoir – any part of Earth or its surroundings with a specified composition or characteristics, such as the atmosphere

reservoir rock – porous, permeable rock that can store and produce fluid resources including, oil, gas, brine, or water

residence time – average period of time during which a substance remains in a reservoir

residuum – insoluble material remaining after intense weathering

resource – concentration of natural material that can be extracted now or in the future; includes reserves and resources

resource curse – see Dutch disease

resource nationalism – the tendency of people and governments to assert control over natural resources located on their territory; usually by restricting development by foreign interests and placing them in government corporations

retained earnings – profits that are retained by a business rather than being distributed as dividends to shareholders

return on investment – profit from an investment expressed as a fraction of the original investment

revenues – income from business operations

reverberatory furnace – a type of furnace used in smelting copper and other ores

rhizosphere – the area of soil surrounding the root of a plant

rift – a linear valley formed by extension of the crust; a graben

riparian rights – the right to use water that is on owned land, although the land owner does not own the water

roasting – heating a rock, mineral or ore, commonly to dry or drive away gases

rock – a consolidated mixture of grains of one or more minerals

rock burst – spontaneous, explosive disintegration of rock exposed in a mine

roll front – arc-shaped, uranium-enriched zone in sandstone

room-and-pillar mining – a type of underground mining involving removal of rooms of ore leaving intervening walls or pillars to support the roof or back

rotary kiln – a type of tilted kiln that rotates causing mineral material to flow from one end to the other

salar – a playa or salt flat

salt dome – salt intrusion that rises from an underlying layer of salt

salt flat – a playa in which evaporites have been deposited

saltwater encroachment – movement of salt water into an aquifer

sand – clastic material between 1/16 and 2 mm in diameter

sandstone – a clastic sedimentary rock consisting of sand-sized grains

saturated – in chemistry, a solution that has dissolved the maximum amount possible of a specific substance; in groundwater refers to porous rock that is filled with water below the water table

scintillometer – a device for measuring particles and energy emitted by radioactive decay

scoria – hardened lava with abundant vesicles, usually basalt

scrubber – device to remove SO_2 and other acid-forming gases produced by combustion of coal and other fossil fuels and by smelting

seam – layer of ore, usually coal or chromite

secondary deposit – mineral deposits that form by weathering of original minerals; commonly applied to metal oxide, carbonate, and sulfate minerals and deposits that form by weathering of sulfide deposits

(but not linked) sedimentary rock – rock formed by deposition of grains of pre-existing rock material (clastic sedimentary rock), precipitation of dissolved material (chemical sedimentary rock) or accumulation of organic matter (organic sedimentary rock)

seismic – (1) refers to movement of natural or anthropogenic shock waves through rock; (2) form of geophysical exploration using artificial shock waves to delineate underground features

severance tax – a tax imposed on the value of minerals removed from the ground

shaft – a vertical or inclined hole used to reach mineral deposits below the surface

shale – a clastic sedimentary rock consisting of fine-grained particles, many of which are clay minerals

share – refers to part interest in a business, usually sold to raise capital for the enterprise (also known as stock in the United States and Canada)

shareholder's equity – total assets of an organization (usually a company) minus its total liabilities; at a more detailed level it is the sum of share capital plus retained earnings minus treasury shares

shatter cone – a conical fracture pattern formed in rocks by passage of shock waves; commonly caused by meteorite impact

shield – the exposed part of a craton

sienna – reddish-yellow natural mineral pigment consisting largely of various mixtures of limonite and hematite

silicate tetrahedron – the basic unit of many silicate minerals; a tetrahedron with a central silicon cation and oxygen anions at each corner

silicon carbide – SiC_4, a synthetic hard substance used as an abrasive

silicosis – a lung disease caused by inhalation of angular fragments rich in silicate rocks or minerals

sill – roughly tabular and horizontal intrusive, igneous rock

silt – clastic material between 1/16 and 1/256 mm in diameter

sink – in geochemistry, a reservoir into which an element, compound, or substance is concentrated

sinter – silica-rich deposit from hot spring waters that reach the surface

skarn – metamorphic rock rich in calcium-bearing silicate minerals, commonly formed at or near intrusive rock contacts by addition of silica to limestone and dolomite; often hosts metal ore minerals

skeletal material – framework for some animals; usually consists of calcite, apatite, or silica in animals and cellulose in plants

skewed – refers to a population that is not normal (mean, mode, and median are not the same)

slag – silicate waste material from smelting of metal ore concentrates

slate – shale that has been metamorphosed to produce a durable rock that splits easily into slabs

slurry pipeline – a pipeline that transports pulverized solid material as a mixture (slurry) with water or some other fluid

smelting – separation of a metal from its ore mineral by pyrometallurgy

smog – a photochemical haze caused by the action of solar ultraviolet radiation on atmosphere polluted with hydrocarbons and oxides of nitrogen especially from automobile exhaust

smoke – vapor and suspended carbon particles resulting from combustion

soapstone – a magnesium-rich rock containing large amounts of the mineral talc [$Mg_3Si_4O_{10}(OH)_2$]

soda ash – Na_2CO_3; a chemical compound prepared from hydrous sodium carbonate minerals such as trona ($NaCO_3.NaHCO_3.2H_2O$)

soft water – water in which soap lathers easily; contains relatively small amounts of dissolved calcium and magnesium

soil – in geology, the upper part of the regolith, which commonly shows compositional zoning reflecting migration of elements during weathering; in engineering, all unconsolidated material above bedrock

solid – substance that resists pressure and/or efforts to change its shape, as contrasted with a liquid or gas

solid solution – a solid material (usually crystalline) that can host two or more different elements or ions in the same crystal position (such as substitution of cadmium for zinc in sphalerite (ZnS))

solubility – degree to which a substance will dissolve, usually in water or an aqueous solution, although can also apply to solid solutions

solute – substance that dissolves in a solvent

solvent – solution in which a solute dissolves, usually applied to aqueous solutions

solvent extraction (SX) – concentration of a desired element (such as gold) from a primary solution by dissolving it in a smaller volume of a secondary solution that is mixed with the primary solution but separates from it because it

is immiscible; first step in solvent extraction-electro winning (SX-EW) process

sour gas – natural gas in which the partial pressure of H_2S is more than 4 ppm at standard temperature and pressure

source – in geochemistry, a reservoir that releases or emits a substance of interest

source rock – rock containing sufficient organic material to produce crude oil and/or natural gas when heated by burial beneath other rocks

sovereign wealth fund – government-owned investment fund investing in real and financial assets

spent fuel – nuclear fuel that can no longer support a fission reaction

stable – (1) refers to minerals or other substances that are present at temperature, pressure, and other conditions under which they do not react to form other minerals or compounds; (2) refers to isotopes that do not undergo spontaneous radioactive decay

stainless steel – corrosion-resistant steel containing 12–50% Cr

standard sample – in chemistry, a sample against which other samples are compared

standard deviation – for any population, the square root of the mean of the squares of the deviations of all samples from the population mean

stannosis – pneumoconiosis caused by inhalation of tin (Sn^{+4}) oxide

statement of income – summary of revenues received by a corporation or other taxable entity, with expenses that are deductible from taxable income

steel – iron-base alloy containing up to 2% C

stock – (1) small, irregularly shaped body of intrusive igneous rock, commonly measuring a few kilometers in diameter; (2) one of the equal parts into which ownership of a corporation can be divided; see also share

stock exchange – public (e.g. open) market in which shares (stock) can be bought and sold

stockpile – material stored for later use, usually by governments

stockwork – system of anastamosing (intersecting) small veins

stope – underground opening or room from which ore has been removed

strata – layers (plural of stratum, a single layer)

strategic metal – a metal that has important applications in defense or related industries and is not available in the country giving it this designation

strategic mineral – mineral material considered necessary to national defense, usually not available domestically

stratiform – having the form of a flat layer or stratum

strip mining – a form of open pit mining in which flat-lying ore bodies are mined in linear zones (strips) and overburden is replaced after mining

stripping ratio – ratio of volume of overburden to volume of mineable ore in an open pit mine

subbituminous coal – low-rank coal

subduction – plate-tectonic process in which a lithospheric plate returns to the mantle at a convergent margin

sub-level caving – type of underground mining in which panels of ore are mined from a series of tunnels known as sub-levels

subsidence – lowering of the land surface caused by collapse of underlying rock; sometimes related to withdrawal of fluids from reservoirs or ore from mines

substitute – in mineral resources, the term refers to either the process by which one commodity takes the place or another, or the process by which one element takes the place of another in a crystal structure; in both cases, the replacing commodity/element has similar properties

sugar – monosaccharide with the general formula CH_2O (most commonly, glucose [$C_6H_{12}O_6$])

sulfate-reducing bacteria – bacteria that derive energy from the conversion of SO_4^{-2} to H_2S or some other form of reduced sulfur

supergene deposit – mineral deposit formed by descending waters; usually referring to mineral deposits formed by weathering; distinguished from hypogene deposits

superphosphate – a fertilizer that contains 16–21% P_2O_5 (see also *triple superphosphate*)

superalloys – steel alloys that retain strength at temperatures above 800 °C

surface rights – ownership or control over surface of a parcel of land

suspended material – particles suspended in a fluid, usually water or air

sweet gas – natural gas containing H_2S with a partial pressure of less that 0.01 bar

swelling clay – clay mineral (usually smectite group) in which the space between individual clay layers increases when water is adsorbed thereby increasing volume

syncrude – "synthetic" oil recovered by processing of tar sand

synroc – substance synthesized by cooling molten material with the composition of average rock; used as a host for nuclear wastes

taconite – see *banded iron formation*

tailings – waste material that remains after processing of pulverized ore to make a concentrate of the desired ore mineral(s)

talus – accumulation of rock debris at the base of a cliff

tar – semi-solid bitumen

tar sand – sand or sandstone in which much of the interstitial pore space is filled with bitumen or tar

tax holiday – period of time during which tax is not imposed

tetraethyllead – organometallic compound $Pb(C_2H_5)_4$ added to some gasoline to prevent premature combustion

thermal coal – coal used for production of heat, usually in electric power plants

thixotropy – pertaining to substances that behave like solids when not stirred but convert to liquids when stirred; desirable property of drilling mud

tidewater land – land between high and low tide levels

timber rights – ownership or control over timber on a parcel of land

tinplate – narrow-guage steel plate with a coating of tin on both sides

torbanite – Australian term for oil shale

toxic – poisonous; detrimental to health (see text for discussion of LD_{50})

trace metal – metal present in very small (trace) amounts

trap – see *hydrocarbon trap*

triple superphosphate (TSP) – fertilizer material containing 44–46% P_2O_5 prepared by treating phosphate minerals with phosphoric acid

tuff – pyroclastic material of approximately sand size

ultramafic – igneous rock consisting largely of magnesium and iron silicates and oxides

umber – purplish-red natural mineral pigment consisting of ferric iron and manganese oxides

unconformity – a surface that represents a break in the sedimentary rock record, caused by erosion or non-deposition of sediments

underground mine – mine that removes ore from beneath the surface through a shaft or adit

unit train – train that carries a single product from one source to a single destination; commonly refers to trains used to haul coal from western US mines to eastern US consumers

unitization – process by which production from an oil or gas field with several owners is apportioned to each owner

unleaded gasoline – see *leaded gasoline*

unsaturated – see *saturated*

unstable – see *stable*

vaporization – process by which a liquid converts to a vapor

vein – planar fracture in rock that has been sealed with minerals precipitated from hydrothermal solutions or cooler water

vesicle – cavity preserved in cooled volcanic rock preserving the form of a bubble of gas that was originally dissolved in the magma, but separated from it during ascent to the surface

vitrinite – remains of wood and bark in coal and other carbon-rich deposits

volatile organic compounds (VOCs) – organic compounds that are easily vaporized, including gasoline components

volcanic rocks – rocks consisting of lava or pyroclastic material

water flooding – a method of enhanced oil recovery involving injection of water into peripheral parts of a field to increase recovery of remaining oil

water rights – ownership or control of water on or below the surface of a parcel of land

water table – top of the zone that is saturated with groundwater

weathering – disintegration and decomposition of rocks at or near Earth's surface

well completion – action taken to bring a oil or gas discovery well into production

well logging – process of lowering measuring devices into exploration drill holes to measure rock characteristics, such as conductivity and radioactivity

wet deposition – precipitation of new compounds formed by reactions among suspended particles, aerosols, and gases in air

working fluid – fluid(s), usually water, that transport heat from a source of thermal energy to turbines in electric power generating plants

REFERENCES

Abandoned Mines.gov 2013. [ONLINE] Available: Abandonedmines. gov

Aboriginal Affairs and Northern Development Canada 2015. A history of treaty-making in Canada. [Online] Available: https:// www.aadnc-aandc.gc.ca/eng/1314977704533/1314977734895.

Abrahamson, D. 1992. Aluminum and global warming. *Nature*, 356, 484.

Abramowski, T., Stoyanova, V. 2012. Deep-sea polymetallic nodules: renewed interest as resources for environmentally sustainable development. *Proceedings of 12th International Multidisciplinary Scientific GeoConference SGEM 2012*, 515–522.

ACS 1997. A national historic landmark chemical landmark: First commercial production of bromine in the United States. American Chemical Society Division of the History and Chemistry and the Office of Public Outreach.

Adams, D. E., Farewell, S.O., Robinson, E., Pack, M. R., Bamesberget, W. L. 1981. Biogenic sulfur strengths. *Environmental Science and Technology*, 15, 51–66.

Adams, S. S., Hite, R. J. 1983. Potash. In Lefond, S. J. (ed.), *Industrial Minerals and Rocks*, 5th edn. New York, NY: Society of Mining Engineers, pp. 1049–1077.

Adams, W. T. 1985. Zirconium and hafnium. *US Bureau of Mines Mineral Facts and Problems*, 675, 941–949.

Adler, T. 1996, Botanical clean-up crews: using plants to tackle polluted water and soil. *Science News*, July 20.

Agency For Toxic Substances 2012. Toxic substances portal – manganese. [Online] Available: http://www.atsdr.cdc.gov/toxprofiles/tp.asp?id=102&tid=23.

Agency For Toxic Substances 2005. Toxicological profile for nickel. [Online] Available: http://www.atsdr.cdc.gov/toxprofiles/tp15.pdf.

Ahmed, A. H., Aria, S., Ikenne, M. 2009. Mineralogy and paragenesis of the Co–Ni arsenide ores of Bou Azzer, anti-Atlas, Morocco. *Economic Geology*, 104, 249–266.

Aieta, E. M., Singley, J. E., Trussel, A. R., Thorbjarnson, K. W., McGuire, M. J. 1987. Radionuclides in drinking water: an overview. *American Water Works Association*, 79, 144–152.

Albers, G., 2015. Aboriginal land claims. Historical Canada. [Online] Available: http://www.thecanadianencyclopedia.ca/en/article/land-claims/.

Al-Chalabi, F. J. 1986. *OPEC at the Crossroads*. Oxford: Pergamon.

Al-Jarallah, M. I., Fazal-ur-Rehman, Musazay, M. S., Aksoy, A. 2004. 2004, Correlation between radon exhalation and radium content in granite samples used as construction material in Saudi Arabia. *Radiation Measurements*, 40, 625–629.

Aldridge, S., Downs, A. J., Downs, T. 2011. *The Group 13 Metals Aluminum, Indium and Thallium: Chemical Patterns and Peculiarities*. Chichester: Wiley.

Alexandratos, N., Bruinsma, J. 2012. World Agriculture Towards 2030/2050. Agriculture Development Economics Division, Food and Agriculture Organization of the United Nations. [Online] Available: www.fao.org/.

Ali, S. H. 2011. Ecological comparison of synthetic versus mined diamonds. Working Paper, Institute for Environmental Diplomacy and Security, University of Vermont.

Allcroft, R. 1956. Copper deficiency disorders in sheep and cattle in Britain. *Journal British Grassland Society*, 11, 182–184.

Amankwah, R. K., Styles, M. T., Nartey, R. S., Al-Hassan, S. 2010. The application of direct smelting of gold concentrates as an alternative to mercury amalgamation in small-scale gold mining in Ghana. *International Journal of Environment and Pollution*, 41, 30–315.

Amaral, P. G. Q., Galembeck, T. M. B., Bonotto, D. M., Artur, A. C. 2012. Uranium distribution and radon exhalation from Brazilian dimension stones. *Applied Radiation and Isotopes*, 70, 808–817.

Ames, J. A., Cutcliffe, W. E. 1983. Cement and cement raw materials. In Lefond, S. J. (ed.), *Industrial Minerals and Rocks*, 5th edn. New York, NY: Society of Mining Engineers, pp. 133–159.

Amini, M. Mueller, K., Abbaspour, K. C., *et al.* 2008a, Statistical modeling of global geogenic fluoride contamination in groundwaters. *Environmental Science and Technology*, 42, 3662–3668.

Amini,M., Abbaspour, K. C., Berg, M., *et al.* 2008b. Statistical modeling of global geogenic arsenic contamination in groundwater. *Environmental Science and Technology*, 42, 3669–3675.

Amos, H. M., Jacob, D. J., Streets, D. G., Sunderland, E. M. 2013. Legacy impacts of all-time anthropogenic emissions on the global mercury cycle. *Global Biogeochemical Cycles*, 27, 410–421.

Ampian, S. G. 1985a. Barite. *US Bureau of Mines Mineral Facts and Problems*, 675, 65–74.

Ampian, S. G. 1985b. Clays. *US Bureau of Mines Mineral Facts and Problems*, 675, 157–170.

Anderson, C. M., Mayes, M., Labelle, R. 2012. Update of occurrence rates for offshore oil spills. BOEM, Bureau of Ocean Energy Management, OCS Report BOEM 2012–069, BSEE 2012–069.

Anderson, M. S., Lakin, H. W., Beeson, K. C., Smith, F. F., Thacker, E. 1961. *Selenium in Agriculture*. Washington, DC: US Government Printing Office,

Andrae, M. O., Crutzen, P. J. 1997. Atmospheric aerosols: biogeochemical sources and role in atmospheric chemistry. *Science*, 276, 1052–1058.

Andreopoulous, S. 2012. Global Economic Forum, London. [Online] Available: www.hpacmag.com/daily_images/1001589773-1001 589774.pdf.

Andrews, A. J., Owsiacki, L., Kerrich, R., Strong, D. F. 1986. The silver deposits at Cobalt and Gowanda, Ontario. I: geology, petrography and whole rock geochemistry. *Canadian Journal of Earth Science*, 23, 1480–1506.

Andrews, M. J., Fuge, M.R. 1986. Cupriferous bogs of the Coed-y-Brenin area, North Wales, and their significance in mineral exploration. *Applied Geochemistry*, 1, 519–525.

Anonymous 2007. 1872 Mining law reform passes House, still faces uphill battle. *Mining Engineering*, 59, 12.

Anstett, T. F., Bleiwas, D. I., Hurdelbrink, R. J. 1985. Tungsten availability – market economy countries. *US Bureau of Mines Information Circular*, 9025.

Anton, M. A. L., Spears, D. A., Somoano, M. C., Tarazona, M. R. M. 2012. Thallium in coal: analysis and environmental implications. *Fuel*, 105, 13–18.

Antonopoulos, I.-S., Perkoulidis, G., Logothetis, D., Karkarian, C. 2014. Ranking municipal solid waste treatment alternative considering sustainability criteria using the analytical hierarchical process tool. *Resources, Conservation and Recycling*, 86, 149–159.

Apodaca, L. E. 2013a. Nitrogen. *US Geological Survey Minerals Yearbook*.

Apodaca, L. E. 2013b. Sulfur. *US Geological Survey Minerals Yearbook*.

Appleyard, F. C. 1983. Gypsum and anhydrite. In Lefond, S. J. (ed.), *Industrial Minerals and Rocks*, 5th edn. New York, NY: Society of Mining Engineers, pp. 775–792.

Applied Ecological Services 2015. Flambeau Copper Mine reclamation. [Online] Available: http://www.appliedeco.com/Projects/FlambeauCopperMine.pdf.

Arioz, O., Kilinc, K., Karasu, B., et al. 2008. A preliminary research on the properties of lightweight expanded clay aggregate. *Journal of Australian Ceramics Society*, 44, 23–30.

Arndt, N. T., Lesher, C. M., Czamanske, G. K. 2005. Mantle-derived magmas and magmatic Ni–Cu–(PGE) deposits. In Hedenquist, J. W., Thompson, J. F. H., Goldfarb, R. J., Richards, J.P. (eds.), *Economic Geology, 100th Anniversary Volume*. Littleton, CO: Society of Economic Geologists, pp. 5–23.

Arora, H. R., Pugh, C. E., Hossner, L. R., Dixon, J. B. 1980. Forms of sulfur in east Texas lignitic coals. *Journal of Environmental Quality*, 9, 383–390.

Arslan, M., Boyhay, M. 1990. A study of the characterization of dustfall. *Atmospheric Environment*, 24A, 2667–2671.

Artiola, J. F. 2014. Biosolids land use in Arizona. The University of Arizona College of Agriculture and Life Sciences, Arizona Cooperative Extension. [Online] Available: http://extension.arizona.edu/sites/extension.arizona.edu/files/pubs/az1426.pdf.

Ashley, S. 2014, World transfer pricing 2014. International tax review. [Online] Available: http://www.internationaltaxreview.com/pdfs/wtp/world-transfer-pricing-2014.pdf.

Ashraf, M., Satapathy, M. 2013. The global quest for light tight oil: myth or reality? Schlumberger Newsroom. [Online] Available: https://www.sbc.slb.com/Our_Ideas/Energy_Perspectives/1st%20Semester13_Content/1st%20Semester%202013_Global.aspx.

Asian Development Bank and Nauru 2015. Fact sheet. [Online] Available: http://www.adb.org/countries/nauru/main.

ASM 2002. Introduction to steel and cast irons. *Metallographer's Guide: Irons and Steels*. Materials Park, OH: ASM International.

Asmiov, I. 1974. *Asimov on Chemistry*. New York, NY: Doubleday and Company.

Aston, R. L. 1993. Surface vs. mineral owners claim near-surface uranium deposit. *Engineering and Mining Journal*, 16RR-16UU.

Attanasi, E., Meyer, R. F. 2010. Natural bitumen and extra-heavy oil. In *Survey of Energy Resources*, 22 edn. London: World Energy Council, pp. 123–140.

Atwood, G. 1975. The strip-mining of western coal. *Scientific American*, 34–39.

Auerbach, S. M., Carrado, K. A., Dutta, P. K., eds. 2003. *Handbook of Zeolite Science and Technology*. Boca Raton, FL: CRC Press.

Aufderheide, A. C., Rapp, G., Wittmers, L. E., et al. Lead exposure in Italy: 800 BC–700 AD. *International Journal of Anthropology*, 9, 9–15.

Austen, I. 2013. A black mound of Canadian oil waste is rising over Detroit. *The New York Times*, May 17.

Austin, G. S., Humphrey, B. 2006. Sodium sulfate resources. In Kogel, J. E., Trivedi, N. C., Barker, N. C., Krukowski, S. T. (eds.), *Industrial Minerals and Rocks*, 7th edn. Littleton, CO: Society for Mining, Metallurgy and Exploration, pp. 879–892.

Austin, G. T. 1991. Gemstones. *US Geological Survey Mineral Yearbook*.

Austin, G. T. 1992. Beyond beauty: high tech uses for gemstones. *Minerals Today*, 6–10.

Australian Coal Association 2015. Coal for jobs. [Online] Available: http://www.australiansforcoal.com.au/coal-4-jobs.html.

Axtman, R. C. 1975. Environmental impact of a geothermal power plant. *Science*, 187, 795–799.

Axtmann, E. V., Luoma, S. N. 1991. Large-scale distribution of metal contamination in the fine-grained sediments of the Clark Fork River, Montana, USA. *Applied Geochemistry*, 6, 75–88.

Aydin, M., Verhulst, K. R., Saltzman, E. S., et al.. 2011, Recent decreases in fossil-fuel emissions of ethane and methane derived from firn air. *Nature*, 476, 198–201.

Babel, M., Schreiber, B. C. 2014. Geochemistry of evaporites and evolution of seawater. In Lollar B. S. (ed.), *Treatise on Geochemistry*, 2nd edn. Amsterdam: Elsevier, vol. 9, pp. 483–548.

Bahn, P. 2008. Killing Lascaux. *Archaeology*, 61, 3.

Bain 2014. The global diamond report 2014. Bain and Company. [Online] Available: http://www.bain.com/Images/BAIN_REPORT_The_Global_Diamond_Report_2014.pdf.

Baird, J. 1991. Soil facts: sulfur as a plant nutrient. *The North Carolina Agricultural Extension Service Publication*, AG-439-15.

Baker, M. 1975. Inactive and abandoned underground mine, water pollution, prevention and control. Washington, DC: Environmental Protection Agency, EPA report 440/9–75-007.

Baldwin, W. L. 1983. *The World Tin Market*. Durham, NC: Duke University Press.

Ballard, R. D., Bischoff, J. L. 1984. Assessment and scientific understanding of hard mineral resources in the EEZ. *US Geological Survey Circular*, 929, 185–208.

Ballhaus, C. G. 1988. Potholes of the Merensky reef at Brakspriut shaft, Rustenburg platinum mines. *Economic Geology*, 83, 1140–1158.

Bardi, U. 2010. Extracting minerals from seawater: an energy analysis. *Sustainability*, 2, 980–992.

Bardossy, G., Aleva, G. J. J. 1990. *Lateritic Bauxites*. Elsevier, Amsterdam.

Barker, J. M. 1983. Sulfur. In Lefond, S. J. (ed.), *Industrial Minerals and Rocks*, 5th edn. New York, NY: Society of Mining Engineers, pp. 1235–1274.

Barley, M. E., Groves, D. I. 1992. Supercontinent cycles and the distribution

Barnes, H. 1983. Irish Wilderness: how the story started. *St. Louis Post Dispatch*, June 20.

Barnes, P. W., Lien, R. 1988. Icebergs rework shelf sediments to 500 m off Antarctica. *Geology*, 16, 1130–1133.

Barnes, S. J., Lightfoot, P. 2005. Formation of magmatic sulfide ore deposits and processes affecting their copper and platinum group element contents. In Hedenquist, J. W., Thompson, J. F. H., Goldfarb, R. J., Richards, J. P. (eds.), *Economic Geology, 100th Anniversary Volume*. Littleton, CO: Society of Economic Geologists, pp. 179–214.

Barnola, J. M. 1987. Vostok ice core provides 160,000-year record of atmospheric CO_2. *Nature*, 329, 410.

Barringer, E. 1992. Researchers ponder the mystery of America's stroke belt in the South. *The New York Times*, July 29.

Barth, H. J. 2002. The 1991 Gulf War Oil Spill: Its ecological effects and recovery rates of intertidal ecosystems at the Saudi Arabian Gulf coast – results of a 10-year monitoring period. Wissenschaftliche Arbeit Im Rahmen des abilitationsverfahrens im Fach Geographie an der Philosophischen Fakultät III der Universität Regensburg. Habilitation Thesis, University of Regensburg.

Bartlett, A. A. 1980a. Forgotten fundamentals of the energy crisis. *Journal of Geological Education*, 28, 4–12.

Bartos, P. J. 2000. The pallacos of Cerro Rico de Potosi, Bolivia: a new deposit type. *Journal of Economic Geology*, 95, 645–654.

Batchelor, B. C. 1979. Geological characteristics of certain coastal and offshore placers as essential guides for tin exploration in Sundaland, Southeast Asia. *Bulletin of the Geological Society of Malaysia*, 11, 283–313.

Battarbee, R. W., Curtis, C. J., Shilland, E. M. 2011. The Round Loch of Glenhead: recovery from acidification, climate change monitoring and future threats. Scottish Natural Heritage Commissioned Report No. 426. [Online] Available: http://www.snh.org.uk/pdfs/publications/commissioned_reports/469.pdf.

Baturin, G. N. 1987. *The Geochemistry of Manganese and Manganese Nodules in the Ocean*. Hingham, MA: Kluwer Academic Publishers.

BBC 2013. Japan extracts gas from methane hydrate in world first. *BBC News Business*, December 3.

Beach, R. H., Bullock, A. M., Heller, K., *et al.* 2000. Lime Production: Industry Profile.

Bear, J. 1972, *Dynamics of Fluids in Porous Media*. New York, NY: Dover Publishing.

Bearden, S. D. 1988. Celestine resources of Mexico. In Zuban, A.-J.W. (ed.), *24th Forum on the Geology of Industrial Minerals*. Columbia, SC: South Carolina Geological Survey.

Beavers, C. 2013. An overview of phosphate mining and reclamation in Florida. M.Sc. Thesis, University of Florida.

Beck, R., Fidora, M. 2008. The impact of sovereign wealth funds on global financial markets. *European Central Bank Occasional Paper Series*, no. 91. [Online] Available: https://www.ecb.europa.eu/pub/pdf/scpops/ecbocp91.pdf.

Bedinger, G. M. 2013. Arsenic. *US Geological Survey Minerals Yearbook.*.

Bekker, A., Slack, J. F., Planavsky, N., *et al.* 2010. Iron formation: the sedimentary product of a complex interplay among mantle, tectonic, oceanic, and biospheric processes. *Economic Geology*, 105, 467–508.

Bekkum, V. H., Flanigen, E. M., Jansen, J. C. 1991. *Introduction to Zeolite Science and Practices*. Elsevier: Amsterdam.

Belanger, D., Brousse, T., Long, J. W. 2008. Manganese oxides: battery materials make the leap to electrochemical capacitors. *Electrochemical Society Interface*, Spring, 49–52.

Belk, S. 2012, Oxy deals may revive oil production in Long Beach and Carson; State Lands Commission to vote on agreement June 7 to spur drilling in tidelands: *Long Beach Business Journal*, June 5.

Bellona Foundation 2010. Environmental Challenges in the Arctic: Norilsk nickel. Oslo, Bellona Foundation.

Benson, S. M., Cole, D. R. 2008. CO_2 sequestration in deep sedimentary formations. *Elements*, 5, 325–331.

Berger, V. I., Singer, D. A., Bliss, J. D., Moring, B. C. 2011. Ni–Co laterite deposits of the world; database and grade tonnage models. *US Geological Survey Open-File Report*, 2011–1058.

Bergenstock, D. J., Deily, M. E., Taylor, I. W. 2006. A cartels' response to cheating: an empirical investigation of the De Beers diamond empire. *Southern Economic Journal*, 73, 173–189.

Berkowitz, N., Brown, R. A. S. 1977. In-situ coal gasification: the Forestberg (Alberta) field test. *CIM Bulletin*, 70, 92–97.

Bernhard, A. 2010. The nitrogen cycle: processes, players and human impact. *Nature Education Knowledge*, 3, 25.

Bernier, L. 1984. Ocean mining activity shifting to exclusive economic zones. *Engineering Mining Journal*, July, 57–62.

Berthelsen, C., Tracy, R. 2014. Senate report: banks had unfair commodities market advantages. *The Wall Street Journal*, November 19.

Beukes, N. J., Gutzmer, J. 2008. Origin and paleoenvironmental significance of major iron formations at the Archean–Paleoproterozoic boundary. In Hagemann, S., Rosiere, C., Gutzmer, J., Beukes, N. (eds.) *Banded Iron Formation-Related High-Grade Ore*. Littleton, CO: Society of Economic Geologists, pp. 5–47.

Bian, Z., Miao, X., Lei, S., *et al.* 2012. The challenges of reusing mining and mineral wastes. *Science*, 337, 702–704.

Bigelow, R. C., Plumlee, G. S. 1995. The Summitville mine and its downstream effects. *US Geological Survey Open-File Report*, 95–23.

Bintania, R., Van de Wal, R., Oerlemans, J. 2005. Modeled atmospheric temperatures and global sea levels of the past millions years. *Nature*, 437, 125–128.

Bird, J. M., Weathers, M.S. 1979. Origin of josephinite. *Geochemical Journal*, 13, 41–55.

Bird, K. J., Houseknecht, D. W. 2008. Arctic National Wildlife Refuge, 1002 area, petroleum assessment, 1998, including economic analysis. *United States Geological Survey Factsheet*, 0028–01.

Bisc, C. J. 1981. Pennsylvania's subsidence control guidelines: should they be adopted by other states? *Mining Engineering*, 33, 1623–1628.

Bishop, A., Bremner, C., Laake, A., *et al.* 2011. Petroleum potential of the Arctic: challenges and solutions. *Oilfield Review*, 22, 36–49.

Blaskett, D. R. 1990. *Lead and its Alloys*. New York, NY: Horwood.

Blatt, H. G., Middleton, G. V., Murray, R. C. 1980. *Origin of Sedimentary Rocks*. Englewood Cliffs, NJ: Prentice-Hall.

Blechman, B. M. 1985. *National Security and Strategic Minerals: An Analysis of US Dependence on Foreign Sources of Cobalt*, Boulder, CO: Westview Press.

Bleimeister, W. C., Brison, R. J. 1960. Beneficiation of rock salt at the Detroit Mine. *Mining Engineering*, 918–922.

Bleiwas, D. I. 2000. Arsenic and old waste. US Geological Survey. [Online] Available: http://minerals.usgs.gov/minerals/mflow/d00-0195/.

Bleiwas, D. I., Gambogi, J. 2013. Preliminary estimates of the quantities of rare-earth elements contained in selected products and in imports of semimanufactured products to the United States, 2010. *US Geological Survey Open-File Report*, 2013–1072.

Bliss, N. W. 1976. Non-bauxite sources of alumina: a survey of Canadian potential. *CIM Bulletin*, 75–87.

BLM 2012. Oil and gas lease utilization onshore and offshore updated report to the president. Bureau of Land Management, US Department of the Interior. [Online] Available: http://www.blm.gov/pgdata/etc/medialib/blm/wy/programs/energy/og/leasing/protests/2012/may/Appeals.Par.42771.File.dat/ExhP.pdf.

BLM 2013a. Public Land Statistics. Bureau of Land Management, US Department of the Interior. [Online] Available: http://www.blm.gov/public_land_statistics/pls12/pls2012.pdf.

BLM 2013b. Oil and gas statistics by year for fiscal years 1988–2013. Bureau of Land Management, US Department of the Interior. [Online] Available: http://www.blm.gov/wo/st/en/prog/energy/oil_and_gas/statistics.html.

BLM 2013c. Interior disburses $14.2 billion in 2013 energy revenues to federal, state, local and tribal goernments. [Online] Avaialable: http://www.blm.gov/or/news/files/11-19-13_FY_2013_Disbursements_Press_Release_Final.pdf.

Blossom, J. W. 1985. Rhenium. *US Bureau of Mines Mineral Facts and Problems*, 675, 665–672.

Blum, A. 2006. Chemical burns. *The New York Times*, November 19.

Blum, J. 2005. 51–49 Senate vote backs Arctic oil drilling. *The Washington Post*, March 17.

Blum, J. D., Popp, B. N., Drazen, J. C., Choy, C. A., Johnson, M. W. 2013 Methylmercury production below the mixed layer in the North Pacific Ocean. *Nature Geoscience*. 6, 879–884.

Bockman, O. C., Kaarstad, O., Lie, O. H., Richards, L. 1990. *Agriculture and Fertilizers*. Oslo: Norsk Hydro.

Boden, T. A., Marland, G., Andres, R. J. 2013. *Global, Regional, and National Fossil-Fuel CO2 Emissions*. Oak Ridge, TN: Carbon Dioxide Information Analysis Center..

BOEM 2011a. Leasing Activities Information. Final Notice of Sale 213, Central Planning Area. [Online] Available: http://www.boem.gov/Oil-and-Gas-Energy-Program/Leasing/Regional-Leasing/Gulf-of-Mexico-Region/Lease-Sales/216-222/fcov222-pdf.aspx.

BOEM 2011b. Assessment of undiscovered technically recoverable oil and gas resources of the nation's outer continental shelf, 2011. *BOEM Fact Sheet*, RED-2011-01-a.

BOEM 2014. *Oil and Gas Lease Sale 233 – Preliminary Bid Recap*. Washington, DC: Bureau of Ocean Energy Management, US Department of the Interior.

Boercker, S. W. 1979. Energy use in the production of primary aluminum. *Materials and Society*, 3, 153–161.

Boggess, W. R., Wixson, B. G. 1978. *Lead in the Environment*. Washington, DC: National Science Foundation.

Bogatuyrev, B. A., Zhukov, V. V. 2009. Bauxite provinces of the world. *Geology of Ore Deposits*, 51, 339–355.

Bogdanich, I. and Purtill, A. 2015. Australia – mining law 2015. International Comparative Legasl Guides. [Online] Available: http://www.iclg.co.uk/practice-areas/mining-law/mining-law-2015/australia.

Bohn, A. 1979. Trace metals in fucoid algae and purple sea urchins near a high Arctic lead/zinc ore deposit. *Marine Pollution Bulletin*, 10, 325–327.

Boice, L. P. 1986. CMA: an alternative to road salt? *Environment*, 28.

Boime, E. I. 2002. Fluid Boundaries: Southern California, Baja California, and the conflict over the Colorado River,1848–1944. Ph.D. dissertaion, University of California, San Diego, CA.

Bolivar, S. L. 1986. An overview of the Prairie Creek diamond-bearing intrusion, Arkansas. *Transactions of the Society of Mining Engineering*, 280, 1988–1994.

Bon, R. L., Gloyn, R. W. and Tabet, D. E. 1995. Utah mineral activity summary. *Utah Geological Survey Circular*, 91.

Bonilla, H. 1978. The War of the Pacific and the national and colonial problem in Peru. *Past and Present*, 81, 92–118.

Booth, C., J. 2003. Groundwater as an environmental constraint of longwall coal mining. *RMZ Materials and Geoenvironment*, 50, 49–52.

Borda, T., Celi, L., Bünemann, E. K., *et al.* 2014. Fertilization strategies affect phosphorus forms and release from soils and suspended solids. *Journal of Environmental Quality*, 43, 1024–1031.

Borgese, E. M. 1985. *The Mines of Neptune: Metals and Minerals from the Sea*. New York, NY: Abrams.

Borrok, D. M., Kesler, S. E., Boer, R. H., Essene, E. J. 1998. The Vergenoeg magnetite–fluorite deposit, South Africa: support for a hydrothermal model for massive iron oxide deposits. *Economic Geology*, 93, 564–586.

Both, R. A., Stumpf, E. F. 1987. Distribution of silver in the Broken Hill orebody. *Economic Geology*, 82, 1037–1043.

Boucher, O. 1995. The sulfate–CCN–cloud albedo effect: a sensitivity study with two general circulation models. *Tellus*, 47B, 281–300.

Boudreau, A. E., Mathex, E. A., McCallum, I. 1986. Halogen geochemistry of the Stillwater and Bushveld Complexes: evidence for transport of the platinum group elements by Cl-rich fluids. *Petrology*, 27, 346–357.

Bourne, H. L. 1989. What's it worth: a review of mineral royalty information. *Mining Engineering*, July, 541–544.

Boutron, C. F., Patterson, C. C. 1983. The occurrence of lead in Antarctic recent snow, firn deposited over the last two centuries and prehistoric ice. *Geochimica et Cosmochimica Acta*, 47, 1355–1368.

Boutron, C. F., Candelone, J.-P., Hong, S. 1994. Past and recent changes in the large-scale tropospheric cycles of lead and other heavy metals as documented in Antarctic and Greenland snow and ice: a review. *Geochimica et Cosmochimica Acta*, 58, 3217–3225.

Bowen, R. 1989. *Geothermal Resources*. Amsterdam: Elsevier.

Bowker, R. P. G. 1990. *Phosphorus Removal From Wastewater*. Park Ridge, NJ: Noyes.

Bowling, D. L. 1988. The geology and genesis of the Apex gallium–geranium deposit. MSc. Thesis, University of Utah.

Boyd, D. R. 2001. Canada vs. the OECD: an environmental comparison. [Online] Available: http://environmentalindicators.com/htdocs/indicators/13nucl.htm

Boyd, F. R., Meyer, H.O.A 1979. *Kimberlites, Diatremes and Diamonds: Their Geology, Petrology and Geochemistry*. Washington, DC: American Geophysical Union.

Boyd, T. L., Lund, J. W., Fresson, D. H. 2011. *Direct Utilization of Geothermal Energy, 2010 Worldwide Review*. Klamath Falls, OR: Geo-Heat Center, Oregon Institute of Technology.

Boyle, R. W. 1968. The geochemistry of silver and its deposits. *Geological Survey of Canada*, 160, 264.

Boynard, A. A., Borbon, A., Leonardis, T., *et al.* 2014, Spatial and seasonal variability of measured anthropogenic non-methane hydrocarbons in urban atmospheres: Implication on emission ratios. *Atmospheric Environment*, 82, 258–267

Boynton, R. S. 1980. *Chemistry and Technology of Lime and Limestone*. New York, NY: John Wiley.

Boynton, R. S., Gutschick, K. A., Freas, R. C., Thompson, J. L. 1983. Lime. In Lefond, S. J. (ed.), *Industrial Minerals and Rocks*, 5th edn. New York, NY: Society of Mining Engineers, pp. 809–831.

Bozic, K. J., Kurtz, S., Lau, E. 2009. The epidemiology of bearing surface usage in total hip arthroplasty in the United States. *Journal of Bone and Joint Surgery (American version)*, 91, 1614–1620.

BP 2012. Annual energy review. British Petroleum. [Online] Available: http://www.bp.com/content/dam/bp/pdf/Statistical-Review-2012/statistical_review_of_world_energy_2012.pdf.

Bradley, D., Mccauley, A. 2013. A preliminary deposit model for lithium–cesium–tantalum (LCT) pegmatites. *US Geological Survey Open-File Report*, 2013-1008.

Brady, N. C. 1990. *The Nature and Property of Soils*. New York, NY: Macmillan.

Branquet, Y., Laumonier, B., Cheilletz, A., Giuliani, G. 1999. Emeralds in the Eastern Cordillera of Colombia: two tectonic settings for one mineralization. *Geology*, 27, 597–600.

Bray, C. J., Spooner, E. T. C. 1983. Sheeted vein Sn–W mineralization and greisenization associated with economic kaolinization, Goonbarrow China clay pit, St. Austell, Cornwall, England: geologic relationships and geochronology. *Economic Geology*, 78, 1064–1089.

Bray, E. L. 2013. Bauxite and alumina. *US Geological Survey Minerals Yearbook*.

Breit, G. N. 1992, Vanadium: resources in fossil fuels. *US Geological Survey Bulletin*, 1877, 1–8.

Brierley, J. A., Brierley, C. L. 2001. Present and future commercial applications of biohydrometallurgy. *Hydrometallurgy*, 59, 233–239.

Britten, R. 2013. Regional metallogeny and genesis of a new deposit type – disseminated awaruite (Ni–Fe alloy) mineralization hosted in the Cache Creek terrane. Society of Economic Geologists Whistler: Geoscience for Discovery, September 24–27, Whistler, BC (Conference abstract).

Brobst, D. A. 1983. Barium minerals. In Lefond, S. J. (ed.), *Industrial Minerals and Rocks*, 5th edn. New York, NY: Society of Mining Engineers, pp. 485–501.

Broecker, W. S. 1983. The ocean. *Scientific American*, 247, 146–161.

Broecker, W. S. 1985. *How to Build a Habitable Planet*, Palisades, NY: Eldigio Press.

Browing, J., Tinker, S. W., Ikonnikova, S., *et al.* 2013. Barnett shale model – 2 (conclusion): Barnett study determines full-field reserves, production forecast. *Oil and Gas Journal*, September.

Brownawell, A. M., Berent, S., Brent, R. L., *et al.* 2005. The potential adverse health effects of dental amalgam. *Toxicological Reviews*, 24, 1–10.

Bubeck, R. C., Diment, W. H., Dick, B. O., Baldick, A. L., Lipton, S. D. 1971. Runoff of deicing salt: effect on Irondequoit Bay, Rochester, New York. *Science*, 172, 1128–1131.

Bues, A. A. 1986. *Geology of Tungsten*. Paris: UNESCO, International Geological Correlation Programme Project 26.

Bui, Q. 2013. U.S. is the world's latest producer of natural gas. Here's what that means. *NPR Planet Money*, November 17.

Buiter, W. H., Purvis, D.D. 1983. Oil, disinflation, and export competitiveness: a model of the "Dutch disease." In Bhandari, J. S., Putnam, B.H. (eds.) *Economic Interdependence and Flexible Exchange Rates*. Cambridge, MA: MIT Press., pp. 221–247.

Bullis, L. H., Mielke, J. E. 1985. *Strategic and Critical Minerals*. Boulder, CO: Westview Press.

Buol, S. W., Hole, F. D., Mccracken R. J. 1989. *Soil Genesis and Classification*, Ames, IA: Iowa State University Press.

Burgess, S., Beilstein, J. 2013. This means war? China's scramble for minerals and resource nationalism in southern Africa. *Contemporary Security Policy*, 34, 120–143.

Burkin, A. R. 1987. *Production of Aluminum and Alumina*. New York, NY: John Wiley.

Burnett, W. M., Ban, S. D. 1989. Changing prospects for natural gas in the United States. *Science*, 244, 305–310.

Burns, P. C., Ewing, R. C., Navrotsky, A. 2012. Nuclear fuel in a reactor accident. *Science*, 335, 1184–1188.

Burns, R. L. 1986. Controls on observed variations in surface geochemical exploration: a case study-the Albion-Scipio trend oilfield. M.Sc. Thesis, Wayne State University, Detroit, MI.

Buryakovsky, L., Eremenko, N. A., Gorfunkel, M. V., Chilingarian, G. V. 2005. *Geology and Geochemistry of Oil and Gas.* Amsterdam: Elsevier.

Bushnell, J., Yihsu, C. 2012. Allocation and leakage in regional cap-and-trade markets for CO_2. *Resource and Energy Economics*, 34, 647–668.

Butt, C. R. M., Cluzel, D. 2013. Serpentinites: essential roles in geodynamics, arc volcanism, sustainable development, and the origin of life. *Elements*, 9, 95–98.

Butterman, W. C., Amey, E.B., III 2005. Mineral commodity profiles: gold. *US Geologic Survey Open-File Report*, 02–303.

Butterman, W. C., Jorgenson, J.D. 2004. Mineral commodity profiles: germanium. *US Geological Survey Open-File Report*, 2004–1218.

Caddey, S. W., Bachman, R. L., Campbell, T. J., Reid, R. R., Ott, R. P. 1991. The Homestake gold mine, an early Proterozoic iron-formation-hosted gold deposit, Lawrence County, South Dakota. *US Geological Survey Bulletin*, 1857.

Callahan, W. H. 1977. The history of the discovery of the zinc deposit at Elmwood, Tennessee, concept and consequence. *Economic Geology*, 72, 1382–1392.

Cammarota, V. A., Jr. 1992. Market transparency in the worldwide minerals trade. *Minerals Today*, 6–12.

Candela, P. A., Piccoli, P. M. 2005. Magmatic processes in the development of porphyry-type ore systems. In Hedenquist, J. W., Thompson, J. F. H., Goldfarb, R. J., Richards, J. P. (eds.), *Economic Geology, 100th Anniversary Volume.* Littleton, CO: Society of Economic Geologists, pp. 25–37.

Canney, F. C., Canon, H. L., Cathrall, J. B., Robinson, K. 1979. Autumn colors, insects, plant disease and prospecting. *Economic Geology*, 74, 1673–1676.

Cannon, W. F., Force, E. R. 1983. Potential for high-grade, shallow-marine manganese deposits in North America. In Shanks, W. C. (ed.), *Unconventional Mineral Deposits.* Littleton, CO: Society of Mining Engineers, pp. 175–189.

Capone, C. A., Jr., Elzinga, K.G. 1987. Technology and energy use before, during and after OPEC: the US Portland cement industry. *Energy Journal*, 8, 93–112.

Cargill, S. M., Root, D. H., Bailey, E. H. 1981. Estimating usable resources from historical industry data. *Economic Geology*, 76, 1081–1095.

Carrington, D. 2014. UK defeats European bid for fracking regulations. *The Economist*, January 14.

Carlin, J. F. 1985a. Bismuth. *US Bureau of Mines Mineral Facts and Problems*, 675, 83–90.

Carlin, J. F. 1985b. Tin. *US Bureau of Mines Mineral Facts and Problems*, 675, 847–858.

Carlin, J. F. 2013. Tin. *US Geological Survey Minerals Yearbook*.

Carlson, C. L., Swisher, J. H. 1987. *Innovative Approaches to Mined Land Reclamation*, Carbondale, IL: Southern Illinois University Press.

Carlson, E. 1986. To salt or not to salt: that is the snow belt states' question. *The Wall Street Journal*, April 22.

Carmalt, S. W., St. John, B. 1986. Giant oil and gas fields. *American Association of Petroleum Geologists Memoir*, 40.

Carn, S. A., Krueger, A. J., Krotkov, N. A., Yang, K., Levelt, P. F. 2007. Sulfur dioxide emissions from Peruvian copper smelters detected by the Ozone Monitoring Instrument. *Geophysical Research Letters*, 34, L09801.

Carpenter, R. H., Carpenter, S. F. 1991. Heavy mineral deposits in the upper coastal plain of North Carolina and Virginia. *Economic Geology*, 86, 1657–1671.

Carpenter, S. B., Kistler, R. B. 2006. Boron and borates. In Lefond, S. J. (ed.), *Industrial Minerals and Rocks*, 5th edn. New York, NY: Society of Mining Engineers, pp. 275–283.

Carr, D. D., Rooney, L. F. 1983. Limestone and dolomite In Lefond, S. J. (ed.), *Industrial Minerals and Rocks*, 5th edn. New York, NY: Society of Mining Engineers.

Carter, C. I. 2013. Queen asserts rights to mine under homes. *The Telegraph*, December 19.

Carter, C., Rausser, G., Smith, A. 2013. *Commodity Storage and the Market Effects of Biofuel Policies.* Davis, CA: UC Davis.

Castilla, J. C. 1983. Environmental impact in sandy beaches of copper mine tailings at Chañaral, Chile. *Marine Pollution Bulletin*, 14, 459–464.

Castor, S. R. 2008. Rare earth deposits of North America. *Resource Geology*, 58, 337–347.

Cathcart, J. B., Sheldon, R. P., Guldrandsen, R. A. 1984. Phosphate rock resources of the US. *United States Geological Survey Circular*, 888.

Cathles, L. M. 1981. Fluid flow and genesis of hydrothermal ore deposits. *Economic Geology*, 75, 424–457.

Cathles, L. M., Adams, J. J. 2005. Fluid flow and petroleum and mineral resources in the upper (< 20-km) continental crust. In Hedenquist, J. W., Thompson, J. F. H., Goldfarb, R. J., Richards, J. P. (eds.), *Economic Geology, 100th Anniversary Volume.* Littleton, CO: Society of Economic Geologists, pp. 77–130.

Cavallo, G., Pandit, M. 2008 Geology and petrography of ochres and white clay deposits in the Rajasthan state, India. *Geoarchaeology and Archaeomineralogy, Proceedings of the International Conference, 29–30 October.* Sofia: Publishing House "St. Ivan Rilski."

Cawthorn, R. G., Barnes, S. J., Ballhaus, C., Malitch, K. N. 2005. Platinum group element, chromium and vanadium deposits in mafic to ultramafic rocks. In Hedenquist, J. W., Thompson, J. F. H., Goldfarb, R. J., Richards, J. P. (eds.), *Economic Geology, 100th Anniversary Volume.* Littleton, CO: Society of Economic Geologists, pp. 215–249.

Cecil, C. B., Dulong, F.T. 1986. Sulfur content of the coal resources of the United States: current status. *Society of Mining Engineers Preprint*, 86–84.

CEQ 1989. *Environmental Trends: Council on Environmental Quality: 23rd Annual Report.* Washington, DC: Executive Office of the President.

CGG 2015. Heavy oil/oil sands. [Online] Available: http://www.cgg.com/default.aspx?cid=3510.

Chaffee, M. A. 1982. A geochemical study of the Kalamazoo porphyry copper deposit. In Titley, S. R. (ed.), *Advances in Geology of the Porphyry Copper Deposits, Southwestern North America*. Tucson, AZ: University of Arizona Press, pp. 211–226.

Champigny, N., Abbot, R. M. 1992. *Understanding the Real World of Environmental Management In The Canadian Mineral Industry*. London: Elsevier.

Chander, S. 1992. *Emerging Process Technologies for a Cleaner Environment*. Littleton, CO: Society for Mining, Metallurgy and Exploration.

Chapman, G. P. 1983. Mica. In Lefond, S. J. (ed.), *Industrial Minerals and Rocks*, 5th edn. New York, NY: Society of Mining Engineers, pp. 915–929.

Chapman, N. A., McKinley, I. G. 1987. *The Geological Disposal of Nuclear Waste*. New York, NY: Wiley.

Chen, B. X. 2014. Apple says supplies don't come from war zones. *The New York Times*, February 13.

Cheney, E. S., Patton, T. C. 1967. Origin of bedrock placer values. *Economic Geology*, 62, 852–853.

Cherkasova, E. F., Ryshenko, B. N. 2011. Natural water contamination under chromite deposits mining. *Mineral Magazine*, 74, 662.

Chester, E. W. 1983. United States Oil Policy and Diplomacy. Westport, CT: Greenwood Press.

Chestnut, A. 1990. Filtration system cleans groundwater at refinery. *Pollution Engineering*, 22, 65–67.

Chiaradia, M., Gulson, B. L., James, M., Jameson, C. W., Johnson, D. 1997. Identification of secondary lead sources in the air of an urban environment. *Atmospheric Environment*, 31, 3511–3521.

Chidester, R. A. 1964. Talc resources of the United States. *US Geological Survey Bulletin*, 1167.

Chilés, J.-P., Delfiner, P. 2012. *Geostatistics: Modeling Spatial Uncertainty*, 2nd edn. New York, NY: Wiley.

Chin, M., Rood, R. B., Lin, S. J., Muller, J. F., Thompson, A. M. 2000. Atmospheric sulfur cycle simulated in the global model GOCART: Model description and global properties. *Journal of Geophysical Research*, 105.

Chorover, J., Kretzschmar, R., Garcia-pichel, F., Sparks, D.L. 2007. Soil biogeochemical processes within the critical zone. *Elements*, 3, 321–326.

Clair, T. A., Hindar, A. 2005. Liming for the mitigation of acid rain effects in freshwater: a review of recent results. *Environmental Reviews*, 13, 91–128.

Clark, L. H., Burrill, G. H. R. 1981. Unconformity-related uranium deposits, Athabasca area, Saskatchewan, and East Alligator Rivers areas, Northern Territory, Australia. *CIM Bulletin*, 75, 91–98.

Clark, S. H. B. 1989. Metallogenic map of zinc, lead and barium deposits and occurrences in Paleozoic sedimentary rocks, east-central United States. *US Geological Survey Miscellaneous Investigation*, I–1773.

Cleetus, R. 2010. Finding common groups in the debate between carbon tax and cap-and-trade policies. *Bulletin of Atomic Scientists*, 67, 19–27.

Cleveland Clinic, 2015. Occupational lung disease. [Online] Available: http://www.clevelandclinicmeded.com/medical pubs/diseasemanagement/pulmonary/occupational-lung-disease/Default.htm.

Cline, J. S., Hofstra, A. H., Muntean, J. L., Tosdal, R. M., Hickey, K. A. 2005. Carlin-type gold deposits in Nevada, USA: critical geologic characteristics and viable models. In Hedenquist, J. W., Thompson, J. F. H., Goldfarb, R. J., Richards, J. P. (eds.), *Economic Geology, 100th Anniversary Volume*. Littleton, CO: Society of Economic Geologists, pp. 451–484.

Cloud, P. E. 1983. The biosphere. *Scientific American*, 249, 176–189.

Clout, J. M. F., Simonson, B. M. 2005 Precambrian iron formations and iron formation-hosted iron ore deposits. In Hedenquist, J. W., Thompson, J. F. H., Goldfarb, R. J., Richards, J.P. (eds.), *Economic Geology, 100th Anniversary Volume*. Littleton, CO: Society of Economic Geologists, pp. 643–679.

Cobalt Development Institute 2013. Production Statistics. [Online] Available: http://www.thecdi.com/cobalt-stats.

Cochran, T. B., Feiveson, H. A., Patterson, W., *et al.* 2010. Fast breeder reactor programs: history and status. A research report of the International Panel on Fissile Materials. International Panel on Fissile Materials. [Online] Available: http://fissilematerials.org/library/rr08.pdf.

Cohen, K. 1984. *Nuclear Power The Resourceful Earth*. Oxford: Basil Blackwell.

Cohen, N. 1991. Regulation of in-place-asbestos-containing material. *Environmental Research*, 55, 97–105.

Comer, J. B. 1974. Genesis of Jamaican bauxite. *Economic Geology*, 69, 1251–1264.

Conger, H. M. 1992. Environmental impact of gold ore processing. In Chander, S. (ed.), *Emerging Process Technologies for a Cleaner Environment*. Littleton, CO: Society for Mining, Metallurgy, and Exploration.

Conradie, A. S. 2008. Gold. In Mabuza, M.(ed.), *South Africa's Mineral Industry 2007/2008*. Pretoria: Directorate, Mineral Economics, Department of Minerals and Energy, pp. 27–31.

Constantine, T. A. 1985. Treating gold and silver mine tailings pond effluents. *Water Pollution Control*, 123.

Contreras-León, G.J., Rodríguez-Satizábal, S-A., Casellanos-romero, C-M., Franco-Herrera, A., Serrano-Gómez, M. 2013. Acute toxicity of drilling muds on *Litopenaeus vannamei* (Boone, 1931) postlarvaeu. *Ciencia, Tecnología y Futuro*, 5, 127–138.

Cook, J. 1985. Nuclear follies. *Forbes*, February 11, 82–100.

Cook, N. G. W. 1982. Ground-water problems in open pit and underground mines. *Geological Society of America Special Paper*, 189, 397–406.

Cook, R. B., Kreis, R. G., Jr, Kingson, J. C., *et al.* 1990. Paleolimnology of McNeary Lake: an acidic lake in northern Michigan. *Journal of Paleolimnology*, 3, 13–34.

Cook, T. A. 2013. Reserve growth of oil and gas fields: investigations and applications. *USGS Scientific Investigations Report*, 2013–5063.

Coolbaugh, M. J. 1967. Special problems of mining in deep potash. *Mining Engineering*, May, 68–73.

Cooper, R. G., Harrison, A. P. 2009. The exposure to and health effects of antimony. *Indian Journal of Occupational and Environmental Medicine*, 13, 3–10.

Copeland, C. 2013. Mountaintop mining: background and current controversies. *Congressional Research Service*, 7–5700.

Coppa, L. V. 1991. Recycled scrap metal trade in jeopardy? *Minerals Today*, 24–28.

Corathers, L. A. 2013a. Manganese. *US Geological Survey Minerals Yearbook*, Advance Release for 2011.

Corathers, L. A. 2013b. Silicon. *US Geological Survey Minerals Yearbook*, Advance Release for 2011.

Corden, W. M., Neary, J. P. 1982. Booming sector and de-industrialization in a small open economy. *The Economic Journal*, 92, 825–848.

Corn, P. S. 2008. The US Geological Survey and wilderness research. *International Journal of Wilderness*, 14, 24–33.

Cornwall, W. 2015, Deepwater Horizon after the oil. *Science*, 348, 22–29.

Correll, D. L. 1998. The role of phosphorus in the eutrophication of receiving waters: a review. *Journal of Environmental Quality*, 27, 261–266.

Cottingham, J. 2012. Arkansas' bromine industry sees rebound. *Arkansas Business*, September 10.

Couch, G. R. 2009. Underground coal gasification. IEA Clean Coal Centre Profile, September. International Energy Agency, London.

Courts, M. J. 2012. Conflict minerals disclosure rule. Report to Congressional Committees. [Online] Available: http://www.gao.gov/assets/600/592458.pdf.

Couturier, G. 1992. Magnesium. *Canadian Minerals Yearbook*, 21.1–27.20.

Cover, M. 2012. Hillary Clinton: Opposition to sea treaty based on 'mythology'. CNSNews.com. [Online] Available: http://cnsnews.com/news/article/hillary-clinton-opposition-sea-treaty-based-mythology [Accessed May 23 2012].

Cowie, R. L. 1994. The epidemiology of tuberculosis in gold miners with silicosis. *American Journal of Respiratory Critical Care Medicine*, 150, 1460–1462.

Cox, D. P., Singer, D. A. 1992. Distribution of gold in porphyry copper deposits. *US Geological Survey Bulletin*, 1877-C, C1–C25.

Cox, D. P., Lindsey, D. A., Singer, D. A., Moring, B. C., Diggles, M.F. 2007. Sediment-hosted copper deposits of the world: models and database. *US Geological Survey Open-File Report*, 03-107.

Cox, S. F. 2005. Coupling between deformation, fluid pressures, and fluid flow in ore-producing hydrothermal systems at depth in the crust. In Hedenquist, J. W., Thompson, J. F. H., Goldfarb, R. J., Richards, J.P. (eds.), *Economic Geology, 100th Anniversary Volume*. Littleton, CO: Society of Economic Geologists, pp. 39–75.

Crabbe, L. 1989. The international gold standard and US monetary policy from World War I to the New Deal. *Federal Reserve Bulletin*, 75, 423–440.

Cramton, P. 2006. How best to auction oil rights. In Humphreys, M., Sachs, J. D., Stiglitz, J. E. (eds.), *Escaping the Resource Curse*. New York, NY: Columbia University Press, pp. 114–151.

Crangle, R. D., Jr. 2012. Gypsum. *United States Geological Survey Minerals Yearbook*.

Crangle, R. D., Jr. 2013. Diatomite. *United States Geological Survey Mineral Commodity Summary*.

Crangle, R. D., Jr. 2014. Boron. *United States Geological Survey Mineral Commodity Summary*.

Crockett, R. N., Sutphin, D. M. 1993. International strategic minerals inventory summary report – niobium (columbium) and tantalum. *US Geological Survey Circular*, 930-M.

Cronan, D. S. 1992. *Marine Minerals in the Exclusive Economic Zones*. London: Chapman and Hall.

Cronquist, C. 2001. *Estimation and Classification of Reserves of Crude Oil, Natural Gas, and Condensate*. Richardson, TX: Society of Petroleum Engineers.

Crookston, J. A., Fitzpatrick, W. D. 1983. Refractories. In Lefond, S. J. (ed.), *Industrial Minerals and Rocks*, 5th edn. New York, NY: Society of Mining Engineers, pp. 373–385.

Crow, P. 1991. The windfall profits tax. *Oil and Gas Journal*, 89, 33.

Crowson, P. 2012. Some observations on copper yields and ore grades. *Resources Policy*, 37, 59–72.

Crozier, S. A. 1992. Overview of environmental regulations and their impact on the United States copper industry. In Chancier, S. (ed.), *Emerging Process Technologies for a Cleaner Environment*. Littleton, CO: Society for Mining, Metallurgy and Exploration, pp. 17–36.

Culhane, P. J. 1981. *Public Lands Politics*. Baltimore, MD: Johns Hopkins University Press.

Cumberlidge, J. T., Chace, F. M. 1967. Geology of the Nickel Mountain mine, Riddle, Oregon. *American Institute Mining, Metal and Petroleum Engineering, Graton-Sales Volume*, 1933–1965.

Cummings, A. B., Given, I. A. 1973. *SME Mining Engineers Handbook*. New York, NY: American Institute of Mining, Metallurgical and Petroleum Engineers.

Cunningham, C. G., Zientek, M. L., Bawiec, W. J., Orris, G. J. 2005. Geology and nonfuel mineral deposits of Latin America and Canada. *US Geological Survey Open-File Report*, 2005-1294B.

Cunningam, W. P., Saigo, B. W. 1992. *Environmental Science: A Global Concern*. Dubuque, IA: William C. Brown.

Dahlkamp, F. J. 1989. Classification scheme for uranium deposits: state of the art review. In *Metallogenesis of Uranium Deposits*. Vienna: International Atomic Energy Agency, pp. 1–32.

Danielou, M. 2012. *Debunking Ten Myths About Phosphate Rock Production: Trends from 1992–2011*. Paris: International Fertilizer Industry Association.

Danielson, A. L. 2013. OPEC: Encyclopedia Britannica. [Online] Available: http://www.britannica.com/EBchecked/topic/454413/OPEC/233528/History

Dankers, A. P. H. 1995. The energy situation in Thai lime industry. *Energy for Sustainable Development*, 1, 36–40.

Danko, P. 2014. World's first full-scale "clean" coal plant opens in Canada. National Geographic. [Online] Available: http://energyblog.nationalgeographic.com/2014/10/02/worlds-first-full-scale-clean-coal-plant-opens-in-canada/.

Darling, T. 2006. *Well Logging and Formation Evaluation*. Amsterdam: Gulf Professional Publishing.

Das, S. 2013. Australia's economic outlook – the Naura solution? *EconoMonitor*, April 2.

Dasgupta, P., Heal, G. M. 1980. *Economic Theory and Exhaustible Resources*, Cambridge: Cambridge University Press.

Davenhall, B. 2012. Geomedicine: geography and personal health. [Online] Available: http://www.esri.com/library/ebooks/geomedicine.pdf.

Davenport, P. H., Hornbrook, E. H. W., Butler, A. J. 1975. Regional lake sediment geochemical survey for zinc mineralization in western Newfoundland. In Elliott, I. L., Fletcher, W. K. (eds.), *Geochemical Exploration 1974*. Amsterdam: Elsevier, pp. 555–578.

Davenport, W. G., King, M., Schlesinger, M., Biswas, A. K. 2002. *Extractive Metallurgy of Copper*, 4th edn. Oxford: Elsevier Science Limited.

David, M. 1977. *Geostatistical Ore Reserve Estimation*. Amsterdam: Elsevier.

Davies, J. C., Mazurek, J. 1998. *Pollution Control in United States: Evaluation the System (Resources for the Future)*. Washington, DC: Routeledge.

Davis, C. J. 1985a. Nitrogen (ammonia). *US Bureau of Mines Mineral Facts and Problems*, 675, 553–562.

Davis, I. J. 1985b. Mica. *US Bureau of Mines Minerals Facts and Problems*, 509–520.

Davy, R., El-Ansary, M. 1986. Geochemical patterns in the laterite profile at the Boddington gold deposit, Western Australia. *Journal of Geochemical Exploration*, 26, 119–144.

De Wit, M. J. 1985. *Minerals and Mining in Antarctica*. Oxford: Clarendon Press.

De Wit, M. J., Kruger, F. J. 1990. *The Economic Potential of the Dufek Complex*. Washington, DC: American Geophysical Union.

Deakin. D. R. 2014. The 10 largest diamonds ever discovered. *The Richest*, February 4. [Online] Available: http://www.therichest.com/rich-list/the-biggest/a-miners-best-friend-the-10-largest-uncut-rough-diamonds/.

DeLong, C. R. 1926. The future demand for bromine. *Industrial and Engineering Chemistry*, 18, 425–428.

Dermatas, D. 2006. Fate and behavior of metal(oid) contaminants in an organic matter-rick shooting range soil: Implications of remediation. *Water, Air & Soil Pollution*, 6, 143–155.

Devuyst, E. A., Robbins, G., Vergunst, R., Tandi, B., Iamarino, P. F. 1991. INCO's cyanide removal technology working well. *Mining Engineering*, February, 1–3.

DeWulf, J., Van Langenhove, H. 1995. Simultaneous determination of C_1 and C_2-halocarbons and monocyclic aromatic hydrocarbons in marine water samples at ng/l concentration levels. *International Journal of Environmental Analytical Chemistry*, 61, 35–46.

DeYoung, J. H. 1981. The Lasky cumulative tonnage–grade relationship: a reexamination. *Journal of Economic Geology*, 76, 1067–1080.

DeYoung, J. H., Jr., Sutphin, D. M., Cannon, W. F. 1984a. International strategic minerals inventory summary report – manganese. *US Geological Survey Circular*, 930-A, 22.

DeYoung, J. H., Jr., Lee, M.P., Lipin, B. R. 1984b. International strategic minerals inventory summary report – chromium. *US Geological Survey Circular*, 930-B, 41.

DeYoung, R. 2014. Some behavioral aspects of energy descent: how a biophysical psychology might help people transition through the lean times ahead. *Frontiers in Psychology*, 5, 1–16.

Dittrich, T., Seifert, T., Gutzmer, J. 2011. Gallium in bauxite deposits. *Mineralogical Magazine*, 75, 765.

Dixit, S. S., Smol, J. P., Kingston, J. C. 1992. Diatoms: powerful indicators of environmental change. *Environmental Science and Technology*, 26, 22–23.

Dobrin, M. B. 1976. *Introduction to Geophysical Prospecting*, New York, NY: McGraw-Hill.

Dockery, D. W., Stone, P. H., 2007. Cardiovascular risks from fine particulate air pollution. *New England Journal of Medicine*, 356, 511–513.

DOD 2014. Strategic and critical materials operations report to congress: operations under the strategic and critical materials stockpiling act during fiscal year 2013. US Department of Defense. [Online] Available: http://www.strategicmaterials.dla.mil/Report%20Library/TAB%20B-FY13%20NDS%20Operations%20Report.pdf.

Department of Defense 2012. Carbon dioxide enhanced oil recovery: untapped domestic energy supply and long term carbon storage solution. [Online] Available: http://www.netl.doe.gov/file%20library/research/oil-gas/small_CO2_EOR_Primer.pdf.

DOE 2015. Enhanced oil recovery. [Online] Available: http://energy.gov/fe/science-innovation/oil-gas-research/enhanced-oil-recovery.

Dogan, A., Dogan, M., Hoskins, J. A. 2008. Erionite series minerals: mineralogical and carcinogenic properties. *Environmental Geochemistry and Health*, 30, 367–381.

Dolley, T. O. 2013. Silica. *US Geological Survey Minerals Yearbook*.

Donovan, E., Donovan, B. L., McKinley, M. A., Cowan, D. M., Paustenbach, D. J. 2012. Evaluation of take home (para-occupational) exposure to asbestos and disease: a review of the literature. *Critical Reviews in Toxicology*, 42, 703–731.

Donville, C. 2013. Rio unit readies $2.4 billion rights offer for mine [Online]. *Bloomberg News*, November 14.

Dorfner, K. 1990, *Ion Exchangers*. Berlin: De Gruyter.

Dotterweich, M. 2013. The history of human-induced soil erosion: geomorphic legacies, early descriptions and research and the development of soil conservation – a global synopsis. *Geomorphology*, 201, 1–34.

Doull, J., Klassen, C. D., Amdur, M. O., eds., 1980. *Casarett and Doull's Toxicology: The Basic Science of Poisons*, New York, NY: Macmillan.

Drew, L. J. 1967. Grid-drilling exploration and its application to the search for petroleum. *Economic Geology*, 62, 698–710.

Drew, L. J. 1990. *Oil and Gas Forecasting*. Oxford: Oxford University Press.

Driscoll, F. G. 1986. *Groundwater and Wells*, St. Paul, MN: Johnson Filtration Systems.

Drummond, I. M. 1987. *The Gold Standard and the International Monetary System 1900–1939*, New York, NY: Macmillan.

Du, M., Wang, C., Hu, X., Zhao, G. 2008. Positive effect of selenium on the immune regulation activity of lingzhi or reishi medicinal mushroom, *Ganoderma lucidum* (W. Curt.: Fr.) P. Karst.

(Aphyllophoromycetideae), proteins in vitro. *International Journal of Medical Mushrooms*, 10, 337–344.

Duderstadt, J. J., Kikuchi, C. 1979. *Nuclear Power*. Ann Arbor, MI: University of Michigan Press.

Duffy, S. 2014. Evolution of the Boundary Water Canoe Area Wilderness: a historical perspective of key controversy. National Wilderness Conference, Albuquerque, October 15-19. [Online] Available: http://www.wilderness50th.org/documents/National%20Wilderness%20Conference%20Program.pdf.

Durrheim, R. J. 2010. Mitigating the risk of rockbursts in the deep hard rock mines of South Africa: 100 years of research. In Brune, J. (ed.), *Extracting the Science: A Century of Mining Research*. Littleton, CO: Society for Mining, Metallurgy and Exploration, pp. 156–171.

Dutta, S., Rajaram, R., Robinson, B. 2005. Mineland reclamation. In V. Rajaram, V., Dutta, S. (eds.), *Sustainable Mining Practices: A Global Perspective*. Leiden: Balkema, ch. 5.

Dyni, J. R. 2010. Oil shale. In Clarke, A. W., Trinnaman, J. A. (eds.), *Survey of Energy Resources*, 22nd edn. Amsterdam: Elsevier, pp. 93–123.

Earman, S., Phillips, F. M., McPherson, B. J. 2005. A revised model for the formation of trona deposits. *Geological Society of America Abstracts with Programs*, 37, no. 7, 467.

Earney, F. C. F. 1990. *Marine Mineral Resources*. London: Routeledge.

Eatough, D. J., Farber, R. J, Watson, J. G. 2000. Second generation chemical mass balance source apportionment of sulfur oxides and sulfate at the Grand Canyon during the Project MOHAVE summer intensive. *Journal Air Waste Management Association*, 50, 759–774.

Ebrahim, Z., Friedrichs, J. 2013. Gas flaring: the burning issue. Resilience.org, September 3. [Online] Available: http://www.resilience.org/stories/2013-09-03/gas-flaring-the-burning-issue.

Economides, M. J., Ungemach, P. O. 1987. *Applied Geothermics*. New York, NY: John Wiley.

Edelstein, D. L. 1985. Arsenic. *US Bureau of Mines Mineral Facts and Problems*, 675, 43–52.

Edelstein, D. L. 2013a. Copper. *US Geological Survey Mineral Yearbook*.

Edelstein, D. L. 2013b. Arsenic. *US Geological Survey Mineral Commodity Summary*.

EIA 1993 US Coal reserves: an update by heat and sulfur content, DOE.EIA-0529(92). [Online] Available: http://large.stanford.edu/publications/coal/references/docs/052992.pdf.

EIA 2012. Annual Energy Outlook. Early Release Overview. US Department of Energy. [Online] Available: http://www.eia.gov/forecasts/archive/aeo12/

EIA 2013b. Annual Energy Outlook. Early Release Overview. US Department of Energy. Available: http://www.eia.gov/forecasts/archive/aeo13/

EIA 2013a. Technically recoverable shale oil and shale gas resources: An assessment of 137 shale formations in 41 countries outside the United States. [Online] Available: http://www.eia.gov/analysis/studies/worldshalegas/.

EIA 2015a. Today in energy. [Online] Available: http://www.eia.gov/todayinenergy/detail.cfm?id=10151.

EIA 2015b. National carbon dioxide emissions from fossil fuel-burning power plants as measured by acid rain program Continuous Emissions Monitoring Systems (CEMS). [Online] Available: http://www.epa.gov/captrade/maps/co2.html

Eiden, R. 1990. The atmosphere: physical properties and climate change. In Hutzinger, O. (ed.), *The Handbook of Environmental Chemistry*. Heidelberg: Springer-Verlag, pp. 147–188.

El-Reedy, M. 2012. *Offshore Structures*. Amsterdam: Gulf Professional Publishing.

Eldridge, T., Boileau, S. 2014. *United Kingdom. Mining Law 2014*, 1st edn. London: Global Legal Group.

Eliot, R. C. (ed.) 1978. *Coal Desulfurization Prior to Combustion*. Park Ridge, NJ: Noyes Data Corp.

Elliot, T.C., Schwieger, R.G., eds. 1985. *The Acid Rain Sourcebook*. New York, NY: McGraw-Hill.

Elliott, J. E. 1992. Tungsten – geology and resources of deposits in southeastern China. *US Geological Survey Bulletin*, 18771, 1–34.

Ellis, R. 1987. Aredor makes the grade. *Mining Magazine*, 206–213.

Elmer, F. L., White, R. W., Powell, R. 2006. Devolatilization of metabasic rocks during greenschist–amphibolite facies metamorphism. *Journal of Metamorphic Geology*, 24, 497–513.

Ely, N. 1964. Minerals titles and concessions. In Robie, E. H. (ed.), *Economics of the Mineral Industries*. New York, NY: American Institute of Mining, Metallurgical and Petroleum Engineers, pp. 81–130.

Emigh, G. D. 1983. Phosphate rock. In Lefond, S. J. (ed.), *Industrial Minerals and Rocks*, 5th edn. New York, NY: Society of Mining Engineers, pp. 1017–1048.

EMJ 1981. Rundle: a pause for reflection and possible downsizing to 125,000 bbl/d. *Engineering and Mining Journal*, 106–110.

EMJ 1982. Exxon shelves the most advanced oil shale project in US. *Engineering and Mining Journal*, May, 33–35.

EMR 1973. *Report on the Administration of the Emergency Gold Mining Assistance Act*. Ottawa: Department of Energy, Mines and Resources.

Emsley, J. 1980. The phosphorus cycle. In Huntzinger, O. (ed.), *The Handbook of Environmental Chemistry*. Heidelberg: Springer-Verlag, pp. 147–167.

Ensign, R. L., Matthews, C. M. 2014. Barrick gold unit is accused of bribery in Africa. *The Wall Street Journal*, June 18.

Environment Canada 2011. Canada's protected areas, 2011. [Online] Available: http://www.ec.gc.ca/indicateurs-indicators/default.asp?lang=en&n=478A1D3D-1.

EPA 2003a. Health effects support document for manganese. [Online] Avaialable: http://www.epa.gov/ogwdw/ccl/pdfs/reg_determine1/support_cc1_magnese_healtheffects.pdf.

EPA 2003b. National emission standards for hazardous air pollutants (NESHAP) for lime manufacturing background information document. Vol II. [Online] Available: http://www.epa.gov/ttnatw01/lime/comments_resp8-20-03.pdf.

EPA 2013. Basic information about arsenic in drinking water. [Online] Available: http://water.epa.gov/drink/contaminants/basicinformation/arsenic.cfm.

EPA 2014. Overview of greenhouse gases. United States Environmental Protection Agency. [Online] Available: http://www.epa.gov/climatechange/ghgemissions/gases.html.

EPA 2015. EPA: National Emissions Inventory, 2015. [Online] Available: http://www.epa.gov/ttnchie1/trends/.

EPA-RFS 2015. Renewable fuel standard (RFS). [Online] Avaialble: http://www.epa.gov/oms/fuels/renewablefuels/.

EPS 2012. 2012 Industry recyling report. [Online] Available: http://www.achfoam.com/2012-industry-recycling-report.aspx.

Ericksen, G. E. 1981. Geology and origin of the Chilean nitrate deposits. *US Geological Survey Professional Paper*, 1188, 37.

Ericksen, G. E. 1983. The Chilean nitrate deposits. *American Scientist*, 71, 366–374.

Erisman, J. W., Sutton, M. A., Galloway, J. N., Klimont, Z., Winiwarter, W. 2008. How a century of ammonia synthesis changed the world. *Nature Geoscience*, 1, 636–639.

Erlich, E. I., Hausel, W. D. 2002. *Diamond Deposits: Origin, Exploration, and History of Discovery*. Littleton, CO: Society for Mining, Metallurgy, and Exploration.

Ehrlich, H. L. 1997, Microbes and metals. *Applied Microbiology Biotechnology*, 48, 687–692.

Esikinazi, B., Chevrier, J., Rauch, S. A., *et al.* 2013. In utero and childhood and polybrominated diphenyl ether (PBDE) exposures and neurodevelopment in the CHAMACOS Study. *Environmental Health Perspectives*, 121, 257–262.

Estes, J. E., Crippen, R. E., Star, J. L. 1985. Natural oil seep detection in the Santa Barbara Channel, California, with shuttle imaging radar. *Geology*, 13, 282–284.

Etacheri, V., Marom, R., Elazari, R., Salitra, G., Aurbach, D. 2011. Challenges in the development of advanced Li-ion batteries: a review. *Energy and Environmental Science*, 4, 3243–3262.

Eugster, H. E., Chou, I. M., 1979. A model for the deposition of Cornwall-type magnetite deposits. *Economic Geology*, 74, 763–774.

Eugster, H. P., Hardie, L. A. 1978. Saline lakes. In Lerman, A. (ed.), *Lakes – Chemistry, Geology and Physics*. Heidelberg: Springer-Verlag, pp. 237–293.

Evangelou, V. P., Zhang, Y. L. 1995. A review: pyrite oxidation mechanisms and acid mine drainage prevention. *Critical Reviews in Environmental Science and Technology*, 25, 141–199.

Evans, D. I., Shoemaker, R. S., Veltman, H., eds. 1979. *International Laterite Symposium*. New Orleans, LA: Society of Mining Engineers of AIME.

Evans, K. A., McCuaig, T. C., Leach, D., Angerer, T., Hagemann, S. G. 2013. Banded iron formation to iron ore: a record of the evolution of Earth environments. *Geology*, 41, 99–102.

Ewing, R. 2006. The nuclear fuel cycle: a role for mineralogy and geochemistry. *Elements*, 2, 331–334.

Ewing, R. C., Murakami, T. 2012. Fukushima Daiichi more than one year later. *Elements*, 8, 181–182.

Ewing, R. C., Ritsema, J. 2011. Underestimating nuclear accident risks: Why are rare events so common? *Bulletin of Atomic Scientists*, May 3.

EY 2013. Resources nationalism update. Mining and Metals, October 13. [Online] Available: http://www.ey.com/Publication/vwLUAssets/EY-M-and-M-Resource-nationalism-update-October-2013/$FILE/EY-M-and-M-Resource-nationalism-update-October-2013.pdf [Accessed March 25 2014].

Eyde, T. H., Holmes, D. H. 2006. Zeolites. In Kogel, J. E., Trivedi, N. C., Barker, N. C., Krukowski, S. T. (eds.), *Industrial Minerals and Rocks*, 7th edn. Littleton, CO: Society for Mining, Metallurgy and Exploration, pp. 1039–1064.

Falkowski, P., Scholes, R. J., Boyle, E., *et al.* 2000, The global carbon cycle: a test of our knowledge of Earth as a system. *Science*, 290, 291–296.

Farnesworth, C. H. 1992. Canada to divide its northern land. *The New York Times*, May 7.

Fay, S. 1982. *Beyond Greed*. New York, NY: Viking.

Fedorinova, Y., Kolesnikova, M. 2012. Potash export grip challenged in China's bond deal. *Bloomberg News*, November 14.

Feldman, S. R. 2006. Salt. In Kogel, J. E., Trivedi, N. C., Barker, N. C., Krukowski, S. T. (eds.), *Industrial Minerals and Rocks*, 7th edn. Littleton, CO: Society for Mining, Metallurgy, and Exploration, pp. 793–813.

Felsenfeld, A. J. 1991. A report on fluorosis in the United States secondary to drinking well water. *Journal of the American Medical Association*, 265, 486–488.

Feng, D., Roberts, H. H. 2010. Pervasive barite deposits at cold seeps from the northern Gulf of Mexico continental slope: geochemical characteristics and formation mechanism. American Geophysical Union, Fall Meeting 2010, Abstract #OS53A-1374.

Fenton, M. D. 2013. Iron and steel. *US Geological Survey Minerals Yearbook*, Advance Release for 2011.

Fergusson, J. E. 1990. *The Heavy Elements: Chemistry, Environmental Impact and Health Effects*. Oxford: Pergamon Press.

Fernández-Martinez, R., Rucandio, I. 2014. Total mercury, organic mercury and mercury fractionation in soil profiles from the Almadén mercury mine area. *Environmental Science: Processes and Impacts*, 2, 333–340.

Ferrero, R. C., Kolak, J. J., Bills, D. J., *et al.* 2013. US Geological Survey Energy and Minerals Science Strategy: a resource life-cycle approach. *US Geological Survey Circular*, 1383-D.

Ferro, S. 2002. Synthesis of diamond. *Journal of Materials Chemistry*, 12, 2843–2855.

Fettweiss, G. B. 1979. *World Coal Resources*. Amsterdam: Elsevier.

Fick, J. Brazil's Petrobras confirms new oil fields near northeast coast. *The Wall Street Journal*, September 27.

Filho, I. A., Lima, P. R. A., Souza, O. M. 1984. Aspects of Geology of the Barreiro Carbonatitic Complex, Araxa, MG, Brazil. In Rodrigues, C. S. and Lima, P. R. A. S. (eds.), *Complexos Brasileira de Metaluria e Mieração*, pp. 21–44.

Filippelli, G. M., Laidlaw, M. A. S., Latimer, J. C., Raftis, R. 2005. Urban lead poisoning and medical geology. *GSA Today*, 15, 4–11.

FIPR 2010. Florida Industrial and Phosphate Research Institute, http://www1.fipr.state.fl.us/.

Fischer, R. P. 1970. Similarities, differences and some genetic problems of the Wyoming and Colorado Plateau types of uranium deposits in sandstone. *Economic Geology*, 65, 778–784.

Fitzgerald, A. M. 2002. Mining agreements: negotiated frameworks in the Australian Mineral sector. Chatswood: Prospect Media. [Online] Available: http://eprints.qut.edu.au/34063/.

Flawn, P. T. 1966. *Mineral Resources*. Chicago, IL: Rand McNally.

Fluker, S. 2009. Wilderness narrative in law: the view from Canada's national parks. Social Science and Research Network, June 26. [Online] Available: http://papers.ssrn.com/sol3/papers.cfm?abstract_id=1531772.

Flynn, A. 2014. Rio Tinto's Mongolia mine hits another snag. *MarketWatch*, June 23.

Foell, E. J., Thiel, H., Schriever, G. 1992. DISCOL: a long-term, large-scale disturbance-recolonization experiments in the abyssal eastern tropical South Pacific ocean. *Mining Engineering*, 44, January, 90–94.

Foley, N., Jaskula, B. 2013. Gallium: a smart metal. US Geological Survey, 2. [Online] Available: http://pubs.usgs.gov/fs/2013/3006/pdf/fs2013-3006.pdf.

Force, E. R., Cannon, W. F. 1988. Depositional model for shallow-marine manganese deposits around black shale basins. *Economic Geology*, 83, 93–117.

Force, E. R. 1991. Geology of titanium-mineral deposits. *Geological Survey of America Special Paper*, 259, 112.

Forstner, U., Wittmann, G. T. W. 1981. *Metal Pollution in the Aquatic Environment*. Berlin: Springer-Verlag.

Fortescue, J. A. C. 1992. Landscape geochemistry: retrospect and prospect – 1990. *Applied Geochemistry*, 7, 1–54.

Foss, P. 1987. *Federal Lands Policy*. New York, NY: Greenwood Press.

Foth, H. D. 1984. *Fundamentals of Soil Science*. New York, NY: John Wiley.

Frakes, L., Bolton, B. 1992. Effects of ocean chemistry, sea level and climate on the formation of primary sedimentary manganese ore deposits. *Economic Geology*, 87, 1207–1217.

Frankham, J. 2010, Final analysis: the use of metal scavengers for recovery of precious, base and heavy metals from waste streams. *Platinum Metals Review*, 54, 200.

Fraser, D. C. 1978. Geophysics of the Montcalm township copper–nickel discovery. *CIM Bulletin*, 99–104.

Freas, R. C., Hayden, J. S., Pryor, C. A., Jr. 2006. Limestone and dolomite. In Kogel, J. E., Trivedi, N. C., Barker, N. C., Krukowski, S.T. (eds.), *Industrial Minerals and Rocks*, 7th edn. Littleton, CO: Society for Mining, Metallurgy and Exploration, pp. 581–597.

Freeman, L. W., Highsmith, R.P. 2014. Supplying society with natural resources: the future of mining – from Agricola to Rachel Carson and beyond. *The Bridge*, 44, 24–32.

Freeze, R. A., Cherry, J.A. 1979. *Groundwater*. Englewood Cliffs, NJ: Prentice-Hall.

Frey, R. W., Howard, J. D., Dorjes, J. 1989. Coastal sequences, eastern Buzzards Bay, Massachusetts, negligible record of an oil spill. *Geology*, 17, 461–465.

Freyssinet, P. H., Butt, C. R. M., Morris, R. C. 2005. Ore-forming processes related to lateritic weathering. In Hedenquist, J. W., Thompson, J. F. H., Goldfarb, R. J., Richards, J. P. (eds.), *Economic Geology, 100th Anniversary Volume*. Littleton, CO: Society of Economic Geologists, pp. 681–722.

Fridell, E., Haeger-Eugensson, M., Moldanova, J., Forsberg, B., Sjoberg, K. 2014. A modeling study of the impact of air quality and health due to emissions from E85 and petro fuelled cars in Sweden. *Atmospheric Environment*, 82, 1–8.

Friederici, P. 2013. WWII-era law keeps Germany hooked on "brown coal" despite renewable shift. InsideClimateNews.org. [Online] Available: http://insideclimatenews.org/news/20131001/ww-ii-era-law-keeps-germany-hooked-brown-coal-despite-renewables-shift.

Frim, R., Ukeles, S. D. 2007. Bromine. *Mining Engineering*, 59, 22–23.

Frimmel, H. E. 2008. Earth's continental crustal gold endowment. *Earth and Planetary Science Letters*, 267, 45–55.

Fukuzaki, N., Tamura, R., Hirano, Y., Mizushima, Y. 1986. Mercury emission from a cement factory and its influence on the environment. *Atmospheric Environment*, 20, 2291–2299.

Fulkerson, W., Judkins, R. R., Sanghavi, M. K. 1990. Energy from fossil fuel. *Scientific American*, 129–135.

Fulton, J. E., III., Miller, R. B. 2006. Fluorspar. In Kogel, J. E., Trivedi, N. C., Barker, N. C., Krukowski, S. T. (eds.), *Industrial Minerals and Rocks*, 7th edn. Littleton, CO: Society or Mining, Metallurgy, and Exploration, pp. 461–473.

Fyfe, W. S. 1978. Fluids in the Earth's crust: their significance in metamorphic, tectonic and chemical transport processes, Amsterdam, Elsevier.

Gahan, C. S., Srichandan, H., Kim, D-J., Akcil, A. 2012. Biohydrometallurgy and biomineral processing technology: a review of its past, present and future. *Research Journal of Recent Sciences*, 10, 85–99.

Galley, A. G., Hannington, M. D., Jonasson, I. R. 2007. Volcanogenic massive sulphide deposits. *Geological Association of Canada, Mineral Deposits Division, Special Publication*, No. 5.

Gambogi, J. 2014. Rare-earth elements. *US Geological Survey Minerals Yearbook*.

Gammons, C. H., Duaime, T. E. 2006. *Long Term Changes in the Limnology and Geochemistry of the Berkeley Pit Lake, Butte, Montana*. Heidelberg: Springer-Verlag.

GAO 2012. Mineral resources: mineral volume, value, and revenue. *Government Accounting Office Report*, GAO-13-45R.

Garbini, S., Schweinfurth, S. P., eds. 1986. A national agenda for coal quality research; symposium proceedings. *US Geological Survey Circular*, 979.

Garbisu, C., Alkora, I. 2001. Phytoextraction: a cost effective plant-based technology for the removal of metals from the environment. *Bioresource Technology*, 77, 229–236.

Gardner, D. 2013. Not all forms of resource nationalism are alike. *Financial Times*, August 18.

Garelick, H., Jones, H., Dybowska, A., Valsami-Jones, E. 2008. Arsenic pollution sources. *Reviews Environmental Contamination and Toxicology*, 197, 17–60.

Garnar, T. W. 1983. Zirconium and hafnium minerals. In Lefond, S. J. (ed.), *Industrial Minerals and Rocks*, 5th edn. New York, NY: Society of Mining Engineers, pp. 1433–1476.

Garnier, V., Giuliani, G., Ohnenstetter, D., *et al.* 2008. Marble-hosted ruby deposits from central and southeast Asia: towards a new genetic model. *Ore Geology Reviews*, 34, 169–191.

Garnish, J. W., ed. 1987. Proceeding of the 1st EEC/US workshop on hot dry rock geothermal energy. *Geothermics*, 16.

Garrett, D. E. 1992. Natural soda ash: occurrences, processing and use. New York, NY: Van Nostrand Reinhold.

Gastech, I. 2007. Viability of underground coal gasification in the "Deep Coals" of the Powder River Basin, Wyoming. Report prepared for the Wyoming Business Council, Business and Industry Division State Energy Office.

Gatti, J. B., Castilgo Queiroz, G., Garcia, E. E. C. 2008. Recycling of aluminum cans in terms of Life-Cycle Inventory (LCI). *International Journal of Life-Cycle Assessment*, 13, 219–225.

Gauthier-Lafaye, F. 2002. 2 billion year old natural analogs for nuclear waste disposal: the natural nuclear fission reactors in Gabon (Africa). *Comptes Rendus Physique*, 3(7–8), 839–849.

Gelb, B. A. 1983. The crude oil windfall profit tax act: context and content: congressional Research Service, Library of Congress, Report 81-270E. [Online] Available: http://energytaxfacts.com/assets/uploads/2013/02/CRS-Windfall-Profits-Tax-Analysis-1981.pdf?97d1a5.

GEMS Monitoring and Assessment Research Centre 1989. *United Nations Environmental Data Report.* Cambridge, MA: Blackwell.

Gentry, D. W., O'Neil, T. J. 1984. *Mineral Investment Analysis.* New York, NY: Society of Mining Engineers.

George, A. C., Hinchliffe, L. 1972. Measurements of uncombined radon daughters in uranium mines. *Health Physics*, 23, 791–803.

George, M. W. 2013. Selenium and tellurium. *US Geological Survey Minerals Yearbook.*

Gershey, E. L., Klein, R. C., Party, E., Wilkerson, A. 1990. *Low-Level Radioactive Waste.* New York, NY: Van Nostrand.

Gerst, M. D. 2008. Revisiting the cumulative grade tonnage relationship for major copper ore types. *Journal of Economic Geology*, 103, 615–628.

Getschow, G., Petzinger, T. 1984. Louisiana marshlands, laced with oil canals, are rapidly vanishing. *The Wall Street Journal*, October 24.

Ghosh, J. A. 1983. *OPEC, The Petroleum Industry and United States Energy Policy.* Westport, CT: Quorum.

Gilchrest, J. D. 1980. *Extractive Metallurgy.* Oxford: Pergamon.

Gilmour, C. C., Podnar, M., Bullock, A. L., *et al.* 2013, Mercury methylation by novel microorganisms from new environments. *Environmental Science and Technology*, 47, 11810–11820.

Girvan, N. 1972. *Copper in Chile.* Woking: Gresham Press.

Glas, J. P. 1988. Protecting the ozone layer: a perspective from industry. In Ausuble, J. H., Sladovich, H. E. (eds.), *Technology and the Environment.* Washington, DC: National Academy Press, pp. 137–155.

Gleeson, T., Wada, Y., Bierkens, M. F. P., Van beek, L. P. H. 2012. Water balance of global aquifers revealed by groundwater footprint. *Nature*, 488, 197–200.

Glenn, G. C., Hathaway, T. K. 1979. Quality control by blind sample analysis. *American Journal of Clinical Pathology*, 72, 156–162.

Godkin, D. 2012. Pushing for potash. *Canadian Mining Journal*, October, 18–23.

Goff, B. H., Weinberg, R., Groves, D. I., Vielreicher, N. M., Fourie, P. J. 2004. The giant Vergenoeg fluorite deposit in a magnetite–fluorite–fayalite REE pipe: a hydrothermally altered carbonatite-related pegmatoid? *Mineralogy and Petrology*, 80, 173–199.

Gold, T., Soter, S. 1980. The deep-earth gas hypothesis. *Scientific American*, 154–164.

Goldberg, I., Hammerbeck, E. C. I., Labuschagne, L. S., Rossouw, C. 1992. International strategic minerals inventory summary report – Vanadium. *US Geological Survey Circular*, 930-K.

Goldfarb, R. J., Baker, T., Dube, B., *et al.* 2005. Distribution, character, and genesis of gold deposits in metamorphic terranes. In Hedenquist, J. W., Thompson, J. F. H., Goldfarb, R. J., Richards, J. P. (eds.), *Economic Geology, 100th Anniversary Volume.* Littleton, CO: Society of Economic Geologists, pp. 407–450.

Goldfarb, R. J., Groves, D. I., Gardoll, S. 2001. Orogenic gold and geologic time: a global synthesis. *Ore Geology Reviews*, 18, 1–75.

Goldfarb, R. J., Bradley, D., Leach, D. L. 2010, Secular variation in economic geology: *Economic Geology*, 105, 459–465.

Goldhaber, S. B. 2003. Trace element risk assessment: essentiality vs. toxicity. *Regulatory Toxicology and Pharmacology*, 38, 232–242.

Goldman, H. B., Reining, D. 1983. Sand and gravel. In Lefond, S. J. (ed.), *Industrial Minerals and Rocks*, 5th edn. New York, NY: Society of Mining Engineers, pp. 1151–1166.

Golightly, J. P. 1981. Nickeliferous laterite deposits. *Economic Geology*, 75, 710–735.

Goodman, C. G. 1992. History of the US Petroleum Depletion Allowance: 1890–1990. *Petroleum Accounting and Financial Management Journal*, 11, 120.

Goold, D., Willis, A., 1997. *The Bre-X Fraud.* Toronto: McClelland & Stewart.

Goolsby, D. A., Battaglin, W. A., Hooper, R. P. 1997. Sources and transport of nitrogen in the Mississippi River Basin. American Farm Bureau Federation Workshop "From the corn belt to the gulf … agriculture and hypoxia in the Mississippi River watershed", Mississippi River Watershed. July 14–15, St. Louis, Missouri.

Goonan, T. 2006. Mercury flow through the mercury-containing lamp sector of the economy of the United States. *United States Geological Survey Scientific Investigations Report*, 2006–5264.

Gorte, R. W., Vincent, C. H., Hanson, L. A., Rosenblum, M. R. 2012. Federal land ownership: overview and data. *Congressional Research Services*, 7–5700 R42346.

Gossen, L. P., Velichkina, L. M. 2006. Environmental problems of the oil-and-gas industry. *Petroleum Chemistry*, 46, 67–72.

Goswami, E., Craven, V., Dahlstrom, D. L., Alexander, D., Mowar, F. 2013. Domestic asbestos exposure: a review of epidemiologic and exposure data. *International Journal of Environmental Research and Public Health*, 10–11, 5629–5670.

Govett, G. J. S., Govett, M. H. 1977. The inequality of the distribution of world mineral supplies. *CIM Bulletin*, 70, 59–68.

Grabowski, R. B., Wetzel, N., Raney, R. G. 1991. Gold. *Minerals Today*, 14–19.

Graedel, T. E., Cao, J. 2010. Metal spectra as indicators of development. *Proceedings of the National Academy of Science*, 107, 20905–20910.

Graney, J. R., Halliday, A. N., Keeler, G. J., *et al.* 1995. Isotopic record of lead pollution in lake sediments from the northeastern United States. *Geochimica et Cosmochimica Acta*, 59, 1715–1728.

Grandfield, J. 2014. *Light Metals*. San Diego, CA: Wiley.

Gray, E. 1982. *The Great Uranium Cartel*. Toronto: McClelland and Stewart.

Graybeal, F. T., Vikre, P. G. 2010. A review of silver-rich mineral deposits and their metallogeny. *Society of Economic Geologists Special Publication*, 15, 85–118.

Grayson, L. E. 1981. *National Oil Companies*. New York, NY: John Wiley, p. 269.

Greasley, D. 1990. Fifty years of coal-mining productivity: the record of the british coal industry before 1939. *The Journal of Economic History*, 50, 877–902.

Great Lakes Protection Fund 2007. Final report: restoring Great Lakes Basin water through the use of Conservation Credits and Integrated Water Balance Analysis System. The Great Lakes Protection Fund Project, #763.

Green, P. S. 2013. Water desalination capacity climbs on power, energy needs. *Bloomberg News*, October 14. [Online] Available: http://www.bloomberg.com/news/articles/2013-10-14/water-desalination-capacity-climbs-on-power-energy-needs.

Green, T. 1981. *World of Diamonds*. London: Weidenfeld and Nicholson, 300.

Greger, M., Landberg, T. 1999. Use of willow in phytoextraction. *International Journal of Phytoremediation*, 1, 115–123.

Griffin, S. O., Jones, K., Tomar, S. L. 2001. An economic evaluation of community water fluoridation. *Journal of Public Health Dentistry*, 61, 78–86.

Grisafe, D. A., Angina, E. E., Smith, S. M. 1988. Leaching characteristics of a high-calcium fly ash as a function of pH: a potential source of selenium toxicity. *Applied Geochemistry*, 3, 601–608.

Groat, L. A., Hart, C. J. R., Lewis, L. L., Neufeld, H. L. D. 2005. Emerald and aquamarine mineralization in Canada. *Geoscience Canada*, 32, 65–76.

Groat, L. A., Giuliani, G., Marshall, D. D., Turner, D. 2008. Emerald deposits and occurrences: a review. *Ore Geology Reviews*, 34, 87–112.

Groffman, A., Peterson, S., Brookins, D. 1992. Removing lead from wastewater using zeolite. *Water Environment and Technology*, 54–59.

Gross, W. H. 1975. New ore discovery and the source of silver–gold veins, Guanajuato, Mexico. *Economic Geology*, 70, 1170–1189.

Grossman, E. 2011. Are flame retardants safe? Growing evidence says 'No'. Yale University Report [Online] Available: http://e360.yale.edu/feature/pbdes_are_flame_retardants_safe_growing_evidence_says_no/2446/.

Grotheer, S. J. 2008. Evaluation of lightweight concrete mixtures for bridge deck and pre-stressed bridge girder applications. M.S. Thesis, Kansas State University.

Groves, D. I., Goldfarb, R. J., Robert, F., Hart, C. J. 2003. Gold deposits in metamorphic belts: overview of current understanding, outstanding problems, future research and exploration significance. *Economic Geology*, 98, 1–29.

Groves, D. I., Bierlein, F. P., Meinert, L. D., Hitzman, M. W. 2010. Iron oxide copper-cold (IOCG) deposits through Earth history: implications for origin, lithospheric setting, and distinction from other epigenetic iron oxide deposits. *Economic Geology*, 105, 641–654.

Groves, S. 2011. UN Convention on the law of the sea erodes US sovereignty over US extended continental shield. *The Heritage Foundation*, Backgrounder #2561 on International Law.

Gruber, P. W., Medina, P. A., Keolian, G. A., *et al.* 2011. Global lithium availability: a constraint for electric vehicles? *Journal of Industrial Ecology*, 15, 760–775.

Guberman, D. E. 2013a. Lead. *US Geological Survey Mineral Yearbook*.

Guberman, D. E. 2013b. Germanium. U.S. Geological Survey Minerals Yearbook Advance Release for 2012, 6.

Guberman, D. E. 2014 Thallium. *US Geological Survey Mineral Commodity Summary*.

Guéguin, M., Cardarelli, F. 2007. Chemistry and mineralogy of titanium-rich slags. Part 1: hemo-ilmenite, sulphate and upgraded titanium slags. *Mineral Processing and Extractive Metallurgy*, 28, 1–58.

Guiliani, G., Silva, L. J. H. C., Couto, P. 1990. Origin of emerald deposits of Brazil. *Mineralium Deposita*, 25, 57–64.

Gunter, M. E. 2009. Asbestos sans mineralogy. *Elements*, 5, 141.

Gupta, C. K., Krishnamurthy, N. 2005. *Extractive Metallurgy of Rare Earths*. Boca Raton, FL: CRC Press.

Gupta, H. K. 1980. *Geothermal Resources: An Energy Alternative*. Amsterdam: Elsevier.

Gura, D. 2010, Toxic red sludge spill from Hungarian aluminum plant an ecological disaster. National Public Radio. [Online] Available: http://www.npr.org/blogs/thetwo-way/2010/10/05/130351938/red-sludge-from-hungarian-aluminum-plant-spill-an-ecological-disaster.

Gurney, J. J., Helmstaedt, H. H., Le Roex, A.P., *et al.* 2005. Diamonds: crustal distribution and formation processes in time and space and an integrated deposit model. In Hedenquist, J. W., Thompson, J. F. H., Goldfarb, R. J., Richards, J. P. (eds.), *Economic Geology, 100th Anniversary Volume*. Littleton, CO: Society of Economic Geologists, pp. 143–177.

Gutentag, E. D., Heimes, F. J., Krothe, N. C., Luckey, R. R., Weeks, J. B. 1984. Geohydrology of the High Plains aquifer in part of Colorado, Kansas, Nebraska, New Mexico, Oklahoma, South Dakota, Texas and Wyoming. *US Geological Survey Professional Paper 1400-B*, 64.

Guthrie, G. D., Jr. 1992. Biological effects of inhaled minerals. *American Mineralogist*, 77, 225–243.

Gutierrez, J., Vallejo, L. E., Lin, J. S., Painter, R. 2010. Impact of longwall mining of coal on highways in southwestern Pennsylvania. *IAHS-AISH Publication*, 339, 120–125.

Gutowski, T. G., Sahni, S., Allwood, J. M., Ashby, M. F., Worrell, E. 2013. The energy required to produce materials: constraints on energy-intensity improvements, parameters of demand. *Philosophical Transactions of the Royal Society*, 371, 2012003.

Gutzmer, J., Buekes, N. J. 2009. The giant Kalahari manganese field: manganese in the twenty-first century: short course notes.

Institute for Geochemical Research, Hungarian Academy of Sciences, 19–28.

Guy-Bray, J. 1989. The mineral resources program of the United Nations Department of Technical Cooperation for Development. *Ontario Geological Survey Special Volume*, 3, 798–801.

Häfele, W. 1990. Energy from nuclear power. *Scientific American*, 263, 136–145.

Hagelüken, C. 2012. Recycling the platinum group metals: a European perspective. *Platinum Metals Review*, 56.

Hagemann, S. G., Rosiere, C., Gutzmer, J., Buekes, N.J. 2008. Introduction to banded iron formation-related high-grade iron ore. *Reviews in Economic Geology*, 15, 1–4.

Haile, P., Hendricks, K., Porter, R. 2010. Recent US offshore oil and gas lease bidding: a progress report. [Online] Available: http://www.econ.yale.edu/~pah29/ijio.pdf.

Halbouty, M. T. 2001. Giant oil and gas fields decade 1990–1999: an introduction. *American Association of Petroleum Geologists Memoir*, 78, 1–13.

Halford, B. 2009. Wallboard woes: odors and corrosion raise concern over drywall imported from China. *Science and Technology*, 50–51.

Halker Consulting 2013. Multi-well site facilities. [Online] Available: http://www.halker.com/projects/multi-well-site-facilities/.

Hamilton, J. M. 1990. Earth science in Canada from a user's viewpoint. *Geoscience Canada*, 16, 213–220.

Hamilton, M. 1992. Water fluoridation: a risk assessment perspective. *Journal of Environmental Health*, 54, 27–31.

Hammond, G. P., Jones, C. I. 2008. Embodied energy and carbon in construction materials. *Proceedings of the Institution of Civil Engineers – Energy*, 161, 87–98.

Hammond, O. H., Baron, R. E. 1976. Synthetic fuels: prices and prospects. *American Scientist*, 64, 407–412.

Hanley, J. J., Mungall, J. E., Pettke, T., Spooner, E. T. C., Bray, C. J. 2008. Fluid and halide melt inclusions of magmatic origin in the ultramafic and lower banded series, Stillwater Complex, Montana, USA. *Journal of Petrology*, 49, 1133–1160

Hannington, M. D., De Ronde, C. E. J., Petersen, S. 2005. Sea-floor tectonics and submarine hydrothermal systems. In Hedenquist, J. W., Thompson, J. F. H., Goldfarb, R. J., Richards, J. P. (eds.), *Economic Geology, 100th Anniversary Volume*. Littleton, CO: Society of Economic Geologists, pp. 111–141.

Hannis, S., Bide, T., Minks, A. 2009. Commodity profiles: cobalt. *British Geological Survey*. [Online] Available: https://www.bgs.ac.uk.

Hanor, J. S. 1987. Origin and migration of subsurface sedimentary brines. *Society of Economic Paleontologists and Mineralogists Short Course*, no. 21, 247.

Hao, Z., Fei, H., Lui, L., Turner, S. 2012. Comprehensive utilization of vanadium–titanium magnetite deposits in China has come to a new level. *Acta Geologica Sinica*, 87, 286–287.

Harben, P. W., Bates, R. L. 1984. *Geology of the Non-Metallics*. New York, NY: Metal Bulletin.

Harben, P. W., Theune, C. 2006. Nitrogen. In Kogel, J. E., Trivedi, N. C., Barker, N. C., Krukowski, S. T. (eds.), *Industrial Minerals and Rocks*, 7th edn. Littleton, CO: Society for Mining, Metallurgy, and Exploration, pp. 671–678.

Harden, B. 2012. Treasure hunt: the battle over Alaska's mega mine. *Frontline*, July 23.

Hargrove, E. C. 1989. *Foundations of Environmental Ethics*. Englewood Cliffs, NJ: Prentice Hall.

Harlow, G. E., Davies, R. 2005. Diamonds. *Elements*, 1, 67–70.

Harper, K. N., Zuckerman, M. K., Harper, M. L., Kingston, J. D., Armelagos, G. J. 2011. The origin and antiquity of syphilis revisited: an appraisal of Old World pre-Columbian evidence for treponemal infection. *American Journal of Physical Anthropology*, 146, Supplement 53, 99–133.

Harr, J. 1996. *A Civil Action*. New York, NY: Vintage.

Harris, D. P. 1979. World uranium resources. *Annual Review of Energy*, 4, 403–432.

Harris, D. P. 1990. *Mineral Exploration Decisions*. New York, NY: Wiley-Interscience.

Harris, J. H., Kesler, S.E. 1996. BLASH vs. GIF theory in mining investment. *Society of Economic Geologists Newsletter*, no. 26, 23–24.

Hart, M. 2013. *Gold: The Race for The World's Most Seductive Metal*. New York, NY: Simon & Schuster.

Hartman, H. L. 1987. *Introductory Mining Engineering*. New York, NY: Wiley-Interscience, 633.

Harvey, C. C., Muray, H. H. 2006. Clays: an overview. In Kogel, J. E., Trivedi, N. C., Barker, N. C., Krukowski, S. T. (eds.), *Industrial Minerals and Rocks*, 7th edn. Littleton, CO: Society for Mining, Metallurgy, and Exploration, pp. 335–342.

Haryett, C. R. 1982. Innovations over the last decade in the Saskatchewan potash industry. *Mining Engineering*, August, 1225–1227.

Hassanpour, M., Shafigh, P., Mahmud, H. B. 2012. Lightweight aggregate concrete fiber reinforcement – a review. *Construction and Building Materials*, 37, 452–461.

Hasselback, D. 2013. Flow-through shares: Canada's quirky tax innovation. Financial Post, March 7. [Online] Available: http://business.financialpost.com/2013/03/07/flow-through-shares-canadas-quirky-tax-innovation/.

Hasselrot, B., Hultberg, H. 1984. Liming of acidified lakes and streams and its consequences for aquatic ecosystems. *Fisheries*, 9, 4–9.

Hatayama, H., Daigo, I., Matsuno, Y., Adachi, Y. 2009. Assessment of the recycling potential of aluminum in Japan, the United States, Europe and China. *Materials Transactions*, 50, 650–656.

Hawtrey, R. G. 1980. *The Gold Standard in Theory and Practice*. Westport, CT: Greenwood Press.

Haynes, B. W., Kramer, G. W. 1983. Characterization of US cement kiln dust. *US Bureau of Mines Information Circular*, 8885.

Hayward, C. L., Reimold, W. U., Gibson, R. L., Robb, L. J. 2005. Witwatersrand Basin, South Africa: evidence for a modified placer origin, and the role of Vredefort impact event. *Geological Society of London Special Publications*, 248, 31–58.

Healy, D. 2012. Hydraulic fracturing or 'fracking': a short summary of current knowledge and potential environmental impacts. Environmental Protection Agency of Ireland. [Online]

Available: http://www.epa.ie/pubs/reports/research/sss/uniaber deen_frackingreport.pdf.

Healy, J. 2013. As oil floods plains towns, crime pours in. *The New York Times*, November 20.

Heaney, P. J., Vicenzi, E. P., De, S. 2005. Strange diamonds: the mysterious origins of carbonado and framesite. *Elements*, 1(2), 85.

Heasler, H. P., Jaworowski, C., Foley, D. 2009. Geothermal systems and monitoring hydrothermal features. In Young, R., Norby, L. (ed.), *Geological Monitoring*. Boulder, CO: Geological Society of America, pp. 105–140.

Heath, G. R. 1978. Deep-sea manganese nodules. *Oceanus*, 21, 60–68.

Heath, G. R. 1981. Ferromanganese nodules of the deep sea. *Economic Geology*, 75, 736–765.

Hedrick, J. B. 1985. Rare earth elements and yttrium. *US Bureau of Mines Mineral Facts and Problems*, 675, 647–664.

Hedrick, J. B. 2014. REE Handbook: the ultimate guide to rare earth elements. [Online] Available: http://www.reehandbook.com/definition.html.

Hedrick, J. B., Templeton, D. A. 1990. Rare-earth minerals and metals. *US Bureau of Mines Minerals Yearbook*.

Heffer, P., Prud'homme, M. 2014. *Fertilizer Outlook 2014–2018*. Paris: International Fertilizer Industry Association.

Heffern, E. L., Coates, D. A. 2004. Geologic history of natural coal-bed fires, Powder River Basin, USA. *International Journal of Coal Geology*, 59, 25–47.

Hein, F.J., Lecki, D., Larter, S., Suter, J. R. 2013. Heavy-oil and oil-sand petroleum systems in Alberta and beyond. *AAPG Studies in Geology*, 64, 1–21.

Heinrich, E. W. 1980. Geological spectrum of glass-sand deposits in eastern Canada and their evaluation. *New York State Museum Bulletin*, 436, 106–112.

Helfrich, L. A., Neves, R. J., Parkhurst, J. 2009. Liming acidified lakes and ponds. *Virginia Cooperative Extension Publication*, 420, 254.

Helman, C. 2013. The world's biggest oil companies 2013. *Forbes*, November 17.

Henderson, G. V., Katzman, H. 1978. Aggregate resources vs. urban development and multiple use planning, San Gabriel valley, California. *Society of Mining Engineers Preprint*, 78-H-32.

Henke, K. R., ed. 2009. *Arsenic: Environmental Chemistry, Health Threats and Waste Treatment*. Hoboken, NJ: Wiley.

Henkels, P. J. 2006. Gypsum plasters and wallboard. In Kogel, J. E., Trivedi, N. C., Barker, N. C., Krukowski, S. T. (eds.), *Industrial Minerals and Rocks*, 7th edn. Littleton, CO: Society for Mining, Metallurgy, and Exploration.

Herkenoff, E. C. 1972. When are we going to mine oil? *Engineering and Mining Journal*, 173, 132–138.

Herrington, R., Maslennikov, V., Zaykov, V., Seravkin, I. 2005. VMS deposits of the South Urals, Russia. *Ore Geology Reviews*, 27, 238–239.

Hesselbach, J., Herrmann, C., eds. 2011. *Globalized Solutions for Sustainability in Manufacturing*. Berlin: Springer-Verlag.

Hessley, R. K., Reasoner, J. W., Riley, J. T. 1986. *Coal Science*. New York, NY: John Wiley and Sons.

Hewett, D. F. 1929. Cycles in metal production. *American Institute of Mining and Metallurgical Engineers Transactions*, 85, 65–92.

Hewitt, C. N. 1998. *Reactive Hydrocarbons in the Atmosphere*. San Diego, CA: Academic Press.

Hight, R. P. 1983. Abrasives. In Lefond, S. J. (ed.), *Industrial Minerals and Rocks*, 5th edn. New York, NY: Society of Mining Engineers, pp. 11–23.

Hill, L. 2013. Cliffs suspends Ontario Ring of Fire chromite project. *Bloomberg News*, November 21.

Hillel, D. 2007. *Soil in the Environment*. New York, NY: Academic Press.

Hilson, G., Christopher, J., Hilson, C. J., Pardie, S. 2007. Improving awareness of mercury pollution in small-scale gold mining communities: challenges and way forward in rural Ghana. *Environmental Research*, 103, 275–287.

Hilts, P. J. 1990. US opens a drive to wipe out lead poisoning in nation's young. *New York Times*, December 19.

Hilyard, J., ed. 2010. *International Petroleum Encyclopedia*. Tulsa, OK: PennWell Publishing Company.

Hiskey, J. B., Devries, F. W. 1992. Environmental considerations for alternates to cyanide processing. In Chancier, S. (ed.), *Emerging Process Technologies for a Cleaner Environment*. Littleton, CO: Society for Mining, Metallurgy and Exploration, pp. 73–80.

Hitzman, M. W., Kirkham, R., Broughton, D., Thorson, J., Selley, D. 2005. The sediment-hosted stratiform copper ore system. In Hedenquist, J. W., Thompson, J. F. H., Goldfarb, R. J., Richards, J. P. (eds.), *Economic Geology, 100th Anniversary Volume*. Littleton, CO: Society of Economic Geologists, pp. 609–642.

Hitzman, M. W., Selley, D., Bull, S. 2010. Formation of sediment-hosted stratiform copper deposits through Earth history. *Economic Geology*, 105, 627–640.

Hodai, K. 2012. Death metal: tin mining in Indonesia. *The Guardian*, November 23.

Hoffert, M. I., Calderia, K., Benford, G., *et al.* 2002. Advanced technology paths to global climate stability: energy for a green-house planet. *Science*, 298, 981–987.

Hohmann, G. W., Ward, S. H. 1981. Electrical methods in mining geophysics. *Economic Geology*, 67, 281–294.

Hohn, M. E. 1998. *Geostatistics and Petroleum Geology*. Dordrecht: Kluwer Academic Publishing.

Holland, H. D. 1972. Granites, solutions and base metal deposits. *Economic Geology*, 67, 281–294.

Holland, H. D. 2005. Sedimentary mineral deposits and the evolution of Earth's near-surface environments. *Economic Geology*, 100, 1489–1509.

Holland, H. D. 2006. The oxygenation of the atmosphere and oceans. *Philosophical Transactions of the Royal Society of London*, 361, 903–915.

Holland, M. 2010. Trans-Alaska pipeline spill toll 5,000 barrels. *Anchorage Daily News*, May 28.

Holloway, J., Roberts, I., Rush, A. 2010. China's steel industry. *Reserve Bank of Australia Bulletin*, December, 19–25.

Holloway, M., Rudd, O. 2013. *Fracking: The Operations and Environmental Consequences of Hydraulic Fracturing*. New York, NY: Wiley–Scrivener Publishing.

Holton, J. R. 1990. Global transport processes in the atmosphere. In Huntzinger, O. (ed.) *The Handbook of Environmental Chemistry*. Heidelberg: Springer-Verlag, vol. 1, pp. 97–145.

Homer, A. W. 2009. Coal mine safety regulation in China and the USA. *Journal of Contemporary Asia*, 39, 424–439.

Honda, Y., Beall, C., Delzell, E. 2002. Mortality among workers at a talc mining and milling facility. *Annals Occupational Hygiene*, 46, 575–585.

Höök, M., Soderbergh, B., Jakobsson, K., Aleklett, K. 2009. The evolution of giant oil field production behavior. *Natural Resources Research*, 18.

Hooke, R. L., Martin-Duque, J. F. 2012. Land transformation by humans: a review. *GSA Today*, 22, 4–10.

Horn, S. 2014. Why ExxonMobil's partnerships with Russia's Rosneft challenge the narrative of US exports as energy weapon. Desmogblog. [Online] Available: http://www.desmogblog.com/2014/03/17/exxonmobil-russia-rosneft-gas-export-weapon.

Horne, R. A. 1978. *The Chemistry of our Environment*, New York, NY: Wiley-Interscience.

Horta-Puga, G., Carriquiry, J. D. 2014. The last two centuries of lead pollution in the southern Gulf of Mexico recorded in the annual bands of the scleractinian coral *Orbicella faveolata*. *Bulletin of Environmental Contaminant Toxicology*, 92, 567–573.

Hossner, L. R. 1988. *Reclamation of Surface Mined Lands*. Boca Raton, FL: CRC Press.

Hostermann, J. W., Patterson, S. H., Good, E. E. 1990. World non-bauxite aluminum resources excluding alunite. *US Geological Survey Professional Paper*, 1076-C, 1–73

Houston, J., Butcher, A., Ehren, P., Evans, K., Godfrey, L. 2011. The evaluation of brine prospects and the requirement for modifications to filing standards. *Economic Geology*, 106, 1225–1239.

Howarth, R. W. 2008. Coastal nitrogen pollution: a review of sources and trends globally and regionally. *Harmful Algae*, 8, 14–20.

Howell, J. W., Cawthorne, J.M. 1987. *Copper in Animals and Man*. Boca Raton, FL: CRC Press.

Hower, M. 2013. Increased corn prices threaten ethanol fuel's long-term viability. *Triple Pundit*, April 19.

Hsu-Kim, H., Kucharzyk, K. H., Zhang, T., DesHusses, M. A. 2013. Mechanisms regulating mercury bioavailability for methylating microorganisms in the aquatic environment: a critical review. *Environmental Science and Technology*, 47, 2441–2456.

Hsü, K. J., Montadert, L., Bernoulli, D., *et al.* 1977. History of the Mediterranean salinity crisis. *Nature*, 267, 399–403.

Hu, R. Z., Qi, H. W., Zhou, M. F., *et al.* 2009. Geological and geochemical constraints on the origin of the giant Lincang coal seam-hosted germanium deposit, Yunnan, SW China: a review. *Ore Geology Reviews*, 36, 221–234.

Hua, Y., Luo, Z., Cheng, S., Rui, Z. 2012. Health risks of organic contaminated soil in and out of service oil refinery site. *Journal of Earth Science*, 23, 121–128.

Hubbert, M. K. 1985. The world's evolving energy system. In Perrine, R. L., Ernst, W.G. (eds.), *Energy: For Ourselves and Our Posterity*. Englewood Cliffs, NJ: Prentice-Hall, pp. 44–100.

Hudson-Edwards, K. A., Jamieson, H.E., Lottermoser, B.G. 2011. Mines wastes: past, present, future. *Elements*, 7, 375–380.

Huffman, T. R. 2000. Exploring the legacy of reserve mining: what does the longest environmental trial in history tell us about the meaning of American environmentalism? *Journal of Policy History*, 12, 339–368.

Hughes, D. 2013. Whither shale oil? Interview with David Hughes. Post Carbon Institute. [Online] Available: http://www.postcarbon.org/whither-shale-oil-interview-with-david-hughes/.

Humphreys, M. H., Sherlock, M. F. 2013. US and world coal production, taxes and incentives. *Congressional Research Service 7–5700*, R43011.

Huncharek, M., Geschwind, J. F., Kupelnick, B. 2003. Perineal application of cosmetic talc and risk of invasive epithelial ovarian cancer: a meta-analysis of 11,933 subjects from sixteen observational studies. *Anticancer Research*, 23, 1955–1960.

Hunt, J. M. 1979. *Petroleum Geochemistry and Geology*. San Francisco, CA: W.H. Freeman.

Hurlbut, C. S., Kammerling, R. C. 1991. *Gemology*. New York, NY: John Wiley.

Hurt, H., III 1981. *Texas Rich: The Hunt Dynasty from the Early Oil Days Through the Silver Crash*. New York, NY: W.W. Norton.

Huston, D. L., Pehrsson, S., Eglington, B. M., Zaw, K., 2010, The geology and metallogeny of volcanic-hosted massive sulfide deposits: variations through geologic time and with tectonic setting. *Economic Geology*, 105, 571–591.

Hutchinson, F. H. 2009. About IGCC Power. Clean-Energy US. Clean-Energy US. [Online] Available: http://www.clean-energy.us/facts/igcc.htm.

Hutchinson, T. C., Meema, K. M. 1987. *Lead, Mercury, Cadmium and Arsenic in the Environment*. New York, NY: John Wiley.

Hylko, J. M., Peltier, R. 2010. The evolution of the ESBWR. *Power Magazine*, October 1.

IAEA 2011. Uranium 2011: Resources, production and demand. [Online] Available: http://www.iea.org/publications/freepublications/publication/weo2011_web.pdf.

IAEA 2013. Nuclear power reactors in the world, 2013 edition. International Atomic Energy Agency. [Online] Available: http://www.iea.org/publications/freepublications/publication/weo2011_web.pdfhylko.

Ibrahim, S. S. 2012. Diatomite ores: origin, characterization and applications. *Journal of International Environmental Application and Science*, 7, 191–199.

Ibrahim, Y. M. 1992. Kuwait struggles with oil damage. *The New York Times*, April 21.

IEA 2012. Coal information 2012. International Energy Agency. [Online] Available: http://www.iea.org/topics/coal/.

IEA 2013 Resources to reserves. International Energy Agency. [Online] Available: http://www.iea.org/publications/freepublications/publication/Resources2013.pdf.

Ilgren, E. B., Brena, M. O., Larragoita, J. C., Navarrete, G. L., Brena, A. F. 2008. A reconnaissance study of a potential emerging Mexican mesothelioma epidemic due to fibrous zeolite exposure. *Indoor and Built Environment*, 17, 496–515.

Inman, M. 2013. The true cost of fossil fuels. *Scientific American*, 59, 58–61.

Institute of Economic Studies, University of Iceland 2009. The effect of power-intensive industrial developments on the Icelandic economy. [Online] Available: https://www.institutenorth.org/assets/images/uploads/files/HHI-skyrslan-a-ensku.pdf.

Institute of Medicine 2000. *DRI Dietary Reference Intakes for Vitamin C, Vitamin E, Selenium and Carotenoids*. Washington, DC: National Academy Press.

Interagency Force on Carbon Capture and Storage 2010. Report. [Online] Available: http://www.epa.gov/climatechange/Downloads/ccs/CCS-Task-Force-Report-2010.pdf.

International Agency for Research on Cancer 2012. *Arsenic, Metals, Fibers, and Dusts: A Review of Human Carcinogens*. Lyon: International Agency for Research on Cancer.

International Copper Study Group 2013. World Copper Factbook for 2013. [Online] Available: http://www.icsg.org/index.php/component/jdownloads/finish/170/1188.

International Magnanese Institute 2013. International Manganese Institute. [Online] Available: http://www.manganese.org/home.

IPCC 2001. *Climate Change 2001: Impacts, Adaptation and Vulnerability. Contribution of Working Group II to the Third Assessment Report of the Intergovernmental Panel on Climate Change*. Cambridge: Cambridge University Press.

IPCC 2007. *Climate Change 2007: Mitigation of Climate Change. Contribution of Working Group III to the Fourth Assessment Report of the Intergovernmental Panel On Climate Change*. Cambridge: Cambridge University Press.

IPCC 2013. *Climate Change 2013: The Physical Science Basis*. Intergovernmental Panel on Climate Change. [Online] Available: http://www.ipcc.ch/report/ar5/wg1/.

IRRI 2001. Rice production in the 21st century. International Rice Research Institute. [Online] Available: http://irri.org/?gclid=CM2GwPW-ssMCFQ6NaQodlYYA1A.

Irvine, N. 1977. Origin of the chromite layers in the Muskox intrusion and other stratiform intrusions: a new interpretation. *Geology*, 5, 273–277.

Irving, P. M. 1991. *Acidic Deposition: State of Science and Technology*. Washington DC: US National Acid Precipitation Assessment Program.

Ishihara, S. 1992. The changing face of economic geology in Japan. *SEG Newsletter*, no. 10, 24–27.

Itokawa, Y., Durlach, J. 1988. *Magnesium in Health and Disease*. London: Libbey.

Izatt, R. M., Izatt, S. R., Bruening, R. L., Izatt, N. E., Moyer, B. A. 2014. Challenges to achievement of metal sustainability in our high-tech society. *Chemical Society Reviews*, 42, 2451–2474.

Jaber, J. O., Sladek, T. A., Mernitz, C., Tarawneh, T. M. 2008. Future policies and strategies for oil shale development in Jordan. *Jordan Journal of Mechanical and Industrial Engineering*, 2, 31–44.

Jacobs, J. 2011. The sustainability of water resources in the Colorado River basin. *The Bridge*, 41, 6–12.

Jacoby, N. H. 1974. *Multinational Oil*. New York, NY: Macmillan.

Jahns, R. H. 1983. Gem materials. In Lefond, S. J. (ed.), *Industrial Minerals and Rocks*, 5th edn. New York, NY: Society of Mining Engineers, pp. 279–338.

James, S. 2011. Black lung disease seen rising in US miners. *Reuters*, May 20.

Jan, J., Roe, L.A. 1983. Iodine. In Lefond, S. J. (ed.), *Industrial Minerals and Rocks*, 5th edn. New York, NY: Society of Mining Engineers, pp. 793–798.

Jasinski, S. M. 2013a. Phosphate rock. *US Geological Survey Minerals Yearbook*.

Jasinski, S. M. 2013b. Potash. *US Geological Survey Minerals Yearbook*.

Jaskula, B. W. 2013a. Beryllium. *US Geological Survey Minerals Yearbook*.

Jaskula, B. W. 2013b. Lithium. *US Geological Survey Minerals Yearbook*.

Javier, L. A., Aquino, N.P. 2012. Nickel-ore exports from Philippines seen rising on Indonesia ban. *Bloomberg News*, April 2. [Online] Available: http://www.bloomberg.com/news/2012-04-01/nickel-ore-exports-from-philippines-seen-rising-on-indonesia-ban.html.

Jennet, J. C., Wixson, B. G., Lowsley, I. H., *et al.* 1977. Transport and distribution from mining, milling and smelting operations in a forest ecosystem. In Boggess, W. R. (ed.), *Lead in the Environment*. Washington, DC: National Science Foundation, RA–770214.

Jensen, J. H., Treckoff, W. E., Anderson, A. P. 1983. Bromine. In Lefond, S. J. (ed.) *Industrial Minerals and Rocks*, 5th edn. New York: Society of Mining Engineers.

Jentleson, B. W. 1986. *Pipeline Politics: The Complex Political Economy of East–West Energy Trade*. Ithaca, NY: Cornell University Press.

Jing, H., Charlet, L. 2013. A review of arsenic presence in China drinking Water. *Journal of Hydrology*, 492, 79–88.

Jogalekar, A. 2013. Who's afraid of nuclear waste?: WIPPing transuranics into shape. *Scientific American*, August 27.

Johany, A. D. 1980. *The Myth of the OPEC Cartel*. New York, NY: John Wiley.

Johnson, C. A., Thornton, I. 1987. Hydrological and chemical factors controlling the concentrations of Fe, Cu, Zn, and As in a river system contaminated by acid mine drainage. *Water Research*, 21, 359–356.

Johnson, D. B. 2003. Chemical and microbiological characteristics of mineral spoils and drainage waters at abandoned coal and metal mines. *Water, Air and Soil Pollution*, 3, 47–66.

Johnson, P. T. J., Townsend, A. R., Cleveland, C. C., *et al.* 2010. Linking environmental nutrient enrichment and disease emergence in humans and wildlife. *Ecological Applications*, 20, 16–29.

Johnson, T. 2007. The return of resource nationalism. Council on Foreign Relations, August 13. [Online] Available: http://www.cfr.org/venezuela/return-resource-nationalism/p13989.

Johnson, T. 2011. Global natural gas potential. Council on Foreign Relations, August 24. [Online] Available: http://www.cfr.org/natural-gas/global-natural-gas-potential/p17946.

Johnson, W., Paone, J. 1982. Land utilization and reclamation in the mining industry, 1930–1980. *US Bureau of Mines, Information Circular*, 8862.

Johnson, W. 1985. Cement. *US Bureau of Mines Mineral Facts and Problems*, 675, 121–131.

Johnson Matthey, 2014. Platinum Today. Market Data Tables. [Online] Available: http://www.platinum.matthey.com/services/market-research/market-data-tables.

Johnston, D.T. 2011, Multiple sulfur isotopes and the evolution of Earth's surface sulfur cycle. *Earth Science Reviews*, 106, 161–183.

Jolly, D., Clark, N. 2012. Labor dispute pits France against ArcelorMittal. *The New York Times*, November 27.

Jolly, J. W. L. 1985. Copper. *US Bureau of Mines Mineral Facts and Problems*, 675, 197–222.

Jolly, J. W. L., Edelstein, D. L. 1992. Copper. *US Bureau of Mines Minerals Yearbook*.

Jones, B. F. 1966. *Geochemical Evolution of Closed Basic Water in the Western Great Basin. Proceedings of the 2nd Symposium on Salt, Volume 1.* Cleveland, OH: Northern Ohio Geological Survey.

Jorgenson, D. B., George, M. W. 2004. Mineral commodity profile: indium. *US Geological Survey Open-File Report*, 2004-1300.

Joshi, S. R. 2005. Comparison of groundwater rights in the United States: lessons for Texas. M.S. Thesis, Texas Tech University.

Jouzel, J. V., Masson-Delmotte, O., Cattani, G., *et al.* 2007. Orbital and millennial Antarctic climate variability over the past 800,000 years. *Science*, 317, 793–797.

Joyce, P., Scannell, T. 1988. *Diamonds in South Africa.* Cape Town: Struik Publishers.

Juhlin, C. 1991. *Scientific Summary Report of the Deep Gas Drilling Project in the Siljan Ring Impact Structure.* Stockholm: Swedish State Power Board.

Juszli, M. P. 1989. Creative distribution techniques can help market penetration of industrial minerals. *Mining Engineering*, 41, 1009–1112.

Kadafa, A. A. 2012. Oil exploration and spillage in the Niger delta of Nigeria. *Civil and Environmental Research*, 2, 38–51.

Kadey, F. L. 1983. Diatomite. In Lefond, S. J. (ed.), *Industrial Minerals and Rocks*, 5th edn. New York, NY: Society of Mining Engineers, pp. 677–689.

Kane, R. E. 2009. Seeking low cost perfection. *Elements*, 5, 169–174.

Kanter, J. 2011 Fancy batteries in electric cars pose recycling challenges. *The New York Times*, August 30.

Kapoor, S., Smalley, G. A., Norman, M. E.1992. Hazardous waste regulations affect refinery wastewater schemes: *Oil and Gas Journal*, 90, p. 45–49.

Karacan 2008. Modeling and prediction of ventilation methane emissions of US longwall mines using supervised artificial neural networks. *International Journal of Coal Geology*, 73, 371–387.

Karr, A. R. 1983. Coal-slurry pipeline measure is rejected by House as railroad lobby succeeds. *The Wall Street Journal*, September 28.

Kasriel, K., Wood, D. 2013. *Upstream Petroleum Fiscal and Valuation Modeling in Excel.* Chichester: Wiley.

Kavlak, G., Graedel, T.E. 2013. Global anthropogenic tellurium cycles for 1940–2010. *Resources, Conservation and Recycling*, 76, 21–26.

Kawahar, K., Kato-Negishi, M. 2011. Link between aluminum and the pathogenesis of alzheimer's disease: the integration of the aluminum and amyloid cascade hypotheses. *International Journal of Alzheimer's Disease*, 17.

Kawamoto, H., Tamaki, W. 2011. Trends in supply of lithium resources and demand of the resources for automobiles. *Science and Technology Trends Quarterly Review*, 39, 51–64.

Kazantzis, G. 2000. Thallium in the environment and health effects. *Environmental Geochemistry and Health*, 22, 275–280.

KCC 1971. Expropriation of the El Teniente copper mine by the Chilean government. *Kennecott Copper Corporation*, 8.

Kean, S. 2010. *The Disappearing Spoon: And Other True Tales of Madness, Love and the History of the World from the Periodic Table of the Elements.* Boston, MA: Little, Brown and Company.

Kearey, P., Brooks, M., Hill, I. 2002. *An Introduction to Geophysical Exploration.* Oxford: Blackwell Publishing.

Kehrig, H. A., Costa, M., Moreira, I., Malm, O. 2001. Methylmercury and total mercury in estuarine organisms from Rio de Janeiro, Brazil. *Environmental Science and Pollution Research International*, 8, 275–279.

Keller, W., Heneberry, J., Gunn, J. M., *et al.* 2004. *Recovery of Acid and Metal-Damaged Lakes Near Sudbury Ontario: Trends and Status.* Ontario: Cooperative Freshwater Ecology Unit, Laurentian University.

Kelly, C. 2012. Aluminum cars, doors, frames – industry's "next frontier": Alcoa. Reuters. [Online] Available: http://www.reuters.com/article/2012/06/13/us-aluminum-alcoa-auto-idUSBRE85C1KW20120613.

Kelting, D. L., Laxson, C. L. 2010. Review of effects and costs of road de-icing with recommendations for winter road management in the Adirondack park. *Adirondack Watershed Institute Report*, #AWI2010-01.

Kemper, K. Minnatullah, K. Foster, S. Tuinhof, A.,Talbi, A. 2005. Arsenic contamination of groundwater in south and east Asian countries: towards a more effective operational response. The World Bank, Technical Report No. 31303, vol. 2.

Kendall, A., Kesler, S. E., Keolian, G. 2008. Geologic vs. geographic constraints on cement resources. *Resources Policy*, 33, 160–167.

Kendall, A., Kesler, S. E., Keolian, G. 2009. Megaquarry versus decentralized mineral production: network analysis of cement production in the Great Lakes regions, USA. *Journal of Transportation Geography*, 18, 322–330.

Kendall, R., Brown, T., Hertherington, L., Chetwyn, C. 2010. Coal. British Geological Survey Commodity Profile. [Online] Available: http://www.bgs.ac.uk/downloads/start.cfm?id=1404.

Kennedy, J., Supreme Court Justice. 1999. Certiorari to the United States Court of Appeals for the Tenth Circuit (No. 98–830) 151 F. 3d 1251 reversed. Cornell University Library publication. [Online] Available: https://www.law.cornell.edu/supct/html/98-830.ZO.html.

Kennedy, J. L. 1984. *Oil and Gas Pipeline Fundamentals.* Tulsa, OK: PennWell Publishing Company.

Kennedy, P. 1990. Nevada flexes regulatory muscle with hefty fine. *Northern Miner*, November 19.

Kent, B. 1994. Groundwater contamination by oil fields and refinery industries. *Environmental Science and Pollution Control Series*, 11, 209–220.

Kerkhoff, B. 2007. *Effects of Substances on Concrete and Guide to Protective Treatments*. Washington, DC: Portland Cement Association.

Kerrich, R., Goldfarb, R. J., Richards, J. 2005. Metallogenic provinces in an evolving geodynamic framework. In Hedenquist, J. W., Thompson, J. F. H., Goldfarb, R. J., Richards, J. P. (eds.), *Economic Geology, 100th Anniversary Volume*. Littleton, CO: Society of Economic Geologists, pp. 1097–1136.

Kesler, S. E. 2005. Ore-forming fluids. *Elements*, 1, 13–18.

Kesler, S. E., Jones, L. M. 1981. Sulfur- and strontium-isotopic geochemistry of celestite, barite and gypsum from the Mesozoic basins of northeastern Mexico. *Chemical Geology*, 31, 211–224.

Kesler, S. E., Wilkinson, B. H. 2008. Earth's copper resources estimated from tectonic diffusion of porphyry copper deposits. *Geology*, 36, 255–258.

Kesler, S. E., Wilkinson, B. H. 2009. Resources of gold in Phanerozoic epithermal deposits. *Economic Geology*, 104, 623–633.

Kesler, S. E., Wilkinson, B. H. 2014. Tectonic-diffusion estimates of global mineral resources: extending method: granitic tin deposits. *Geological Society of London Special Publication*, 393.

Kesler, S. E., Chryssoulis, S. L., Simon, A. G. 2002. Gold in porphyry copper deposits: its abundance and fate. *Ore Geology Reviews*, 21, 103–124.

Kesler, S. E., Gruber, P. W. Medina, P. A., *et al.* 2012. Global lithium resources: relative importance of pegmatite, bring and other deposits. *Ore Geology Reviews*, 48, 55–69.

Kesler, T. L. 1939. Sienna (ocher) deposits of the Cartersville district, Georgia. *Economic Geology*, 34, 324–341.

Kesler, T. L. 1950. Geology and mineral deposits of the Cartersville district, Georgia. *US Geological Survey Professional Paper*, 224, 97.

Kesler, T. L. 1956. Environment and origin of the Cretaceous kaolin deposits of Georgia and South Carolina. *Economic Geology*, 51, 541–554.

Kesler, T. L. 1970. Hydrothermal kaolinization in Michoacán, Mexico. *Clays and Clay Minerals*, 18, 121–124.

Kessler, G. 1987. *Nuclear Fission Reactors*. New York, NY: Springer-Verlag.

Kestenbaum, R. 1991. The tin men: a chronicle of crisis. *Metal Bulletin PLC*, 260.

Kharecha, P. A., Hansen, J. E. 2013. Prevented mortality and greenhouse gas emissions from historical and projected nuclear power. *Environmental Science and Technology*, 47, 4889–4895.

King, K. 2008. Oil field resource growth. GEP-03 Geological basis for estimating the world's petroleum resources: challenges and uncertainties. International Geological Congress, Oslo.

Kingsnorth, J. D. 2009. The rare earths market: can supply meet demand? IMCOA. [Online] Available: http://www.usmagnetic materials.com/documents/Rare-Earths-Presentation.pdf

Kirk, J., Ruiz, J., Chelsey, J., Titley, S. 2002. The origin of gold in South Africa. *Scientific American*, 91, 534–541.

Kirkley, M. B., Gurney, J. J., Levinson, A. A. 1992. Age origin and emplacement of diamonds: a review of scientific advances in the last decade. *CIM Bulletin*, 85, 46–59.

Kitman, J. L. 2000. The secret history of lead. *The Nation*, March 20.

Kizny, T. 2004. *Gulag: Life and Death inside the Soviet Concentration Camps*. Ontario: Firefly Books.

Klemme, H. D. 1980. Types of petroliferous basins. In Foster, N. H., Beaumont E. A. (eds.), *Geologic Basins*. Tulsa, OK: American Association of Petroleum Geologists, pp. 87–101.

Klitgord, K. D., Watinds, J. S. 1984. Geologic studies related to oil and gas development in the EEZ. *US Geological Survey Circular*, 929, 107–184.

Knoop, P. A., Owen, R. W., Morgan, C. L. 1998. Regional variability in ferromanganese nodule composition: northeastern tropical Pacific Ocean. *Marine Geology*, 147, 1–12.

Kobayashi, J. 1971. Relation between the 'itai-itai' disease and the pollution of river water by cadmium from a mine. Advances in Water Pollution Research, Proceedings of the 5th international conference, San Francisco.

Koederitz, I. E., Harvey, A. H., Honarpour, M. 1989. *Introduction to Petroleum Reservoir Analysis*. Houston, TX: Gulf Publishing.

Kogel, J. E., Trivedi, N. C., Barker, N. C., Krukowski, S. T. 2006. *Industrial Minerals and Rocks*, 7th edn. Littleton, CO: Society for Mining, Metallurgy, and Exploration.

Kohl, W. L., ed. 1991. *After the Oil Price Collapse: OPEC, the United States and the World Oil Market*. Baltimore, MD: Johns Hopkins Press.

Kolata, G. 1992. New Alzheimer's study questions link to metal. *The New York Times*, November 10.

Kullgren, I. K. 2014. Do cities really save $38 for every $1 they spend on fluoridation? Politifact.com, August 23. [Online] Available: http://www.politifact.com/oregon/statements/2012/aug/23/nick-fish/do-cities-really-save-38-every-1-they-spend-fluori/.

Koljonen, T., Gustavsson, N., Noras, P., Tanskanen, H. 1989. Geochemical atlas of Finland: preliminary aspects. *Journal of Geochemical Exploration*, 32, 231–242.

Komnenic, A. 2013. Copper costs up, grades down. *Metal Bulletin PLC*, July 23.

Kondolf, G. M. 1997. Hungry water: effects of dams and gravel mining on river channels. *Environmental Management*, 21, 533–551.

Konhauser, K. O., Pecoits, E., Lalonde, S. V., *et al.* 2009. Oceanic nickel depletion and a methanogen famine before the Great Oxidation Event. *Nature*, 458, 750–754.

Konikow, L. F. 2013. Groundwater depletion in the United States (1900–2008). *US Geological Survey Scientific Investigations Report* 2013–5079, 63.

Kordosky, G. A. 1992. Copper solvent extraction: the state of the art. *Journal of the Minerals, Metals and Materials Society*, 4, 40–45.

Kornze, N. G. 2013. Statement of Neil G. Kornze, Principal Deputy Direction Bureau of Land Management Before the House Appropriations Committee, Hearing of the FY 2014 Budget Request for the Bureau of Land Management. [Online] Available: http://www.doi.gov/budget/appropriations/2014/upload/testimony_INTH_NK20130507.pdf.

Koschmann A. H., Bergendahl, M. H. 1968. Principal gold-producing districts of the United States. *US Geological Survey Professional Paper*, 610.

Kostic, D. E. 2013b. Sodium sulfate. *US Geological Survey Minerals Commodity Summary*, 2.

Kostick, D. S. 2013c. Soda ash. *US Geological Survey Minerals Yearbook*.

Kostick, D. S. 2013a. Salt. *US Geological Survey Minerals Yearbook*.

Krabbenhoft, D., Schuster, P. 2002. *US Geological Survey Fact Sheet*, FS-051-02. [Online] Available: http://toxics.usgs.gov/pubs/FS-051-02/.

Kramer, A.E., 2006. Evraz, a Russian steel maker, reaches deal to buy Oregon Steel Mills. *International Herald Tribune*, October 20.

Kramer, D. A. 1985. Magnesium. *US Bureau of Mines Mineral Facts and Problems*, 675, 471–482.

Kramer, D. A., Plunkert, P. A. 1991. Lightweight materials for new cars. *Minerals Today*, August, 8–15.

Krauskopf, K. B. 1988. *Radioactive Waste Disposal*. London: Chapman and Hall.

Krauskopf, K. B., Bird, D. 1994. *Introduction to Geochemistry*. New York, NY: McGraw-Hill.

Krauss, C. 2014. Potential crackdown on Russia risks also punishing western oil companies. *The New York Times*, March 27.

Krauss, U. H., Saam, H. G., Schmidt, H. W. 1984. International strategic minerals inventory summary report – phosphate. *US Geological Survey Circular*, 930-C.

Krauss, U. H., Schmidt, H. W., Taylor, H. A., Jr., Suthin, D.M. 1988. International strategic minerals inventory: natural graphite. *US Geological Survey Circular*, 930-H.

Krebs, W., Brombacher, C., Bosshard, P. P., Bachofen, R., Brandl, H. 1997. Microbial recovery of metals from solids. *FEMS Microbiology Reviews*, 20, 605–617.

Kretzchmar, G. L., Kirchner, A., Sharifzyanova, L. 2010. Resource nationalism – limits to foreign direct investment. *The Energy Journal*, 31, 27–52.

Krishnan, F. R., Hellwig, G. V. 1982. Trace emissions from coal and oil combustions. *Environmental Progress and Sustainable Energy*, 1, 290–295.

Krukowski, S. T., Johnson, K. S. 2006. Iodine. In Kogel, J. E., Trivedi, N. C., Barker, N. C., Krukowski, S. T. (eds.), *Industrial Minerals and Rocks*, 7th edn. Littleton, CO: Society for Mining, Metallurgy, and Exploration, pp. 541–555.

Kuck, P. H. 1985. Vanadium. *US Bureau of Mines Mineral Facts and Problems*, 675, 895–916.

Kuck, P. 2013. Nickel. *US Geological Survey Minerals Yearbook*.

Kuleshov, V. N. 2011. Manganese deposits: communication 1, genetic models of manganese ore formation. *Lithology and Mineral Resources*, 46, 473–493.

Kump, L. R. 2008. The role of seafloor hydrothermal systems in the evolution of seawater composition during the Phanerozoic. *Geophysical Monograph Series*, 178, 275–283.

Kvenvolden, K. S. 1988. Methane hydrate: a major reservoir of carbon in the shallow geosphere. *Chemical Geology*, 71, 41–51.

Kvenvolden, K. S., Ginsburg, G. D., Soloviev, V. A. 1993. Worldwide distribution of subaquatic gas hydrates. *Marine Letters*, 13, 32–40.

Lang, R. D. 1986. Development of Australia's first major diamond discovery. *Mining Engineering*, 38, 13–17.

Langer, W. H. 1988. Natural aggregates of the conterminous United States. *US Geological Survey Bulletin*, 1594, 33.

Langer, W. 2006. Aggregates. In Kogel, J. E., Trivedi, N. C., Barker, N. C., Krukowski, S. T. (eds.), *Industrial Minerals and Rocks*, 7th edn. Littleton, CO: Society for Mining, Metallurgy, and Exploration, pp. 159–169.

Langton, J. 2008. Bakken Formation: will it fuel Canada's oil industry? *CBC News*, January 27.

Lantzy, R. J., MacKenzie, F. T. 1979 Atmospheric trace metals: global cycles and assessment of man's impact. *Geochimica et Cosmochimica Acta*, 43, 511–525.

Larsen, K. G. 1977. Sedimentology of the Bonne Terre formation, southeast Missouri. *Economic Geology*, 72, 408–419.

Lavelle, M. 2013. Space view of natural gas flaring darkened by budget woes. *National Geographic*, October 10.

Lavoie, R. A., Jardine, T. D., Chumchal, M. M., Kidd, K. A., Campbell, L. M. 2013. Biomagnification of mercury in aquatic food webs: a world wide meta analysis. *Environmental Science and Technology*, 47, 13385–13394.

Laznicka, P. 2006. *Giant Metallic Deposits*. Berlin: Springer.

Le Quéré, G. P. P., Andres, R. J. Andrew, R. M., *et al.* 2013. Global Carbon Budget 2013. *Earth System Science Data*, 6, 235–263.

Leach, D. L., Sangster, D. F., Kelley, K. D. 2005 Sediment-hosted lead-zinc deposits: a global perspective. In Hedenquist, J. W., Thompson, J. F. H., Goldfarb, R. J., Richards, J. P. (eds.), *Economic Geology, 100th Anniversary Volume*. Littleton, CO: Society of Economic Geologists, pp. 561–607.

Leachman, W. D. 1988. Helium. *US Bureau of Mines Minerals Yearbook*.

LeCompte, C. 2013. Fertilizer plants spring up to take advantage of the US's cheap natural gas. *Scientific American*, April 25.

Lee, C. O., Bartlett, Z. W., Feierabend, R. H. 1960. The Grand Isle mine. *Mining Engineering*, June, 578–584.

Lee, J-E., Oliveria, R., Dawson, T., Fung, I. 2005. Root functioning modifies seasonal climate. *Proceedings of the National Academy of Sciences of United States of America*, 102:17576–17581.

Lefebvre, B. 2014. Shale-oil boom spurs refining binge. *The Wall Street Journal*, March 2.

Leffler, W. L., Pattarozzi, R., Sterling, G. 2003. *Deepwater Petroleum Exploration and Production: A Nontechnical Guide*. Tulsa, OK: PennWell Publishing Company.

Leffler, W. L. 2012. *Petroleum Refining in Nontechnical Language*. Tulsa, OK: PennWell Publishing Company.

Lefond, S. J., Jacoby, C.H. 1983. Salt. In Lefond, S. J. (ed.), *Industrial Minerals and Rocks*, 5th edn. New York, NY: Society of Mining Engineers, pp. 1119–1150.

Lehmann, B. 1990. *Metallogeny of Tin*. Heidelberg: Springer-Verlag.

Leney, G. W. 1964. Geophysical exploration for iron ore. *Transactions of Society of Mining Engineering*, 231, 355–372.

Lessmark, O., Thornelof, E. 1986. Liming in Sweden. *Water, Air and Soil Pollution*, 31, 809–815.

Levinsohn, B. 2014. US Steel: tariff decision prompts meltdown, undermines bull. *Wall Street Journal*, February 19.

Levinson, A. A. 1983. *Introduction to Exploration Geochemistry*, 2nd edn. Calgary: Applied Publishing.

Lewan, M. D. 1984. Factors controlling the proportionality of vanadium to nickel in crude oils. *Geochimica et Cosmochimica Acta*, 48, 2231–2238.

Lewis, A. E. 1985. Oil from shale: the potential, the problems, and a plan for development. In Perrine, R. L., Ernst, W.G. (ed.), *Energy for Ourselves and our Posterity*. Englewood Cliffs, NJ: Prentice-Hall, pp. 101–122.

Lewis, J., Benedict, C. P. 1981. Icebergs on the Grand Banks: oil and gas considerations. *World Oil*, January, 109–111.

Li, R. X., Xiao, C. D., Sneed, S. B., Yan, M. 2012, A continuous 293-year record of volcanic events in an ice core from Lambert glacier basin, East Antarctica. *Antarctic Science*, 24, 293–298.

Li, C., Cornett, J. 2011. Increased zinc concentrations in Canadian Arctic air. *Atmospheric Pollution Research*, 2, 45–48.

Li, V. C., Fischer, G., Lepech, M. 2004. Crack-resistant concrete material for transportation construction. Transportation Research Board 83rd Annual Meeting, Washington, DC. [Online] Available: http://hdl.handle.net/2027.42/84828.

Likens, G. E., Driscoll, C.T., Buso, D.C. 1996. Long-term effects of acid rain: response and recovery of a forest ecosystem. *Science*, 272: 244–246.

Likens, G. E., Buso, D. C., Butler, T. J. 2005. Long-term relationships between SO_2 and NO_x emissions and SO_2^{2-} and NO_3 concentration in bulk deposition at the Hubbard Brook Experimental Forest, NH. *Journal of Environmental Monitoring*, 7, 964–968.

Lim, S. R., Kang, D., Ogunseitan, O. A., Schoenung, J.M 2013. Potential environmental impacts from the metals in incandescent, compact fluorescent lamp (CFL), and light-emitting diode (LED) bulbs. *Environmental Science and Technology*, 47, 1040–1047.

Lin, H. 2010. Earth's critical zone and hydropedology: concepts, characteristics and advances. *Hydrology Earth Systems Science*, 14, 25–45.

Link, P. K. 1982. *Basic Petroleum Geology*. Tulsa, OK: Oil and Gas Consultants International.

Link, W. K. 1952. Significance of oil and gas seeps in world oil exploration. *American Association of Petroleum Geologists Bulletin*, 36, 1505–1529.

Lipin, B. R. 1982. Low-grade chromium resources. *Society of Mining Engineers Reprint*, 82–62, 17.

Liu, G. G. 2007. China's coal supply/demand and their impact on international coal market. AAA Minerals International. [Online] Available: www.aaamineral.com/coal/ppt/Aruba%20George.ppt.

Liu, W., Yang, J., Xiao, B. 2009, Application of Bayer red mud for iron recovery and building material production from aluminosilicate residues. *Journal of Hazardous Materials*, 161, 474–478.

Loebenstein, J. R. 1985. Platinum-group metals. *US Bureau of Mines Mineral Facts and Problems*, 675, 595–615.

Loferski, P. J. 2012. Platinum-group metals. *US Geological Survey Minerals Yearbook*.

Logue, O.T., 1991. Safety and pneumoconioses: abrasive blasting and protective respiratory equipment. *Materials Performance*, 30, September, 32–37; December, 5–6.

Lohr, S. 1985. Settlement is imposed in tin crisis. *The New York Times*, March 8.

London, D. 2008. Pegmatites. *Canadian Mineralogist Special Publication*, 10.

Longhurst, J. W. S. 1989. *Acid Deposition: Sources, Effects and Controls*, London: British Library Science Reference and Information Service.

Lovelock, J. 1971. Atmospheric fluorine compounds as indicators of air movements, *Nature*, 230, 379.

Lowell, J. D. 1968. Geology of the Kalamazoo orebody, San Manuel district, Arizona. *Economic Geology*, 63, 645–654.

Lowell, J. D., Guilbert, J. M. 1970. Lateral and vertical alteration – mineralization zoning in porphyry ore deposits. *Economic Geology*, 65, 373–408.

Lowers, H., Meeker, G. 2002. Tabulation of asbestos-related terminology. *US Geological Survey Open-File Report*, 02-0458.

Lund, J. W., Bertani, R. 2010. Worldwide geothermal utilization 2010. *Geothermal Resources Council Transactions*. 34, 195–198.

Lüthi, D. M., Le Floch, B., Bereiter, T., *et al.* 2008. High-resolution carbon dioxide concentration record years before present. *Nature*, 453, 379–382.

Lutz, B., Bernhardt, E. S., Schlesinger, W. 2013. The environmental price tag on a ton of mountaintop removal coal. *PloS One*, E73203.

Luxbacher, G. W., Sell, D. P., Skaggs, G. L., McPhee, G. T. 1992. *The Clear Air Act Amendments of 1990 – An Eastern Coal Producer's View*. Littleton, CO: Society for Mining, Metallurgy and Exploration, Preprint 92–229.

Lyday, P. A. 1999 *Iodine and Iodine Compounds*. New York, NY: Wiley-VCH.

Lyday, P. A. 1985. Iodine. *US Bureau of Mines Mineral Facts and Problems*, 675, 377–384.

Lynd, L. E. 1985. Titanium. *US Bureau of Mines Mineral Facts and Problems*, 675, 859–880.

Lyons, W. 2009. *Working Guide to Reservoir Engineering*. Amsterdam: Gulf Professional Publishing.

Lyons, T. W., Reinhard, C. T., Planavsky, N. J. 2014. The rise of oxygen in Earth's early ocean and atmosphere. *Nature*, 506, 307–315.

MacAvoy, P. W. 1988. *Explaining Metal Prices*. Boston, MA: Kluwer Academic Publishers.

Machel, H. G. 2004. Concepts and models of dolomitization: a critical appraisal. *Geological Society of London Special Publication*, 235, 7–63.

Mackay, A. L., Terrones, H. 1991. Diamond from graphite. *Nature*, 352, 762.

Macleod, A. 2014. China's gold demand. [Online] [Online] Available: http://news.goldseek.com/GoldSeek/1392393720.php.

Madigan, M. T., Martinko, J. M, Bender, K. S., *et al.* 2014. *Brock Biology of Microorganisms*, 14th edn. Boston, MA: Benjamin Cummings.

Madlool, N. A., Saidur, R., Hossain, M. S., Rahim, N. A. 2011. A critical review on energy use and savings in the cement industries. *Renewable and Sustainable Energy Reviews*. 15, 2042–2060.

Madongo, I. 2014. SA mining: tough questions beyond BRICs. *Financial Times*, February 13.

Magon, K. M. 1990. The asbestos dilemma: make the workplace safe without endangering workers. *Safety and Health*, 142, 50–53.

Maher, K., McGinty, T. 2013. Coal's decline hits hardest in the mines of Kentucky. *The Wall Street Journal*, November 26.

Mahoney, M. C., Naasca, P. C., Burnett, W. S. 1991. Bone cancer incidence rates in New York state: time trends and fluoridated drinking water. *American Journal of Public Health*, 81, 475–479.

Maier, W. D. 2005. Platinum-group element (PGE) deposits and occurrences: mineralisation styles, genetic concepts, and exploration criteria. *Journal of African Earth Sciences*, 41, 165–191.

Malinconico, L. L. 1987. On the variation of SO_2 emissions from volcanoes. *Journal of Volcanology and Geothermal Research*, 33, 231–237.

Malm, E. 2013. Federal mineral royalty disbursements to states and the effects of sequestration. *Tax Foundation Fiscal Fact*, No. 371.

Malm, W. C. 1999. *Introduction to Visibility*. Fort Collins, CO: Colorado State University, Cooperative Institute for Research in the Atmosphere.

Manahan, S. E. 1990. *Environmental Chemistry*, New York, NY: Lewis.

Mann, C. C. 2011. *1493: Uncovering the New World Columbus Created*. New York, NY: Knopf.

Mann, E. L. 1983. Asbestos. In Lefond, S. J. (ed.), *Industrial Minerals and Rocks*, 5th edn. New York, NY: Society of Mining Engineers, pp. 435–449.

Mann, P., Horn, M., Cross, I. 2007. Emerging trends from 69 giant oil and gas fields discovered from 2000–2006. American Association of Petroleum Geologists, Annual Meeting, Long Beach, CA.

Mannion, J. E. 1983. Sodium carbonate deposits. In Lefond, S. J. (ed.), *Industrial Minerals and Rocks*, 5th edn. New York, NY: Society of Mining Engineers, pp. 1187–1206.

Marcak, H., Mutke, G. 2013. Seismic activation of tectonic stresses by mining. *Journal of Seismology*, 17, 1139–1148.

Mares, D. R. 2010 Resource nationalism and energy security in Latin America.. James A. Baker III Institute for Public Policy, Rice University. [Online] Available: http://pages.ucsd.edu/~dmares/MaresResourceNationalismWorkPaper.pdf.

Mari, M., Domingo, J. L. 2010. Toxic emissions from crematories: a review. *Environment International*, 36, 131–137.

Mark, C. 2007. Multiple seam longwall mining in the United States: lessons for ground control. Online] Available: http://www.cdc.gov/niosh/mining/UserFiles/works/pdfs/mslmi.pdf

Markandya, A., Wilkinson, P. 2007. Electricity generation and health. *The Lancet*, 370(9591), 979–990.

Market Wired 2012. World lime market to reach 79.1 billion tonnes in 2016. Market Wired, August 29. [Online] Available: http://www.marketwired.com/press-release/world-lime-market-to-reach-791-billion-tonnes-in-2016-1695377.htm.

Marr, I. L., Cresser, M. S. 1983. *Environmental Chemical Analysis*, New York, NY: Chapman and Hall.

Marshall, E. 1989. Gasoline: the unclean fuel? *Science*, 246, 199–201.

Martin, M. 2013. Bakken: home of the next big oil boom? *CNB News*, March 23.

Mason, R. 2013. *Trace Metals in Aquatic Systems*. Chichester: Wiley-Blackwell.

Mathur, A., Morris, A. C. 2014. Distributional effects of a carbon tax in broader US fiscal reform. *Energy Policy*, 66, 326.

Matthews, R. G. 2010. China steps up its steelmaking game. *The Wall Street Journal*, October 8.

Maugeri, L. 2013. *The Shale Oil Boom: A US Phenomenon*. Cambridge, MA: Harvard Kennedy School, Belfer Center for Science and International Affairs.

Maynard, J. B. 1983. *Geochemistry of Sedimentary Ore Deposits*. Berlin: Springer-Verlag.

Maynard, J. B., Okita, O. M. 1991. Bedded barite deposits in the United States, Canada, Germany, and China: two major types based on tectonic setting. *Economic Geology*, 86, 364–376.

May, P. W. 2000. Diamond thin films: a 21st-century material. *Philosophical Transactions of the Royal Society A*, 348, 473–495.

Mays, G. C. 1992. *Durability of Concrete Structures: Investigation, Repair, Protection*. London: E&F Spon.

Mayuga, M. N. 1970. Geology and development of California's giant – Wilmington oil field. In Halbouty, M. T. (ed.), *Geology of Giant Petroleum Fields*. Tulsa, OK: American Association of Petroleum Geologists, pp. 158–184.

McCarl, H. N. 1985. Lightweight aggregate. In Lefond, S. J. (ed.), *Industrial Minerals and Rocks*, 5th edn. New York, NY: Society of Mining Engineers, pp. 81–92.

McConnell, D. 1935. The Chilean nitrate industry. *The Journal of Political Economy*, 43, 506–529.

McConville, L. B. 1975. The Athabasca tar sands. *Mining Engineering*, 27, 19–25.

McDonald, T. A. 2002. A perspective on the potential health risks of PBDEs. *Chemosphere*, 46, 745–755.

McDonough, W., Braungart, M. 2013. *The Up-Cycle: Beyond Sustainability – Designing for Abundance*. New York, NY: North Point Press.

McFarlin, R. F. 1992. Florida phosphate and the environment: practices, problems and emerging technologies. In Chander, S. (ed.), *Emerging Process Technologies for a Cleaner Environment*. Littleton, CO: Society for Mining, Metallurgy and Exploration, pp. 29–39.

McGarrity, J. 2013. Britain seen putting off subsea coal gasification projects. *Reuters*, October 1.

McKeith, T. D., Schodde, R. C., Baltis, E. J. 2010. Gold discovery trends. *Society of Economic Geologists Newsletter*, 81.

McKelvey, V. E. 1960. Relations of reserves of the elements to their crustal abundance. *American Journal of Science*, 258A, 234–241.

McLelland, J. M., Selleck, B. W. 2011. Megacrystic Gore Mountain-type garnets in the Adirondack Highlands: age, origins and tectonic implications. *Geosphere*, 7, 1194–1208.

McNamee, K. 1990. Canada's endangered spaces: preserving the Canadian wilderness. In Lime, D. W. (ed.), *Managing American's Enduring Wilderness Resource*. University of

Minnesota, St. Paul, MN: Minnesota Extension Service, pp. 425–443.

McWhitert, C., McMahon, D. 2013. Spotted again in America: textile jobs. *The Wall Street Journal*, December 22.

Meadows, D. H., Meadows, D.C., Randers, J., Behrens, W.W., III. 1972. *The Limits to Growth*, Universe Books, New York.

Megaw, P. K. W., Ruiz, J., Titley, S. R. 1988. High temperature carbonate-hosted Ag–Pb–Zn(Cu) deposits of northern Mexico. *Economic Geology*, 83, 1856–1885.

Megill, R. E., Wightman, R. B., 1984. The ubiquitous overbid. *American Association of Petroleum Geologists Bulleitn*, 68, 417–425.

Melcher, F., Obertur, T., Rammalmeir, D. 2006. Geochemical and mineralogical distribution of germanium in the Khusib Springs Cu–Zn–Pb–Ag sulfide deposit, Otavi Mountain Land, Namibia. *Ore Geology Reviews*, 28, 32–56.

Meng, Y.-F., Yan, C.-S., Lai, J., *et al.* Enhanced optical properties of chemical vapor deposited single crystal diamond by low-pressure/high-temperature annealing. *Proceedings of the National Academy of Sciences*, 105, 17620–17625.

Menzie, W. D., Barry, J. J., Bleiwas, D. I., *et al.* 2010. The global flow of aluminum from 2006 through 2010. *US Geology Survey Open-File Report*, 2010–1256.

Meriläinen, J. 1986. *Diatoms and Lake Acidity*. Dordrecht, Kluwer Academic Publishers.

Merritt, K. A., Amirbahman, A. 2009. Mercury methylation dynamics in estuarine and coastal marine environments, a critical review. *Earth Science Reviews*, 96, 54–66.

Metlay, D., Sarewitz, D. 2012. Decision strategies for addressing complex "messy" problems. *The Bridge*, 42, 6–16.

Meyer, F. M. 2004. Availability of bauxite reserves. *Natural Resources Research*, 13, 161–172.

Meyer, H. O. A. 1985. Genesis of a diamond: a mantle saga. *American Mineralogist*, 70, 344–355.

Meyers, R. A. E. 1981. *Coal Handbook*. New York, NY: Marcel Dekker.

Michalski, G., Bohlke, J. K., Thiemens, M. H. 1994. Long term atmospheric deposition as the source of nitrate and other salts in the Atacama Desert, Chile: new evidence from mass-independent oxygen isotopic compositions. *Geochimica et Cosmochimica Acta*, 68, 403–408.

Mikami, H. M. 1983. Chromite. In LeFond, S. J. (ed.), *Industrial Minerals and Rocks*, 5th edn. New York, NY: Society of Mining Engineers, pp. 567–584.

Mikdashi, A. 1986. *Transnational Oil: Issues, Policies and Perspectives*. New York, NY: St. Martin's Press.

Mikesell, R. F. 1986. *Stockpiling Strategic Materials: An Evaluation of the National Program*. Washington, DC: American Enterprise Institute for Public Policy Research.

Mikesell, R. F. 1987. *Nonfuel Minerals: Foreign Dependency and National Security*. Ann Arbor, MI: University of Michigan Press.

Mikesell, R. F., Whitney, J. W. 1987. *The World Mining Industry*. London: Allen and Unwin.

Mikesell, R. F. 1988. *The Global Copper Industry*. New York, NY: Croom Helm.

Miller, M. M. 2011. Lime in the United States 1960 to 2009. *US Geological Survey Mineral Industry Surveys*, 17.

Miller, M. M. 2013. Fluorspar. *US Geological Survey Minerals Yearbook*.

Miller, M. M. 2013. Barite. *US Geological Survey Minerals Yearbook*.

Miller, R. W., Pesaran, P. 1980b. Effects of drilling fluids on soil and plants: II. Complete drilling fluid mixtures. *Environment Quality*, 9, 552–558.

Miller, R. W., Honarvar, S., Hunsaker, B. 1980a. Effects of drilling fluids on soils and plants: I. *Environmental Quality*, 9, 547–551.

Miller, S. L., Urey, H. C. 1959. Organic compounds synthesis on the primitive Earth. *Science*, 130, 245–251.

Milliman, J. D., Syvitski, P. M. 1992. Geomorphic/tectonic control of sediment discharge to the ocean: the importance of small mountainous rivers. *Journal of Geology*, 100, 525–544.

Mills, H. N. 1983. Glass raw materials. In Lefond, S. J. (ed.), *Industrial Minerals and Rocks*, 5th edn. New York, NY: Society of Mining Engineers, pp. 339–351.

Mills, P. 2006. Barium minerals. In Kogel, J. E., Trivedi, N. C., Barker, N. C., Krukowski, S. T. (eds.), *Industrial Minerals and Rocks*, 7th edn. Littleton, CO: Society for Mining, Metallurgy, and Exploration, pp. 219–226.

Miner, R., Upton, B. 2002. Methods for estimating greenhouse gas emissions from lime kilns at Kraft pulp mills. *Energy*, 27, 729–738.

Mining Watch Canada 2012. Introduction to the legal framework for mining in Canada. [Online] Available: http://www.miningwatch.ca/publications/introduction-legal-framework-mining-canada#Q7.

Minter, W. E. L. 2006. The sedimentary setting of Witwatersrand placer mineral deposits in an Archean atmosphere. *GSA Memoirs*, 198, 105–119.

Mishra, C. P., Sheng-Fogg, C. D., Christiansen, R. G., Lemons, J. F., Jr., Degiacomo, D. L. 1985. Cobalt availability market economy countries. *US Bureau of Mines Information Circular*, 9012.

Mitchell, A. W. 1984. Barite in the western Ouachita Mountains, Arkansas. *Arkansas Geological Commission Guidebook*, 84–2, 124–131.

Mitchell, B. R. 1988. *British Historical Statistics*. Cambridge: Cambridge University Press.

Mitchell, P. 2009. Taxation and investment issues in mining. Extractive Industries Transparency Initiative. [Online] Available: http://eiti.org/files/MINING%20Compressed.pdf.

Mitchell, R. H. 1991. *Petrology of Lamproites*. New York, NY: Plenum Press.

Miyamoto, C., Ketterings, Q., Cherney, J., Kilcer, T. 2008. Nitrogen fixation. *Cornell University, Cooperative Extension Fact Sheet*, 39.

Moller, P., Cerny, P., Saupe, F., eds. 1989. *Lanthanides, Tantalum and Niobium*, Heidelberg: Springer-Verlag.

Montes, J., Ilif, L., Luhnow, D. 2013. Mexico congress passes historic energy bill. *The Wall Street Journal*, December 12.

Moore, C., Marshall, R. I. 1991. *Steelmaking*. London: Institute of Metals.

Moore, J. W. 1991. *Inorganic Contaminants of Surface Water. Research and Monitoring Properties*. New York, NY: Springer Verlag.

Moore, P. 2012. Potash mining: booming pink gold. *International Mining*, July, 58–62.

Moorehouse, D. F. 1997. The intricate puzzle of oil and gas reserves growth. US Energy Information Administration/Natural Gas Monthly, July.

Morell, J. B. 1992. *The Law of the Sea: An Historical Analysis of the 1982 Treaty and its Rejection by the United States*. Jefferson, NC: McFarland.

Morgenstern, R., Tirpak, D. 1990. Environmental Protection Agency. *EPA Journal*, 16, no. 2.

Morrow, H. 2011. *Proceedings of the Eighth International Cadmium Conference, Kunming China, November 10–13*. Brussels: International Cadmium Association.

Morse, D. E. 1985. Salt. *US Bureau of Mines Mineral Facts and Problems*, 675, 679–688.

Mosier, D. L., Singer, D. A., Moring, B. C., Galloway, J. P. 2012. Podiform chromite deposits – database and grade and tonnage models. US Geological Survey Scientific Investigation Report 2012-5157. [Online] Avaialable: http://pubs.usgs.gov/sir/2012/5157/.

Moskalyk, R. R. 2003. Gallium: the backbone of the electronics industry. *Minerals Engineering*, 16, 921–928.

Mossman, B. T., Bignon, J., Corn, M., Seaton, A., Gee, J. B. L. 1989. Asbestos: scientific developments and implications for public policy. *Science*, 247, 294–301.

Mouawad, J., Meier, B. 2010. Risk-taking rises as oil rigs in Gulf drill deeper. *The New York Times*, August 29.

Mowatt, T. 1992. Turning Bogota into the emerald city. *The New York Times*, June 22.

Muffler, L. J. P. 1977. Technical analysis of geothermal resources. *Ecology Law Quarterly*, 6, no. 2, 253–270.

Muhn, J., Stuart. *Opportunity and Challenge: The Story of BLM*. Washington, DC: US Department of the Interior, Bureau of Land Management.

Muntean, J., Cline, J., Simon, A., Longo, A. 2011. Magmatic–hydrothermal origin of Nevada's Carlin-type gold deposits. *Nature Geoscience*, 4, 122–127.

Murakami, H., Ishihara, S. 2013 Trace elements of indium-bearing sphalerite from tin-polymetallic deposits in Bolivia, China and Japan: a femto-second LA-ICPMS study. *Ore Geology Reviews*, 53, 223–243.

Murphy, G. F., Brown, R. E. 1985. Silicon. *US Bureau of Mines Mineral Facts and Problems*, 675, 713–728.

Murphy, K. 2011. What's lurking in your countertop? *The New York Times*, July 24.

Murray, R. L. 1989. *Understanding Radioactive Waste*. Columbus, OH: Battelle Press.

Murtaugh, K. A. 2006. Analysis of sustainable water supply options for Kuwait. Ph.D. thesis, Massachusetts Institute of Technology.

Nable, R. O., Banuelos, G. S., Paull, J. G. 1997. Boron toxicity. *Plant and Soil*, 193, 181–198.

Naicker, K., Cukrowska, E., McArthy, T. S. 2003. Acid mine drainage from gold mining activities in Johannesburg, South Africa and environs. *Environmental Pollution*, 122, 29–40.

Naldrett, A. J. 1989. *Magmatic Sulfide Deposits*. Oxford: Oxford University Press.

Naldrett, A. J. 2003. From impact for riches: evolution of geological understanding as seen at Sudbury, Canada. *GSA Today*, 13, 4–9.

Naldrett, A. J. 2010. Secular variation of magmatic sulfide deposits and their source magmas. *Economic Geology*, 105, 669–688.

NAPAP 1991. *National Acid Precipitation Assessment Program, 1990 Integrated Assessment Report*. Washington, DC: National Acid Prevention Assessment Program.

Nash, J. T., Granger, H. C., Adams, S. S. 1981. Geology and concepts of genesis of important types of uranium deposits. *Economic Geology*, 75, 63–116.

Naslund, H. R. 1983. The effect of oxygen fugacity on liquid immiscibility in iron-bearing silicate melts. *American Journal of Science*, 283, 1034–1059.

National Lime Association 2014. Using lime for flue gas treatment. [Online] Available: http://www.lime.org/documents/publications/free_downloads/fgd-final-2000.pdf.

National Mining Association, 2015, Value and rank of nonfuel mineral production in the US for 2014. [Online] Available: http://www.nma.org/pdf/m_value_rank.pdf.

Natural Resources Canada 2014. Tables on the structure and rates of main taxes. [Online] Available: http://www.nrcan.gc.ca/mining-materials/taxation/mining-taxation-regime/8890.

National Research Council 2007. *Colorado River Basin Water Management*. Washington, DC: National Academic Press.

National Response Team Report 2011. On scene coordinator report Deepwater Horizon Oil Spill. [Online] Available: http://www.uscg.mil/foia/docs/dwh/fosc_dwh_report.pdf

NEA 1981. *The Environmental and Biological Behaviour of Plutonium and Some Other Transuranium Elements*. Paris: Organization for Economic Cooperation and Development, Nuclear Energy Agency.

Needleman, H. L. 2004. Lead poisoning. *Annual Review of Medicine*, 55, 209–222.

Neff, K. M., McKelvie, S., Ayers, R. C., Jr. 2000. Environmental impacts of synthetic based drilling fluids. US Department of the Interior, Minerals Management Service, Gulf of Mexico, OCS Region, New Orleans, LA, OCS Study. MMS 2000–064.

Nehring, R. 1982. Prospects for conventional world oil reserves. *Annual Review of Energy*, 7, 175–200.

Newell, R. G., Rogers, K. 2003, The US experience with the phase-down of lead in gasoline. Resources for the Future [Online]. Available: http://web.mit.edu/ckolstad/www/Newell.pdf.

Ng, A., Patterson, C. 1981. Natural concentrations of lead in ancient Arctic and Antarctic ice. *Geochimica et Cosmochimica Acta*, 45, 2109–2121.

Ngo, C., Natowitz, J. 2009. Our Energy Future: Resources, Alternatives and the Environment. Hoboken, NJ: Wiley.

Nicoletopolous, V. 2013. Spanish mining: taxation and regulation. *Industrial Minerals*, 32–36.

Niec, M. 1986. Geology of sulfur deposits. In *Fertilizer Minerals in Asia and the Pacific. Volume 1. Mineral Concentrations and Hydrocarbon Accumulations in the ESCAP Region*. Bangkok, Thailand: United Nations, Economic Social Commission for Asia and the Pacific (UNESCO), pp. 84–92.

Nielsen, L. 2011. Classifications of countries based on their level of development: how it is done and how it could be done [Online]. International Monetary Fund. Available: http://www.imf.org/external/pubs/cat/longres.aspx?sk=24628.

NIOSH 2013. National Institute for Occupational Safety and Health (NIOSH), criteria for a recommended standard: occupational exposure to hexavalent chromium. *NIOSH Publication*, No. 2013–128, January.

NRMCA 2012. Concrete CO_2 fact sheet. National Ready Mix Concrete Association Publication number 2PCO2. [Online] Available: http://www.nrmca.org/sustainability/CONCRETE%20CO2%20FACT%20SHEET%20FEB%202012.pdf.

Norgate, T. E., Jahanshahi, S., Rankin, W. J. 2007. Assessing the environmental impact of metal production processes. *Journal of Cleaner Production*, 15, 838–848.

Norman, N.C. 1998. *Chemistry of Arsenic, Antimony, and Bismuth*. London: Blackie Academic & Professional.

Northey, S., Mohr, S., Mudd, G. M., Weng, Z., Guiro, D. 2014 Modeling future copper grade decline based on a detailed assessment of copper resources and mining. *Resources, Conservation and Recycling*, 83, 190–201.

Norton, D. 1978. Sourcelines, source regions and pathlines for fluids in hydrothermal systems related to cooling plutons. *Economic Geology*, 73, 21–28.

Notholt, A. J. G., Sheldon, R. P., Davidson, D. F. 1989. *Phosphate Deposits of the World*. Cambridge: Cambridge University Press.

Novelline, R. 1997. *Fundamentals of Radiology*, 5th edn. Cambridge, MA: Harvard University Press.

NRC 1978. *Arsenic*. Washington, DC: National Research Council, National Academy of Science.

NRC 1990. *Surface Coal Mining Effects on Groundwater Recharge*. Washington, DC: National Academy Press.

NRC 2015. Location of projected new nuclear power reactors, Nuclear Regulatory Commission (NRC). [Online] Available: http://www.nrc.gov/reactors/new-reactors/col/new-reactor-map.html.

NRDC 2011. Arctic wildlife refuge: why trash an American treasure for a tiny percentage of our needs? [Online] Available: http://www.nrdc.org/land/wilderness/arctic.asp.

Nriagu, J. O. 1978. *The Biogeochemistry of Lead in the Environment*. Amsterdam: Elsevier.

Nriagu, J. O. 1979. *Copper in the Environment*. New York, NY: John Wiley.

Nriagu, J. O. 1983. *Lead and Lead Poisoning in Antiquity*. New York, NY: Wiley Interscience.

Nriagu, J. O. 1987. *Zinc in the Environment*. New York, NY: Wiley Interscience.

Nriagu, J. O. 1989. A global assessment of natural sources of atmospheric trace metals. *Nature*, 338, 47–49.

Nriagu, J. O. 1990a. The rise and fall of leaded gasoline. *The Science of the Total Environment*, 92, 13–28.

Nriagu, J. O. 1990b. Global metal pollution. *Environment*, 32, no. 7, 7–33.

Nriagu, J. O. 1992. Human lead exposure. In Needleman, H. L. (ed.), *Human Lead Exposure*. Boca Raton, FL: CRC Press, pp. 3–21.

Nriagu, J. O., Rao, S. S. 1987. Response of lake sediments to changes in trace metal emission from the smelters at Sudbury, Ontario. *Environmental Pollution*, 44, 211–218.

Nriagu, J. O., Pacyna, J. M. 1988. Quantitative assessment of worldwide contamination of air, water and soils by trace metals. *Nature*, 333, 134–139.

Nriagu, J. O., Pfeiffer, W. C., Malm, O., Mierle, G. 1992. Mercury pollution in Brazil. *Nature*, 356, 389.

Nuclear Energy Institute 2015. World nuclear generation and capacity. [Online] Available: http://www.nei.org/Knowledge-Center/Nuclear-Statistics/World-Statistics/World-Nuclear-Generation-and-Capacity.

Nwoke, C. 1987. *Third World Mineral and Global Pricing*. London: Zed Books.

O'Neil, T. J. 1974. The minerals depletion allowance, parts I and II. *Mining Engineering, October and November*, 61–64 and 39–42.

Oancea, D. 2011. Mining in Mexico. InfoMine. [Online] Available: http://www.infomine.com/countries/SOIR/mexico/welcome.asp?i=mexico-soir-2.

Ober, J. A. 1992. Strontium. *US Bureau of Mines Minerals Yearbook*.

Ober, J. A. 2002. Materials flow of sulfur. *US Geological Survey Open-File Report*, 02-298.

Ober, J. A. 2006a. Strontium minerals. In Kogel, J. E., Trivedi, N. C., Barker, N. C., Krukowski, S. T. (eds.), *Industrial Minerals and Rocks*, 7th edn. Littleton, CO: Society for Mining, Metallurgy, and Exploration, pp. 971–986.

Ober, J. 2006b. Sulfur. In Kogel, J. E., Trivedi, N. C., Barker, N. C., Krukowski, S. T. (eds.), *Industrial Minerals and Rocks*, 7th edn. Littleton, CO: Society for Mining, Metallurgy, and Exploration, pp. 935–970.

Ober, J. A. 2013. Bromine. *US Geological Survey Minerals Yearbook*.

Ober, J. A. 2014. Strontium. *US Geological Survey Mineral Commodity Summary*.

Obregon, M. J., Escobar del Rey, F., Morreale de Escobar, G. 2005. The effects of iodine deficiency on thyroid hormone deiodination. *Thyroid*, 15, 917–929.

OECD 1983. *Aluminum Industry, Energy Aspects for Structural Change*. Paris: Organization for Economic Cooperation and Development.

OECD 2010. Global forum on environment focusing on sustainable materials management, 2010, materials case study 2: aluminum. [Online] Available: http://www.oecd.org/environment/waste/46194971.pdf.

Office of Surface Mining 2015. Reclaiming abandoned mine lands. [Online] Available: http://www.osmre.gov/programs/aml.shtm.

OGJ 2013. MGM gives up McKenzie delta conventional licenses. *Oil & Gas Journal*, May 30. [Online] Available: http://www.ogj.com/articles/2013/05/northwest-territories-mgm-relinquishes-licenses.html.

OGJ 2015. Sharp drop expected in global E&P spending in 2015, study says. *Oil & Gas Journal*, January 8. [Online] Available:

http://www.ogj.com/articles/2015/01/sharp-drop-expected-in-global-e-p-spending-in-2015-study-says.html.

Ohle, E. L., Bates, R. L. 1981. Geology, geologists, and mineral exploration. *Society of Economic Geologists*, 75, 766–774.

Oldfield, F. 2005. *Environmental Change: Key Issues and Alternative Approaches*. Cambridge, Cambridge University Press.

Olson, D. K. 1986. Michigan silver: native silver occurrences in the copper mines of Upper Michigan. *Mineralogical Record*, 17, 37–48.

Olson, D. W. 2013. Industrial diamond. *US Geological Survey Minerals Yearbook*.

Olson, D. W. 2014. Gemstones. *US Geological Survey Minerals Yearbook*.

Olson, K. W., Klecky, J.W. 1989. Severance tax stability. *National Tax Journal*, 42, 69–78.

Olson, R. H. 1983. Zeolites. In Lefond, S. J. (ed.), *Industrial Minerals and Rocks*, 5th edn. New York, NY: Society of Mining Engineers, pp. 1391–1422.

ONRR 2013. US Department of the Interior, New Release November 19. [Online] Available: http://www.onrr.gov/about/pdfdocs/20131119a.pdf.

Optima 1985. Diamond trading over fifty years. *Optima*, 32(1), 40–48.

Orris, G. J. 1992. Barite: a comparison of grades and tonnages for bedded barite deposits with and without associated base-metals sulfides. *US Geological Survey Bulletin*, 1877, B1–B12.

Osborne, D. G. 1988. *Coal Preparation Technology*. Lancaster: Kluwer Academic Publishers.

Osterman, J. W. 1990. Evaluating the impact of municipal water fluoridation on the aquatic environment. *American Journal of Public Health*, 80, 1230–1235.

OTA 1978. *Management of Fuel and Nonfuel Minerals in Federal Lands: Current Status and Issues*. Washington, DC: Office of Technology Assessment, US Congress.

OTA 1987. *Marine Minerals: Exploring Our New Ocean Frontier*. Washington, DC: Congress of the United States, Office of Technology Assessment.

OTA 1988. Copper: technology and competitiveness. *US Government Printing Office, Office of Technology Assessment*, OTA-E–367.

Owens, M. 2012. Federal judge ruling distinguishes coalbed methane as separate from the coal it's found in. [Online] Available: http://www.tricities.com/news/article_a318d3ab-6a1c-5adf-94c0-df31c7fb725a.html.

Pacyna, J. M., Pacyna, E. G. 2001. An assessment of global and regional emissions of trace metals to the atmosphere from anthropogenic sources worldwide. *Environmental Reviews*, 2001, 9, 269–298

Palencia, C. M. 1985. Molybdenum availability – market economy countries. *US Bureau of Mines Information Circular*, 9044, 30.

Paone, R. M. 1992. *Strategic Nonfuel Minerals and Western Security*. New York, NY: University Press of America.

Papke, K.G. 1984. Barite in Nevada. *Nevada Bureau of Mines and Geology Bulletin*, 95.

Papp, J. F. 1985. Chromium. *US Bureau of Mines Mineral Facts and Problems*, 675, 139–156.

Papp, J. F. 2007. Chromium – a national mineral commodity perspective. [Online] Available: http://pubs.usgs.gov/of/2007/1167/.

Papp, J. F. 2013a. Chromium. *US Geological Survey Minerals Yearbook*, Advance Release for 2011.

Papp, J.F. 2013b. Niobium. *US Geological Survey Minerals Yearbook*.

Papp, J. F. 2013c. Tantalum. *US Geological Survey Mineral Commodity Summary*.

Parks Canada 2009. Parks Canada, National Parks System Plan 2009. [Online] Available: http://www.pc.gc.ca/docs/v-g/nation/nation1.aspx.

Parra, F. 2010. *Oil Politics: A Modern History of Petroleum*. London: I. B.Tauris.

Parrenin, F., Barnola, J.-M., Beer, J. *et al.* 2007. The EDC3 chronology for the EPICA Dome C ice core. *Climate of the Past*, 3, 483–497.

Parsons, R. B. 1981. Earned depletion and the mining industry. *CIM Bulletin*, 74, no. 829, 132–138.

Paruchuri Y., Siuniak A., Johnson N., *et al.* 2010. Occupational and environmental mercury exposure among small-scale gold miners in the Talensi–Nabdam district of Ghana's Upper East region. *Science of the Total Environment*, 408, 6079–6085.

Patrick, L. 2006. Lead toxicity, a review of the literature. Part I: exposures, evaluations, and treatment. *Alternative Medicine Review: a Journal of Clinical Therapeutic*, 11, 2–22.

Patterson, S. H. 1977. Aluminum from bauxite: are there alternatives? *American Scientist*, 65, 345–351.

Patterson, S. H., Kurtz, H. F., Olson, J. C., Neeley, C. L. 1986. World's bauxite resources. *US Geological Survey Professional Paper*, 1076-B, 1–151.

Patterson, S. M., Murray, H. 1983. Clays. In Lefond, S. J. (ed.), *Industrial Minerals and Rocks*, 5th edn. New York, NY: Society of Mining Engineers, pp. 585–599.

Peabody, C. E., Einaudi, M. T. 1992. Origin of petroleum and mercury in the Culver-Baer cinnabar deposit, Mayacmas district, California. *Economic Geology*, 87, 1078–1103.

Peacy, J. G., Davenport, W.G. 1979. *The Iron Blast Furnace, Theory and Practice*. Oxford: Pergamon.

Pearson, D. G., Davies, G. R., Nixon, P. H. 1989. Graphitized diamonds from a peridotite massif in Morocco and implications for anomalous diamond occurrences. *Nature*, 338, 60–62.

Peck, M. C., ed. 1988. *The World Aluminum Industry in a Changing Era*. Washington, DC: Resources for the Future.

Peeling, G. R., Kendall, G., Shinya, W., Keyes, R. 1992. *Canadian Policy Perspective on Environmental Aspects of Minerals and Metals*. London: Elsevier.

Pelham, L. 1985. Fluorspar. *US Bureau of Mines Mineral Facts and Problems*, 675, 277–290.

Pemberton, M. 2012. Alaska's oil pipeline makes 35-year mark. *Bloomberg BusinessWeek*, June 21.

Perez-Fodich, A., Reich, M., Alvarez, F., *et al.* 2014. Climate change and tectonic uplift triggered the formation of the Atacama Desert's giant nitrate deposits. *Geology*, 42, 251–254.

Peréz, L., Medina-ramón, M., Künzli, N., *et al.* 2009. Size fractionate particulate matter, vehicle traffic, and case-specific daily mortality in Barcelona (Spain). *Environmental Science and Technology*, 43, 4707–4713.

Perkins, G., Sahajwallaa, V. 2007. Modeling of heat and mass transport phenomena and chemical reaction in underground coal gasification. *Chemical Engineering Research and Design*, 85, 329–343.

Perrino, C. 2010. Atmospheric particulate matter. Proceedings of a CISB Minisymposium, March, 2010.

Perry, H. 1974. The gasification of coal. *Scientific American*, 230, 19–25.

Persistence Market Research 2014. Bromine market: global industry analysis and forecast 2014–2020. [Online] Available: http://www.persistencemarketresearch.com/market-research/bromine-market.asp.

Peters, G. P., Minx, J. C., Weber, C. L., Edenhofer, O. 2011. Growth in emission transfers via international trade from 1990–2008. *Proceedings of the National Academy of Sciences*, 108, 8903–8908.

Petersen, U. 1971. Laterite and bauxite formation. *Economic Geology*, 66, 1070–1072.

Petersen, U., Maxwell, R. S. 1979. Historical mineral production and price trends. *Mining Engineering*, January, 25–34.

Peterson, C. H., Rice, S. D., Short, J. W., *et al.* 2003. Long-term ecosystem response to the Exon Valdez oil spill. *Science*, 302, 2082–2086.

Petkof, B. 1985. Beryllium. *US Bureau of Mines Mineral Facts and Problems*, 75–82.

Phillips, G. N., Law, J. D. M. 2000. Witwatersrand gold fields: geology, genesis, and exploration. *Society of Economic Geologists, Gold in 2000*, 439–495.

Phillips, T. A. 1977. An economic evaluation of a process to separate raw urban refuse into its metal, mineral and energy components. *US Bureau of Mines Information Circular*, 8732.

Pickard, A. 2012. Vale moves forward on Victor-Capre project. Northern Life CA, June 15. [Online] Available: http://www.northernlife.ca/news/localNews/2012/06/15-vale-victor-capre-sudbury.aspx.

Pirrone, N., Cinnirella, S., Feng, X., *et al.* 2010. Global emissions to the atmosphere from anthropogenic and natural sources. *Atmospheric Chemistry and Physics*, 10, 5951–5964.

Pitcairn, I. K., Teagle, D. A. H., Craw, D., *et al.* 2006. Sources of metals and fluids in orogenic gold deposits: insights from the Otago and Alpine schists, New Zealand. *Economic Geology*, 101, 1525–1546.

Pitfield, P., Brown, T., Gunn, G., Rayner, D. 2011. Commodity profiles: tungsten. British Geological Survey. {online} Avaialbale: www.mineralsUK.com.

Planavsky, N. ., Asael, D., Hofmann, A., *et al.* 2014, Evidence for oxygenic photosynthesis half a billion years before the Great Oxidation Event. *Nature Geoscience*, 7, 283–286.

PLS 1988. Public Land Statistics. Bureau of Land Management, US Department of Interior. [Online] Available: http://www.blm.gov/public_land_statistics/.

Plumlee, G. S., Goldhaber, M. B., Rowan, E. L. 1995. The potential role of magmatic gases in the genesis of Illinois–Kentucky fluorspar deposits; implications from chemical reaction path modeling. *Economic Geology*, 89, 1860–1882.

Plummer, B. 2013. Everything you need to know about the EPA's carbon limits for new power plants. *Washington Post*, September 20.

Plunkert, P. A. 1985a. Antimony. *US Bureau of Mines Mineral Facts and Problems*, 675, 33–42.

Plunkert, P. A. 1985b. Cadmium. *US Bureau of Mines Mineral Facts and Problems*, 675, 111–120.

Plunkert, P. A. 1985c. Germanium. *US Bureau of Mines Mineral Facts and Problems*, 675, 317–323.

Plunkert, P. A. 1985d. Thallium. *US Bureau of Mines Mineral Facts and Problems*, 675, 829–834.

Pocock, S. J., Shaper, A. G., Powell, P. 1985. The British regional heart study: cardiovascular disease and water quality. In Caladrese, E. J., Tuthill, R. W., Condie, L. (eds.), *Inorganics in Drinking Water and Cardiovascular Diseases*. Princeton, NJ: Princeton Science Publishing, pp. 11–127.

Podolsky, G., Keller, D. P. 1994. Pigment: iron oxide. In Carr, D. D. (ed.), *Industrial Minerals and Rocks*, 6th edn. Littleton, CO: Society for Mining, Metallurgy, and Exploration, pp. 765–781.

Polamski, J., Fluchler, H., Blasner, P. 1982. Accumulation of airborne fluoride in soils. *Journal of Environmental Quality*, 11, 451–461.

Polinares 2012. Fact sheet: lithium. [Online] Available: http://www.polinares.eu/docs/d2-1/polinares_wp2_annex2_factsheet4_v1_10.pdf.

Polmar, N., White, M. 2012. *Project Azorian: The CIA and the Raising of the K-129*. Annapolis, MD: US Naval Institute.

Polmear, I. J. 1981. *Light Alloys*. Metals Park, OH: American Society of Metals.

Polyak, D. E. 2015. Iodine. *US Geological Survey Mineral Yearbook*.

Polyak, D. E. 2013a. Molybdenum. *US Geological Survey Minerals Yearbook*.

Polyak, D. E. 2013b. Vanadium. *US Geological Survey Minerals Yearbook*.

Poonian, C. 2003. The effects of the 1991 Gulf War on the marine and coastal environment of the Arabian Gulf: impact, recovery and future prospects. M.Sc. Thesis, King's College, London.

Popkin, R. 1985. Fighting waste from gold mining. *EPA Journal*, 11.

Porcella, D. 1990, Mercury in the environment. *EPRI Journal*, April/May, pp. 46–48.

Porter, J. 2014. "Extremely high" levels of PCBs, lead at Raleigh scrap recycling business. *Indy Week*, April 9.

Potter, M. J. 2006. Feldspars. In Kogel, J. E., Trivedi, N. C., Barker, N. C., Krukowski, S. T. (eds.), *Industrial Minerals and Rocks*, 7th edn. Littleton, CO: Society for Mining, Metallurgy, and Exploration, pp. 451–460.

Powell, J. D. 1988. Origin and influence of coal mine drainage on streams of the United States. *Environmental Geology and Water Science*, 11, 141–152.

Power, W. R. 1983. Dimension and cut stone. In Lefond, S. J. (ed.), *Industrial Minerals and Rocks*, 5th edn. New York, NY: Society of Mining Engineers.

Prain, R. 1975. *Copper: The Anatomy of an Industry*. London: Mining Journal Books.

Prather, B. E. 1991. Petroleum geology of the Upper Jurassic and Lower Cretaceous, Baltimore Canyon Trough, Western North Atlantic Ocean. *American Association of Petroleum Geologists Bulletin*, 75, 258–277.

Presser, T., Barnes, I. 1985. Dissolved constituents including selenium in water in the vicinity of Kesterton National Wildlife Refuge and West Grassland, Fresno and Merced Counties, California. US Geological Survey Water Resources Investigations Report, 85–4220.

Pressler, J. W. 1985. Gem stones. *US Bureau of Mines Mineral Facts and Problems*, 675, 305–316.

Pressler, J. W., Pelham, L. 1985. Lime, calcium and calcium compounds. *US Bureau of Mines Mineral Facts and Problems*, 675, 453–460.

Pretorius, D. A. 1981. Gold, geld, gilt: future supply and demand. *Economic Geology*, 76, 2032–2042.

Price, L., Hasanbeigi, A., Aden, N. 2011. A comparison of Iron and Steel Production Energy Intensity in China and the US. ACEEE Summer Study on Energy Efficiency in Industry. [Online] Available: http://escholarship.org/uc/item/8mb2d99v#page-1.

Prieto, C. 1973. *Mining in the New World*. New York, NY: McGraw-Hill.

Prindle, D. F. 1981. *Petroleum Politics and the Texas Railroad Commission*, Austin, TX: University of Texas Press.

Prior, D. B., Doyle, E. H., Kaluza, M. J. 1989. Evidence for sediment eruption on deep sea floor, Gulf of Mexico. *Science*, 243, 517–520.

Property Saskatchewan 2014. Using new technology to overcome ancient potash geology in Saskatchewan. [Online] Available: http://prosperitysaskatchewan.wordpress.com/2012/06/21/using-new-technology-to-overcome-ancient-potash-geology-in-saskatchewan/.sauet

Prospero, J.M. 1999, Long-range transport of mineral dust in the global atmosphere: impact of African dust on the environment of the southeastern United States. *Proceedings of the National Academy of Science*, 96, 3396–3403.

Prud'homme, M., Krukowski, S. T. 2006. Potash. In Kogel, J. E., Trivedi, N. C., Barker, N. C., Krukowski, S. T. (eds.), Industrial Minerals and Rocks, 7th edn. Littleton, CO: Society for Mining, Metallurgy and Exploration, pp. 723–742.

Prugger, F. F., Prugger, A. A. 1991. Water problems in Saskatchewan potash mining: what can be learned from them? *Canadian Institution of Mining and Metallurgy Bulletin*, 84, 58–65.

Pura, R. 1986. Malaysia's tin scheme stuns the industry. *The Wall Street Journal*, September 25.

Quantin, C., Becquer, T., Rouiller, J. H., Berthelin, J. 2001. Oxide weathering and trace metal release by bacterial reduction in a New Caledonia Ferralsol. *Biogeochemistry*, 53, 323–340.

Qian, W. 2013. Vast deposits of flammable ice found. *CRI English China Daily*, December 18.

Radetsky, M. 1985. *State Mineral Enterprises in Developing Countries: Their Impact on International Mineral Markets*. Baltimore, MD: Johns Hopkins University Press.

Rafelski, J. and Jones, S. E., 1987. Cold nuclear fusion. *Scientific American*, 257, 84–89.

Raisbeck, M. F., Siemion, R. S., Smith, M. A. 2006. Modest copper supplementation blocks molybdenosis in cattle. *Journal Veterinary Diagnostic Investigations*, 18, 566–582.

Rampino, M. R., Self, S., Stothers, R. B. 1988. Nuclear winters. *Annual Review of Earth and Planetary Science*, 16, 73–99.

Randall, R. W. 1972. *Real del Monte, A British Mining Venture in Mexico*, Austin, TX: University of Texas Press.

Rankin, J. 2012. Energy use in metal production. [Online] Available: https://publications.csiro.au/rpr/download?pid=csiro:EP12183&dsid=DS3.

Ratner, M., Tiemann, M. 2014. An overview of unconventional oil and natural gas: resources and federal actions. Congressional Research Service. [Online] Available: http://fas.org/sgp/crs/misc/R43148.pdf.

Rauch, J. N., Pacyna, J. M. 2009, Earth's global Ag, Al, Cr, Cu, Fe, Ni, Pb, and Zn cycles. *Global Biogeochemical Cycles*, 23, GB2001, DOI:10.1029/2008GB003376.

Ravenscroft, P. 2007. Predicting the global distribution of arsenic pollution in groundwater. Royal Geographical Society Annual International Conference. [Online] Available: http://www.geog.cam.ac.uk/research/projects/arsenic/symposium/S1.2_P_Ravenscroft.pdf.

Reading, H. G., ed. 1986. *Sedimentary Environments and Facies*. Oxford: Blackwell.

Reckling, K., Hoy, R. B., Lefond, S. J. 1983. Diamonds. In Lefond, S. J. (ed.), *Industrial Minerals and Rocks*, 5th edn. New York, NY: Society of Mining Engineers, pp. 653–671.

Red Book 1991. *Annual, Uranium Resources, Production and Demand*. Paris: Organization for Economic Cooperation and Development.

Reed, S., Kramer, A. 2013. Chevron and Ukraine set shale gas deal. *The New York Times*, Novemeber 5.

Retallack, G. 2008. Cool-climate or warm-spike lateritic bauxites at high latitudes? *Journal of Geology*, 116, 761–778.

Reuters 2011. Flue gas desulfurization – future key lime market driver according to Merchant Research and Consulting Ltd. *Reuters*, June 8.

Reuters 2013. Dubai gold trade hurt by new Indian import tariffs.

Rice, D. R. 1986. Oil and gas assessment: methods and applications. *American Association of Petroleum Geologists, Studies in Geology*, 21.

Rich, R. A., Holland, H. D., Petersen, U. 1977. *Hydrothermal Uranium Deposits*. Amsterdam: Elsevier.

Richards, H. G., Savage, D., Andrews, J. N. 1992. Granite–water reactions in an experimental hot dry rock geothermal reservoir, Rosemanowes test site, Cornwall, UK. *Applied Geochemistry*, 7, 193–222.

Richards, M., Mumin, A. H. 2013. Magmatic–hydrothermal processes within an evolving Earth: iron oxide–copper–gold and porphyry Cu±Mo±Au deposits. *Geology*, 41, 767–770.

Riding, A. 1991. Accord bans oil exploration in the Antarctic for 50 years. *The New York Times*, October 5.

Riordan, P. H. 1981. *Geology of Asbestos Deposits*. New York, NY: Society of Mining Engineers.

Robb, L. J., Meyer, F. M. 1991. A contribution to recent debate concerning epigenetic versus syngenetic mineralization processes in the Witwatersrand basin. *Economic Geology*, 86, 396–401.

Robelius, F. 2007. Giant oil fields – the highway to oil: giant oil fields and their importance for future oil production. Ph.D. Dissertation, Uppsala University.

Roberts, J. Oil and the Iraq war of 2003. *International Research Center for Energy and Economic Development, Occasional Paper*, 37.

Robertson, A. H. F., Hudson, J. D. 1973. Cyprus umbers: chemical precipitates on a Tethyan ocean ridge. *Earth and Planetary Science Letters*, 18, 93–101.

Robinson, D. 2012. The nickel laterite challenge. [Online] Available: http://www.miningaustralia.com.au/features/the-nickel-laterite-challenge.

Robinson, G. R., Jr. 2004. Trends in availability of aggregate. *Mining Engineering*, 56, 17–24.

Robson, A. 2014. Australia's carbon tax: an economic evaluation. *Economic Affairs*, 34, 35–45.

Rodbell, D. T., Delman, E. M., Abbott, M. B., Besonen, M. T, Tapia, P.M. 2014. The heavy metal contamination of Lake Junin National Reserve, Peru: an unintended consequence of the juxtaposition of hydroelectricity and mining. *GSA Today*, DOI: 10.1130/GSATG200A.1.

Roddy, D., Gonzalez, G. 2010. Underground coal gasification (UGC) with carbon capture and storage (CCS). In Hester, R. E., Harrison, R. M. (eds.), *Issues in Environmental Science and Technology*. Cambridge: Royal Society of Chemistry, pp. 102–125.

Rodell, M., Velicogna, I., Famiglietti, J. S. 2009. Satellite-based estimates of groundwater depletion in India. *Nature*, 460, 999–1002.

Rodriguez, A. L., Stulberg, A., Kim, D., *et al.* 2013. The great debate: mining in Latin America. Journal of International Affairs, Columbia University. [Online] Available: http://jia.sipa.columbia.edu/online-articles/great-debate-mining-in-latin-america/.

Roebroeks, W., Siera, M. J., Nielsena, T. K., *et al.* 2012. Use of red ochre by early Neanderthals. *Proceedings of the National Academy of Sciences*, 109, 1889–1894.

Rogers, C. E., Tomita, A. V., Trowbridge, P. R., *et al.* 1997. Hair analysis does not support hypothesized arsenic and chromium exposure from drinking water in Woburn, Massachusetts. *Environmental Health Perspectives*, 105, 1090–1097.

Rogich, D. G., Matos, G. R. 2008. The global flows of metals and minerals. *US Geological Survey Open-File Report*, 2008-1355.

Rohan 2012. Bromine market worth 776,844.91 tons by 2018. Market Watch, *The Wall Street Journal*.

Romero, R. A. 1991. Aluminum toxicity. *Environmental Science Technology*, 25, 1658f.

Romero, S., Thomas, L. 2014. Brazil's star, Petrobras, is hobbled by scandal and stagnation. *The New York Times*, April 15.

Root, D. H., Attansai, E. D. 1992. Oil field growth in the United States – how much is left in the barrel? *US Geological Survey Circular*, 1074.

Rose, A. W., Hawkes, H. E., Webb, J. S. 1979. *Geochemistry in Mineral Exploration*, New York, NY: Academic Press.

Rosner, D., Markowitz, G. 2006. Deadly dust: silicosis and the politics of twentieth century America. Ann Arbor, MI: University of Michigan Press.

Ross, M. 1987. Minerals and health: the asbestos problem. In Peirce, H. W. (ed.), *21st Forum on Geology of Industrial Minerals*. Tuscon, AZ: Arizona Bureau of Geology and Mineral Technology, Special Paper 4, pp. 83–89.

Ross, T. 2013. EU plan for fracking law threatens UKs shale gas boom. *The Telegraph*, December 15.

Rossman, G. R. 2009. The geochemistry of gems and its relevance to gemology: different traces, different prices. *Elements*, 5, 159–162.

Rossman, K. J., Chisholm, W., Hong, S., Candelone, J. -P, Boutron, C. F. 1997. Lead from Carthaginian and Roman Spanish mines isotopically identified in Greenland ice dated from 600 BC to 300 AD. *Environmental Science and Technology*, 31, 3413–3416.

Rossman, M. D., Preuss, O. P., Powers, M. B. 1991. *Beryllium: Biomedical and Environmental Aspects*. Baltimore, MD: Williams and Wilkins.

Rothschild, B. M. 2005. History of syphilis. *Infectious Diseases Society of America*, 40, 1454–1463.

Round, F. E., Crawford, R. M., Mann, D. G., Repak, A. J. 1990. The diatoms. *American Scientist*, 79, March/April, 174.

Rowan, E. L., Engle, M. A., Kirby, C. S., Kraemer, T. F. 2011. Radium content of oil- and gas-field produce waters in the northern Appalachian basin (USA): summary and discussion of data. *US Geological Survey Scientific Investigations Report*, 2011–5135.

Rowe, G. W. 1988. Well contaminated by water softener regeneration discharge water. *Journal of Environmental Health*, 50, 272–276.

Rudnick, R. L., Gao, S. 2003. Composition of the continental crust. *Treatise on Geochemistry*, 3, 1–64.

Ruiz., K., Kesler, S. E., Jones, L. M., Sutter, J. F. 1985. Geology and geochemistry of the Las Cuevas fluorite deposits, San Luis Potosi, Mexico. *Economic Geology*, 80, 1200–1209.

Russel, A. 1989. Pyrites as a sulphur source, down but not out. *Industrial Minerals*, 261, 41–52.

Russell, O. L. 1981. Oil shale. *Mining Engineering*, 33, 29–54.

Rybach, L., Muffler, L. J. P. 1981. *Geothermal Systems: Principals and Case Histories*. New York, NY: John Wiley.

Rytuba, J. J. 2003. Mercury from mineral deposits and potential environmental impact. *Environmental Geology*, 43, 326–338.

Rytuba, J. J., John, D. A., Foster, A. L., Ludington, S. D., Kotlyar, B. 2004. Hydrothermal enrichment of gallium in zones of advanced argillic alteration; examples from the Paradise Peak and McDermitt ore deposits, Nevada. *US Geological Survey Bulletin*, 2209C, 1.

Rytuba, J. L., Heropoulos, C. 1992. Mercury: an important by-product of epithermal gold systems. *US Geological Survey Bulletin*, 1187D, 1–23.

SACSIS. 2014. The nationalisation debate: South Africa's mineral resources equivalent to one million rand per citizen. [Online] Available: http://sacsis.org.za/site/article/1924.

Sainsbury, C. L. 1969. Tin resources of the world. *US Geological Survey Bulletin*, 1301, 55.

Salt Institute 2013. Safe and sustainable snowfighting. Snowfighter's handbook. [Online] Available: http://www.saltinstitute.org/wp-content/uploads/2013/07/Snowfighters_HB_2012.pdf.

Salter, R. S. 1987. *International Symposium on Gold Metallurgy*, Elmsford, NY: Pergamon.

Saltman, J. 2014. *Drinking Water: A History*. New York, NY: Overlook Press.

Salzarulo, W. P. 1997. An analysis of the importance of the percentage depletion allowance for a construction aggregates company. *Oil and Gas Tax Quarterly*, 45, 445–466.

Sampson, A. 1975. The Seven Sisters. New York, NY: Bantam.

Samuel, C. R., Liu, X. 2009. *Advanced Drilling Engineering*. Houston, TX: Gulf Publishing.

Sanchez-Mejorada, R. V. 2000. Mining law in Mexico. *Mineral Resources Engineering*, 9, 129–139.

Sander, W. 2007. Sierra foothills companies find gold in gravel mining. Sacramento Business Journal. [Online] Available: http://www.bizjournals.com/sacramento/stories/2007/09/24/focus2.html?page=all.

Sandrea, R., Sandrea, I. 2010. Deepwater crude oil output: how large will the uptick be. *Oil and Gas Journal*, 108.41.

Sankaran, K. K. 2006. Enhancing innovation and competitiveness through investments in fundamental research. [Online] Available: http://enhancinginnovation.wustl.edu/Sankaran.pdf.

Sauer, J. R. 1988. *Brazil – Paradise of Gemstones*. Rio de Janeiro: Sauer and Co.

Sauerwein, K. 2004. Less mining, more business. *Los Angeles Times*, June 21.

Saulnier, G. J., Goddard, K. E. 1982. Usage of mathematical models to predict impacts of mining energy minerals on the hydrologic system in northwestern Colorado. *Mining Engineering*, 34, 284–292.

Sautter, C., Poletti, S., Zhang, P., Gruissem, W. 2006. Biofortification of essential nutritional compounds and trace elements in rice and cassava. *Proceedings of the Nutrition Society*, 65, 153–159.

Sawkins, F. J. 1989. Anorogenic felsic magmatism, rift sedimentation, and giant Proterozoic Pb–Zn deposits. *Geology*, 17, 657–660.

Sawkins, F. J. 1990. *Metal Deposits in Relation to Plate Tectonics*. New York, Springer-Verlag.

Scanlon, B. R., Faunt, C. C., Longuevergne, L., *et al.* 2012. Groundwater depletion and sustainability of irrigation in the US High Plains and Central Valley. *Proceedings of the National Academy of Sciences*, 109, 9320–9325.

Schell, D. 1985. Road salt contaminates well, cases health hazard. *Journal of Environmental Health*, 47, 202–203.

Schenck, G. H. K., Torries, T. F., Jr. 1985. Aggregate-crushed stone. In Lefond, S. J. (ed.), *Industrial Minerals and Rocks*, 5th edn. New York, NY: Society of Mining Engineers, pp. 60–80.

Schernikau, L. 2010. *Economics of the International Coal Trade. The Renaissance Of Steam*. Berlin: Springer.

Schilling, P., Jr. 2013 Hard-fought United States vs. Reserve Mining chanted environmentalism. Minn Post. [Online] Available: https://www.minnpost.com/minnesota-history/2013/02/hard-fought-united-states-vs-reserve-mining-changed-environmentalism.

Schmalz, R. F. 1969. Deep-water evaporite deposits: a genetic model. *American Association of Petroleum Geologists Bulletin*, 53, 798–823.

Schmalz, R. F. 1991. The Mediterranean salinity crisis: alternative hypothesis. *Carbonates and Evaporites*, 6, 121–126.

Schmidt, M. G. 1989. *Common Heritage or Common Burden? The United States Position on the Development of a Regime for Mining in the Law of the Sea Convention*, Oxford: Clarendon Press.

Schneider, K. 1990. US wrestles with gap in radiation exposure rules. *The New York Times*, December 26.

Schnoor, J. L. 2014, Re-emergence of emerging contaminants. *Environmental Science and Technology*, DOI: 10.1021/es504256j.

Schobert, H. H. 1987. *Coal*. Washington, DC: American Chemical Society.

Schoenfeld, G. 1990. Rad storm rising. The Atlantic. [Online] Available: http://www.theatlantic.com/magazine/archive/1990/12/rad-storm-rising/306380/

Schreier, H. 1989. *Asbestos in the Natural Environment*. Amsterdam: Elsevier.

Schreiber, B. C., Babel, M., Lugli, S. 2007. Introduction and overview. *Journal of Geological Society, Special Publication*, 285, 1–13.

Schulte, R. F., Taylor, R. D., Piatak, N. M., Seal, R. R., II 2012. Stratiform chromite deposit model. *US Geological Survey Scientific Investigations Report*, 2010–5070-E, xiv.

Schulz, T. L. 2005. Westinghouse AP1000 advanced passive plant. *Nuclear Engineering and Design*, 236, 1547–1557.

Schumann, W. 2006. *Gemstones of the World*. New York, NY: Sterling.

Schwarz-Schampera, U., Herzig, P. M. 2002. *Indium*. Berlin: Springer.

Schwela, U. 2011. *TIC Statistics and Transport Project*. Lasne: Tantalum–Niobium International Study Center.

Seddon, M. 2012. World tungsten report. [Online] Available: http://blog.metal-pages.com/wp-content/uploads/2012/09/World Tungsten_201209.pdf.

Sedgwick, J. 1991. Strong but sensitive. *Atlantic*, April.

Self, S. J., Reddy, B. V., Rosen, M. A. 2012. Review of underground coal gasification technologies and carbon capture. *International Journal of Energy and Environmental Engineering*, 3, 16.

Selfin, Y., Snook, R., Gupta, H. 2011. The impact of sovereign wealth funds on economic success., PriceWaterhouseCoopers, October. [Online] Available: http://www.pwc.co.uk/en_UK/uk/assets/pdf/the-impact-of-sovereign-wealth-funds-on-economic-success.pdf.

Seltenrich, N. 2013. Incineration versus recycling: in Europe, a debate over trash. Environment 360. [Online] Available: http://e360.yale.edu/feature/incineration_versus_recycling__in_europe_a_debate_over_trash/2686/.

Semkin, R. G., Kramer, J. R. 1976. Sediment geochemistry of Sudbury area lakes. *Canadian Mineralogist*, 14, 73–90.

Sengupta, P. 2013. Potential health impacts of hard water. *International Journal of Preventive Medicine*, 4, 866–875.

Seramur, K., Cowan, E., Lane, J. D. 2012. Transport and deposition of coal fly ash in the Watts Barr reservoir system from the

Kingston, Tennessee ash storage cell failure in 2008. *Abstracts with Programs – Geological Society of America*, 44, 464–465.

Sessitsch, A., Kuffner, M., Kidd, P., *et al.* 2013. The role of plant-associated bacteria in the mobilization and phytoextraction of trace elements in contaminated soils. *Soil Biology and Biochemistry*, 60, 182–194.

Severinghaus, N., Jr. 1983. Fillers, filters and absorbents. In Lefond, S. J. (ed.), *Industrial Minerals and Rocks*, 5th edn. New York, NY: Society of Mining Engineers, pp. 204–235.

Shabecoff, P. 1990. Bush cuts back areas off coasts open for drilling. *The New York Times*, June 27.

Shannon, S. S., JR., 1983. Rare earths and thorium. In Lefond, S. J. (ed.), *Industrial Minerals and Rocks*, 5th edn. New York, NY: Society of Mining Engineers, pp. 231–267.

Shaw, R., Goodenough, K., Gunn, G., Brown, T., Rayner, D. 2011. Commodity profiles: niobium–tantalum. *British Geological Survey*. [Online} Available: www.mineralsUK.com.

Shawe, D. R. 1981. US Geological Survey workshop on nonfuel mineral-resource appraisal of wilderness and CUSMAP areas. *US Geological Survey Circular*, 848.

Shedd, K. B. 2013a. Cobalt. *US Geological Survey Minerals Yearbook*.

Shedd, K. B. 2013b. Tungsten. *US Geological Survey Minerals Yearbook*.

Sheng, J. 2001. Main metallogenic characteristics of tungsten deposits in China. *Proceedings of the Sixth Biennial SGA-SEG Meeting*, 6, 1141–1144.

Shephard, M., Golden, D., Komai, R., Marasky, T. 1985. Utility solid waste: Managing the by-products of coal combustion. *Electric Power Research Institute Journal*, 10, no. 8, 20–35.

Shigley, J. E., Dirlam, D. M., Schmetzer, K., Jobbins, E. A. 1990. Gem review. *Gems and Gemology*, 26, 4–12.

Shirey, S. B., Shigley, J. E. 2013. Recent advances in understanding the geology of diamonds. *Gems and Gemology*, 49, 188–222 .

Shogren, E. 2010. Dishes still dirty? Blame phosphate-free detergent. National Public Radio. [Online] Avaialable: http://www.npr.org/2010/12/15/132072122/it-s-not-your-fault-your-dishes-are-still-dirty.

Shulman, P. A. 2011. The making of a tax break: the oil depletion allowance, scientific taxation, and natural resources policy in the early twentieth century. *Journal of Policy History*, 23, 281–322.

Sibson, R. H., Moore, J., McM., Rankin, A. H. 1975. Seismic pumping: a hydrothermal fluid transport mechanism. *Journal of the Geological Society of London*, 131, 653–659.

Sibson, R. H., Robert, F., Poulsen, K. H. 1988. High-angle reverse faults, fluid-pressure cycling, and mesothermal gold-quartz deposits. *Geology*, 16, 551–555.

Siegel, A. 1960. Iron oxide pigments. In Gillson, J. L. (ed.), *Industrial Minerals and Rocks*, 3rd edn. New York, NY: Society of Mining Engineers, pp. 585–593.

Sievert, S. M., Kiene, R. P., Schulz-vogt, H.N. 2007. The sulfur cycle. *Oeanography*, 20, 117–123.

Sigel, A., Sigel, H., Sigel, K. O. 2013. *Cadmium: From Toxicity to Essentiality*. New York, NY: Springer.

Sillitoe, R. H. 2005. Supergene oxidized and enriched porphyry copper and related deposits. In Hedenquist, J. W., Thompson, J. F. H., Goldfarb, R. J., Richards, J. P. (eds.), *Economic Geology, 100th Anniversary Volume*. Littleton, CO: Society of Economic Geologists, pp. 723–768.

Sillitoe, R. H. 2010. Porphyry copper systems. *Economic Geology*, 105, 3–41.

Sillitoe, R. H., Hedenquist, J. H. 2003. Linkages between volcanotectonic settings, ore-fluid compositions and epithermal precious metal deposits. *Society of Economic Geologists Special Publication*, 10, 315–343.

Sillitoe, R. H., Halls, C., Grant, J. N. 1975. Porphyry tin deposits in Bolivia. *Economic Geology*, 70, 913–927.

Silverstein, K. 2013. Secret oil company memos on pollution in Louisiana. *Harper's Blog*, October 11.

Simandl, G. J. 2009. Fluorspar market and selected fluorite-bearing deposits, British Columbia, Canada. *Geofile*, 2009–03.

Simmons, M. 2002. The world's giant oilfields. White Paper from January 9, 2002. [Online] Available: http://www.firstenercastfinancial.com/pdfs/3221.pdf.

Simmons, S. F., White, N. C., John, D. A. 2005. Geologic characteristics of epithermal precious and base metal deposits. In Hedenquist, J. W., Thompson, J. F. H., Goldfarb, R. J., Richards, J. P. (eds.), *Economic Geology, 100th Anniversary Volume*. Littleton, CO: Society of Economic Geologists, pp. 485–522.

Simms, R., Moolji, S. 2011. In the belly of the machine: indigenous mining experience in Panama. Report Prepared for McGill University, Centro de Incidencia Ambiental (CIAM). Washington, DC: The Smithsonian Institute and the Comarca Ngobe-Bugle.

Simon, A. C., Ripley, E. M. 2011. The role of magmatic sulfur in the formation of ore deposits. *Reviews in Mineralogy and Geochemistry*, 73, 513–578.

Simon, A. C., Pettke, T. 2009. Platinum solubility and partitioning in a felsic-melt-vapor–brine assemblage. *Geochimica et Cosmochimica Acta*, 73, 438–454.

Simon, W. 1991. Mexican silver mining. *Engineering Mining Journal*, 192, 18–22.

Simpson, I. J., Anderson, M. P. S., Meinardi, S., *et al.* 2012. Long-term decline of global atmospheric ethane concentrations and implications for methane. *Nature*, 488, 490–494.

Sinclair, W. D., Jonasson, I. R., Kirkham, R. V., Soregaroli, A. E. 2009. Rhenium and other platinum-group metals in porphyry deposits. *Geological Survey of Canada Open-File Report*, 6181.

Sinclair, A. J., Blackwell, G. H. 2006. *Applied Mineral Inventory Estimation*. Cambridge: Cambridge University Press.

Sindeeva, N. D. 1964. *Mineralogy and Types of Deposits of Selenium and Tellurium*. New York, NY: John Wiley Interscience.

Singer, D. A., Drew, L. J. 1976. The area of influence of an exploratory drill hole. *Economic Geology*, 71, 642–647.

Singer, D. A. 1977. Long-term adequacy of metal resources. *Resources Policy*, 3, 127–133.

Singer, D. A., Ovenshine, A. T. 1979. Assessing metallic resources in Alaska. *American Scientist*, 67, 582–589.

Singer, D. A., Kouda, R. 1988. Integrating spatial and frequency information in the search for Kuroko deposits of the Hokuroko district, Japan. *Economic Geology*, 83, 18–29.

Singer, D. A., Menzie, W. D. 2010. *Quantitative Mineral Resource Assessments: An Integrated Approach*, New York, NY: Oxford University Press.

Singer, D. A. 2013. The lognormal distribution of metal resources in mineral deposits. *Ore Geology Reviews*, 55, 80–86.

Singhal, R. K., Kolada, R. 1988. Surface mining of oil sands in Canada: development in productivity improvement. *Mining Engineering*, 38, 267–275.

Sinkankas, J. 1989. *Emerald and Other Beryls*. Boulder, CO: Geoscience Press.

Skinner, B. J. 1975. A second iron age ahead? *American Scientist*, 64, 258–269.

Skinner, H. C. W., Ross, M., Fronde, C. 1988. *Asbestos and Other Fibrous Minerals*. New York, NY: Oxford University Press.

Slack, J. F., Causey, J. D., Eppinger, R. G., *et al.* 2010. Co–Cu–Au deposits in metasedimentary rocks – a preliminary report. *US Geological Survey Open-File Report*, 2010-1212.

Slansky, M. 1986. *Geology of Sedimentary Phosphates*. Amsterdam: Elsevier.

Slichter, L. B. 1960. The need of a new philosophy of prospecting. *Mining Engineering*, 14, 570–576.

Smith, C. L., Ficklin, W. H., Thompson, J. M. 1987. Concentrations of arsenic, antimony, and boron in steam and steam condensate at the Geysers, California. *Journal of Volcanology and Geothermal Research*, 32, 329–341.

Smith, D. 2013. South African gold miners strike over 'slave wages in white man's economy'. *The Guardian*, September 3.

Smith, D. B. 2015, Soil geochemisty landscapes of the conterminous United States: US Geological Survey [Online] Available: http://minerals.cr.usgs.gov/projects/soil_geochemical_landscapes/index.html.

Smith, G. R. 1990. Tungsten. *US Bureau of Mines Minerals Yearbook*.

Smith, J. V., Dawson, J. S. 1985. Carbonado: diamond aggregates from early impacts of crustal rocks? *Geology*, 13, 342–343.

Smith, R. D., Campbell, J. A., Felix, W. D. 1980. Atmospheric trace element pollutants from coal combustion. *Mining Engineering*, 32, 1603–1610.

Smith, S. J., Van Aardenne, J. Klimont, Z., *et al.* 2011. Anthropogenic sulfur dioxide emissions: 1850–2005. *Atmospheric Chemistry and Physics*, 11, 1101–1116.

Smolders, A. J. P., Lamers, L. P. M., Lucassen, E. C. H. T., van der Velde, G., Roelofs, J. G. M. 2006. Internal eutrophication: how it works and what to do about it: a review. *Chemistry and Ecology*, 22, 93–111.

Sokoloski, A. A. 1996. Explanation of discovery. [Online] Available: http://www.blm.gov/iemtap/discovery.html.

Song, D., Wang, M., Ahang, J., Zheng, C. 2008. Contents and occurrence of cadmium in the coals from Guizhou Province, China. *Annals of the New York Academy of Sciences*, 1140, 274–281.

Sparks, D. 2002. *Environmental Soil Chemistry*. San Diego, CA: Academic Press.

Spencer, S. M., Rupert, F. R., Yon, J. W. 1990. Fuller's earth deposits in Florida and southwestern Georgia. In Zupan, A. W., Maybin, A. H., III (eds.), *Proceedings of 24th Forum on the Geology of Industrial Minerals*, Columbia, SC: South Carolina Geological Survey, pp. 121–127.

St. Petersburg Times 2005. EVRAZ Raises $422M in soft London IPO. *St. Petersburg Times*, June 3. [Online] Available: https://www.yumpu.com/en/document/view/10730721/prosecutors-rule-papers-use-of-word-zhid-the-st-petersburg-/5.

Standish-lee, P., Loboschefsky, E., Beuhler, M. 2005. The future of water: identifying and developing effective methods for managing water in arid and semi-arid regions [Online]. Available: http://acwi.gov/swrr/Rpt_Pubs/wef_session107/107_0130.pdf.

Stanley, G. G. 1988. *The Extractive Metallurgy of Gold in South Africa*. Johannesburg: South African Institute of Mining and Metallurgy, Chamber of Mines of South Africa.

State of Norway, 2014. Map: areas where critical loads are exceeded. [Online] Available: http://www.environment.no/Topics/Air-pollution/Acid-rain/Map-Areas-were-critical-loads-are-exceeded/.

Stauffer, H. C. 1981. Oil shale, tar sands, and related materials. *American Chemical Society Symposium Series*, 163.

Steel, T., 2014. Worldwide nonferrous metals explorations budgets down 25% in 2014. *SNL Metals and Mining Research*, October 21. [Online] Available: https://www.snl.com/InteractiveX/Article.aspx?cdid=A-29544169-11817.

Steinnes, E. 1987. Impact of long-range atmospheric transport of heavy metals to the terrestrial environment in Norway. In Hutchinson, T. C., Meema, K. M. (eds.), *Lead, Mercury, Cadmium and Arsenic in the Environment*. New York, NY: John Wiley.

Stevens, W. K. 1989. To treat the attack of acid rain, add limestone to water and wait. *New York Times*, January 31.

Stewart, H. J., Webb, T. C., Shaw, C. S. J. 2007. Correlating geologic properties with durability of construction aggregate. *Atlantic Geology*, 41, 28–29.

Stewart, M. G., Wang, X., Nguyen, M.N. 2011. Climate change impact and risks of concrete infrastructure deterioration. *Engineering Structures*, 33(4), 1326–1337.

Stinson, T. F. 1982. State severance taxes and the federal system. *Mining Engineering*, April, 382–386.

Stoiber, R. E., Williams, S. N., Huebert, B. 1987. Annual contribution of sulfur dioxide to the atmosphere by volcanoes. *Journal of Volcanology and Geothermal Research*, 33, 1–8.

Stowasser, W. F. 1985. Phosphate rock. *US Bureau of Mines Mineral Facts and Problems*, 675, 579–594.

Stowe, C. W. 1987. *Evolution of Chromium Ore Fields*. New York, NY: Van Nostrand Reinhold.

Streets, D. G., Devane, M. K., Lu, Z., *et al.* 2011. All-time releases of mercury to the atmosphere form human activities. *Environmental Science and Technology*, 45, 10485–10491.

Strong, D. F., Fryer, B. J., Kerrich, R. 1984. Genesis of the St. Lawrence fluorspar deposits as indicated by fluid inclusion, rare earth element and isotopic data. *Economic Geology*, 79, 1142–1496.

Struble, L, Livesey, P., Del Strother, P., Bye, G. C. 2011. *Portland Cement*, 3rd edn. London: Institute of Civil Engineers (ICE Publishing).

Sturges, W. T., Harrison, R. M. 1986a. Bromine: lead ratios in airborne particles from urban and rural sites. *Atmospheric Environment*, 20, 577–588.

Sturges, W. T., Harrison, R. M. 1986b. Bromine in marine aerosols and the origin, nature and quantity of natural atmospheric bromine. *Atmospheric Environment*, 20, 1485–1496.

Styles, M. T., Amankwah, R. K., Al-Hassan, S., Nartet, R. S. 2010. Identification and testing of a method for mercury-free gold processing for artisanal and small-scale gold miners in Ghana. *International Journal of Environment and Pollution*, 41, 289–303.

Sullivan, D. E. 2006. Materials in use in US interstate highways. *US Geological Survey Fact Sheet*, 2006–3127. [Online] Available: http://pubs.usgs.gov/fs/2006/3127/2006-3127.pdf.

Sullivan, J., Gaines, L. 2012. Status of life-cycle inventories for batteries. *Energy Conversion and Management*, 58, 134–148.

Sulphur Institute 2014. An introduction to sulfur. [Online] Available: http://www.sulphurinstitute.org/learnmore/sulphur101.cfm.

Suttill, K. R. 1988. A fabulous silver porphyry: Cerro Rico de Potosi. *Engineering Mining Journal*, 189, March, 50–53.

Sverdrup, H. U., Ragnarsdottir, K. V., Koca, D. 2014a. On modeling the global copper mining rates, market supply, copper price and the end of copper reserves. *Resources, Conservation and Recycling*, 84, 158–174.

Sverdrup, H. U., Koca, D., Ragnarsdottir, K. V. 2014b. Investigating the sustainability of the global silver supply, reserves, stocks in society and market price using different approaches. *Resources, Conservation and Recycling*, 86, 121–140.

Swain, E. B., Jakus, P. M., Rice, G., *et al.* 2007. Socioeconomic consequences of mercury use and pollution. *AMBIO: A Journal of Human Environment*, 36, 45–61.

Syvitski, J. P. M., Harvey, N., Wolanski, E., *et al.* 2005, Dynamics of the coastal zone. In Crossland, C. J. Kremer, H. H. Lindeboom, H. J. Crossland, J. I. M., Le Tissier, M.D.A. (eds.), *Coastal Change and the Anthropocene: The Land–Ocean Interactions in the Coastal Zone Project of the International Geosphere–Biosphere Programme*. Berlin: Springer, pp. 39–94.

Szatmari, P. 1992. Role of modern climate and hydrology in world oil preservation. *Geology*, 20, 1143–1146.

Tang, Y. S., Saling, J. H. 1990. *Radioactive Waste Management*. New York, NY: Hemisphere Publishing.

Tanner, A. O. 2014. Kyanite and related minerals. *United States Geological Survey Minerals Yearbook*.

Tanner, A. O. 2014. Feldspar. *US Geological Survey Mineral Commodity Summary*.

Tanner, A. O. 2014. Iron oxide pigments. *US Geological Survey Mineral Commodity Summary*.

Taugbol, G., ed. 1986, The Norwegian Monitoring Programme for Long-Range Transported Air Pollutants. Results 1980–1984, Norwegian State Pollution Control Authority Report 230/86.

Taylor, B. E., Wheeler, M. C., Nordstrom, D. K. 1984. Stable isotope geochemistry of acid mind drainage: experimental oxidation of pyrite. *Geochimica et Cosmochimica Acta*, 48, 2269–2678.

Taylor, H. A., Jr. 1985. Dimension stone. *US Bureau of Mines Mineral Facts and Problems*, 675, 769–776.

Taylor, H. A., Jr. 1991. Chicago's Amoco building marble to be recycled. *Minerals Today*, March, 27.

Taylor, J., Yokell, M. 1979. *Yellowcake: The International Cartel*. New York, NY: Pergamon Press.

Trefis Team 2012. Shell's record Transocean deal shows the importance of ultra-deepwater. *Forbes*, February 10.

Tepordei, V. V. 1999. Natural aggregates: foundation of America's future. *US Geological Survey Fact Sheet*, 144–97. [Online] Available: http://minerals.usgs.gov/minerals/pubs/commodity/aggregates/fs14497.pdf.

Terry, R. C. 1974. *Road Salt, Drinking Water, and Safety: Improving Public Policy and Practice*. Cambridge, MA: Ballinger.

Thoburn, J. 1981. *Multinationals, Mining And Development: A Study of the Tin Industry*. Westmead: Gower Publishing.

Thomas, C. R., Kelley, T. R. 2010. A brief review of silicosis in the United States. *Environmental Health Insights*, 4, 21–26.

Thompkins, R. W. 1982. Radiation in uranium mines. *CIM Bulletin*, September, 149–159.

Thompson, B. 2013. Industry fired back at AP ethanol story. *AG Web*, November 12.

Thompson, I., Boutilier, R.G. 2011. Social license to operate. In Darling, P. (ed.), *SME Mining Engineering Handbook*. Denver, CO: Society of Mining Metallurgy and Exploration, pp. 1779–1796.

Thompson, L. K., Sidhu, S. S., Roberts, B. A. 1979. Fluoride accumulations in soil and vegetation in the vicinity of a phosphorus plant. *Environmental Pollution*, 18, 221–234.

Thompson Reuters GMFS 2013. Gold survey, 2013. [Online] Available: http://share.thomsonreuters.com/PR/Misc/GFMS/GoldSurvey2013Update1.pdf.

Thompson Reuters GFMS 2014. World silver survey 2014. [Online] Available: https://www.silverinstitute.org/site/wp-content/uploads/2011/06/WSS2014Summary.pdf.

Thornton, I., ed. 1983. *Applied Environmental Geochemistry*, New York, NY: Academic Press.

Tian, H. Z., Lu, L., Cheng, K., *et al.* 2012. Anthropogenic atmospheric nickel emissions and its distribution characteristics in China. *Science of the Total Environment*, 417–418.

Tian, Y., Zhu, Q., Geng, Y. 2013. An analysis of energy-related greenhouse gas emission in the Chinese iron and steel industry. *Energy Policy*, 56, 352–361.

Tierney, S. F. 2012. *Why Coal Plants Retire*. Boston, MA: Analysis Group, Inc.

Tissot, B. P., Welte, D. H. 1978. *Petroleum Formation and Occurrence*. Heidelberg: Springer-Verlag.

Titus, J. T. 2011. Was lead-free solder worth the effort? *ECN Magazine*, December 28.

Tivey, M. K. 2007. Generation of seafloor hydrothermal vent fluids and associated mineral deposits. *Oceanography*, 20, 50–65.

Tolcin, A. C. 2013a. Lead. *US Geological Survey Mineral Commodity Summary*.

Tolcin, A. C. 2013b. Cadmium. *US Geological Survey Minerals Yearbook*.

Tollinsky, N. 2013. Emission reduction program keeps Vale in compliance with SO$_2$ emissions. *Sudbury Mining Solutions Journal*, August 13.

Tollinsky, N., 2015. Sudbury's Superstack outlives its usefulness. *Sudbury Mining Solutions Journal*, February 25

Toth, L. M., Malinauskas, A.P., Eidam, G. R., Burton, H. M. 1986. The Three Mile Island accident: diagnosis and prognosis. *American Chemical Society*, 293, 193–211.

Towler, B. F. 2014. Nuclear energy. In *The Future of Energy*. London: Academic Press, pp. 135–159.

Towner, R. R., Gray, J. M., Porter, L. M. 1988. International strategic minerals inventory summary report: titanium. *US Geological Survey Circular*, 930-G.

Transportation Research Board 2007. *Guidelines for Selection of Snow and Ice Control Materials to Mitigate Environmental Impacts: NCHRP Report 577*. Washington, DC: Transportation Research Board.

Trasande, L., Landrigan, P. J., Schechter, C. 2005. Public health and economic consequences of methyl mercury toxicity to the developing brain. *Environmental Health Perspectives*, 113, 590–596.

Tsan, Y. T., Peng, K. Y., Hung, D .Z., Hu, W. H., Yang, D. Y. 2005. Case report: the clinical toxicity of dimethylamine borane. *Environmental Health Perspectives*, 113, 1784–1786.

Tschakert, P., Singha, K. 2007. Contaminated identities: mercury and marginalization in Ghana's artisanal mining sector. *Geoforum*, 38, 1304–1321.

Tseng, W. P. 1977. Effects and doe–response relationships of skin cancer and Blackfoot disease with arsenic. *Environmental Health Perspectives*, 19, 109–119.

Tuck, C. A. 2013. Cesium. *US Geological Survey Minerals Commodity Summary*.

Tuck, C. A., Virta, R.L. 2013. Iron ore. *US Geological Survey Minerals Yearbook*.

Turneaure, F. S. 1971. The Bolivian tin–silver province. *Economic Geology*, 66, 212–225.

Ullman, F. D. 1991. Cement. *Mining Engineering*, 44, 553–554.

UMW, 2013. United Mine Workers of American. Health and Safety on the job. [Online] Available: http://www.umwa.org/?q=content/black-lung.

UNEP, 2013. Global mercury assessment 2013: sources, emissions, releases and environmental transport. UNEP Chemicals Branch, Geneva. [Online] Available: http://www.unep.org/PDF/PressReleases/GlobalMercuryAssessment2013.pdf.

United Nations 1982. United Nations convention on the law of the sea. Montego Bay, Jamaica: United Nations. [Online] Available: http://www.un.org/Depts/los/convention_agreements/convention_overview_convention.htm.

US Department of Health and Human Services 1999. Toxicological profile for mercury. [Online] Available: http://www.atsdr.cdc.gov/toxprofiles/tp46.pdf.

US DOI. 2008. Inventory of onshore federal oil and natural gas resources and restrictions to their development, phase III inventory – onshore United States. [Online] Available: http://www.blm.gov/pgdata/etc/medialib/blm/wo/MINERALS__REALTY__AND_RESOURCE_PROTECTION_/energy/EPCA_Text_PDF.Par.18155.File.dat/Executive%20Summary%20text.pdf.

US Energy Information Administration 2014. Annual Energy Outlook 2014. [Online] Available: http://www.eia.gov/forecasts/aeo/mt_liquidfuels.cfm.

US Forest Service 1995. Anatomy of a mine from prospect to production. US Forest Service General Technical Report INT-GTR-35. [Online] Available: http://www.fs.fed.us/geology/anatomy_mine.pdf.

USGS 1988. National energy resource issues. *USGS Bulletin*, 1850.

USGS 1989. Focus: assessment of petroleum potential in the Arctic National Wildlife Refuge, Alaska. *USGS Professional Paper*, 1850, 26–28.

USGS 1999. The oil and gas resources potential of the Arctic National Wildlife Refuge 1002 Area, Alaska. *US Geological Survey Open-File Report*, 98–34.

USGS 2013. Supporting data for the US Geological Survey 2012 world assessment of undiscovered oil and gas resources. [Online] Available: http://pubs.usgs.gov/dds/dds-069/dds-069-ff/.

USGS 2014. The world's water. The USGS Water Science School. US Geological Survey. US Department of Interior. [Online] Available: https://water.usgs.gov/edu/earthwherewater.html.

USGS 2015. Gas hydrates prime: US Geological Survey Gas Hydrates Project. http://woodshole.er.usgs.gov/project-pages/hydrates/primer.html.

Vallero, D., Brasier, C. 2008. *Sustainable Design: The Science of Sustainability and Green Engineering*. Hoboken, NJ: Wiley.

van der Vyver, J. D. 2002. Nationalization of mineral rights in South Africa. *De Jure*, 45, 126–142.

Van Oss, H. 2014. Cement. *US Geological Survey Minerals Yearbook*.

van Rensburg, W. C. J. 1986. *Strategic Minerals*. Englewood Cliffs, NJ: Prentice-Hall.

van Wijnbergen, S. 1984. The "Dutch disease": a disease after all? *The Economic Journal*, 94, 373.

van Zyl, A. A. 1988. De Beers' 100. *Geobulletin*, 31, 24–28.

VanLandingham, S. L. 1985. *Gemology of World Gem Deposits*. New York, NY: Van Nostrand Reinhold.

Vann, A. 2012. Energy projects on federal lands: leasing and authorization. Congressional Research Service 7–5700, 20. [Online] Available: http://www.fas.org/sgp/crs/misc/R40806.pdf.

Velasco, P., Gurmendi, A. C. 1988. *The Mineral Industries of Southern South America, 1988 International Review*. Washington, DC: USBM.

Velders, G. J. M., Ravishankara, A. R., Miller, M. K., *et al.* 2012. Preserving Montreal protocol climate benefits by limiting HFCs. *Science*, 335, 922–923.

Verplanck, P. L., van Gosen, B. S. 2011. Carbonatite and alkaline intrusion-related rare earth element deposits: a deposit model. *US Geological Survey Open-File Report*, 2011-1256.

Victor, D. G., Morgan, M. G., Steinbruner, J., Ricke, K. 2009. The geoengineering option: a last resort against global warming? *Foreign Affairs*, 88, 64–76.

Victor, D. G., Hults, D. R., Thurber, M. C. 2012. *Oil and Governance: State Owned Enterprises and the World Energy Supply*. Cambridge: Cambridge University Press.

Vigrass, L. 1979. Final well report, U or R Regina 3–8–17–19 (W.2nd. Mer.) Saskatchewan. *Earth Physics Branch Open File*, 79–9.

Viljayan, S., Melnyk, A. J., Singh, R. D., Nutall, K. 1989. Rare earths: their mining processing and growing industrial usage. *Mining Engineering*, 41, 13–18.

Virta, R. L. 2002. Geology, mineralogy, mining. *US Geological Survey Open-File Report*, 02-149.

Virta, R. L. 2006. Asbestos. In Kogel, J. E., Trivedi, N. C., Barker, N. C., Krukowski, S. T. (eds.), *Industrial Minerals and Rocks*, 7th edn. Littleton, CO: Society for Mining, Metallurgy and Exploration, pp. 195–206.

Virta, R. L. 2014. Zeolites. *US Geological Survey Mineral Commodity Summary*.

Virta, R. L. 2013a. Clay and shale. *US Geological Survey Minerals Yearbook*.

Virta, R. L. 2013b. Talc and pyrophyllite. *US Geological Survey Minerals Yearbook*.

Visconti, L. F., Fernandes, E. S. 2013. The International comparative legal guide to: Mining Law 2014. [Online] Avaialable: http://www.iclg.co.uk/practice-areas/mining-law/mining-law-2015/brazil.

Visschedijk, A. H. J., Denier van der Gon, H. A. C., Hulskotte, J. H. J., Quass, U. 2013. Anthropogenic vanadium emission to air and ambient air concentrations in north-west Europe. *E3S Web of Conferences*, 1, 03004. [Online] Available http://www.e3s-conferences.org/articles/e3sconf/pdf/2013/01/e3sconf_ichm13_03004.pdf.

Vogely, W. A., ed. 1985. *Economics of the Mineral Industries*. New York, NY: American Institute of Mining, Metallurgical and Petroleum Engineers.

Vourvopoulos, G. 1987. Proceedings of the international conference on trace elements in coal. *Journal of Coal Quality*, 8, 18–27.

Wada, Y., Beek, L. P. H., Van Kempen, C. M., *et al.* 2010. Global depletion of groundwater resources. *Geophysical Research Letters*, 37, 20402L.

Wald, M. 2012. Another step to smaller reactors. *New York Times*, November 28.

Wald, M. L. 2013. Deal advances development of a smaller nuclear reactor. *The New York Times*, February 20.

Wald, M. W. 2014. Nuclear waste solution seen in desert salt beds. *The New York Times*, February 9.

Waldman, J. 2015. *Rust: The Longest War*. New York, NY: Simon and Schuster.

Waldrop, M. W. 2014, Plasma physics: the fusion upstarts. *Nature*, 511, 398–400.

Wallace, R. E., Kraemer, T. F., Taylor, R. E., Wesselman, J. B. 1979. Assessment of geopressurized–geothermal resources in the northern Gulf of Mexico basin. *US Geological Survey*, 790, 132–155.

Walston, R. E. 1986. Western water law. *Natural Resources and Environment*, 1, no. 4, 6–8, 48–52.

Walters, A., Lusty, P. 2011. Rare earth elements, Commodity Profile. British Geological Survey. [Online] Available: https://www.bgs.ac.uk/mineralsuk/statistics/mineralProfiles.html.

Wanty, R. B., Johnson, S. L., Briggs, P. H. 1991. Radon-222 and its parent radionuclides in groundwater from two study areas in New Jersey and Maryland, USA. *Applied Geochemistry*, 6, 305–318.

Ward, C. R. 1984. *Coal Geology and Coal Technology*. Palo Alto, CA: Blackwell Scientific Publications.

Wargo, J. G. 1989. *In situ* leaching of disseminated gold deposits: geological factors. *Mining Engineering*, 40, 973–975.

Warneck, P. 1988. *Chemistry of the Natural Atmosphere*. New York, NY: Academic Press.

Warren, J. K. 2006. *Evaporites: Sediments, Resources and Hydrocarbons*. Berlin: Springer.

Warren, H. V. 1989. Geology, trace elements health. *Social Science and Medicine*, 29, 285.

Warren, J. K. 1989. *Evaporite Sedimentology*. New York, NY: Prentice Hall.

Wassel, R. 2012. Lessons from the Macondo well blowout in the Gulf of Mexico. *The Bridge*, 42, 46–53.

Watson, J. G., Zhu, T., Chow, J. C., *et al.* 2002. Recepter modeling application framework for particle source appointment. *Chemosphere*, 49, 1093–1136.

Wayne, L. 2009. Chinese drywall linked to corrosion. *The New York Times*, November 23.

Wearden, G. 2010. BHP Billiton urged to pull Potash Corp out of Western Sahara. *The Guardian*, August 22.

Weatherly, D. K., Henley, R. W. 2013. Flash vaporization during earthquakes evidenced by gold deposits. *Nature Geoscience*, 6, 294–298.

Wei, W. Q., Abnet, C. C., Qiao, Y. L., *et al.* 2004, Prospective study of serum selenium concentrations and esophageal and gastric cardia cancer, heart disease, stroke, and total death. *American Journal of Clinical Nutrition*, 79, 80–85.

Weisman, J. 2005. House budget bills face senate fight. *The Washington Post*, December 20.

Weisman, W. I., McIlveen, S. 1983. Sodium sulfate deposits. In Lefond, S. J. (ed.), *Industrial Minerals and Rocks*, 5th edn. New York, NY: Society of Mining Engineers, pp. 1207–1224.

Wellmer, F. W. 1989. *Economic Evaluations in Exploration*. Heidelberg: Springer-Verlag.

Wellmer, F. W. 1992. The concept of lead time. *Minerals Industry International*, March, 39–40.

Wellmer, F. W., Becker-Platen, J. D. 2002. Sustainable development and the exploitation of mineral and energy resources: a review. *International Journal of Earth Sciences (Geologische Rundschau)*, 91, 723–745.

Wellstead, J. 2012. The future of phosphate is in Morocco. [Online] Available: http://potashinvestingnews.com/6590-future-phosphate-morocco-ocp-fertilizer-production-flsmidth-agrium-potashcorp.html.

Wentzler, T. H., Aplan, F.F. 1992. Neutralization reactions between acid mine waters and limestone. page range?

Werner, M., Lochner, R. 1998. Growth and application of undoped and doped diamond films. *Reports Progress in Physics*, 61, 1665.

Wessel, G. R., Wimberly, B. H. 1992. *Native Sulfur: Developments in Geology and Exploration*. Littleton, CO: Society for Mining, Metallurgy and Exploration.

West, E. G. 1982. *Copper and its Alloys*. New York, NY: Halsted Press.

Westall, J., Stumm, W. 1980. The hydrosphere. In Huntzinger, O. (ed.), *The Handbook of Environmental Chemistry*. Heidelberg: Springer-Verlag, vol. 1, part A, pp. 17–50.

Wheatcroft, G. 1985. *The Randlords*. London, Weidenfeld and Nicholson.

Whewell, T. 2008. China to seal $9bn DR Congo Deal. *BBC News*, April 14. [Online] Available: http://news.bbc.co.uk/2/hi/programmes/newsnight/7343060.stm.

White, D. E. 1974. Diverse origins of hydrothermal ore fluids. *Economic Geology*, 75, 382–493.

White, D. E., Muffler, L. J. P., Truesdell, A. H. 1971. Vapor-domination hydrothermal systems compared with hot-water systems. *Economic Geology*, 66, 75–97.

White, J. A. L. 1968. Native sulfur deposits associated with volcanic activity. *Mining Engineering*, June, 47–54.

White, N. C., Hedenquist, J. W. 1995. Epithermal gold deposits: styles, characteristics and exploration. *Society of Economic Geology Newsletter*, 23, 9–13.

White, W. H., Bookstrom, A. A., Kamilli, R. J., Ganster, *et al.* 1981. Character and origin of Climax-type molybdenum deposits. *Economic Geology*, 75, 270–316.

Whitehead, D. 1968. *The Dow Story*. New York, NY: McGraw-Hill.

Whitehouse, D. 2012. *Glass: A Short History*. Washington, DC: Smithsonian Books.

Whitney, E. 2005. Claiming public lands. National Public Radio. [Online] Available: http://www.loe.org/shows/shows.html?programID=05-P13-00043#feature4

WHO 1980. Environmental health criteria: tin and organo-tin compounds. World Health Organization. [Online] Available: http://www.inchem.org/documents/ehc/ehc/ehc015.htm.

WHO 1981a. Environmental health criteria: manganese. World Health Organization. [Online] Available: http://www.inchem.org/documents/ehc/ehc/ehc017.htm.

WHO 1981b. Environmental health criteria: chromium. World Health Organization. [Online] Available: http://www.inchem.org/documents/ehc/ehc/ehc61.htm.

WHO 1981c. Environmental health criteria: arsenic. World Health Organization. [Online] Available: http://www.inchem.org/documents/ehc/ehc/ehc018.htm

WHO 1982. Environmental health criteria: titanium. World Health Organization. [Online] Available: http://www.inchem.org/documents/ehc/ehc/ehc24.htm.

WHO 1984. Environmental health criteria: fluorine and fluorides. World Health Organization. [Online] Available: http://www.inchem.org/documents/ehc/ehc/ehc36.htm.

WHO 1988. Environmental health criteria: vanadium. World Health Organization. [Online] Available: http://www.inchem.org/documents/ehc/ehc/ehc81.htm.

WHO 1989a. Environmental health criteria: lead. World Health Organization. [Online] Available: http://www.inchem.org/documents/ehc/ehc/ehc85.htm

WHO 1990a. Environmental health criteria: beryllium. World Health Organization. [Online] Available: http://www.inchem.org/documents/ehc/ehc/ehc106.htm.

WHO 1990b. Environmental health criteria: methylmercury. World Health Organization. [Online] Available: http://www.inchem.org/documents/ehc/ehc/ehc101.htm

Wicken, O. M., Duncan, L. R. 1983. Magnesite and related minerals. In LeFond, S. J. (ed.), *Industrial Minerals and Rocks*, 5th edn. New York, NY: Society of Mining Engineers, pp. 881–896.

Wilburn, D. R. 2008. Material use in the United States: selected case studies for cadmium, cobalt, lithium and nickel in rechargeable batteries. *US Geological Survey Scientific Investigations Report*, 2008–5141.

Wilburn, D. R. 2011. Cobalt mineral exploration and supply from 1995 through 2013. *US Geological Survey Scientific Investigations Report*, 2011–5084, 16.

Wilkinson, B. H., McElroy, B. J. 2007. The impact of humans on continental erosion and sedimentation. *Geological Society of America Bulletin*, 119, 140–156.

Wilkinson, B. H., Kesler, S. E. 2009. Quantitative identification of metallogenic deposits and provinces: applications to Phanerozoic porphyry copper deposits. *Economic Geology*, 104, 607–622.

Will, R. 2004. Shifting trends in fluorocarbons. *Industrial Minerals*, 441, 22.

Willett, J. C. 2013. Mica. *US Geological Survey Mineral Yearbook*.

Williams, B. 1991. *US Petroleum Strategies in the Decade of the Environment*. Tulsa, OK: PennWell Publishing Company.

Williams, O. R., Driver, N. E., Ponce, S. L. 1990. Managing water resources in wilderness areas. *US Forest Service General Technical Report*, SE–66.

Williams, S. J. 1992. Sand and gravel: an enormous offshore resource within the US exclusive economic zone. *US Geological Survey Bulletin*, 1877H, 1–10.

Wilson, F. 1985. Mineral wealth, rural poverty. *Optima*, 33, no.1, 23–37.

Wilson, J. D. 2011. Resource nationalism or resource liberalism? *Australian Journal of International Affairs*, 65, 283.

Wilson-Jones, M. 2003. *Principles of Roman Architecture*. New Haven, CT: Yale University Press.

Winchester, J. W. 1980. Transport processes in air. In Hutzinger, O. (ed.), *The Handbook of Environmental Chemistry*. Heidelberg: Springer-Verlag, vol. 2, Part A, pp. 19–30.

Winterhalder, K. 1988. Trigger factors initiating natural revegetation processes on barren, acid, metal-toxic soil near Sudbury, Ontario, smelters. *US Bureau of Mines Information Circular*, 9184, 118–124.

Woodall, R. 1984. Success in mineral exploration. *Geoscience Canada*, 11, 41–46, 83–90, 127–132.

Woodall, R. 1988. The role of mineral exploration toward the year 2000. Geophysics, 7, no. 2, 35–37.

Woodall, R. 1992. Challenge of minerals exploration in the 1990s. *Mining Engineering*, 44, 679–683.

Woollins, J. D., Laitinen, R. 2011. *Selenium and Tellurium Chemistry.* New York, NY: Springer.

World Alumnum Institute 2013. The International Aluminum Institute's report on the aluminum industry's global perfluorocarbon gas emissions reduction programme – results of the 2003 anode effect survey. [Online] Available: http://www.world-aluminium.org/media/filer_public/2013/01/15/fl0000134.pdf

World Bank 2011. Overview of state ownership in the global minerals industry. Raw Minerals Group. [Online] Available: http://siteresources.worldbank.org/INTOGMC/Resources/GlobalMiningIndustry-Overview.pdf.

World Bank 2013. Mining: sector results profile. [Online] Available: http://www.worldbank.org/en/results/2013/04/14/mining-results-profile.

World Bank 2015. Electricity production from coal sources (% of total). [Online] Available: http://data.worldbank.org/indicator/EG.ELC.COAL.ZS.

World Coal Association 2014. Coal statistics. [Online] Available: http://www.worldcoal.org/resources/coal-statistics/.

World Energy Council 2010. Survey of energy resources: focus on shale gas. [Online] Available: http://www.worldenergy.org/wp-content/uploads/2012/10/PUB_shale_gas_update_2010_WEC.

World Nuclear Association 2015a. Supply of uranium. [Online] Available: http://www.world-nuclear.org/info/Nuclear-Fuel-Cycle/Uranium-Resources/Supply-of-Uranium/.

World Nuclear Association 2015b. World nuclear power reactors and uranium requirements. [Online] Available: http://www.world-nuclear.org/info/Facts-and-Figures/World-Nuclear-Power-Reactors-and-Uranium-Requirements/.

Worstall, T. 2012. What will Apple do when indium runs out in 2017? Forbes. [Online] Available: http://www.forbes.com/sites/timworstall/2012/03/09/what-will-apple-do-when-indium-runs-out-in-2017/.

WyoFile 2013. Interior disburses $14.2 billion in 2013 energy revenues to federal, state, local and tribal goernments. *WyoFile*, Novemeber 19. [Online] Available: wyofile.com

Yaacob, S. E., Ahmad, S. 2014. Return to gold-based monetary system: analysis based on gold price and inflation. *Asian Social Science*, 10, 18–28.

Yen, T. F., Chilingarian, G. V. 1979. Asphalthenes and asphalts, 1. In Yen, T. F. and Chilingarian, G. V. (eds) *Developments in Petroleum Science*. Amsterdam: Elsevier, pp. 1–6.

Yergin, D. 1990. *The Prize: The Epic Quest for Oil, Money and Power*, New York, NY: Simon and Schuster.

Yergin, D. 2011. *The Quest: Energy, Security, and the Remaking of the Modern World*. New York, NY: Penguin Press.

Young, G. M. 1992. Late Proterozoic stratigraphy and the Canada–Australia connection. *Geology*, 20, 215–218.

Young, R. S. 1979. *Cobalt in Biology and Biochemistry*. New York, NY: Academic Press.

Younger, S. 2013. Sand rush: fracking book spurs rush on Wisconsin silica. *National Geographic*, July 4. [Online] Available: http://news.nationalgeographic.com/news/energy/2013/07/130703-wisconsin-fracking-sand-rush/.

Youngson, J. H., Craw, D. 1993. Gold nuggets growth during tectonically induced sedimentary recycling, Central Otago, New Zealand. *Sedimentary Geology*, 84, 71–88.

Younos, T. 2005. The economics of desalination. *Journal of Contemporary Water Research and Education*, 132, 39–45.

Yudovich, Y. E., Ketris, M. P. 2004. Arsenic in coal: a review. *International Journal of Coal Geology*, 61, 141–196.

Yudovich, Y. E., Ketris, M. P. 2006. Selenium in coal: a review. *International Journal of Coal Geology*, 67, 112–126.

Yurth, C. 2012. Dine politicians applaud uranium ban. *Navajo Times*, January 13.

Zaihua, L., Daoxian, Y. 1991. Effect of coal mine waters of variable pH on spring water quality: a case study. *Environmental Geology and Water Science*, 17, 219–225.

Zarzycki, J. 1991. *Glasses and the Vitreous State*. Cambridge: Cambridge University Press.

Zdunczyk, M. 1991. Aggregate imports: is it achievable? *Mining Engineering*, 43, 145.

Zemba, S., Ames, M., Green, L., Botelho, M. J., Gossman, D. 2011. Emissions of metals and polychlorinated dibenzo(p)dioxins and furans (PCDD/Fs) from Portland cement manufacturing plants: inter-kiln variability and dependence on fuel types. *Science of the Total Environment*, 409, 4198–4205.

Zeng, Q., Liu, J., Qin, K., *et al.* 2013. Types, characteristics and time–space distribution of molybdenum deposits in China. *International Geology Review*, 55, 1311–1358.

Zenhder, A. J. B., Zinder, S. H. 1980. The sulfur cycle. In Hutzinger, O. (ed.), *The Handbook of Environmental Chemistry*, Heidelberg: Springer-Verlag, vol. 1, Part A, pp. 105–146.

Zernike, K. 2012. In New Jersey, a battle over fluoridation bill, and the facts. *The New York Times*, March 2.

Zhao, C., Qin, S., Yang, Y., Li, Y., Lin, M. 2009 Concentration of gallium in the Permo-Carboniferous coals of China. *Energy Exploration and Exploitation*, 27, 333–343.

Zhang, P, Wiegel, R., Hassan, E.-S. 2007, Phosphate rock. In Kogel, J. E., Trivedi, N. C., Barker, J. M, Krukowski, S.T., (eds.), *Industrial Minerals and Rocks*, 7th edn. Littleton, CO: Society for Mining, Metallurgy, and Exploration, pp, 703–722.

Zhang, H. H., Yuan, H. X., Hu, Y. G., *et al.* 2006, Spatial distribution and vertical variation of arsenic in Guangdon soil profiles, China. *Environmental Pollution*, 144, 492–499.

Zhou, T. F., Fan, T., Yuan, F., Wu, M. A., Hou, M. J. 2005. Xiangquan: the world's first reported sediment-hosted thallium-only deposit in northeastern margin of the Yangtze Block, eastern China. *Proceedings of the Biennial GSA (Society for Geology Applied to Mineral Deposits) Meeting*, 8, 515–517.

Zoltai, S.C. 1988. Distribution of base metals in peat near a smelter at Flin Flon, Manitoba. *Water, Air and Soil Pollution*, 37, 217–228.

Zumberge, H. 1979. Mineral resources and geopolitics in Antarctica. *American Scientist*, 65, 345–349.

INDEX

uranium mining, 159
vanadium, 202
zinc, 255
dysprosium, 9, 236, 350

Eagle, copper–nickel mine, Michigan, 93, 191
earnings, 107, 108, 150
earthquake, 128, 149, 167, 264
    Fukushima nuclear incident, 165
    hydraulic fracturing, 155
    orogenic gold deposits, 261
    Yellowstone, 171
East Tennessee
    Mississippi Valley-type deposit, 226
East Texas
    oil field, 55
    tight gas reservoir, 154
economic rent, 119, 120
Economic Stabilization Act of 1970, 118
Economic Stabilization Act of 2008, 128
EDTA, 36
earnings, 107, 108, 150
erbium, 236
evaporite, 16, 18, 21, 31, 201, 246, 290, 294, 310,
    317, 325
    deposits, 292–294
Egypt, 138, 198, 218, 227, 234, 246, 261, 270,
    279, 294, 306, 320, 321
El Indio, 259
El Salvador, 169
El Teniente, copper deposit, Chile, 218, 221, 261
electric arc furnace, 182
electric charge, 53
electric power generation, 49, 53, 54, 122, 129,
    131, 132, 134, 162, 164, 168
    CO_2 emissions, 45
electromagnetic exploration, 61
electrum, 258, 272, 273, 359
element, 3, 9, 11, 14, 29, 32
Elk Basin oil field, Alberta, 138
Elliot Lake, uranium deposits, Ontario, 159, 251
embargo, 110, 116, 118, 150, 173, 192, 197, 347
emerald, 239, 257, 278, 285–286
Emergency Gold Mining Assitance Act.
    Canada, 105
emery, 338, 340
eminent domain, 92
emissions, 6, 34, 35, 40, 41, 43, 45, 76, 78, 115,
    223, 227, 321, 351
    acid-rain, 45
    aluminum production, 210
    arsenic, 238, 239
    cement production, 316
    CO_2, 46
    coal combustion, 131, 133
    cobalt production, 199
    copper production, 223
    ethanol, 144
    fluorine, 308
    geothermal power, 170

lime production, 291
manganese, 187
mercury, 249, 251
methane, 128, 351
nickel smelting, 191
nuclear power, 166
particulate, 42
power plants, 134
radon, 41, 321
selenium, 254
SO_2, 42
steel production, 182
sulfur, 38
Emperor, gold mine, Fiji, 254
employment, 85, 100, 105, 112, 115, 130, 135,
    138, 182, 267, 302, 314
Endicott, oil field, Alaska, 70, 154
energy
    metal production, 75
    nuclear, 31
    production, CO_2 emissions, 46
    value of production, 7
energy policy, 150, 166, 174
Energy Policy Act, 144
Energy Policy and Conservation Act of
    1975, 347
England, 135, 156, 172, 218, 229, 307, 315,
    328, 334
    groundwater law, 82
English mining law, 80
enhanced oil recovery, 139, 173
environmental assessment, 48, 86
environmental impact statement, 71
enzyme, 37, 191, 215, 227
epilimnion, 34
epithermal, 24–25
    antimony, 236
    gold deposits, 258–260
    silver deposits, 272–275
Equator Principles, 105
equilibrium, 29
    economic, 116
    groundwater, 19
erionite, 333, 337
erosion, 14, 17, 23, 27, 28, 41, 123, 137, 195, 208,
    209, 216, 231, 264, 282, 318, 322, 323, 341
Escondida, copper deposit, Chile, 68, 69,
    221, 223
estimation methods, reserves or resources, 342
ethane, 34, 60, 61, 77, 136
ethanol, 144–145, 306
Ethiopia, 278, 301
European Union, 10, 46, 92, 94, 115, 150, 232,
    348, 349
europium, 9, 360
eutrophication, 298, 300, 304, 348
evaporate, 16, 18, 21, 31, 201, 246, 290, 294, 310,
    317, 325
    salt deposits, 292–294
evaporite, 27

basins, 294, 295, 300
deposit, 245
deposits, 16, 21, 191, 292, 299–300, 305,
    309–311, 322, 325, 326
Excelsior diamond, 283
excise tax, 111, 115, 128
Exclusive Economic Zone, 186
exhalative deposit, 24
Exotica, copper deposit, Chile, 221
exploration
    drilling, 344
    drilling data, 344
    frontier, 151
    methods, 55–79
    oil, 95–100
explosions
    in mines, 69, 128, 264
    in nuclear reactors, 164
expropriation, 102
extender, 292, 322, 327–329, 337, 361
extends, 231
external cost, 109
extraction
    mining and pumping methods, 67–71
extralateral rights, 85
extrapolation technique
    toxity, 52

Fairbanks, Trans-Alaska pipeline, 149
Fairbanks, Alaska, gold deposits, 262
Far Southeast, copper deposit, Philippines, 261
feasibility analysis, 65–66
Federal Land Policy and Management Act of
    1976, 84
feeder zone
    VMS deposits, 219, 221
feldspar, 241, 259, 264, 279, 315, 323, 326, 329,
    332, 337, 338, 355, 356, 361
ferroalloys
    ferrochromium, 175
    ferromanganese, 186
    ferronickel, 191
    ferrosilicon, 197
    tungsten, 254
fertilizer, xi, 1, 6, 42, 95, 144, 237, 240, 288–290,
    296–305, 313, 347–348, 362
felsic, 12–13, 25, 27, 172, 190, 193, 204, 208, 218,
    226, 239, 248, 259, 321, 355, 356
FGD
    flue-gas desulfurization, 54
fiber
    asbestos, 331
fiberglass, 317
Fiji, 254
financing, 66, 92, 111
Finland, 2, 41, 168, 199, 278
fire
    mine, 72
    oil blowout, 66
fish, 40, 42, 45, 55, 133, 151, 249, 267, 293, 301